NEUROSCIENCE

NEUROSCIENCE

Edited by

DALE PURVES

GEORGE J. AUGUSTINE

DAVID FITZPATRICK

LAWRENCE C. KATZ

ANTHONY-SAMUEL LaMANTIA

JAMES O. McNAMARA

Department of Neurobiology
Duke University Medical Center

SINAUER ASSOCIATES, INC.
PUBLISHERS
SUNDERLAND, MASSACHUSETTS

THE COVER

The background in the cover illustration is a pattern of orientation columns in the primary visual cortex revealed by optical imaging. (Courtesy of David Fitzpatrick; see Chapter 11.)

Library of Congress Cataloging-in-Publication Data

Neuroscience / edited by Dale Purves . . . [et al.].
                     p.   cm.
        Includes bibliographical references and index.
        ISBN 0-87893-747-1 (alk. paper)
        1. Neurosciences. 2. Neurophysiology. 3. Neurology. I. Purves, Dale.
        [DNLM: 1. Nervous systems—physiology. 2. Neurochemistry. WL 102
        N4996 1997]
        QP355.2.N487   1997
        612.8—dc20
        DNLM/DLC
        for Library of Congress   96-43031
                                CIP

Printed in U.S.A.
5 4 3 2 1

Contributors

George J. Augustine, Ph.D.

Nell B. Cant, Ph.D.

John H. Casseday, Ph.D.

Gillian Einstein, Ph.D.

David Fitzpatrick, Ph.D.

Lawrence C. Katz, Ph.D.

Anthony-Samuel LaMantia, Ph.D.

Julie Kauer, Ph.D.

Donald C. Lo, Ph.D.

James O. McNamara, M.D.

Richard D. Mooney, Ph.D.

Miguel A. L. Nicolelis, M.D., Ph.D.

Dale Purves, M.D.

Peter H. Reinhart, Ph.D.

Sidney A. Simon, Ph.D.

All the contributors are associated with the Department of Neurobiology at Duke University Medical Center.

Contents in Brief

Contents

**Unit III
Movement and Its Central
Control**

16. Descending Control of Spinal Cord Circuitry 311

17. Modulation of Movement by the Basal Ganglia and Cerebellum 329

18. Mechanisms of Motor Modulation 345

Preface

Whether judged in molecular, cellular, systemic, or behavioral terms, the human nervous system is a stupendous piece of biological machinery. Given its accomplishments—all the artifacts of human culture, for instance—there is good reason for wanting to understand how the brain and the rest of the nervous system works. The debilitating and costly effects of neurological and psychiatric disease add a further sense of urgency to this quest. The aim of this book is to highlight the intellectual challenges and excitement—and the uncertainties—of what many see as the last great frontier of biological science. The information presented should serve as a starting point for pre-med students, medical students, graduate students in the neurosciences, and others who simply want to understand how the human nervous system operates. Like any other great challenge, neuroscience should be, and is, full of dissension and considerable fun. Both these ingredients have gone into the construction of this book; we hope they will be conveyed in equal measure to readers at all levels.

Acknowledgments

We are grateful to numerous colleagues who provided helpful criticisms and suggestions as we were writing this book. We particularly wish to thank David Amaral, Gary Banker, Ursula Bellugi, Dan Blazer, Robert Burke, John Chapin, Milt Charlton, Michael Davis, Robert Desimone, Allison Doupe, Sascha du Lac, Jens Eilers, Peter Eimas, Everett Ellinwood, Robert Erickson, Howard Fields, Bob Fremeau, Michela Gallagher, Steve George, Pat Goldman-Rakic, Mike Haglund, Zach Hall, Bill Henson, John Heuser, Jonathan Horton, Alan Humphrey, Jon Kaas, Herb Killackey, Len Kitzes, Arthur Lander, Story Landis, Darrell Lewis, Alan Light, Steve Lisberger, Arthur Loewy, Eve Marder, Robert McCarley, Jim McIlwain, Ron Oppenheim, Scott Pomeroy, Rodney Radtke, Louis Reichardt, Steve Roper, John Rubenstein, David Rubin, Josh Sanes, Cliff Saper, Lynn Selemon, Carla Shatz, Larry Squire, John Staddon, Peter Strick, Warren Strittmatter, Joe Takahashi, and Christina Williams. It is understood, of course, that any remaining errors are in no way attributable to our critics. We also thank the first-year students at Duke University Medical School, who from 1991 through 1996 read and criticized earlier drafts of this text that were used in introducing them to neuroscience. Finally, we owe special thanks to Polly Garner, Dana Hall, Robert Reynolds, and Cathy Wooton, who labored long and hard to put the manuscript together, and to Andy Sinauer and John Woolsey and their staffs for turning the manuscript into such a lovely book.

CHAPTER **1**

THE ORGANIZATION OF THE NERVOUS SYSTEM

■ OVERVIEW

Perhaps the major reason that neuroscience is such an exciting field is the wealth of fundamental questions about the human brain (and the rest of the nervous system) that remain unanswered. Understanding the many functions of this remarkable organ entails unraveling the interconnections of large numbers of nerve cells that are organized into myriad systems and subsystems. Adding to the challenge is the fact that a specialized vocabulary has arisen to describe the structures of the nervous system. Despite these conceptual and semantic difficulties, comprehending the brain and the rest of the nervous system is feasible—and greatly facilitated—if some basic facts about neural organization and the relevant terminology are understood at the outset.

■ THE NERVOUS SYSTEM IS MADE UP OF CELLS

The fact that cells are the basic elements of living organisms was recognized early in the nineteenth century. It was not until well into the twentieth century, however, that neuroscientists agreed that nervous tissue, like all other organs, is made up of these fundamental units. The extraordinary shapes of individual nerve cells and the great extent of some of their branches obscured their resemblance to the cells of other tissues, leading some biologists to conclude that each nerve cell was connected to its neighbors by a reticulum of protoplasmic links. This "reticular theory" of nerve cell communication eventually fell from favor, primarily as a result of the work of the Spanish neuroanatomist Santiago Ramón y Cajal. Based on light microscopic examination of nervous tissue stained with silver salts (see Figure 1.2), Cajal argued persuasively that nerve cells are discrete entities, and that they communicate with one another by means of specialized contacts called synapses. The advent of biological electron microscopy in the 1950s finally established beyond any doubt that nerve cells are indeed functionally independent units.

The nineteenth-century histological studies of Cajal, Camillo Golgi, and a host of successors led to the consensus that the cells of the nervous system can be divided into two broad categories: **nerve cells** (or **neurons**) and a variety of **supporting cells**. Nerve cells are specialized for electrical signaling over long distances, and understanding this process represents one of the more dramatic success stories in modern biology (the subject of Unit 1). Supporting cells, in contrast, are not capable of electrical signaling. In the central nervous system (the brain and spinal cord), these supporting cells consist mostly of **neuroglial cells**. Although the cells of the human nervous system are in many ways similar to those of other organs, they are unusual in their extraordinary numbers (the human brain is estimated to contain 100 billion neurons and several times as many supporting cells), their rich functional diversity, and the ability of nerve cells to form the intricate ensembles, or circuits, on which sensation, perception, and behavior ultimately depend.

■ NERVE CELLS

In most respects, the structure of neurons resembles that of other cells. Thus, each nerve cell has a **cell body** containing a **nucleus**, **endoplasmic reticulum**, **ribosomes**, **Golgi apparatus**, **mitochondria**, and other organelles that are essential to the function of all cells (Figure 1.1). Nonetheless, nerve cells are highly specialized for intercellular communication, a fact reflected in their morphology, in the molecular specialization of their membranes, and in the functional intricacies of the synaptic contacts between them.

The specialization of neurons for signaling is most apparent in their bizarre and fascinating geometries. The most salient morphological feature is the elaboration of the **dendrites** (also called dendritic branches or dendritic processes) that arise from the neuronal cell body. Most neurons have multiple dendrites, which are typically short and highly branched. The dendrites (together with the cell body) provide sites for the synaptic contacts made by the terminals of other nerve cells and can thus be regarded as specialized for receiving information.

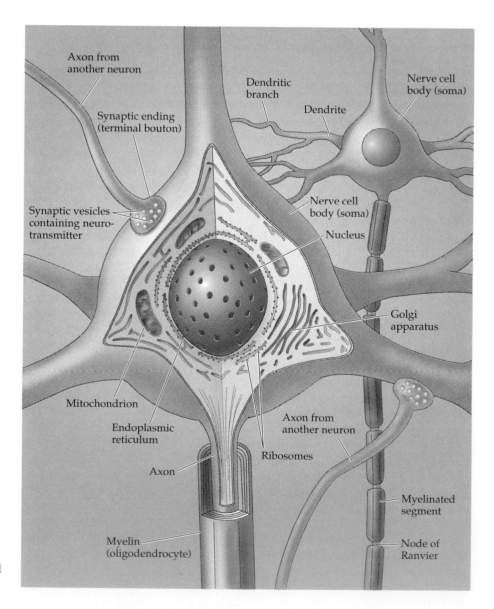

Figure 1.1 Diagram of a prototypical nerve cell and its component parts.

The spectrum of neuronal geometries ranges from a small minority of cells that lack dendrites altogether to neurons with dendritic arborizations that rival the complexity of a mature tree (Figure 1.2). The number of inputs that a particular neuron receives depends on the complexity of its dendritic arbor: nerve cells that lack dendrites are innervated by just one or a few other nerve cells, whereas those with increasingly elaborate dendrites are innervated by a commensurately larger number of other neurons. Since a fundamental purpose of nerve cells is to integrate information from other neurons, the number of inputs received by each nerve cell (which in the human nervous system ranges from 1 to about 100,000) is an especially important determinant of neuronal function.

The information from the inputs that impinge on the dendrites is "read out" at the origin of the **axon**, the portion of the nerve cell specialized for signal conduction (see Figure 1.1). The axon is a unique extension from the neuronal cell body that may travel a few hundred micrometers or much farther, depending on the type of neuron and the size of the species. The majority of

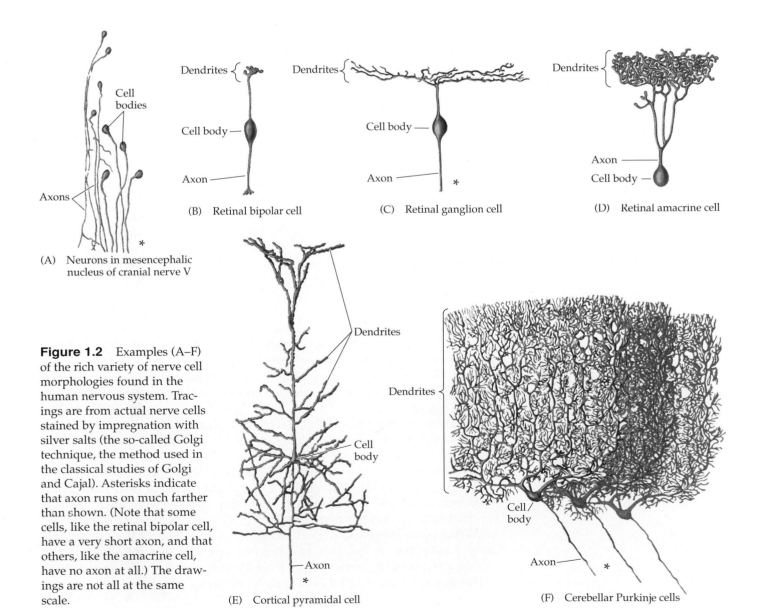

Figure 1.2 Examples (A–F) of the rich variety of nerve cell morphologies found in the human nervous system. Tracings are from actual nerve cells stained by impregnation with silver salts (the so-called Golgi technique, the method used in the classical studies of Golgi and Cajal). Asterisks indicate that axon runs on much farther than shown. (Note that some cells, like the retinal bipolar cell, have a very short axon, and that others, like the amacrine cell, have no axon at all.) The drawings are not all at the same scale.

(A) Neurons in mesencephalic nucleus of cranial nerve V

Cell bodies

Axons

(B) Retinal bipolar cell

Dendrites
Cell body
Axon

(C) Retinal ganglion cell

Dendrites
Cell body
Axon

(D) Retinal amacrine cell

Dendrites
Axon
Cell body

(E) Cortical pyramidal cell

Dendrites
Cell body
Axon

(F) Cerebellar Purkinje cells

Dendrites
Cell/body
Axon

nerve cells in the human brain have axons only a few millimeters long, and a few have no axons at all (see, for example, the retinal amacrine cell in Figure 1.2D; *amacrine* means lacking a long process). Such neurons transmit information locally. In contrast, the axons that run from the human spinal cord to the foot are about a meter long. The axonal mechanism that carries signals over such distances is called the **action potential,** a self-regenerating electrical wave that propagates from its point of initiation at the cell body (called the axon hillock) to the terminus of the axon.

The information encoded by action potentials is passed on to the next cell in the pathway by means of **synaptic transmission**. Accordingly, axon terminals are highly specialized to convey this information to target cells—which include other neurons in the brain, spinal cord, autonomic ganglia, and muscles and glands throughout the body. These terminal specializations are called **synaptic endings** (or **terminal boutons**), and the contacts they make with the target cells are called **chemical synapses** (see Figure 1.1; another type of contact called electrical synapses is described in Chapter 5). A single neuron can receive many thousands of synaptic endings, and can contact as many as a thousand other cells. Each synaptic ending contains secretory organelles called **synaptic vesicles**, as well as membrane specializations that promote vesicle fusion and exocytosis. The release of **neurotransmitters** from synaptic vesicles modifies the electrical properties of the target cell, thus generating a signal in the contacted (postsynaptic) cell. The postsynaptic cells are activated by virtue of **neurotransmitter receptors** on their surfaces, which specifically bind and react to the neurotransmitters released by the presynaptic cells.

■ NEUROGLIAL CELLS

Neuroglial cells—usually referred to simply as **glial cells** or **glia**—are quite different from nerve cells. The major distinction is that glia do not participate directly in electrical signaling, although some of their supportive functions help maintain the signaling abilities of neurons. Glia are more numerous than nerve cells in the brain, outnumbering them by a ratio of perhaps 3 to 1. Although glial cells can also be complex in form, they are generally smaller than neurons, and lack axons and dendrites. The term *glia* (from the Greek word meaning "glue") reflects the nineteenth-century presumption that these cells held the nervous system together in some way. The word has survived, despite the lack of any evidence that binding nerve cells together is among the many functions of glial cells. Glial roles that *are* reasonably well established include maintaining the ionic milieu of nerve cells, modulating the rate of nerve signal propagation, modulating synaptic action by controlling the uptake of neurotransmitters, and aiding in recovery from neural injury.

There are three types of glial cells in the central nervous system: astrocytes, oligodendrocytes, and microglial cells (Figure 1.3). **Astrocytes**, which are restricted to the brain and spinal cord, have elaborate local processes that give these cells a starlike appearance. The major function of astrocytes is to maintain, in a variety of ways, an appropriate chemical environment for neuronal signaling (see also Box D). **Oligodendrocytes**, which are also restricted to the central nervous system, lay down a laminated wrapping called **myelin** around some, but not all, axons (see Figure 1.1). Myelin has important effects on the speed of action potential conduction (see Chapter 4). In the peripheral nervous system, the cells that elaborate myelin are called **Schwann cells**. **Microglial cells**, smaller cells derived from hematopoietic

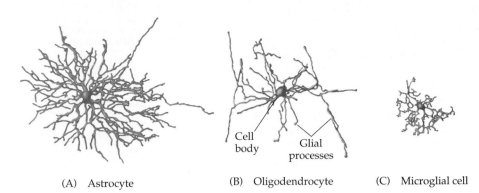

Figure 1.3 Neuroglial cells. Tracings of an astrocyte, an oligodendrocyte, and a microglial cell. Like neurons, glia can be visualized by impregnation with silver salts. The tracings are at approximately the same scale. (After Jones and Cowan, 1983.)

(A) Astrocyte   (B) Oligodendrocyte   (C) Microglial cell

Cell body   Glial processes

stem cells, share many properties with tissue macrophages. Indeed, some neurobiologists prefer to categorize microglia as a type of macrophage. Whatever the correct designation, microglia proliferate following injury to the nervous system and presumably help repair neural damage.

It should be apparent from this brief account that the various glial cells are grouped together for primarily historical reasons. In fact, glia comprise diverse cell types with entirely different functions; their common denominator is simply that they do not conduct neural signals. Moreover, the list of glial functions given here is generally thought to be incomplete. For example, many neurobiologists suspect that glial cells participate in neuronal remodeling (see Unit 4).

■ NEURAL CIRCUITS AND SYSTEMS

Neurons do not function in isolation; they are organized into ensembles called **circuits** that process specific kinds of information. Although the arrangement of neural circuitry is extremely varied, some features are characteristic of all such ensembles. Neural connections are typically made in a dense tangle of axons terminals, dendrites, synapses, and glial cell processes that are together called the **neuropil** (the suffix *pil* comes from the Greek word *pilos*, meaning "felt") (Figure 1.4). Nerve cells that carry information toward the central nervous system (or further centrally within the spinal cord and brain) are called **afferent neurons**; nerve cells that carry information away from the brain or spinal cord (or away from the circuit in question) are called **efferent neurons**. Nerve cells that only participate in the local aspects of a circuit are called **interneurons**. These three classes—afferent neurons, efferent neurons, and interneurons—are the constituents of all neural circuits. The simple spinal reflex circuit in Figure 1.5 illustrates this terminology.

Processing circuits are combined in turn into **systems** that serve broader functions (the visual system or the auditory system, for example). A quite general functional distinction in the nervous system is between the sensory and motor systems. The sensory component of the nervous system includes all those cells, circuits, and subsystems that provide information

Figure 1.4 The business of neural circuits is carried out in a dense matrix of axons, dendrites, and their connections that occupy the space between the nerve cell bodies. This complex is called the neuropil, shown here in an electron micrograph of the cerebral cortex. (From Peters et al., 1970.)

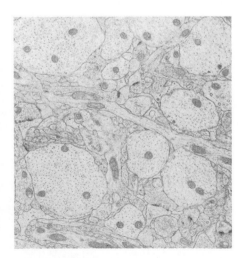

Figure 1.5 A simple reflex circuit, the knee-jerk response (more formally, the myotatic reflex), illustrates several points about the functional organization of neural circuits. Stimulation of peripheral sensors (a muscle stretch receptor in this case) initiates action potentials that travel centrally along the afferent axons of the sensory neurons. This information stimulates spinal motor neurons by means of synaptic contacts. The action potentials generated in motor neurons travel peripherally in efferent axons, giving rise to muscle contraction and an observable behavioral response. One of the purposes of this particular reflex is to help maintain an upright posture in the face of unexpected changes.

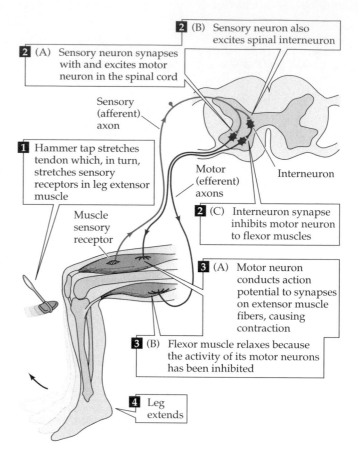

2 (A) Sensory neuron synapses with and excites motor neuron in the spinal cord

2 (B) Sensory neuron also excites spinal interneuron

Sensory (afferent) axon

1 Hammer tap stretches tendon which, in turn, stretches sensory receptors in leg extensor muscle

Motor (efferent) axons

Interneuron

Muscle sensory receptor

2 (C) Interneuron synapse inhibits motor neuron to flexor muscles

3 (A) Motor neuron conducts action potential to synapses on extensor muscle fibers, causing contraction

3 (B) Flexor muscle relaxes because the activity of its motor neurons has been inhibited

4 Leg extends

about the environment, both external and internal. The motor system consists of all those elements that respond to such information by generating movement.

■ THE BASIC ANATOMICAL SUBDIVISIONS OF THE NERVOUS SYSTEM

Neuroscientists and physicians have traditionally divided the nervous system structurally into central and peripheral components (Figure 1.6). The **central nervous system** comprises the brain (cerebrum, cerebellum, and brainstem) and the spinal cord. The **peripheral nervous system** includes sensory neurons, which connect the brain and spinal cord to sensory receptors, as well as motor neurons, which connect brain and spinal cord to muscles and glands. The elements of the peripheral nervous system devoted to motor function are further categorized into somatic and autonomic divisions. The **somatic division** innervates skeletal muscles; the **autonomic division** (Box A) innervates smooth muscles, cardiac muscle, and glands.

The peripheral nervous system has two major anatomical components: **ganglia**, which are accumulations of nerve cell bodies and supporting cells, and **nerves**, which are bundles of nerve cell axons and their supporting cells. The sensory axons and sensory ganglia of the peripheral nervous system gather information about events at the surface of the body (as well as within it) and relay this information to the central nervous system. The **sensory ganglia** lie adjacent to either the spinal cord (in which case they are referred

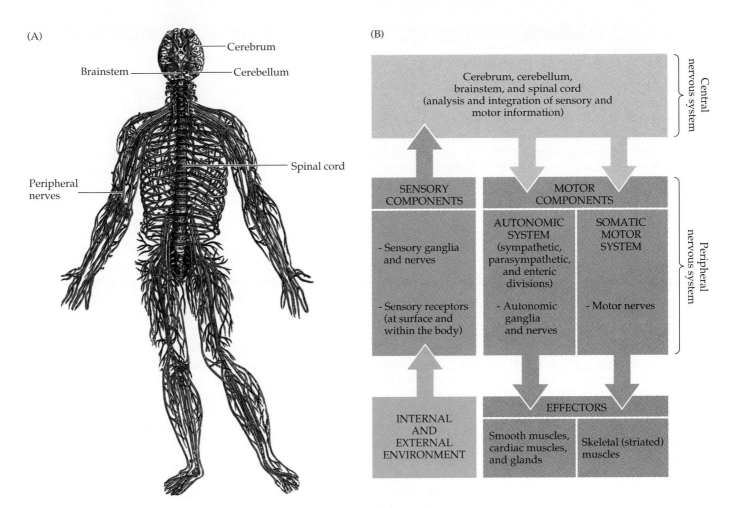

Figure 1.6 The major components of the nervous system and their functional relationships. (A) Overall structure of the nervous system, showing its central and peripheral components. The brain is turned upward to show the brainstem and cerebellum; notice that the spinal cord is hidden by the vertebrae. (B) Diagram of the major components of the central and peripheral nervous systems and their functional relationships. Stimuli from the environment convey information to processing circuits within the brain and spinal cord, which in turn interpret their significance and send signals to peripheral effectors that move the body or adjust the workings of its internal organs.

to as **dorsal root ganglia**; see Figure 1.13) or the brainstem (in which case they are called **cranial nerve ganglia**). The nerve cells in sensory ganglia send axons to the periphery that terminate in (or contact) specialized receptors that transduce information about a wide variety of stimuli. The central processes of these sensory cells enter the spinal cord or brainstem. The cells and circuits of the spinal cord and brainstem in turn relay the information received from the periphery to various regions of the brain. The motor axons of the peripheral nervous system connect motor neurons in the brainstem and spinal cord to the muscles and viscera of the body. Somatic motor neurons innervate striated muscles and thus control all skeletal movements and consequently most overt, voluntary behaviors. Autonomic motor neurons in the brainstem and spinal cord form synapses with motor neurons that lie in the **autonomic ganglia** (see Box A). The neurons of the autonomic ganglia

Box A
THE AUTONOMIC NERVOUS SYSTEM

The autonomic nervous system controls involuntary (visceral) functions and has three divisions. The sympathetic and parasympathetic divisions consist of two-neuron chains that connect the central nervous system with the smooth muscles and glands of the viscera, blood vessels, and skin. The enteric division is a largely independent system that lies in the walls of the gastrointestinal tract and controls many digestive functions. The sympathetic system organizes the involuntary responses that anticipate maximal exertion (in the extreme, the so-called "fight-or-flight" reaction). Conversely, the parasympathetic system organizes the involuntary responses that generally reflect visceral function in a state of relaxation.

Sympathetic and parasympathetic ganglia are innervated by preganglionic neurons in the spinal cord. Sympathetic preganglionic axons arise from neurons in the thoracic and upper lumbar spinal cord. The preganglionic neurons that innervate the head and thoracic organs are in the upper and middle thoracic segments, and those that innervate the abdominal and pelvic organs are in the lower thoracic and upper lumbar segments. The parasympathetic preganglionic axons arise from neurons in the brainstem and sacral spinal cord. Many organs—including the salivary glands, heart, bladder, and sex organs—receive inputs from both the sympathetic and parasympathetic systems. Other targets receive only sympathetic innervation.

These include the sweat glands, the adrenal medulla, the piloerector muscles of the skin, and most blood vessels. The neurons innervated by the preganglionic sympathetic axons are for the most part found in the sympathetic chain ganglia, whereas the parasympathetic motor neurons are located in ganglia within the organs they control. (The term *ganglion* simply means a cluster of nerve cells, along the course of a peripheral nerve.)

The enteric nervous system, although it receives sympathetic and parasympathetic innervation, acts to some degree independently of the rest of the autonomic system. A rich intrinsic circuitry of sensory neurons, interneurons, and motor neurons interconnects different levels of the gut and coordinates activity along its length. Indeed, the enteric system is said to contain more neurons than the entire spinal cord! Abetted by sympathetic and parasympathetic influences, the enteric system governs gut motility, secretion, and the transfer of substances across the gut epithelium.

Sensory inputs from the viscera modulate autonomic activity. Like other primary sensory neurons, the relevant cell bodies lie in dorsal root and cranial nerve ganglia; the visceral sensory axons that enter the spinal cord terminate mainly in the intermediate gray matter, near the preganglionic neurons of the thoracolumnar and sacral cord. Those that enter the brainstem in cranial nerves VII, IX, and X terminate in the nucleus

of the solitary tract, which participates in many important autonomic reflexes. Sensory fibers that travel in the sympathetic nerves convey visceral sensations, usually pain. Other fibers, including most of those that travel in the parasympathetic nerves to the nucleus of the solitary tract, convey information that does not reach consciousness, but which is nonetheless important for integration of autonomic reflexes. Examples include the axons innervating arterial baroreceptors and chemoreceptors. In addition to mediating the function of the body's glands and visceral muscles, the autonomic nervous system has provided researchers with a set of relatively accessible pathways and peripheral preparations that have greatly stimulated neurobiological research for more than a century.

References

JANSEN, A. S. P., X. V. NGUYEN, V. KARPITSKIY, T. C. METTENLEITER AND A. D. LOEWY (1995) Central command neurons of the sympathetic nervous system: basis of the fight or flight response. Science 270: 644–646.

LOEWY, A. D. AND K. M. SPYER (eds.). (1990) *Central Regulation of Autonomic Functions.* New York: Oxford.

PICK, J. (1970) *The Autonomic Nervous System: Morphological, Comparative, Clinical and Surgical Aspects.* Philadelphia: J.B. Lippincott Company.

The autonomic nervous system. (The enteric division is not shown.) ▶

innervate smooth muscle, glands, and cardiac muscle, thus controlling most involuntary behavior.

The central nervous system is usually considered to include seven basic parts: the **spinal cord**, the **medulla**, the **pons**, the **cerebellum**, the **midbrain**, the **diencephalon**, and the **cerebral hemispheres** (Figure 1.7). The medulla, pons, and midbrain are collectively called the **brainstem**; the diencephalon

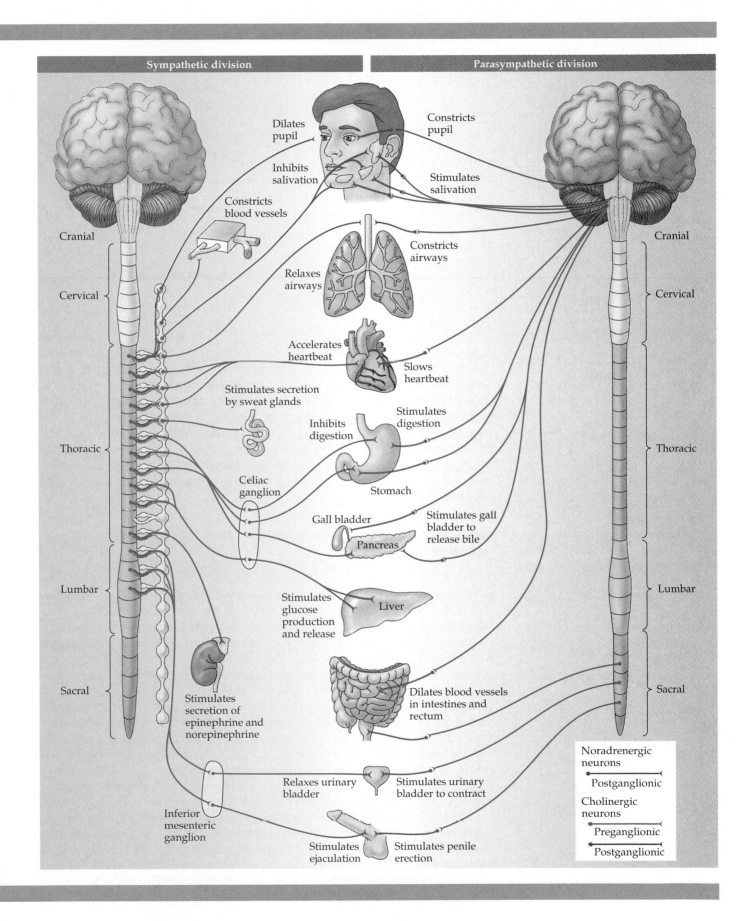

Figure 1.7 A midsagittal view of the adult brain showing the major divisions of the central nervous system.

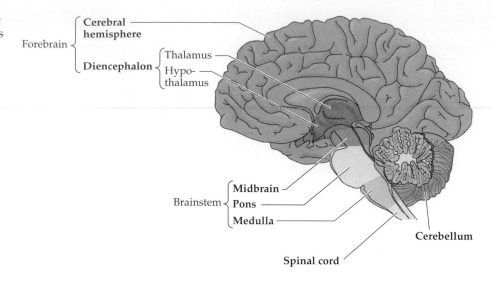

and cerebral hemispheres are collectively called the **forebrain**. These divisions of the adult brain are related to embryonic subdivisions that are apparent at the earliest stages of neural development; indeed, some knowledge of how the brain is formed helps greatly in understanding its adult structure (and nomenclature) (see Chapter 20). Although this categorization of the central nervous system is somewhat artificial (neurons often span the boundaries of these subdivisions, and most neural functions depend on more than one of these components), the division of the central nervous system into its major parts provides an anatomical framework essential to any discussion of function.

■ THE EXTERNAL ANATOMY OF THE CEREBRAL HEMISPHERES

When the human brain is viewed from the side (Figure 1.8A), three major structures are visible: the brainstem, the cerebellum, and the cerebral hemispheres. The latter are so large that they hide the rest of the brain's subdivisions from view. In addition to their large size (the cerebral hemispheres represent 85% of the brain by weight), an obvious feature of the hemispheres is their highly convoluted surface. The ridges are known as **gyri** (singular: gyrus), and the valleys are called **sulci** (singular: sulcus) or, if they are especially deep, **fissures**. The entire convoluted surface of the hemispheres comprises a laminated rind of neurons and supporting cells about 2 mm thick called the **cerebral cortex**. The reasons for cerebral sulcation are not entirely clear, but the infolding of the brain obviously allows a great deal more cortical surface area (2.2 m^2 on average) to exist within the confines of the cranium. A sulcus or fissure often corresponds to a boundary between two functionally distinct areas; thus, the mechanism of sulcation probably involves the differential growth of distinct cortical regions.

Each hemisphere is conventionally divided into four **lobes** (Figure 1.8B) named for the bones of the skull that overlie them, the **frontal, parietal, temporal**, and **occipital lobes**. The frontal lobe is the most anterior and is separated from the parietal lobe by the **central sulcus**. A particularly important feature of the frontal lobe is the **precentral gyrus**. (The prefix *pre*, when

(A)

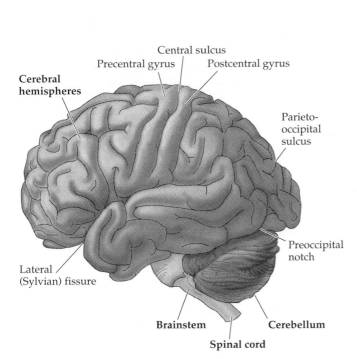

(B)

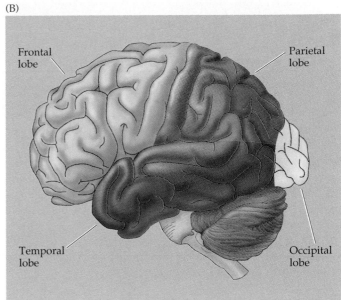

(C)

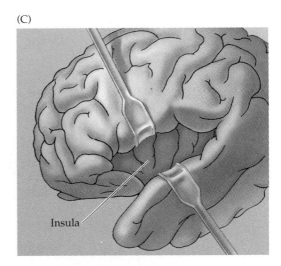

Figure 1.8 Lateral view of the human brain. (A) Some of the major sulci and gyri. (B) The four lobes of the brain. (C) The banks of the lateral, or Sylvian, fissure have been pulled apart to expose the insula.

used anatomically, refers to something that is in front of or anterior to something else; Box B reviews other important anatomical terms) The cortex of the precentral gyrus is referred to as the **motor cortex** because it contains neurons whose axons project to the motor neurons in the brainstem and spinal cord that innervate the skeletal (striated) muscles.

The temporal lobe extends almost as far anteriorly as the frontal lobe but is located inferior to it, the two lobes being separated by the **lateral (or Sylvian) fissure**. The superior aspect of the temporal lobe contains cortex concerned with audition. The parietal lobe lies posterior to the central sulcus and superior to the lateral fissure. The **postcentral gyrus**, the most anterior gyrus in the parietal lobe, harbors cortex that is concerned with somatic (bodily) sensation; this area is therefore referred to as the **somatic sensory**

Box B
SOME ANATOMICAL TERMINOLOGY

The terms used to specify location in the central nervous system are the same ones used for gross anatomical description of the rest of the body. Nonetheless, because some anatomical terms refer to the long axis of the body, which is straight, whereas others refer to the long axis of the central nervous system, which has a bend in it, the descriptive anatomy of the nervous system can be confusing. A consideration of Figures A and B should help clarify the standard terms for location.

Another set of important terms concerns the planes that allow discussion of a histological section or a tomographic image (Figure C; see Boxes C and E for examples). Sections taken in the plane dividing the two hemispheres are called sagittal, and can be further categorized as median and paramedian according to whether the section is near the midline (median or midsagittal) or more lateral (paramedian). The plane of the image in Figure 1.10, for instance, is in the median sagittal plane. Histological

sections or tomograms taken in the plane of the face are called frontal or coronal. Horizontal sections refer to the anterior/posterior plane that passes through both ears. Different terms are usually used to refer to sections of the spinal cord. The plane of section orthogonal to the length of the cord is called transverse, whereas sections in the axis of the cord are called longitudinal.

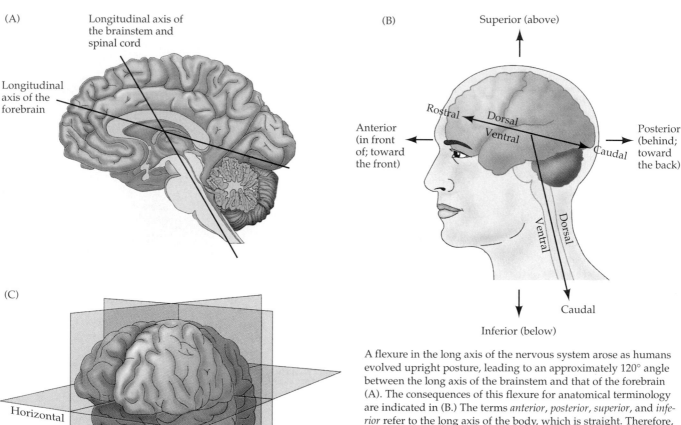

(A) Longitudinal axis of the brainstem and spinal cord
Longitudinal axis of the forebrain

(B) Superior (above)
Rostral
Dorsal
Ventral
Anterior (in front of; toward the front)
Posterior (behind; toward the back)
Caudal
Ventral
Dorsal
Caudal
Inferior (below)

(C) Horizontal
Coronal
Sagittal

A flexure in the long axis of the nervous system arose as humans evolved upright posture, leading to an approximately 120° angle between the long axis of the brainstem and that of the forebrain (A). The consequences of this flexure for anatomical terminology are indicated in (B.) The terms *anterior*, *posterior*, *superior*, and *inferior* refer to the long axis of the body, which is straight. Therefore, these terms indicate the same direction for both the forebrain and the brainstem. In contrast, the terms *dorsal*, *ventral*, *rostral*, and *caudal* refer to the long axis of the central nervous system. The dorsal direction is toward the back for the brainstem and spinal cord, but toward the top of the head for the forebrain. The opposite direction is ventral. The rostral direction is toward the top of the head for the brainstem and spinal cord, but toward the face for the forebrain. The opposite direction is caudal. (C) The major planes of section used in cutting or imaging the brain.

cortex. The boundary between the parietal lobe and the occipital lobe, the most posterior of the hemispheric lobes, is somewhat arbitrary (a line from the parieto-occipital sulcus to the preoccipital notch). The occipital lobe, only a small part of which is apparent from the lateral surface of the brain, is concerned with vision. In addition to their role in primary and sensory processing, each lobe of the cerebral hemispheres has characteristic cognitive functions (Chapter 24). Thus the frontal lobe is critical in planning behavior, the parietal lobe in attending to important stimuli, the temporal lobe in recognizing objects and faces, and the occipital lobe in a variety of visual analyses.

Part of the lateral surface of the hemisphere, the **insula**, is hidden beneath the frontal and temporal lobes, and can be seen only if these two lobes are pulled apart or removed (see Figure 1.8C). Despite its name, the insular cortex is not an island; it is buried simply because of the relatively greater growth of the rest of the hemisphere around it. The insular cortex is largely concerned with visceral and autonomic function, including taste.

Other important external features of the brain can only be seen from its ventral surface (Figure 1.9). Extending along the inferior surface of the frontal lobe near the midline are the **olfactory tracts**, which arise from enlargements at their anterior ends called the **olfactory bulbs**. The olfactory bulbs receive input from the rootlets that make up the first **cranial nerve** (cranial nerve I is therefore called the olfactory nerve). Many of the other cranial nerves (II–XII), which connect the brain to sensory organs and muscles of the head, are also visible in this view. The major functions of the cranial nerves are summarized in Table 1.1. On the inferior surface of the temporal lobe, the parahippocampal gyrus conceals the **hippocampus** (see Figure 1.14), a highly specialized cortical structure that is folded into the medial temporal lobe; the hippocampus figures importantly in memory (see Chapters 23 and 29).

When the brain is divided in the midsagittal plane (see Box B), all of its major subdivisions are visible on the cut surface (Figures 1.7 and 1.10). In this view, the cerebral hemispheres, because of their great size, are still the most obvious structures. The frontal lobe of each hemisphere extends forward from the central sulcus, the medial end of which can just be seen. The **parieto-occipital sulcus**, running from the superior to the inferior aspect of the hemisphere, separates the parietal and occipital lobes. The **calcarine sulcus** divides the medial surface of the occipital lobe, running at very nearly a right angle from the parieto-occipital sulcus and marking the location of the primary visual cortex. A long, roughly horizontal sulcus, the **cingulate sulcus**, extends across the medial surface of the frontal and parietal lobes. The prominent gyrus below it, the **cingulate gyrus**, along with the cortex adjacent to it, is known as the **limbic lobe** (although the use of the term "lobe" here is less precise than for the four lobes of the hemisphere described above). The limbic lobe (*limbic* means border or edge), which wraps around the corpus callosum (see Figure 1.14), and the subcortical areas connected to it, are referred to as the **limbic system**. These limbic structures are important in the regulation of visceral motor activity and emotional expression, among other functions (see Chapter 27).

■ THE EXTERNAL ANATOMY OF THE DIENCEPHALON AND BRAINSTEM

The overall structure of the diencephalon and brainstem are perhaps best appreciated in midsagittal view (see Figure 1.10). From this perspective, the diencephalon can be seen to consist of three parts arrayed dorsoventrally. The **dorsal thalamus**, the largest component of the diencephalon, comprises a

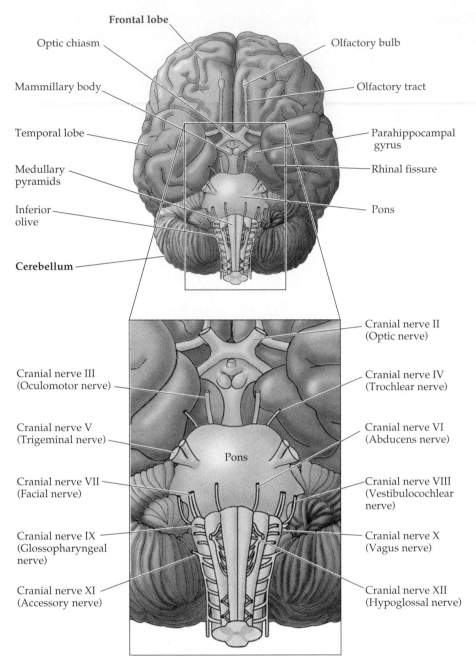

Figure 1.9 Ventral view of the human brain, indicating the major features visible from this perspective.

number of subdivisions, all of which relay information to the cerebral cortex from other parts of the brain. The underlying **subthalamus** is concerned with control of motor functions. The **hypothalamus**, a small but especially crucial part of the diencephalon, is devoted to the control of homeostatic and reproductive functions; recall that the hypothalamus is intimately related, both structurally and functionally, to the pituitary gland, a critical endocrine organ whose posterior part is attached to the hypothalamus by the pituitary stalk (or infundibulum). The midbrain lies caudal to the thalamus, with the supe-

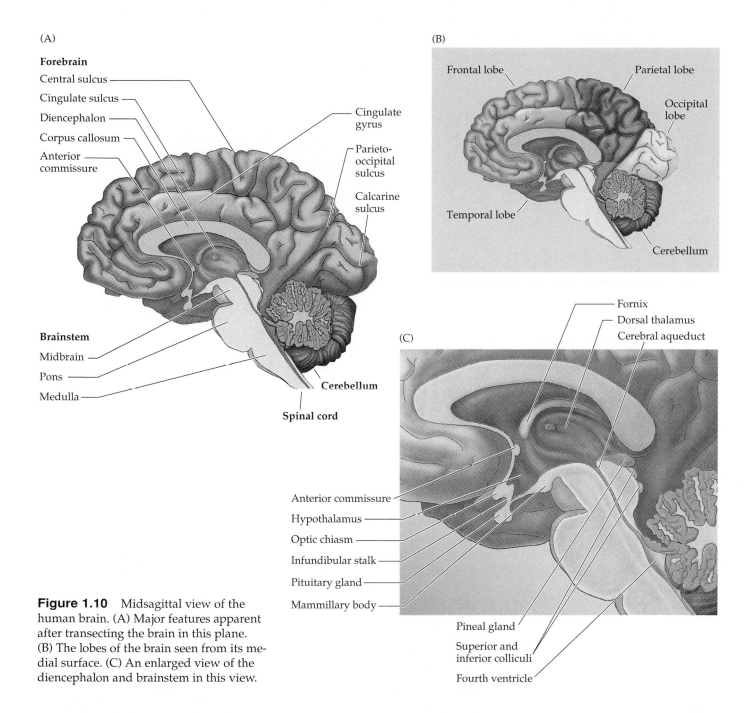

Figure 1.10 Midsagittal view of the human brain. (A) Major features apparent after transecting the brain in this plane. (B) The lobes of the brain seen from its medial surface. (C) An enlarged view of the diencephalon and brainstem in this view.

rior and inferior colliculi on the dorsal surface of the midbrain forming its **tectum** (meaning "roof"; see also Figure 1.12). The other prominent external features of the midbrain—the cerebral peduncles—are not seen in the midsagittal view because they do not reach the midline. The pons is caudal to the midbrain, with the cerebellum lying over the pons just beneath the cerebral hemispheres. The major function of the cerebellum is coordination of motor activity, posture, and equilibrium. Like the cerebrum, the cerebellar surface is covered by a thin cortex that is thrown into folds (in this case called **folia**). Caudal to the pons and cerebellum is the medulla, which merges into the spinal cord.

Figure 1.11 The ventral surface of the brainstem. The functions of the cranial nerves indicated here are summarized in Table 1.1.

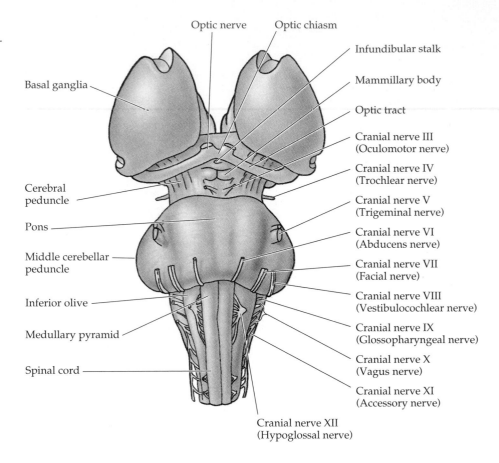

Other aspects of the diencephalon and brainstem are best seen when the hemispheres are removed altogether. The component of the diencephalon that can be seen from the surface in a ventral view (Figure 1.11) is the hypothalamus, which is bounded rostrally by the **optic chiasm** (formed by the crossing of axons in cranial nerve II, the optic nerve) and includes the **mammillary bodies** caudally. The **infundibular stalk**, which is located between these structures, connects the hypothalamus to the **pituitary gland**. Clinically, the proximity of the optic chiasm and the pituitary gland is important because pituitary tumors often give rise to a unique combination of visual and endocrine signs and symptoms.

The midbrain lies caudal to the diencephalon, being the most rostral of the three components of the brainstem. The ventral surface of the midbrain is characterized by two large fiber bundles, the **cerebral peduncles**, which contain the axons traveling between the cerebral cortex and the brainstem and spinal cord. Cranial nerve III (the oculomotor nerve) arises from the midbrain between the two cerebral peduncles (a region called the interpeduncular fossa). The other cranial nerve that arises from the midbrain, cranial nerve IV (the trochlear nerve), is the only cranial nerve to exit the brain dorsally.

The pons, the part of the brainstem that lies caudal to the midbrain, is evident in ventral view as a massive enlargement on the surface of the brainstem. The pons is so named because it is made up of neurons with transversely oriented axons that run across the base of the brainstem (*pons* means "bridge"). The pontine axons extend into the middle cerebellar peduncles, one of three sets of fiber bundles that attach the pons to the cerebellum (see

TABLE 1.1
The Cranial Nerves and Their Primary Functions

Cranial nerve	Name	Major function
I	Olfactory nerve	Sense of smell
II	Optic nerve	Vision
III	Oculomotor nerve	Eye movements; pupillary constriction and accommodation; muscles of eyelid
IV	Trochlear nerve	Eye movements
V	Trigeminal nerve	Somatic sensation from face, mouth, cornea; muscles of mastication
VI	Abducens nerve	Eye movements
VII	Facial nerve	Controls the muscles of facial expression; taste from anterior tongue; lacrimal and salivary glands
VIII	Auditory/vestibular nerve	Hearing; sense of balance
IX	Glossopharyngeal nerve	Sensation from pharynx; taste from posterior tongue; carotid baroceptors
X	Vagus nerve	Autonomic functions of gut; sensation from pharynx; muscles of vocal cords; swallowing
XI	Accessory nerve	Shoulder and neck muscles
XII	Hypoglossal nerve	Movements of tongue

Figures 1.9, 1.11, and 1.12). Cranial nerve V (the trigeminal nerve), which is easily recognized because of its large size, is the only cranial nerve that emerges from the pons.

Caudal to the pons is the third component of the brainstem, the medulla. In ventral view the medulla is characterized by two longitudinal prominences near the midline, the **medullary pyramids**. The pyramids contain axons that arise in the precentral gyrus (the motor cortex) and project to the spinal cord. (More rostrally, these same axons are part of the cerebral peduncles that lie on the ventral surface of the midbrain.) Lateral to the pyramids are the **inferior olives**; these protrusions from the brainstem are also involved in motor control. Cranial nerves VI–X and XII arise from the medulla or from the junction of the pons and medulla (cranial nerve XI arises from the spinal cord).

The dorsal surface of the diencephalon and brainstem is hidden in the intact brain by the cerebellum and posterior portions of the cerebral hemispheres. When these structures are removed, several important features can be seen (Figure 1.12). The dorsal surface of the midbrain is characterized by four "hills," or **colliculi**. The neurons of the **superior colliculus** are important in the control of eye movements; those of the **inferior colliculus** process auditory information. Cranial nerve IV emerges from the brainstem just caudal to the inferior colliculus. On the dorsal surface of the pons, where the cerebellum has been cut away, the three large fiber bundles, or peduncles that attach the cerebellum to the rest of the brain are evident. The **middle cerebellar peduncle**, which is the largest of these fiber bundles, lies between the **superior** and **inferior peduncles**. As might be expected, all three peduncles contain axons that carry information to or from the cerebellum. The dorsal surface of the medulla consists of a thin layer of connective tissue derived from pial and ependymal cells and associated blood vessels. This layer covers the fourth ventricle, part of the fluid-filled ventricular system that lies at the core of the brain (see page 25).

Figure 1.12 The dorsal surface of the diencephalon and brainstem.

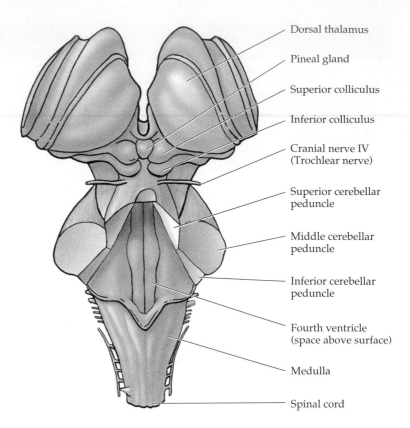

Dorsal thalamus

Pineal gland

Superior colliculus

Inferior colliculus

Cranial nerve IV (Trochlear nerve)

Superior cerebellar peduncle

Middle cerebellar peduncle

Inferior cerebellar peduncle

Fourth ventricle (space above surface)

Medulla

Spinal cord

■ THE EXTERNAL ANATOMY OF THE SPINAL CORD

The spinal cord lies in the vertebral canal and, in the adult, extends from the first cervical vertebra to about the level of the twelfth thoracic vertebra (Figure 1.13). The vertebral column and the spinal cord are divided into **cervical, thoracic, lumbar, sacral**, and **coccygeal** regions. The peripheral nerves that innervate most of the body arise from the spinal cord as 31 pairs of segmental spinal nerves (Figure 1.13A, B). Sensory information carried by the afferent axons in the peripheral nerves enters the cord via the **dorsal roots**; motor commands carried by the efferent axons leave the cord via the **ventral roots** (Figure 1.13C). Once the roots join, sensory and motor axons (with some exceptions) travel together in the segmental spinal nerves. The cervical region of the cord gives rise to eight cervical nerves (C1–C8), the thoracic to twelve thoracic nerves (T1–T12), the lumbar to five lumbar nerves (L1–L5), the sacral to five sacral nerves (S1–S5), and the coccygeal to one coccygeal nerve. The segmental nerves leave the vertebral column through the intervertebral foramina that lie just rostral to the respectively numbered vertebral body. Because the spinal cord is considerably shorter than the vertebral column, the lumbar and sacral nerves run for some distance in the vertebral canal before emerging, thus forming a collection of nerve roots known as the **cauda equina**. Important clinical procedures are the collection of cerebrospinal fluid from the space surrounding these nerves for analysis (called "lumbar puncture") and the introduction of local anesthetics to produce spinal anesthesia; at this level the risk of damage to the spinal cord from a poorly placed needle is minimized.

Two regions of the spinal cord are enlarged to accommodate the greater number of nerve cells and connections needed to process information related

(A)

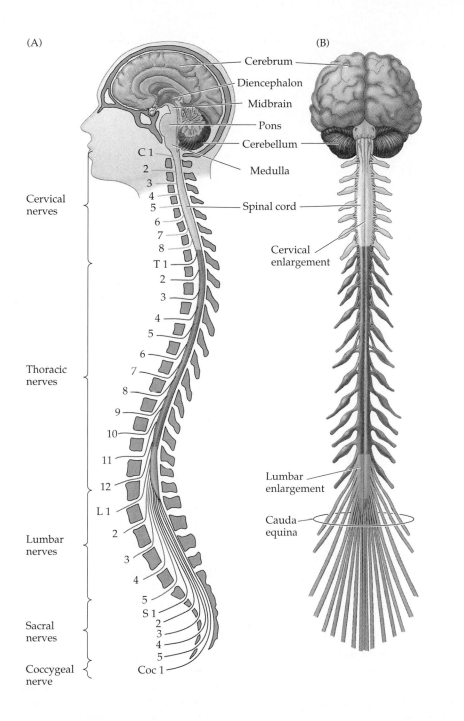

Cervical nerves

Thoracic nerves

Lumbar nerves

Sacral nerves

Coccygeal nerve

C 1
2
3
4
5
6
7
8
T 1
2
3
4
5
6
7
8
9
10
11
12
L 1
2
3
4
5
S 1
2
3
4
5
Coc 1

Cerebrum
Diencephalon
Midbrain
Pons
Cerebellum
Medulla
Spinal cord

(B)

Cervical enlargement

Lumbar enlargement

Cauda equina

(C)

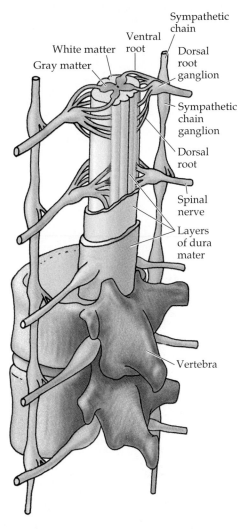

Gray matter
White matter
Ventral root
Sympathetic chain
Dorsal root ganglion
Sympathetic chain ganglion
Dorsal root
Spinal nerve
Layers of dura mater
Vertebra

Figure 1.13 The external organization of the spinal cord. (A) A lateral view indicating the position of the spinal cord segments with reference to the vertebrae. (Note that the position of the brackets on the left side of the figure refers to the vertebrae, not the spinal segments.) (B) The spinal cord in ventral view, indicating the emergence of the segmental nerves and the cervical and lumbar enlargements. (C) Diagram of several cord segments, showing the external anatomy in more detail.

to the upper and lower limbs (see Figure 1.13). The expansion at the level of the nerves to the arms is called the **cervical enlargement** and includes spinal segments C5–T1; the expansion at the level of the nerves to the legs is called the **lumbar enlargement** and includes spinal segments L2–S3.

■ THE INTERNAL ANATOMY OF THE CEREBRAL HEMISPHERES AND DIENCEPHALON

When the brain is dissected, sectioned, or observed with noninvasive imaging techniques (Box C), many deeper structures are apparent (Figures 1.14,

Figure 1.14 Major internal structures of the brain, shown after the upper half of the left hemisphere has been cut away.

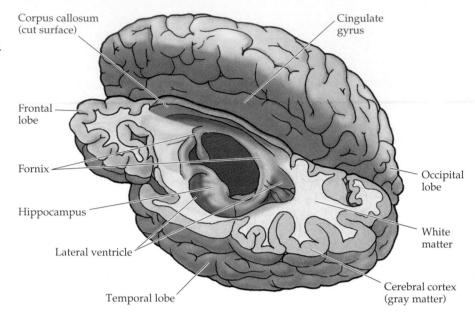

Corpus callosum (cut surface)

Cingulate gyrus

Frontal lobe

Fornix

Hippocampus

Lateral ventricle

Temporal lobe

Occipital lobe

White matter

Cerebral cortex (gray matter)

1.15, and 1.16). In sections through the forebrain, the cerebral cortex is evident as a thin layer of neural tissue that covers the entire cerebrum. Most cerebral cortex is made up of six layers, and is referred to as **neocortex**. Phylogenetically older cortex (called paleocortex) with fewer cell layers occurs on the inferior and medial aspect of the temporal lobe, being separated from neocortex by the rhinal fissure (see Figure 1.9). Cortex with even fewer layers (three), referred to as archicortex, is found in the hippocampus. The hippocampal cortex is folded into the medial aspect of the temporal lobe, and is therefore visible only in dissected brains or in sections (Figures 1.14 and 1.15).

Cortex—whether cerebral, cerebellar, or hippocampal—is made up of neuronal cell bodies, their dendrites, the terminal arborizations of axons, and glial cells. Macroscopically, this collection of cells and processes is referred to as **gray matter** (recall that microscopically the tangle between the neurons is called neuropil; see Figure 1.4). Axons entering the cortex, as well as those leaving it, form the so-called **white matter**, which makes up a large part of the subcortical tissue of the hemispheres.

A number of key structures are more deeply embedded within the hemispheres. The largest of these are the **basal ganglia** (Figure 1.15). The neurons of these large nuclei receive input from the cerebral cortex and participate in the organization and guidance of complex motor functions. (The term *ganglia* does not usually refer to clusters of neurons within the central nervous system, which are called **nuclei**; the usage here is an exception.) The basal ganglia surround the diencephalon rostrally and laterally and include three main substructures: the **caudate**, the **putamen**, and the **globus pallidus**. The caudate and putamen are structurally and functionally quite similar and are collectively called the **striatum**. In the base of the forebrain, ventral to the basal ganglia, are several clusters of nerve cells known as the **basal forebrain nuclei**. These nuclei are relatively small in humans, but are of particular interest because they are specifically affected in Alzheimer's disease. Another nucleus buried in the hemisphere is the **amygdala**, which lies in front of the hippocampus in the anterior pole of the temporal lobe. The amygdala is an important participant in the control of emotional behavior.

(A)

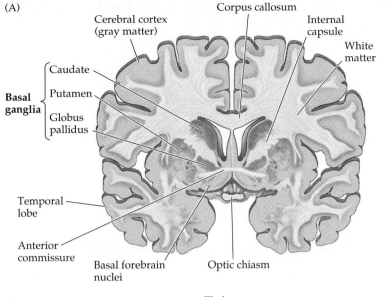

(C)

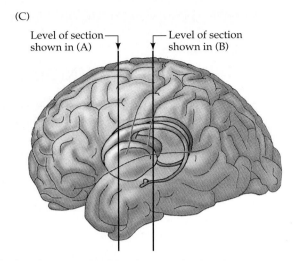

(B)

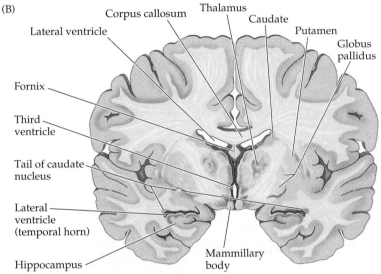

Figure 1.15 Internal structures of the brain as seen in coronal section. (A) This plane of section runs through the basal ganglia. (B) A somewhat more posterior plane of section that includes the thalamus. (C) A transparent view of the basal ganglia showing the approximate location of the sections in (A) and (B). Notice that because the caudate nucleus has a tail that arcs into the temporal lobe, it may appear twice in the same section. The same is true of several other brain structures, including the lateral ventricles.

Different parts of the two hemispheres are interconnected by three large bundles of axons—the corpus callosum, anterior commissure, and fornix (see Figures 1.14 and 1.15). The **corpus callosum** and **anterior commissure** link the cortex on the two sides of the brain, whereas the **fornix** interconnects the hippocampus and the hypothalamus within each hemisphere. Axons descending from (and ascending to) the cerebral cortex assemble into another large fiber bundle called the **internal capsule**. The internal capsule lies just lateral to the diencephalon (forming a "capsule" around it), and many of its axons arise from or terminate in the dorsal thalamus. Other axons descending from the cortex in the internal capsule continue past the diencephalon to enter the cerebral peduncles of the midbrain. Thus, the internal capsule is the major pathway linking the cerebral cortex to the rest of the brain and spinal cord. Strokes or other injury to this structure interrupt the flow of ascending and descending nerve traffic, usually with devastating consequences.

Box C
BRAIN IMAGING TECHNIQUES

Until the early 1970s, the only technique available for imaging the structure of a living brain was conventional X-ray technology. X-ray images are formed on photographic film by the differential absorption of this relatively high-energy radiation by bone and soft tissue, such that bone appears white and soft tissues dark. (The high energy explains the danger of X-radiation, which can cause cellular damage, including gene mutation.) Although X-rays do a good job of delineating the bones of the skull, they do not reveal very much about the brain. Radiologists therefore developed several tricks to make the cerebral ventricles visible; for example, if air is injected into the ventricles to displace the cerebrospinal fluid, the ventricles appear darker on X-rays. But this technique, known as pneumoencephalography, was difficult, painful, and risky and is now obsolete. Radiopaque contrast materials can also be injected into the arterial circulation to reveal the cerebral vasculature and associated pathologies in considerable detail;

however there are also risks associated with such angiograms, including allergic reactions to the injected material and complications from the injection procedure itself. Finally, conventional X-ray films provide only a two-dimensional view of brain structure or vasculature, making localization of lesions uncertain. As a result of these several problems, there was strong motivation in the 1960s and 1970s to discover better ways to image the brain.

A major advance was the development of computerized tomography (CT). CT uses a movable X-ray tube that is rotated around the patient's head. Opposite the tube (i.e., on the other side of the patient's head) are X-ray detectors far more sensitive than conventional film, thereby allowing much shorter exposure times (and less risk of radiation damage). Rather than acquiring a single image, as in conventional X-ray pictures, a CT scan gathers intensity information from many positions around the patient's head. These data are entered

into a matrix, and the radiodensity at each point in the three-dimensional space of the head is calculated. With a sufficiently narrow X-ray beam, sensitive detectors, and digital signal processing techniques, small differences in radiodensity can be converted into an image. Since the information is gathered for the full volume of the head, the computed matrix contains information about the entire brain. It is therefore possible to generate "slices," or tomograms (*tomo* means "cut" or "slice") of various planes through the brain, visualizing internal structures at any desired level. CT scans readily distinguish gray matter and white matter, differentiate the ventricles quite well, and show up many other brain structures with a spatial resolution of several millimeters.

Although computerized tomography opened a new era in brain imaging, it now competes with a technique called magnetic resonance imaging, or MRI. MRI provides strikingly detailed images of the structure of the brain with a reso-

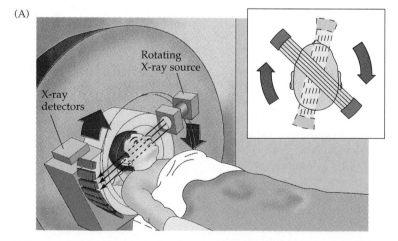

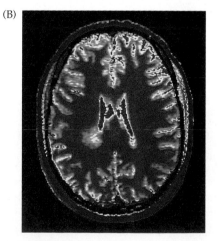

(A) In computerized tomography, the X-ray source and detectors are moved around the patient's head. This approach generates a matrix of intersecting points that have been obtained from several directions. The signal at each point can then be computed, allowing reconstruction of a "slice" through the brain that preserves three-dimensional relationships. (B) This CT scan shows a horizontal section of a normal adult brain.

lution of under 1 mm. Moreover, variants of this approach can give information about regional blood flow and the metabolic or biochemical state of selected brain regions (see Box E). In certain research applications (referred to as MR microscopy), this approach can provide even higher resolution. Like CT, MRI is entirely noninvasive; subjects are simply exposed to a strong magnetic field that is harmless (although accidents can happen if unsecured ferromagnetic objects are left in the vicinity). These features have made MRI the technique of choice for diagnostic studies or research when high resolution of the intact brain is required.

How nuclear magnetic resonance produces an image is more difficult to explain than X-ray imaging. Magnetic resonance derives from the interaction of a magnet and a magnetic field. Consider, for example, the earth's magnetic field and a compass magnet. At rest, the compass needle points north. If the needle is tapped, however, it will swing back and forth (oscillate) at a frequency directly proportional to the magnetic field

strength. The needle will continue to oscillate until friction dissipates the energy imparted by the tap and the needle again points north. Since the oscillation frequency is proportional to magnetic field strength, knowledge about the spatial variation of the field could in principle be used to detect (and make an image of) the location of the needle on the earth's surface (albeit crudely). In MRI, atomic nuclei—principally hydrogen—act as the compass needle, and a strong magnet plays the role of the earth's magnetic field. If all the atomic nuclei are aligned by the magnetic field and then "tapped" with a brief radiofrequency pulse, they emit energy in an oscillatory fashion as they return to the alignment imposed by the field. By using sensitive detectors of the radio frequencies emitted by the oscillating nuclei, together with computer techniques similar to those used in CT scanning, it is possible to construct extraordinarily detailed images of the brain. Moreover, unlike CT scans, views can be obtained from any angle. Since many brain struc-

tures are best seen in a particular plane, the ability to make "slices" from any point of view is a big advantage.

The resolution of MRI depends primarily on the strength of the magnetic field; currently, most clinical machines have field strengths of 1.5 Tesla, which provide millimeter resolution. The higher field strength magnets (3–4 Tesla) now beginning to be used, together with the recognition of the paramagnetic properties of hemoglobin, are ushering in another new era in brain imaging (see Box E).

References

CORMACK, A. M. (1980) Early two-dimensional reconstruction and recent topics stemming from it. Science 209: 1482–1486.

HOUNSFIELD, G. N. (1980) Computed medical imaging. Science 210: 22–28.

OLDENDORF, W. AND W OLDENDORF JR. (1988) *Basics of Magnetic Resonance Imaging*. Boston, MA: Kluwer Academic Publishers.

SCHILD, H. (1990) *MRI Made Easy (…Well, Almost)*. Berlin: H. Heeneman.

STARK, D. D. AND W. G. BRADLEY (1988) *Magnetic Resonance Imaging*. St. Louis: Mosby.

(C)

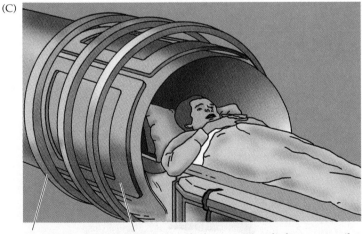

Magnetic coils that produce a static magnetic field in the long axis of the patient

Magnetic coils that produce a static field perpendicular to the long axis

A radiofrequency coil specifically designed for the head or other body part (not shown) perturbs the static fields to generate an MRI

(D)

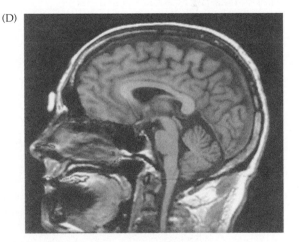

(C) Diagram of the machine used to obtain clinical MR images. (D) An MR image taken in the midsagittal plane. Note the extraordinary clarity with which all major brain components can be seen (compare with Figure 1.10).

■ THE INTERNAL ANATOMY OF THE BRAINSTEM AND SPINAL CORD

Like the forebrain, the brainstem and spinal cord consist of white matter and gray matter. In the spinal cord, the arrangement is relatively simple (Figure 1.16). The interior of the cord is formed by gray matter surrounded by white matter. The pathways of the white matter are subdivided into dorsal (or posterior), lateral, and ventral (or anterior) columns, each of which contains axon bundles related to specific functions. The **lateral columns** include axons that travel from the cerebral cortex to contact spinal motor neurons, thus forming the major caudal extension of the medullary pyramids. The **dorsal columns** carry ascending sensory information from somatic mechanoreceptors, and the **ventral (and ventrolateral) columns** carry both ascending pain and temperature information and descending motor information. The gray matter of the spinal cord is divided into dorsal and ventral (or posterior and anterior) "horns." The neurons of the **dorsal horn** receive sensory information that enters the spinal cord via the dorsal roots of the spinal nerves. The **ventral horn** contains the cell bodies of motor neurons that send axons out the ventral roots to terminate on striated muscles. Thus, a general rule of spinal cord organization is that the neurons that process sensory information lie dorsally, whereas the somatic motor neurons lie ventrally. The preganglionic neurons of the autonomic nervous system (see Box A) lie in an intermediate zone between the dorsal and ventral horns.

Although the internal organization of the brainstem is more complex than that of the spinal cord, some of the same principles of organization apply to both structures (Figure 1.17). The sensory and motor cell groups in the brainstem are known collectively as the **cranial nerve nuclei**. A relatively small core region of gray matter in the brainstem (called the **tegmentum**) contains these nuclei. As in the case of the spinal cord, the primary sensory

Figure 1.16 Internal structure of the spinal cord. (A) Cross sections of the cord at three different levels, showing the characteristic arrangement of gray and white matter in the cervical, thoracic, and lumbar cord. (B) Diagram of the internal structure of the spinal cord in cross section, showing the position of the major white matter columns and other features.

(A)

(B)

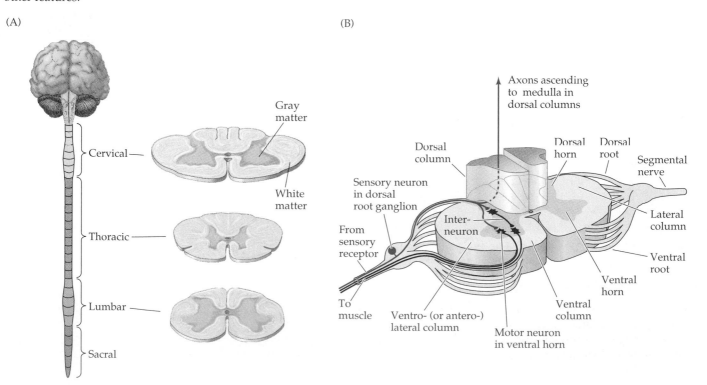

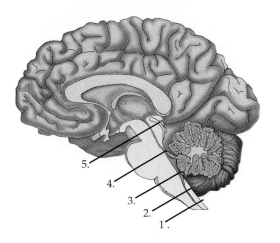

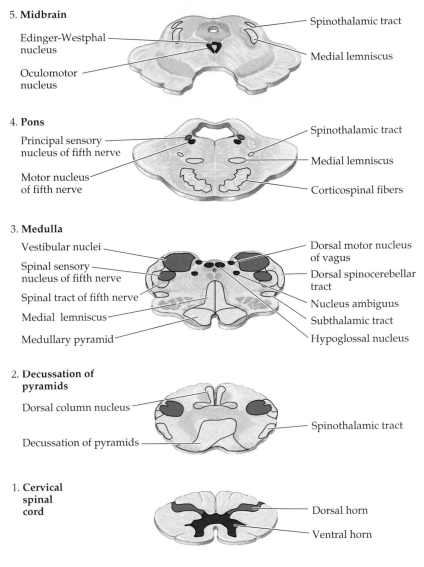

5. Midbrain

Edinger-Westphal nucleus

Oculomotor nucleus

Spinothalamic tract

Medial lemniscus

4. Pons

Principal sensory nucleus of fifth nerve

Motor nucleus of fifth nerve

Spinothalamic tract

Medial lemniscus

Corticospinal fibers

3. Medulla

Vestibular nuclei

Spinal sensory nucleus of fifth nerve

Spinal tract of fifth nerve

Medial lemniscus

Medullary pyramid

Dorsal motor nucleus of vagus

Dorsal spinocerebellar tract

Nucleus ambiguus

Subthalamic tract

Hypoglossal nucleus

2. Decussation of pyramids

Dorsal column nucleus

Decussation of pyramids

Spinothalamic tract

1. Cervical spinal cord

Dorsal horn

Ventral horn

Figure 1.17 The internal structure of the brainstem. Sections through the decussation of the pyramids in the area of transition between spinal cord and medulla (2), medulla (3), pons (4), and midbrain (5). A section through the cervical spinal cord (1) is shown for comparison. The diagram on the left indicates the level at which each section was taken. Areas that receive sensory inputs are shown in blue; areas that contain motor neurons are shown in red. Tracts of white matter in the brainstem that carry information between the spinal cord, brainstem, and forebrain are indicated in outline. Some of these are continuous with columns of white matter in the spinal cord. Sections are not to scale.

neurons are located in ganglia associated with the cranial nerves (just as dorsal root ganglia are associated with segmental nerves). The nuclei within the brainstem that receive sensory input are located separately from those that give rise to motor output; in general, the sensory nuclei are found laterally in the brainstem, whereas the motor nuclei are located more medially (corresponding to the dorsal and ventral horns of gray matter in the spinal cord). The segregation of sensory and motor functions in the brainstem is clinically important; as a result of this arrangement, specific signs and symptoms can often indicate the precise location of a brainstem lesion.

■ THE VENTRICULAR SYSTEM

The ventricular system is a series of interconnected, fluid-filled spaces in the core of the forebrain and brainstem (Figure 1.18A). The presence of a ventricular space in each subdivision of the brain reflects the fact that the ventricles are the adult derivative of the lumen of the embryonic neural tube (see Figure 1.18B and Chapter 20). Except for the lateral ventricles, all these

Figure 1.18 The ventricular system. (A) Location of the ventricles as seen in a transparent left lateral view. (B) Table showing the ventricular spaces associated with each of the major subdivisions of the brain. (See Chapter 20 for a detailed account of brain development.)

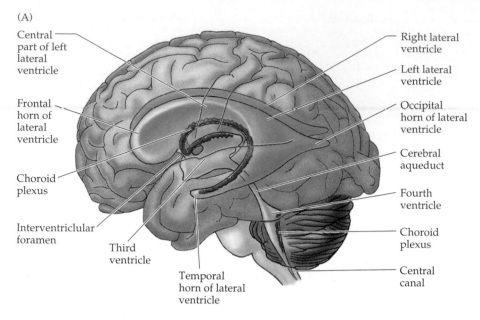

(A)

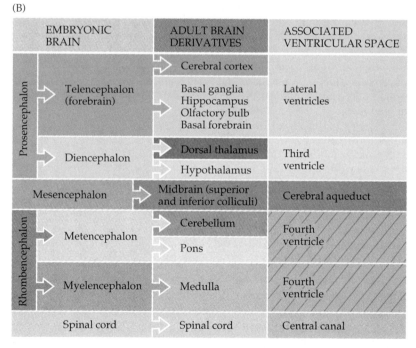

(B)

	EMBRYONIC BRAIN	ADULT BRAIN DERIVATIVES	ASSOCIATED VENTRICULAR SPACE
Prosencephalon	Telencephalon (forebrain)	Cerebral cortex	Lateral ventricles
		Basal ganglia Hippocampus Olfactory bulb Basal forebrain	
	Diencephalon	Dorsal thalamus	Third ventricle
		Hypothalamus	
	Mesencephalon	Midbrain (superior and inferior colliculi)	Cerebral aqueduct
Rhombencephalon	Metencephalon	Cerebellum	Fourth ventricle
		Pons	
	Myelencephalon	Medulla	Fourth ventricle
	Spinal cord	Spinal cord	Central canal

spaces can be seen on the medial surface of a brain cut in the midsagittal plane (see Figure 1.10). The **third ventricle** forms a narrow space between the right and left thalamus, and communicates with the **lateral ventricle** through a small opening at the anterior end of the third ventricle (the interventricular foramen). In midsagittal views of the brain, the lateral ventricle is hidden behind the septum pellucidum, a thin layer of tissue that forms its medial wall. The third ventricle is continuous caudally with the **cerebral aqueduct** which runs though the midbrain. At its caudal end, the aqueduct opens into the **fourth ventricle**, a larger space in the dorsal pons and medulla. The fourth ventricle narrows caudally to form the central canal of the spinal cord. **Cerebrospinal fluid** is produced by the **choroid plexus**, a network of specialized secretory tissue in the lateral, third, and fourth ventricles (Figure 1.19). The cerebrospinal fluid percolates through the ventricular system and flows into the subarachnoid space through perforations in the thin covering of the fourth ventricle; it is eventually absorbed by specialized structures called arachnoid villi (or granulations; see Figure 1.19) and returned to the general circulation.

■ THE MENINGES

Although the brain is often considered in isolation, it lies within the cranial cavity, which both supports and protects it. This space is conventionally divided into three regions: the anterior, middle, and posterior cranial fossae. Beneath the bony shell of the cranium lie three protective tissue layers, which also extend down the brainstem and the spinal cord. Together these layers are called the **meninges** (Figure 1.19). The outermost layer of the meninges is called the **dura mater** because it is thick and tough. The middle layer is called the **arachnoid mater** because of spiderlike processes that extend from it toward the third layer, the **pia mater**, a thin, delicate layer of cells that closely invests the surface of the brain. The space between the arachnoid and pia (the **subarachnoid space**) is filled with cerebrospinal fluid, which helps to cushion the brain within the cranial cavity. Since the pia closely adheres to the brain as its surface curves and folds, whereas the arachnoid does not, there are places—referred to as **cisterns**—where the subarachnoid space is especially large. The major arteries supplying the brain course through the subarachnoid space where they give rise to branches that penetrate the substance of the hemispheres. Not surprisingly, then, the subarachnoid space is a frequent site of bleeding following trauma. A collection of blood in this space is referred to as a subarachnoid hemorrhage.

■ THE BLOOD SUPPLY OF THE BRAIN AND SPINAL CORD

The brain receives blood from two sources: the **internal carotid arteries**, which arise at the point in the neck where the common carotid arteries bifurcate, and the **vertebral arteries**, which arise in the chest as branches of the subclavian arteries (Figure 1.20). The right and left vertebral arteries come together at the level of the pons on the ventral surface of the brainstem to form the midline **basilar artery**, which in turn joins the blood supply from the internal carotids in an arterial ring at the base of the brain called the **circle of Willis** (Figure 1.20A). The posterior cerebral arteries arise at this confluence. Conjoining the two major sources in this arterial anastomosis improves the chances of any region of the brain continuing to receive blood if one of the major arteries becomes occluded. The major branches that arise from the internal carotid artery—the **anterior** and **middle cerebral arteries**—form the **anterior circulation** that supplies the forebrain (Figure 1.20B). Each of these arteries gives rise to branches that supply the cortex, and branches that penetrate the basal surface of the brain, supplying deep structures such as the basal ganglia, thalamus, and internal capsule.

The **posterior circulation** of the brain supplies the brainstem, and is made up of arterial branches arising from the posterior cerebral, basilar, and vertebral arteries. The pattern of arterial distribution is similar for all the subdivisions of the brainstem: midline arteries supply medial structures, lateral arteries supply the lateral brainstem, and dorsal arteries supply dorsal brainstem structures and the cerebellum (Figure 1.21). Among the most important

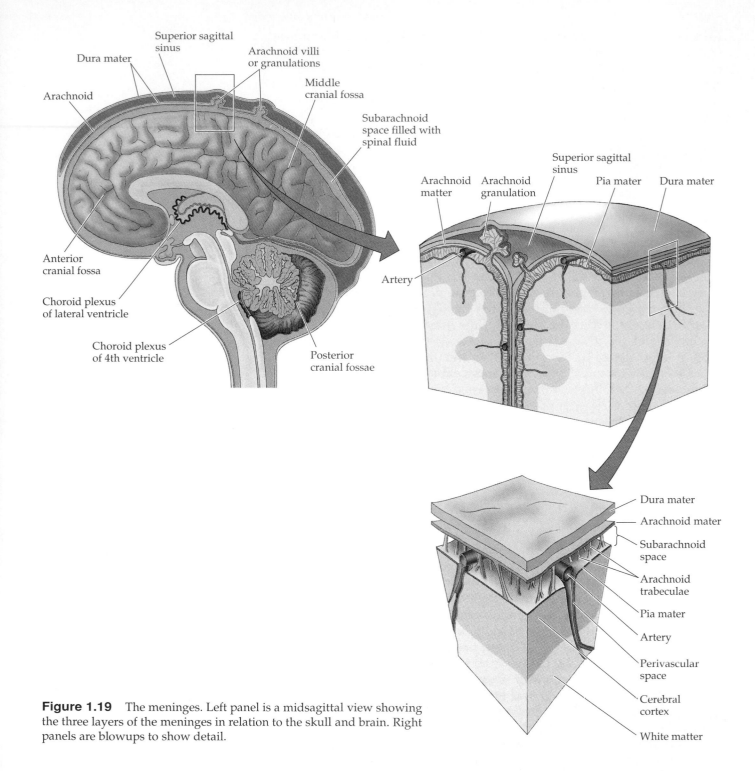

Figure 1.19 The meninges. Left panel is a midsagittal view showing the three layers of the meninges in relation to the skull and brain. Right panels are blowups to show detail.

dorsal arteries are the posterior inferior cerebellar artery and the anterior inferior cerebellar artery, which supply several cranial nerve nuclei in the dorsolateral pons, and portions of the cerebellum and cerebellar peduncles, respectively. These and the arterial branches that penetrate the brainstem from its ventral and lateral surfaces are especially common sites of occlusion.

The spinal cord is supplied by the vertebral arteries and approximately ten **medullary arteries** that arise from segmental branches of the aorta; these

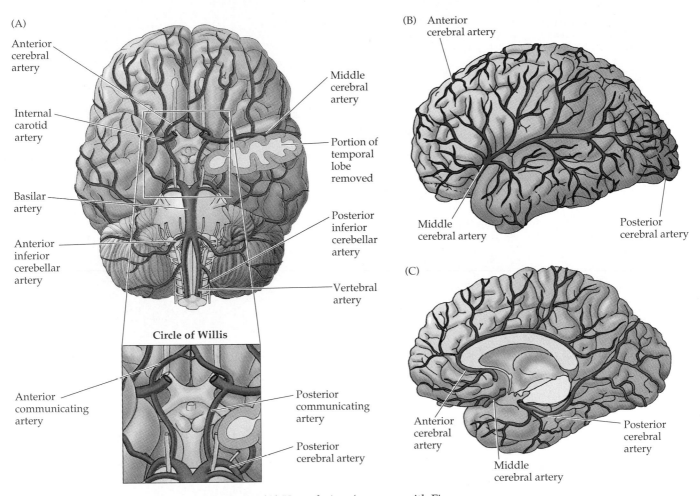

Figure 1.20 The major arteries of the brain. (A) Ventral view (compare with Figure 1.9). The blowup of the boxed area shows the circle of Willis. (B) Lateral and (C) midsagittal views showing anterior, middle, and posterior cerebral arteries.

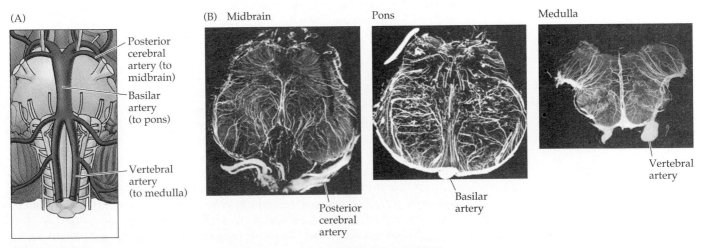

Figure 1.21 Blood supply of the three subdivisions of the brainstem. (A) Diagram of major supply. (B) Photomicrographs of brainstem sections after injection of a contrast agent into the vascular system. The brainstem is supplied by midline, lateral, and dorsal perforating arteries. (From Hassler, 1967.)

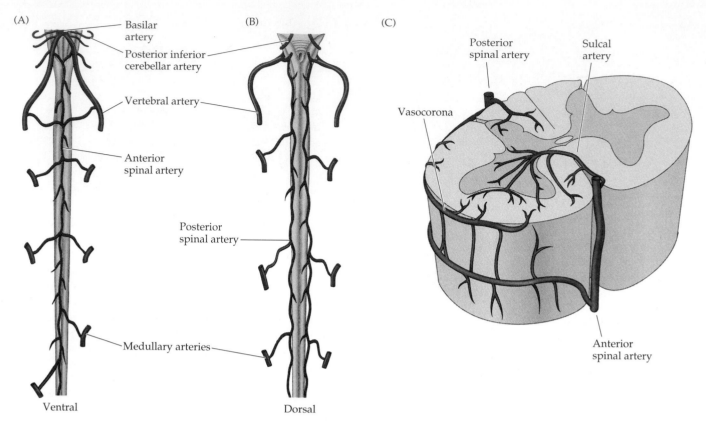

(A)

Basilar artery

Posterior inferior cerebellar artery

Vertebral artery

Anterior spinal artery

Medullary arteries

Ventral

(B)

Posterior spinal artery

Dorsal

(C)

Posterior spinal artery

Sulcal artery

Vasocorona

Anterior spinal artery

Figure 1.22 Blood supply of the spinal cord. (A) View of the ventral (anterior) surface of the spinal cord. At the level of the medulla, the vertebral arteries give off branches that merge to form the anterior spinal artery. Approximately 10 to 12 segmental arteries (which arise from various branches of the aorta) join the anterior spinal artery along its course. These segmental arteries are known as medullary arteries. (B) The vertebral arteries (or the posterior inferior cerebellar artery) give rise to paired posterior spinal arteries that run along the dorsal (posterior) surface of the spinal cord. (C) Cross section through the spinal cord, illustrating the distribution of the anterior and posterior spinal arteries. The anterior spinal arteries give rise to numerous sulcal branches that supply the anterior two-thirds of the spinal cord. The posterior spinal arteries supply much of the dorsal horn and the dorsal columns. A network of vessels known as the vasocorona connects these two sources of supply and sends branches into the white matter around the margin of the spinal cord.

join to form the anterior and posterior spinal arteries (Figure 1.22). If any of the medullary arteries are obstructed or damaged (during abdominal surgery, for example), the blood supply to specific parts of the spinal cord may be compromised. The pattern of the resulting neurological damage depends on whether the supply to the posterior or anterior artery is interrupted. As might be expected from the arrangement of ascending and descending pathways in the spinal cord, loss of the posterior supply leads to loss of mechanosensory function, whereas loss of the anterior supply leads to motor deficits and diminished pain and temperature sensation.

The blood supply of the brain is particularly significant because neurons are so sensitive to oxygen deprivation (as well as to altered concentrations of other substances; Box D). As a result of its high metabolic rate, brain tissue deprived of oxygen and glucose, even briefly, is at high risk. Strokes (the death or dysfunction of brain tissue due to vascular disease) often follow

Box D
THE BLOOD-BRAIN BARRIER

The interface between the walls of capillaries and the surrounding tissue is important throughout the body, as it keeps vascular and extravascular concentrations of ions and molecules at appropriate levels in these two compartments. In the brain, this interface is especially significant and has been accorded an alliterative name, "the blood-brain barrier." The special properties of the blood-brain barrier were first observed by the nineteenth-century bacteriologist Paul Ehrlich, who noted that intravenously injected dyes leaked out of capillaries in most regions of the body to stain the surrounding tissues; the brain, however, remained unstained. Ehrlich wrongly concluded that the brain had a low affinity for the dyes; it remained for his student, Edwin Goldmann, to show that such dyes do not traverse the specialized walls of brain capillaries.

The restriction of large molecules like Ehrlich's dyes (and many smaller molecules) to the vascular space is the result of tight junctions between neighboring capillary endothelial cells in the brain. Such junctions are not found in capillaries elsewhere in the body, where the spaces between adjacent endothelial cells allow much more ionic and molecular traffic. The structure of tight junctions was first demonstrated in the 1960s by Tom Reese, Morris Karnovsky, and Milton Brightman. Using electron microscopy after the injection of electron-dense intravascular agents such as lanthanum salts, they showed that the close apposition of the endothelial cell membranes prevented such ions from passing. Substances that traverse the walls of brain capillaries must move *through* the endothelial cell membranes. Accordingly, molecular entry into the brain should be determined by an agent's solubility in lipids, the major constituent of cell membranes. Nevertheless, many ions and molecules not readily soluble in lipids *do* move quite readily from the vascular space into brain tissue. A molecule like glucose, the primary source of metabolic energy for neurons and glial cells, is an obvious example. This paradox is explained by the presence of specific transporters for glucose and other critical molecules and ions.

In addition to tight junctions, astrocytic "end feet" (the terminal regions of astrocytic processes) surround the outside of capillary endothelial cells. The reason for this endothelial–glial allegiance is unclear, but may reflect an influence of astrocytes on the formation and maintenance of the blood-brain barrier.

The brain, more than any other organ, must be carefully shielded from abnormal variations in its ionic milieu, as well as from the potentially toxic molecules that find their way into the vascular space by ingestion, infection, or other means. The blood-brain barrier is thus important for protection and homeostasis. It also presents a significant problem for the delivery of drugs to the brain. Large (or lipid-insoluble) molecules can be introduced to the brain, but only by transiently disrupting the blood-brain barrier with hyperosmotic agents like mannitol.

References

BRIGHTMAN, M. W. AND T. S. REESE (1969) Junctions between intimately opposed cell membranes in the vertebrate brain. J. Cell Biol. 40: 648–677.

SCHMIDLEY, J. W. AND E. F. MAAS (1990) Cerebrospinal fluid, blood-brain barrier and brain edema. In *Neurobiology of Disease*, A. L. Pearlman and R.C. Collins (eds.). New York: Oxford University Press, Chapter 19, pp. 380–398.

REESE, T. S. AND M. J. KARNOVSKY (1967) Fine structural localization of a blood–brain barrier to exogenous peroxidase. J. Cell Biol. 34: 207–217.

(A)

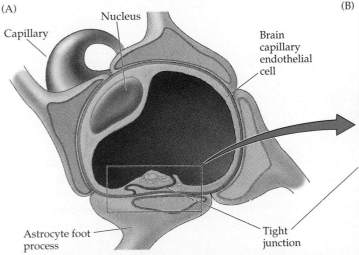

Capillary

Nucleus

Brain capillary endothelial cell

Astrocyte foot process

Tight junction

(B)

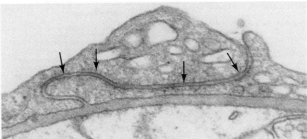

The cellular basis of the blood-brain barrier. (A) Diagram of a brain capillary in cross section and reconstructed views, showing endothelial tight junctions and the investment of the capillary by astrocytic end feet. (B) Electron micrograph of boxed area in (A), showing the appearance of tight junctions between neighboring endothelial cells (arrows). (A after Goldstein, Goldstein and Betz, 1986; B from Peters et al., 1991.)

BOX E
IMAGING TECHNIQUES BASED ON BLOOD FLOW: PET, SPECT, AND *f*MRI

The most informative brain-imaging techniques now rely on detecting small changes in blood flow to visualize active areas of the brain. The brain utilizes a remarkably large fraction of the body's energy resources (about 20% of circulating glucose is consumed by the brain). Not surprisingly, at any given moment the most active nerve cells use more glucose and oxygen than relatively quiescent neurons. To meet the increased metabolic demands of particularly active neurons, the local flow of blood to the relevant brain area increases. Detecting and mapping these local changes in cerebral blood flow form the basis for three widely used functional brain imaging techniques: positron emission tomography (PET), single-photon emission computerized tomography (SPECT), and functional magnetic resonance imaging (*f*MRI). Because these techniques reveal patterns of activity in the intact brain, they have greatly enhanced the ability to understand both normal brain function and abnormal brain states associated with a variety of pathologies.

In PET scanning, unstable positron-emitting isotopes are synthesized in a cyclotron by bombarding nitrogen, carbon, oxygen, or fluorine with protons. Examples of the isotopes used include ^{15}O (half-life, 2 min), ^{18}F (110 min), and ^{11}C (20 min). These probes can be incorporated into many different reagents (including water, precursor molecules of specific neurotransmitters, or glucose) and used to analyze specific aspects of brain function. When the radiolabeled compounds are injected into the bloodstream, they distribute according to the physiological state of the brain. Thus, labeled oxygen and glucose accumulate in more metabolically active areas, and labeled transmitter probes are taken up selectively by appropriate regions. As the unstable isotope decays, the extra proton breaks down into a neutron and a positron. The emitted positrons travel several millimeters, on average, until they collide with an electron. The collision of a positron with an electron destroys both particles, emitting two gamma rays from the site of the collision in directions that are exactly 180° apart. Gamma ray detectors placed around the head are therefore arranged to register a "hit" only when two opposite detectors (i.e., 180° apart) react simultaneously. By reconstructing the sites of the positron-electron collisions, the location of active regions can be imaged. The mean free path of the positrons in brain tissue limits the resolution of PET scanning to about 4 mm. Nonetheless, PET images can be superimposed on MRI images from the same subject (see Box C), providing detailed information about specific brain areas involved in a wide variety of functions. The elegance and

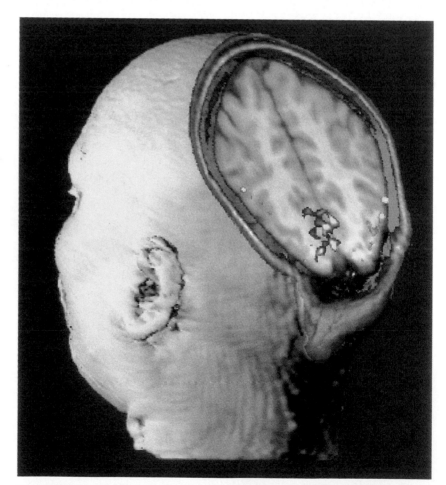

Example of functional magnetic resonance imaging. Regional changes in cerebral blood flow were measured during visual stimulation; the area of activated visual cortex (color) was then mapped onto the brain, a section of which is shown at the appropriate level in the head. (From Belliveau et al., 1991.)

power of this technique are evident in Figures 24.6 and 25.6.

SPECT imaging is an outgrowth of older techniques for measuring regional cerebral blood flow. A radiolabeled compound with a relatively short half-life (for example, ^{133}Xe) is inhaled or injected into the circulation (in the latter case, ^{123}I-labeled iodoamphetamine is used); the probes bind to red blood cells and are carried throughout the body. As the label undergoes radioactive decay, it emits high-energy photons. The rate of clearance of the probes was initially detected using an array of sodium iodide photon detectors placed around the head. More recent approaches have used a gamma camera that can be rapidly moved around the head to collect photons from many different angles, thus permitting a more accurate three-dimensional image. The information gathered using SPECT can also be combined with structural information from other imaging techniques, such as CT scans and MRI scans, to provide better localization of the active areas. A limitation of SPECT scanning is its relatively low resolution (about 8 mm). Although this level is not sufficient to resolve the finer features of the brain, it reveals the major areas involved in normal processing or disease. SPECT imaging is neither as flexible nor as accurate as PET imaging, but it is much simpler, primarily because the radiolabeled probes are commercially available and do not require an on-site cyclotron (as does the synthesis of PET probes).

A variant of MRI, called functional MRI (fMRI), now offers perhaps the best approach to analyzing the brain at work. fMRI is based on the fact that oxyhemoglobin (the oxygen-carrying form of hemoglobin) has a different magnetic resonance signal than deoxyhemoglobin (the oxygen-depleted form of hemoglobin). Brain areas activated by a specific task (e.g., the occipital cortex during visual behavior; see accompanying figure) utilize more oxygen. Initially, this activity decreases the levels of oxyhemoglobin and increases levels of deoxyhemoglobin. Within seconds, the brain microvasculature responds to this local oxygen depletion by increasing the flow of oxygen-rich blood to the active area. The local response leads to an increase in the oxyhemoglobin-to-deoxyhemoglobin ratio, which forms the basis for the fMRI signal. Thus, unlike PET or SPECT, fMRI uses signals intrinsic to the brain rather than signals originating from exogenous, radioactive probes; consequently, repeated observations can be made on the same individual, providing a major advantage over other imaging methods. fMRI also offers superior spatial localization (currently a few millimeters), as well as good temporal resolution (on the order of seconds under optimal circumstances, compared to minutes for other functional imaging techniques). As a result of these advantages, fMRI has emerged as the technology of choice for probing both the normal and abnormal functional architecture of the human brain.

References

BELLIVEAU, J. W., D. N. KENNEDY JR., R. C. MCKINSTRY, B. R BUCHBINDER, R. M. WEISSKOFF, M. S. COHEN, J. M. VEVEA AND B. R. ROSEN (1991) Functional mapping of the human visual cortex by magnetic resonance imaging. Science 254: 716–719.

COHEN, M. S. AND S. Y. BOOKHEIMER (1994) Localization of brain function using magnetic resonance imaging. Trends Neurosci. 17: 268–277.

KWONG, K. K., J. W. BELLIVEAU, D. A. CHESLER, I. E. GOLDBERG, R. M. WEISSKOFF, B. P. PONCELOT, D. N. KENNEDY, B. E. HOPPEL, M. S. COHEN AND R. TURNER (1992) Dynamic magnetic resonance imaging of human brain activity during primary sensory stimulation. Proc. Natl. Acad. Sci. USA 89: 5675–5679.

OGAWA, S., D. W. TANK, R. MENON, J. M. ELLERMAN, S. G. KIM, H. MERKLE AND K. UGURBIL. (1992) Intrinsic signal changes accompanying sensory stimulation: functional brain mapping with magnetic resonance imaging. Proc. Natl. Acad. Sci. USA 89: 5951–5955.

PETERSEN, S. E., P. T. FOX, A. Z. SNYDER AND M. E. RAICHLE (1990) Activation of extrastriate and frontal cortical areas by visual words and word-like stimuli. Science 249: 1041–1044.

RAICHLE, M. E. (1994) Images of the mind: studies with modern imaging techniques. Annu. Rev. Psychol. 45: 333–356.

RAICHLE, M. E. AND M. I. POSNER (1994) Images of Mind. New York: Scientific American Library.

the occlusion of (or hemorrhage from) the brain's arteries. Historically, studies of the functional consequences of strokes have provided information about the location of various brain functions. The location of language functions in the left hemisphere, for instance, was discovered in this way in the latter part of the nineteenth century. Now, noninvasive imaging techniques based on blood flow (Box E) have largely supplanted the correlation of clinical signs and symptoms with the location of tissue damage observed at autopsy, allowing researchers and clinicians to observe the function of different brain regions in normal subjects as well as in living patients.

■ SUMMARY

Although the human brain is often discussed as if it were a single organ, it comprises a large number of systems and subsystems. The various types of neurons in these systems are assembled into richly interconnected circuits that relay and process the electrical signals that are the currency of all neural functions. Knowledge about the structural organization of the brain provides an essential first step toward unraveling its many purposes. The human nervous system, like that of all vertebrates, comprises a central nervous system, which consists of the brain and spinal cord, and a peripheral nervous system, which consists of peripheral nerves (and their ganglia) extending to a wide array of targets. Sensory components of the peripheral nervous system supply information to the central nervous system about the internal and external environment. The integrated effects of central processing are eventually translated into action by the motor components of the central and peripheral nervous systems. Different brain regions mediate an enormous range of functions, including sensory perception, cognition, language, sleep, emotion, sexuality, and memory, to name but a few. The wealth of structural information described in this chapter provides a framework for understanding these phenomena.

Additional Reading

BRODAL, P. (1992) *The Central Nervous System: Structure and Function.* New York: Oxford University Press.

CARPENTER, M. B. AND J. SUTIN (1983) *Human Neuroanatomy,* 8th Ed. Baltimore: Williams and Wilkins.

ENGLAND, M. A. AND J. WAKELY (1991) *Color Atlas of the Brain and Spinal Cord: An Introduction to Normal Neuroanatomy.* St. Louis: Mosby Year Book.

HAINES, D. E. (1995) *Neuroanatomy: An Atlas of Structures, Sections, and Systems,* 2nd Ed. Baltimore: Urban and Schwarzenberg.

MARTIN, J. H. (1996) *Neuroanatomy: Text and Atlas.* 22nd Ed. Stamford, CT: Appleton & Lange.

NETTER, F. H. (1983) *The CIBA Collection of Medical Illustrations,* Vols. I and II. A. Brass and R. V. Dingle (eds.). Summit, NJ: CIBA Pharmaceutical Co.

PETERS, A., S. L. PALAY AND H. DE F. WEBSTER (1991) *The Fine Structure of the Nervous System: Neurons and Their Supporting Cells,* 3rd Ed. New York: Oxford University Press.

RAMÓN Y CAJAL, S. New York: (1984) *The Neuron and the Glial Cell.* (Transl. by J. de la Torre and W. C. Gibson.) Springfield, IL: Charles C. Thomas.

RAMÓN Y CAJAL, S. (1990) *New Ideas on the Structure of the Nervous System in Man and Vertebrates.* (Transl. by N. Swanson and L. W. Swanson.) Cambridge, MA: MIT Press.

WAXMAN, S. G. AND J. DEGROOT (1995) *Correlative Neuroanatomy,* 22nd Ed. Norwalk, CT: Appleton and Lange.

I
NEURAL SIGNALING

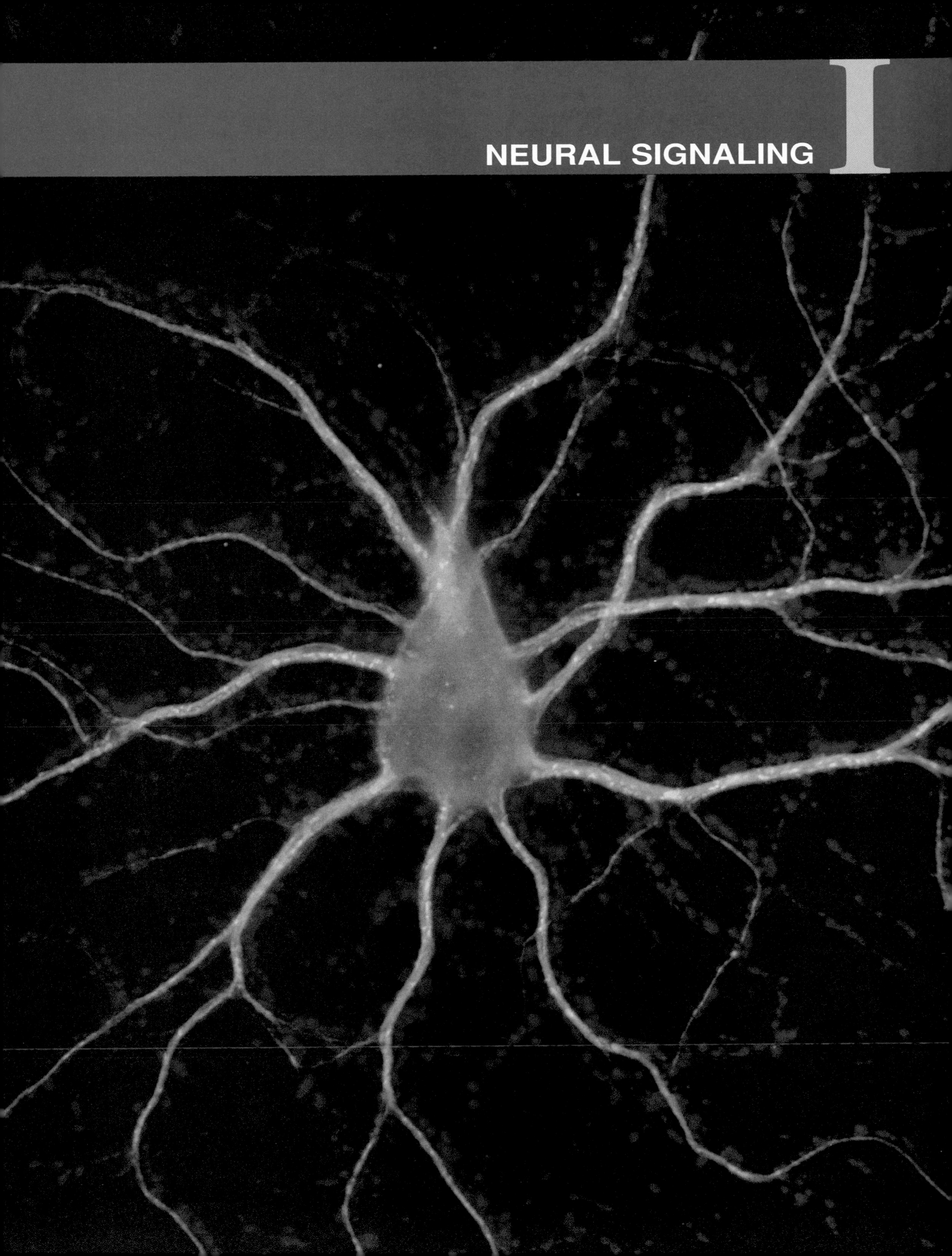

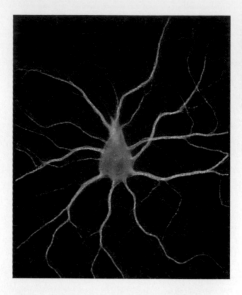

UNIT I
NEURAL SIGNALING

Cultured hippocampal neuron stained for the dendritic protein MAP2 (green fluorescence), and for a synaptic vesicle protein, synaptotagmin (orange-red fluorescence). The cell body and dendrites are visualized by MAP2 immunoreactivity. Orange-red dots represent synaptic vesicle-filled presynaptic nerve terminals originating from neurons not visible in this field. (Courtesy of Olaf Mundigl and Pietro DeCamilli.)

The primary purpose of the brain is to acquire, coordinate, and disseminate information about the body and its environment. To perform this task, neurons have evolved a sophisticated means of generating electrical signals. This unit describes these signals and explains how they are produced and what they mean. It also explains how one type of electrical signal, the action potential, allows information to travel along the length of a nerve cell and how other types of signals—both electrical and chemical—are generated at synaptic connections between nerve cells. Synapses permit information transfer by interconnecting the multitude of neurons in the nervous system. These two types of signaling mechanisms—action potentials and synaptic signals—are the basis for all the information-processing capabilities of the brain.

The cellular and molecular processes that give neurons their unique signaling abilities are also targets for disease processes that can compromise the function of the nervous system. Therefore, knowledge of the cellular and molecular biology of neurons is fundamental to understanding brain pathology. An increasing number of diseases of the nervous system are beginning to be understood as discrete lesions of neuronal signaling molecules. Such information has stimulated novel pharmacological and molecular biological approaches to diagnosing and treating these disorders.

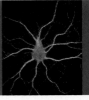

CHAPTER 2

■ OVERVIEW

Nerve cells generate electrical signals that transmit information. Although neurons are not intrinsically good conductors of electricity, they have evolved elaborate mechanisms for generating electrical signals based on the flow of ions across their plasma membranes. Ordinarily, neurons generate a negative potential, called the resting membrane potential, that can be measured by intracellular recording. The action potential abolishes the negative resting potential and makes the transmembrane potential transiently positive. Action potentials are propagated along the length of axons and are the fundamental electrical signal of neurons. Generation of both the resting potential and the action potential can be understood in terms of the nerve cell's selective permeability to different ions and the normal distribution of these ions across the cell membrane.

ELECTRICAL SIGNALS OF NERVE CELLS

■ ELECTRICAL POTENTIALS ACROSS NERVE CELL MEMBRANES

Because electrical signals are the basis of information transfer in the nervous system, it is essential to understand how these signals arise. The use of electrical signals—as in sending electricity over wires to provide power or information—presents a fundamental problem for neurons: neuronal axons, which can be quite long (remember that a spinal motor neuron can extend for a meter or more), are not good electrical conductors. Although neurons and wires are both capable of passively conducting electricity, the electrical properties of neurons compare poorly to even the most ordinary wire. To compensate for this deficiency, neurons have evolved a "booster system" that allows them to conduct electrical signals over great distances despite their intrinsically poor electrical characteristics. The electrical signals produced by this booster system are called **action potentials**, also referred to as spikes or impulses.

The best way to observe an action potential is to use an intracellular microelectrode to record directly the electrical potential across the neuronal plasma membrane (Figure 2.1). A typical microelectrode is a piece of glass tubing pulled to a very fine point (with an opening of less than 1 μm diameter) and filled with a good electrical conductor, such as a concentrated salt solution. This conductive core can then be connected to a voltmeter, such as an oscilloscope, to record the transmembrane voltage of the nerve cell. When a microelectrode is inserted through the membrane of the neuron, it records a negative potential, implying that the cell has a means of generating a constant voltage across its membrane when it is at rest. This voltage, called the **resting membrane potential**, depends on the type of neuron being examined, but is always a fraction of a volt (typically −40 to −90 mV).

Action potentials represent transient changes in the resting membrane potential of neurons. One way to elicit an action potential is to pass electrical current across the membrane of the neuron. In normal circumstances, this current would be generated by another neuron, at the synapse between two nerve cells, or by the transduction of an external stimulus in sensory neurons

(A)

(B)

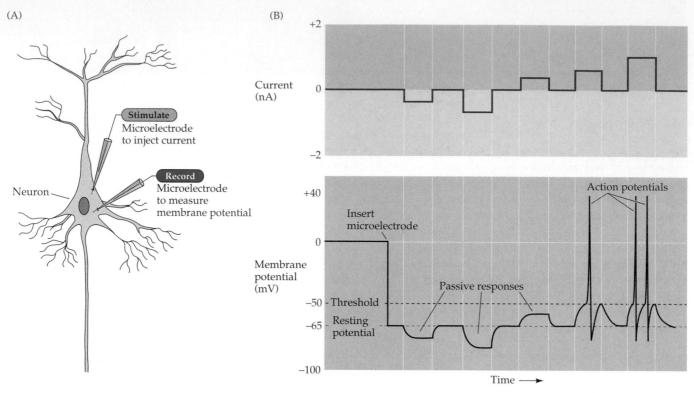

Figure 2.1 Recording passive and active electrical signals in a nerve cell. (A) Two microelectrodes are inserted into a neuron; one of these measures membrane potential while the other injects current into the neuron. (B) Inserting the voltage-measuring microelectrode into the neuron reveals a negative potential, the resting membrane potential. Injecting current through the current-passing microelectrode alters the neuronal membrane potential. Hyperpolarizing current pulses produce only passive changes in the membrane potential. Depolarizing currents that cause the membrane potential to meet or exceed threshold evoke action potentials. Action potentials are active responses in the sense that they are generated by changes in the permeability of the neuronal membrane.

(see Unit 2). In the laboratory, however, it is easy to produce a suitable electrical current by inserting a second microelectrode into the neuron and then connecting the electrode to a battery. If the current delivered in this way is such as to make the membrane potential more negative (**hyperpolarization**), nothing very dramatic happens. The membrane potential simply changes in proportion to the magnitude of the injected current. Such hyperpolarizing responses do not require any unique property of neurons and are therefore called passive electrical responses. A much more interesting phenomenon is seen if current of the opposite polarity is delivered, so that the membrane potential of the nerve cell becomes more positive than the resting potential (**depolarization**). In this case, at a certain level of membrane potential, called the **threshold potential**, an action potential occurs (see Figure 2.1B).

The action potential, which is an active response generated by the neuron, appears on an oscilloscope as a brief (about 1 ms) change from negative to positive in the neuronal membrane potential. Importantly, the amplitude of the action potential is independent of the magnitude of the current used to evoke it; that is, larger currents do not elicit larger action potentials. The action potentials of a given neuron are therefore said to be all-or-none, because they occur fully or not at all. If the amplitude or duration of the stimulus current is increased sufficiently, multiple action potentials occur. Thus, the intensity of a stimulus is encoded in the frequency of action potentials rather than in their amplitude.

This chapter addresses the basic question of how nerve cells can generate electricity by the flow of ions across the neuronal membrane. Chapter 3 explores the means by which action potentials are produced and demonstrates how these signals solve the problem of long-distance electrical conduction within nerve cells. Chapter 4 examines the properties of membrane molecules responsible for producing action potentials. Finally, Chapters 5–7 consider how electrical signals are transmitted between nerve cells by synaptic contacts.

■ HOW IONIC MOVEMENTS PRODUCE ELECTRICAL SIGNALS

Electrical potentials are generated across the membranes of neurons—and, indeed, all cells—because (1) there are *differences in the concentrations* of specific ions across nerve cell membranes, and (2) the membranes are *selectively permeable* to some of these ions. These two facts depend in turn on two different kinds of proteins in the cell membrane (Figure 2.2). The ion concentration gradients are established by proteins known as **ion pumps**, which, as their name suggests, actively move ions into or out of cells against their concentration gradients. The selective permeability of membranes is due largely to **ion channels**, proteins that allow only certain kinds of ions to cross the membrane in the direction of their concentration gradients. Thus, channels and pumps basically work against each other, and in so doing they generate cellular electricity.

To appreciate the role of ion gradients and selective permeability in generating a membrane potential, consider a simple system in which an imaginary membrane separates two compartments containing solutions of

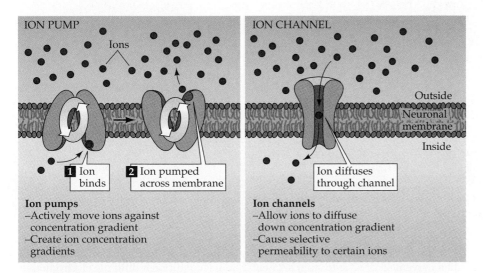

Figure 2.2 Ion pumps and ion channels are responsible for ionic movements across neuronal membranes. Pumps create ion concentration differences by actively transporting ions against their chemical gradients. Channels take advantage of these concentration gradients, allowing selected ions to move, via diffusion, down their chemical gradients.

ions. First, take the case of a membrane that is permeable only to potassium ions (K^+). If the concentration of K^+ on each side of this membrane is equal, then no electrical potential across it will be measured (Figure 2.3A). However, if the concentration of K^+ is not the same on the two sides, then an electrical potential will be generated. For example, if the concentration of K^+ on one side of the membrane (compartment 1) is 10 times higher than the K^+ concentration on the other side (compartment 2), then the electrical potential of compartment 1 will be negative relative to compartment 2 (Figure 2.3B). This difference in electrical potential is generated because the potassium ions flow down their concentration gradient and take their electrical charge (one positive charge per ion) with them as they go. Because neuronal membranes contain pumps that accumulate K^+ in the cell cytoplasm and because potassium-permeable channels in the plasma membrane allow a transmembrane flow of K^+, an analogous situation exists in living nerve cells. A continual resting efflux of K^+ is therefore responsible for the resting membrane potential.

In the artificial case just described, an equilibrium will quickly be reached. As K^+ moves from compartment 1 to compartment 2 (the initial conditions in Figure 2.3B), a potential is generated that tends to impede further flow of K^+. This impediment results from the fact that the potential gradient across the membrane repels the positive potassium ions as they attempt to move across the membrane. That is, as compartment 2 becomes

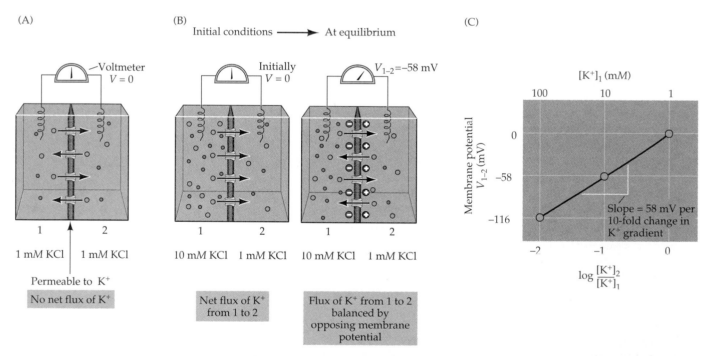

Figure 2.3 Electrochemical equilibrium. (A) A membrane permeable only to K^+ separates compartments 1 and 2, which contain the indicated concentrations of KCl. (B) Increasing the KCl concentration in compartment 1 to 10 mM initially causes a small movement of K^+ into compartment 2 (initial conditions) until the electromotive force acting on K^+ balances the concentration gradient, and the net movement of K^+ becomes zero (at equilibrium). (C) The relationship between the transmembrane concentration gradient ($[K^+]_2/[K^+]_1$) and the membrane potential. As predicted by the Nernst equation, this relationship is linear when plotted on semilogarithmic coordinates, with a slope of 58 mV per tenfold difference in the concentration gradient.

positive relative to compartment 1, this positivity makes compartment 2 less attractive to the positively charged K^+. The net movement (or flux) of K^+ will stop at the point (at equilibrium in Figure 2.3B) where the potential change across the membrane (the relative positivity of compartment 2) exactly offsets the concentration gradient (the 10× excess of K^+ in compartment 1). At this **electrochemical equilibrium**, there is an exact balance between two opposing forces: (1) the concentration gradient that causes K^+ to move from compartment 1 to compartment 2, taking along positive charge, and (2) an opposing electrical gradient that increasingly tends to stop K^+ from moving across the membrane (Figure 2.3B). The number of ions that needs to flow to generate this electrical potential is *very* small ($\approx 10^{-12}$ moles of K^+/cm^2 of membrane). This last fact is significant in two ways. First, it means that the concentrations of permeant ions on each side of the membrane remain essentially constant, even after the flow of ions has generated the potential. Second, the tiny fluxes of ions required to establish the membrane potential do not disrupt chemical electroneutrality because each ion has an oppositely charged counter ion (chloride, Cl^-, in the example shown in Figure 2.3) to maintain the neutrality of the solutions on each side of the membrane. The concentration of K^+ remains equal to the concentration of Cl^- in the solutions in compartments 1 and 2, such that the separation of charge that creates the potential difference is restricted to the immediate vicinity of the membrane.

■ THE FORCES THAT CREATE MEMBRANE POTENTIALS

The electrical potential generated across the membrane at electrochemical equilibrium, the **equilibrium potential**, can be predicted by a simple formula called the **Nernst equation**. This relationship is generally expressed as

$$E_X = \frac{RT}{zF} \ln \frac{[X]_2}{[X]_1}$$

where E_X is the equilibrium potential for any ion X, R is the gas constant, T is the absolute temperature (in Kelvin units), z is the valence (electrical charge) of the permeant ion, and F is the Faraday constant (the amount of electrical charge contained in a mole of a univalent ion). The brackets indicate the concentrations of ion X on each side of the membrane and the symbol ln indicates the natural logarithm of the concentration gradient. Because it is easier to perform calculations and experiments using base 10 logarithms at room temperature, this relationship is usually simplified to

$$E_X = \frac{58}{z} \log \frac{[X]_2}{[X]_1}$$

where log indicates the base 10 logarithm of the concentration ratio. Thus, for the example in Figure 2.3B, the potential across the membrane at electrochemical equilibrium is

$$E_K = \frac{58}{z} \log \frac{[K]_2}{[K]_1} = 58 \log \frac{1}{10} = -58 \text{ mV}$$

The equilibrium potential is conventionally defined in terms of the potential difference between the reference compartment, side 2 in Figure 2.3, and side 1. This relationship holds equally well for biological systems: the outside of the cell, which is low in K^+ relative to the cell interior, is the conventional reference point (defined as zero potential); thus, an inside-negative potential is measured across the K⁺-permeable neuronal membrane.

For a simple hypothetical system with only one permeant ion species, the Nernst equation allows the electrical potential across the membrane at equilibrium to be predicted exactly. For example, if the concentration of K⁺ on side 1 is increased to 100 mM, the membrane potential will be −116 mV. More generally, if the membrane potential is plotted against the logarithm of the K⁺ concentration gradient ($[K]_2/[K]_1$), the Nerst equation predicts that this relationship will be linear with a slope of 58 mV (actually 58/z) per ten-fold change in the K⁺ gradient (Figure 2.3C).

To reinforce and extend the concept of electrochemical equilibrium, consider some additional experiments on the influence of ionic species and ionic permeability that could be performed on the simple model system in Figure 2.3. For instance, what would happen to the electrical potential across the membrane (the potential of side 1 relative to side 2) if the potassium on side 2 were replaced with 10 mM sodium (Na⁺) and the K⁺ in compartment 1 were replaced by 1 mM Na⁺? No potential would be generated, because no Na⁺ could flow across the membrane (which was defined as being permeable only to K⁺). However, if under these ionic conditions (10 times more Na⁺ in compartment 2) the K⁺-permeable membrane were magically replaced with a membrane permeable only to Na⁺, a potential of +58 mV would be measured at equilibrium. If 10 mM calcium (Ca²⁺) were present in compartment 2 and 1 mM Ca²⁺ in compartment 1, and a Ca²⁺-selective membrane separated the two sides, what would happen to the membrane potential? A potential of +29 mV would develop, because the valence of calcium is 2. Finally, what would the membrane potential be if 10 mM Cl⁻ were present in compartment 1 and 1 mM Cl⁻ were present in compartment 2, with the two sides separated by a Cl⁻-permeable membrane? Because the valence of this anion is −1, the potential would be +58 mV.

The balance of chemical and electrical forces at equilibrium means that the electrical potential can determine ionic fluxes across the membrane, just as the ionic gradient can determine the membrane potential. To examine the influence of membrane potential on ionic flux, imagine connecting a battery across the two sides of the membrane in Figure 2.3 to change the electrical potential across the membrane without changing the distribution of ions on the two sides. As long as the battery is off things will be just as before, with the flow of K⁺ from compartment 1 to compartment 2 causing a negative membrane potential. However, if the battery is used to make compartment 1 more negative relative to compartment 2, there will be less K⁺ flux, because the negative potential will tend to keep K⁺ in compartment 1. How negative will side 1 need to be before there is no net flux of K⁺? The answer is −58 mV, the voltage needed to counter the tenfold difference in K⁺ concentrations on the two sides of the membrane. If compartment 1 is made *more* negative than −58 mV, then K⁺ will actually flow from compartment 2 into compartment 1, because the positive ions will be attracted to the more negative potential of compartment 1. Thus, in some circumstances the electrical potential can overcome an ionic concentration gradient.

These simple examples demonstrate that the magnitude and direction of ion fluxes depend on the electrical potential imposed across the membrane, as well as on the concentration gradients. The ability to alter ion flux experimentally by changing either the potential imposed on the membrane or the transmembrane concentration gradient for an ion provides a convenient means of studying ion fluxes across the plasma membranes of neurons, as will be evident in many of the experiments described in the following chapters.

■ ELECTROCHEMICAL EQUILIBRIUM IN A MULTI-ION ENVIRONMENT

Now consider a somewhat more complex situation in which Na^+ and K^+ are unequally distributed across the membrane, as in Figure 2.4A. Imagine, for example, what would happen if 10 mM K^+ and 1 mM Na^+ were present in compartment 1, and 1 mM K^+ and 10 mM Na in compartment 2. If the membrane were permeable only to K^+, the membrane potential would be −58 mV; if the membrane were permeable only to Na^+, the potential would be +58 mV. But what would the potential be if the membrane were permeable to *both* K^+ and Na^+? In this case, the potential depends on the *relative* permeability of the membrane to K^+ and Na^+. If it is more permeable to K^+, the potential approaches −58 mV, and if it is more permeable to Na^+, the potential is closer to +58 mV. Since there is no permeability term in the Nernst equation, which only considers the simple case of a single permeant ion species, a more elaborate equation is needed that takes into account both the concentration gradients of the permeant ions and the relative permeability of the membrane to each permeant species.

Such an equation was developed by David Goldman in 1943. For the case most relevant to neurons, in which K^+, Na^+, and Cl^- are the primary permeant ions, the **Goldman equation** is written

$$V = 58 \log \frac{P_K[K]_2 + P_{Na}[Na]_2 + P_{Cl}[Cl]_1}{P_K[K]_1 + P_{Na}[Na]_1 + P_{Cl}[Cl]_2}$$

where V is the voltage across the membrane (again, compartment 1 relative to the reference compartment 2) and P indicates the permeability of the

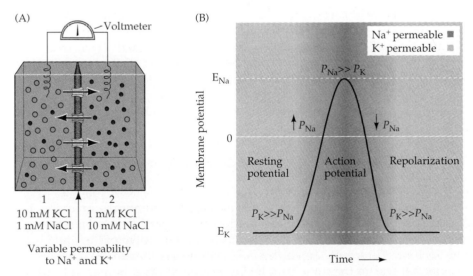

(A) Voltmeter

1   2
10 mM KCl   1 mM KCl
1 mM NaCl   10 mM NaCl

Variable permeability to Na^+ and K^+

(B)
Na^+ permeable ■
K^+ permeable ▨

$P_{Na} \gg P_K$

E_{Na}

Membrane potential

0

$\uparrow P_{Na}$ $\downarrow P_{Na}$

Resting potential   Action potential   Repolarization

$P_K \gg P_{Na}$ $P_K \gg P_{Na}$

E_K

Time ⟶

Figure 2.4 Resting and action potentials entail permeabilities to different ions. (A) Hypothetical situation in which a membrane variably permeable to Na^+ and K^+ separates two compartments that contain both ions. (B) Schematic representation of the membrane ionic permeabilities associated with resting and action potentials. At rest, neuronal membranes are more permeable to K^+ (yellow) than to Na^+ (red); accordingly, the resting membrane potential is negative and approaches E_K. During an action potential, the membrane becomes very permeable to Na^+ (red); thus the membrane potential becomes positive and approaches E_{Na}. The rise in Na permeability is transient, however, so that the membrane again becomes primarily permeable to K^+ (yellow), causing the potential to return to its negative resting value.

membrane to each ion of interest. The Goldman equation is thus an extended version of the Nernst equation that takes into account the relative permeabilities of each of the ions involved. The relationship between the two equations becomes obvious in the situation where the membrane is permeable only to one ion, say, K^+; in this case, the Goldman expression collapses back to the simpler Nernst equation. In this context, it is important to note that the valence factor (z) in the Nernst equation has been eliminated; this is why the concentrations of negatively charged chloride ions, Cl^-, have been inverted relative to the concentrations of the positively charged ions [remember that $-\log (A/B) = \log (B/A)$].

If the membrane in Figure 2.4A is permeable to only K^+ and Na^+, the terms involving Cl^- drop out because P_{Cl} is 0. In this case, solution of the Goldman equation yields a potential of -58 mV when only K^+ is permeant, $+58$ mV when only Na^+ is permeant, and some intermediate value if both ions are permeant. For example, if K^+ and Na^+ are equally permeant, then the potential would be 0 mV.

It is particularly pertinent to ask what would happen if the membrane started out being permeable to K^+, then temporarily switched to become most permeable to Na^+. In this circumstance, the membrane potential would start out at a negative level, become positive while the Na^+ permeability remained high, and fall back to a negative level as the Na^+ permeability decreased again. As it turns out, this last case essentially describes what goes on in a neuron during the generation of an action potential. In the resting state, P_K of the neuronal plasma membrane is much higher than P_{Na}; since, as a result of the action of ion pumps, there is always more K^+ inside the cell than outside (Table 2.1), the resting potential is negative (Figure 2.4B). As the membrane potential is depolarized (by synaptic action, for example), P_{Na} increases. The transient increase in Na^+ permeability causes the membrane potential to become even more positive, because Na^+ rushes in (remember that there is much more Na^+ outside a neuron than inside, again as a result of ion pumps). Because of this positive feedback loop, an action potential occurs. This rise in Na^+ permeability during the action potential is transient. As the membrane permeability to K^+ is restored, the membrane potential quickly returns to its resting level.

Armed with this appreciation of a few electrochemical principles, it will be much easier to understand the more detailed account that follows of how neurons generate resting and action potentials.

■ THE IONIC BASIS OF THE RESTING MEMBRANE POTENTIAL

The action of ion pumps creates substantial transmembrane gradients for most ions. Table 2.1 summarizes the ion concentrations measured in an exceptionally large nerve cell found in the nervous system of the squid (Box A). Such measurements are the basis for stating that there is much more K^+ inside the neuron than out, and much more Na^+ outside than in. Similar concentration gradients occur in the neurons of most animals, including humans. However, because the ionic strength of mammalian blood is lower than that of sea-dwelling animals such as squid, the concentrations of each ion in mammals are several times lower. These pump-dependent concentration gradients are the source of the resting neuronal membrane potential and, indirectly, the action potential.

Once the ion concentration gradients across various neuronal membranes are known, the Nernst equation can be used to calculate that the

TABLE 2.1
Extracellular and Intracellular Ion Concentrations

Ion	Concentration (mM)	
	Intracellular	*Extracellular*
Squid neuron		
Potassium (K^+)	400	20
Sodium (Na^+)	50	440
Chloride (Cl^-)	40–150	560
Calcium (Ca^{2+})	0.0001	10
Mammalian neuron		
Potassium (K^+)	140	5
Sodium (Na^+)	5–15	145
Chloride (Cl^-)	4–30	110
Calcium (Ca^{2+})	0.0001	1–2

equilibrium potential for K$^+$ is more negative than that of any other major ion. Since the resting membrane potential of the squid neuron is approximately −65 mV, K$^+$ is the ion that is closest to electrochemical equilibrium when the cell is at rest. This fact implies that the resting membrane is more permeable to K$^+$ than to the other ions listed in Table 2.1.

It is possible to test this guess, as Alan Hodgkin and Bernard Katz did in 1949, by asking what happens to the resting membrane potential as the concentration of K$^+$ outside the neuron is altered. If the resting membrane is permeable only to K$^+$, then the Goldman equation (or even the simpler Nernst equation) predicts that the membrane potential will vary in proportion to the logarithm of the K$^+$ concentration gradient across the membrane. Assuming that the internal K$^+$ concentration is unchanged during the experiment, a plot of membrane potential against the logarithm of the external K$^+$ concentration will yield a straight line with a slope of 58 mV per tenfold change in external K$^+$ concentration at room temperature (see Figure 2.3C). (The slope becomes about 61 mV at mammalian body temperatures.)

When Hodgkin and Katz performed this experiment on a living squid neuron, they found that the resting membrane potential did indeed change when the external K$^+$ concentration was modified, becoming less negative as external K$^+$ concentration was raised (Figure 2.5A). When the external K$^+$ concentration was raised high enough to equal the concentration of K$^+$ inside the neuron, thus making the K$^+$ equilibrium potential 0 mV, the resting membrane potential was also approximately 0 mV. In short, the resting membrane potential varied with the logarithm of the K$^+$ concentration, with a maximal slope that approached the predicted 58 mV per tenfold change in K$^+$ concentration (Figure 2.5B). The value obtained was not exactly 58 mV because other ions, such as Cl$^-$ and Na$^+$, are also slightly permeable, and thus influence the resting potential to a small degree. The contribution of these other ions is particularly evident at low external K$^+$ levels, precisely as predicted by the Goldman equation. In general, however, manipulation of

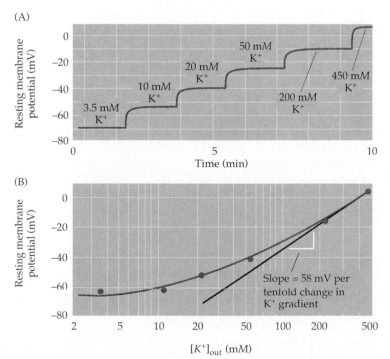

Figure 2.5 Experimental evidence that the resting membrane potential of a squid giant axon is determined by the K$^+$ concentration gradient across the membrane. (A) Increasing the external K$^+$ concentration makes the resting membrane potential more positive. (B) Relationship between resting membrane potential and external K$^+$ concentration, plotted on a semilogarithmic scale. The straight line represents a slope of 58 mV per tenfold change in concentration, as given by the Nernst equation. (After Hodgkin and Katz, 1949.)

Box A
THE REMARKABLE GIANT NERVE CELLS OF SQUID

Many of the initial insights into how ion concentration gradients and changes in membrane permeability produce electrical signals came from experiments performed on the extraordinarily large nerve cells of the squid. The axons of these nerve cells can be up to 1 mm in diameter—100 to 1000 times larger than mammalian axons. Squid axons are large enough to allow experiments that would be impossible on most other nerve cells. For example, it is not difficult to insert simple wire electrodes inside these giant axons and make reliable electrical measurements. The relative ease of this approach yielded the first intracellular recordings of action potentials from nerve cells and, as will be discussed in the next chapter, the first experimental measurements of the ionic currents that produce action potentials.

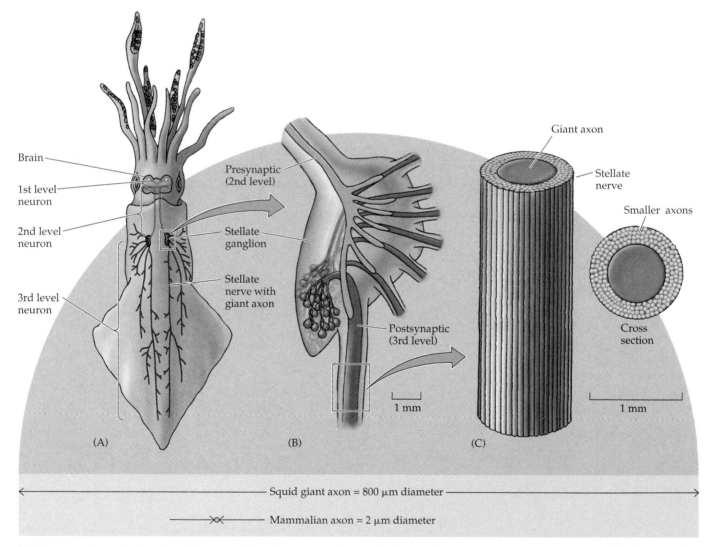

Brain

1st level neuron

2nd level neuron

3rd level neuron

(A)

Presynaptic (2nd level)

Stellate ganglion

Stellate nerve with giant axon

Postsynaptic (3rd level)

1 mm

(B)

Giant axon

Stellate nerve

Smaller axons

Cross section

1 mm

(C)

Squid giant axon = 800 μm diameter

Mammalian axon = 2 μm diameter

(A) Diagram of a squid, showing the location of its giant nerve cells. Different colors indicate the neuronal components of the escape circuitry. The first- and second-level neurons originate in the brain, while the third-level neurons are in the stellate ganglion and innervate muscle cells of the mantle. (B) Giant synapses within the stellate ganglion. The second-level neuron forms a series of fingerlike processes, each of which makes an extraordinarily large synapse with a single third-level neuron. (C) Structure of a giant axon of a third-level neuron lying within its nerve. The difference in the diameters of a squid giant axon and a mammalian axon are shown below.

It also is practical to extrude the cytoplasm from giant axons and measure its ionic composition (see Table 2.1). In addition, some giant nerve cells form synaptic contacts with other giant nerve cells, producing very large synapses that have been extraordinarily valuable in understanding the fundamental mechanisms of synaptic transmission (see Chapter 5).

Giant neurons evidently evolved in squid because they enhanced survival. These neurons participate in a simple neural circuit that activates the contraction of the mantle muscle, producing a jet propulsion effect that allows the squid to move away from predators at a remarkably fast speed. As discussed in Chapter 3, larger axonal diameter allows faster conduction of action potentials. Thus, squid presumably have these huge nerve cells to escape more successfully from their numerous enemies. Today—nearly 60 years after their discovery by John Z. Young at University College, London—the giant nerve cells of squid remain useful experimental systems for probing basic neuronal functions.

References

LLINÁS, R. (1982) Calcium in synaptic transmission. Sci. Am. 247(4): 56–65.

YOUNG, J. Z. (1939) Fused neurons and synaptic contacts in the giant nerve fibres of cephalopods. Phil. Trans. R. Soc. Lond. B 229: 465–503.

the external concentrations of these other ions has only a small effect, emphasizing that K$^+$ permeability is the primary source of the resting membrane potential.

In summary, Hodgkin and Katz showed that the inside-negative resting potential of neurons arises because (1) the membrane of the resting neuron is more permeable to K$^+$ than to any of the other ions present, and (2) there is more K$^+$ inside the neuron than outside. The selective permeability to K$^+$ is caused by K$^+$-permeable membrane channels that are open in resting neurons, and the large K$^+$ concentration gradient is produced by membrane pumps that selectively accumulate K$^+$ within neurons. (The structure and function of these channels and pumps are described in Chapter 4.) Many subsequent studies on neurons have confirmed the general validity of these principles.

■ THE IONIC BASIS OF ACTION POTENTIALS

What causes the membrane potential of a neuron to depolarize during an action potential? Although a general answer to this question has been given (increased permeability to Na$^+$), it is well worth reasoning out in more detail. Given the data presented in Table 2.1, one can use the Nernst equation to calculate that the equilibrium potential for Na$^+$ (E_{Na}) in neurons, and indeed in most cells, is positive. Thus, if the membrane were to become highly permeable to Na$^+$, the membrane potential would approach E_{Na}. Based on these considerations, Hodgkin and Katz hypothesized that the action potential arises because the neuronal membrane becomes temporarily permeable to Na$^+$.

Taking advantage of the same style of ion substitution experiment they used to assess the resting potential, Hodgkin and Katz tested the role of Na$^+$ in generating the action potential by asking what happens to the action potential when Na$^+$ is removed from the external medium. They found that lowering the external Na$^+$ concentration reduces both the rate of rise of the action potential and its peak amplitude (Figure 2.6A–C). Indeed, when they examined this Na$^+$ dependence quantitatively, they found a more-or-less linear relationship between the amplitude of the action potential and the logarithm of the external Na$^+$ concentration (Figure 2.6D). The slope of this

Figure 2.6 The role of sodium in the generation of an action potential in a squid giant axon. (A) An action potential evoked with the normal ion concentrations inside and outside the cell. (B) The amplitude and rate of rise of the action potential diminish when external sodium concentration is reduced to one-third of normal, but (C) recover when the Na^+ is replaced. (D) While the amplitude of the action potential is quite sensitive to the external concentration of Na^+, the resting membrane potential (E) is little affected by changing the concentration of this ion. (After Hodgkin and Katz, 1949.)

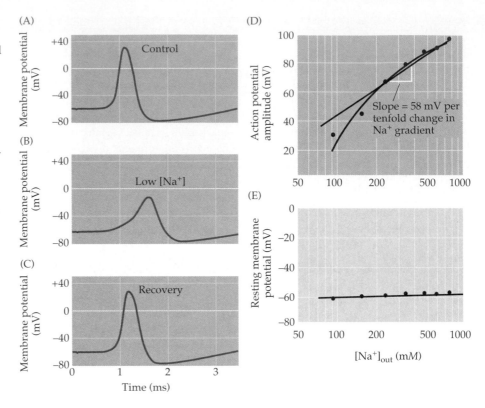

relationship approached a value of 58 mV per tenfold change in Na^+ concentration, as expected for a membrane selectively permeable to Na^+. In contrast, lowering Na^+ concentration had very little effect on the resting membrane potential (Figure 2.6E). Thus, while the resting neuronal membrane is only slightly permeable to Na^+, the membrane becomes extraordinarily permeable to Na^+ during the **rising** and **overshoot phases** of the action potential. (Box B provides an explanation of action potential nomenclature.) This temporary increase in Na^+ permeability results from the opening of Na^+-selective channels that are essentially closed in the resting state. Recall further that membrane pumps maintain a large electrochemical gradient for Na^+, which is in much higher concentration outside the neuron than in. Thus, when the Na^+ channels open, Na^+ flows into the neuron, causing the membrane potential to depolarize and approach E_{Na}.

The time that the membrane potential lingers near E_{Na} (about +58 mV) during the overshoot of an action potential is brief because the increased membrane permeability to Na^+ is short-lived. The membrane potential rapidly repolarizes, and after this falling phase there is a transient **undershoot**. As indicated in the following chapter, these latter events in the action potential cycle are largely due to an increase in the K^+ permeability of the membrane. During the undershoot, the membrane potential is transiently hyperpolarized because K^+ permeability is even greater than at rest. The action potential ends when this phase of enhanced K^+ permeability subsides and the membrane potential returns to its normal resting level.

The ion substitution experiments carried out by Hodgkin and Katz provide convincing evidence that the resting membrane potential results from a high resting membrane permeability to K^+, whereas depolarization during an action potential results from a transient rise in membrane Na^+ permeability. Although these experiments identified the ions that flow dur-

Box B
ACTION POTENTIAL FORM AND NOMENCLATURE

The action potential of the squid giant axon has a characteristic shape, or waveform, with a number of different phases (Figure A). During the rising phase, the membrane potential rapidly depolarizes. In fact, action potentials cause the membrane potential to depolarize so much that the membrane potential transiently becomes positive with respect to the external medium, producing an overshoot. The overshoot of the action potential gives way to a falling phase in which the membrane potential rapidly repolarizes. Repolarization takes the membrane potential to levels even more negative than the resting membrane potential for a short time; this brief period of hyperpolarization is called the undershoot.

Although the waveform of the squid action potential is typical, the details of the action potential form vary widely from neuron to neuron in different animals. In myelinated axons of vertebrate motor neurons (Figure B), the action potential is virtually indistinguishable from that of the squid axon. However, the action potential recorded in the cell body of this same motor neuron (Figure C) looks rather different. Thus, the action potential waveform can vary even within the same neuron. More

complex action potentials are seen in other central neurons. For example, action potentials recorded from the cell bodies of neurons in the mammalian inferior olive (a region of the brainstem involved in motor control) last tens of milliseconds (Figure D). These action potentials exhibit a pronounced plateau during their falling phase, and their undershoot lasts even longer than that of the motor neuron. One of the most dramatic types of action potentials occurs in the cell bodies of cerebellar Purkinje neurons (Figure E). These potentials have several complex phases that result from the summation of multiple, discrete action potentials.

The variety of action potential waveforms could mean that each type of neuron has a different mechanism of action potential production. Fortunately, however, these diverse waveforms all result from relatively minor variations in the scheme used by the squid giant axon. For example, plateaus in the repolarization phase result from the presence of ion channels that are permeable to Ca^{2+}, and long-lasting undershoots result from the presence of extra types of K^+ channels. The complex action potential of the Purkinje cell results from these extra features

plus the fact that different types of action potentials are generated in various parts of the Purkinje neuron—cell body, dendrites, and axons—and are summed together in recordings from the cell body. Thus, the lessons learned from the squid axon are applicable to, and indeed essential for, understanding action potential generation in all neurons.

References

BARRETT, E. F. AND J. N. BARRETT (1976) Separation of two voltage-sensitive potassium currents, and demonstration of a tetrodotoxin-resistant calcium current in frog motoneurones. J. Physiol. (London) 255: 737–774.

DODGE, F. A. AND B. FRANKENHAEUSER (1958) Membrane currents in isolated frog nerve fibre under voltage clamp conditions. J. Physiol. (London) 143: 76–90.

HODGKIN, A. L. AND A. F. HUXLEY (1939) Action potentials recorded from inside a nerve fibre. Nature 144: 710–711.

LLINÁS, R. AND M. SUGIMORI (1980) Electrophysiological properties of in vitro Purkinje cell dendrites in mammalian cerebellar slices. J. Physiol. (London) 305: 197–213.

LLINÁS, R. AND Y. YAROM (1981) Electrophysiology of mammalian inferior olivary neurones in vitro. Different types of voltage-dependent ionic conductances. J. Physiol. (London) 315: 549–567.

(A) 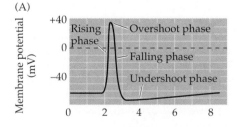 (B) 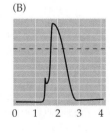 (C) 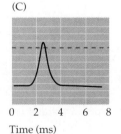 (D) 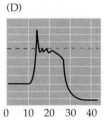 (E)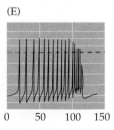

Time (ms)

(A) The phases of an action potential of the squid giant axon. (B) Action potential recorded from a myelinated axon of a frog motor neuron. (C) Action potential recorded from the cell body of a frog motor neuron. The action potential is smaller and the undershoot prolonged in comparison to the action potential recorded from the axon of this same neuron (B). (D) Action potential recorded from the cell body of a neuron from the inferior olive of a guinea pig. This action potential has a pronounced plateau during its falling phase. (E) Action potential recorded from the cell body of a Purkinje neuron in the cerebellum of a guinea pig. (A after Hodgkin and Huxley, 1939; B after Dodge and Frankenhaeuser, 1958; C after Barrett and Barrett, 1976; D after Llinás and Yarom, 1981; E after Llinás and Sugimori, 1980.)

ing an action potential, they did not establish *how* the neuronal membrane is able to change its ionic permeability to generate the action potential, or what mechanisms trigger this critical change. The next chapter addresses this issue, with the surprising conclusion that the neuronal membrane potential itself affects the membrane permeability. In short, the membrane potential changes the membrane permeability to generate action potentials.

■ SUMMARY

Nerve cells generate electrical signals to convey information over substantial distances and to transmit it to other cells by means of synapses. The action potential—the signal that conveys information along nerve cell axons—ultimately depends on the resting electrical potential across the neuronal membrane. A difference in electrical potential is generated whenever ions flow across cell membranes. Such ion fluxes occur when the membranes are permeable to one or more ion species, and when there is an electrochemical gradient that favors ion flow. At rest, a negative membrane potential (the resting potential) results from a net efflux of K^+ across neuronal membranes that are predominantly permeable to K^+. An action potential occurs when a transient rise in Na^+ permeability allows a net flow of Na^+ across a membrane that is now predominantly permeable to Na^+. The brief rise in membrane Na^+ permeability is followed by a secondary, transient rise in membrane K^+ permeability that repolarizes the neuronal membrane and produces a brief undershoot of the action potential. As a result, the membrane is depolarized in an all-or-none fashion. When these active permeability changes subside, the membrane potential returns to its resting level because of the high resting membrane permeability to K^+.

Additional Reading

Reviews

HODGKIN, A. L. (1951) The ionic basis of electrical activity in nerve and muscle. Biol. Rev. 26: 339–409.

HODGKIN, A. L. (1958) The Croonian Lecture: Ionic movements and electrical activity in giant nerve fibres. Proc. R. Soc. Lond. (B) 148: 1–37.

Important Original Papers

BAKER, P. F., A. L. HODGKIN AND T. I. SHAW (1962) Replacement of the axoplasm of giant nerve fibres with artificial solutions. J. Physiol. (London) 164: 330–354.

COLE, K. S. AND H. J. CURTIS (1939) Electric impedence of the squid giant axon during activity. J. Gen. Physiol. 22: 649–670.

GOLDMAN, D. E. (1943) Potential, impedance, and rectification in membranes. J. Gen. Physiol. 27: 37–60.

HODGKIN, A. L. AND P. HOROWICZ (1959) The influence of potassium and chloride ions on the membrane potential of single muscle fibres. J. Physiol. (London) 148: 127–160.

HODGKIN, A. L. AND B. KATZ (1949) The effect of sodium ions on the electrical activity of the giant axon of the squid. J. Physiol. (London) 108: 37–77.

HODGKIN, A. L. AND R. D. KEYNES (1953) The mobility and diffusion coefficient of potassium in giant axons from *Sepia*. J. Physiol. (London) 119: 513–528.

HODGKIN, A. L. AND R. D. KEYNES (1955) Active transport of cations in giant axons from *Sepia* and *Loligo*. J. Physiol. (London) 128: 28–60.

KEYNES, R. D. (1951) The ionic movements during nevous activity. J. Physiol. (London) 114: 119–150.

Books

HODGKIN, A. L. (1967) *The Conduction of the Nervous Impulse*. Springfield, IL: Charles C. Thomas.

HODGKIN, A. L. (1992) *Chance and Design*. Cambridge: Cambridge University Press.

JUNGE, D. (1992) *Nerve and Muscle Excitation*, 3rd Ed.. Sunderland, MA: Sinauer Associates.

KATZ, B. (1966) *Nerve, Muscle, and Synapse*. New York: McGraw-Hill.

■ OVERVIEW

The action potential, the primary electrical signal generated by nerve cells, reflects changes in membrane permeability to specific ions. Contemporary understanding of membrane permeability is based on evidence obtained by the voltage clamp technique, which permits detailed characterization of permeability changes as a function of membrane potential and time. For most types of axons, these changes consist of a rapid and transient rise in sodium permeability, followed by a slower but more sustained rise in potassium permeability. Both permeabilities are voltage-dependent, increasing as the membrane potential depolarizes. The kinetics and voltage dependence of Na^+ and K^+ permeabilities provide a complete explanation of action potential generation. Depolarizing the membrane potential to the threshold level causes a rapid, self-sustaining increase in Na^+ permeability that produces the rising phase of the action potential; however, the Na^+ permeability increase is short-lived and is followed by a slower increase in K^+ permeability that restores the membrane potential to its usual negative resting level. A mathematical model that describes the behavior of these permeabilities predicts virtually all of the observed properties of action potentials. This same ionic mechanism also permits action potentials to be propagated along the length of neuronal axons, explaining how electrical signals are conveyed throughout the nervous system.

■ IONIC CURRENTS ACROSS NERVE CELL MEMBRANES

The previous chapter introduced the idea that nerve cells generate electrical potentials by virtue of a membrane that is differentially permeable to various ion species. In particular, a transient increase in the permeability of the neuronal membrane to Na^+ initiates the action potential. This chapter considers exactly how this increase in Na^+ permeability occurs. A key to understanding this phenomenon is the observation that action potentials occur *only* when the neuronal membrane potential becomes more positive than a certain threshold level. This relationship suggests that the mechanism responsible for the increase in Na^+ permeability is sensitive to the membrane potential. Therefore, if one could understand how a change in membrane potential activates Na^+ permeability, it should be possible to explain how action potentials are generated.

The fact that the Na^+ permeability that generates the membrane potential change is itself potential-sensitive presents both conceptual and practical obstacles to studying the mechanism of the action potential. A practical problem is the difficulty of systematically varying the membrane potential to study the permeability change, because such changes in membrane potential will produce an action potential, which will cause further, uncontrolled, changes in the membrane potential. Historically, then, it was not really possible to understand action potentials until a technique was developed that allowed experimenters to control membrane potential and simultaneously measure the underlying permeability changes. This technique, the **voltage**

VOLTAGE-DEPENDENT MEMBRANE PERMEABILITY

clamp method (Box A), provides all the information needed to define the ionic permeability of the membrane at any level of membrane potential.

In the late 1940s, Alan Hodgkin and Andrew Huxley used the voltage clamp technique to work out the permeability changes underlying the action potential. They again chose to use the giant neuron of the squid because its large size (up to 1 mm in diameter; see Box A in Chapter 2) allowed insertion

Box A
THE VOLTAGE CLAMP METHOD

Breakthroughs in scientific research often rely on the development of new technologies. In the case of the action potential, detailed understanding came only after of the invention of the voltage clamp technique by Kenneth Cole in the 1940s. This device is called a voltage clamp because it controls, or clamps, membrane potential (or voltage) at any level desired by the experimenter. The method measures the membrane potential with a microelectrode (or other type of electrode) placed inside the cell (1) electronically comparing this voltage to the voltage to be maintained, the command voltage (2). The clamp circuitry then passes a current back into the cell

though another intracellular electrode (3). This electronic feedback circuit holds the membrane potential at the desired level, even in the face of permeability changes that would normally alter the membrane potential (such as those generated during the action potential). Most importantly, the device permits the simultaneous measurement of the current needed to keep the cell at a given voltage (4). Therefore, the voltage clamp technique can indicate how membrane potential influences ionic current flow across the membrane. This information gave Hodgkin and Huxley the key insights that led to their model for action potential generation.

Today, the voltage clamp method remains widely used to study ionic currents in neurons and other cells. The most popular contemporary version of this approach is the patch clamp technique, a method that can be applied to virtually any cell and has a resolution high enough to measure the minute electrical currents flowing through single ion channels (see Box A in Chapter 4).

References

COLE, K. S. (1968) *Membranes, Ions and Impulses: A Chapter of Classical Biophysics.* Berkeley, CA: University of California Press.

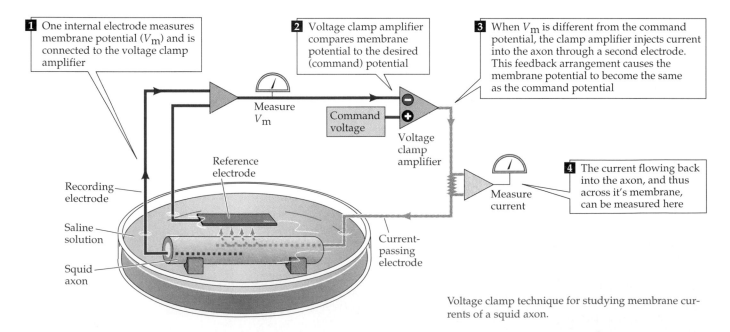

1 One internal electrode measures membrane potential (V_m) and is connected to the voltage clamp amplifier

2 Voltage clamp amplifier compares membrane potential to the desired (command) potential

3 When V_m is different from the command potential, the clamp amplifier injects current into the axon through a second electrode. This feedback arrangement causes the membrane potential to become the same as the command potential

Measure V_m

Command voltage

Voltage clamp amplifier

Reference electrode

Recording electrode

Saline solution

Squid axon

Current-passing electrode

Measure current

4 The current flowing back into the axon, and thus across it's membrane, can be measured here

Voltage clamp technique for studying membrane currents of a squid axon.

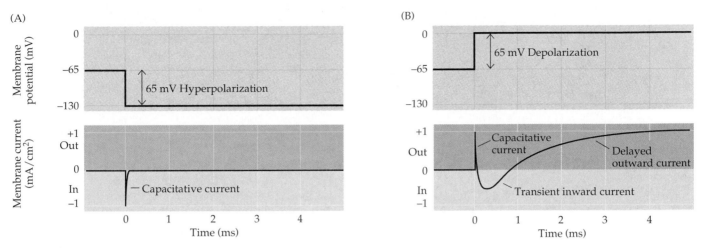

Figure 3.1 Current flow across a squid axon membrane during a voltage clamp experiment. (A) A 65-mV hyperpolarization of the membrane potential produces only a very brief capacitive current. (B) A 65-mV depolarization of the membrane potential also produces a brief capacitive current, but this is followed by a longer-lasting but transient phase of inward current and a delayed but sustained outward current. (After Hodgkin et al., 1952.)

of the electrodes necessary for voltage clamping. Hodgkin and Huxley were the first investigators to test directly the hypothesis that potential-sensitive Na^+ and K^+ permeability changes are both necessary and sufficient for the production of action potentials.

Hodgkin and Huxley's first goal was to determine whether neuronal membranes do, in fact, have voltage-dependent permeabilities. To address this issue, they asked whether ionic currents flow across the membrane when its potential is changed. The result of one such experiment is shown in Figure 3.1. Figure 3.1A illustrates the currents produced by a squid axon when its membrane potential, V_m, is hyperpolarized from the resting level of −65 mV to −130 mV. The initial response of the axon results from the redistribution of charge across the axonal membrane. This capacitative current is nearly instantaneous, ending within a fraction of a millisecond. Aside from this brief event, very little current flows when the membrane is hyperpolarized. However, when the membrane potential is depolarized from −65 mV to 0 mV, the response is quite different (Figure 3.1B). Following the capacitive current, the axon produces a rapidly rising inward ionic current (*inward* refers to a positive charge entering the cell—that is, cations in or anions out), which gives way to a more slowly rising, delayed outward current. The fact that membrane depolarization elicits these ionic currents establishes that the membrane permeability of axons is indeed voltage-dependent.

■ TWO TYPES OF VOLTAGE-DEPENDENT IONIC CURRENT

The results shown in Figure 3.1 demonstrate that the ionic permeability of neuronal membranes is voltage-sensitive, but the experiments do not identify how many types of permeability exist, or which ions are involved. As discussed in Chapter 2, varying the potential across a membrane makes it possible to deduce the equilibrium potential for the ionic fluxes through the membrane, and thus to identify the ions that are flowing. Since the voltage

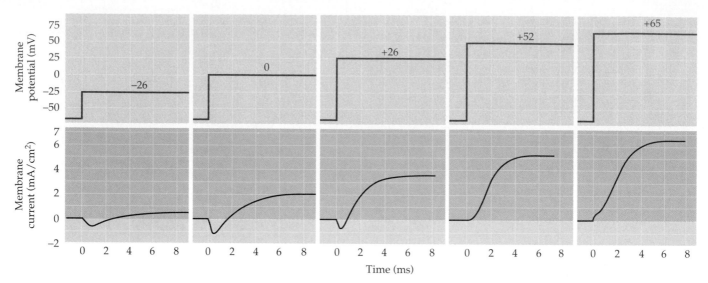

Figure 3.2 Current produced by membrane depolarizations to several different potentials. The early current first increases, then decreases in magnitude as the depolarization increases; note that this current is reversed in polarity at potentials more positive than about +55 mV. In contrast, the late current increases monotonically with increasing depolarization. (After Hodgkin et al., 1952.)

clamp method allows the membrane potential to be changed while ionic currents are being measured, it was a straightforward matter for Hodgkin and Huxley to determine ionic permeability by examining how the properties of the early inward and late outward currents changed as the membrane potential was varied (Figure 3.2). As already noted, no appreciable ionic currents flow at membrane potentials more negative than the resting potential. At more positive potentials, however, the currents not only flow but change in magnitude. The early current has a U-shaped dependence on membrane potential, increasing over a range of depolarizations up to approximately 0 mV but decreasing as the potential is depolarized further. In contrast, the late current increases monotonically with increasingly positive membrane potentials. These different responses to membrane potential can be seen more clearly when the magnitudes of the two current components are plotted as a function of membrane potential, as in Figure 3.3.

The voltage sensitivity of the early inward current gives an important clue about the nature of the ions carrying the current, namely, that no current flows when the membrane potential is clamped at +52 mV. For the squid neurons studied by Hodgkin and Huxley, the external Na^+ concentration is 440 mM, and the internal Na^+ concentration is 50 mM. For this concentration gradient, the Nernst equation predicts that the equilibrium potential for Na^+ should be +55 mV. Recall further from Chapter 2 that at the Na^+

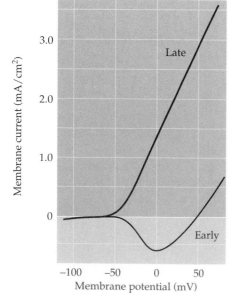

Figure 3.3 Relationship between current amplitude and membrane potential, taken from experiments such as the one shown in Figure 3.2. Whereas the late outward current increases steeply with increasing depolarization, the early inward current first increases in magnitude, but then decreases and reverses to outward current at about +55 mV (the sodium equilibrium potential). (After Hodgkin et al., 1952.)

equilibrium potential, there is no net flux of Na$^+$ across the membrane, even if the membrane is highly permeable to Na$^+$. Thus, the experimental observation that no current flows at the membrane potential where Na$^+$ cannot flow is a strong indication that the early inward current is carried by entry of Na$^+$ into the axon.

An even more compelling way to test whether Na$^+$ carries the early inward current is to examine the behavior of this current after removing external Na$^+$. Removing the Na$^+$ outside the axon makes E_{Na} negative; if the permeability to Na$^+$ is increased under these conditions, current should flow outward as Na$^+$ leaves the neuron, due to the reversed concentration gradient. When Hodgkin and Huxley performed such an experiment, they obtained the result shown in Figure 3.4. Removing external Na$^+$ caused the early inward current to reverse its polarity and become an outward current at a membrane potential that gave rise to an inward current when external Na$^+$ was present. This result demonstrates convincingly that the early inward current measured when Na$^+$ is present in the external medium must be due to Na$^+$ entering the neuron.

Notice that removal of external Na$^+$ in the experiment shown in Figure 3.4 has little effect on the outward current that flows after the neuron has been kept at a depolarized membrane voltage for several milliseconds. This result shows that the late outward current must be due to the flow of an ion other than Na$^+$. Several lines of evidence presented by Hodgkin, Huxley, and others showed that this late outward current is caused by K$^+$ exiting the neuron. Perhaps the most compelling demonstration of K$^+$ involvement is that the amount of K$^+$ efflux from the neuron, measured by loading the neuron with radioactive K$^+$, is closely correlated with the magnitude of the late outward current.

Taken together, these voltage clamp experiments show that changing the membrane potential to a level more positive than the resting potential produces two effects: an early influx of Na$^+$ into the neuron, followed by a delayed efflux of K$^+$. The early influx of Na$^+$ produces a transient inward current, whereas the delayed efflux of K$^+$ produces a sustained outward current. The differences in the time course and ion selectivity of the two fluxes suggest that two different ionic permeability mechanisms are activated by changes in membrane potential. Further evidence that there are two distinct mechanisms has come from pharmacological studies of drugs that specifically affect these two currents (Figure 3.5). **Tetrodotoxin**, an alkaloid neurotoxin found in certain puffer fish, tropical frogs, and salamanders, blocks the Na$^+$ current without affecting the K$^+$ current. Conversely, **tetraethylammonium ions** block K$^+$ currents without affecting Na$^+$ currents. The differential sensitivity of Na$^+$ and K$^+$ currents to these drugs provides strong additional evidence that Na$^+$ and K$^+$ flow through independent permeability pathways. As discussed in Chapter 4, it is now known that these pathways are ion channels that are selectively permeable to either Na$^+$ or K$^+$. In fact, tetrodotoxin, tetraethylammonium, and other drugs that interact with specific types of ion channels have been extraordinarily useful tools in characterizing these channel molecules (Box B).

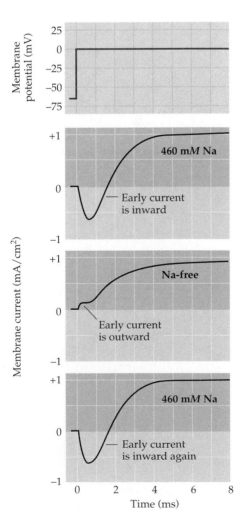

Figure 3.4 Dependence of the early inward current on sodium. In the presence of normal external concentrations of Na$^+$, depolarization of a squid axon to 0 mV produces an inward initial current. However, removal of external Na$^+$ causes the initial inward current to become outward, an effect that is reversed by restoration of external Na$^+$. (After Hodgkin and Huxley, 1952a.)

Figure 3.5 Pharmacological separation of Na⁺ and K⁺ currents into sodium and potassium components. Panel (1) shows the curent that flows when the membrane potential of a squid axon is depolarized to −10 mV in control conditions. (2) Treatment with tetrodotoxin causes the early Na⁺ currents to disappear but spares the late K⁺ currents. (3) Addition of tetraethylammonium blocks the K⁺ currents without affecting the Na⁺ currents. (After Moore et al., 1967 and Armstrong and Binstock, 1965.)

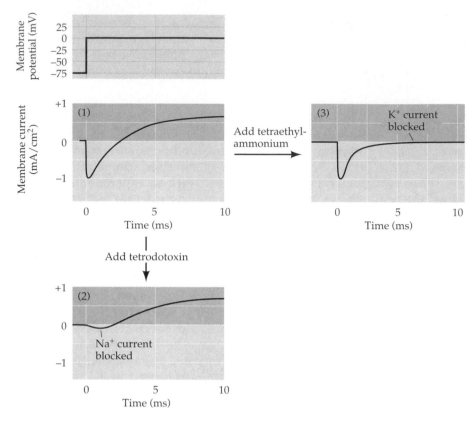

■ TWO VOLTAGE-DEPENDENT MEMBRANE CONDUCTANCES

The next goal Hodgkin and Huxley set for themselves was to describe Na⁺ and K⁺ permeability changes mathematically. To do so, they assumed that the ionic currents are due to a change in **membrane conductance**, defined as the reciprocal of the membrane resistance; thus, membrane conductance is closely related (although not identical) to membrane permeability. When evaluating ionic movements from an electrical standpoint, it is convenient to describe them in terms of ionic conductances rather than ionic permeabilities. For present purposes, permeability and conductance can be considered synonymous. If membrane conductance (g) obeys Ohm's Law (which states that voltage is equal to the product of current and resistance), then the ionic current that flows during an increase in membrane conductance is given by

$$I_{ion} = g_{ion}\left(V_m - E_{ion}\right)$$

where I_{ion} is the ionic current, V_m is the membrane potential, and E_{ion} is the equilibrium potential for the ion flowing through the conductance, g_{ion}. The difference between V_m and E_{ion} is the electrochemical driving force acting on the ion.

Hodgkin and Huxley used this simple relationship to calculate the dependence of Na⁺ and K⁺ conductances on time and membrane potential. Specifically, they measured the difference between currents recorded in the presence and absence of external Na⁺ (as shown in Figure 3.4) to determine separately the currents carried by Na⁺ and K⁺—I_{Na} and I_K. Since they already knew V_m, E_{Na}, and E_K, they could calculate g_{Na} and g_K. Hodgkin and Huxley were able to draw two fundamental conclusions about these conductances. The first conclusion is that both conductances are voltage-dependent—that is,

Box B
TOXINS THAT POISON ION CHANNELS

Given the importance of Na$^+$ and K$^+$ channels for neuronal excitation, it is not surprising that a number of organisms have evolved channel-specific toxins as mechanisms for self-defense or for capturing prey. A rich collection of natural toxins selectively target the ion channels of neurons and other cells. These toxins are valuable not only for survival, but for studying the function of cellular ion channels. The best-known channel toxin is tetrodotoxin, which is produced by certain puffer fish and other animals. Tetrodotoxin produces a potent and specific obstruction of the Na$^+$ channels responsible for action potential generation, thereby paralyzing the animals unfortunate enough to ingest it. Saxitoxin, a chemical homologue of tetrodotoxin produced by dinoflagellates, has a similar action on Na$^+$ channels. The potentially lethal effects of eating shellfish that have ingested these "red tide"

dinoflagellates are due to the potent neuronal actions of saxitoxin.

Scorpions paralyze their prey by injecting a potent mix of peptide toxins that also affect ion channels. Among these are the α-toxins, which slow the inactivation of Na$^+$ channels (Figure A1); exposure of neurons to these toxins prolongs the action potential (Figure A2), thereby scrambling information flow within the nervous system of the soon-to-be-devoured victim. Other peptides in scorpion venom, called β-toxins, shift the voltage dependence of Na$^+$ channel activation (Figure B). These toxins cause Na$^+$ channels to open at potentials much more negative than normal, disrupting action potential generation. Some alkaloid toxins combine these actions, both removing inactivation *and* shifting activation of Na$^+$ channels. One such toxin is batrachotoxin, produced by a species of frog; some tribes of South American

Indians use this poison on their arrow tips. A number of plants produce similar toxins, including aconitine, from buttercups; veratridine, from lilies; and a number of insecticidal toxins produced by plants such as chrysanthemums and rhododendrons.

Potassium channels have also been targeted by toxin-producing organisms. Peptide toxins affecting K$^+$ channels include dendrotoxin, from wasps; apamin, from bees; and charybdotoxin, yet another toxin produced by scorpions. All of these toxins block K$^+$ channels as their primary action; no toxin is known to affect the activation or inactivation of these channels, although such agents may simply be awaiting discovery.

References

CAHALAN, M. (1975) Modification of sodium channel gating in frog myelinated nerve fibers by *Centruroides sculpturatus* scorpion venom. J. Physiol. (Lond.) 244: 511–534.

SCHMIDT, O. AND H. SCHMIDT (1972) Influence of calcium ions on the ionic currents of nodes of Ranvier treated with scorpion venom. Pflügers Arch. 333: 51-61.

STANSFELD, C. E., S. J. MARSH, D. N. PARCEJ, J. O. DOLLY, AND D. A. BROWN (1987) Mast cell degranulating peptide and dendrotoxin selectively inhibit a fast-activating potassium current and bind to common neuronal proteins. Neurosci. 23: 893–902.

(A)

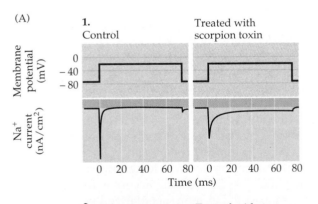

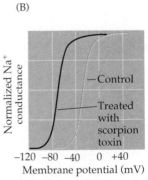

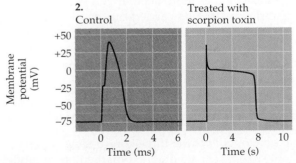

(A) Effects of toxin treatment on frog axons. (1) α-Toxin from the scorpion *Leiurus quinquestriatus* prolongs Na$^+$ currents recorded with the voltage clamp method (see Box A). (2) As a result of the increased Na$^+$ current, α-toxin greatly prolongs the duration of the axonal action potential—note the change in time scale after treating with toxin. (B) Treatment of a frog axon with β-toxin from another scorpion, *Centruroides sculpturatus*, shifts the activation of Na$^+$ channels, so that Na$^+$ conductance begins to increase at potentials much more negative than usual. (A after Schmidt and Schmidt, 1972; B after Cahalan, 1975.)

Figure 3.6 Depolarization increases Na⁺ and K⁺ conductances of the squid giant axon. The peak magnitude of Na⁺ conductance and steady-state value of K⁺ conductance both increase steeply as the membrane potential is depolarized. (After Hodgkin and Huxley, 1952b.)

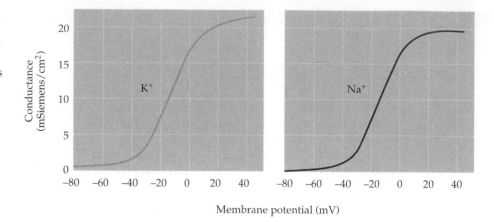

both the Na⁺ and K⁺ conductances increase progressively as the neuron is depolarized. Figure 3.6 illustrates this relationship by plotting each conductance versus the membrane potential. Note the similar voltage dependence for each conductance; both are sigmoidal functions of membrane potential, and are quite small at negative potentials, maximal at very positive potentials, and exquisitely dependent on membrane voltage at intermediate potentials. The observation that these conductances are sensitive to changes in membrane potential shows that the mechanism underlying the conductances somehow "senses" the voltage across the membrane.

The second conclusion derived from the calculations of Hodgkin and Huxley is that the Na⁺ and K⁺ conductances also change over time. For example, both the Na⁺ and K⁺ conductances require some time to **activate**, or turn on. In particular, the K⁺ conductance has a pronounced delay, requiring several milliseconds to reach its maximum; the Na⁺ conductance reaches its maximum more rapidly (Figure 3.7). The more rapid activation of the Na⁺ conductance allows the resulting inward Na⁺ current to precede the delayed outward K⁺ current (Figure 3.7A, B). Although the Na⁺ conductance rises rapidly, it quickly declines, even though the membrane potential is kept at a depolarized level. This fact shows that depolarization not only causes the Na⁺ conductance to activate, but also causes it to decrease over time, or **inactivate**. The K⁺ conductance of the squid axon does not inactivate in this way; thus, while the Na⁺ and K⁺ conductances share the property of time-dependent activation, only the Na⁺ conductance inactivates. (Inactivating K⁺ conductances have since been discovered in other types of nerve cells; see Chapter 4.) The time courses of the Na⁺ and K⁺ conductances are also voltage-dependent, with the speed of both activation and inactivation increasing at more depolarized potentials (see Figure 3.7C, D). This finding accounts for more rapid courses of membrane currents measured at more depolarized potentials.

All told, the voltage clamp experiments carried out by Hodgkin and Huxley showed that the ionic currents that flow when the neuronal membrane is depolarized are due to three different voltage-sensitive processes: (1) activation of Na⁺ conductance, (2) activation of K⁺ conductance, and (3) inactivation of Na⁺ conductance.

■ RECONSTRUCTION OF THE ACTION POTENTIAL

From the experimental measurements of these three conductance changes, Hodgkin and Huxley were able to construct a detailed mathematical model of these processes to determine whether conductance changes to Na⁺ and K⁺

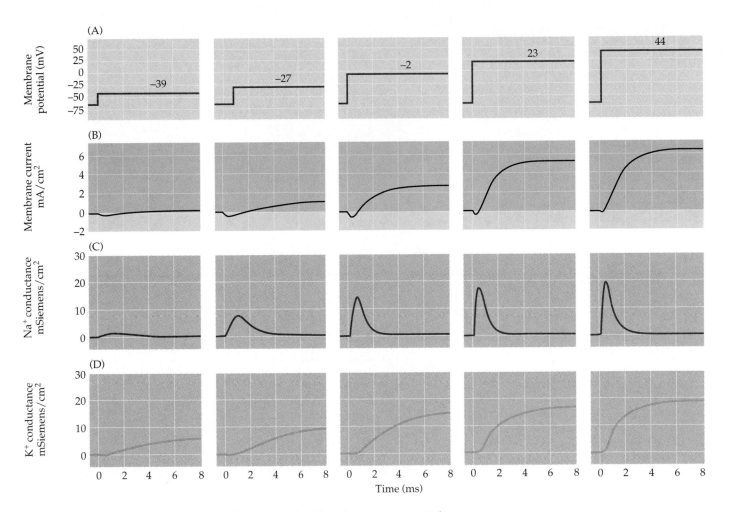

Figure 3.7 Membrane conductance changes underlying the action potential are time- and voltage-dependent. Depolarizations to various membrane potentials (A) elicit different membrane currents (B). Below are shown the Na+ (C) and K+ (D) conductances calculated from these currents. Both peak Na+ conductance and steady-state K+ conductance increase as the membrane potential becomes more positive. In addition, the activation of both conductances, as well as the rate of inactivation of the Na+ conductance, occur more rapidly with larger depolarizations. (After Hodgkin and Huxley, 1952b.)

alone are sufficient to produce an action potential. Using this information, they could in fact generate the form and time course of the action potential with remarkable accuracy (Figure 3.8A). Further, the Hodgkin-Huxley model predicted other features of action potential behavior in the squid axon, such as how the delay before action potential generation changes in response to stimulating currents of different intensities (Figure 3.8B). Figure 3.8A shows a reconstructed action potential, together with the time courses of the underlying Na+ and K+ conductances. The coincidence of the initial increase in Na+ conductance with the rapid rising phase of the action potential demonstrates that a selective increase in Na+ conductance is responsible for action potential initiation. The increase in Na+ conductance causes Na+ to enter the neuron, thus depolarizing the membrane potential, which approaches E_{Na}. The rate of depolarization subsequently falls both because the electrochemical driving force on Na+ decreases and because the Na+ conduc-

Figure 3.8 Mathematical reconstruction of the action potential. (A) Reconstruction of an action potential (black curve) together with the underlying changes in Na⁺ (red curve) and K⁺ (yellow curve) conductance. The size and time course of the action potential were calculated using only the properties of g_{Na} and g_K measured in voltage clamp experiments. Real action potentials evoked by brief current pulses of different intensities (B) are remarkably similar to those generated by the mathematical model (C). (After Hodgkin and Huxley, 1952d.)

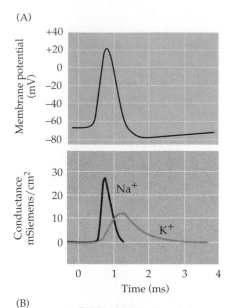

Figure 3.9 Feedback cycles responsible for membrane potential changes during an action potential. Membrane depolarization rapidly activates a positive feedback cycle fueled by the voltage-dependent activation of Na⁺ conductance. This phenomenon is followed by the slower activation of a negative feedback loop as depolarization activates a K⁺ conductance, which helps to repolarize the membrane potential and terminate the action potential.

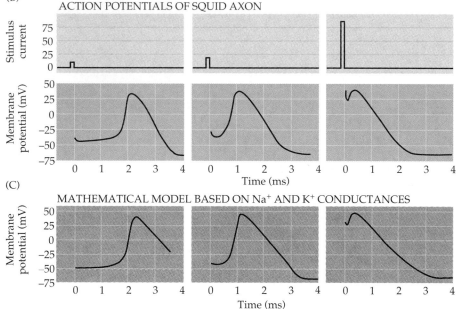

(B) ACTION POTENTIALS OF SQUID AXON

(C) MATHEMATICAL MODEL BASED ON Na⁺ AND K⁺ CONDUCTANCES

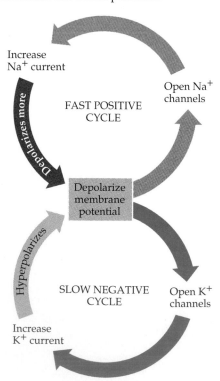

tance inactivates. At the same time, depolarization slowly activates the voltage-dependent K⁺ conductance, causing K⁺ to leave the cell and repolarizing the membrane potential toward E_K. Because the K⁺ conductance becomes temporarily higher than it is in the resting condition, the membrane potential actually becomes briefly more negative than the normal resting potential (the undershoot). The hyperpolarization of the membrane potential causes the voltage-dependent K⁺ conductance (and any Na⁺ conductance not inactivated) to turn off, allowing the membrane potential to return to its resting level.

This mechanism of action potential generation represents a positive feedback loop: activating the voltage-dependent Na⁺ conductance increases Na⁺ entry into the neuron, which makes the membrane potential depolarize, which leads to the activation of still more Na⁺ conductance, more Na⁺ entry, and still further depolarization (Figure 3.9). Positive feedback continues unabated until Na⁺ conductance inactivation and K⁺ conductance activation restore the membrane potential to the resting level. Because this positive feedback loop, once initiated, is sustained by the intrinsic properties of the neuron—namely, the voltage dependence of the ionic conductances—the action potential is self-supporting, or **regenerative**. This regenerative quality explains why action potentials exhibit all-or-none behavior (see Figure 2.1), and why they have a well-defined threshold (Box C).

Hodgkin and Huxley's reconstruction of the action potential and all its features shows that the properties of the voltage-sensitive Na^+ and K^+ conductances, together with the electrochemical driving forces created by ion pumps, are sufficient to explain action potentials. Their use of both empirical

Box C
THRESHOLD

An important—and potentially puzzling—property of the action potential is its initiation at a particular membrane potential, called threshold. Indeed, action potentials never occur without a depolarizing stimulus that brings the membrane to this level. The depolarizing "trigger" can be one of several events: a synaptic input; a receptor potential generated by specialized receptor organs; the endogenous pacemaker activity of cells that generate action potentials spontaneously; or the local current that mediates the spread of the action potential down the axon (see next section).

Why the action potential "takes off" at a particular level of depolarization can be understood by comparing the underlying events to a chemical explosion (A). As shown in the figure, exogenous heat (analogous to the depolarizing trigger event) stimulates an exothermic chemical reaction, which produces more heat, which further enhances the reaction (B). As a result of this positive feedback loop, the rate of the reaction builds up exponentially—the definition of an explosion. In any such process, however, there is a threshold, that is, a point up to which heat can be supplied without resulting in an explosion. The threshold for the chemical explosion diagrammed here is the point at which the amount of heat supplied exogenously is just equal to the

amount of heat that can be dissipated by the circumstances of the reaction.

The threshold of action potential initiation is, in principle, quite similar (C). There is a range of "subthreshold" depolarization, within which the rate of increased sodium entry is less than the rate of potassium exit (remember that the membrane at rest is highly permeable to K^+, which therefore flows out as the membrane is depolarized). The point at which Na^+ inflow just equals K^+ outflow represents an unstable equilibrium analogous to the ignition point of an explosive mixture. The behavior of the membrane at threshold reflects this instability: the membrane potential may linger at the threshold level for a vari-

able period before either returning to the resting level or flaring up into a full-blown action potential. In theory at least, if there is a net internal gain of a single sodium ion, an action potential occurs; conversely, the net loss of a single potassium ion leads to repolarization. A precise definition of threshold, therefore, is that value of membrane potential, in moving toward zero from the resting potential, at which the current carried by Na^+ entering the neuron is exactly equal to the K^+ current that is flowing out. Once the triggering event depolarizes the membrane beyond this point, the positive feedback loop of Na^+ entry on membrane potential closes and the action potential "fires."

(A)   (B)   (C)

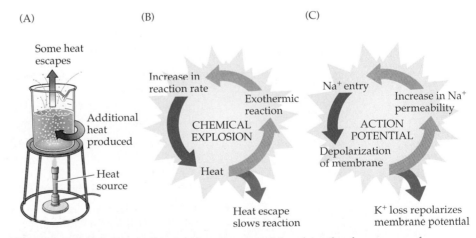

A positive feedback loop underlying the action potential explains the phenomenon of threshold.

and theoretical methods brought an unprecedented level of rigor to a long-standing problem, setting a standard of proof that is achieved only rarely in biological research.

■ LONG-DISTANCE SIGNALING BY MEANS OF ACTION POTENTIALS

The voltage-dependent mechanisms of action potential generation also explain the long-distance transmission of these electrical signals. Recall from Chapter 2 that neurons are relatively poor conductors of electricity, at least compared to a wire. Current conduction by wires, and by neurons in the absence of action potentials, is called **passive current flow**. The passive electrical properties of a nerve cell axon can be determined by measuring the voltage change resulting from a current pulse passed across the axonal membrane (Figure 3.10A). If this current pulse is not large enough to generate action potentials, the magnitude of the potential change that results decays exponentially with increasing distance from the site of current injection (Figure 3.10B). Typically, the potential falls to a small fraction of its ini-

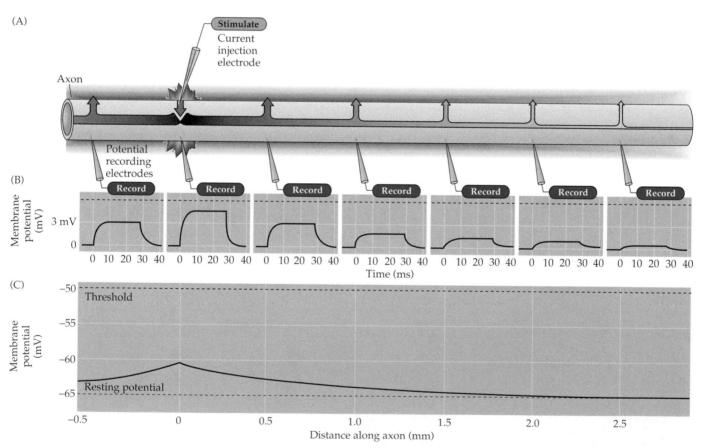

Figure 3.10 Passive current flow in an axon. (A) Experimental arrangement for examining the local flow of electrical current in an axon. A current-passing electrode produces a subthreshold change in membrane potential, which spreads passively along the axon. (B) Potential responses recorded at the positions indicated by microelectrodes. With increasing distance from the site of current injection, the amplitude of the potential change is attenuated. (C) Relationship between the amplitude of potential responses and distance. (After Hodgkin and Rushton, 1938.)

tial value at a distance no more than a couple of millimeters away from the site of injection (Figure 3.10C). The progressive decrease in the amplitude of the induced potential change occurs because the injected current leaks across the axonal membrane; accordingly, less current is available to change the membrane potential further along the axon. Thus, the leakiness of the axonal membrane prevents effective passive transmission of electrical signals in all but the shortest axons (those 1 mm or less in length).

If the experiment shown in Figure 3.10 is repeated with a depolarizing current pulse sufficiently large to produce an action potential, the result is dramatically different (Figure 3.11). In this case, an action potential occurs without decrement along the entire length of the axon, which may be a distance of a meter or more. Thus, action potentials somehow circumvent the inherent leakiness of neurons.

How are action potentials capable of traversing great distances along such a poor passive conductor? The answer is in part provided by the observation that the amplitude of the action potentials recorded at different distances is constant. This all-or-none behavior indicates that more than simple passive flow of current must be involved in action potential propagation. A second clue comes from examination of the time of occurrence of the action potentials recorded at different distances from the site of stimulation: action potentials occur later and later at greater distances along the axon (Figure 3.11B). Thus, the action potential has a measurable rate of transmission,

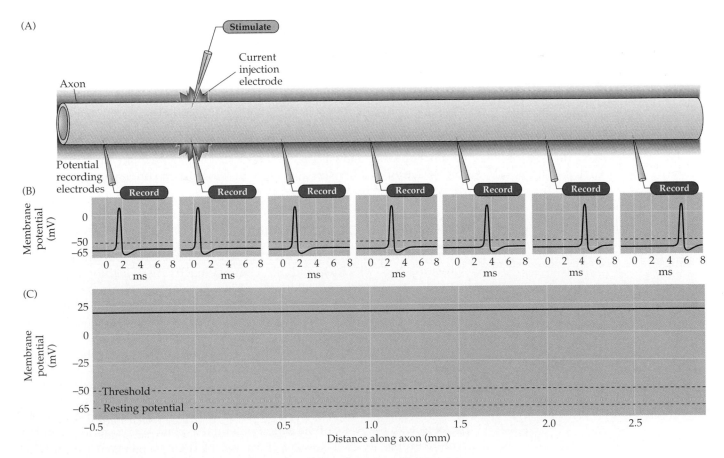

Figure 3.11 Propagation of an action potential produces the same voltage changes at all intervals along the length of an axon, albeit with an increasing delay.

Figure 3.12 Action potential conduction requires both active and passive current flow. Depolarization at one point along an axon opens Na⁺ channels locally (1) and produces an action potential in this region (A) of the axon (time point 1). The resulting inward current flows passively along the axon (2), depolarizing the adjacent region (B) of the axon. At a later time (time point 2), the depolarization of the adjacent membrane has opened Na⁺ channels in region B, resulting in the initiation of the action potential at this site and additional inward current that again spreads passively to an adjacent region (C) farther along the axon (3). At a still later time (time point 3), the action potential has propagated even farther. This cycle continues along the full length of the axon (5). Note that as the action potential spreads, the membrane potential repolarizes due to K⁺ channel opening and Na⁺ channel inactivation, leaving a "wake" of refractoriness behind the action potential that prevents its backward propagation (4).

called the **conduction velocity**. The delay in the arrival of the action potential at successively more distant points along the axon differs from the case shown in Figure 3.10, in which the electrical changes produced by passive current flow occur at more or less the same time at the successive points.

The mechanism of action potential propagation is easy to grasp once one understands how action potentials are generated and how current passively flows along an axon (Figure 3.12). A depolarizing stimulus—usually a synaptic signal or a receptor potential in an intact neuron, or an injected current pulse in an experiment—locally depolarizes the axon, thus opening the voltage-sensitive Na⁺ channels in that region. The opening of Na⁺ channels causes inward movement of Na⁺, and the resultant depolarization of the membrane potential generates an action potential at that site. Some of the local current generated by the action potential will then flow passively down the axon for the same reason that subthreshold currents spread along the axon (see Figure 3.10). This passive current flow depolarizes the membrane potential in the adjacent region of the axon, thus opening the Na⁺ channels in the neighboring membrane. The local depolarization produces an action potential in this region, which then spreads again in a continuing cycle until the end of the axon is reached. Thus, the regenerative properties of Na⁺ channel opening allow action potentials to propagate in an all-or-none fashion by acting as a booster at each point along the axon, ensuring the long-distance transmission of electrical signals.

■ THE REFRACTORY PERIOD

The depolarization that produces Na⁺ channel opening also causes slower activation of K⁺ channels and Na⁺ channel inactivation, leading to repolarization of the membrane potential as the action potential sweeps along the length of an axon (see Figure 3.12). In its wake, the action potential leaves the Na⁺ channels inactivated and K⁺ channels activated for a brief time. These transitory changes make it harder for the axon to produce subsequent action potentials during this interval, which is called the **refractory period**. Thus, the refractory period limits the number of action potentials that a given nerve cell can produce per unit time; as might be expected, different types of neurons have different maximum rates of action potential firing due to different types and densities of ion channels. The refractoriness of the membrane in the wake of the action potential explains why action potentials do not propagate back toward the point of their initiation as they travel along an axon.

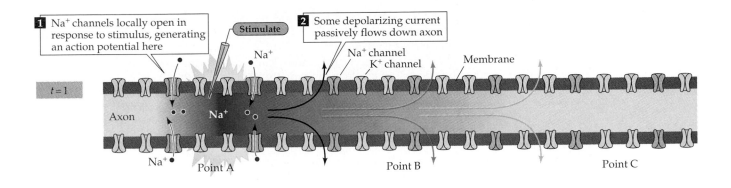

1 Na⁺ channels locally open in response to stimulus, generating an action potential here

Stimulate

2 Some depolarizing current passively flows down axon

Na⁺ channel
K⁺ channel

Membrane

t = 1

Axon

Na⁺

Na⁺

Point A

Point B

Point C

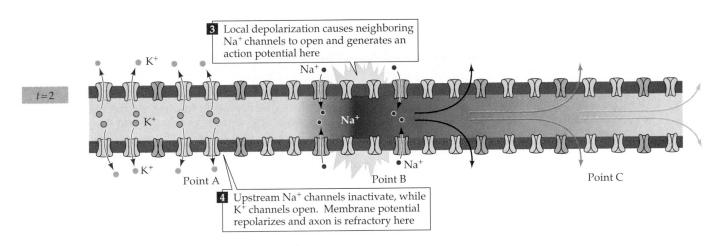

3 Local depolarization causes neighboring Na⁺ channels to open and generates an action potential here

t = 2

K⁺

K⁺

K⁺

Na⁺

Na⁺

Na⁺

Point A

Point B

Point C

4 Upstream Na⁺ channels inactivate, while K⁺ channels open. Membrane potential repolarizes and axon is refractory here

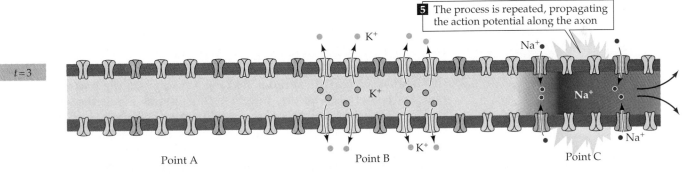

5 The process is repeated, propagating the action potential along the axon

t = 3

K⁺

Na⁺

K⁺

Na⁺

Na⁺

K⁺

Point A

Point B

Point C

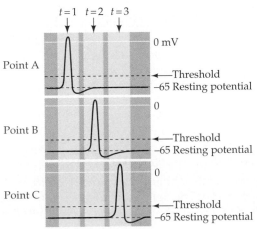

t = 1   t = 2   t = 3

Point A

0 mV
Threshold
−65 Resting potential

Point B

0
Threshold
−65 Resting potential

Point C

0
Threshold
−65 Resting potential

Figure 3.13 Saltatory action potential conduction along a myelinated axon. (A) Diagram of a myelinated axon. (B) Local current in response to action potential initiation at a particular site flows locally, as in Figure 3.12. However, the presence of myelin prevents the local current from leaking across the internodal membrane; it therefore flows farther along the axon than it would in the absence of myelin. Moreover, voltage-gated Na^+ channels are present only at the nodes of Ranvier. This arrangement means that the generation of active, voltage-gated currents need only occur at these unmyelinated regions. The result is a greatly enhanced velocity of action potential conduction.

■ INCREASED CONDUCTION VELOCITY AS A RESULT OF MYELINATION

Action potential propagation requires the coordinated action of two forms of current flow—the passive flow of current as well as active currents flowing through voltage-dependent ion channels (Figure 3.12). As a result, the rate of action potential propagation is determined by both of these phenomena. One way of improving passive current flow is to increase the diameter of an axon, which effectively decreases the internal resistance to passive current flow. The consequent increase in action potential conduction velocity presumably explains why invertebrates such as squid have evolved giant axons.

Another strategy to improve the passive flow of electrical current is to insulate the axonal membrane, reducing the ability of current to leak out of the axon, thereby increasing the distance along the axon that a given local current can flow passively. This strategy has resulted in the evolution of **myelination**, a process by which oligodendrocytes in the central nervous system (and Schwann cells in the peripheral nervous system) wrap the axon in **myelin**, which comprises multiple layers of closely opposed glial membrane (Figure 3.13; see also Chapter 1). By acting as an electrical insulator, myelin greatly speeds up action potential conduction. (Unmyelinated axon conduction velocities range from about 0.5 to 10 m/s, whereas myelinated axons can conduct at velocities up to 150 m/s.) The major reason underlying this marked increase in speed is that the time-consuming process of action potential generation occurs only at certain points along the axon, called **nodes of Ranvier**, where there is a gap in the myelin wrapping. If the entire surface of an axon were insulated, there would be no place for current to flow out of the axon and action potentials could not be generated. As it happens, an action potential generated at one node of Ranvier elicits current that flows passively within the myelinated segment until the next node is reached. This local current flow then generates an action potential in the neighboring segment, and the cycle is repeated along the length of the axon. Because current flows across the neuronal membrane only at the nodes (see Figure 3.13), this type of propagation is called **saltatory**, meaning that the action potential jumps from node to node.

■ SUMMARY

The action potential and all its complex properties can be explained by time- and voltage-dependent changes in the Na^+ and K^+ permeabilities of neuronal membranes. This conclusion derives from evidence obtained by a device called the voltage clamp. The voltage clamp technique is an electronic feedback method that allows control of neuronal membrane potential and, simultaneously, direct measurement of the voltage-dependent fluxes of Na^+ and K^+ that produce the action potential. Voltage clamp experiments show

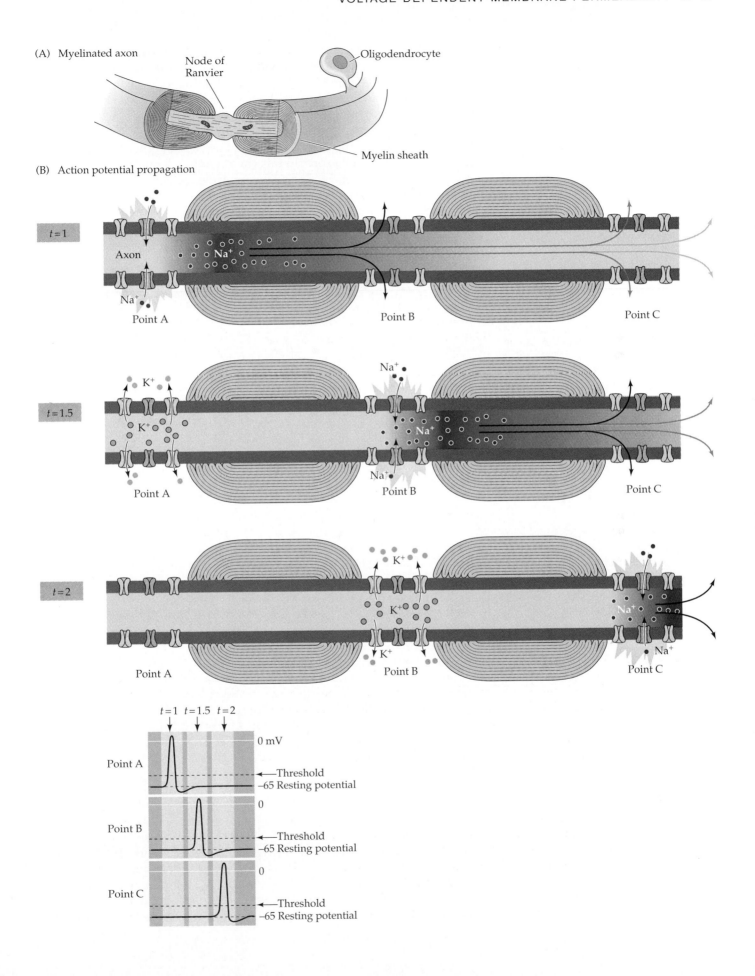

(A) Myelinated axon

Node of Ranvier

Oligodendrocyte

Myelin sheath

(B) Action potential propagation

t = 1

Axon

Na⁺

Point A

Na⁺

Point B

Point C

t = 1.5

K⁺

K⁺

Point A

Na⁺

Na⁺

Na⁺

Point B

Point C

t = 2

Point A

K⁺

K⁺

K⁺

Point B

Na⁺

Na⁺

Point C

t = 1   t = 1.5   t = 2

Point A

0 mV

Threshold
−65 Resting potential

Point B

0

Threshold
−65 Resting potential

Point C

0

Threshold
−65 Resting potential

that a transient rise in Na$^+$ conductance activates rapidly and then inactivates during a sustained depolarization of the membrane potential. Such experiments also demonstrate a rise in K$^+$ conductance that activates in a delayed fashion and, in contrast to the Na$^+$ conductance, does not inactivate. Mathematical modeling of the properties of these conductances indicates that they, and they alone, are responsible for the production of all-or-none action potentials. By virtue of local current flow, action potentials can propagate along the nerve cell axons. In this way, action potentials compensate for the relatively poor passive electrical properties of nerve cells and enable neural signaling over long distances.

Additional Reading

Reviews

RALL, W. (1977) Core conductor theory and cable properties of neurons. In *Handbook of Physiology,* Section 1: *The Nervous System,* Vol. 1: *Cellular Biology of Neurons.* E. R. Kandel (ed.). Bethesda: American Physiological Society, pp. 39–98.

Important Original Papers

ARMSTRONG, C. M. AND L. BINSTOCK (1965) Anomalous rectification in the squid giant axon injected with tetraethylammonium chloride. J. Gen. Physiol. 48: 859–872.

HODGKIN, A. L. AND A. F. HUXLEY (1952a) Currents carried by sodium and potassium ions through the membrane of the giant axon of *Loligo.* J. Physiol. 116: 449–472.

HODGKIN, A. L. AND A. F. HUXLEY (1952b) The components of membrane conductance in the giant axon of *Loligo.* J. Physiol. 116: 473–496.

HODGKIN, A. L. AND A. F. HUXLEY (1952c) The dual effect of membrane potential on sodium conductance in the giant axon of *Loligo.* J. Physiol. 116: 497–506.

HODGKIN, A. L. AND A. F. HUXLEY (1952d) A quantitative description of membrane current and its application to conduction and excitation in nerve. J. Physiol. 116: 507–544.

HODGKIN, A. L. AND W. A. H. RUSHTON (1938) The electrical constants of a crustacean nerve fibre. Proc. R. Soc. Lond. 133: 444–479.

HODGKIN, A. L., A. F. HUXLEY AND B. KATZ (1952) Measurements of current–voltage relations in the membrane of the giant axon of *Loligo.* J. Physiol. 116: 424–448.

MOORE, J. W., M. P. BLAUSTEIN, N. C. ANDERSON AND T. NARAHASHI (1967) Basis of tetrodotoxin's selectivity in blockage of squid axons. J. Gen. Physiol. 50: 1401–1411.

Books

HILLE, B. (1992) *Ionic Channels of Excitable Membranes,* 2nd Ed. Sunderland, MA: Sinauer Associates.

JUNGE, D. (1992) *Nerve and Muscle Excitation,* 3rd Ed. Sunderland, MA: Sinauer Associates.

■ OVERVIEW

The generation of electrical signals in neurons requires both selective membrane permeability and specific ion concentration gradients across the plasma membrane. The membrane proteins that give rise to these two essential conditions are called ion channels and pumps, respectively. Ion channels, as the phrase implies, have pores that allow particular ions to diffuse across the neuronal membrane. Some channels also have specialized domains that sense the electrical potential across the membrane. Such channels open or close in response to the level of membrane potential, allowing the membrane permeability to be voltage-sensitive. Many types of voltage-sensitive ion channels have been identified, and this diversity generates a wide spectrum of electrical characteristics among neuron types. Pumps are membrane proteins that produce and maintain ion concentration gradients. The Na$^+$ pump, which regulates the intracellular concentrations of both Na$^+$ and K$^+$ by hydrolyzing ATP to fuel the translocation of these ions across the plasma membrane, is the best-studied example. Other pumps produce concentration gradients for a variety of other ions. From the perspective of neural signaling, pumps and channels are complementary: pumps create the concentration gradients that impel ions to diffuse through channels, thus generating electrical signals.

■ ION CHANNELS INVOLVED IN ACTION POTENTIAL PROPAGATION

Although Hodgkin and Huxley had no knowledge of the physical nature of the conductance mechanisms underlying action potentials, they nonetheless proposed that nerve cell membranes contain channels that allow ions to permeate selectively, passing from one side of the membrane to the other (see Chapter 3). Based on the ionic conductances and currents determined by voltage clamp experiments, the postulated channels had to have several properties. First, since the ionic currents are quite large, the channels had to be capable of moving ions across the membrane at high rates. Second, because the ionic currents depend on the electrochemical gradient across the membrane, channels had to make use of these gradients to move ions. Third, because Na$^+$ and K$^+$ flow across the membrane independently of each other, different channel types had to be capable of discriminating between Na$^+$ and K$^+$, allowing only one of these ions to flow across the membrane under certain conditions. Finally, since the conductances are voltage-dependent, they had to be able to sense the voltage across the membrane, opening only when the voltage reaches appropriate levels. While this concept of channels was highly speculative in the 1950s, later experimental work established beyond any doubt that voltage-sensitive ion channels do explain all the ionic conductance phenomena described in Chapter 3.

   The first direct evidence for the presence of voltage-sensitive, ion-selective channels in nerve cell membranes came from measurements of the ionic currents flowing through individual ion channels. The voltage-clamp apparatus used by Hodgkin and Huxley could only resolve the *aggregate* current

Box A
THE PATCH CLAMP METHOD

A wealth of new information about ion channels resulted from the invention of the patch clamp method in the 1970s. This technique is based on a very simple idea. A glass pipette with a very small opening is used to make tight contact with a tiny area, or patch, of neuronal membrane. After the application of a small amount of suction to the back of the pipette, the connection between pipette and membrane becomes so strong that no ions can flow between the pipette and the membrane. Thus, all the ions that flow when a single ion channel opens must flow into the pipette. The resulting electrical current, though small, can be measured with an ultra-sensitive electronic amplifier connected to the pipette. Based on the geometry involved, this arrangement usually is called the cell-attached patch clamp method. As with the conventional voltage clamp method, the patch clamp method allows experimental control of the membrane potential to characterize the voltage dependence of membrane currents.

Although the ability to record currents flowing through single ion channels is an important advantage of the cell-attached patch clamp method, it has turned out that minor technical modifications yield still other advantages. For example, if the membrane patch within the pipette is disrupted by briefly applying strong suction, the interior of the pipette becomes continuous with the cytoplasm of the cell. This arrangement allows measurements of electrical potentials and currents from the entire cell, and is therefore called the whole-cell method. The whole-cell configuration also allows diffusional exchange between the pipette and the cytoplasm, producing a convenient way to inject substances into the interior of a "patched" cell.

Various configurations in patch clamp measurements of ionic currents.

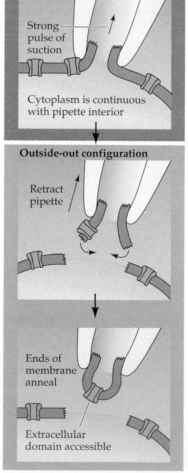

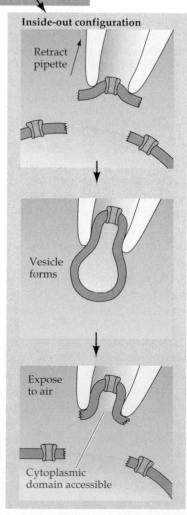

Two other variants of the patch clamp method originate from the finding that once a tight seal has formed between the membrane and the glass pipette, small pieces of membrane can be pulled away from the cell without disrupting the seal; this yields a preparation that is free of the complications imposed by the rest of the cell. Simply retracting a pipette that is in the cell-attached configuration causes a small vesicle of membrane to remain attached to the pipette. By exposing the tip of the pipette to air, the vesicle opens to yield a small patch of membrane with its (former) intracellular surface exposed. This arrangement, called the inside-out patch configuration, allows the measurement of single-channel currents with the added benefit of making it possible to change the medium to which the intracellular surface of the membrane is exposed. Thus, the inside-out configuration is particularly valuable when studying the influence of intracellular regulators on ion channel function. Alternatively, if the pipette is retracted while it is in the whole-cell configuration, a membrane patch is produced that has its extracellular surface exposed. This arrangement, called the outside-out configuration, is optimal for studying how channel activity is influenced by extracellular signals, such as neurotransmitters (see Chapter 7). This range of possible configurations makes the patch clamp method an unusually versatile technique for studies of ion channel function.

References

HAMILL, O. P., A. MARTY, E. NEHER, B. SAKMANN AND F. J. SIGWORTH (1981) Improved patch-clamp techniques for high-resolution current recording from cells and cell-free membrane patches. Pflügers Arch. 391: 85–100.

resulting from the flow of ions through many thousands of channels. A technique capable of measuring the currents flowing through single channels was devised by Erwin Neher and Bert Sakmann in 1976. This remarkable technique, called patch clamping (Box A), effectively revolutionized the study of membrane currents. In particular, the patch clamp method provided the means to test Hodgkin and Huxley's proposals about the characteristics of ion channels.

Currents flowing through Na^+ channels are best examined in experimental circumstances that prevent the flow of current through other membrane channels, K^+ channels in particular. Under such conditions, depolarizing the potential (Figure 4.1A) across a patch of membrane from a nerve cell that possesses voltage-dependent Na^+ conductances causes tiny inward currents to flow, but only occasionally (Figure 4.1B). The size of these currents is minuscule—approximately 1 pA (10^{-12} ampere), orders of magnitude smaller than the Na^+ currents measured by voltage clamping an entire axon. The currents flowing through single channels are called **microscopic currents** to distinguish them from the **macroscopic currents** flowing through a large number of channels in a large region of the membrane. Although microscopic currents are certainly small, a current of 1 pA nonetheless reflects the flow of thousands of ions per millisecond. Thus, a single channel can let many ions pass through the membrane very rapidly.

Several observations prove that the microscopic currents shown in Figure 4.1B are due to the opening of single, voltage-activated Na^+ channels. First, the currents are carried by Na^+ because they are directed inward at potentials more negative than E_{Na}, reverse their polarity at E_{Na}, and are outward at more positive potentials. This behavior exactly parallels that of the macroscopic Na^+ currents described in Chapter 3. Second, the channels have a time course of opening and closing that matches the kinetics of macroscopic Na^+ currents. This correspondence is difficult to appreciate in the measurement of microscopic currents flowing through a single open channel, because individual channels open and close in a stochastic (random) manner, as can be seen in the individual traces shown in Figure 4.1B. However, repeated depolarization of the membrane potential causes each Na^+

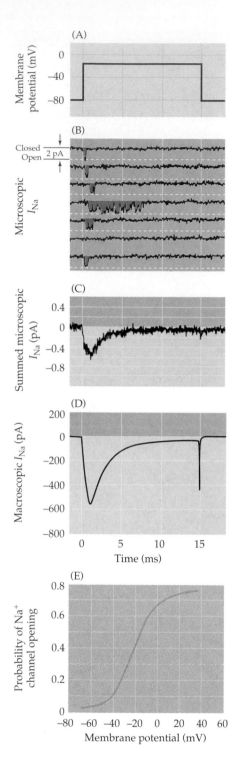

Figure 4.1 Patch clamp measurements of ionic currents flowing through single Na⁺ channels in a squid giant axon. In these experiments, Cs⁺ was applied to the axon to block voltage-gated K⁺ channels. Depolarizing voltage pulses (A) applied to a patch of membrane containing a single Na⁺ channel result in brief currents (B; downward deflections) in the seven successive recordings of membrane currents (I_{Na}). (C) The sum of many such current records shows that most channels open in the initial 1–2 ms, after which the probability of channel openings diminishes because of channel inactivation. (D) A macroscopic current measured from another axon shows the close correlation between the time courses of microscopic and macroscopic Na⁺ currents. (E) The probability of an Na⁺ channel opening depends on the membrane potential, increasing as the membrane is depolarized. (B and C after Bezanilla and Correa, 1995; D after Vandenburg and Bezanilla, 1991; E after Correa and Bezanilla, 1994.)

channel to open and close many times. When the current responses to a large number of such stimuli are averaged together, the collective response has a time course that looks much like the macroscopic Na⁺ current. In particular, the channels open mostly at the beginning of a prolonged depolarization, showing that they subsequently inactivate, as predicted from the macroscopic Na⁺ current (compare Figures 4.1C and 4.1D). Third, the opening and closing of the channels are voltage-dependent; thus, the channels are closed at −80 mV and open when the membrane potential is depolarized. In fact, the probability that any given channel will be open varies with membrane potential (Figure 4.1E), again as predicted from the macroscopic Na⁺ conductance (see Figure 3.6). Finally, tetrodotoxin, which blocks the macroscopic Na⁺ current, also blocks microscopic Na⁺ currents. Taken together, these results show that the macroscopic Na⁺ current measured by Hodgkin and Huxley does indeed arise from the aggregate effect of many thousands of microscopic Na⁺ currents, each representing the opening of a single voltage-sensitive Na⁺ channel.

Patch clamp experiments have also revealed the properties of the channels responsible for the macroscopic K⁺ currents measured by Hodgkin and Huxley. When the membrane potential is depolarized (Figure 4.2A), microscopic outward currents (Figure 4.2B) can be observed under conditions that block Na⁺ channels. These microscopic outward currents exhibit all the features expected for currents flowing through axonal K⁺ channels. Thus, neither the microscopic currents (Figure 4.2C) nor their macroscopic counterparts (Figure 4.2D) inactivate during brief depolarizations. Moreover, the single-channel currents are sensitive to ionic manipulations and drugs that affect the macroscopic K⁺ currents and, like the macroscopic K⁺ currents, are voltage-dependent (Figure 4.2E). This evidence shows that macroscopic K⁺ currents arise from the opening of many voltage-sensitive K⁺ channels.

The evidence illustrated in Figures 4.1 and 4.2 shows that neuronal membranes contain at least two types of channels—one selectively permeable to Na⁺ and a second selectively permeable to K⁺. Both these channel types are **voltage-gated**, meaning that their opening is influenced by membrane potential (Figure 4.3). For both, depolarization increases the probability of opening. Depolarization also slowly inactivates the Na⁺ channel (but not the K⁺ channel), causing it to pass into a nonconducting state. In summary, patch clamping has allowed direct observation of microscopic ionic currents flowing through single ion channels, thus confirming that such channels are responsible for the macroscopic currents and conductances that underlie the action potential.

Figure 4.2 Patch clamp measurements of ionic currents flowing through single K⁺ channels in a squid giant axon. In these experiments, tetrodotoxin was applied to the axon to block voltage-gated Na⁺ channels. Depolarizing voltage pulses (A) applied to a patch of membrane containing a single K⁺ channel results in brief currents (B; upward deflections) whenever the channel opens. (C) The sum of 300 such current records shows that most channels open with a delay, but remain open for the duration of the depolarization. (D) A macroscopic current measured from another axon shows the correlation between the time courses of microscopic and macroscopic K⁺ currents. (E) The probability of a K⁺ channel opening depends on the membrane potential, increasing as the membrane is depolarized. (B and C after Augustine and Bezanilla, in Hille 1992; D after Augustine and Bezanilla, 1990; E after Perozo et al., 1991.)

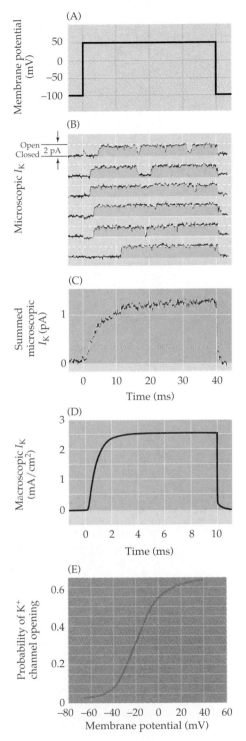

■ OTHER TYPES OF ION CHANNELS

The patch clamp method has been used to demonstrate that a large number of other ion channel types exist, not only in axonal membranes but also in the membranes of neuronal cell bodies, dendrites, and presynaptic terminals. This diversity of ion channel types has been separated into three large groups: voltage-gated, extracellular ligand-gated, and intracellular ligand-gated channels.

Voltage-gated channels specifically permeable to the four major physiological ions (Na⁺, K⁺, Ca²⁺, and Cl⁻) have now been characterized. As

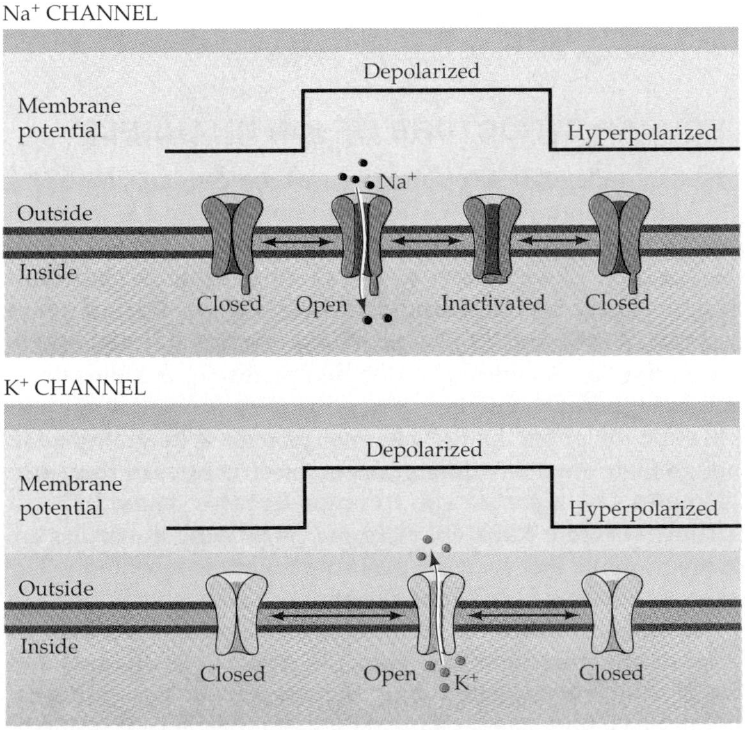

Na⁺ CHANNEL

K⁺ CHANNEL

Figure 4.3 Functional states of voltage-gated Na⁺ and K⁺ channels. Both channels are closed when the membrane potential is hyperpolarized and open when the potential is depolarized. Na⁺ channels also inactivate during prolonged depolarization, whereas many types of K⁺ channels do not.

discussed in the preceding section, Na$^+$ channels are important in the initiation and propagation of action potentials—a role that can also be played by Ca^{2+} channels in some neurons. Ca^{2+} channels also play an essential part during the release of neurotransmitters at synapses (see Chapter 5); it is not surprising, therefore, that at a number of Ca^{2+} channels subtypes have been described. However, the largest and most diverse class of voltage-gated ion channels are the K$^+$ channels. Dozens of K$^+$ channel types have been described, all varying in their activation/inactivation properties (see Figure 4.5). Some of these activate only at hyperpolarized potentials and help to set the resting membrane potential, whereas others can change the frequency of action potential trains or alter the duration of individual action potentials.

Extracellular ligand-gated channels are generally insensitive to changes in the membrane potential, but are activated by binding specific extracellular ligands such as neurotransmitters. These channels are essential for synaptic transmission and other forms of cell-cell signaling. As discussed more fully in Chapter 7, these ion channels contain two functional domains within a single protein entity: an extracellular ligand-binding domain, and an ion channel. While most voltage-gated ion channels allow only one type of ion to permeate, many extracellular ligand-gated ion channels are less selective and often pass two or even three species of ions. As implied by their name, intracellular ligand-gated channels also contain ligand-binding and ion channel domains, but in this case the ligand-binding domain is located on the intracellular face of the molecule. Intracellular ligands include Ca^{2+}, Na$^+$, and ATP.

The variety of ion channels present in different populations of neurons, and even at different sites within a single neuron, endows each cell type (or cell domain) with particular electrical signaling properties.

■ THE MOLECULAR STRUCTURE OF ION CHANNELS

Profound advances in understanding voltage-gated ion channels have also come from molecular genetic studies. Genes encoding Na$^+$ and K$^+$ channel proteins, as well as many others, have been identified and cloned. From these genes, the amino acid sequences of the channel proteins and their higher-order structures have been deduced. Expression of ion channel genes in defined experimental systems has also provided insights into the operation of ion channels (Box B). As a result, a detailed picture of channel structure and function has gradually emerged.

The general structure of Na$^+$ and K$^+$ channel proteins is diagrammed in Figure 4.4. Although there are many differences in the structures of these two channel types (Figure 4.4A), there are also common features shared by most voltage-gated channels (Figure 4.4B). For example, all of these molecules are **integral membrane proteins** that span the plasma membrane repeatedly. The membrane-spanning domains form a **central pore** through which ions can diffuse (Figure 4.4C,D). All voltage-sensitive channels also possess a domain of amino acids called the **pore loop** that lines the pore of the channel and accounts for the pore's ability to allow only certain ions to pass through. Another feature of all voltage-gated channels is a voltage sensor, a membrane-spanning segment that includes a series of positively charged amino acids (Figure 4.4B). These charged residues sense the electrical potential across the membrane and permit the pore of the channel to open.

Despite these shared features, a number of structural properties distinguish various voltage-gated channels. As expected, the amino acid composition of the pore loop differs among channel types, allowing each channel

Box B
EXPRESSION OF ION CHANNELS IN *XENOPUS* OOCYTES

The transition from sequencing an ion channel gene to understanding channel function is a challenging problem. To tackle this problem, it is essential to have an experimental system in which the gene product can be expressed efficiently, and in which the function of the resulting channel can be studied with methods such as the patch clamp technique. Ideally, the vehicle for expression should be readily available, have few endogenous channels, and be large enough to permit mRNA and DNA to be microinjected with ease. Oocytes (immature eggs) from the clawed African frog, *Xenopus laevis* (Figure A), fulfill all these demands. These huge cells (approximately 1 mm in diameter; Figure B) are easily harvested from the female *Xenopus*. Work performed in the 1970s by John Gurdon, a developmental biologist, showed that injection of exogenous mRNA into frog oocytes causes them to synthesize foreign protein in prodigious quantities. In the early 1980s, Ricardo Miledi, Eric Barnard, and other neurobiologists demonstrated that *Xenopus* oocytes could express exogenous ion channels, and that physiological methods could be used to study the ionic currents generated by the newly synthesized channels (Figure C).

As a result of these pioneering studies, heterologous expression experiments have now become a standard way of studying ion channels. The approach has been especially valuable in deciphering the relationship between channel structure and function. In such experiments, precise mutations (often affecting a single nucleotide) are made in the part of the channel gene that encodes a structure of interest; the resulting channel proteins are then expressed in oocytes to assess the functional consequences of the mutation.

The ability to combine molecular and physiological methods in a single cell system has made *Xenopus* oocytes a powerful experimental tool. Indeed, this system has been as valuable to contemporary studies of voltage-gated ion channels as the squid axon was to such studies in the 1950s and 1960s.

References

GUNDERSEN, C. B., R. MILEDI AND I. PARKER (1984) Slowly inactivating potassium channels induced in *Xenopus* oocytes by messenger ribonucleic acid from *Torpedo* brain. J. Physiol. (Lond.) 353: 231–248.

GURDON, J. B., C. D. LANE, H. R. WOODLAND AND G. MARBAIX (1971) Use of frog eggs and oocytes for the study of messenger RNA and its translation in living cells. Nature 233:177–182

SUMIKAWA, K., M. HOUGHTON, J. S. EMTAGE, B. M. RICHARDS AND E. A. BARNARD (1981) Active multi-subunit ACh receptor assembled by translation of heterologous mRNA in *Xenopus* oocytes. Nature 292: 862–864.

(A)

(B)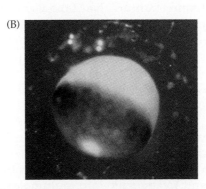

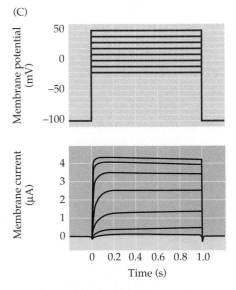

(A) The clawed African frog, *Xenopus laevis*. (B) A single oocyte from *Xenopus*. (C) Results of a voltage clamp experiment showing K⁺ currents produced following injection of K⁺ channel mRNA into an oocyte. (B courtesy of J. Green, Dana-Farber Cancer Institute; C after Gundersen et al., 1984.)

type to conduct a specific ion. Some channels also have sequences of amino acids that plug the channel pore during prolonged depolarization, giving rise to channel inactivation. As noted, different types of channels are selectively sensitive to various toxins (see Box B in Chapter 3); toxin sensitivity is

Figure 4.4 Structure of voltage-gated channels. (A) Topology of the principal subunits of voltage-gated Na⁺ and K⁺ channels. Repeating motifs of Na⁺ channels are labeled I, II, III, and IV. (B) The generalized motif of voltage-gated channels. In addition to six membrane-spanning domains, voltage-sensitive channels contain a pore loop responsible for ion conduction and selectivity, a charged domain for sensing the membrane potential, and other regions for channel inactivation, intracellular regulation, and association with other channel subunits. (C) The three-dimensional arrangement of the membrane-spanning domains of a voltage-gated channel. (D) Organization of channel proteins in membranes. In channels with four repeating motifs, the motifs are organized into a symmetrical structure with a central pore for ion flow. Channels with single repeats form similar structures by assembly into tetramers. (A after Catterall, 1988; D after Hille, 1992.)

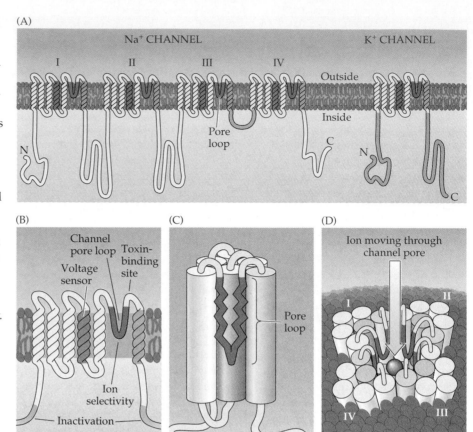

conferred by unique amino acid residues that allow a specific toxin to bind the channels. Finally, the number of membrane-spanning domains varies for different channel subunits; whereas most K⁺ channel subunits span the membrane six times, Na⁺ (and Ca²⁺) channel subunits cross the membrane 24 times (see Figure 4.4A). Evidently an entire channel can be produced by single Na⁺ and Ca²⁺ channel subunits, whereas four K⁺ channel subunits must aggregate to form a single functional ion channel. The variety of known K⁺ channel subunits can be mixed in different ways to yield a remarkably large number of functionally distinct K⁺ channels (see the next section).

In short, ion channels are integral membrane proteins with characteristic features that allow them to conduct ions and, in many cases, to sense the transmembrane potential. Specializations of these channels further allow them to discriminate among different ion species, to bind various neurotoxins, and to assemble into multimolecular aggregates. Importantly, several diseases produce subtle changes in channel structure and action potential generation, with potentially disastrous consequences (Box C).

■ THE DIVERSITY OF VOLTAGE-GATED CHANNELS

A surprising fact that has emerged from molecular studies of ion channels is the large number of genes that code for voltage-activated and ligand-activated channels. For example, at least six different Na⁺ channel genes have been identified. This finding was unexpected because Na⁺ channels from many different cell types have similar functional properties, as though they

Figure 4.5 Diverse properties of four different types of voltage-gated K⁺ channels found in the fruitfly *Drosophila*. Messenger RNA transcribed from each of four different K⁺ channel genes, termed *Shaker, Shal, Shab,* and *Shaw,* was injected into *Xenopus* oocytes (see Box B). Voltage clamp measurements of currents flowing through each type of expressed K⁺ channel showed that they vary markedly in their activation and inactivation properties. (After Wei et al., 1990.)

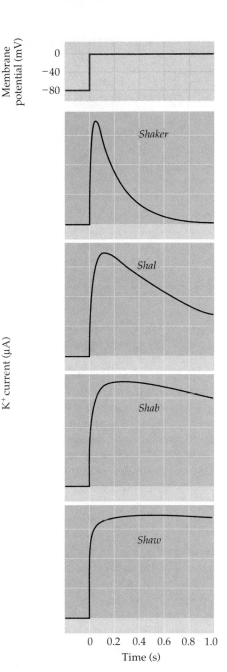

were all the product of a single gene. It is now clear, however, that the several Na⁺ channel genes produce proteins that differ in their structure, function, and expression in specific tissues. A similar number of Ca²⁺ channel genes have been identified, although this particular diversity is consistent with the wide range of functional properties exhibited by Ca²⁺ channels. Most remarkable of all is the diversity of K⁺ channels. The number of known genes for K⁺ channel subunits is expanding very rapidly; six major K⁺ channel families have been identified and the total number of known K⁺ channel subunit genes now stands at more than 25. Because each K⁺ channel consists of four subunits, the products of these genes can combine in many ways to yield a very large number of functionally distinct channel types. The various K⁺ channels differ primarily in their rates of inactivation; some take minutes to inactivate, as in the case of squid axon K⁺ channels studied by Hodgkin and Huxley (see Chapter 3), whereas others inactivate within milliseconds, as is typical of Na⁺ channels (Figure 4.5). Other types of K⁺ channels are sensitive to the concentration of intracellular modulators, such as Ca²⁺ or ATP. The result of this channel diversity is an extraordinary range of electrical characteristics in different types of neurons.

Other mechanisms also contribute to the diversity of ion channels. For example, genes often contain one or more sites for splicing, and can thus create multiple forms of channel subunits from a single gene. Further, most voltage-gated channel subunits associate with other proteins, called accessory subunits, that alter the properties of the channels. Finally, channel proteins can undergo posttranslational modifications, such as phosphorylation by protein kinases (see Chapter 7), which can produce still other changes in their functional characteristics. Thus, although the basic electrical signal of the nervous system—the action potential—is relatively stereotyped, the proteins responsible for generating this and other signals exhibit a rich diversity, the significance of which is only beginning to be understood.

■ PUMPS CREATE AND MAINTAIN ION GRADIENTS

Up to this point, the discussion of electrical signaling has largely taken for granted the remarkable fact that nerve cell membranes can generate large concentration gradients for selected ions. None of the ions of particular physiological importance (Na⁺, K⁺, Cl⁻, Ca²⁺, and H⁺) are in electrochemical equilibrium. Because channels produce electrical effects by allowing these ions to diffuse down their electrochemical gradients, there would be a gradual dissipation of concentration gradients unless nerve cells could restore ions displaced during electrical current flow. The essential tasks of generating and maintaining ionic concentration gradients are carried out by a group of plasma membrane proteins known as ion pumps.

Several kinds of ion pump have been identified. A single protein is responsible for sustaining the chemical gradients of both Na⁺ and K⁺; this molecule is called the **Na⁺ pump** (or, more properly, the **Na⁺/K⁺ pump**;

Box C
GENETIC DISEASES THAT AFFECT VOLTAGE-GATED ION CHANNELS

Given the importance of voltage-gated ion channels, it is easy to imagine that mutations in the genes encoding them would have serious consequences. Indeed, several genetic diseases are known to result from small but critical alterations in ion channel genes. The best-characterized ion channel diseases are those affecting skeletal muscle cells. In these disorders, alterations in ion channel proteins produce either myotonia (muscle stiffness due to excessive electrical excitability) or paralysis (due to insufficient muscle excitability).

Certain forms of myotonia are due to defects in Cl⁻ channels. Patients suffering from such diseases have muscle stiffness because their muscle cells spontaneously produce action potentials, resulting in excessive muscle tone (see Chapter 16). Physiological analysis indicates that the muscle cells in these patients have a reduced Cl⁻ permeability, and genetic studies have shown that this defect is due to one or more mutations in a Cl⁻ channel gene (Figure A). Reducing the Cl⁻ permeability of the muscle cell makes it easier for Na⁺ channel currents to excite the muscle cell, leading to the myotonic hyperexcitability.

Other inherited muscle diseases arise from Ca²⁺ channel defects. Two different point mutations in a Ca²⁺ channel gene result in muscle paralysis (Figure B). Both mutations reduce macroscopic Ca²⁺ channel currents, one by decreasing the number of functional Ca²⁺ channels, the other by causing Ca²⁺ channels to inactivate at much more negative potentials than usual. As Ca²⁺ channels do not play an important role in action potential production in skeletal muscle cells, it is not yet clear why these mutations cause muscle paralysis.

Still other muscle cell diseases

(A) Cl⁻ CHANNEL

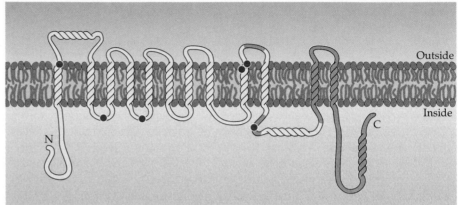

(B) Ca²⁺ CHANNEL

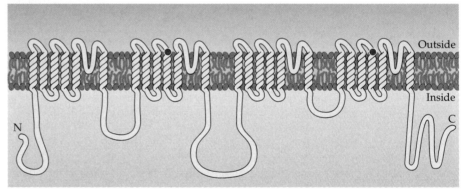

(C) Na⁺ CHANNEL

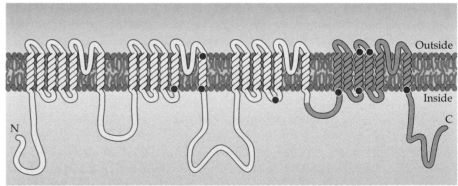

Genetic mutations in (A) Cl⁻ channels, (B) Ca²⁺ channels, and (C) Na⁺ channels that result in muscle cell diseases. Red regions indicate the sites of these mutations; the red circles indicate mutations in single amino acids, while red shading shows regions that are deleted in certain diseases. (After Hoffmann et al., 1995.)

result from mutations in a single Na⁺ channel gene (Figure C). Even though these diseases all target the same gene, some produce myotonia whereas others yield paralysis. Even more surprising is the observation that all of the mutations studied thus far have the same general effect: slowing the inactivation of Na⁺ currents (Figure D). How can slowing Na⁺ channel inactivation produce either myotonia or paralysis? During myotonia, there is a small depolarization of the muscle cell resting membrane potential. Because the mutant Na⁺ channels are less prone to inactivate in response to depolarizations, the sustained inward current that results can cause the muscle cell to repeatedly fire action potentials and produce stiffness. In contrast, paralysis is thought to result from a somewhat larger depolarization of the muscle cell resting potential. In this case, the depolarization does inactivate a sufficient number of Na⁺ channels to prevent action potential production, thus causing paralysis rather than stiffness.

A number of cardiac disorders have recently been linked to mutations in ion channels. Long QT syndromes (LQT) are inherited heart disorders characterized by prolonged ventricular action

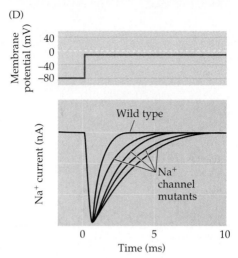

(D) Mutations in Na⁺ channels slow the rate of inactivation of Na⁺ currents. (After Barchi, 1995.)

potentials that result in arrhythmias, seizures, and sudden death. These disorders are associated with mutations at any one of three chromosomal locations, and the affected genes at all three of these sites have turned out to code for ion channels. LQT2 is caused by mutations of a K⁺ channel gene called *HERG*, and LQT3 is cause by a mutated Na⁺ channel gene. LQT1 is caused by mutations in another gene that appears to encode a K⁺ channel. The *HERG* mutation is located

within the K⁺ channel pore and results in a decreased current flow. This in turn prolongs the repolarization phase of the cardiac action potential. The Na⁺ channel-linked mutations produce an inactivation-resistant Na⁺ current, which also acts to prolong the repolarization phase of the action potential. The recent advances in the molecular genetics of LQT have allowed the development of tests to identify the exact location of the genetic defect in patients and, hence, to the most appropriate drug treatments.

Single-channel recording techniques (Box A) in conjunction with molecular biological approaches (Box B) represent powerful means of identifying genetic diseases that affect ion channels. While most successes to date have been with diseases affecting peripheral sites, these approaches also promise to unravel ion channel-mediated defects within the CNS.

References

BARCHI, R. L. (1995) Molecular pathology of the skeletal muscle sodium channel. Annu. Rev. Physiol. 57: 355–385.

HOFFMAN, E. P., F. LEHMANN-HORN AND R. RUDEL (1995) Overexcited or inactive: Ion channels in muscle disease. Cell 80: 681–686.

Figure 4.6A). In contrast, two different pumps share the important job of keeping intracellular Ca²⁺ concentrations low. These proteins are called the Ca²⁺ pump and the Na⁺/Ca²⁺ exchanger (Figure 4.6B, C). Several mechanisms exist for regulating intracellular Cl⁻ levels; the best understood of these swaps intracellular Cl⁻ for another extracellular anion, bicarbonate (Figure 4.6D). Still other pumps, such as an Na⁺/H⁺ exchanger, keep intracellular pH at appropriate levels (Figure 4.6E).

While the specific job of each pump differs, all are faced with the task of translocating ions against their electrochemical gradients. Moving ions uphill requires the consumption of energy, and the pumps considered here can be sorted into two classes based on their energy sources. Some pumps acquire energy directly from the hydrolysis of ATP. These are called **ATPase pumps**; examples include the Na⁺/K⁺ pump and the Ca²⁺ pump. Other pumps use the electrochemical gradients of other ions, often Na⁺, as their

Figure 4.6 Examples of ion pumps found in cell membranes. Some of these pumps are powered by the hydrolysis of ATP (ATPase pumps), whereas others use the electrochemical gradients of co-transported ions as a source of energy (ion exchange pumps).

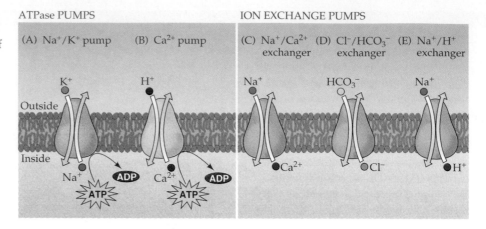

energy source. These pumps all carry one or more ions *against* their electrochemical gradient while simultaneously taking another ion, such as Na^+, *down* its gradient. Because at least two species of ions are involved in such transactions, these pumps are usually called **ion exchange pumps**. Examples of ion exchange pumps include the Na^+/Ca^{2+} exchanger and the Na^+/H^+ exchanger. Although the electrochemical gradient of Na^+ (or other counter ions) is the proximate source of energy for ion exchange pumps, these gradients ultimately depend on the hydrolysis of ATP by ATPase pumps like the Na^+/K^+ pump.

■ FUNCTIONAL PROPERTIES OF THE Na⁺/K⁺ PUMP

Richard Keynes in the 1950s was the first to use radioactive Na^+ to demonstrate the pump-dependent efflux of Na^+ from a neuron. Keynes and his collaborators found that disrupting the supply of ATP in a squid axon by treatment with metabolic poisons stops this efflux (Figure 4.7A). Other conditions that lower intracellular ATP, such as cooling, also prevent Na^+ efflux. These experiments showed that removing intracellular Na^+ requires cellular metabolism. Further studies with radioactive K^+ demonstrated that Na^+ efflux is associated with simultaneous, ATP-dependent influx of K^+. These opposing fluxes of Na^+ and K^+ are operationally inseparable: removal of external K^+ greatly reduces Na^+ efflux (Figure 4.7A). Such movements of Na^+ and K^+ implicated an ATP-hydrolyzing Na^+/K^+ pump in the generation of transmembrane gradients of Na^+ and K^+. The exact mechanism responsible for these fluxes of Na^+ and K^+ is still not entirely clear, but it is thought that the pump alternately shuttles these ions across the membranes in a cycle fueled by the transfer of a phosphate group from ATP to the pump protein (Figure 4.7B).

Quantitative studies of the movements of Na^+ and K^+ indicate that these two ions are not pumped at identical rates; the K^+ influx is only about two-thirds the Na^+ efflux. Thus, the pump transports these ions in a ratio of 2 K^+ brought into the cell for every 3 Na^+ removed (see Figure 4.7B). This stoichiometry causes the net loss of one positively charged ion from inside of the cell during each round of pumping, meaning that the pump generates an electrical current that can hyperpolarize the membrane potential. For this reason, the Na^+/K^+ pump is said to be **electrogenic**. Because pumps act much more slowly than ion channels, the current produced by the Na^+/K^+ pump is quite small. For example, in the squid axon, the current generated

Figure 4.7 Ionic movements due to the Na$^+$/K$^+$ pump. (A) Measurement of radioactive Na$^+$ efflux from a squid giant axon. This efflux depends upon external K$^+$ and intracellular ATP. (B) A model for the movement of ions by the Na$^+$/K$^+$ pump. Uphill movements of Na$^+$ and K$^+$ are driven by ATP, which phosphorylates the pump. These fluxes are asymmetrical, with 3 Na$^+$ carried out for every 2 K$^+$ brought in. (A after Hodgkin and Keynes, 1955; B after Lingrel et al., 1994.)

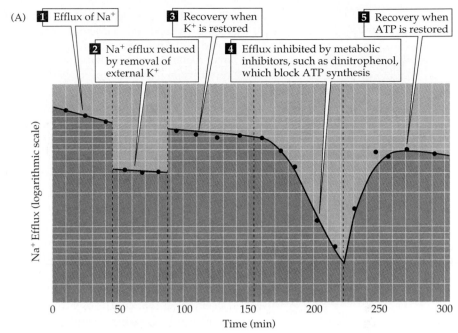

(A)

1 Efflux of Na$^+$

2 Na$^+$ efflux reduced by removal of external K$^+$

3 Recovery when K$^+$ is restored

4 Efflux inhibited by metabolic inhibitors, such as dinitrophenol, which block ATP synthesis

5 Recovery when ATP is restored

Na$^+$ Efflux (logarithmic scale)

Time (min)

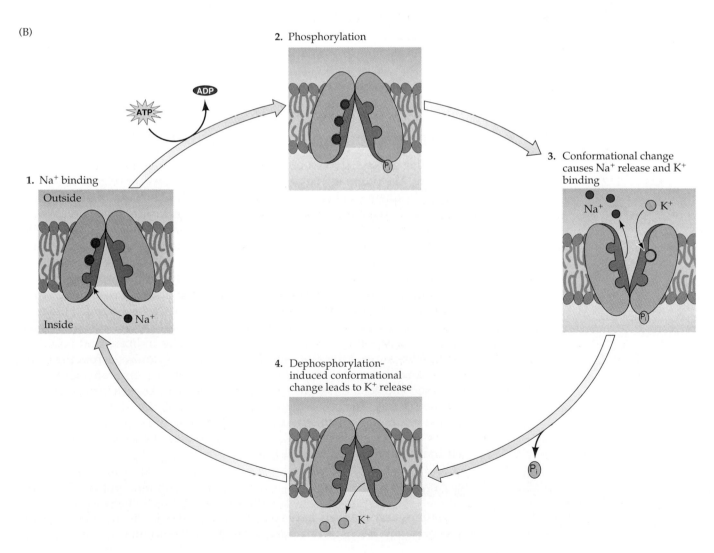

(B)

2. Phosphorylation

1. Na$^+$ binding
Outside
Inside
Na$^+$

ATP
ADP

3. Conformational change causes Na$^+$ release and K$^+$ binding
Na$^+$
K$^+$

4. Dephosphorylation-induced conformational change leads to K$^+$ release
K$^+$

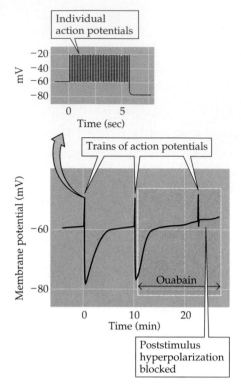

Individual action potentials

mV
−20
−40
−60
−80

Time (sec)
0   5

Trains of action potentials

Membrane potential (mV)
−60
−80

Ouabain

Time (min)
0   10   20

Poststimulus hyperpolarization blocked

Figure 4.8 The electrogenic transport of ions by the Na^+/K^+ pump can influence membrane potential. Measurements of the membrane potential of a small unmyelinated axon show that a train of action potentials (inset) is followed by a long-lasting hyperpolarization. This hyperpolarization is blocked by ouabain, indicating that it results from the activity of the Na^+/K^+ pump. (After Rang and Ritchie, 1968.)

by the pump is less than 1% of the current flowing through voltage-gated Na^+ channels.

Although the electrical current generated by the activity of the Na^+/K^+ pump is small, under some circumstances the pump can significantly influence neuronal membrane potentials. For instance, prolonged stimulation of small unmyelinated axons produces a substantial hyperpolarization of their membrane potential (Figure 4.8). During the period of stimulation, Na^+ enters through voltage-gated channels and accumulates within the axons. As the pump removes this extra Na^+, the resulting current generates a long-lasting hyperpolarization. Support for this interpretation comes from the observation that conditions that block the Na^+/K^+ pump—for example, treatment with ouabain, a plant glycoside that specifically inhibits the pump—prevent the hyperpolarization. The electrical contribution of the Na^+/K^+ pump is particularly significant in these small-diameter axons because their large surface-to-volume ratio causes intracellular Na^+ concentration to rise to higher levels than it would in other cells. Nonetheless, it is important to emphasize that, in most circumstances, the Na^+/K^+ pump has no direct role in generating the action potential and has relatively little effect on the resting potential.

■ THE MOLECULAR STRUCTURE OF THE Na^+/K^+ PUMP

Molecular genetic studies of the Na^+/K^+ pump have begun to define its structure. The pump is a large, integral membrane protein (Figure 4.9A). The primary sequence indicates that the protein spans the membrane 10 times, with most of the molecule found on the cytoplasmic side of the membrane.

Although a detailed account of the functional domains of the Na^+/K^+ pump is not yet available, some parts of the amino acid sequence have identified functions (Figure 4.9B). One intracellular domain of the protein is required for ATP binding and hydrolysis and the amino acid phosphorylated by ATP has been identified. Another extracellular domain may represent the binding site for ouabain, the toxin that blocks the pump. However, the sites involved in the most critical function of the pump—the movement of Na^+ and K^+—remain unknown. Kinetic studies indicate that both ions bind to the pump at the same site. Because these ions move across the membrane, it is likely that this site traverses the plasma membrane; and, since both Na^+ and K^+ are positively charged, it is also likely that the site has a negative charge. The observation that removing one negatively charged glutamate residue in a membrane-spanning domain of the protein (orange in Figure 4.9B) greatly reduces K^+ binding provides at least a hint about the ion-translocating domain of the pump.

■ SUMMARY

Ion pumps and channels have complementary functions. The primary purpose of pumps is to generate transmembrane concentration gradients, which

(A)

(B)

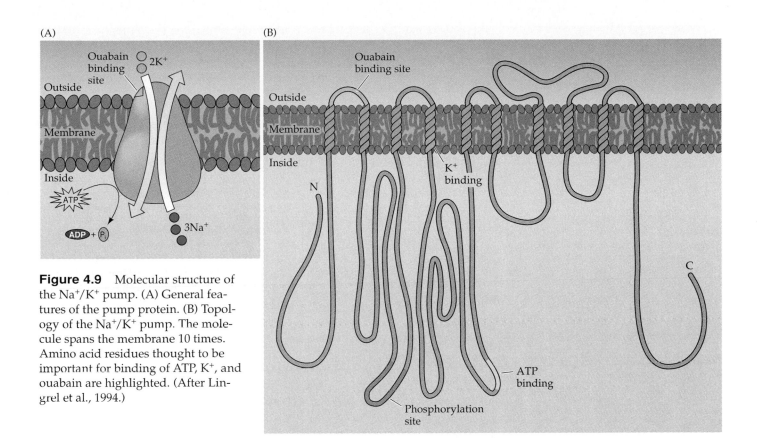

Figure 4.9 Molecular structure of the Na$^+$/K$^+$ pump. (A) General features of the pump protein. (B) Topology of the Na$^+$/K$^+$ pump. The molecule spans the membrane 10 times. Amino acid residues thought to be important for binding of ATP, K$^+$, and ouabain are highlighted. (After Lingrel et al., 1994.)

are then exploited by ion channels to generate electrical signals. Ion channels are responsible for the voltage-dependent conductances of nerve cell membranes. The channels underlying the action potential are integral membrane proteins that open or close their pores in response to the membrane potential, allowing ions to diffuse across the membrane. The flow of ions through single open channels can be detected as tiny electrical currents; the synchronous opening of many such channels generates the macroscopic currents that produce action potentials. Molecular studies in a variety of animal species show that such voltage-gated channels have conserved the structures that are responsible for features such as ion permeation and voltage sensing, as well as the features that specify ion selectivity and toxin sensitivity. The large number of ion channel genes creates channels with a wide range of functional characteristics, thus allowing different types of neurons to have distinctive electrical properties. Pump proteins are quite different. The cellular energy needed for ion movement against a concentration gradient is provided either by the hydrolysis of ATP, or by the electrochemical gradient of co-transported ions. The Na$^+$/K$^+$ pump produces and maintains the transmembrane gradients of Na$^+$ and K$^+$, while other pumps are responsible for the electrochemical gradients for other physiologically important ions, such as Cl$^-$, Ca^{2+}, and H$^+$. Together, pumps and channels provide a remarkably complete explanation of neuronal signaling.

Additional Reading

Reviews

BEZANILLA, F. AND A. M. CORREA (1995) Single-channel properties and gating of Na⁺ and K⁺ channels in the squid giant axon. In *Cephalopod Neurobiology*, N. J. Abbott, R. Williamson and L. Maddock (eds.). New York: Oxford University Press, pp. 131–151.

CATTERALL, W. A. (1988) Structure and function of voltage-sensitive ion channels. Science 242: 50–61.

LINGREL, J. B., J. VAN HUYSSE, W. O'BRIEN, E. JEWELL-MOTZ, R. ASKEW AND P. SCHULTHEIS (1994) Structure–function studies of the Na,K-ATPase. Kidney Internat. 45: S32–S39.

SKOU, J. C. (1988) Overview: The Na,K pump. Meth. Enzymol. 156: 1–25.

Important Original Papers

AUGUSTINE, C. K. AND F. BEZANILLA (1990) Phosphorylation modulates potassium conductance and gating current of perfused giant axons of squid. J. Gen. Physiol. 95: 245–271.

CORREA, A. M. AND F. BEZANILLA (1994) Gating of the squid sodium channel at positive potentials. II. Single channels reveal two open states. Biophys. J. 66: 1864–1878.

HODGKIN, A. L. AND R. D. KEYNES (1955) Active transport of cations in giant axons from *Sepia* and *Loligo*. J. Physiol. 128: 28–60.

LLANO, I., C. K. WEBB AND F. BEZANILLA (1988) Potassium conductance of squid giant axon. Single-channel studies. J. Gen. Physiol. 92: 179–196.

PEROZO, E., D. S. JONG AND F. BEZANILLA (1991) Single-channel studies of the phosphorylation of K⁺ channels in the squid giant axon. II. Nonstationary conditions. J. Gen Physiol. 98: 19–34.

RANG, H. P. AND J. M. RITCHIE (1968) On the electrogenic sodium pump in mammalian non-myelinated nerve fibres and its activation by various external cations. J. Physiol. 196: 183–221.

SIGWORTH, F. J. AND E. NEHER (1980) Single Na⁺ channel currents observed in cultured rat muscle cells. Nature 287: 447–449.

THOMAS, R. C. (1969) Membrane current and intracellular sodium changes in a snail neurone during extrusion of injected sodium. J. Physiol. 201: 495–514.

VANDERBERG, C. A. AND F. BEZANILLA (1991) A sodium channel model based on single channel, macroscopic ionic, and gating currents in the squid giant axon. Biophys. J. 60: 1511–1533.

WEI, A. M., A. COVARRUBIAS, A. BUTLER, K. BAKER, M. PAK AND L. SALKOFF (1990) K⁺ current diversity is produced by an extended gene family conserved in *Drosophila* and mouse. Science 248: 599–603.

Books

HILLE, B. (1992) *Ionic Channels of Excitable Membranes*, 2nd Ed. Sunderland, MA: Sinauer Associates.

JUNGE, D. (1992) *Nerve and Muscle Excitation*, 3rd Ed. Sunderland, MA: Sinauer Associates.

NICHOLLS, D. G. (1994) *Proteins, Transmitters and Synapses*. Oxford: Blackwell Scientific Publications.

CHAPTER 5

■ OVERVIEW

The human brain contains some 100 billion neurons, each with the ability to influence many other cells. A highly efficient mechanism is needed to enable communication among this astronomical number of elements. Such communication is made possible by synapses, the functional contacts between neurons. Although there are many synaptic subtypes within the brain, they can be divided into two general classes: electrical synapses and chemical synapses. Electrical synapses permit direct, passive flow of electrical current from one neuron to another. The current flows through gap junctions, which are specialized membrane channels that connect the two cells. Chemical synapses enable communication via the secretion of neurotransmitters; in this case, chemical agents released by the presynaptic neurons produce secondary current flow in postsynaptic neurons by activating specific receptor molecules. The secretion of neurotransmitters is triggered by voltage-gated Ca^{2+} channels, which elevate Ca^{2+} within the presynaptic terminal. The rise in Ca^{2+} concentration causes synaptic vesicles—presynaptic organelles that store neurotransmitters—to fuse with the plasma membrane and release their contents into the space between the pre- and postsynaptic cells. Although it is not yet understood exactly how Ca^{2+} triggers exocytosis, proteins on the surface of the synaptic vesicle and elsewhere in the presynaptic terminal evidently mediate this process.

■ ELECTRICAL SYNAPSES

Although they are a distinct minority, electrical synapses are found in all nervous systems, including the human brain. The generalized structure of an electrical synapse, as seen with an electron microscope, is shown schematically in Figure 5.1A. The membranes of the two communicating neurons are unusually close together, and are linked by a special kind of intercellular contact called a **gap junction**. Gap junctions contain precisely aligned, paired channels in the membrane of each neuron, such that each channel pair forms a pore (Figure 5.2A). The pore of a gap junction channel is much larger than the pores of the voltage-gated ion channels described in the previous chapter. As a result, a variety of substances can simply diffuse between the cytoplasm of the the pre- and postsynaptic neurons. Substances that can be exchanged via gap junctions include ions, as well as much larger molecules with molecular weights as great as several hundred daltons; the latter include ATP and other important intracellular metabolites such as second messengers.

Electrical synapses work by allowing ionic current to flow passively through the gap junction pores from one neuron to another. The usual source of the current is the potential difference generated locally by the action potential (see Chapter 3). The "upstream" neuron, which is the source of current, is termed **presynaptic**, and the "downstream" neuron into which this current flows is termed **postsynaptic**. This arrangement has a number of interesting consequences. One is that transmission can be bidirectional; that is, current can flow in either direction across the gap junction, depending

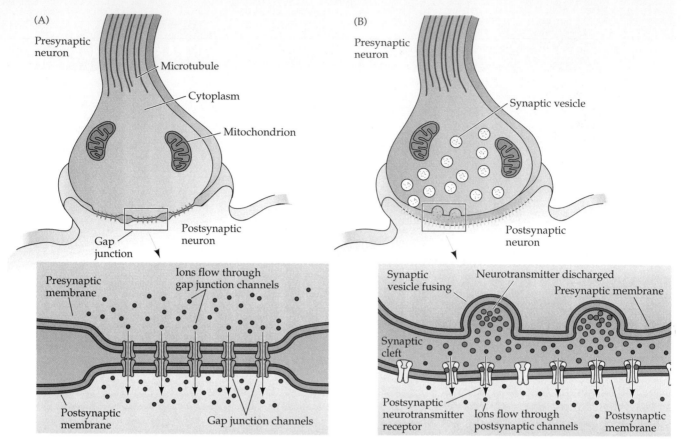

Figure 5.1 Electrical and chemical synapses differ fundamentally in their transmission mechanisms. (A) At electrical synapses, gap junctions between pre- and postsynaptic membranes permit current to flow passively through intercellular channels (blowup). This current flow changes the postsynaptic membrane potential, initiating (or in some instances inhibiting) the generation of postsynaptic action potentials. (B) At chemical synapses, there is no intercellular continuity, and thus no direct flow of current from pre- to postsynaptic cell. Synaptic current flows across the postsynaptic membrane only in response to the secretion of neurotransmitters, which opens or closes postsynaptic ion channels after binding to receptor molecules (blowup).

on which member of the coupled pair is invaded by an action potential (although some types of gap junctions have special features that render their transmission unidirectional). Another important feature of the electrical synapse is that transmission is extraordinarily fast: because passive current flow across the gap junction is virtually instantaneous, communication can occur quickly.

The first example of electrical transmission was discovered in the crayfish nervous system (Figure 5.2B), where electrical synapses connect many of the neurons in a circuit that allows the crayfish to escape from its predators with a minimum of delay. A more general purpose of electrical synapses is to synchronize electrical activity among populations of neurons. For example, certain hormone-secreting neurons within the hypothalamus are connected by electrical synapses. This arrangement ensures that all cells fire action potentials at about the same time to maximize a burst of hormone secretion into the blood. In principle, the fact that gap junction pores are large enough

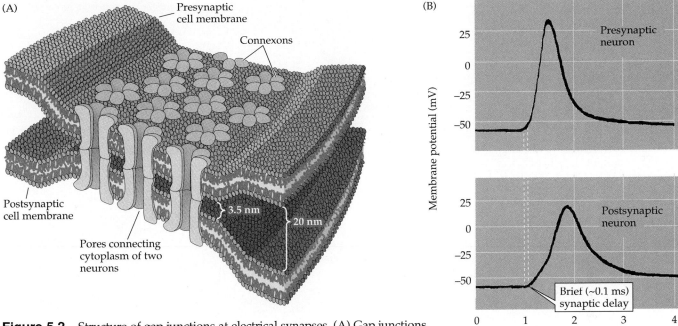

Figure 5.2 Structure of gap junctions at electrical synapses. (A) Gap junctions consist of hexameric complexes formed by the coming together of subunits called connexons, which are present in both the pre- and postsynaptic membranes. The pores of the channels connect to one another, creating electrical continuity between the two cells. (B) Rapid transmission of signals at an electrical synapse in the crayfish. An action potential in the presynaptic neuron causes the postsynaptic neuron to be depolarized within a fraction of a millisecond. (B after Furshpan and Potter, 1959.)

to allow molecules such as ATP and second messengers to diffuse intercellularly also permits electrical synapses to coordinate the metabolism of coupled neurons.

■ CHEMICAL SYNAPSES

The generalized structure of a chemical synapse is shown schematically in Figure 5.1B. The separation between the pre- and postsynaptic neurons is substantially greater at chemical synapses than at electrical synapses and is called the **synaptic cleft**. The key feature of all chemical synapses is the presence of small, membrane-bounded organelles called **synaptic vesicles** within the presynaptic terminal. These spherical organelles are filled with one or more **neurotransmitters**, the chemical signals secreted from the presynaptic neuron. The use of such chemical agents as messengers between the communicating neurons gives this type of synapse its name. There are many kinds of neurotransmitters (see Chapter 6); perhaps the best studied is acetylcholine, the transmitter employed at peripheral neuromuscular synapses, in autonomic ganglia, and at some central synapses.

Transmission at chemical synapses is based on the elaborate sequence of events depicted in Figure 5.3. The process is initiated when an action potential invades the terminal of the presynaptic neuron. The change in membrane potential associated with the arrival of the action potential causes the opening of voltage-gated calcium channels in the presynaptic membrane. Because of the steep concentration gradient of Ca^{2+} across the presynaptic

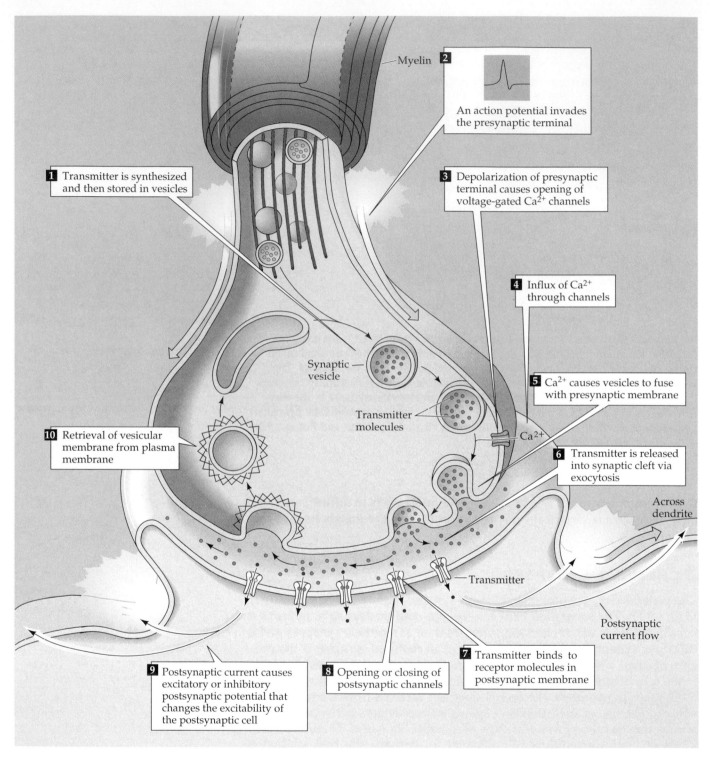

Figure 5.3 Sequence of events involved in transmission at a typical chemical synapse.

membrane (the external Ca^{2+} concentration is approximately 10^{-3} M, whereas the internal Ca^{2+} concentration is approximately 10^{-7} M), the opening of these channels causes a rapid influx of Ca^{2+} into the presynaptic terminal. Calcium influx causes the Ca^{2+} concentration of the cytoplasm in the terminal to rise from its normally low level to a much higher value. Elevation of the presynaptic Ca^{2+} concentration allows synaptic vesicles to fuse with the plasma membrane of the presynaptic neuron, since this process is Ca^{2+}-

dependent. The fusion of synaptic vesicles causes their contents, most importantly neurotransmitters, to be released into the synaptic cleft. Following this exocytosis, transmitters diffuse throughout the synaptic cleft and bind to specific receptors on the membrane of the postsynaptic neuron. The binding of neurotransmitter to the receptors causes channels in the postsynaptic membrane to open (or sometimes to close). The resulting neurotransmitter-induced current flow alters the membrane potential of the postsynaptic neuron, increasing or decreasing the probability that the neuron will fire an action potential.

■ QUANTAL TRANSMISSION AT NEUROMUSCULAR SYNAPSES

Much of the evidence leading to present understanding of chemical synaptic transmission was obtained from experiments on neuromuscular junctions, the synapses between spinal motor neurons and skeletal muscle cells. Because of their simplicity, large size, and peripheral location, these synapses are particularly amenable to experimental analysis. Such synapses occur at specializations called **end plates** because of the saucerlike appearance of the site on the muscle fiber where the presynaptic axon elaborates its terminals (Figure 5.4A). Most of the pioneering work on neuromuscular transmission was performed in the laboratory of Bernard Katz at University College, London during the 1950s and 1960s. Katz has been widely recognized for his remarkable contributions to understanding synaptic transmission as more recent experiments have verified the applicability of events at the neuromuscular junction to transmission at chemical synapses throughout the nervous system.

When an intracellular microelectrode records the membrane potential of a muscle cell, an action potential in the presynaptic motor neuron can be seen to elicit a transient depolarization of the postsynaptic muscle cell. This change in membrane potential, called an **end plate potential (EPP)**, is nor-

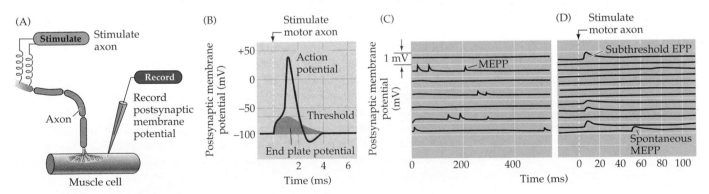

Figure 5.4 Synaptic transmission at neuromuscular junctions. (A) Experimental arrangement. The axon of the motor neuron innervating the muscle fiber is stimulated with an extracellular electrode, while an intracellular microelectrode is inserted into the postsynaptic muscle cell to record its electrical responses. (B) End plate potentials (EPPs) evoked by stimulation of a motor neuron are normally above threshold and therefore produce an action potential in the postsynaptic muscle cell. (C) Spontaneous miniature EPPs (MEPPs) occur in the absence of presynaptic stimulation. (D) When the neuromuscular junction is bathed in a solution that has a low concentration of Ca^{2+}, stimulating the motor neuron evokes EPPs whose amplitudes are reduced to about the size of MEPPS. (After Fatt and Katz, 1952.)

mally large enough to bring the membrane potential of the muscle cell well above the threshold for producing a postsynaptic action potential (Figure 5.4B). The postsynaptic action potential, triggered by the EPP, causes the muscle fiber to contract. One of Katz's seminal findings, in studies carried out with Paul Fatt in 1951, was that spontaneous changes in muscle cell membrane potential occur even in the absence of stimulation of the presynaptic motor neuron (Figure 5.4C). These changes have the same shape as EPPs but are much smaller (typically less than 1 mV in amplitude). Both EPPs and these small, spontaneous events are sensitive to pharmacological agents that block postsynaptic acetylcholine receptors, such as curare (see Chapter 7). These and other parallels between EPPs and the spontaneously occurring depolarizations led Katz and his colleagues to call the spontaneous events **miniature end plate potentials**, or **MEPPs**.

The relationship between the full-blown end plate potential and MEPPs was clarified by careful analysis of the amplitudes of EPPs. The EPP provides a convenient electrical assay of neurotransmitter secretion from the motor neuron terminal, as long as precautions are taken to prevent muscle contraction from dislodging the microelectrode used to detect the EPP. The usual means of eliminating muscle contractions is either to lower Ca^{2+} concentration in the extracellular medium, or to partially block the postsynaptic transmitter receptors with curare. As expected from the scheme illustrated in Figure 5.3, lowering the Ca^{2+} concentration reduces neurotransmitter secretion, thus reducing the magnitude of the EPP below the threshold for postsynaptic action potential production. Under such conditions, repetitive stimulation of the motor neuron produces EPPs that fluctuate in amplitude (Figure 5.4D). These fluctuations give considerable insight into the mechanisms responsible for neurotransmitter release. In particular, the evoked response in low Ca^{2+} appears to result from the release of unit amounts of neurotransmitter by the presynaptic nerve terminal. Thus, the amplitude of the smallest evoked response is strikingly similar to the size of single MEPPs (compare Figures 5.4C and D). In accord with this similarity, increments in the EPP response (Figure 5.5A) occur in units about the size of single MEPPs (Figure 5.5B). These "quantal" fluctuations in the amplitude of EPPs suggest that EPPs are made up of individual units, each equivalent to a MEPP.

The idea that EPPs represent the simultaneous release of multiple MEPP-like units can be tested statistically. A model based on the independent occurrence of unitary events (Poisson statistics) predicts what the distribution of EPP amplitudes should look like during a large number of trials of motor neuron stimulation. The experimentally determined distribution of EPP amplitudes is consistent with that expected if transmitter release from the motor neuron is indeed quantal (smooth curve in Figure 5.5B). Such analyses confirm the idea that transmitter release does indeed occur in discrete packets, each equivalent to a MEPP. Thus, a presynaptic action potential causes a postsynaptic EPP because it synchronizes the release of many transmitter quanta.

■ RELEASE OF TRANSMITTERS FROM SYNAPTIC VESICLES

The discovery of the quantal release of packets of neurotransmitter immediately raised the question of how such quanta are formed and discharged into the synaptic cleft. At about the time Katz and his colleagues discovered quantal release of neurotransmitter by means of electrophysiology, electron microscopy revealed, for the first time, the presence of synaptic vesicles in presynaptic terminals. Putting these two discoveries together, Katz and oth-

(A)

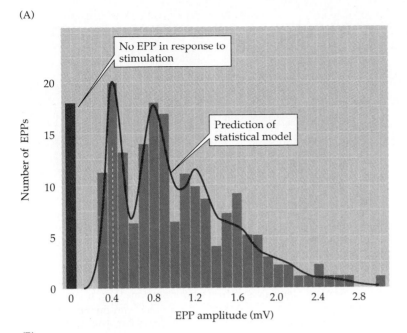

(B)

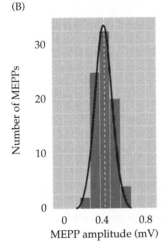

Figure 5.5 Quantized distribution of EPP amplitudes evoked in a low Ca^{2+} solution. Peaks of EPP amplitudes (A) occur in integer multiples of the mean amplitude of MEPPs, whose amplitude distribution is shown in (B). The leftmost bar in the EPP amplitude distribution shows trials in which presynaptic stimulation failed to elicit an EPP in the muscle cell. The red curve indicates the prediction of a statistical model based on the assumption that the EPPs result from the independent release of multiple MEPP-like quanta. The observed match, including the predicted number of failures, supports this interpretation. (After Boyd and Martin, 1955.)

ers proposed that synaptic vesicles loaded with transmitter were the source of the electrically recorded quanta. Subsequent biochemical studies showed that the vesicles are indeed repositories of transmitter. The concentration of the neurotransmitter acetylcholine in synaptic vesicles of motor neurons, for instance, is on the order of 100 mM; given the diameter of a single synaptic vesicle, this concentration translates into approximately 10,000 molecules of neurotransmitter per vesicle.

Synaptic vesicles secrete their contents by fusing with the plasma membrane of the presynaptic terminal. To prove that fusion actually occurs, it is necessary to show that each fused vesicle causes a single quantal event to be recorded postsynaptically. This challenge was met by correlating measurements of vesicle fusion with the quantal content of EPPs at the neuromuscular junction. The number of vesicles that fused with the presynaptic plasma membrane was measured by electron microscopy in terminals that had been treated with a drug (4-aminopyridine, or 4-AP) that enhanced the number of vesicle fusion events produced by single action potentials (Figure 5.6A). Parallel electrical measurements were made of the quantal content of the EPPs

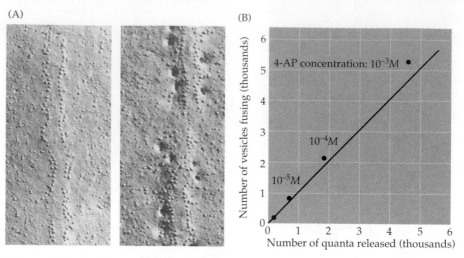

(A)

(B)

Figure 5.6 Relationship of synaptic vesicle exocytosis and quantal transmitter release. (A) Freeze-fracture electron microscopy was used to visualize the fusion of synaptic vesicles in presynaptic terminals of frog motor neurons. *Left*, image of the plasma membrane of an unstimulated presynaptic terminal. *Right*, image of the plasma membrane of a terminal stimulated by an action potential. Stimulation causes the appearance of dimple-like structures that represent the fusion of synaptic vesicles with the presynaptic membrane. The view is as if looking down on the release sites from inside the presynaptic terminal. (B) Comparison of the number of observed vesicle fusions to the number of quanta released by a presynaptic action potential. Transmitter release was varied by using a drug (4-AP) that affects the duration of the presynaptic action potential, thus changing the amount of calcium that enters during the action potential. The diagonal line is the 1:1 relationship expected if each vesicle that opened released a single quantum of transmitter. (From Heuser et al., 1979.)

elicited in this way. A comparison of the number of synaptic vesicle fusions observed with the electron microscope and the number of quanta released at the very same synapse showed a good correlation between these two measures (Figure 5.6B). This experimental result, obtained in the 1970s by John Heuser, Tom Reese, and their colleagues, remains one of the strongest lines of support for the idea that a quantum of transmitter release is due to a synaptic vesicle fusing with the presynaptic membrane. More recent evidence, based on other means of measuring vesicle fusion, has left no doubt about the validity of this interpretation.

■ LOCAL RECYCLING OF SYNAPTIC VESICLES

The fusion of synaptic vesicles causes new membrane to be added to the plasma membrane of the presynaptic terminal, but the addition is not permanent. Although a bout of exocytosis can dramatically increase the surface area of presynaptic terminals, this extra membrane is removed within a few minutes. Heuser and Reese performed another important set of experiments that showed the fused vesicle membrane is retrieved and taken back into the cytoplasm of the nerve terminal (a process called endocytosis). The experiments, again carried out at the neuromuscular junction, were based on filling the synaptic cleft with horseradish peroxidase (HRP), an enzyme that produces a dense product that can be seen with the electron microscope. In these circumstances, endocytosis can be visualized by the uptake of HRP into the nerve terminal (Figure 5.7A). To activate endocytosis, the presynaptic terminal was

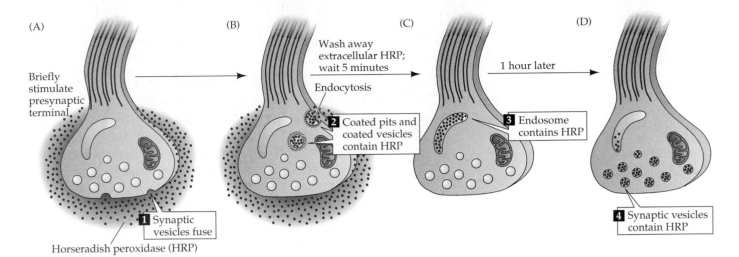

(A) Briefly stimulate presynaptic terminal.

1 Synaptic vesicles fuse

Horseradish peroxidase (HRP)

(B) Wash away extracellular HRP; wait 5 minutes

Endocytosis

2 Coated pits and coated vesicles contain HRP

(C) 1 hour later

3 Endosome contains HRP

(D)

4 Synaptic vesicles contain HRP

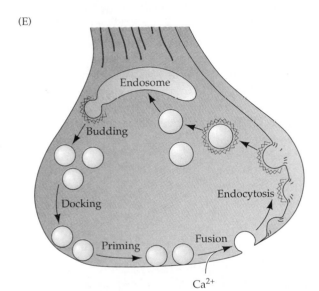

(E)

Endosome

Budding

Docking

Priming

Fusion

Endocytosis

Ca²⁺

Figure 5.7 Local recycling of synaptic vesicles in presynaptic terminals. (A) Horseradish peroxidase (HRP) introduced into the synaptic cleft was used to follow the fate of membrane retrieved from the presynaptic plasma membrane. Stimulation of endocytosis by presynaptic action potentials causes HRP to be taken up into the presynaptic terminals via a pathway that includes (B) coated vesicles and (C) endosomes. (D) Eventually, the HRP is found in newly formed synaptic vesicles. (E) Interpretation of the results shown in A–D. Calcium-regulated fusion of vesicles with the presynaptic membrane is followed by endocytotic retrieval of vesicular membrane via coated vesicles and endosomes, and subsequent reformation of new synaptic vesicles. (After Heuser and Reese, 1973.)

stimulated with a train of action potentials, and the subsequent fate of the HRP was followed by electron microscopy. Immediately following stimulation, the HRP was found within special endocytotic organelles called coated vesicles (Figure 5.7B). A few minutes later, however, the coated vesicles had disappeared and the HRP was found in a different organelle, the endosome (Figure 5.7C). Finally, approximately an hour after stimulating the terminal, the HRP reaction product appeared inside synaptic vesicles (Figure 5.7D).

These observations indicate that synaptic vesicle membrane is recycled within the presynaptic terminal via the sequence summarized in Figure 5.7E. In this sequence, called the **synaptic vesicle cycle**, the retrieved vesicular membrane passes through diverse intracellular compartments (coated vesicles and endosomes) and is eventually used to make new synaptic vesicles. The newly made vesicles are refilled with neurotransmitter, docked at the presynaptic plasma membrane and primed to participate in exocytosis once again. Vesicles are *originally* produced by processing through the endoplasmic reticulum and Golgi apparatus in the neuronal cell body. Because of the long distance between the cell body and the presynaptic terminal in most neurons, transport of vesicles from the soma would not permit rapid replenishment of synaptic vesicles during continuous neural activity. Thus, local recycling is well suited to the peculiar anatomy of neurons, giving nerve terminals the means to provide a continual supply of synaptic vesicles.

■ THE ROLE OF CALCIUM IN TRANSMITTER SECRETION

Lowering the concentration of Ca^{2+} outside a presynaptic motor nerve terminal reduces the size of the EPP (compare Figures 5.4B and D). Measurement of the number of transmitter quanta released under such conditions shows that the reason the EPP gets smaller is that lowering Ca^{2+} concentration decreases the number of vesicles that fuse with the plasma membrane of the terminal. An important insight into how Ca^{2+} regulates the fusion of synaptic vesicles was the discovery that presynaptic terminals have voltage-sensitive Ca^{2+} channels in their plasma membranes. The first indication of presynaptic Ca^{2+} channels derived from the observation by Katz and Ricardo Miledi that presynaptic terminals treated with tetrodotoxin (to block Na^+ channels; see Chapter 3) could still produce an action potential. The explanation for this surprising finding was that current was still flowing through Ca^{2+} channels, substituting for the current ordinarily carried by Na^+ channels. Subsequent voltage clamp experiments, performed at a giant presynaptic terminal of the squid (Figure 5.8A), confirmed the presence of voltage-gated Ca^{2+} channels in the presynaptic terminal (Figure 5.8B). Blockade of these Ca^{2+} channels with drugs inhibits transmitter release, confirming that

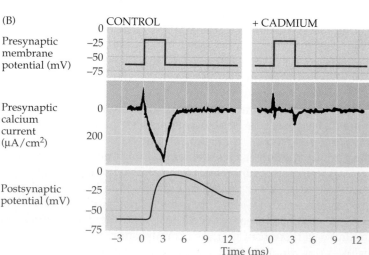

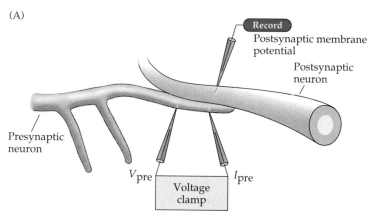

Figure 5.8 Currents flowing through the calcium channels of a presynaptic terminal cause transmitter release. (A) The voltage clamp method was used to measure currents flowing across the presynaptic membrane when the membrane potential was depolarized. (B) Pharmacological agents that block currents flowing through Na^+ and K^+ channels reveal a remaining inward current flowing through Ca^{2+} channels. This influx of calcium triggers transmitter secretion, as indicated by a change in the postsynaptic membrane potential. Treatment of the same presynaptic terminal with cadmium, a calcium channel blocker, eliminates both the presynaptic calcium current and the postsynaptic response. (After Augustine and Eckert, 1984.)

(A)

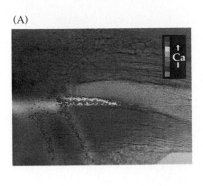

(B)

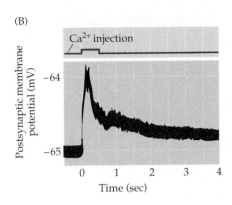

(C)

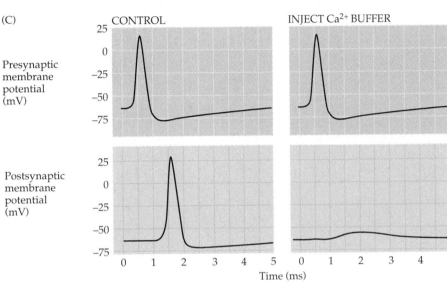

Figure 5.9 Further evidence that calcium entry triggers transmitter release from presynaptic terminals. (A) Fluorescence microscopy measurements of presynaptic Ca^{2+} concentration at the squid giant synapse. A train of presynaptic action potentials causes a rise in Ca^{2+} concentration, as revealed by a dye (fura-2) that fluoresces more strongly when the Ca^{2+} concentration increases. (B) Microinjection of Ca^{2+} into a squid giant presynaptic terminal triggers transmitter release, measured as a depolarization of the postsynaptic membrane potential. (C) Microinjection of BAPTA, a Ca^{2+} chelator, into a squid giant presynaptic terminal prevents transmitter release. (A from Smith et al., 1993; B after Miledi, 1971; C after Adler et al., 1991.)

the channels are directly involved. Thus, presynaptic action potentials open voltage-gated Ca^{2+} channels, and the resulting influx of Ca^{2+} somehow triggers transmitter release.

The fact that Ca^{2+} entry into presynaptic terminals causes a rise in the concentration of Ca^{2+} within the terminal has been directly documented by microscopic imaging of terminals filled with Ca^{2+}-sensitive fluorescent dyes (Figure 5.9A). The consequences of the rise in presynaptic Ca^{2+} concentration for neurotransmitter release has been shown in two ways. First, microinjection of Ca^{2+} directly into presynaptic terminals triggers transmitter release in the absence of presynaptic action potentials (Figure 5.9B). Second, presynaptic microinjection of Ca^{2+} chelators (chemicals that bind Ca^{2+} and keep its concentration buffered at low levels) prevents presynaptic action potentials from causing transmitter secretion (Figure 5.9C). These results show that a rise in presynaptic Ca^{2+} concentration is both necessary and sufficient for transmitter release.

■ MOLECULAR MECHANISMS OF TRANSMITTER SECRETION

Precisely how an increase in presynaptic Ca^{2+} concentration goes on to trigger vesicle fusion and neurotransmitter release is not well understood. However, some important clues have come from molecular studies that have

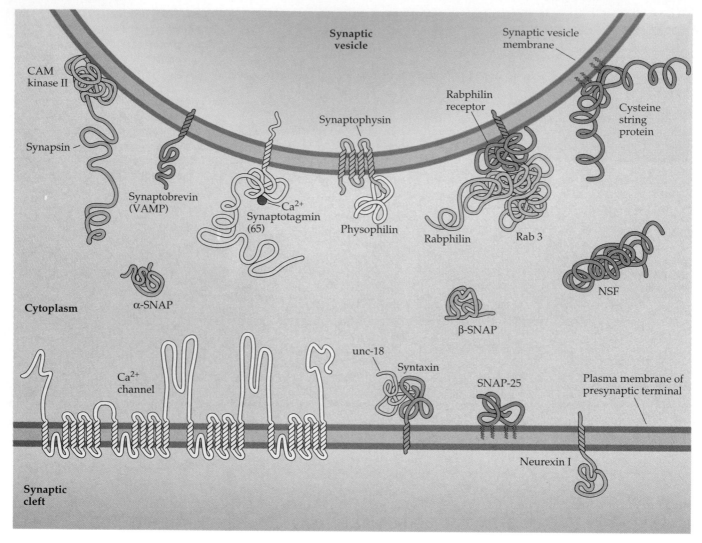

Figure 5.10 Presynaptic proteins potentially involved in neurotransmitter release. (After Jessell and Kandel, 1993.)

identified and characterized a number of proteins found on synaptic vesicles and presynaptic plasma membranes. Some of these presynaptic proteins are represented in Figure 5.10. Although the parts played by these molecules in transmitter secretion are not yet clear, the properties of the proteins give some indication of their functions. For example, synaptotagmin, a protein found in the membrane of synaptic vesicles, is capable of binding Ca^{2+}. This observation suggests that synaptotagmin acts as a Ca^{2+} sensor, signaling the elevation of Ca^{2+} within the terminal and triggering vesicle fusion. In support of this idea are genetic experiments in which synaptotagmin genes have been altered in mice, fruit flies, and other experimental animals. Such mutations greatly impair Ca^{2+}-dependent neurotransmitter release and are in most cases lethal. How Ca^{2+} binding to synaptotagmin could lead to exocytosis is not yet known. Moreover, studies of toxins that affect transmitter release implicate several other presynaptic proteins in transmitter exocytosis (Box A).

It is likely that a complex cascade of proteins, acting in concert, allow neurons to secrete transmitters. Some of these proteins are probably involved

Box A
TOXINS THAT AFFECT TRANSMITTER RELEASE

Several important insights about the molecular basis of neurotransmitter secretion have come from analysis of the actions of certain presynaptic neurotoxins. One family of such agents is the clostridial toxins, extremely potent bacterial proteins that destroy synaptic transmission by blocking transmitter release. These toxins are responsible for botulism and tetanus, two serious human disorders caused by anaerobic bacterial growth in food or infected tissue, respectively. Clever and patient biochemical work has shown that these toxins are highly specific proteases that cleave selected presynaptic proteins. Tetanus toxin and botulinum toxin (types B, D, and F) specifically cleave the vesicle membrane protein, synaptobrevin (see Figure 5.10). Other botulinum toxins are proteases that cleave syntaxin and SNAP-25, proteins found on the presynaptic plasma membrane. These observations argue that destruc-

tion of the presynaptic proteins is the basis for the actions of the toxins on neurotransmitter release. The evidence also implies that these three synaptic proteins are somehow important in the process of vesicle–plasma membrane fusion, probably by forming the SNARE complex (see text).

Another toxin that targets neurotransmitter release is α-latrotoxin, a protein found in the venom of the female black widow spider. Application of this molecule to neuromuscular synapses causes a massive discharge of synaptic vesicles, even when Ca^{2+} is absent from the extracellular medium. While it is not yet clear how this toxin triggers Ca^{2+}-independent exocytosis, α-latrotoxin binds to neurexins, a group of integral membrane proteins found in presynaptic terminals (see Figure 5.10). Because the neurexins bind to synaptotagmin, a vesicular Ca^{2+}-binding protein known to be important in exocytosis, this interac-

tion may allow α-latrotoxin to bypass the usual Ca^{2+} requirement for triggering vesicle fusion.

Still other toxins produced by snakes, spiders, snails, and other predatory animals are known to affect transmitter release, but their sites of action have yet to be identified. Based on the precedents described here, it is likely that these poisons will continue to provide valuable tools for elucidating the molecular basis of neurotransmitter release.

References

MONTECUCCO, C. AND G. SCHIAVO (1994) Mechanism of action of tetanus and botulinum neurotoxins. Mol. Microbiol. 13: 1–8.

PETRENKO, A. G., M. S. PERIN, B. A. DARLETOV, Y. A. USHKARYOV, M. GEPPERT AND T. C. SÜDHOF (1991) Binding of synaptotagmin to the α-latrotoxin receptor implicates both in synaptic vesicle exocytosis. Nature 353: 65–68.

in other types of membrane fusion events common to all cells. For example, two proteins known to be important for the fusion of vesicles with membranes of the Golgi apparatus, termed NSF (NEM-sensitive fusion protein) and SNAPs (soluble NSF-attachment proteins), also appear to be involved in the fusion of synaptic vesicles with the presynaptic membrane. These two proteins work together with the membrane-associated SNARE (SNAP receptor) proteins (synaptobrevin, syntaxin and SNAP-25; see Box A) to form a macromolecular complex that plays an essential but poorly understood role in release of neurotransmitters. Although the detailed mechanisms responsible for transmitter secretion are not known, the recent application of molecular biological approaches has greatly energized the study of this long-standing problem.

■ SUMMARY

Synapses communicate the information carried by action potentials from one neuron to another. The cellular mechanisms that underlie synaptic transmission are closely related to the mechanisms that generate resting and action potentials, namely ion flow through ion channels in the neuronal membrane. In the case of electrical synapses, these channels are gap junctions. Direct but passive flow of current through the gap junctions is the basis for transmission at these synapses. In the case of chemical synapses, channels with smaller and more selective pores are activated by the binding of neurotransmitters to postsynaptic receptors. Transmitter agents are released from presynaptic terminals in quanta after the arrival of an action potential, reflecting their storage within synaptic vesicles. Vesicles discharge their contents when presynaptic depolarization opens voltage-gated calcium channels, allowing Ca^{2+} to accumulate within the presynaptic terminal. How calcium triggers neurotransmitter release is not yet established, but a number of key proteins found within the presynaptic terminal are apparently involved.

Additional Reading

Reviews

AUGUSTINE, G. J., M. P. CHARLTON AND S. J. SMITH (1987) Calcium action in synaptic transmitter release. Annu. Rev. Neurosci. 10: 633–693.

JESSELL, T. M. AND E. R. KANDEL (1993) Synaptic transmission: A bidirectional and self-modifiable form of cell–cell communication. Cell 72/Neuron 10: (Supplement): 1–30.

ROTHMAN, J. E. (1994) Mechanisms of intracellular protein transport. Nature 372: 55–63.

SÜDHOF, T. C. (1995) The synaptic vesicle cycle: A cascade of protein-protein interactions. Nature 375: 645–653.

Important Original Papers

ADLER, E., M. ADLER, G. J. AUGUSTINE, M. P. CHARLTON AND S. N. DUFFY (1991) Alien intracellular calcium chelators attenuate neurotransmitter release at the squid giant synapse. J. Neurosci. 11: 1496–1507.

BOYD, I. A. AND A. R. MARTIN (1955) The end-plate potential in mammalian muscle. J. Physiol. (Lond.) 132: 74–91.

DEL CASTILLO, J. AND B. KATZ (1954) Quantal components of the end plate potential. J. Physiol. (Lond.) 124: 560–573.

FATT, P. AND B. KATZ (1951) An analysis of the end plate potential recorded with an intracellular electrode. J. Physiol. (Lond.) 115: 320–370.

FATT, P. AND B. KATZ (1952) Spontaneous subthreshold activity at motor nerve endings. J. Physiol. (Lond.) 117: 109–128.

FURSHPAN, E. J. AND D. D. POTTER (1959) Transmission at the giant motor synapses of the crayfish. J. Physiol. (Lond.) 145: 289–325.

HEUSER, J. E. AND T. S. REESE (1973) Evidence for recycling of synaptic vesicle membrane during transmitter release at the frog neuromuscular junction. J. Cell Biol. 57: 315–344.

HEUSER, J. E., T. S. REESE, M. J. DENNIS, Y. JAN, L. JAN AND L. EVANS (1979) Synaptic vesicle exocytosis captured by quick freezing and correlated with quantal transmitter release. J. Cell Biol. 81: 275–300.

MILEDI, R. (1973) Transmitter release induced by injection of calcium ions into nerve terminals. Proc. R. Soc. Lond. B 183: 421–425.

SMITH, S. J., J. BUCHANAN, L. R. OSSES, M. P. CHARLTON AND G. J. AUGUSTINE (1993) The spatial distribution of calcium signals in squid presynaptic terminals. J. Physiol. (Lond.) 472: 573–593.

Books

HALL, Z. (1992) An Introduction to Molecular Neurobiology. Sunderland, MA: Sinauer Associates.

KATZ, B. (1966) Nerve, Muscle, and Synapse. New York: McGraw-Hill.

KATZ, B. (1969) The Release of Neural Transmitter Substances. Liverpool: Liverpool University Press.

NICHOLLS, D. G. (1994) Proteins, Transmitters, and Synapses. Oxford: Blackwell.

CHAPTER 6

NEURO-TRANSMITTERS

■ OVERVIEW

For the most part, neurons in the human brain communicate with each other by releasing chemical messengers called neurotransmitters (electrical synapses are present, but in the distinct minority). The utility cycle of all neurotransmitter molecules is similar: (1) they are synthesized and packaged into vesicles in the presynaptic cell; (2) they are released from the presynaptic cell and bind to receptors on one or more postsynaptic cells; and (3) once released into the synaptic cleft, they are rapidly removed or degraded. The total number of neurotransmitters is not known, but is likely to be well over 100. Despite this diversity, these agents can be classified into two broad categories: small-molecule neurotransmitters and neuropeptides. In general, small-molecule neurotransmitters mediate rapid reactions, whereas neuropeptides tend to modulate slower, ongoing brain functions. Abnormal transmitter functions cause a wide range of neurological and psychiatric disorders; as a result altering the actions of neurotransmitters by pharmacological or other means is central to many modern therapeutic strategies.

■ WHAT DEFINES A NEUROTRANSMITTER?

As broadly described in the preceding chapter, neurotransmitters are chemical signals released from presynaptic nerve terminals into the synaptic cleft. The subsequent binding of neurotransmitters to specific receptors on postsynaptic neurons (or other cell classes) then briefly changes the electrical properties of the target cells.

The notion that electrical information can be transferred from one neuron to the next by means of chemical signaling was the subject of intense debate through the first half of the twentieth century. The initial experiment supporting this idea was performed in 1926 by German physiologist Otto Loewi. Acting on an idea that allegedly came to him in the middle of the night, Loewi rushed to his laboratory and performed a simple but conclusive test. He isolated and perfused the hearts of two frogs, monitoring the rates at which they were beating. He then electrically stimulated the vagus nerve attached to one heart, which caused a slowing of the heartbeat. The perfusate flowing through this heart was collected and transferred to the second heart. Even though the second heart had not been stimulated, its beat also slowed. This result demonstrated that the vagus nerve regulates the heartbeat by releasing a chemical that accumulated in the perfusate. Originally referred to as "vagus substance," this agent was later shown to be **acetylcholine (ACh)**, which is now the most thoroughly studied neurotransmitter. ACh acts not only in the heart but at a variety of postsynaptic targets in the central and peripheral nervous systems, preeminently at the neuromuscular junction (see Chapter 5).

Over the years, a number of formal criteria have emerged that definitively identify a substance as a neurotransmitter (Box A). Identifying the neurotransmitter active at any particular synapse remains a difficult undertaking, and for many synapses (particularly in the brain), the nature of the neurotransmitter is not yet known. Substances that have not met all the criteria outlined in Box A are referred to as putative neurotransmitters.

Box A
CRITERIA THAT DEFINE A SUBSTANCE AS A NEUROTRANSMITTER

A number of standard criteria are used to confirm that a suspected agent is indeed a neurotransmitter at a given synapse.

1. *The substance must be present within the presynaptic neuron.* Clearly, a chemical cannot be secreted from a presynaptic neuron unless it is present there. Because elaborate biochemical pathways are required to produce neurotransmitters, showing that the enzymes and precursors required to synthesize the substance are present in presynaptic neurons provides additional evidence that the substance is used as a transmitter. (Note, however, that since the transmitters glutamate, glycine, and aspartate are also needed for protein synthesis and other metabolic reactions in all neurons, their presence is *not* sufficient evidence to establish them as neurotransmitters.)

2. *The substance must be released in response to presynaptic depolarization, which must occur in a Ca^{2+}-dependent manner.* Another essential criterion for identifying a neurotransmitter is to demonstrate that it is released from the presynaptic neuron in response to presynaptic electrical activity, and that this release requires Ca^{2+} influx into the presynaptic terminal. Meeting this criterion is technically challenging, not only because it may be difficult to selectively stimulate the presynaptic neurons, but also because enzymes and transporters efficiently remove the secreted neurotransmitters.

3. *Specific receptors for the substance must be present on the postsynaptic cell.* A neurotransmitter cannot act on its target unless specific receptors for the transmitter are present in the postsynaptic membrane. One way to demonstrate receptors is to show that application of

exogenous transmitter mimics the postsynaptic effect of presynaptic stimulation. A more rigorous way to demonstrate receptors is to show that agonists and antagonists that alter the normal postsynaptic response have the same effect when the substance in question is applied exogenously. High-resolution histological methods can also be used to show that specific receptors are present in the postsynaptic membrane (by detection of radioactively labeled receptor antibodies, for example).

Fulfilling these criteria establishes unambiguously that a given substance is used as a transmitter at a given synapse. Practical difficulties, however, have prevented these standards from being applied at many types of synapses. It is for this reason that so many substances must still be referred to as "putative neurotransmitters."

(1)

(2)

(3)

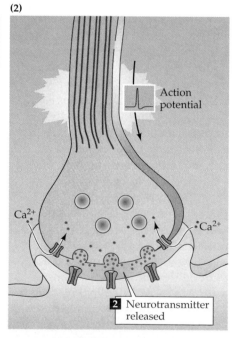

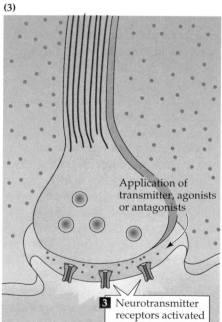

Demonstrating the identity of a neurotransmitter at a synapse requires showing (1) its presence, (2) its release, and (3) the postsynaptic presence of specific receptors.

(A)

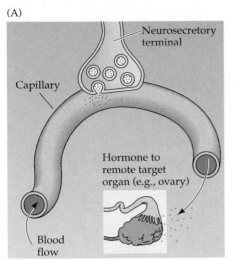

(B)

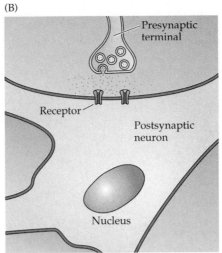

(C)

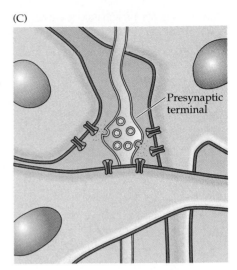

Figure 6.1 Localization of neurotransmitter action. (A) Hormones are released into the general circulation to affect target organs at distant sites; some hormones also act as neurotransmitters. Neurotransmitters in general, however, act either locally (B), by altering the electrical excitability of a small region of a single postsynaptic cell, or more diffusely (C), by altering the electrical excitability of a few postsynaptic cells.

The special characteristics of neurotransmitters are made clearer by comparing their actions to another type of chemical signal, the hormones secreted by the endocrine system. Hormones typically influence target cells far removed from the hormone-secreting cell (Figure 6.1A). This "action at a distance" is achieved by the release of hormones into the bloodstream. In contrast, the distance over which neurotransmitters act is always much less. At many synapses, transmitters bind only to receptors on the postsynaptic cell that directly underlies the presynaptic terminal (Figure 6.1B and Chapter 5); in such cases, the transmitter acts over distances less than a micrometer. At other synapses, neurotransmitters diffuse locally to alter the electrical properties of multiple postsynaptic (and sometimes presynaptic) cells in the vicinity of the presynaptic release sites (Figure 6.1C).

While the distinction between neurotransmitters and hormones is generally clear-cut, a substance can act as a neurotransmitter in one region of the brain while serving as a hormone elsewhere. For example, **vasopressin** and **oxytocin**, two peptide hormones that are released into the circulation from the posterior pituitary, also function as neurotransmitters at a number of central synapses. A number of other peptides also serve as both hormones and neurotransmitters.

■ TWO MAJOR CATEGORIES OF NEUROTRANSMITTERS

By the 1950s, the list of neurotransmitters had expanded (by the criteria described in Box A) to include three amines—epinephrine, dopamine, and serotonin—in addition to ACh. Over the following decade, four amino acids—glutamate, aspartate, γ-aminobutyric acid (GABA), and glycine—were also shown to be neurotransmitters. Subsequently, other small molecules, including norepinephrine and histamine, were identified as transmitters (Figure

SMALL-MOLECULE NEUROTRANSMITTERS

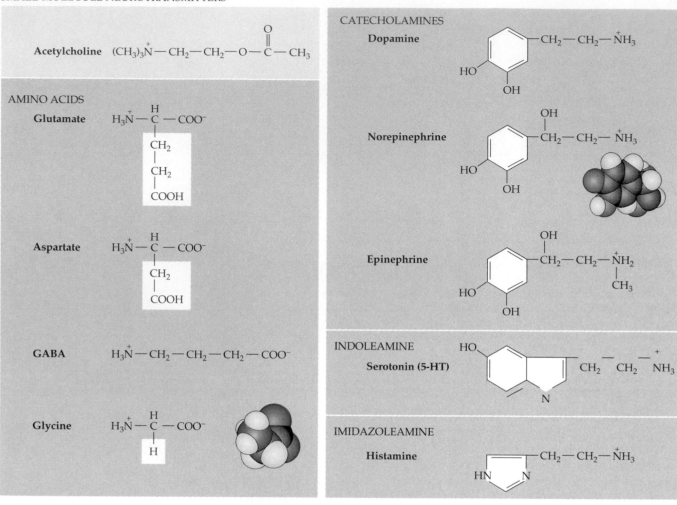

PEPTIDE NEUROTRANSMITTERS (more than 100 peptides, usually 3–30 amino acids long)

Example: **Methionine enkephalin** (Tyr–Gly–Gly–Phe–Met)

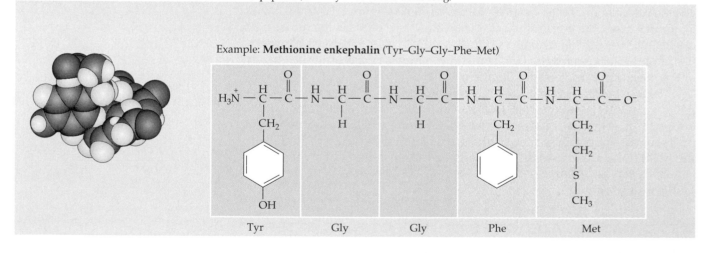

6.2), and considerable evidence now suggests that several purines (such as ATP, adenosine, and AMP) should be added to the list. The most recent class of molecules now known to be transmitters are a large number of polypeptides; since the 1970s, more than 100 such molecules have been shown to meet at least some of the criteria for a neurotransmitter.

◀ **Figure 6.2** There are 10 well-established small-molecule neurotransmitters, and more than 100 peptide neurotransmitters. Small-molecule transmitters can be subdivided into acetylcholine, the amino acids, and the biogenic amines. The catecholamines, so named because they all share the catechol moiety (i.e., the hydroxylated benzene ring), make up a distinctive subgroup within the biogenic amines. Serotonin and histamine contain an indole ring and an imidazole ring, respectively. Size differences between the small-molecule neurotransmitters and the peptide neurotransmitters are indicated by the space-filling models for glycine, norepinephrine, and methionine enkephalin. (Carbon atoms are black, nitrogen atoms blue, and oxygen atoms red.)

For purposes of discussion, it is useful to separate this variety of agents into two broad categories based on size (Figures 6.2 and 6.3). **Neuropeptides** are relatively large transmitter molecules composed of 3 to 36 amino acids. Individual amino acids, such as glutamate and GABA, as well as acetylcholine, serotonin, and histamine, are much smaller than neuropeptides and are therefore called **small-molecule neurotransmitters**. Within the category of small-molecule neurotransmitters, the **biogenic amines** (dopamine, norepinephrine, epinephrine, serotonin, and histamine) are often discussed

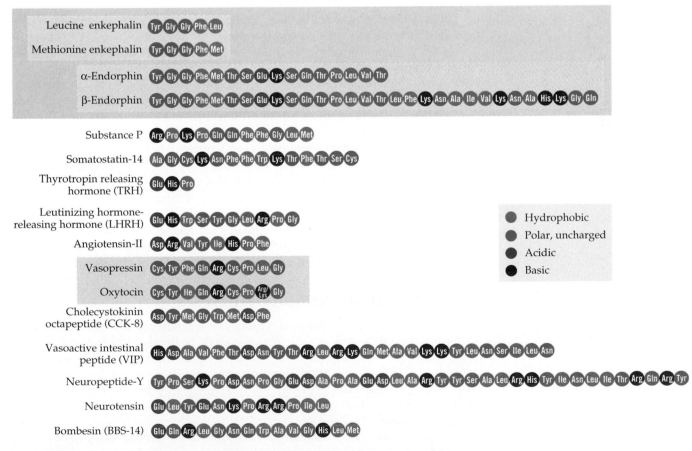

Figure 6.3 Neuropeptides vary in length, but usually contain between 3 and 36 amino acids. Note that one peptide can include the sequence of other neuroactive peptides within it. For example, β-endorphin also contains α-endorphin and methionine enkephalin (met-enkephalin).

separately because their chemical properties and postsynaptic actions are distinct from the other neurotransmitters in this group.

■ NEURONS MAY RELEASE MORE THAN ONE TRANSMITTER

Until recently, it was believed that a given neuron produced only a single type of neurotransmitter. There is now convincing evidence, however, that many neurons contain and release two or more different neurotransmitters. There are numerous examples of different peptides in the same terminal, as well as cases in which two small-molecule neurotransmitters are found within the same neuron, or in which a peptide neurotransmitter is found along with a small-molecule neurotransmitter. When more than one transmitter is present within a nerve terminal, the molecules are called **co-transmitters**. Because each class of transmitter is usually packaged in a separate population of synaptic vesicles, co-transmitters often are segregated within a presynaptic terminal (although there are also instances in which two or more co-transmitters are present in the same synaptic vesicle).

The presence of co-transmitters lends considerable versatility to synaptic transmission. In particular, if a presynaptic terminal packages co-transmitters in different types of vesicles, then these transmitters need not be released simultaneously. In fact, co-transmitter release varies with the frequency of presynaptic stimulation: empirically, low-frequency stimulation often releases only small neurotransmitters, whereas high-frequency stimulation is required to release neuropeptides from the same presynaptic terminals (Figure 6.4). In this way, the presence of co-transmitters allows the chemical signaling properties of a synapse to change according to the level of presynaptic activity.

The differential release of co-transmitters is probably based on the distribution of Ca^{2+} and vesicles in presynaptic terminals. Typically, a presynaptic terminal packages small-molecule co-transmitters into relatively small synaptic vesicles (often with a clear core), some of which are docked at the plasma membrane; in contrast, peptide co-transmitters are contained within large dense-core synaptic vesicles that are farther away from the plasma membrane. At low firing frequencies, the concentration of Ca^{2+} may increase only in the vicinity of presynaptic Ca^{2+} channels, limiting release to small-molecule transmitters because of the selective fusion of small vesicles located immediately adjacent to the channels. High-frequency stimulation increases the Ca^{2+} concentration more evenly throughout the presynaptic terminal, thereby inducing the release of neuropeptides from the larger, more distant vesicles.

■ NEUROTRANSMITTER SYNTHESIS

Effective synaptic transmission requires close control of the concentration of neurotransmitters within the synaptic cleft. Neurons have therefore developed a sophisticated ability to regulate the synthesis, packaging, release, and degradation (or removal) of neurotransmitters (Figure 6.5A). In general, each of these processes is specific for a particular transmitter and requires a number of enzymes that are found only in neurons that use the transmitter at their synapses.

As a rule, the synthesis of small-molecule neurotransmitters occurs locally within presynaptic terminals (Figure 6.5B). The enzymes needed for transmitter synthesis are transported to the nerve terminal cytoplasm at a rate of 0.5 to 5 millimeters a day by a mechanism known as **slow axonal**

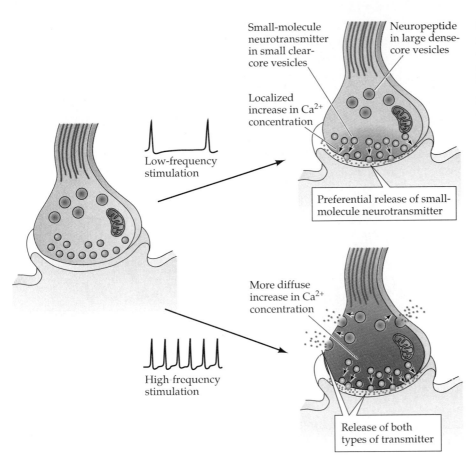

Small-molecule neurotransmitter in small clear-core vesicles

Neuropeptide in large dense-core vesicles

Localized increase in Ca²⁺ concentration

Low-frequency stimulation

Preferential release of small-molecule neurotransmitter

More diffuse increase in Ca²⁺ concentration

High-frequency stimulation

Release of both types of transmitter

Figure 6.4 Differential release of neuropeptide and small-molecule co-transmitters. Low-frequency stimulation preferentially raises the Ca²⁺ concentration close to the membrane, causing the release of transmitter from small clear-core vesicles docked at presynaptic specializations. High-frequency stimulation leads to a more general increase in Ca²⁺, causing the release of peptide neurotransmitters from large dense-core vesicles as well as small-molecule neurotransmitters from small clear-core vesicles.

transport. The precursor molecules used by these enzymes are usually taken into the nerve terminal by transport proteins found in the plasma membrane of the terminal. The synthetic enzymes generate a cytoplasmic pool of free neurotransmitter that must then be loaded into synaptic vesicles by vesicular membrane transport proteins (see Chapter 5).

The mechanisms responsible for the synthesis and packaging of peptide transmitters are fundamentally different from those of the small-molecule neurotransmitters (Figure 6.5C). Peptide-secreting neurons, like other cells, carry out gene transcription in their cell bodies. Transcription often results in the synthesis of polypeptides that are much larger than the final, "mature" peptide. Processing these polypeptides, called **pre-propeptides** (or pre-pro-proteins), takes place by a sequence of reactions in a number of intracellular organelles. Pre-propeptides are synthesized in the rough endoplasmic reticulum, where the signal sequence of amino acids—that is, the sequence indicating that the peptide is to be secreted—is removed. The remaining polypeptide, called a **propeptide** (or proprotein), then traverses the Golgi apparatus and is packaged into vesicles in the *trans*-Golgi network. The final stages of peptide neurotransmitter processing occur after packaging into vesicles, and involve proteolytic cleavage, modification of the ends of the peptide, glycosylation, phosphorylation, and disulfide bond formation.

In general, neuropeptide synthesis is much like the synthesis of proteins secreted from non-neuronal cells (pancreatic enzymes, for instance). A major difference, however, is that the neuronal axon often presents a very long distance between the site of a peptide's synthesis and its secretion. The

(A) LIFE CYCLE OF NEUROTRANSMITTER

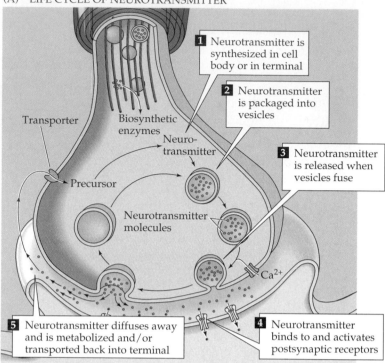

1 Neurotransmitter is synthesized in cell body or in terminal

2 Neurotransmitter is packaged into vesicles

3 Neurotransmitter is released when vesicles fuse

4 Neurotransmitter binds to and activates postsynaptic receptors

5 Neurotransmitter diffuses away and is metabolized and/or transported back into terminal

Transporter
Biosynthetic enzymes
Neurotransmitter
Precursor
Neurotransmitter molecules
Ca²⁺

(B) SMALL-MOLECULE TRANSMITTERS

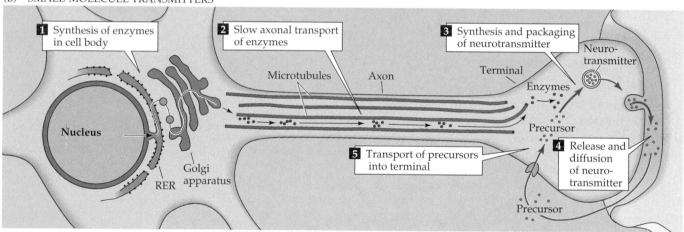

1 Synthesis of enzymes in cell body

2 Slow axonal transport of enzymes

3 Synthesis and packaging of neurotransmitter

4 Release and diffusion of neurotransmitter

5 Transport of precursors into terminal

Nucleus
Golgi apparatus
RER
Microtubules
Axon
Terminal
Enzymes
Neurotransmitter
Precursor
Precursor

(C) PEPTIDE TRANSMITTERS

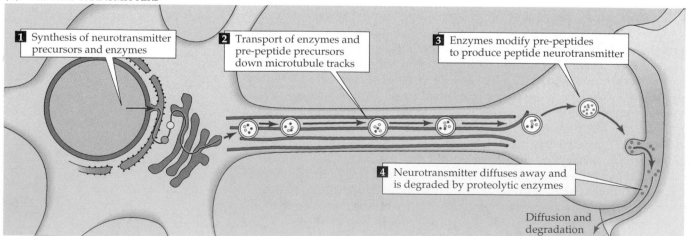

1 Synthesis of neurotransmitter precursors and enzymes

2 Transport of enzymes and pre-peptide precursors down microtubule tracks

3 Enzymes modify pre-peptides to produce peptide neurotransmitter

4 Neurotransmitter diffuses away and is degraded by proteolytic enzymes

Diffusion and degradation

◀ **Figure 6.5** The synthesis, packaging, secretion, and removal of neurotransmitters. (A) The life history of transmitter agents entails (1) neurotransmitter synthesis, (2) packaging into vesicles, (3) fusion of vesicles resulting in neurotransmitter release, and (4) neurotransmitter removal. In some cases, the neurotransmitter and/or a breakdown product is reused for neurotransmitter synthesis (5). (B) The life cycle of small-molecule neurotransmitters unfolds entirely at nerve terminals (red arrows). Precursors are taken up into the terminals by specific transporters, and neurotransmitter synthesis and packaging take place within the nerve endings. After vesicle fusion and release the neurotransmitter diffuses away and, depending on the transmitter, may also be enzymatically degraded. The reuptake of the neurotransmitter (or its metabolites) starts another cycle of synthesis, packaging, release, and removal. The enzymes necessary for neurotransmitter synthesis are made in the cell body of the presynaptic cell and are transported down the axon by slow axonal transport. (C) Peptide neurotransmitters, as well as the enzymes that modify their precursors, are synthesized in the cell body. Enzymes and propeptides are packaged into vesicles in the Golgi apparatus. During fast axonal transport of these vesicles to the nerve terminals, the enzymes modify the propeptides to produce one or more neurotransmitter peptides. After vesicle fusion and exocytosis, the peptides diffuse away and are degraded by proteolytic enzymes.

peptide-filled vesicles must therefore be transported along the axon to the synaptic terminal. The mechanism responsible for such movement, known as **fast axonal transport**, carries vesicles at rates up to 400 mm/day along cytoskeletal elements called microtubules. Microtubules are long, cylindrical filaments, 25 nm in diameter, that are present throughout neurons and other cells. Peptide-containing vesicles are moved along these microtubule "tracks" by ATP-requiring "motor" proteins such as kinesin.

■ THE PACKAGING OF NEUROTRANSMITTERS

Following their synthesis, neurotransmitters are stored within synaptic vesicles. The nature of these vesicles varies for different transmitters. Some of the small-molecule neurotransmitters—acetylcholine and the amino acid transmitters—are packaged in small vesicles 40–60 nm in diameter, the centers of which appear clear in electron micrographs; accordingly, these vesicles are referred to as small clear-core vesicles (Figure 6.6A). Neurotransmitters are

(A)

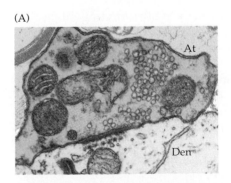

(B)

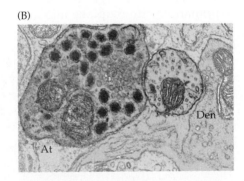

Figure 6.6 Different types of synaptic vesicles. (A) Small clear-core vesicles at a synapse(s) between an axon terminal (At) and a dendritic spine (Den) in the central nervous system. Such vesicles typically contain small-molecule neurotransmitters. (B) Large dense-core vesicles in another type of central axon terminal (At) synapsing onto a dendrite (Den). Such vesicles typically contain neuropeptides (or in some cases biogenic amines). (From Peters, Palay, and Webster, 1991.)

concentrated in synaptic vesicles by transporter proteins in an energy-requiring mechanism. Neuropeptides, in contrast, are packaged into larger synaptic vesicles that range from 90 to 250 nm in diameter and, with appropriate fixation, appear electron-dense in electron micrographs—hence, these are referred to as large dense-core vesicles (Figure 6.6B). The biogenic amine neurotransmitters are packaged into at least two types of vesicles that are different from either the small clear-core vesicles or the large dense-core vesicles. Vesicles containing biogenic amines can be either small (40–60 nm diameter) dense-core vesicles, or larger (60–120 nm diameter), irregularly shaped, dense-core vesicles, depending on the particular class of neuron.

■ NEUROTRANSMITTER RELEASE AND REMOVAL

Once filled with transmitter molecules, vesicles associate with the presynaptic membrane and fuse with it in response to Ca^{2+} influx (see Chapter 5). The mechanisms of exocytotic release are similar for all transmitters, although there are differences in the speed of this process. In general, small-molecule transmitters are secreted more rapidly than peptides. For example, while secretion of ACh from motor neurons requires only a fraction of a millisecond, many neuroendocrine cells, such as those in the hypothalamus, require high-frequency bursts of action potentials for many seconds to release peptide hormones from their nerve terminals. These differences in the rate of transmitter release make neurotransmission relatively rapid at synapses employing small-molecule transmitters and slower at synapses that use peptides. Such differences in the rate of release probably arise from spatial differences in vesicle localization and presynaptic Ca^{2+} signaling, as illustrated in Figure 6.4. Thus, the small clear-core vesicles used to store small-molecule transmitters are often docked at active zones (specialized regions of the presynaptic membrane; see Chapter 5), whereas the large dense-core vesicles used to store peptides are not (compare Figures 6.6A and B). Biogenic amines are packaged into small vesicles that dock at active zones in some neurons, while in others they are packaged and released much like peptides.

When the transmitter has been secreted into the synaptic cleft, it binds to specific receptors on the postsynaptic cell, thereby generating a postsynaptic electrical signal (see Chapter 7). The transmitter must then be removed rapidly to enable the postsynaptic cell to engage in another cycle of neurotransmitter release, binding, and signal generation. The mechanisms by which neurotransmitters are removed vary, but always involve diffusion in combination with reuptake into nerve terminals or surrounding glial cells, degradation by specific enzymes, or in some cases a combination of these. For most of the small-molecule neurotransmitters there are transporters that remove the transmitters (or their metabolites) from the synaptic cleft, ultimately delivering these molecules back to the presynaptic terminal (see Figure 6.5A).

The major distinctions among the transmitters presented in the following sections are summarized in Table 6.1. Not surprisingly, the particulars of the processes of synthesis, packaging, release and removal differ for each neurotransmitter.

■ ACETYLCHOLINE

Acetylcholine, Loewi's "vagus substance," is the neurotransmitter at neuromuscular junctions, at synapses in sympathetic and parasympathetic ganglia of the peripheral autonomic nervous system, and at many sites within the central nervous system. Two major cholinergic neuronal groups are the basal

TABLE 6.1
Properties of Some of the Major Neurotransmitters

Neurotransmitter	Postsynaptic effect[a]	Precursor(s)	Rate-limiting step in synthesis	Removal mechanism	Type of vesicle
ACh	Excitatory	Choline + acetyl CoA	CAT	AChEase	Small, clear
Glutamate	Excitatory	Glutamine	Glutaminase	Transporters	Small, clear
GABA	Inhibitory	Glutamate	GAD	Transporters	Small, clear
Glycine	Inhibitory	Serine	Phosphoserine	Transporters	Small, clear
Catecholamines (epinephrine, norepinephrine, dopamine)	Excitatory	Tyrosine	Tyrosine hydroxylase	Transporters, MAO, COMT	Small dense-core, or large irregular dense-core
Serotonin (5-HT)	Excitatory	Tryptophan	Tryptophan hydroxylase	Transporters, MAO	Large, dense-core
Histamine	Excitatory	Histidine	Histidine decarboxylase	Transporters	Large, dense-core
ATP	Excitatory	ADP	Mitochondrial oxidative phosphorylation; glycolysis	Hydrolysis to AMP and adenosine	Small, clear
Neuropeptides	Excitatory and inhibitory	Amino acids (protein synthesis)	Synthesis and transport	Proteases	Large, dense-core

[a]The most common postsynaptic effect is indicated; recall that the same transmitter can elicit postsynaptic excitation *or* inhibition depending on the nature of the ion channels affected by transmitter binding (see Chapters 5 and 7).

forebrain nuclear complex and the cholinergic nuclei of the brainstem tegmentum. ACh may also act in some pain and chemosensory pathways. Whereas a great deal is known about the function of cholinergic transmission at the neuromuscular junction and at ganglionic synapses, the role of ACh in the central nervous system is not as well understood.

Acetylcholine is synthesized in nerve terminals from acetyl coenzyme A (acetyl CoA) and choline, in a reaction catalyzed by choline acetyltransferase (CAT; Figure 6.7). The presence of CAT in a neuron is thus a strong indication that ACh is used as a transmitter. In contrast to most other small-molecule neurotransmitters, the postsynaptic actions of ACh are not terminated by reuptake, but by a powerful hydrolytic enzyme, acetylcholinesterase (AChE). This enzyme is concentrated in the synaptic cleft, ensuring a rapid decrease in ACh concentration after its release from the presynaptic terminal. AChE has a very high catalytic activity (5000 molecules of ACh per AChE molecule per second) and hydrolyzes ACh into acetate and choline. Cholinergic nerve terminals contain a high-affinity, Na^+-dependent transporter that takes up the choline produced by ACh hydrolysis.

Among the many interesting drugs that interact with cholinergic enzymes are the organophosphates. These compounds include mustard gas (a chemical widely used in World War I), numerous insecticides, and Sarin, the agent recently made notorious by a group of Japanese terrorists. Organophosphates can be lethal to humans (and insects) because they inhibit AChE, causing ACh to accumulate at cholinergic synapses. This build-up of ACh depolarizes the postsynaptic cell and renders it refractory to subsequent ACh release, causing neuromuscular paralysis.

■ GLUTAMATE

Glutamate is generally conceded to be the most important transmitter for normal brain function. Nearly all excitatory neurons in the central nervous

Figure 6.7 Acetylcholine metabolism in cholinergic nerve terminals. The synthesis of acetylcholine from choline and acetyl CoA requires choline acetyltransferase. Acetyl CoA is derived from pyruvate generated by glycolysis, while choline is transported into the terminals via a Na^+-dependent transporter. After release, acetylcholine is rapidly metabolized by acetylcholinesterase and choline is transported back into the terminal.

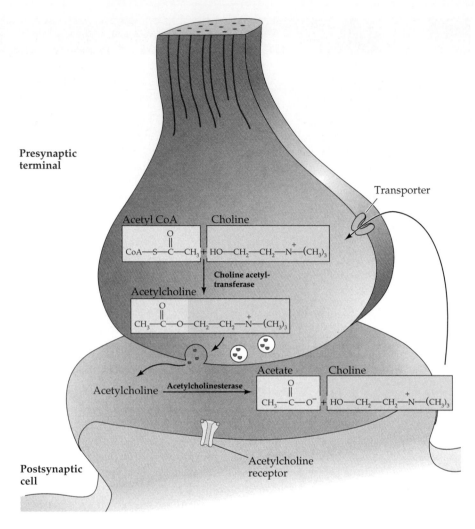

system are glutamatergic, and it is estimated that over half of all brain synapses release this agent. Glutamate plays an especially important role in clinical neurology because elevated concentrations of extracellular glutamate, released as a result of neural injury, are highly toxic to neurons (Box B).

The most prevalent glutamate precursor in synaptic terminals is glutamine. Glutamine is released by glial cells and, within presynaptic terminals, is metabolized to glutamate by the mitochondrial enzyme glutaminase (Figure 6.8). Following its packaging and release, glutamate is removed from the synaptic cleft by high-affinity glutamate transporters present in both glial cells and presynaptic terminals. Glial cells contain the enzyme glutamine synthetase, which converts glutamate into glutamine; glutamine is then transported out of the glial cells and into terminals. In this way, synaptic terminals work together with glial cells to maintain an adequate supply of the neurotransmitter. This synthetic pathway is referred to as the **glutamate-glutamine cycle**.

■ GABA AND GLYCINE

Most inhibitory neurons in the brain and spinal cord use either γ-aminobutyric acid (GABA) or glycine as a neurotransmitter. Like glutamate, GABA

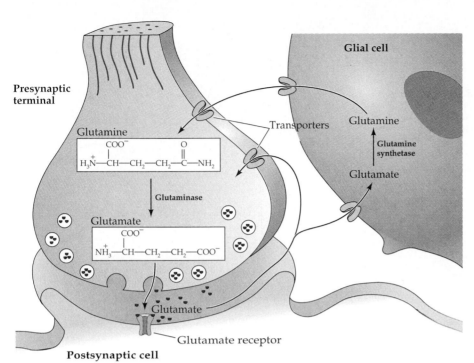

Figure 6.8 Glutamate synthesis and cycling between neurons and glia. The action of glutamate released into the synaptic cleft is terminated by uptake into neurons and surrounding glial cells via specific transporters. Within the nerve terminal, the glutamine released by glial cells is converted back to glutamate.

was identified in brain tissue during the 1950s and the details of its synthesis and degradation were worked out shortly thereafter. David Curtis and John Watkins were the first to show that GABA inhibits the ability of mammalian neurons to fire action potentials. Remarkably, as many as one-third of the synapses in the brain appear to use GABA as their neurotransmitter. Unlike glutamate, GABA is not an essential metabolite, nor is it incorporated into protein. Thus, the presence of GABA in neurons and terminals is a good initial indication that the cells use GABA as a neurotransmitter. GABA is most commonly found in local-circuit interneurons, although the Purkinje cells of the cerebellum provide an example of a GABAergic projection neuron (see Chapter 18).

GABA is synthesized from glutamate by the enzyme glutamic acid decarboxylase (GAD), which is found almost exclusively in GABAergic neurons (Figure 6.9A). GAD requires a cofactor, pyridoxal phosphate, for activity. Because pyridoxal phosphate is derived from vitamin B_6, a dietary deficiency of B_6 can lead to diminished GABA synthesis. The significance of this fact became clear after a disastrous series of infant deaths was linked to the omission of vitamin B_6 from infant formula. The lack of B_6 resulted in a large reduction in the GABA content of the brain; the subsequent loss of synaptic inhibition caused seizures that in some cases were fatal.

The mechanism of GABA removal is similar to that for glutamate; both neurons and glia contain high-affinity transporters for GABA (see Figure 6.9A). Most GABA is eventually converted to succinate, which is metabolized further in the tricarboxylic acid cycle that mediates cellular ATP synthesis. The enzymes required for this degradation, GABA aminotransferase and succinic semialdehyde dehydrogenase, are both mitochondrial enzymes. Inhibition of GABA breakdown causes a rise in tissue GABA content and an increase in the activity of inhibitory neurons. Because epileptic seizures can arise from a decrease in neuronal inhibition, a GABA aminotransferase inhibitor, sodium dipropylacetate, is widely used as an anticonvulsant. Drugs

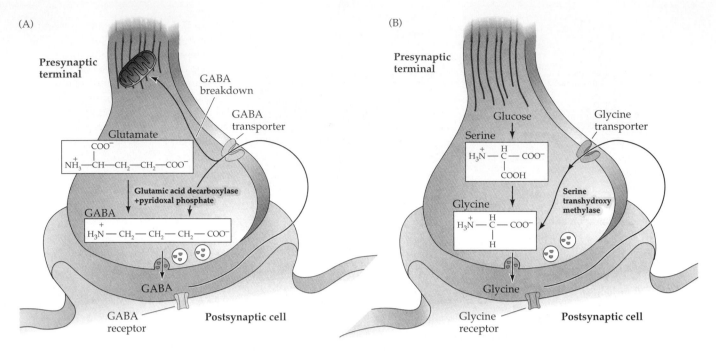

Figure 6.9 Synthesis, release, and reuptake of the inhibitory neurotransmitters GABA and glycine. (A) GABA is synthesized from glutamate by the enzyme glutamic acid decarboxylase, which requires pyridoxal phosphate. (B) Glycine can be synthesized by a number of metabolic pathways; in the brain, the major precursor is serine. High-affinity transporters return both GABA and glycine to the synaptic terminals for reuse.

that act as agonists or as modulators on postsynaptic GABA receptors, such as barbiturates, are also used clinically to treat epilepsy, and are effective sedatives and anesthetics.

The distribution of the neutral amino acid glycine in the central nervous system is more localized than that of GABA. Glycine inhibits the firing of spinal cord and brainstem motor neurons but has only a weak effect on neurons of the cerebral cortex. About half of the inhibitory synapses in the spinal cord use glycine; most of the others use GABA. Glycine is synthesized from serine by the mitochondrial isoform of serine hydroxymethyltransferase (Figure 6.9B). Once released from the presynaptic cell, glycine is rapidly removed from the synaptic cleft by specific membrane transporters. Mutations in the genes coding for some of these enzymes result in hyperglycinemia, a devastating neonatal disease characterized by lethargy, seizures, and mental retardation.

■ THE BIOGENIC AMINES

There are five established biogenic amine neurotransmitters: the three **catecholamines**—norepinephrine (**noradrenaline**), **epinephrine** (**adrenaline**), and **dopamine**—and **histamine** and **serotonin** (see Figure 6.2). Some aspects of the synthesis and degradation of the amine neurotransmitters still are not well defined, but many of the properties of these processes fall somewhere between those of the other small-molecule neurotransmitters and those of the neuropeptides. Drugs that interfere with biogenic amine metabolism are

Box B
EXCITOTOXICITY IN ACUTE NEURONAL INJURY

Excitotoxicity refers to the ability of glutamate and related compounds to destroy neurons through prolonged activation of the processes that underlie excitatory synaptic transmission. Normally, the concentration of glutamate released into the synaptic cleft rises to high levels (approximately 1 mM), but remains at this concentration for some milliseconds only. If abnormally high concentrations of glutamate accumulate in the extracellular space, the excessive activation of neuronal glutamate receptors can literally excite neurons to death.

The phenomenon of excitotoxicity was discovered in 1957, when D. R. Lucas and J. P. Newhouse serendipitously found that feeding sodium glutamate to infant mice destroys neurons in the retina. Roughly a decade later, J. W. Olney extended this discovery by showing that regions of glutamate-induced neuronal loss occurred throughout the brain. The damage inflicted by glutamate was evidently restricted to the postsynaptic cells: the dendrites of the target neurons were grossly swollen, while the postsynaptic terminals were spared. Olney also examined the relative potency of glutamate analogues and found that their neurotoxic actions paralleled their ability to activate postsynaptic glutamate receptors. Furthermore, glutamate receptor antagonists were

effective in blocking the neurotoxic effects of glutamate. He therefore postulated that glutamate destroyed neurons by a mechanism similar to that employed at excitatory glutamatergic synapses, and coined the term *excitotoxin* to reflect this effect.

Evidence that excitotoxicity is an important cause of neuronal damage after injury has come from studying the consequences of oxygen deprivation. The most common cause of reduced blood flow to the brain (ischemia) is occlusion of a cerebral blood vessel (a stroke). The idea that synaptic activity might contribute to ischemic injury emerged from the observation that concentrations of glutamate and aspartate in the extracellular space increase during ischemia. Further, microinjection of glutamate receptor antagonists in experimental animals protects neurons from ischemia-induced damage. Together, these findings suggest that extracellular accumulation of glutamate during ischemia activates glutamate receptors and triggers a chain of events culminating in neuronal death. Presumably, the reduced supply of oxygen elevates extracellular glutamate levels by slowing the energy-dependent uptake of glutamate at synapses. Excitotoxic mechanisms have now been implicated in other acute forms of neuronal insult, including hypo-

glycemia, traumatic injury, and status epilepticus (repeated intense seizures).

The excitotoxicity hypothesis, if correct, has profound implications for treating many neurological disorders. For example, blockade of glutamate receptors after onset of the insult could rescue neurons otherwise destined to die. Clinical trials are now evaluating the protective effects of such antagonists in the immediate aftermath of a stroke. This approach could ameliorate some of the devastating consequences of strokes, trauma, and repeated seizures.

References

CHOI, D. W. (1988) Glutamate neurotoxicity and diseases of the nervous system. Neuron 1: 623–634.

LUCAS, D. R. AND J. P. NEWHOUSE (1957) The toxic effect of sodium l-glutamate on the inner layers of the retina. Arch. Ophthalmol. 58: 193–201.

OLNEY, J. W. (1969). Brain lesions, obesity and other disturbances in mice treated with monosodium glutamate. Science 164: 719–721.

OLNEY, J. W. (1971) Glutamate-induced neuronal necrosis in the infant mouse hypothalamus: An electron microscopic study. J. Neuropathol. Exp. Neurol. 30: 75–90.

ROTHMAN, S. M. (1983) Synaptic activity mediates death of hypoxic neurons. Science 220: 536–537.

especially important as treatments for a variety of clinical disorders (Box C).

All the catecholamines are derived from a common precursor, the amino acid tyrosine (Figure 6.10). The first step in catecholamine synthesis is catalyzed by tyrosine hydroxylase and results in the synthesis of dihydroxyphenylalanine (DOPA). Because tyrosine hydroxylase is rate-limiting for the synthesis of all three transmitters, its presence is a valuable criterion for identifying catecholaminergic neurons.

• *Dopamine* is produced by the action of DOPA decarboxylase on DOPA (see Figure 6.10). Although present in several brain regions, the major

Box C
BIOGENIC AMINE NEUROTRANSMITTERS AND PSYCHIATRIC DISORDERS

The regulation of the biogenic amine neurotransmitters is altered in a variety of psychiatric disorders. Indeed, most psychotropic drugs (defined as drugs that alter behavior, mood, or perception) selectively affect one or more steps in the synthesis, packaging, or degradation of biogenic amines. Sorting out how these drugs work has been extremely useful in beginning to understand the molecular mechanisms underlying some of these diseases.

Based on their effects on humans, psychotherapeutic drugs can be divided into several broad categories: antipsychotics, antianxiety drugs, antidepressants, and stimulants. Antipsychotic drugs are used to ameliorate psychoses such as schizophrenia. Reserpine, the first of these drugs, was developed in the 1950s and initially was used as an anti-hypertensive agent. Reserpine blocks the uptake of norepinephrine into synaptic vesicles and therefore depletes the transmitter at aminergic terminals, diminishing the ability of the sympathetic nervous system to cause vasoconstriction. A major side effect in hypertensive patients treated with reserpine—behavioral depression—suggested the possibility of using it as an antipsychotic agent in patients suffering from agitation and pathological anxiety. (Its ability to cause depression in mentally healthy individuals also suggested that aminergic transmitters are involved in mood disorders; see Box D in Chapter 27.)

Although reserpine is no longer used as an antipsychotic agent, its initial success stimulated the development of antipsychotic drugs such as chlorpromazine, haloperidol, and benperidol, which over the last several decades have radically changed the approach to treating psychotic disorders. Prior to the discovery of these drugs, psychotic patients were typically hospitalized for long periods, sometimes indefinitely. Modern antipsychotic drugs now allow most patients to be treated on an outpatient basis after a brief hospital stay. The clinical effectiveness of these drugs is correlated with their ability to block brain dopamine receptors, implying that excessive dopamine release is responsible for some types of psychotic illness. A great deal of effort continues to be expended on developing more effective antipsychotic drugs with fewer side effects, and on discovering the site of action of these medications.

The second category of psychotherapeutic drugs is the antianxiety agents. Anxiety disorders are estimated to afflict between 10 and 35% of the population, making them the most common psychiatric disorders. The two major forms of pathological anxiety—panic attacks and generalized anxiety disorder—both respond to drugs that affect aminergic transmission. The agents used to treat panic disorders included inhibitors of the enzyme monoamine oxidase (MAO inhibitors, or MAOIs) required for the catabolism of the amine neurotransmitters, and blockers of serotonin receptors. The most effective drugs in treating generalized anxiety disorder have been benzodiazepines, such as chlordiazepoxide (Librium®), and diazepam (Valium®). In contrast to most other psychotherapeutic drugs, these agents increase the efficacy of transmission at $GABA_A$ synapses rather than acting at aminergic synapses.

Antidepressants and stimulants also affect aminergic transmission. A large number of drugs are used clinically to treat depressive disorders, and in the laboratory to explore the neurotransmitter abnormalities underlying these common conditions. The three major classes of antidepressants—MAOI, tricyclic antidepressants, and serotonin uptake blockers such as fluoxetine (Prozac®) and trazodone—all influence various aspects of aminergic transmission. MAOIs such as phenelzine block the breakdown of amines, whereas the tricyclic antidepressants such as desipramine block the reuptake of norepinephrine and other amines. The popular antidepressant fluoxetine (Prozac®) selectively blocks the reuptake of serotonin without affecting the reuptake of catecholamines. Stimulants such as amphetamine are also used to treat some depressive disorders. Amphetamine stimulates the release of norepinephrine from nerve terminals; the transient "high" resulting from taking amphetamine is presumably the emotional opposite of the depression that sometimes follows reserpine-induced norepinephrine depletion.

Despite the relatively small number of aminergic neurons in the brain, this litany of pharmacological actions emphasizes that these neurons are critically important in the maintenance of mental health.

References

Brunello, N. C., C. Masotto, L., Steardo, R. Markstein and G. Racagni. (1995) New insights into the biology of schizophrenia through the mechanism of action of clozapine. Neurpsychopharmacol. 13: 177–213.

Carlsson, A. (1993) Thirty years of dopamine research. Adv. Neurol. 60: 1–10.

Fibiger, H. C. (1995) Neurobiology of depression: Focus on dopamine. Adv. Biochem. Psychopharmacol. 49: 1–17.

Perry, P. J. (1995) Clinical use of the newer antipsychotic drugs. Am. J. Health Syst. Pharm. 52: S9–S14.

Seiden, L. S., K. E. Sabol and G. A. Ricaurte. (1993) Amphetamine: Effects on catecholamine systems and behavior. Annu. Rev. Pharmacol. Toxicol. 33: 639–677.

dopamine-containing area of the brain is the substantia nigra, which plays an essential role in the control of body movements (see Chapters 17 and 18). In Parkinson's disease, the dopaminergic neurons of the substantia nigra degenerate, leading to a characteristic motor dysfunction (see Box B in Chapter 18). Because dopamine does not readily cross the blood-brain barrier, the disease can be treated by administering DOPA together with drugs that prevent catecholamine breakdown.

  • *Norepinephrine* synthesis requires dopamine β-hydroxylase, which catalyzes the production of norepinephrine from dopamine (see Figure 6.10). Neurons that synthesize norepinephrine are largely restricted to the locus coeruleus, a brainstem nucleus that projects diffusely to the midbrain and telencephalon. These neurons are especially important in modulating sleep and wakefulness (Chapter 26).

  • *Epinephrine* is present at much lower levels in the brain than any of the other catecholamines. The enzyme that synthesizes epinephrine, phenylethanolamine-*N*-methyltransferase (see Figure 6.10), is present only in adrenaline-secreting neurons. Sensitive methods that identify epinephrine have confirmed the existence of epinephrine-containing neurons in the central nervous system and shown them to be located in two groups in the rostral medulla. The function of these epinephrine-containing neurons in the brain is not known.

  All three catecholamines are removed by reuptake into terminals, or into surrounding glial cells, by an Na⁺-dependent transporter. The two major enzymes involved in the catabolism of catecholamines are monoamine oxidase (MAO) and catechol *O*-methyltransferase (COMT), both of which are present within catecholaminergic nerve terminals and are the targets of numerous psychotropic drugs.

  • *Histamine* has long been known to be released from mast cells and platelets in response to allergic reactions or tissue damage. Only recently, however, has this amine been implicated as a neurotransmitter. Histamine is produced from the amino acid histidine by a histidine decarboxylase (Figure 6.11A). High concentrations of histamine and histamine decarboxylase are found in the hypothalamus, from whence histaminergic neurons send sparse but widespread projections to almost all regions of the brain and spinal cord. Their function remains uncertain.

  • *Serotonin*, or *5-hydroxytryptamine* (5-HT), is also synthesized from one of the common amino acids—in this case, tryptophan. An essential dietary requirement, tryptophan is taken up into neurons by a plasma membrane transporter and hydroxylated in a reaction catalyzed by the enzyme tryptophan-5-hydroxylase (Figure 6.11B). As in the case of the catecholamines, this reaction is the rate-limiting step for 5-HT synthesis. Serotonin is located in discrete groups of neurons in the raphe regions of the pons and upper brainstem; these cells send widespread projections to the telencephalon and diencephalon and have also been implicated in the regulation of sleep and wakefulness (see Chapter 26).

■ ATP AND OTHER PURINES

All synaptic vesicles contain ATP, which is co-released with one or more "classical" neurotransmitters. This observation raises the possibility that ATP could also act as a co-transmitter. It has been known since the 1920s that the extracellular application of ATP (or its breakdown products AMP and adenosine) to neurons can elicit electrical responses. Since then, the idea that

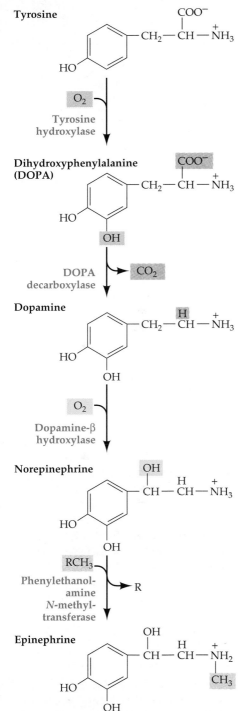

Figure 6.10 The biosynthetic pathway for the catecholamine neurotransmitters. The amino acid tyrosine is the precursor for all three catecholamines. The first step in this reaction pathway, catalyzed by tyrosine hydroxylase, is rate-limiting.

(A)

Histidine

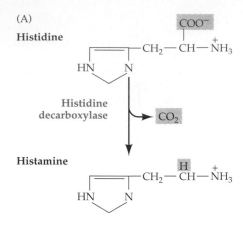

Histidine
decarboxylase → CO_2

Histamine

(B)

Tryptophan

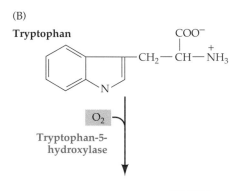

O_2

Tryptophan-5-
hydroxylase

5-Hydroxytryptophan

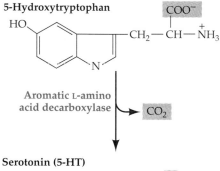

Aromatic L-amino
acid decarboxylase → CO_2

Serotonin (5-HT)

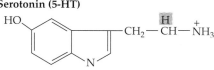

Figure 6.11 Synthesis of histamine and serotonin. (A) Histamine is synthesized from the amino acid histidine. (B) Serotonin is derived from the amino acid tryptophan by a two-step process that requires the enzymes tryptophan-5-hydroxylase and a decarboxylase.

these purines (so named because of the characteristic purine ring) are neurotransmitters at some synapses has received considerable experimental support. There is now strong evidence that ATP acts as an excitatory neurotransmitter in the periphery. Postsynaptic actions of ATP have also been demonstrated in the central nervous system, specifically at dorsal horn neurons and in a subset of hippocampal neurons.

Purines act on a large and diverse family of receptors, many of which have recently been cloned. Whether or not purines play a role in synaptic transmission depends on the presence and/or distribution of purinergic receptors near the sites of release. These receptors have been separated into two major families: the P1 receptors, activated predominantly by ATP and ADP; and the P2 receptors, activated predominantly by AMP and adenosine. These receptors can be either ion channels or G-protein-coupled receptors (see Chapter 7), and their activation may subtly shape postsynaptic responses to classical neurotransmitters. Alternatively, if purinergic receptors are located presynaptically, their activation could modulate neurotransmitter release. In any event, it seems likely that excitatory synaptic transmission mediated by purinergic receptors is widespread in the mammalian brain.

■ PEPTIDE NEUROTRANSMITTERS

Many peptides are well known as hormones in endocrine cells, including neurons in the neuroendocrine regions of the brain such as the hypothalamus and pituitary. Advances in the ability to detect and isolate these molecules have now shown that peptides may also act as neurotransmitters, often being co-released with small-molecule neurotransmitters.

The biological activity of the peptide neurotransmitters depends on the sequence of their amino acids (see, for example, Figure 6.3). Propeptide precursors are often many times larger than their active peptide products and can give rise to more than one species of neuropeptide (Figure 6.12). Since each of these peptide products can be separately contained in synaptic vesicles, transmission based on peptides often elicits complex postsynaptic responses. Peptide transmitters have been implicated in modulating emotions (Chapter 27), and some, such as substance P and the opioid peptides, are involved in the perception of pain (Chapter 9). Still other peptides, such as melanocyte-stimulating hormone, adrenocorticotropin, and β-endorphin, regulate complex responses to stress.

• *Substance P.* The study of neuropeptides began more than 60 years ago with the accidental discovery of substance P, a powerful hypotensive agent. (The name derives from the fact that this molecule was an unidentified component of *p*owder extracts from brain and intestine.) Substance P is an 11-amino acid peptide (see Figure 6.3) present in high concentrations in the human hippocampus and neocortex. It is also released from C fibers, the small-diameter afferents in peripheral nerves that convey information about pain and temperature (as well as postganglionic autonomic signals). Substance P is a sensory neurotransmitter in the spinal cord, where its release can be inhibited by opioid peptides released from spinal cord interneurons, resulting in the suppression of pain. The protease responsible for the inactivation of substance P is associated with synaptic membranes.

• *Opioid peptides.* Morphine has long been known to be an especially effective analgesic. This and other opiate drugs affect the perception of pain by interacting with specific receptors expressed at a number of sites in the central and peripheral nervous systems. The endogenous opioid peptides were discovered during a search for endogenous compounds that mimicked the actions of morphine. It was hoped that such compounds would be anal-

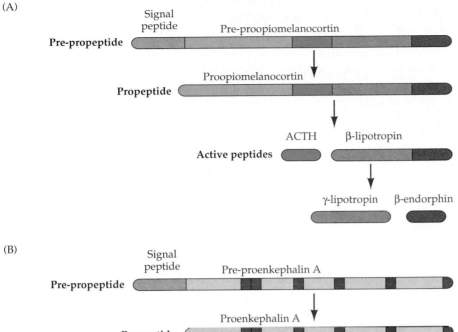

(A)

Pre-propeptide — Signal peptide — Pre-proopiomelanocortin

Propeptide — Proopiomelanocortin

Active peptides — ACTH — β-lipotropin

γ-lipotropin — β-endorphin

(B)

Pre-propeptide — Signal peptide — Pre-proenkephalin A

Propeptide — Proenkephalin A

Active peptides

Met-enkephalin   Leu-enkephalin   Met-enkephalin

Figure 6.12 Proteolytic processing of the pre-propeptides, pre-proopiomelanocortin (A) and pre-proenkephalin A (B). For each pre-propeptide, the signal sequence is indicated in orange at the left; the locations of active peptide products are darkly shaded. The maturation of the pre-propeptides involves cleaving the signal sequence and other proteolytic processing. Such processing can result in a number of different neuroactive peptides such as ACTH, γ-lipotropin, and β-endorphin (A), or multiple copies of the same peptide, such as met-enkephalin (B).

gesics, and that their understanding would shed light on addiction to morphine and other narcotics. The endogenous ligands of the opioid receptors have now been identified as a family of more than 20 opioid peptides grouped into three classes: the endorphins, the enkephalins, and the dynorphins, each class being liberated from an inactive pre-propeptide. These precursors are the product of three distinct genes: pre-proopiomelanocortin, pre-proenkephalin A, and pre-prodynorphin. Opioid precursor processing is tissue-specific due to the differential expression of the processing enzymes. The pro-opiomelanocortin precursor also contains the sequences for several nonopioid neuropeptides, such as the stress hormone adrenocorticotropic hormone (ACTH) and α-, β-, and γ-melanocyte stimulating hormone (MSH) (Figure 6.12).

Opioid peptides are widely distributed throughout the brain. In general, these peptides tend to be depressants. When injected intracerebrally, they act as analgesics, and have been implicated in the mechanisms underlying acupuncture-induced analgesia (see Chapter 9). Unfortunately, the repeated administration of endorphins leads to tolerance and addiction. Although the results of opioid research have not yet provided a complete understanding of narcotic addiction, a solid basis of knowledge has been established that promises ultimately to solve this extraordinary social and medical problem.

■ NITRIC OXIDE: NEW MECHANISMS OF NEUROTRANSMITTER ACTION

Ten years ago nitric oxide (NO) was thought about primarily as a toxic gas relevant to air pollution. The discovery, in 1987, that NO is a signaling molecule that modulates vascular tone has sparked tremendous interest in the biological effects of this molecule, which is now recognized as a novel chemical messenger for a number of cell types, including neurons. Nitric oxide is certainly not a classical neurotransmitter in the central nervous system; it is a short-lived radical that interacts with surrounding neurons by diffusing across membranes, rather than being released by exocytosis and interacting with membrane-bound receptors. In this sense, NO represents a significant departure from all neurotransmitter mechanisms characterized to date.

Nitric oxide is synthesized from L-arginine following stimulation of the enzyme nitric oxide synthase (NOS). Neuronal-type nitric oxide synthase (nNOS) is a widely distributed calmodulin-regulated enzyme and is coupled to a variety of neurotransmitter systems in the brain and in peripheral tissues. Once generated, NO can diffuse locally and interact with target molecules such as guanylyl cyclase, the enzyme catalyzing cGMP synthesis. NO and cGMP together comprise an especially wide-ranging signal transduction system that may also play an important role in neurological diseases. An emerging hypothesis is that the balance between nitric oxide and superoxide generation is a critical factor in the etiology of some neurodegenerative diseases. Nitric oxide defies current classification schemes in that it can function both as a neurotransmitter and a second messenger. In any event, it demonstrates that, despite a century-long analysis of neurotransmitter mechanisms, this field still has some surprises in store.

■ SUMMARY

The large number of neurotransmitters in the nervous system can be divided into two broad classes: small-molecule transmitters and neuropeptides. Neurotransmitters are synthesized from defined precursors by regulated enzymatic pathways, packaged into one of a variety of vesicle types, and released into the synaptic cleft in a Ca^{2+}-dependent manner. Many synapses release more than one type of neurotransmitter, and multiple transmitters are sometimes packaged in the same synaptic vesicle. The postsynaptic effects of neurotransmitters are terminated by the degradation of the transmitter in the synaptic cleft, by transport of the transmitter back into cells, or by diffusion out of the synaptic cleft. Glutamate is the major excitatory neurotransmitter in the brain, whereas GABA and glycine are the major inhibitory neurotransmitters. The actions of these small-molecule neurotransmitters are typically faster than those of the neuropeptides. Thus, the small-molecule transmitters usually mediate synaptic transmission when a speedy response is essential, whereas the neuropeptide transmitters, as well as the biogenic amines and some other small-molecule neurotransmitters, can regulate or modulate ongoing activity in the brain or in peripheral target tissues. The enormous importance of drugs that influence transmitter actions in the treatment of neurological and psychiatric disorders guarantees that a steady stream of new information in this field will be forthcoming.

Additional Reading

Reviews

BECKER, C.-D. (1995) Glycine receptors: Molecular heterogeneity and implications for disease. Neuroscientist 1: 130–141.

CARLSSON, A. (1987) Perspectives on the discovery of central monoaminergic neurotransmission. Annu. Rev. Neurosci. 10: 19–40.

DAWSON, T. M. AND S. H. SNYDER (1994) Gases as biological messengers: Nitric oxide and carbon monoxide in the brain. J. Neurosci. 14: 5147–5159.

EMSON, P. C. (1979) Peptides as neurotransmitter candidates in the CNS. Prog. Neurobiol. 13: 61–116

HÖKFELT, T. D., K. MILLHORN, SEROOGY, Y. TSURUO, S. CECCATELLI, B. LINDH, B. MEISTER, T. MELANDER, M. SCHALLING, T. BARTFAI AND L. TERENIUS (1987) Coexistence of peptides with classical neurotransmitters. Experientia Suppl. 56:154-179.

INESTROSSA, N. C. AND A. PERELMAN (1989) Distribution and anchoring of molecular forms of acetylcholinesterase. Trends Pharmacol. Sci. 10: 325–329.

JUNG, L. H. AND R. H. SCHELLER (1991) Peptide processing and targeting in the neuronal secretory pathway. Science 251: 1330–1335.

MELDRUM, B. AND J. GARTHWAITE (1990) Glutamate neurotoxicity may underlie slowly progressive degenerative diseases such as Huntington's Disease and Alzheimer's Disease. Trends Pharmacol. Sci. 11: 379.

SCHWARTZ, J. C., J. M. ARRANG, M. GARBARG, H. POLLARD AND M. RUAT (1991) Histaminergic transmission in the mammalian brain. Physiol. Rev. 71(1): 1–51.

SHAFQAT, S., M. VELAZ-FAIRCLOTH, A. GUANANO-FERRAZ AND R. T. FREMEAU (1993) Molecular characterization of neurotransmitter transporters. Molec. Endocrinol. 7: 1517–1529.

TUCEK, S., J. RICNY AND V. DOLEZAL (1990) Advances in the biology of cholinergic neurons. Adv. Neurol. 51: 109–115.

Important Original Papers

CHEN, Z. P., A. LEVY, ET AL. (1995) Nucleotides as extracellular signalling molecules. J. Neuroendicrinol. 7(2): 83–96.

CURTIS, D. R., J. W. PHILLIS AND J. C. WATKINS (1959) Chemical excitation of spinal neurons. Nature 183: 611–612.

DALE, H. H., W. FELDBERG AND M. VOGT (1936) Release of acetylcholine at voluntary motor nerve endings. J. Physiol. 86: 353–380.

HÖKFELT, T., O. JOHANSSON, A. LJUNGDAHL, J. M. LUNDBERG AND M. SCHULTZBERG (1980) Peptidergic neurons. Nature 284: 515–521.

HUGHES, J., T. W. SMITH, H. W. KOSTERLITZ, L. A. FOTHERGILL, B. A. MORGAN AND H. R. MORRIS (1975) Identification of two related pentapeptides from the brain with potent opiate agonist activity. Nature 258: 577–580.

KUPFERMANN, I. (1991) Functional studies of cotransmission. Physiol. Rev. 71: 683–732.

LOEWI, O. (1921) Über humorale übertragbarheit der herznervenwirkung. Pflügers Arch. 189: 239–242.

SOSSIN, W. S., A. SWEET-CORDERO AND R. H. SCHELLER (1990) Dale's hypothesis revisited: Different neuropeptides derived from a common prohormone are targeted to different processes. Proc. Natl. Acad. Sci. U.S.A. 87: 4845–4548.

Books

BRADFORD, H. F. (1986) Chemical Neurobiology. New York: W. H. Freeman.

COOPER, J. R., F. E. BLOOM AND R. H. ROTH (1991) The Biochemical Basis of Neuropharmacology. New York: Oxford University Press.

HALL, Z. (1992) An Introduction to Molecular Neurobiology. Sunderland, MA: Sinauer Associates, Chapters 3–7.

NICHOLLS, D. G. (1994) Proteins, Transmitters, and Synapses. Boston: Blackwell Scientific.

CHAPTER 7

NEURO-TRANSMITTER RECEPTORS AND THEIR EFFECTS

■ OVERVIEW

Neurotransmitters evoke postsynaptic electrical responses by binding to members of a diverse group of proteins called neurotransmitter receptors, which in turn give rise to electrical signals by opening or closing ion channels in the postsynaptic membrane. Whether the postsynaptic actions of a particular neurotransmitter are excitatory or inhibitory is determined by the class of ion channel affected by the transmitter. There are two major classes of receptors: those in which the receptor molecule is also an ion channel, and those in which the receptor and ion channel are separate molecules. The former are called ligand-gated ion channels, and they give rise to fast postsynaptic responses that typically last only a few milliseconds. The latter are called metabotropic receptors, and they produce slower postsynaptic effects that may endure much longer. Metabotropic receptors affect postsynaptic ion channels indirectly by activating transducer molecules called G-proteins. G-proteins may alter the properties of ion channels directly, or may activate intracellular second-messenger pathways that then modulate the channels.

■ NEUROTRANSMITTER RECEPTORS ALTER POSTSYNAPTIC MEMBRANE PERMEABILITY

In 1907, the British physiologist John N. Langley introduced the concept of **receptor molecules** to explain the specific and potent actions of certain chemicals on muscle and nerve cells. This concept has now been successfully invoked to account for the ability of neurotransmitters, hormones, and drugs to alter the functional properties of neurons. While it has been clear since Langley's day that receptors are important for synaptic transmission, their molecular identity and mechanism of action remained a mystery until quite recently. It is now known that neurotransmitter receptors are proteins embedded in the postsynaptic plasma membrane where they play a dual role in synaptic transmission. First, receptors have domains that extend into the synaptic cleft and bind neurotransmitters that are released into this space by the presynaptic neuron. Second, the binding of neurotransmitters to these domains opens or closes ion channels in the postsynaptic membrane (Figure 7.1). Typically, the resulting ion fluxes change the membrane potential of the postsynaptic cell, thus mediating the transfer of electrical information across the synapse.

■ PRINCIPLES DERIVED FROM STUDIES OF THE NEUROMUSCULAR JUNCTION

A particularly useful system for understanding neurotransmitter receptors has been the experimentally accessible neuromuscular junction. The binding of the neurotransmitter acetylcholine (ACh) to postsynaptic receptors opens ion channels in the muscle fiber membrane. This effect can be demonstrated directly by using the patch clamp method to measure the minute postsynaptic currents that flow when ACh binds to its receptors (see Box A in Chapter 4). Exposure of the extracellular surface of a patch of postsynaptic membrane to ACh causes single-channel currents to flow for a few milliseconds (Figure

Figure 7.1 Receptors that mediate the postsynaptic actions of neurotransmitters have two functions. First, receptors detect the presence of neurotransmitters in the synaptic cleft by specific binding. Second, receptors alter the ionic permeability of the postsynaptic membrane by virtue of being coupled, directly or indirectly, to ion channels in the postsynaptic membrane. Opening or closing these channels as a result of transmitter binding causes ionic currents to flow, thus changing the postsynaptic membrane potential.

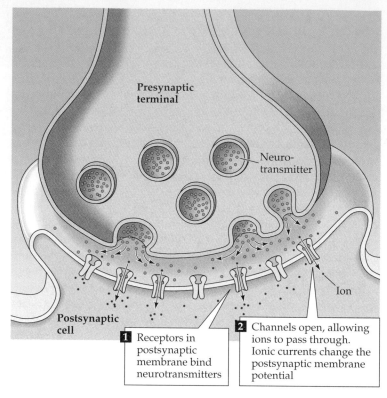

Presynaptic terminal

Neurotransmitter

Ion

Postsynaptic cell

1 Receptors in postsynaptic membrane bind neurotransmitters

2 Channels open, allowing ions to pass through. Ionic currents change the postsynaptic membrane potential

Figure 7.2 Activation of ACh receptors at neuromuscular synapses. (A) Patch clamp measurement of single ACh receptor currents from a patch of membrane removed from the postsynaptic muscle cell. When ACh is applied to the extracellular surface of the membrane, the repeated opening of a single channel can be seen as inward currents (downward deflections). (B) Synchronized opening of many ACh-activated channels at a synapse. (1) If a single channel is examined during the release of ACh from the presynaptic terminal, the channel opens transiently. (2) If several channels are examined together, ACh release opens the channels almost synchronously. (3) The opening of many postsynaptic channels sum to produce a macroscopic EPC. (C) The inward EPC depolarizes the postsynaptic muscle cell, giving rise to an EPP.

(A) Patch clamp measurement of single ACh receptor current

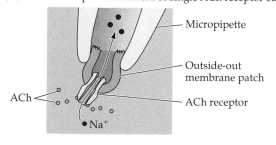

Micropipette

Outside-out membrane patch

ACh receptor

ACh

Na$^+$

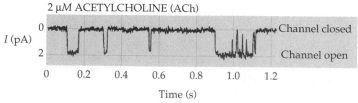

2 μM ACETYLCHOLINE (ACh)

I (pA)

Channel closed

Channel open

Time (s)

(B) Currents produced by:

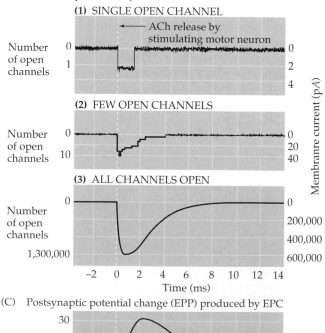

(1) SINGLE OPEN CHANNEL

ACh release by stimulating motor neuron

Number of open channels

(2) FEW OPEN CHANNELS

Number of open channels

(3) ALL CHANNELS OPEN

Number of open channels

Membrane current (pA)

Time (ms)

(C) Postsynaptic potential change (EPP) produced by EPC

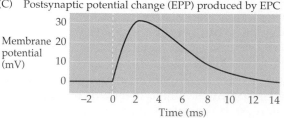

Membrane potential (mV)

Time (ms)

7.2A). Even if ACh is continually applied such that the transmitter is nearly always bound to receptors, the channels open and close repeatedly. Thus, ACh opens ion channels stochastically when it binds to receptors in the postsynaptic membrane.

The electrical consequences of exogenous ACh action on a membrane patch are greatly multiplied when an action potential in a presynaptic motor neuron causes the transient release of several million molecules of ACh into the synaptic cleft. In this case, the transmitter molecules bind to many thousands of ACh receptors packed in a dense array on the postsynaptic surface, transiently opening a very large number of postsynaptic ion channels. Each time two ACh molecules bind to a receptor, a single channel opens briefly (Figure 7.2B, 1). The opening of a large number of channels is synchronized by the secreted ACh (Figure 7.2B, 2). As discussed in Chapter 3, the resulting macroscopic current is called the **end plate current**, or **EPC**. Because the current flowing during the EPC is normally inward, it causes the postsynaptic membrane potential to depolarize. This depolarizing change in potential—the **end plate potential** (**EPP**; Figure 7.2C)—typically triggers a postsynaptic action potential due to the opening of voltage-activated ion channels (see Chapter 5). In the experiment shown in Figure 7.2C, no action potential is generated due to presence of TTX, a Na^+ channel blocker.

The ions that flow when ACh is bound to its postsynaptic receptors can be identified using the principles of ion permeation that were introduced in Chapters 2–4. In particular, the identity of the ions that are carrying the current can be determined by knowing the membrane potential at which no current flows in response to transmitter binding. When the potential of the postsynaptic muscle cell is controlled by the voltage clamp method (Figure 7.3A; see Box A in Chapter 3), the magnitude of the membrane potential clearly affects the amplitude and polarity of EPCs (Figure 7.3B). Thus, when the postsynaptic membrane potential is made more negative than the resting potential, the amplitude of the EPC becomes larger, whereas this current is reduced when the membrane potential is made more positive. At approximately 0 mV, no EPC is detected, and at even more positive potentials, the current reverses its polarity, becoming outward rather than inward (Figure 7.3C). The potential where the EPC reverses is called the **reversal potential**.

The magnitude of the EPC at any membrane potential is given by the product of the ionic conductance activated by ACh (g_{ACh}) and the electrochemical driving force on the ions flowing through ligand-gated channels. The driving force is simply the difference between the postsynaptic membrane potential, V_m, and the reversal potential for the EPC, E_{rev}. Thus, the value of the EPC is given by the relationship

$$I = gV$$

where the units of I are amperes, the units of g are Siemens, and the units of V are volts. More specifically,

$$EPC = g_{ACh}(V_m - E_{rev})$$

This relationship predicts that the EPC will be a large inward current at potentials more negative than E_{rev} because the electrochemical driving force, $V_m - E_{rev}$, is a large negative number; conversely, the EPC becomes smaller at potentials approaching E_{rev} because the driving force is reduced. At potentials more positive than E_{rev}, the EPC is outward because the dri-

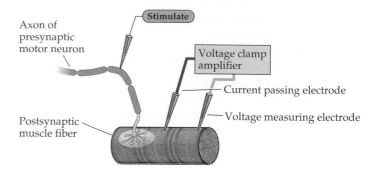

(A)   Scheme for voltage clamping postsynaptic muscle fiber

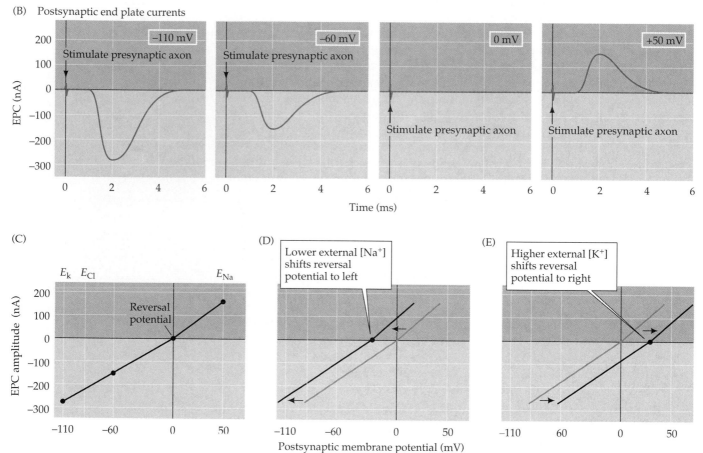

Figure 7.3   The influence of the postsynaptic membrane potential on end plate currents. (A) A postsynaptic muscle fiber is voltage clamped using two electrodes, while the presynaptic neuron is electrically stimulated to cause the release of ACh from presynaptic terminals. This experimental arrangement allows the recording of macroscopic EPCs produced by ACh. (B) Amplitude and time course of EPCs when the postsynaptic cell is clamped at four different membrane potentials. (C) The relationship between the peak amplitude of EPCs and postsynaptic membrane potential is nearly linear, with a reversal potential (the voltage at which the direction of the current changes from inward to outward) close to 0 mV. Also indicated on this graph are the equilibrium potentials of Na^+, K^+, and Cl^- ions. Lowering the external Na^+ concentration causes EPCs to reverse at more negative potentials (D), while raising the external K^+ concentration makes the reversal potential more positive (E). (Modified from Takeuchi and Takeuchi, 1960.)

ving force is reversed in direction (that is, positive). Since the channels opened by ACh are not sensitive to membrane voltage, g_{ACh} should depend only on the number of channels opened by ACh (which depends in turn on the concentration of ACh in the synaptic cleft). Thus, the magnitude and polarity of the postsynaptic membrane potential determines the direction and amplitude of the EPC solely by altering the driving force on ions flowing through the receptor channels opened by ACh.

At the reversal potential, there is no net driving force on the ions that permeate the receptor-activated channel. Thus, the identity of the ions that flow during the EPC can be deduced by observing how the reversal potential of the EPC compares to the equilibrium potential for various ion species. For example, if ACh were to open an ion channel permeable only to K^+, then the reversal potential of the EPC would be at the equilibrium potential for K^+, which for a muscle cell is close to -100 mV. If the ACh-activated channels were permeable only to Na^+, then the reversal potential of the current would be approximately $+60$ mV, the Na^+ equilibrium potential; if these channels were permeable only to Cl^-, then the reversal potential would be somewhere between -40 and -70 mV. By this reasoning, ACh-activated channels cannot be permeable to only one of these ions, because the reversal potential of the EPC is not near the equilibrium potential for any of them (Figure 7.3C). However, if these channels were about equally permeable to Na^+ and K^+, then the reversal potential of the EPC would be about halfway between $+60$ mV and -100 mV.

The fact that EPCs reverse at approximately 0 mV (the actual value is -5 to -15mV) is therefore consistent with the idea that ACh-activated ion channels are permeable to both Na^+ and K^+. This implication has been tested by experiments in which the extracellular concentration of these two ions is altered. As expected, the magnitude and reversal potential of the EPC are changed by altering the concentration gradient of each ion. Lowering the external Na^+ concentration, which makes E_{Na} more negative, produces a negative shift in E_{rev} (Figure 7.3D), whereas elevating external K^+ concentration, which makes E_K more positive, causes E_{rev} to shift to a more positive potential (Figure 7.3E). Such experiments confirm that the ACh-activated ion channels are in fact permeable to both Na^+ and K^+.

Even though the channels opened by the binding of ACh to its receptors are permeable to both Na^+ and K^+, the EPC is generated primarily by Na^+ influx (Figure 7.4). If the membrane potential is kept at E_K, the EPC arises entirely from an influx of Na^+ because at this potential there is no driving force on K^+ (Figure 7.4A). At the usual mucle fiber resting membrane potential of -90 mV, there is a small driving force on K^+, but a much greater one on Na^+. Thus, during the EPC, much more Na^+ flows into the muscle cell than K^+ flows out; it is the net influx of positively charged Na^+ that constitutes the inward current measured as the EPC (Figure 7.4B). At the reversal potential of 0 mV, Na^+ influx and K^+ efflux are exactly balanced, so no current flows during the opening of channels by ACh binding (Figure 7.4C). At potentials more positive than E_{rev} the balance reverses; for example, at E_{Na} there is no influx of Na^+ and a large efflux of K^+ because of the large driving force on this ion (Figure 7.4D). Even more positive potentials cause efflux of both Na^+ and K^+ and produce an even larger outward EPC.

Were it possible to measure the end plate potential at the same time as the end plate current (the voltage clamp technique prevents this by keeping membrane potential constant), the EPP would be seen to vary in parallel with the amplitude and polarity of the EPC (Figures 7.4E, F). At the usual

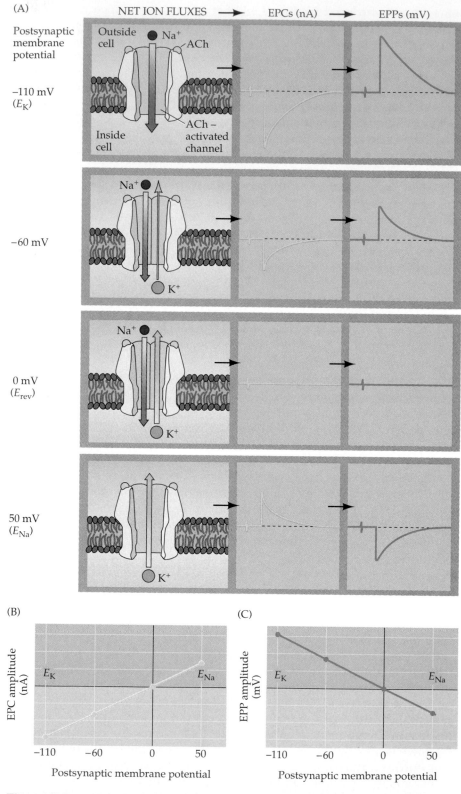

Figure 7.4 Na$^+$ and K$^+$ movements during EPCs and EPPs. (A) Each of the post-synaptic potentials (V_{post}) indicated at the left results in different relative fluxes of Na$^+$ and K$^+$ (ion fluxes). These ion fluxes determine the amplitude and polarity of the EPCs, which in turn determine the EPPs. (B) Voltage dependence of the EPCs and EPPs (C) are shown below.

postsynaptic resting membrane potential of −90 mV, the large inward EPC causes the postsynaptic membrane potential to become more depolarized (see Figure 7.4B). However, at 0 mV, the EPP reverses its polarity, and at more positive potentials, the EPP is hyperpolarizing (see Figure 7.4D). Thus, the polarity and magnitude of the EPP depend on the electrochemical driving force that determines the polarity and magnitude of the EPC. EPPs will depolarize when the membrane potential is more negative than E_{rev}, and hyperpolarize when the membrane potential is more positive than E_{rev}. The general rule, then, is that the action of a transmitter drives the postsynaptic membrane potential toward E_{rev}.

Although this discussion has focused on the neuromuscular junction, similar mechanisms generate postsynaptic responses at all synapses. A general principle that emerges is that transmitter binding to postsynaptic receptors produces a postsynaptic conductance change as ion channels are opened—or sometimes closed (see the next section). The postsynaptic conductance is increased if—as at the neuromuscular junction—channels are opened, and decreased if channels are closed. This conductance change generates an electrical current, the **postsynaptic current (PSC)**, which in turn changes the postsynaptic membrane potential to produce a **postsynaptic potential (PSP)**. As in the case of the EPP at the neuromuscular junction, the PSP will be depolarizing if its reversal potential is more positive than the postsynaptic membrane potential and hyperpolarizing if its reversal potential is more negative.

PSPs are the ultimate outcome of excitatory and most inhibitory synaptic transmission, concluding a sequence of electrical and chemical events that begins with the invasion of an action potential into the terminals of a presynaptic neuron. In many ways, the events that produce PSPs at synapses are similar to those that generate action potentials in axons; in both cases conductances produced by ion channels opening or closing lead to ionic current flow that changes the membrane potential.

■ EXCITATORY AND INHIBITORY POSTSYNAPTIC POTENTIALS

Postsynaptic potentials alter the probability that an action potential will be produced in the postsynaptic cell. At the neuromuscular junction, synaptic action only increases the probability that an action potential will occur in the postsynaptic muscle cell. At many other synapses, PSPs can actually decrease the probability that the postsynaptic cell will generate an action potential. PSPs are called **excitatory (EPSPs)** if they increase the likelihood of a postsynaptic action potential occurring, and **inhibitory (IPSPs)** if they decrease this likelihood. Given that most neurons receive inputs from both inhibitory and excitatory synapses, it is important to understand more precisely the mechanisms that determine whether a particular synapse excites or inhibits its postsynaptic partner.

The principles of excitation have just been described for the neuromuscular junction, and by extension for all excitatory synapses. The principles of postsynaptic inhibition are much the same as for excitation. In both cases, neurotransmitters binding to receptors open or close ion channels in the postsynaptic cell. Whether a postsynaptic response is an EPSP or an IPSP depends on the type of channel that is coupled to the receptor. In fact, the only factor that distinguishes postsynaptic excitation from inhibition is the reversal potential of the PSP in relation to the threshold voltage for generating action potentials in the postsynaptic cell.

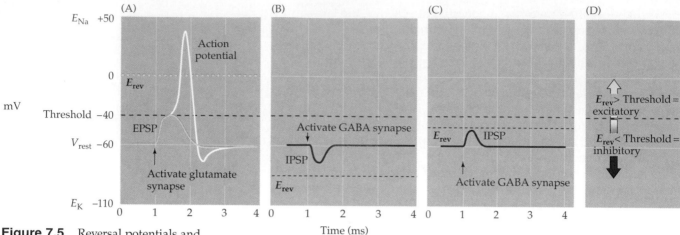

Figure 7.5 Reversal potentials and threshold potentials determine postsynaptic excitation and inhibition. (A) If the reversal potential for a PSP (0 mV) is more positive than the action potential threshold (−40 mV), the effect of a transmitter is excitatory, and it generates EPSPs. (B) If the reversal potential for a PSP is more negative than the action potential threshold, the transmitter is inhibitory and may generate IPSPs. (C) IPSPs can nonetheless depolarize the postsynaptic cell if their reversal potential is between the resting potential and the action potential threshold. (D) The general rule of postsynaptic action is: If the reversal potential is more positive than threshold, excitation results; inhibition occurs if the reversal potential is more negative than threshold.

Consider a neuronal synapse that uses glutamate as the transmitter. Such synapses often have receptors that, like the ACh receptors at neuromuscular synapses, open ion channels that are nonselectively permeable to cations. When such glutamate receptors are activated, both Na^+ and K^+ flow across the postsynaptic membrane. The reversal potential for the resulting EPSP is approximately 0 mV, whereas the resting potential of neurons is approximately −60 mV. The EPSP will depolarize the postsynaptic membrane potential, bringing it toward 0 mV. For the particular neuron shown in Figure 7.5A, the action potential threshold voltage is −40 mV. Thus, the EPSP increases the probability that the postsynaptic neuron will produce an action potential, defining this synapse as excitatory.

As an example of inhibitory postsynaptic action, consider a neuronal synapse that uses GABA as its transmitter. At such synapses, the GABA receptors open channels that are selectively permeable to Cl^-. When these channels open, negatively charged chloride ions can flow across the membrane. Assume that the postsynaptic neuron has a resting potential of −60 mV and an action potential threshold of −40 mV, just as in the previous example. If E_{Cl} is −70 mV, transmitter release at this synapse will inhibit the postsynaptic cell, because E_{Cl} is more negative than the action potential threshold. In this case, the electrochemical driving force ($V_m − E_{rev}$) causes Cl^- to flow into the cell, generating an outward PSC (because Cl^- is negatively charged) and consequently a hyperpolarizing IPSP (Figure 7.5B). Because E_{Cl} is more negative than the action potential threshold, the conductance change arising from the binding of GABA keeps the postsynaptic membrane potential more negative than threshold, thereby reducing the probability that the postsynaptic cell will fire an action potential.

However, not all inhibitory synapses produce hyperpolarizing IPSPs. For example, in the neuron just described, if E_{Cl} were −50 mV instead of −70 mV, then the synapse would still be inhibitory because the reversal potential of the IPSP is still more negative than the action potential threshold (−40 mV). Because the electrochemical driving force now causes Cl^- to flow out of the cell, the IPSP is actually depolarizing (Figure 7.5C). Nonetheless, this depolarizing IPSP still inhibits the postsynaptic cell because the cell's membrane potential is kept more negative than the threshold potential for action potential initiation. Thus, while EPSPs always depolarize the postsynaptic cell, IPSPs can hyperpolarize or depolarize; indeed, an inhibitory conductance charge may produce no potential change at all!

Although the particulars of postsynaptic action can be complex, a simple rule distinguishes postsynaptic excitation from inhibition: An EPSP has a reversal potential more positive than the action potential threshold, whereas an IPSP has a reversal potential more negative than threshold (Figure 7.5D). Intuitively, this rule can be understood by realizing that an EPSP will tend to depolarize the membrane potential so that it exceeds threshold, whereas an IPSP will always act to keep the membrane potential more negative than the threshold potential.

■ SUMMATION OF SYNAPTIC POTENTIALS

The postsynaptic effects of most synapses in the brain are not nearly as large as those at the neuromuscular junction; indeed, PSPs due to the activity of individual synapses are well below the threshold for generating postsynaptic action potentials, and may be only a fraction of a millivolt. How, then, can the brain transmit information from presynaptic to postsynaptic neurons if most central synaptic effects are subthreshold? The answer is that most neurons are innervated by thousands of synapses, and the PSPs produced by each active synapse can *sum together*—in space and in time—to determine the behavior of the postsynaptic neuron.

Consider the simplified case of a neuron that is innervated by two excitatory synapses, each generating a subthreshold EPSP, and an inhibitory synapse that produces an IPSP (Figure 7.6A). While activation of either one of the excitatory synapses alone (E1 or E2 in Figure 7.6B) produces a subthreshold EPSP, activation of both excitatory synapses at about the same time causes the two EPSPs to sum together. If the sum of the two EPSPs (E1 + E2) depolarizes the postsynaptic neuron sufficiently to reach the threshold potential, a postsynaptic action potential results. Such **summation** thus allows subthreshold EPSPs to influence action potential production. Likewise, an IPSP generated by an inhibitory synapse (I) can sum (algebraically speaking) with a subthreshold EPSP to reduce its amplitude (E1 + I) or can sum with suprathreshold EPSPs, such as the one produced by the sum of E1 + E2, to prevent the postsynaptic neuron from reaching threshold (E1 + I + E2).

Figure 7.6 Summation of postsynaptic potentials. (A) A microelectrode records the postsynaptic potentials produced by the activity of two excitatory synapses (E1 and E2) and an inhibitory synapse (I). (B) Electrical responses to synaptic activation. Stimulating either excitatory synapse (E1 or E2) produces a subthreshold EPSP, whereas stimulating both synapses at the same time (E1 + E2) produces a suprathreshold EPSP that evokes a postsynaptic action potential. Activation of the inhibitory synapse alone (I) results in a hyperpolarizing IPSP. Summing this IPSP with the EPSP produced by one excitatory synapse (E1 + I) reduces the amplitude of the EPSP, while summing it with the suprathreshold EPSP produced by activating synapses E1 and E2 keeps the postsynaptic neuron below threshold, so that no action potential is evoked.

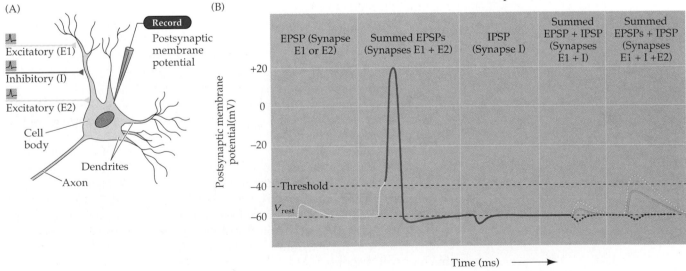

In short, the summation of EPSPs and IPSPs by a postsynaptic neuron permits the cell to integrate the electrical information provided by all the inhibitory and excitatory synapses acting on it at any moment. Whether the sum of active synaptic inputs results in the production of an action potential depends on the balance between excitation and inhibition. If the EPSPs are of sufficient amplitude to raise the membrane potential above threshold, then the postsynaptic cell will produce an action potential. Conversely, if inhibition prevails, then the postsynaptic cell will remain silent. Of course, the balance between EPSPs and IPSPs changes continually over time, depending on the number of excitatory and inhibitory synapses active at a given moment and on the magnitude of the current at each synapse. Summation is therefore an ionic tug-of-war between excitatory and inhibitory synaptic influences; the outcome of the contest determines whether or not a postsynaptic neuron becomes an active element in the neural circuit it belongs to.

■ TWO FAMILIES OF POSTSYNAPTIC RECEPTORS

Neurotransmitter receptors open or close postsynaptic ion channels. This feat is accomplished in two different ways by two broad families of receptor proteins. The receptors in one family—called **ionotropic receptors**—are linked directly to ion channels (the Greek *tropos* means to move in response to a stimulus). Ionotropic receptors contain two functional domains: an extracellular site that binds neurotransmitters, and a membrane-spanning domain that forms an ion channel (Figure 7.7A). These receptors combine both transmitter-binding and channel functions into a single molecular entity and are, therefore, also called **ligand-gated ion channels** to reflect this concatenation. Ionotropic receptors are multimers and are usually made up of five individual protein subunits, each of which spans the plasma membrane and contributes to the pore of the ion channel.

The second family of neurotransmitter receptors are the **metabotropic receptors** (so called because the eventual movement of ions through a channel depends on one or more metabolic steps). These receptors do not have ion channels as part of their structure; instead, they affect channels by the activation of intermediate molecules called **G-proteins** (Figure 7.7B). For this reason, metabotropic receptors are also called **G-protein coupled receptors**. Metabotropic receptors are monomeric proteins with an extracellular domain that contains a neurotransmitter binding site, and an intracellular domain that binds to G-proteins. Neurotransmitter binding to metabotropic receptors activates G-proteins, which then dissociate from the receptor and interact directly with ion channels or bind to other effector proteins, such as enzymes, that then change the conductance of the ion channels (see Figure 7.7B). Thus, G-proteins can be thought of as transducers that couple neurotransmitter binding to the regulation of postsynaptic ion channels.

The two families of postsynaptic receptors give rise to PSPs with very different time courses, producing postsynaptic actions that range from less than a millisecond to minutes, hours, or even days. Ligand-gated ion channel receptors generally mediate rapid postsynaptic effects. Examples are the EPP produced at neuromuscular synapses by ACh (see Figures 7.2 and 7.3), EPSPs produced at certain glutamatergic synapses (see Figure 7.5A), and IPSPs produced at certain GABAergic synapses (see Figure 7.5B). In all three cases, the PSPs arise within a millisecond or two of an action potential invading the presynaptic terminal and last for only up to tens of milliseconds. In contrast, the activation of metabotropic receptors typically produces much slower responses. At synapses where G-proteins interact directly with

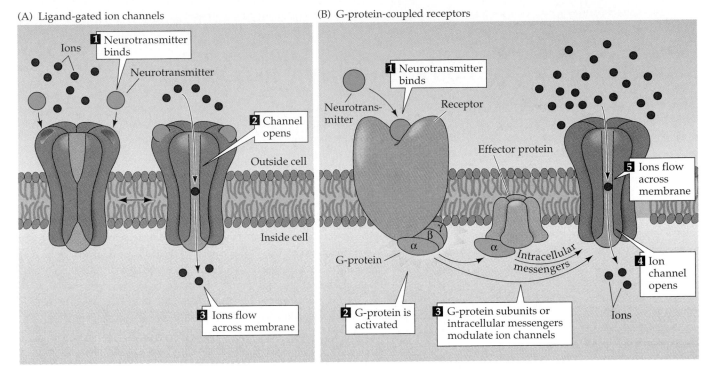

Figure 7.7 A neurotransmitter can affect the activity of a postsynaptic cell via two different types of receptor proteins: ligand-gated ion channels and metabotropic receptors. (A) Ligand-gated ion channels combine receptor and channel functions in a single protein complex. (B) Metabotropic receptors activate the G-proteins, which can modulate ion channels directly or indirectly through intracellular effector enzymes and second messengers.

ion channels, PSPs may last for up to hundreds of milliseconds. When G-proteins modulate ion channels indirectly by activating one or more effector proteins, PSPs may last minutes, hours, or even longer. Importantly, a given transmitter may activate both metabotropic receptors and ligand-gated ion channels to produce both fast and slow PSPs at a single synapse.

A property shared by many metabotropic and ionotropic receptors is that they undergo **desensitization**. Receptor desensitization is defined as a decrease in the amplitude of postsynaptic responses in the continued presence of the transmitter. This phenomenon is usually observed during the repeated application of transmitters, as might occur during a train of action potentials. Individual receptor subtypes desensitize at different rates, and in some subtypes, desensitization can occur during a single PSP. Hence, receptor desensitization may also influence the duration of PSPs.

Perhaps the most important principle to keep in mind is that the response elicited by a given neurotransmitter is determined by the postsynaptic complement of receptors and channels. Exactly how postsynaptic responses are produced by different types of neurotransmitter receptors is considered in the following sections.

■ LIGAND-GATED ION CHANNELS: FAST-ACTING CHOLINERGIC RECEPTORS

Ligand-gated ion channels can produce rapid excitatory or inhibitory PSPs, depending on the types of ions that permeate the ion channel domain of the

Box A
NEUROTOXINS THAT ACT ON POSTSYNAPTIC RECEPTORS

Poisonous plants and animals are widespread in nature. The toxins they produce have been used for a variety of purposes, including hunting, healing, mind-altering, and, more recently, research. Many of the toxins present in poisonous organisms have potent actions on the nervous system, often interfering with synaptic transmission by targeting neurotransmitter receptors. The poisons found in some organisms contain a single type of toxin, whereas others contain a mixture of tens or even hundreds of toxins.

Given the central role of ACh receptors in mediating muscle contraction at neuromuscular junctions, it is not surprising that a large number of natural toxins interfere with transmission at this synapse. In fact, the classification of nicotinic and muscarinic ACh receptors is based on the sensitivity of these receptors to the toxic plant alkaloids nicotine and muscarine, which activate nicotinic and muscarinic ACh receptors, respectively. Nicotine is derived from the dried leaves of the tobacco plant *Nicotinia tabacum*, and muscarine is from the poisonous red mushroom *Amanita muscaria*. Both toxins are stimulants that produce nausea, vomiting, mental confusion, and convulsions. Muscarine poisoning can also lead to circulatory collapse, coma, and death.

The poison α-bungarotoxin, one of many peptides that together make up the venom of the banded krait, *Bungarus multicinctus* (Figure A), blocks transmission at neuromuscular junctions and is used by the snake to paralyze its prey. This 74-amino-acid toxin blocks neuromuscular transmission by irreversibly binding to nicotinic ACh receptors, thus preventing ACh from opening postsynaptic ion channels. Paralysis ensues because skeletal muscles can no longer be activated by motor neurons. As a result of its specificity and its high affinity for nicotinic ACh receptors, α-bungarotoxin has contributed greatly to understanding the ACh receptor molecule. Other snake toxins that block nicotinic ACh receptors are cobra α-neurotoxin and the sea snake peptide erabutoxin. The same strategy these

(A)

(B)

(C)

(A) The banded krait *Bungarus multicinctus*. (B) A marine cone snail (*Conus* sp.) uses venomous darts to kill a small fish. (C) Betel nuts, *Areca catechu*, growing in Malaysia. (A, Robert Zappalorti/Photo Researchers, Inc.; B, Zoya Maslak and Baldomera Olivera, University of Utah; C, Fletcher Baylis/Photo Researchers, Inc.)

receptor. Some insight into the speed with which ligand-gated ion channels respond to neurotransmitters can be gained by examining their molecular structure.

The best-studied ligand-gated ion channel is the nicotinic ACh receptor (nAChR), so named because nicotine also binds to these receptors. The pri-

snakes use to paralyze prey was adopted by South American Indians who used curare, a mixture of plant toxins from *Chondodendron tomentosum*, as an arrowhead poison to immobilize their quarry. Curare also blocks nicotinic ACh receptors; the active agent is the alkaloid δ-tubocurarine.

Another interesting class of animal toxins that selectively block nicotinic ACh and other receptors are peptides produced by fish-hunting marine cone snails (Figure B). These colorful snails kill small fish by "shooting" venomous darts into them. The venom contains hundreds of peptides, known as the conotoxins, many of which target proteins that are important in synaptic transmission. There are conotoxin peptides that block Ca^{2+} channels, Na^+ channels, glutamate receptors, and ACh receptors. The array of physiological responses produced by these peptides all serve to immobilize any prey unfortunate enough to encounter the cone snail. Many other organisms, including other mollusks, corals, worms, and frogs, also utilize toxins containing specific blockers of ACh receptors.

Other natural toxins have mind- or behavior-altering effects and have been used by some cultures for thousands of years. Two examples are plant alkaloid toxins that block muscarinic ACh receptors: atropine from deadly nightshade (belladona), and scopolamine from henbane. Because these plants grow wild in many parts of the world, exposure is not unusual. Poisoning by either toxin can lead to an altered state of consciousness, coma, and death.

Another postsynaptic neurotoxin that, like nicotine, is used as a social drug is found in the seeds from the betel nut, *Areca catechu* (Figure C). Betel nut chewing, although unknown in the United States, is practiced by up to 25% of the population in India, Bangladesh, Ceylon, Malaysia, and the Philippines. Chewing these nuts produces a euphoria caused by arecoline, an alkaloid agonist of nicotinic ACh receptors. Like nicotine, arecoline is an addictive central nervous system stimulant.

Many other neurotoxins alter transmission at noncholinergic synapses. For example, amino acids found in certain mushrooms, algae, and seeds are potent glutamate receptor agonists. The excitotoxic amino acids kainate, from the red alga *Digenea simplex*, and quisqualate, from the seed of *Quisqualis indica*, are used to separate two families of non-NMDA glutamate receptors (see text). Other neurotoxic amino acid activators of glutamate receptors include ibotenic acid and acromelic acid, both found in mushrooms, and domoate, which occurs in algae, seaweed, and mussels. Another large group of peptide neurotoxins block glutamate receptors. These include the α-agatoxins from the funnel web spider, NSTX-3 from the orb weaver spider, Joro toxin from the Joro spider, β-philanthotoxin from wasp venom, as well as the many cone snail toxins.

All of these toxins discussed so far target excitatory synapses. The inhibitory GABA and glycine receptors, however, have not been overlooked by the exigencies of survival. Strychnine, an alkaloid extracted from the seeds of *Strychnos nux-vomica*, is the only drug known to have specific actions on trans-

mission at glycinergic synapses. Because the toxin blocks glycine receptors, strychnine poisoning causes overactivity in the spinal cord and brainstem, leading to seizures. Strychnine is used commercially as a poison for rodents. Neurotoxins that block $GABA_A$ receptors include plant alkaloids such as bicuculline from Dutchman's breeches and picrotoxin from *Anamerta cocculus*. Dieldrin, a very stable commercial insecticide, also blocks these receptors. These agents are, like strychnine, powerful central nervous system stimulants. Muscimol, a mushroom toxin that is a powerful depressant as well as a hallucinogen, activates $GABA_A$ receptors. A synthetic analogue of GABA, baclofen, is a $GABA_B$ agonist that reduces EPSPs in some brainstem neurons, and is used clinically to reduce the frequency and severity of muscle spasms.

Chemical warfare between species has thus given rise to a staggering array of molecules that target synapses throughout the nervous system. Although these toxins are designed to defeat normal synaptic transmission, they have also provided a set of powerful tools to understand postsynaptic mechanisms.

References

ADAMS, M. E. AND B. M. OLIVERA (1994) Neurotoxins: Overview of an emerging research technology. Trends Neurosci. 17: 151–155.

HUCHO, F. AND Y. OVCHINNIKOV (1990) *Toxins as Tools in Neurochemistry.* Berlin: Walter de Gruyer.

MYERS, R. A., L. J. CRUZ, J. E. RIVIER AND B. M. OLIVERA (1993). *Conus* peptides as chemical probes for receptors and ion channels. Chem. Rev. 93: 1923–1926.

mary reason this receptor is so well understood is its accessibility at the vertebrate neuromuscular junction. Another reason is that a number of biological toxins specifically block these receptors (Box A). The availability of these highly specific ligands—α-bungarotoxin in particular—provided a means to isolate and purify the nAChR. This pioneering work paved the way to cloning

(A)

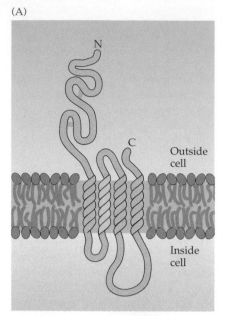

(B)

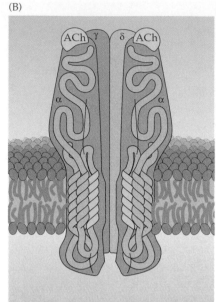

(C)

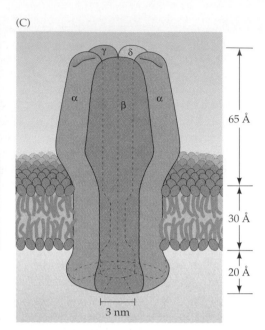

(D)

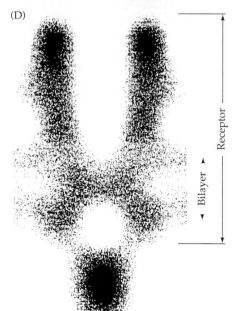

Figure 7.8 The structure of the ACh receptor/channel. (A) Each receptor subunit crosses the membrane four times. (B) Five such subunits come together to form a complex structure containing 20 transmembrane domains that surround a central pore. (C) The openings at either end of the channel are very large—approximately 3 nm in diameter; even the narrowest region of the pore is approximately 0.6 nm in diameter. By comparison, the diameter of Na^+ or K^+ is less than 0.3 nm. (D) An electron micrograph of the ACh receptor, showing the position and size of the protein with respect to the membrane. (From Toyoshima and Unwin, 1990.)

and sequencing the genes encoding the subunits of the nAChR. Based on these molecular studies, the nAChR of muscle cells is now known to be a large protein complex consisting of five subunits arranged around a central pore (Figure 7.8). The pentamer contains two α subunits, which bind ACh and other ligands, such as nicotine and α-bungarotoxin. At the neuromuscular junction, these subunits are combined with three other subunits—β, γ, and δ—in the ratio $2\alpha{:}\beta{:}\gamma{:}\delta$. The intimate association of the binding site and the channel presumably accounts for the rapid response of such receptors.

Neuronal nAChRs differ from those of muscle in that they contain only two receptor subunit types (α and β); these are present in a ratio of 3 α subunits to 2 β subunits in the pentamer. Most neuronal nAChR lack sensitivity to α-bungarotoxin and differ in several other functional properties from their neuromuscular counterparts.

In summary, the nicotinic ACh receptor is a pentameric complex comprising five individual protein subunits. In each of these subunits, there are four transmembrane domains that make up the ion channel portion of the receptor, together with a long extracellular region that makes up the ACh-binding domain. This general structure is retained in all of the ligand-gated ion channels at fast-acting synapses (Figure 7.9). Thus, the nicotinic receptor has served as a paradigm for studies of other ligand-gated ion channels; it has also led to a much deeper appreciation of several neuromuscular diseases (Box B).

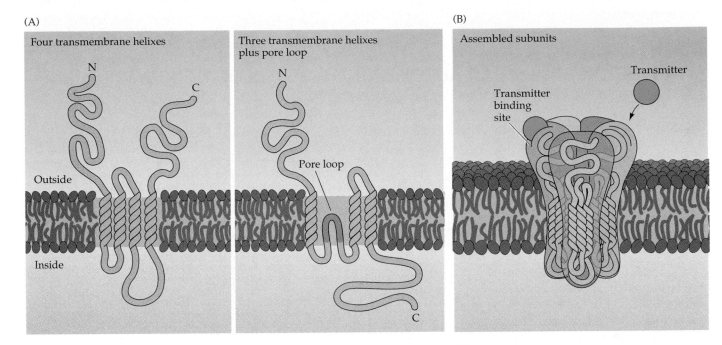

(A)

Four transmembrane helixes

Three transmembrane helixes plus pore loop

(B)

Assembled subunits

Figure 7.9 The general transmembrane architecture of ligand-gated receptors. (A) One of the five subunits of a complete receptor. The long N-terminal region forms the ligand-binding site, while the remainder of the protein spans the membrane either 4 times (left) or 3 times (right). (B) Assembly of five subunits into a complete receptor. (C) Known subunits of ionotropic glutamate (AMPA and NMDA), GABA and glycine receptors.

(C)

Receptor	AMPA	NMDA	GABA	Glycine
Subunits (combination of 5 required for each receptor type)	Glu R1	R1	α	α1
	Glu R2	R2A	β	α2
	Glu R3	R2B	γ	α3
	Glu R4	R2C	δ	α4
	Glu R5	R2D	ρ	β
	Glu R6			

■ LIGAND-GATED ION CHANNELS: FAST-ACTING GLUTAMATE RECEPTORS

Three types of glutamate receptors have been identified. Two of these are ligand-gated ion channels called, respectively, **NMDA receptors** and **AMPA/kainate receptors**. They are named after the agonists that activate them: NMDA (N-methyl-D-aspartate), AMPA (α-amino-3-hydroxyl-5-methyl-4-isoxazole-propionate), and kainic acid. AMPA/kainate receptors are also referred to as **non-NMDA receptors**. The third type of glutamate receptor is the metabotropic glutamate receptor mGluR, which modulates postsynaptic ion channels by activating G-proteins (see page 137). The NMDA and AMPA/kainate receptors are usually more selective for monovalent than dioalvent cations.

AMPA/kainate receptors are responsible for the most rapid EPSPs produced by glutamate. Like other ligand-gated channel receptors, AMPA/kainate receptors are formed from several protein subunits that can co-assemble in different ways to produce a wide range of receptors types (see Figure 7.9). The central pore formed by these subunits allows the passage of Na^+ and K^+, and in some cases small amounts of Ca^{2+}. Thus, the postsynaptic

Box B
MYASTHENIA GRAVIS: AN AUTOIMMUNE DISEASE OF NEUROMUSCULAR SYNAPSES

Myasthenia gravis, a disease that interferes with transmission between motor neurons and skeletal muscle, afflicts approximately 1 of every 200,000 people. Originally described by Thomas Willis in 1685, the hallmark of this disease is muscle weakness, particularly during sustained activity. Although the course of myasthenia gravis is quite variable, the disease commonly affects muscles controlling the eyelids (resulting in drooping of the eyelids, or ptosis) and eye movements (resulting in double vision). Muscles controlling facial expression, chewing, swallowing, and speaking are other common targets of the disease.

An important indication of the cause of myasthenia gravis came from the clinical observation that the muscle weakness improves following treatment with inhibitors of acetylcholinesterase, the enzyme that normally degrades acetylcholine at the neuromuscular junction. Studies of human muscle obtained by biopsy from myasthenic patients showed that both end plate potentials (EPPs) and miniature end plate potentials (MEPPs) are much smaller than normal. Because both the frequency of MEPPs and the quantal content of EPPs are normal, it appeared that myasthenia gravis selectively affects the properties of postsynaptic muscle cells. The structure of neuromuscular junctions is also altered in myasthenia; changes include a widening of the synaptic cleft and a reduction in the number of nicotinic acetylcholine receptors in the postsynaptic membrane.

A chance observation led to the discovery of the underlying cause of these changes. Jim Patrick and Jon Lindstrom, then working at the Salk Institute, were attempting to raise antibodies to nicotinic acetylcholine receptors by

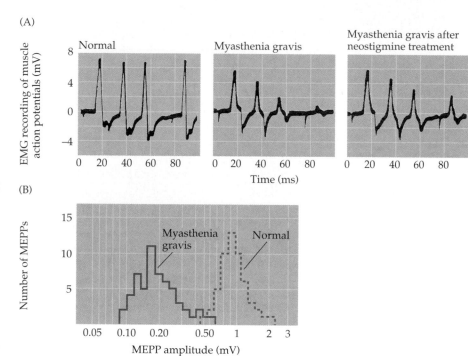

(A) Myasthenia gravis reduces the efficiency of neuromuscular transmission. Electromyographs (EMGs) show muscle responses elicited by stimulating the motor nerves. In normal individuals, each stimulus in a train evokes the same contractile response. Transmission rapidly fatigues in myasthenic patients, but can be partially restored by administration of the acetylcholinesterase inhibitor neostigmine. (B) Distribution of MEPP amplitudes in muscle fibers from myasthenic patients (solid line) and controls (dashed line). The smaller size of MEPPs in myasthenics is due to a diminished number of postsynaptic receptors. (A after Harvey et al., 1941; B after Elmqvist et al., 1964.)

immunizing rabbits with the receptors. Unexpectedly, the immunized rabbits developed muscle weakness that improved after treatment with acetylcholinesterase inhibitors. Subsequent work showed that the blood of myasthenic patients contains antibodies directed against the acetylcholine receptor and that these antibodies are present at neuromuscular synapses. Further, injecting the serum of myasthenic patients (which carries circulating antibodies) into mice produces myasthenic effects in the mice. Taken together, these findings indicate that myasthenia gravis

is an autoimmune disease that targets nicotinic acetylcholine receptors. The immune response reduces the number of functional receptors at the neuromuscular junction, reducing the efficiency of synaptic transmission; muscle weakness occurs because motor neurons are less capable of exciting the postsynaptic muscle cells to contract. This etiology also explains why cholinesterase inhibitors alleviate the signs and symptoms of myasthenia: the inhibitors increase the concentration of acetylcholine in the synaptic cleft, allowing more effective activation of those postsynaptic receptors not yet destroyed by the immune system.

Despite extraordinary progress in understanding this disease, it is not yet clear what triggers the immune system to produce an autoimmune response to acetylcholine receptors.

References

DRACHMAN, D. B. (1994) Myasthenia gravis. New Eng. J. Med. 330: 1797–1810.

ELMQVIST, D., W. W. HOFMANN, J. KUGELBERG AND D. M. J. QUASTEL (1964) An electrophysiological investigation of neuromuscular transmission in myasthenia gravis. J. Physiol. (Lond.) 174: 417–434.

PATRICK, J. AND J. LINDSTROM (1973) Autoimmune response to acetylcholine receptor. Science 180: 871–872.

actions of AMPA/kainate receptors are quite similar to those produced by the nAChRs at the neuromuscular synapse.

The NMDA subfamily of glutamate receptors also form multisubunit, cation-selective ion channels similar to most other ligand-gated ion channel receptors. However, these receptors have some unique functional properties that make them especially interesting. Perhaps most significant is the fact that NMDA receptor ion channels allow the entry of Ca^{2+} in addition to monovalent cations such as Na^+ and K^+. As a result, EPSPs produced by NMDA receptors can increase the concentration of Ca^{2+} within the postsynaptic neuron; the Ca^{2+} concentration change can then act as a second messenger to activate intracellular signaling cascades. Other unique properties of NMDA receptors are that opening the channel requires the presence of a co-agonist (the amino acid glycine), and that extracellular Mg^{2+} blocks the channel at hyperpolarized, but not depolarized, voltages. Hence, NMDA receptors allow the passage of cations only when the Mg^{2+} block is removed by the depolarization of the postsynaptic cell, either by inputs from many presynaptic neurons or by the repetitive firing of the presynaptic cell. These properties are widely thought to be the basis for some forms of information storage in the adult brain, as described in Chapter 23. There are at least five forms of NMDA receptor subunits (NMDA-R1, and NMDA-R2A through NMDA-R2D); different synapses have distinct combinations of these subunits, producing a variety of NMDA receptor subtypes (Figure 7.9C).

While some glutamatergic synapses have only AMPA/kainate receptors, many have both AMPA/kainate and NMDA receptors. An antagonist of NMDA receptors, APV (2-amino-5-phosphono-valerate), is often used to differentiate between the two receptor types. The use of this drug has also revealed differences between the EPSPs produced by NMDA and AMPA/kainate receptors, such as the fact that the EPSPs produced by NMDA receptors are slower and longer-lasting than the EPSPs produced by AMPA/kainate receptors.

■ LIGAND-GATED ION CHANNELS: GABA AND GLYCINE RECEPTORS

Inhibitory synapses employing GABA as their transmitter utilize two types of receptors, called $GABA_A$ and $GABA_B$. $GABA_A$ receptors are ligand-gated ion channels, while $GABA_B$ receptors are metabotropic receptors (see below). Some inhibitory synapses use only one type of GABA receptor, whereas others have both types. $GABA_A$ receptors are inhibitory because their associated channels are permeable to Cl^-; the flow of the negatively charged chloride ions inhibits postsynaptic cells since the reversal potential for Cl⁻ is more negative than the threshold for neuronal firing.

$GABA_A$ receptors are similar in structure to other ionotropic receptors, such as nACh receptors and NMDA and AMPA/kainate glutamate receptors. They are pentamers assembled from a combination of five types of subunits ($\alpha, \beta, \gamma, \delta$, and ρ) that are expressed in neurons. As a result of the diversity of subunits, as well as their variable stoichiometry, the function of $GABA_A$ receptors differs widely among neuronal types. The receptors for glycine are also ligand-gated Cl⁻ channels, their general structure mirroring that of other ionotropic receptors (see Figure 7.9C).

GABA-activated ion channels are the targets of numerous drugs of clinical importance. Benzodiazepines, such as diazepam (Valium®) and chlordiazepoxide (Librium®), are tranquilizing drugs that bind to the α and β subunits of $GABA_A$ receptors. Barbiturates, such as phenobarbital, are hypnotics that bind to the γ subunits of these receptors. Thus the actions of these drugs depends on the subunit composition of GABA receptors at a given synapse. Alcohol also affects GABA receptors and probably produces some of intoxication's familiar behavioral manifestations by interfering with inhibitory synaptic transmission.

■ OTHER LIGAND-GATED ION CHANNELS

Other ligand-gated ion channel receptors include serotonin (5-HT) receptors and purinergic receptors. Of the three major classes of 5-HT receptors, the great majority belong to the 5-HT1 and 5-HT2 subclasses. Only the 5-HT3 receptors are ligand-gated ion channels; the others are metabotropic receptors (see next section). Like ionotropic glutamate receptors, 5-HT3 receptors are nonselective cation channels, show desensitization in the continued presence of neurotransmitter, and allow the entry of Ca2+ in some cell types.

Of the numerous types of purinergic receptors now known to exist, one subclass, referred to as the P_{2x} receptors, belongs to the family of ligand-gated ion channels. Surprisingly, the genes encoding these channels predict a subunit transmembrane arrangement different than other ligand-gated ion channels in that there appear to be only two transmembrane domains. This receptor also forms a nonselective cation channel with significant permeability to Ca^{2+}.

■ METABOTROPIC RECEPTORS AND THE ACTIVATION OF G-PROTEINS

Metabotropic receptors differ fundamentally from ionotropic receptors in that they affect ion channels via the activation of G-proteins (see Figure 7.7B). To date, over 100 different metabotropic receptors have been characterized. These include multiple types of small-molecule receptors for glutamate (mGluR1–6); epinephrine and norepinephrine ($\alpha 1, \alpha 2, \beta 1, \beta 2$); acetylcholine (m1–m5); dopamine (D1–D5); and GABA ($GABA_B$), as well as receptors for all of the neuropeptides. All metabotropic neurotransmitter receptors are part of a superfamily of G-protein-coupled receptors that share a common molecular architecture. Unlike ligand-gated ion channels, which are multimeric proteins, metabotropic receptors consist of a single protein molecule that has a characteristic structure comprising seven membrane-spanning domains (Figure 7.10A). At least four of these domains contribute to the extracellular binding site for neurotransmitters, while two intracellular regions make up the site that binds G-proteins.

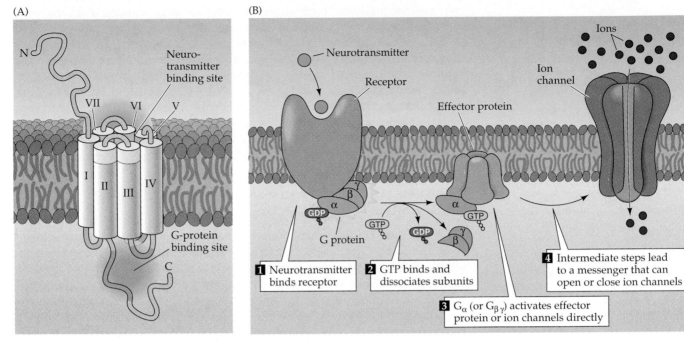

(A)

N

Neuro-transmitter binding site

VII   VI   V

I   II   III   IV

G-protein binding site

C

(B)

Neurotransmitter

Receptor

Effector protein

Ion channel

Ions

β
α
GDP

G protein

γ

α
GTP

GTP   GDP   β
γ

1 Neurotransmitter binds receptor

2 GTP binds and dissociates subunits

3 G_α (or $G_{\beta\gamma}$) activates effector protein or ion channels directly

4 Intermediate steps lead to a messenger that can open or close ion channels

Figure 7.10 Structure and function of metabotropic receptors. (A) The transmembrane architecture of metabotropic receptors. These monomeric proteins contain seven transmembrane domains. Portions of domains II, III, VI and VII (shaded) make up the neurotransmitter-binding region. G-proteins bind to both the loop between domains V and VI and to portions of the C-terminal region (shaded). (B) Mechanism of activation of metabotropic receptors. The binding of neurotransmitters leads to the association of GTP with the α subunit. The resulting dissociation of the G-protein trimer produces both an activated α subunit that is bound to GTP, and a dimer of β and δ subunits. Either of these activated complexes can produce postsynaptic electrical responses, although only the action of the α subunit is illustrated.

The G-proteins activated by metabotropic receptors consist of three protein subunits (α, β, and γ) and are often referred to as the **heterotrimeric G-proteins**. G-proteins are so named because the α subunit binds guanine nucleotides, such as GTP and GDP. A large number of different α, β, and γ subunits have been isolated, allowing a bewildering number of G-protein permutations. Despite this diversity, G-proteins exist in two functional conformations: bound to GTP or bound to GDP. When GDP is bound to the α subunit, the α subunit binds to the β and γ subunits to form an inactive trimer; however, when GTP is bound to the α subunit, the α subunit dissociates from the $\beta\gamma$ complex. An exchange of GDP for GTP is triggered by the binding of neurotransmitters to metabotropic receptors (Figure 7.10B). This switch of nucleotides activates the G-protein, allowing both the GTP-bound α subunit and the free $\beta\gamma$ complex to bind to downstream effector molecules and thereby mediate a variety of postsynaptic responses.

■ DIRECT MODULATION OF ION CHANNELS BY G-PROTEINS

In some instances, the activation of G-proteins by metabotropic receptors allows binding of activated G-protein subunits directly to ion channels to produce PSPs. For example, a number of neuronal types, as well as heart muscle cells, have metabotropic receptors that bind ACh. Because these receptors are also activated by the ACh receptor agonist muscarine, they are usually called **muscarinic receptors**. Activation of muscarinic ACh receptors can open K^+ channels, thereby inhibiting neuronal firing rates and slowing the beat of heart muscle cells. Because this effect is mimicked by exposure of the intracellular component of these channels to $\beta\gamma$ subunits, the inhibitory response is believed to be the result of the opening of K^+ channels following direct G-protein binding. The activation of α subunits can also lead to the rapid closing of voltage-gated Ca^{2+} and Na^+ channels. Because these latter channels are involved in generating action potentials, closing them makes it more difficult for postsynaptic cells to fire.

Even though direct interaction between G-proteins and ion channels is the most efficient means by which metabotropic receptors can generate post-synaptic conductance changes, the onset of such effects still requires tens of milliseconds and may last for seconds or even minutes. Thus, even the fastest actions of metabotropic receptors are much slower than the actions of ligand-gated channels. The comparative slowness of metabotropic receptor actions reflects the fact that multiple proteins need to bind to each other sequentially in order to produce the final physiological response.

■ INDIRECT MODULATION BY INTRACELLULAR MESSENGER PATHWAYS

The elicitation of postsynaptic effects is slowed even more when metabotropic receptors are coupled to intracellular **effector enzymes**. The best-characterized effector enzymes are adenylyl cyclase and phospholipase C. More recently, phospholipase A_2 has also been implicated as an effector for G-proteins. Each of these effector enzymes generates intracellular second messengers that lead to complex biochemical signaling cascades (Figure 7.11). Since the various cascades are activated by specific G-protein subunits, the pathways activated by a particular metabotropic receptor are determined by the combination of subunits associated with the receptor.

Adenylyl cyclase, an enzyme found in nearly all cell types, is activated when bound by a particular stimulatory G-protein (G_s). This enzyme catalyzes the production of the ubiquitous second messenger, **cyclic AMP** (**cAMP**). In some cells, cAMP or a related nucleotide, cyclic GMP (cGMP), can directly bind to ion channels and open or close them. More typically,

Figure 7.11 The three major G-protein-mediated signal pathways that act through enzymatic effectors share a number of properties. Neurotransmitter binding to a seven-transmembrane receptor leads to activation of a G-protein. The activated G-proteins associated with each receptor type can then: (1) activate adenylyl cyclase to generate cAMP and thereby activate protein kinase A (PKA); (2) activate phospholipase C, leading to the generation of two messengers, IP_3 and diacylglycerol (DAG). DAG is a potent activator of protein kinase C (PKC); (3) activate phospholipase A_2, leading to the generation of arachidonic acid. This compound is rapidly metabolized into a large number of short-lived intermediates. The end products of each of these pathways can then open or close ion channels.

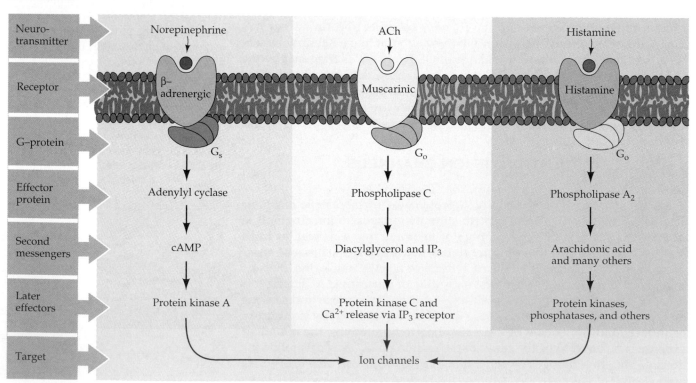

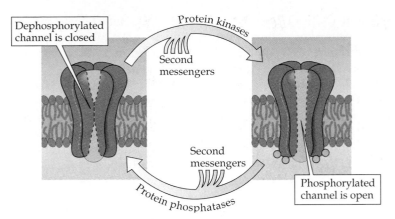

Figure 7.12 Ion channels can be modulated by protein phosphorylation. Second messengers often lead to the activation of protein kinases. These enzymes can then directly phosphorylate ion channels, as shown here, or can phosphorylate other proteins closely associated with ion channels. Protein phosphorylation leads to an increase in the probability of ion channels opening. Dephosphorylation is catalyzed by protein phosphatases, and these enzymes, too, can be regulated by second messengers.

cAMP activates protein kinases, such as protein kinase A (PKA), that phosphorylate target proteins such as ion channels by transferring phosphate groups from ATP to the proteins (Figure 7.12). Phosphorylation then regulates the opening or closing of the postsynaptic ion channels to produce PSPs. For example, recent evidence has shown that AMPA and kainate glutamate receptors are modulated by protein phosphorylation. Phosphorylation by PKA increases the amplitude of neurotransmitter-induced EPSPs, making the postsynaptic cell more excitable. Similarly, the direct phosphorylation of Na^+ channels has been shown to alter the size of Na^+ currents. In this instance, however, phosphorylation by PKA decreases the size of this current, tending to make the postsynaptic cell less excitable.

The activation of **phospholipase C** produces two second messengers: **inositol trisphosphate (IP$_3$)** and **diacylglycerol**. IP$_3$ can release Ca^{2+} from intracellular stores, which then leads to an activation of Ca^{2+}-activated K^+ channels and Ca^{2+}-activated Cl^- channels. Diacylglycerol activates protein kinase C, an enzyme that regulates the gating of many different types of ion channels. The G-protein-mediated activation of **phospholipase A$_2$** results in a complex cascade of biochemical reactions initiated by the release of arachidonic acid from the cell membrane. Many of the intermediates in this cascade are unstable and have not been well characterized in the brain. Nevertheless, a number of ion channels can be modulated by arachidonic acid or its metabolites.

The coupling of metabotropic receptors to ion channels via effector enzymes generates a wide range of postsynaptic actions. Because many second-messenger molecules can be produced by a single effector enzyme, this mechanism of coupling also permits substantial amplification of postsynaptic action. In general, these intermediate effector enzymes slow postsynaptic responses because of the additional time it takes the enzymes to generate their products.

■ POSTSYNAPTIC RESPONSES INVOLVING GENE EXPRESSION

The postsynaptic responses discussed thus far generally occur on a time scale of milliseconds to minutes or sometimes hours. The slowest and most persistent postsynaptic actions of neurotransmitters do not necessarily generate PSPs; instead, they alter the expression of genes that can indirectly influence the electrical activity of postsynaptic cells for much longer periods. Altered expression of genes can be achieved by the activation of second messengers,

Figure 7.13 Neurotransmitters can give rise to prolonged postsynaptic charges by modulating gene expression. The activation of G-protein-coupled receptors leads to the production of a second messenger such as cAMP. Virtually all of the responses of cAMP are mediated by protein kinase A (PKA). The kinase can phosphorylate response elements such as CREB, which in turn modulate the transcription of target genes. The affected genes may code for proteins involved in the synthesis of neurotransmitters, ion channel modulators, or ion channels themselves.

which in turn leads to the phosphorylation of proteins that regulate gene transcription. The phosphorylation of proteins such as the cAMP response element binding protein (CREB) allows the transcription of specific gene products to be modulated (Figure 7.13; see also Chapter 20).

A good example of such modulation occurs in some aminergic neurons. Repeated stimulation of these cells leads to an increase in the rate of norepinephrine synthesis. This effect is mediated by presynaptic metabotropic receptors that activate adenylyl cyclase. The subsequent increase in protein kinase A activity results in the phosphorylation of transcriptional regulators in the cell nucleus. Phosphorylation and activation of regulators such as CREB determines whether or not an adjacent gene is transcribed. In this case, the phosphorylation of CREB leads to an increase in the synthesis of tyrosine hydroxylase, the rate-limiting enzyme in catecholamine synthesis

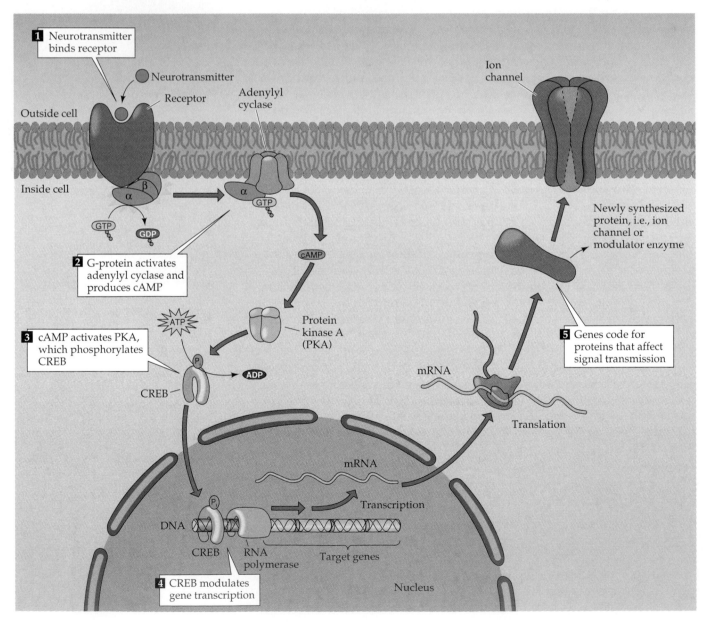

(see Chapter 6). In other cells, transcriptional regulators may lead to the synthesis of ion channels or other proteins that modulate channels.

In short, neurotransmitters can alter the electrical activity of cells not only by regulating the state of ion channels, but also by regulating the synthesis of channels or related proteins. Such changes first appear on a time scale of minutes to hours and can persist for weeks, months, or longer, and may therefore be involved in the long-term encoding of neural information (see Chapter 23).

■ SUMMARY

Neurotransmitter receptors are proteins embedded in the postsynaptic plasma membrane. Receptors translate chemical signals into electrical signals by binding neurotransmitter molecules secreted by presynaptic neurons, which leads in turn to opening or closing postsynaptic ion channels. The postsynaptic currents produced by the synchronous opening or closing of the ion channels changes the membrane potential of the postsynaptic cell. Potential changes that increase the probability of firing an action potential are excitatory, whereas those that decrease the probability of generating an action potential are inhibitory. Because postsynaptic neurons are usually innervated by many different inputs, the integrated effect of all EPSPs and IPSPs produced in a postsynaptic cell at any moment determines whether or not the cell fires an action potential. Two broadly different families of neurotransmitter receptors have evolved to carry out the postsynaptic signaling actions of neurotransmitters. Ligand-gated ion channels combine the neurotransmitter receptor and ion channel in one molecular entity and therefore give rise to rapid postsynaptic electrical responses. Metabotropic receptors regulate the activity of postsynaptic ion channels indirectly, via G-proteins, and induce slower and longer-lasting electrical responses. The faster effects of metabotropic receptors—which are still slower than ligand-gated effects—occur when G-proteins themselves activate ion channels. Slower metabotropic effects involve the activation of intracellular effector enzymes that modulate the phosphorylation of target proteins and/or gene transcription. The postsynaptic response at a given synapse is determined by the combination of receptor subtypes, G-protein subtypes, and ion channels that are expressed in the postsynaptic cell. Because each of these features can vary both within and among neurons, a tremendous diversity of transmitter-mediated postsynaptic effects is possible.

Additional Reading

Reviews

BROWN, A. M. AND L. BIRNBAUMER (1990) Ionic channels and their regulation by G-proteins. Annu. Rev. Physiol. 52: 197.

CHANGEUX, J.-P. (1993) Chemical signaling in the brain. Sci. Am. 269(5): 58–62.

DINGLEDINE, R. (1991) Molecular properties of AMPA/kainate receptors. Trends Pharmacol. Sci. 12: 360.

HILLE, B. (1994) Modulation of ion channel function by G-protein-coupled receptors. Trends Neurosci. 17: 531–535.

LEVITAN, I. B. (1994) Modulation of ion channels by protein phosphorylation and dephosphorylation. Annu. Rev. Physiol. 56: 193–212.

NAKANISHI, S. (1992) Molecular diversity of glutamate receptors and implication for brain function. Science 258: 597.

Important Original Papers

BADING, H., D. D. GINTY AND M. E. GREENBERG (1993) Regulation of gene expression in hippocampal neurons by distinct calcium signaling pathways. Science 260: 181–186.

HARRIS, B. A., J. D. ROBISHAW, S. M. MUMBY AND A. G. GILMAN (1985) Molecular cloning of complementary DNA for the alpha subunit of the G protein that stimulates adenylate cyclase. Science 229: 1274–1277.

HOLLMANN, M., C. MARON AND S. HEINEMANN (1994) N-glycosylation site tagging suggests a three transmembrane domain topology for the glutamate receptor GluR1. Neuron 13: 1331–1343.

HOLZ, G. G. I., S. G. RANE AND K. DUNLAP (1986) GTP-binding proteins mediate transmitter inhibition of voltage-dependent calcium channels. Nature 319: 670–672.

MATTERA, R., M. P. GRAZIANO, A. YATANI, Z. ZHOU, R. GRAF, J. CODINA, L. BIRNBAUMER, A. G. GILMAN AND A. M. BROWN (1989) Splice variants of the alpha subunit of the G-protein G_s activate both adenylyl cyclase and calcium channels. Science 243: 804–807.

UNWIN, N. (1995) Acetylcholine receptor channels imaged in the open state. Nature 373: 37–43.

WICKMAN, K., J. INIGUEZ-LLULI, P. DAVENPORT, R. A. TAUSSIG, G. B. KRAPIVINSKY, M. E. LINDER, A. G. GILMAN AND D. E. CLAPHAM (1994) Recombinant $G_{\beta\gamma}$ activates the muscarinic-gated atrial potassium channel I_{KACh}. Nature 368: 255–257.

Books

HALL, Z. (1992) *An Introduction to Molecular Neurobiology*. Sunderland, MA: Sinauer Associates, Chapters 3–7.

HILLE, B. (1992) *Ionic Channels of Excitable Membranes*. Sunderland, MA: Sinauer Associates, Chapters 1–8, 16–20.

NICHOLLS, D. G. (1994) *Proteins, Transmitters, and Synapses*. Boston: Blackwell Scientific.

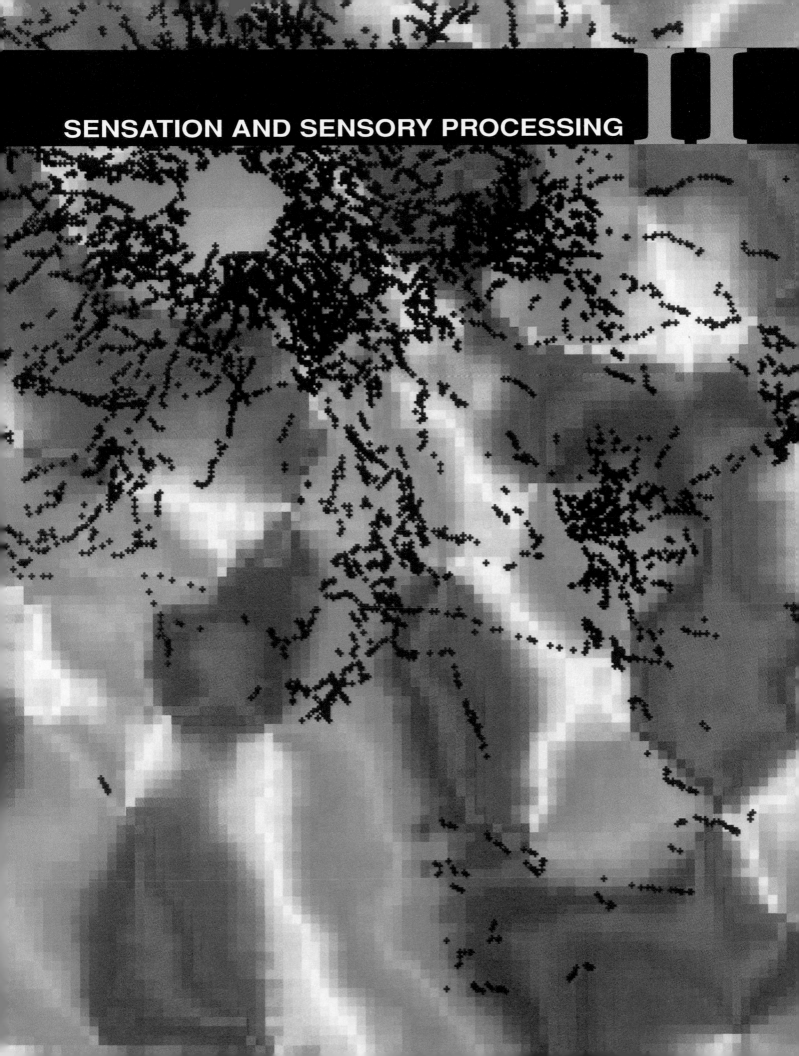

SENSATION AND SENSORY PROCESSING

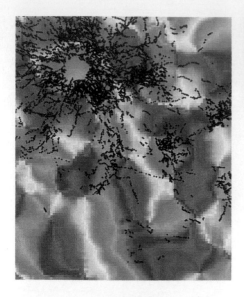

UNIT II

SENSATION AND SENSORY PROCESSING

Surface view of the visual cortex, obtained by optical imaging of a metabolic signal that reveals a partitioning of the cortical sheet into units related to orientation; the position of a single labeled neuron is also indicated. (Courtesy of David Fitzpatrick.)

Sensation entails the ability to transduce, encode, and perceive information about the outside world, and a good deal of the brain is devoted to this task. Although the basic senses—somatic sensation, vision, audition, vestibular sensation and the chemical senses—are very different from one another, a few fundamental rules govern the way the brain handles each of these diverse modalities. Highly specialized nerve cells called receptors convert the energy associated with mechanical forces, photons, sound waves, head movements, and odorant molecules or ingested chemicals into neural signals that convey information about the stimulus to the brain. These afferent sensory signals activate central neurons that interpret both the qualitative and quantitative nature of the stimulus. The central circuitry for sensory processing is typically organized into maps that further determine interactions within and among the major categories of sensation.

The clinical practice of medicine often requires an assessment of sensory deficits to infer the nature and location of a particular neurological problems. Knowledge of where and how the different sensory modalities are transduced, relayed, represented, and processed is therefore essential to understanding and treating a wide variety of diseases. Accordingly, these chapters on the neurobiology of sensation also serve to introduce the importance of structure/function relationships in clinical neurology.

CHAPTER 8

THE SOMATIC SENSORY SYSTEM

■ OVERVIEW

The somatic sensory system has two major components: a subsystem for the detection of mechanical stimuli such as light touch, vibration, and pressure; and a subsystem for the detection of painful stimuli and temperature. Together, these two subsystems give animals the ability to identify the shapes and textures of objects, to monitor the internal and external forces acting on the body at any moment, and to detect potentially harmful situations. Mechanosensory processing of external stimuli is initiated by the activation of a diverse population of cutaneous and subcutaneous receptors at the body surface that relays information to the central nervous system for interpretation, and ultimately action. Additional receptors located in muscles, joints, and other deep structures, monitor mechanical forces that are generated internally; these are called proprioceptors. Mechanosensory information is carried to the brain by several ascending pathways that run in parallel through the spinal cord, brainstem, and thalamus to reach the primary somatic sensory cortex in the postcentral gyrus of the parietal lobe. The primary somatic sensory cortex projects to higher-order association cortices and back to the subcortical structures involved in mechanosensory information processing. This chapter focuses on the mechanosensory subsystem; the pain and temperature subsystem is taken up in the following chapter.

■ CUTANEOUS AND SUBCUTANEOUS SOMATIC SENSORY RECEPTORS

The specialized sensory receptors in the cutaneous and subcutaneous tissues are dauntingly diverse (Table 8.1). They include free nerve endings in the skin, nerve endings associated with specializations that act as amplifiers or filters, and sensory terminals associated with specialized transducing cells that influence the ending by virtue of synapse-like contacts. Based on function, this variety of receptors can be divided into three groups: **mechanoreceptors**, **nociceptors**, and **thermoceptors**. On the basis of their morphology, the receptors near the body surface can also be divided into **free** and **encapsulated** types. Nociceptor and thermoceptor specializations are referred to as **free nerve endings** because the unmyelinated terminal branches of these neurons ramify widely in the upper regions of the dermis and epidermis; their role in somatic sensation is discussed in Chapter 9. Most other cutaneous receptors show some degree of encapsulation, which determines the nature of the stimuli to which they respond.

Despite their variety, all somatic sensory receptors work in fundamentally the same way: stimuli applied to the skin deform or otherwise change the nerve endings, which in turn affects the ionic permeability of the receptor membrane. Changes in permeability generate a depolarizing current in the nerve ending, thus creating a **receptor** (or **generator**) **potential** that in turn triggers action potentials (see Unit I). This overall process, in which the energy of a stimulus is converted into an electrical signal in the sensory neuron, is called **sensory transduction**, and is the critical first step in all sensory perceptions.

Figure 8.1 General organization of the somatic sensory system. (A) Mechanosensory information about the body reaches the brain by way of a three-neuron relay (shown in blue). The first synapse is made by the dorsal root ganglion cells onto neurons in the brainstem nuclei (the local spinal branches are not shown here). The axons of these second-order neurons synapse on third-order neurons of the ventral posterior nuclear complex of the thalamus, which in turn send their axons to the somatic sensory cortex. Information about pain and temperature takes a different course (shown in yellow; the anterolateral system) and is discussed in the following chapter. (B) Lateral view of the left cerebral hemisphere, illustrating the approximate location of the somatic sensory cortex in the anterior parietal lobe, just posterior to the central sulcus.

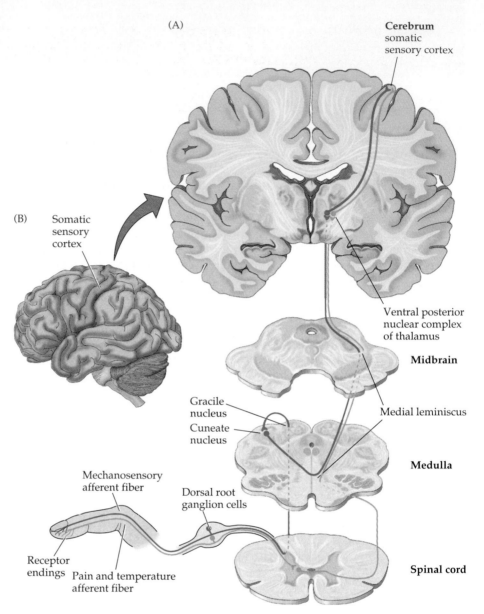

The *quality* of a perceived mechanosensory (or any other) stimulus is determined by the properties of the relevant receptors and the location of their central targets (Figure 8.1). The *strength* of the stimulus is conveyed by the rate of action potential discharge, although this relationship is nonlinear and sometimes quite complex. In addition, some receptors fire rapidly when a stimulus is first presented and then fall silent in the presence of continued stimulation, whereas others generate a sustained discharge in the presence of an ongoing stimulus (Figure 8.2). The usefulness of having some receptors that adapt quickly and others that adapt more slowly is to provide information about both the *dynamic* and *static* qualities of a stimulus. Receptors that initially fire in the presence of a stimulus and then become quiescent are particularly effective in conveying information about change; receptors that go on firing convey information about the persistence of a stimulus. Accordingly, somatic sensory receptors and the neurons that give rise to them are

TABLE 8.1
The Major Classes of Somatic Sensory Receptors

Receptor type	Anatomical characteristics	Associated axons[a]	Location	Function	Rate of adaptation	Threshold of activation
Free nerve endings	Minimally specialized nerve endings	C, Aδ	All skin	Pain, temperature, crude touch	Slow	High
Meissner's corpuscles	Encapsulated; between dermal papillae	Aβ	Principally glabrous skin	Touch, pressure (dynamic)	Rapid	Low
Pacinian corpuscles	Encapsulated; onion-like covering	Aβ	Subcutaneous tissue, interosseous membranes, viscera	Deep pressure, vibration (dynamic)	Rapid	Low
Merkel's disks	Encapsulated; associated with peptide-releasing cells	Aβ	All skin, hair follicles	Touch, pressure (static)	Slow	Low
Ruffini's corpuscles	Encapsulated; oriented along stretch lines	Aβ	All skin	Stretching of skin	Slow	Low
Muscle spindles	Highly specialized (see Figure 8.5 and Chapter 15)	Ia and II	Muscles	Muscle length	Both slow and rapid	Low
Golgi tendon organs	Highly specialized (see Chapter 15)	Ib	Tendons	Muscle tension	Slow	Low
Joint receptors	Minimally specialized	—	Joints	Joint position	Rapid	Low

[a]In the 1920s and 1930s, there was a virtual cottage industry classifying axons according to their conduction velocity. Three main categories were discerned, called A, B, and C. A comprises the largest and fastest axons, C the smallest and slowest. Mechanoreceptor axons generally fall into category A. The A group is further broken down into subgroups designated α (the fastest), β, and δ (the slowest). To make matters even more confusing, muscle afferent axons are usually classified into four additional groups—I (the fastest), II, III, and IV (the slowest)—with subgroups designated by lower case roman letters!

usually classified into rapidly or slowly adapting types (see Table 8.1). **Rapidly adapting**, or **phasic**, receptors respond maximally but briefly to stimuli; their response decreases if the stimulus is maintained. Conversely, **slowly adapting**, or **tonic**, receptors keep firing as long as the stimulus lasts.

■ MECHANORECEPTORS SPECIALIZED TO RECEIVE TACTILE INFORMATION

Four major types of encapsulated mechanoreceptors are specialized to provide information to the central nervous system about touch, pressure, vibration, and skin tension: Meissner's corpuscles, Pacinian corpuscles, Merkel's disks, and Ruffini's corpuscles (Figure 8.3 and Table 8.1). These receptors are referred to collectively as **low-threshold** (or high-sensitivity) mechanoreceptors because even weak mechanical stimulation of the skin induces them to produce action potentials. All low-threshold mechanoreceptors are innervated by myelinated fibers (Aβ, 6–12 μm in diameter; see Table 8.1), ensuring the rapid central transmission of tactile information.

Meissner's corpuscles are elongated receptors formed by a connective tissue capsule that comprises several lamellae of Schwann cells. The center of the capsule contains one or more afferent nerve fibers that generate rapidly adapting action potentials following minimal skin depression. Meissner's corpuscles lie between the dermal papillae just beneath the epidermis of the fingers, palms, and soles. They are the most common mechanoreceptors of glabrous (smooth, hairless) skin (the fingertips, for instance), and rapidly adapting afferent fibers innervating Meissner's corpuscles account

Figure 8.2 Slowly adapting mechanoreceptors continue responding to a stimulus, whereas rapidly adapting mechanoreceptors respond only at the onset (and often the offset) of stimulation. These functional differences allow the mechanoreceptors to provide information about both the static (via slowly adapting receptors) and dynamic (via rapidly adapting receptors) qualities of a stimulus.

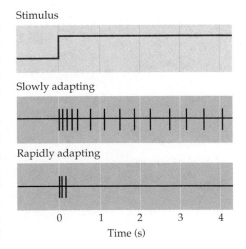

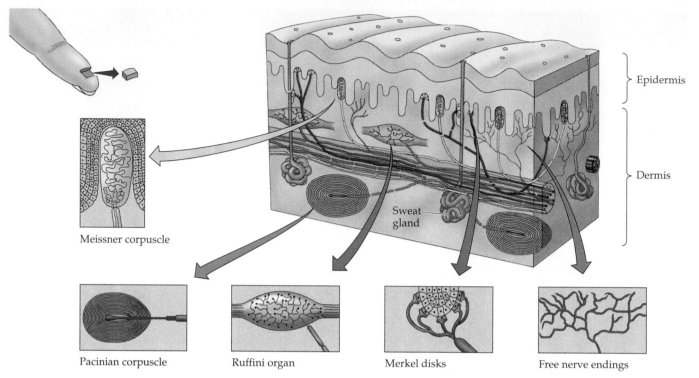

Meissner corpuscle

Pacinian corpuscle   Ruffini organ   Merkel disks   Free nerve endings

Epidermis

Dermis

Sweat gland

Figure 8.3 The skin harbors a variety of morphologically distinct mechanoreceptors. This diagram represents the smooth, hairless (also called glabrous) skin of the fingertip. The major characteristics of the various receptor types are summarized in Table 8.1. (After Darian-Smith, 1984.)

for about 40% of the sensory innervation of the human hand. Meissner's corpuscles are particularly efficient in transducing responses from the relatively low-frequency vibrations (30–50 Hz) that occur when textured objects are moved across the skin.

Pacinian corpuscles are large encapsulated endings located in the subcutaneous tissue (and more deeply in interosseous membranes and gut mesenteries). These receptors differ from Meissner's corpuscles in their morphology, distribution, and response threshold. The Pacinian corpuscle has an onionlike capsule in which an inner core of membrane lamellae is separated from an outer lamella by a fluid-filled space. One or more rapidly adapting afferent axons lie at the center of this structure. The capsule acts as a filter that allows only transient disturbances at high frequencies (250–350 Hz) to activate the nerve endings. Pacinian corpuscles adapt more rapidly than Meissner's corpuscles and have a lower response threshold. These attributes suggest that Pacinian corpuscles are involved in the discrimination of fine surface textures or other moving stimuli that produce high-frequency vibration of the skin. In corroboration of this supposition, stimulation of Pacinian corpuscle afferent fibers in humans induces a sensation of vibration or tickle. They make up 10–15% of the cutaneous receptors in the hand. Pacinian corpuscles located in interosseous membranes probably detect vibrations transmitted to the skeleton. Structurally similar endings found in the bills of ducks and geese and in the legs of cranes and herons detect vibrations in water; such endings in the wings of soaring birds detect vibrations produced by air currents. Because they are rapidly adapting, both Meissner's and

Pacinian corpuscles provide information primarily about the dynamic qualities of mechanical stimuli.

Slowly adapting cutaneous mechanoreceptors include **Merkel's disks** and **Ruffini's corpuscles** (see Figure 8.3 and Table 8.1). Merkel's disks are located in the epidermis, where they are precisely aligned with the papillae that lie beneath the dermal ridges. They account for about 25% of the mechanoreceptors of the hand, and are particularly dense in the fingertips, lips, and external genitalia. The slowly adapting nerve fiber associated with each Merkel's disk enlarges into a saucer-shaped ending that is closely applied to another specialized cell containing vesicles that apparently release peptides that modulate the nerve terminal. Selective stimulation of these receptors in humans produces a sensation of light pressure. These several properties have led to the supposition that Merkel's disks play a major role in the static discrimination of shapes, edges, and rough textures.

Ruffini's corpuscles, although structurally similar to other tactile receptors, are not as well understood. These elongated, spindle-shaped capsular specializations are located deep in the skin, as well as in ligaments and tendons. The long axis of the corpuscle is usually oriented parallel to the stretch lines in skin; thus, Ruffini's corpuscles are particularly sensitive to stretching produced by digit or limb movements. They account for about 20% of the receptors in the human hand, and do not elicit any particular tactile sensation when stimulated electrically. Although there is still some question as to their function, they probably respond primarily to internally generated stimuli (see the section on proprioception below).

■ DIFFERENCES IN MECHANOSENSORY DISCRIMINATION ACROSS THE BODY SURFACE

The accuracy with which tactile stimuli can be sensed varies from one region of the body to another. Figure 8.4 illustrates the results of an experiment in which variation in tactile ability across the body surface was measured by **two-point discrimination**. This test measures the minimal interstimulus distance required to perceive two simultaneously applied skin indentations as distinct (the points of a pair of calipers, for example). Such stimuli applied to the skin of the fingertips are discretely perceived if they are only 2 mm apart. In contrast, the same stimuli applied to the forearm are not perceived as distinct until they are at least 40 mm apart! This marked regional difference in tactile ability is explained by the fact that the encapsulated mechanoreceptors that respond to the stimuli are three to four times more numerous in the fingertips than in other areas of the hand, and many times more dense than in the forearm. Equally important are the differences in the size of the neuronal receptive fields. The **receptive field** of a somatic sensory neuron is the region of the skin within which a tactile stimulus evokes a sensory response in the cell or its axon. Analysis of the human hand shows that the receptive fields of mechanosensory neurons are 1–2 mm in diameter on the fingertips but 5–10 mm on the palms. The receptive fields on the arm are larger still. The importance of receptive field size is easy to envision. If, for instance, the receptive fields of all cutaneous receptor neurons covered the entire digital pad, it would be difficult or impossible to discriminate two spatially separate stimuli applied to the fingertip.

Receptor density and receptive field sizes in different regions are not the whole story, however. Psychophysical analysis of tactile performance under diverse circumstances suggests that something more than the cutaneous periphery is needed to explain variations in tactile perception. For

Figure 8.4 Variation in the sensitivity of tactile discrimination as a function of location on the body surface, measured here by two-point discrimination. (After Weinstein, 1969.)

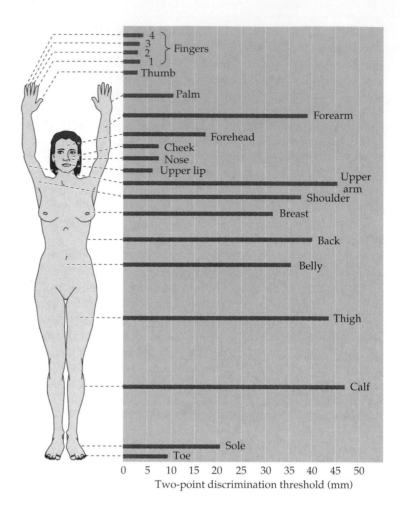

instance, sensory thresholds in two-point discrimination tests vary with practice, fatigue, and stress. The significance of stimuli is also important; even though we spend most of the day wearing clothes, we usually ignore the tactile stimulation that they produce. Some aspect of the mechanosensory system allows us to filter out this information and pay attention to it only when necessary. The fascinating phenomenon of "phantom limb" sensations after amputation (see Box A in Chapter 9) provides further evidence that tactile perception is not fully explained by the peripheral information that travels centrally. The central nervous system clearly plays an active role in determining our perception of the mechanical forces that act on us.

■ MECHANORECEPTORS SPECIALIZED FOR PROPRIOCEPTION

Whereas cutaneous mechanoreceptors provide information derived from external stimuli, another major class of receptors provides information about mechanical forces arising from the body itself. These are called **proprioceptors**, meaning "receptors for self." The purpose of proprioceptors is primarily to give detailed and continuous information about the position of the limbs and other body parts in space. Low-threshold mechanoreceptors, including muscle spindles, Golgi tendon organs, and joint receptors, provide this kind of sensory information, which is essential to the accurate performance of complex movements. (Particularly important is information about the position and motion of the head; in this case, proprioceptors are aided by the highly specialized vestibular system, which is considered separately in Chapter 13.)

Much present knowledge about proprioception derives from studies of **muscle spindles**, which are found in all but a few striated (skeletal) muscles. Muscle spindles consist of from four to eight specialized **intrafusal muscle fibers** surrounded by a capsule of connective tissue. The intrafusal fibers are distributed among the ordinary (extrafusal) fibers of skeletal muscle in a parallel arrangement (Figure 8.5). In the largest of the several intrafusal fibers, the nuclei are collected in an expanded region in the center of the fiber called a bag; hence the name nuclear bag fibers. The nuclei in the remaining two to six smaller intrafusal fibers are lined up single file, such that these fibers are called nuclear chain fibers. Myelinated sensory axons belonging to group Ia innervate muscle spindles by encircling the middle portion of both types of intrafusal fibers (see Figure 8.5 and Table 8.1). The Ia axon terminal is known as the **primary sensory ending** of the spindle. Secondary innervation is provided by group II axons that innervate the nuclear chain fibers and give off a minor branch to the nuclear bag fibers. The intrafusal muscle fibers can contract when commanded to do so by motor axons derived from a pool of specialized motor neurons in the spinal cord (called **gamma motor neurons**). The major function of muscle spindles is to provide information about muscle length (that is, the degree to which they are being stretched). A detailed account of how these important receptors function during movement is given in Chapters 15 and 16.

The density of spindles in human muscles varies. Large muscles that generate coarse movements have relatively few spindles; in contrast, extraocular muscles and the intrinsic muscles of the hand and neck are richly supplied with spindles, reflecting the importance of accurate eye movements, the need to manipulate objects with great finesse, and the continuous demand for precise positioning of the head. This relationship is consistent with the generalization that the sensory-motor apparatus at all levels of the nervous system is much richer for muscles of the hand, head, speech organs, and other parts of the body that are used to perform important and demanding tasks. Spindles are lacking altogether in a few muscles, such as those of

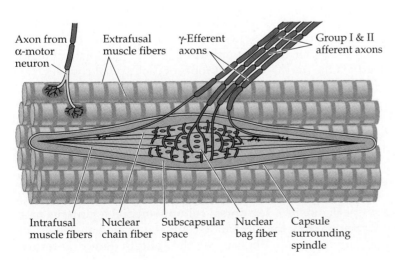

Axon from
α-motor
neuron

Extrafusal
muscle fibers

γ-Efferent
axons

Group I & II
afferent axons

Intrafusal
muscle fibers

Nuclear
chain fiber

Subscapsular
space

Nuclear
bag fiber

Capsule
surrounding
spindle

Figure 8.5 A muscle spindle and several extrafusal muscle fibers. See text for description. (After Matthews, 1964.)

the internal ear, which apparently do not require the kind of feedback that these receptors provide.

Whereas muscle spindles are specialized to signal changes in muscle length, low-threshold mechanoreceptors in tendons inform the central nervous system about changes in muscle tension. These mechanoreceptors, called **Golgi tendon organs**, are innervated by branches of group Ib afferents and are distributed among the collagen fibers that form the tendons (see Chapter 15).

Finally, rapidly adapting mechanoreceptors in and around joints gather dynamic information about limb position and joint movement. These **joint receptors** remain poorly understood.

■ ACTIVE TACTILE EXPLORATION

Tactile discrimination normally entails the active exploration of things in the environment. In humans, this is accomplished by using the hands to grasp and manipulate objects, or by moving the fingers across a surface so that a sequence of contacts between the skin and the object of interest is established. Psychophysical evidence indicates that relative movement between the skin and a surface is the single most important requirement for accurate discrimination of texture by humans. Animal experiments confirm the dependence of tactile discrimination on active exploration. Rats, for instance, discriminate texture by rhythmic movements of their facial whiskers that brush across surfaces. Such active touching, or **haptics**, involves the interpretation of complex spatiotemporal patterns of stimuli that are likely to activate many classes of mechanoreceptors. Haptics also requires dynamic interactions between motor and sensory signals, probably inducing sensory responses in central neurons that differ from the responses of the same cells during passive stimulation of the skin.

■ THE MAJOR MECHANOSENSORY PATHWAY: THE DORSAL COLUMN–MEDIAL LEMNISCUS SYSTEM

The action potentials generated by tactile and other mechanosensory stimuli are transmitted to the spinal cord by afferent sensory axons traveling in the peripheral nerves (Box A). The neuronal cell bodies that give rise to these first-order axons are located in the **dorsal root** (or **sensory**) **ganglia**; one such ganglion is associated with each segmental spinal nerve (see Figure 8.1). Dorsal root ganglion cells are also known as **first-order neurons** because they initiate the sensory process. The ganglion cells thus give rise to long peripheral axons that end in the somatic receptor specializations already described, as well as shorter central axons that reach the dorsolateral region of the spinal cord via the **dorsal** (**sensory**) **roots** of each spinal cord segment (see Chapter 1). The large myelinated fibers that innervate low-threshold mechanoreceptors are derived from the largest neurons in these ganglia, whereas the smaller ganglion cells give rise to the afferent nerve fibers that end in the high-threshold nociceptors and thermoceptors.

Depending on whether they belong to the mechanosensory system or to the pain and temperature system, the first order axons carrying information from somatic receptors have different patterns of termination in the spinal cord and define distinct somatic sensory pathways within the central nervous system (see Figure 8.1). The **dorsal column–medial lemniscus pathway** carries information from the mechanoreceptors that mediate tactile discrimination and proprioception (Figure 8.6); the **spinothalamic (anterolat-**

Box A
DERMATOMES

Each dorsal root ganglion and associated spinal nerve arises from an iterated series of embryonic tissue masses called somites. This fact of development explains the overall segmental arrangement of somatic nerves in the adult. The skin region innervated by the sensory axons of a single dorsal root and its related spinal nerve is called a dermatome. In humans, the area of each dermatome has been defined in studies of patients in whom specific dorsal roots are affected (as in herpes zoster, or "shingles") or surgically interrupted (for pain relief). Such studies show that dermatomal maps vary among individuals. Dermatomes also overlap substantially, so that injury to an individual dorsal root does not lead to complete loss of sensation in the relevant skin region. Moreover, dermatomal overlap is more extensive for touch, pressure, and vibration than for pain and temperature. From a clinical point of view, this latter fact means that testing for pain sensation provides a more precise assessment of a segmental nerve injury than does testing for responses to touch, pressure, or vibration. Finally, the segmental distribution of proprioceptors does not follow the dermatomal map, but is more closely allied with the pattern of muscle innervation. Despite these caveats, knowledge of dermatomes is essential in the clinical evaluation of neurological patients, particularly in determining the level of a spinal lesion.

Reference

HAYMAKER, W. AND B. WOODHALL (1967) *Peripheral Nerve Injuries: Principles of Diagnosis.* Philadelphia: W.B. Saunders.

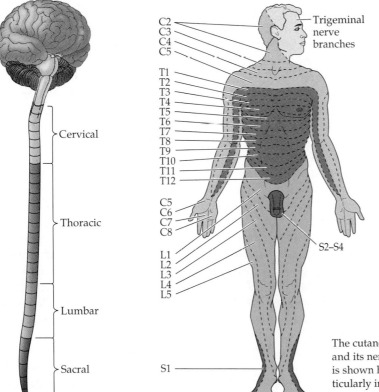

The cutaneous innervation arising from a single dorsal root ganglion and its nerve is called a dermatome. The full set of sensory dermatomes is shown here for a typical adult. Knowledge of this arrangement is particularly important in defining the location of suspected spinal (and other) lesions. The numbers refer to the spinal nerves.

eral) **pathway** mediates pain and temperature sensation (see Chapter 9). This anatomical difference is one of the major reasons that pain and temperature sensation is treated separately.

   Upon entering the spinal cord, the first-order axons carrying information from mechanoreceptors bifurcate into ascending and descending branches, which in turn send collateral branches to several spinal segments. Some collateral branches penetrate the dorsal horn of the cord and synapse on neurons located mainly in a region called Rexed's laminae III–V (named after the neuroanatomist who first studied the details of this part of the spinal cord). These synapses mediate, among other things, segmental reflexes such as the "knee-jerk" reflex described in Chapter 1, and are further considered in Chapters 15 and 16. The major branch of the incoming axons, however, ascends ipsilaterally through the **dorsal columns** (also called the posterior funiculi) of the cord, all the way to the lower medulla, where it terminates by contacting **second-order neurons** in the **gracile** and **cuneate nuclei** (together referred to as the **dorsal column nuclei**; see Figures 8.1 and 8.6A). Axons in the dorsal columns are topographically organized such that the fibers that convey information from lower limbs are in the medial subdivision of the dorsal columns, called the **gracile tract**. The lateral subdivision, called the **cuneate tract**, contains axons conveying information from the upper limbs, trunk, and neck. The dorsal columns in humans account for more than a third of the cross-sectional area of the spinal cord.

Interestingly, lesions limited to the dorsal columns of the spinal cord in both humans and monkeys have only a modest effect on the performance of simple tactile tasks, although they impede the ability to detect the direction and speed of tactile stimuli, and proprioception. Dorsal column lesions may also reduce a patient's ability to initiate active movements related to tactile exploration. For instance, such individuals have difficulty recognizing the direction of lines or identifying numbers and letters drawn on the skin. The relatively mild deficit that follows dorsal column lesions is presumably explained by the fact that some axons responsible for cutaneous mechanoreception also run in the spinothalamic (pain and temperature) pathway described in Chapter 9.

The second-order relay neurons in the dorsal column nuclei send their axons to the somatic sensory portion of the thalamus (see Figure 8.6A). These axons project in the dorsal portion of each side of the lower brainstem, where they form the **internal arcuate tract**. The internal arcuate axons subsequently cross the midline to form a large tract that is elongated dorsoventrally, called the **medial lemniscus**. (The crossing of these fibers is called the decussation of the medial lemniscus; the word *lemniscus*, incidentally, means "ribbon.") In a section through the medulla, such as the one shown in Figure 8.6A, the medial lemniscal axons carrying information from the lower limbs are located ventrally, whereas the axons related to the upper limbs are located dorsally. As the medial lemniscus ascends through the pons and midbrain, it rotates 90° laterally, so that the upper body is eventually represented in the medial portion of the tract, and the lower body in the lateral portion. The axons of the medial lemniscus thus reach the **ventral posterior lateral (VPL) nucleus** of the thalamus, whose cells are the **third-order neurons** of the dorsal column-medial lemniscus system.

■ THE TRIGEMINAL PORTION OF THE MECHANOSENSORY SYSTEM

To make matters even more complicated, tactile and proprioceptive information from the face is conveyed from the periphery to the thalamus by a different route. The mechanosensory pathways described in the preceding section carry somatic information from the upper and lower body and from the posterior third of the head. Information derived from the face is transmitted to the central nervous system by the **trigeminal somatic sensory system** (Figure 8.6B). Low-threshold mechanoreception in the face is mediated by first-order neurons located in the trigeminal (cranial nerve V) ganglion. The peripheral processes of these neurons form the three main subdivisions of the **trigeminal nerve** (the **ophthalmic**, **maxillary**, and **mandibular branches**),

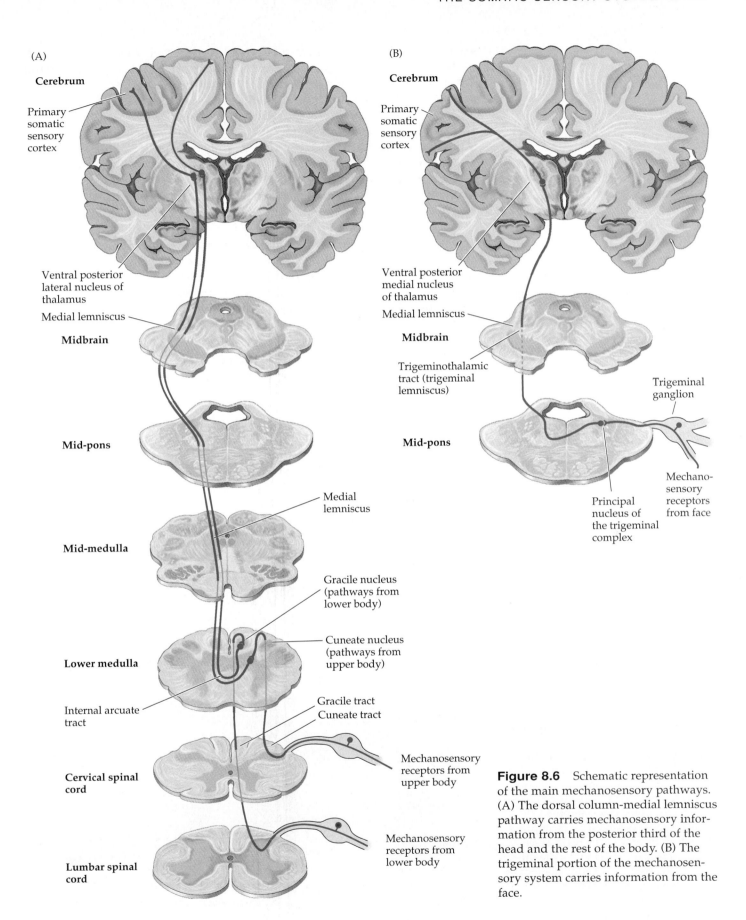

(A)

Cerebrum

Primary somatic sensory cortex

Ventral posterior lateral nucleus of thalamus

Medial lemniscus

Midbrain

Mid-pons

Medial lemniscus

Mid-medulla

Gracile nucleus (pathways from lower body)

Cuneate nucleus (pathways from upper body)

Lower medulla

Internal arcuate tract

Gracile tract
Cuneate tract

Cervical spinal cord

Mechanosensory receptors from upper body

Lumbar spinal cord

Mechanosensory receptors from lower body

(B)

Cerebrum

Primary somatic sensory cortex

Ventral posterior medial nucleus of thalamus

Medial lemniscus

Midbrain

Trigeminothalamic tract (trigeminal lemniscus)

Trigeminal ganglion

Mid-pons

Mechano-sensory receptors from face

Principal nucleus of the trigeminal complex

Figure 8.6 Schematic representation of the main mechanosensory pathways. (A) The dorsal column-medial lemniscus pathway carries mechanosensory information from the posterior third of the head and the rest of the body. (B) The trigeminal portion of the mechanosensory system carries information from the face.

each of which innervates a well-defined territory on the face and head, including the teeth and the mucosa of the oral and nasal cavities. The central processes of trigeminal ganglion cells form the sensory roots of the trigeminal nerve, and enter the brain stem at the level of the pons to terminate on subdivisions of the **trigeminal brainstem complex.**

The trigeminal brainstem complex has two major components: the **principal nucleus** (responsible for processing mechanosensory stimuli), and the **spinal nucleus** (responsible for processing painful and thermal stimuli). Thus, most of the axons carrying information from low-threshold cutaneous mechanoreceptors in the face terminate in the principal nucleus. In effect, this nucleus corresponds to the dorsal column nuclei that relay mechanosensory information from the rest of the body. The spinal nucleus corresponds to a portion of the spinal cord that contains the second-order neurons in the pain and temperature system for the rest of the body (see Chapter 9). The second-order neurons of the trigeminal brainstem nuclei give off axons that cross the midline and ascend to the **ventral posterior medial nucleus** of the thalamus through the trigeminothalamic tract (also called the trigeminal lemniscus).

■ THE SOMATIC SENSORY COMPONENTS OF THE THALAMUS

Each of the several ascending somatic sensory pathways originating in the spinal cord and brainstem converge on the thalamus (Figure 8.7). As noted, the ventral posterior (VP) complex of the thalamus, which comprises a lateral and a medial nucleus, is the main target of these ascending pathways. The more laterally located ventral posterior lateral (VPL) nucleus receives projections from the medial lemniscus carrying all somatosensory information from the body and posterior head, whereas the more medially located ventral posterior medial (VPM) nucleus receives axons from the trigeminal lemniscus (that is, mechanosensory and nociceptive information from the

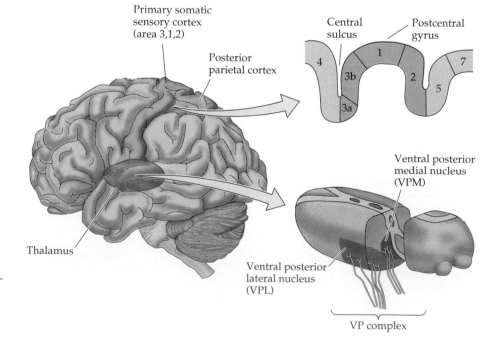

Figure 8.7 Diagram of the somatic sensory portions of the thalamus and their cortical targets in the postcentral gyrus. The VP complex comprises the VPM, which relays somatic sensory information carried by the trigeminal system (from the face), and the VPL, which relays somatic sensory information from the rest of the body. Inset above shows organization of the primary somatosensory cortex in the postcentral gyrus, shown here in a section cutting across the gyrus. (After Brodal, 1992; Jones et al., 1982.)

face). Accordingly, the VP complex contains a complete representation of the somatic sensory periphery.

■ THE SOMATIC SENSORY CORTEX

The axons arising from neurons in the VP complex of the thalamus project to cortical neurons located primarily in layer IV of the **somatic sensory cortex** (Figure 8.7; see Box A in Chapter 24 for a description of cortical lamination). The somatic sensory cortex in humans, which is located in the parietal lobe, comprises four distinct regions, or fields, known as **Brodmann's areas 3a, 3b, 1,** and **2**. Although area 3b is generally known as the **primary somatic sensory cortex** (also called SI), all four areas are involved in processing tactile information. Experiments carried out in nonhuman primates indicate that neurons in areas 3b and 1 respond primarily to cutaneous stimuli, whereas neurons in 3a respond mainly to stimulation of proprioceptors; area 2 neurons process both tactile and proprioceptive stimuli. Mapping studies in humans and other primates show further that each of these four cortical areas contains a separate and complete representation of the body. In these **somatotopic maps**, the foot, leg, trunk, forelimbs, and face are represented in a medial to lateral arrangement (Figures 8.8 and 8.9).

Although the topographic organization of the several somatic sensory areas is similar, the functional properties of the neurons in each region are

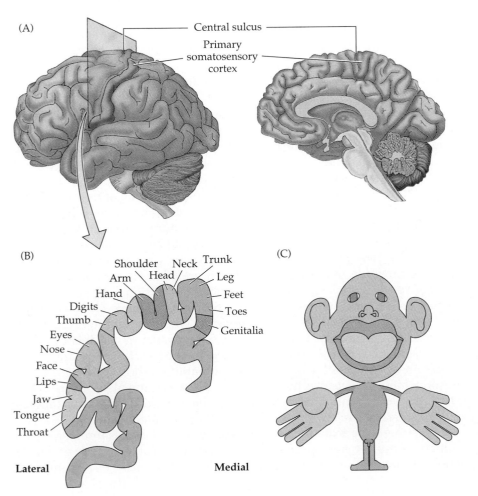

Figure 8.8 Somatotopic order in the human primary somatic sensory cortex. (A) Diagram showing the approximate region of the human cortex from which electrical activity is recorded following mechanosensory stimulation of different parts of the body. The patients in the study were undergoing neurosurgical procedures for which such mapping was required. Although modern imaging methods are now refining these classical data, the human somatotopic map first defined in the 1930s has remained generally valid. (B) Diagram along the plane in (A) showing the somatotopic representation of body parts from medial to lateral. (C) Cartoon of the homunculus constructed on the basis of such mapping. Note that the amount of somatic sensory cortex devoted to the hands and face is much larger than the relative amount of body surface in these regions. A similar disproportion is apparent in the primary motor cortex, for much the same reasons (see Chapter 16). (After Penfield et al., 1953; Corsi, 1991.)

Figure 8.9 The primary somatic sensory map in the owl monkey based, as in Figure 8.8, on the electrical responsiveness of the cortex to peripheral stimulation. Much more detailed mapping is possible in experimental animals than in neurosurgical patients. The enlargement on the right shows areas 3b and 1, which process most cutaneous mechanosensory information. The arrangement is generally similar to that determined in humans. (After Kaas, 1983.)

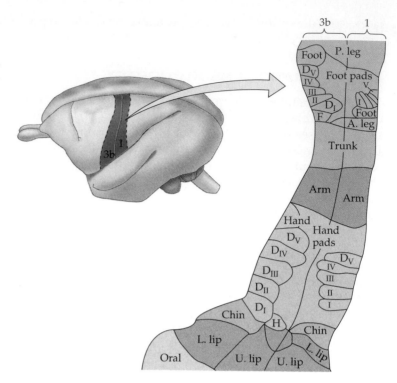

distinct (Box B). For instance, the neuronal receptive fields are relatively simple in area 3b, where responses can be elicited by stimulation of a single finger; in areas 1 and 2, the majority of the receptive fields are multidigit. Furthermore, neurons in area 1 respond preferentially to particular directions of skin stimulation, whereas many area 2 neurons require complex stimuli for their activation (such as a particular shape). Lesions restricted to area 3b produce a severe deficit in both texture and shape discrimination. In contrast, damage confined to area 1 affects the ability of monkeys to perform accurate texture discrimination. Area 2 lesions tend to produce deficits in finger coordination and in shape and size discrimination.

A curious feature of cortical maps, recognized soon after their discovery, is their failure to represent the body in actual proportion. When neurosurgeons determined the representation of the human body in the primary sensory (and motor) cortex, the homunculus (literally, "little man") defined by such mapping procedures had a grossly enlarged face and hands compared to the torso and proximal limbs (Figure 8.8C). These anomalies presumably arise because manipulation, facial expression, and speaking are extraordinarily important for humans, requiring more brain circuitry to govern them. Such distortions are also apparent when topographical maps are compared across species. In the rat brain, for example, an inordinate amount of the somatic sensory cortex is devoted to representing the large facial whiskers (which provide a substantial somatic sensory input for rats and mice), while raccoons overrepresent their paws and the platypus its bill. In short, the sensory input (or motor output) that is particularly significant to a given species gets relatively greater cortical representation.

As might be expected, the representation of each subregion of the somatic sensory map is proportional to the density and complexity of the relevant circuitry at subcortical levels. Thus, in humans, the number of

Box B
PATTERNS OF ORGANIZATION WITHIN THE SENSORY CORTICES: BRAIN MODULES

A variety of observations over the last 40 years have made it clear that in many parts of the brain an iterated substructure exists within somatotopic (and other topographical) cortical maps. This substructure takes the form of units called modules, each module involving hundreds or thousands of nerve cells in repeating patterns. The advantages of these iterated patterns for brain function remain largely mysterious; for the neurobiologist, however, such patterns have provided important clues about cortical connectivity and the mechanisms by which neural activity influences brain development (see Chapters 21 and 22).

The observation that the somatic sensory cortex comprises elementary units of vertically linked cells was first noted in the 1920s by the Spanish neuroanatomist Rafael Lorente de Nó, based on his studies of the rat. The potential importance of cortical modularity remained largely unexplored until the 1950s, when electrophysiological experiments indicated an arrangement of repeating units in the cat brain. Vernon Mountcastle, a neurophysiologist at Johns Hopkins, found that vertical microelectrode penetrations in the primary somatosensory cortex of cats and monkeys encountered cells that responded to the same sort of mechanical stimulus presented at the same location on the body surface. Soon after Mountcastle's pioneering work, David Hubel and Torsten Wiesel discovered a similar arrangement in the primary visual cortex of the cat. Taken together, these and other observations led Mountcastle to the more general view that "the elementary pattern of organization of the cerebral cortex is a vertically oriented column or cylinder of cells capable of input-output functions of considerable complexity." Since these discoveries in the late 1950s

and early 1960s, the view that modular circuits represent a fundamental feature of the mammalian cerebral cortex has gained wide acceptance, and many such entities have now been described.

This wealth of evidence for patterned circuits has led many neuroscientists to conclude that modules are a fundamental feature of the cerebral cortex, essential for perception, cognition, and perhaps even consciousness. Despite the prevalence of modules, some problems with the view that modular units are universally important in cortical function are apparent. First, although modular circuits of a given class are readily seen in the brains of some species, they have not been found in other, sometimes closely related, animals. Second, not all regions of the mammalian cortex are organized in a modular fashion. And third, no consistent function of modules has been discerned, much effort and speculation notwithstanding. This salient feature of the organization of the somatic sensory cortex and other cortical (and some subcortical) regions therefore remains a tantalizing puzzle.

References

HUBEL, D. H. (1988) *Eye, Brain, and Vision.* Scientific American Library. New York: W.H. Freeman.

MOUNTCASTLE, V. B. (1957) Modality and topographic properties of single neurons of cat's somatic sensory cortex. J. Neurophysiol 20: 408–434.

PURVES, D., D. RIDDLE AND A. LAMANTIA (1992) Iterated patterns of brain circuitry (or how the cortex gets its spots). Trends Neurosci. 15: 362–368.

WOOLSEY, T. A. AND H. VAN DER LOOS (1970) The structural organization of layer IV in the somatosensory region (SI) of mouse cerebral cortex. The description of a cortical field composed of discrete cytoarchitectonic units. Brain Res. 17: 205–242.

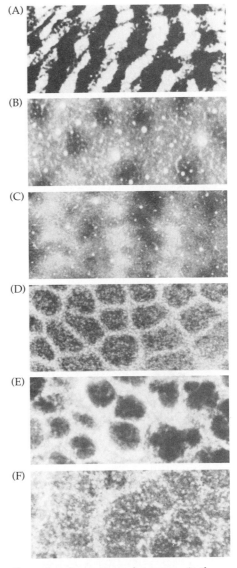

Examples of repeating substructures in the mammalian brain. (A) Ocular dominance columns in layer IV in the primary visual cortex (V1) of a rhesus monkey. (B) Repeating units called "blobs" in layers II and III in V1 of a squirrel monkey. (C) Stripes in layers II and III in V2 of a squirrel monkey. (D) Barrels in layer IV in primary somatic sensory cortex of a rat. (E) Glomeruli in the olfactory bulb of a mouse. (F) Iterated units called "barreloids" in the thalamus of a rat. These examples indicate that modular organization is commonplace in the brain. (From Purves et al., 1992.)

mechanosensory receptors is much higher in the fingertips than in the rest of the body, and the cervical spinal cord is enlarged to accommodate the extra circuitry related to the hand and upper limb.

■ HIGHER-ORDER CORTICAL REPRESENTATIONS

Somatic sensory information is distributed from the primary somatic sensory cortex to "higher-order" cortical fields, as well as to subcortical structures. One of these higher-order cortical centers, the adjacent secondary somatosensory cortex (sometimes called SII), receives convergent projections from SI and sends projections in turn to limbic structures such as the amygdala and hippocampus. This latter pathway is believed to play an important role in tactile learning and memory. Neurons in motor cortical areas also receive tactile information from the anterior parietal cortex and, in turn, provide feedback projections to several cortical somatic sensory regions. Such integration of sensory and motor information is considered in Chapters 19 and 24, where the role of these "association" regions of the cerebral cortex are discussed.

Finally, it is worth calling attention to a fundamental but often neglected feature of the somatic sensory system: the presence of massive descending projections. These pathways originate in sensory cortical fields and run to the thalamus, brainstem, and spinal cord. Indeed, descending projections from the somatosensory cortex outnumber ascending somatic sensory pathways! Although their physiological role is not well understood, it is generally assumed (with some experimental support) that descending projections modulate the ascending flow of sensory information at the level of the thalamus and brainstem.

■ SUMMARY

A major component of the somatic sensory system responds to and interprets mechanical stimuli that reach the body surface or are generated within the body (the pain and temperature components of this diverse system are dealt with in Chapter 9). These essential tasks are performed by neurons distributed across several brain structures and connected by ascending and descending pathways. Transmission of mechanosensory information from the periphery to the brain begins with a variety of receptor types that initiate action potentials. This activity is conveyed centrally via a chain of neurons, referred to as the first-, second-, and third-order cells. First-order neurons are located in the dorsal root and cranial nerve ganglia. Second-order neurons are located in brainstem nuclei. Third-order neurons are found in the thalamus, from whence they project to the cerebral cortex. These afferent pathways are topographically arranged throughout the system, the amount of cortical and subcortical space allocated to various body parts being proportional to the density of peripheral receptors. Studies in experimental animals show that specific cortical regions correspond to each functional submodality; area 3b, for example, processes information from low-threshold cutaneous receptors, and area 3a from proprioceptors. Thus, at least two broad criteria operate in the organization of the somatic sensory system: modality and somatotopy. The end result of this complex interaction is a unified perceptual representation of the body and its ongoing interaction with the outside world.

Additional Reading

Reviews

CHAPIN, J. K. (1987) Modulation of cutaneous sensory transmission during movement: possible mechanisms and biological significance. In *Higher Brain Function: Recent Explorations of the Brain's Emergent Properties.* S. P. Wise (ed.). New York: John Wiley & Sons, pp. 181–209.

DARIAN-SMITH, I. (1984) The sense of touch: performance and peripheral neural processes. In *Handbook of Physiology: The Nervous System,* Volume III. J. M. Brookhart and V. B. Mountcastle (eds.). Bethesda, MD: American Physiological Society, pp. 739–788.

JOHANSSON, R. S. AND A. B. VALLBO (1983) Tactile sensory coding in the glabrous skin of the human. Trends Neurosci. 6: 27–32.

KAAS, J. H. (1990) Somatosensory system. In *The Human Nervous System.* G Paxinos (ed.). San Diego, CA: Academic Press, pp. 813–844.

KAAS, J. H. (1993) The functional organization of somatosensory cortex in primates. Ann. Anat. 175: 509–518.

MOUNTCASTLE, V. B. (1975) The view from within: Pathways to the study of perception.
Johns Hopkins Med. J. 136: 109–131.

MOUNTCASTLE, V. B. (1984) Central nervous mechanisms in mechanoreceptive sensibility. In *Handbook of Physiology,* Section 1: *The Nervous System,* Vol. 3: *Sensory Processes.* L. M. Brookhart and S. R. Geiger (eds.). American Physiological Society and Williams & Wilkens, pp. 789–878.

WOOLSEY, C. (1958) Organization of somatic sensory and motor areas of the cerebral cortex. In *Biological and Biochemical Bases of Behavior.* H. F. Harlow and C. N. Woolsey (eds.). Madison, WI: University of Wisconsin Press, pp. 63–82.

Important Original Papers

ADRIAN, E. D. AND Y. ZOTTERMAN (1926) The impulses produced by sensory nerve endings. Part II. The response of a single end organ. J. Physiol. 61: 151–171.

JOHANSSON, R. S. (1978) Tactile sensibility of the human hand: receptive field characteristics of mechanoreceptive units in the glabrous skin. J. Physiol. (Lond.) 281: 101–123.

JOHNSON, K. O. AND G. D. LAMB (1981) Neural mechanisms of spatial tactile discrimination: Neural patterns evoked by braille-like dot patterns in the monkey. J. Physiol. (London) 310: 117–144.

JONES, E. G. AND D. P. FRIEDMAN (1982) Projection pattern of functional components of thalamic ventrobasal complex on monkey somatosensory cortex. J. Neurophysiol. 48: 521–544.

JONES, E. G. AND T. P. S. POWELL (1969) Connexions of the somatic sensory cortex of the rhesus monkey. I. Ipsilateral connexions. Brain 92: 477–502.

LaMOTTE, R. H. AND M. A. SRINIVASAN (1987) Tactile discrimination of shape: Responses of rapidly adapting mechanoreceptive afferents to a step stroked across the monkey fingerpad. J. Neurosci. 7: 1672–1681.

SUR, M. (1980) Receptive fields of neurons in areas 3b and 1 of somatosensory cortex in monkeys. Brain Res. 198: 465–471.

WALL, P. D. AND W. NOORDENHOS (1977) Sensory functions which remain in man after complete transection of dorsal columns. Brain 100: 641–653.

CHAPTER 9

■ OVERVIEW

A natural assumption is that the sensation of pain arises from excessive stimulation of the same receptors that generate other somatic sensations (discussed in Chapter 8). This is not the case. Although similar in some ways to the sensory processing of routine mechanical stimulation, nociception (*noci* is derived from the Latin for "hurt") depends on specifically dedicated receptors and pathways. Since alerting the brain to the dangers implied by noxious stimuli differs substantially from informing it about more ordinary somatic sensory stimuli, it makes good sense that a special subsystem be devoted to the perception of potentially threatening circumstances. The importance of pain in clinical practice, as well as the many aspects of pain sensation that remain imperfectly understood, continue to make nociception an extremely active area of research.

■ NOCICEPTORS

The relatively unspecialized nerve cell endings that initiate the sensation of pain are called **nociceptors**. Like other cutaneous and subcutaneous receptors, they transduce a variety of stimuli into receptor potentials and action potentials. Moreover, nociceptors, like other somatic sensory receptors, arise from cell bodies in dorsal root ganglia (or in the trigeminal ganglion) that send one axonal process to the periphery and the other into the spinal cord or brainstem.

Because peripheral nociceptive axons terminate in unspecialized "free endings," it is conventional to categorize nociceptors according to the properties of the axons associated with them. As described in the previous chapter, the somatic sensory receptors responsible for the perception of innocuous mechanical stimuli are associated with myelinated axons that have relatively rapid conduction velocities (see Table 8.1). The axons associated with nociceptors, in contrast, conduct relatively slowly, being only lightly myelinated or, more commonly, unmyelinated. Accordingly, axons conveying information about pain fall into either the Aδ group of myelinated axons, which conduct at about 20 m/s, or into the C fiber group of unmyelinated axons, which conduct at velocities generally less than 2 m/s. Thus, even though the conduction of all nociceptive information is relatively slow, there are fast and slow pain pathways.

In general, the faster-conducting Aδ nociceptors respond either to dangerously intense mechanical or to mechanothermal stimuli and have receptive fields that consist of clusters of sensitive spots. Other unmyelinated nociceptors tend to respond to thermal, mechanical, and chemical stimuli, and are therefore said to be polymodal. In short, there are three major classes of nociceptors in the skin: **Aδ mechanosensitive nociceptors**, **Aδ mechanothermal nociceptors**, and **polymodal nociceptors** associated with C fibers. The receptive fields of all pain-sensitive neurons are relatively large, particularly at the level of the thalamus and cortex, presumably because the detection of pain is more important than its precise localization.

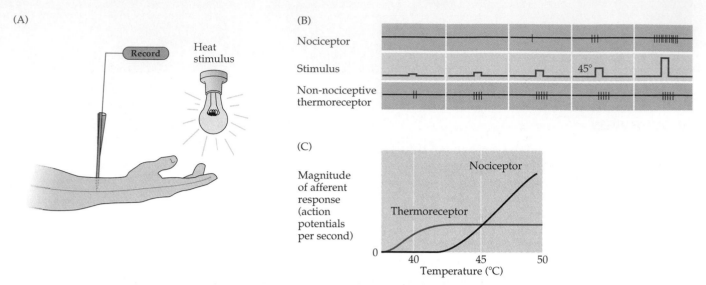

Figure 9.1 Experimental demonstration that nociception involves specialized neurons, not simply excessive discharge of the neurons that respond to normal stimulus intensities. (A) Arrangement for transcutaneous nerve fiber recording. (B) In the painful stimulus range, thermoreceptors fire action potentials at the same rate as at lower temperatures; the number and frequency of action potentials in the nociceptive fiber, however, continue to increase. (Note that 45°C is the approximate threshold for pain in this situation.) (C) Summary of results. (After Fields, 1987.)

Studies carried out in both humans and experimental animals demonstrated some time ago that the rapidly conducting axons that subserve somatic sensory sensation are not involved in the transmission of pain. An experiment of this sort is illustrated in Figure 9.1. The peripheral axons responsive to nonpainful mechanical or thermal stimuli do not discharge at a greater rate when painful stimuli are delivered to the same spot on the skin surface. The nociceptive axons, on the other hand, begin to discharge only when the stimulus (a thermal one in the example in Figure 9.1) reaches high levels; at this same stimulus intensity, other thermoreceptors discharge at a rate no different from the maximum rate already achieved within the non-painful temperature range, indicating that there are both nociceptive and nonnociceptive thermoreceptors. Equally important, direct stimulation of the large-diameter somatic sensory afferents at any frequency in humans does not produce sensations that are described as painful. In contrast, the smaller-diameter, more slowly conducting Aδ and C fibers are active when painful stimuli are delivered; and when stimulated electrically in human subjects, they produce pain.

■ THE PERCEPTION OF PAIN

How do the different classes of nociceptors lead to the perception of pain? As mentioned, one way of determining the answer has been to stimulate different nociceptors in human volunteers while noting the sensations reported. In general, two categories of pain perception have been described: a sharp **first pain** and a more delayed (and longer-lasting) sensation that is generally called **second pain** (Figure 9.2A). Stimulation of the large, rapidly conducting Aα and Aβ axons in peripheral nerves does not elicit the sensation of pain. When the stimulus intensity is raised to a level that activates a subset

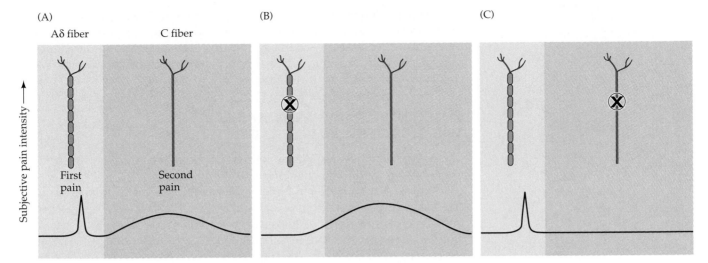

Figure 9.2 Pain can be separated into an early perception of sharp pain and a later perception of a duller, burning quality. (A) First and second pain are carried by different axons, as can be shown by (B) the selective blockade of the more rapidly conducting myelinated axons that carry the sensation of first pain, or (C) blockade of the more slowly conducting C fibers that carry the sensation of second pain. (After Fields, 1990.)

of Aδ fibers, however, a tingling sensation or, if the stimulation is intense enough, a feeling of sharp pain is reported. If the stimulus intensity is increased still further, so that the small-diameter, slowly conducting C fiber axons are brought into play, then a duller, longer-lasting sensation of pain is experienced. It is also possible to selectively anesthetize C fibers and Aδ fibers; in general, these selective blocking experiments confirm that the Aδ fibers are responsible for first pain, and that C fibers are responsible for a duller, longer-lasting pain (Figure 9.2B,C).

■ HYPERALGESIA AND SENSITIZATION

Painful stimuli occurring outside the laboratory are usually associated with tissue damage (such as cuts, scrapes, and bruises). Most people have experienced the phenomenon of **hyperalgesia**, defined as enhanced sensitivity and responsivity to stimulation of the area around damaged tissue. In the region surrounding an injury, stimuli that would not normally cause pain are perceived as painful, and stimuli that would ordinarily be painful are significantly more so (therefore, *hyper*algesia). The cause of this phenomenon is the **sensitization** of nociceptors by various substances released when tissue is damaged (Table 9.1). Evidently, the release of bradykinin, histamine, prostaglandins, and other agents from the site of injury enhances the responsiveness of nociceptive endings. Electrical activity in the nociceptors themselves also stimulates the local release of chemical substances (such as substance P) that cause vasodilatation, swelling, and the release of histamine from mast cells. Injury and pain are thus intertwined in a complex cascade of local signals. The involvement of these substances in the production of pain has also provided clues about how some analgesics may work, suggesting strategies for relief. Aspirin, for example, probably acts by inhibiting cyclooxygenase, an enzyme important in the biosynthesis of prostaglandins.

**TABLE 9.1
Substances Released Following
Tissue Damage**

Substance	Source
Potassium	Damaged cells
Serotonin	Platelets
Bradykinin	Plasma
Histamine	Mast cells
Prostaglandins	Damaged cells
Leukotrienes	Damaged cells
Substance P	Primary afferent fibers

Source: Modified from Fields, 1987.

The presumed purpose of the complex chemical signaling arising from local damage is not only to protect the injured area (as a result of the painful perceptions produced by ordinary stimuli close to the site of damage), but also to promote healing and protect against infection by means of local effects such as increased blood flow.

■ CENTRAL PAIN PATHWAYS: THE SPINOTHALAMIC TRACT

The pathways that carry information about noxious stimuli to the brain, as might be expected for such an important and multifaceted system, are complex. The approach here will be to emphasize the major pathways (summarized in Figure 9.3), omitting some of the subsidiary routes. Because projections from non-nociceptive temperature-sensitive neurons follow the same anatomical route, they are included here, even though they are not part of the pain system.

As in the case of the other sensory neurons located in dorsal root ganglia, the central axons of nociceptive nerve cells enter the spinal cord via the dorsal roots (Figure 9.3A). Axons carrying information from pain and temperature receptors are generally found in the most lateral division of the dorsal roots, but the cell bodies are not discretely localized within the ganglia. When these axons reach the dorsal horn, they branch into ascending and descending collaterals, forming what is called the **dorsolateral tract of Lissauer**. Axons in Lissauer's tract run up and down for one or two spinal cord segments before they penetrate the gray matter of the dorsal horn. Once within the dorsal horn, the axons give off branches that contact neurons located in several of Rexed's laminae (remember that these laminae are the descriptive divisions of the spinal gray matter in cross section; see Figure 15.6). Both Aδ and C fibers send branches to innervate neurons in lamina I (called the marginal zone) and lamina II (called the substantia gelatinosa).

Information from lamina II is transmitted to second-order projection neurons in laminae IV, V, and VI. These cells also receive direct innervation from branches of the first order neurons. The axons of the second-order neurons in laminae IV–VI (collectively known as the **nucleus proprius**) cross the midline and ascend all the way to the brainstem and thalamus in the anterolateral (ventrolateral) quadrant of the contralateral half of the spinal cord. These fibers, together with axons from second-order lamina I neurons, form the **spinothalamic tract**, the major ascending pathway for information concerning pain and temperature. This overall pathway is also referred to as the **anterolateral system**, much as the mechanosensory pathway is referred to as the dorsal column–medial lemniscus system.

The location of the spinothalamic tract is particularly important clinically because of the characteristic sensory deficits that follow cord injury. Since the mechanosensory pathway ascends ipsilaterally in the spinal cord, a unilateral spinal lesion will produce sensory loss of touch, pressure, vibration, and proprioception below the lesion on the same side. The pathways for pain and temperature, however, cross the midline to ascend on the opposite side of the cord. Therefore, diminished sensation of pain below the lesion will be observed on the side opposite the mechanosensory loss (and the lesion). This pattern is referred to as a **dissociated sensory loss**, and helps define the level of the lesion (Figure 9.4).

As is the case of the mechanosensory pathway, noxious and thermal sensations from the face follow a separate route to the thalamus (see Figure 9.3B). First-order axons originating from the trigeminal ganglion cells carry

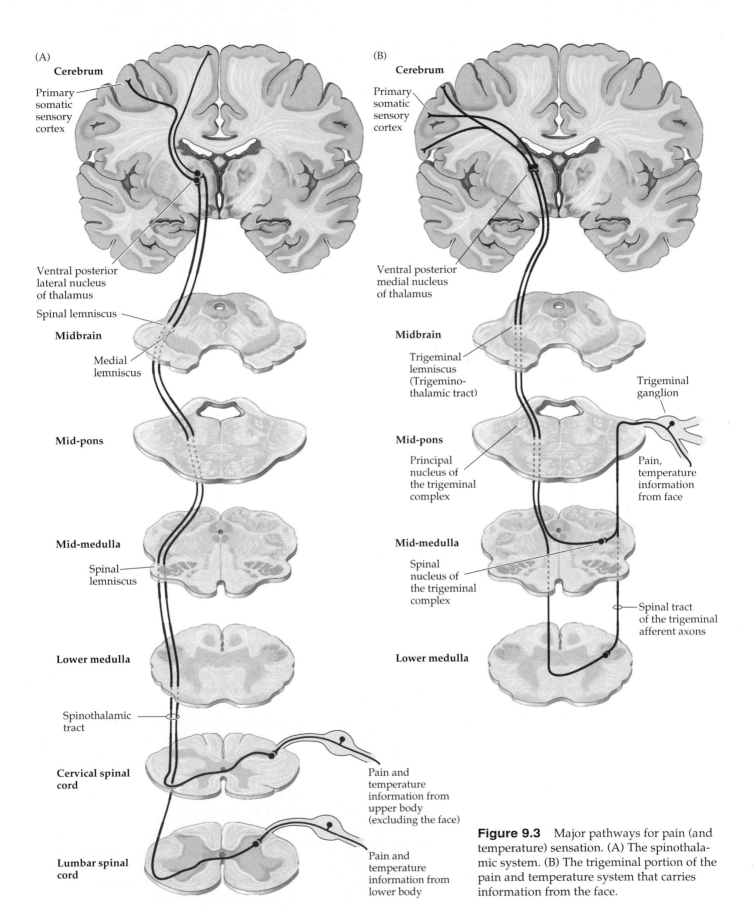

(A) Cerebrum

Primary somatic sensory cortex

Ventral posterior lateral nucleus of thalamus

Spinal lemniscus

Midbrain

Medial lemniscus

Mid-pons

Mid-medulla

Spinal lemniscus

Lower medulla

Spinothalamic tract

Cervical spinal cord

Pain and temperature information from upper body (excluding the face)

Lumbar spinal cord

Pain and temperature information from lower body

(B) Cerebrum

Primary somatic sensory cortex

Ventral posterior medial nucleus of thalamus

Midbrain

Trigeminal lemniscus (Trigemino-thalamic tract)

Trigeminal ganglion

Mid-pons

Principal nucleus of the trigeminal complex

Pain, temperature information from face

Mid-medulla

Spinal nucleus of the trigeminal complex

Spinal tract of the trigeminal afferent axons

Lower medulla

Figure 9.3 Major pathways for pain (and temperature) sensation. (A) The spinothalamic system. (B) The trigeminal portion of the pain and temperature system that carries information from the face.

Figure 9.4 Pattern of "dissociated" sensory loss following a spinal cord hemisection at tenth thoracic level on the left side. This pattern, together with motor weakness on the same side as the lesion, is sometimes referred to as the Brown-Séquard's syndrome.

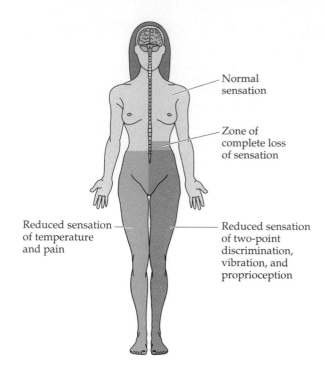

Normal sensation

Zone of complete loss of sensation

Reduced sensation of temperature and pain

Reduced sensation of two-point discrimination, vibration, and proprioception

information from facial nociceptors and thermoreceptors to the central nervous system. After entering the pons, these small myelinated and unmyelinated trigeminal fibers *descend* to the medulla, forming the **spinal trigeminal tract** (or spinal tract of V), and terminate in two subdivisions of the spinal trigeminal complex: the pars interpolaris and pars caudalis. Axons from the second order neurons in these two trigeminal nuclei, like their counterparts in the spinal cord, cross the midline and ascend to the contralateral thalamus in the **trigeminal lemniscus** (also called the trigemino-thalamic tract).

The complexity of the pain pathways (recall that several minor routes are omitted in this account) often makes the origin of a pain difficult to assess (Box A). For the same reason, chronic pain is very difficult to treat. Such pain can arise from inflammation (as in neuritis), injury to nerve endings and scar formation (as in the pain that can follow surgical amputation; Box B), invasive disease such as cancer, and a variety of other causes. Injuries to the central nervous system structures that process nociceptive information can also lead to excruciating pain. The common denominator of conditions that lead to chronic pain is irritation of nociceptive endings, axons, or processing circuits causing abnormal activity in the pain system. Surgical interruption of a particular tract to abolish chronic pain is not usually effective; the pain, although initially alleviated, tends to return. Indeed, there is often no successful treatment for these unfortunate patients.

■ THE NOCICEPTIVE COMPONENTS OF THE THALAMUS AND CORTEX

In the thalamus, the major target nuclei of the ascending pain and temperature axons are, like the targets for mechanosensory axons, in the ventral posterior (VP) complex. The ventral posterior medial (VPM) and ventral posterior lateral (VPL) nuclei receive the bulk of these axons. Neurons in the VPM nucleus receive nociceptive information from the face; neurons in the VPL

Box A
REFERRED PAIN

Surprisingly, there are few, if any, spinal neurons specialized solely for the appreciation of visceral pain. Obviously we recognize such pain, but it is sensed via dorsal horn neurons that are also concerned with cutaneous sensation. The result is that a disorder of an internal organ is sometimes perceived as pain in a cutaneous receptive field. A patient may therefore present to the physician with the complaint of pain at a site other than its actual source, a phenomenon called referred pain. The most common clinical example is anginal pain (pain arising from heart muscle that is not being perfused with sufficient blood) referred to the upper chest wall, with radiation into the left arm and hand. Other important examples are gallbladder pain referred to the scapular region, esophogeal pain referred to the chest wall, ureteral pain (e.g., from passing a kidney stone) referred to the lower abdominal wall, bladder pain referred to the perineum, and the pain from an inflamed appendix referred to the anterior abdominal wall around the umbilicus. Understanding these phenomena can lead to an astute diagnosis when others are stumped.

References

CAPPS, J. A. AND G. H. COLEMAN (1932) An Experimental and Clinical Study of Pain in the Pleura, Pericardium, and Peritoneum. New York: Macmillan.

HEAD, H. (1893) On disturbances of sensation with special reference to the pain of visceral disease. Brain 16: 1–32.

KELLGREW, J. H. (1939–1942) On the distribution of pain arising from deep somatic structures with charts of segmental pain areas. Clin. Sci. 4: 35–46.

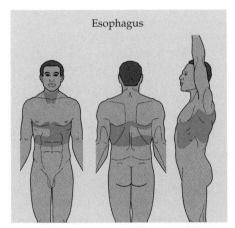
Esophagus

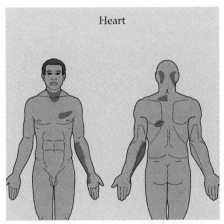
Heart

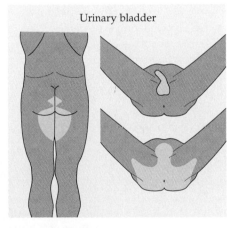
Urinary bladder

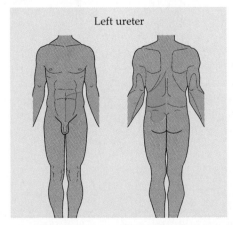
Left ureter

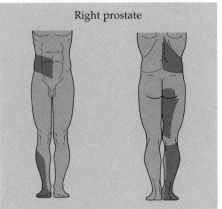
Right prostate

Examples of pain arising from a visceral disorder that is referred to a cutaneous region (color).

Box B
PHANTOM LIMBS AND PHANTOM PAIN

Following the amputation of an extremity, nearly all patients experience an illusion that the missing limb is still present. Although this sensation usually diminishes over time, it persists in some degree throughout the amputee's life and can often be reactivated by injury to the stump or other perturbations. Such phantom sensations are not limited to amputated limbs; phantom breasts following mastectomy, phantom genitalia following castration, and phantoms of the entire lower body following spinal cord transection have all been reported. Phantoms are also common after local nerve block for surgery. During recovery from brachial plexus anesthesia, for example, it is common for the patient to experience a phantom arm, perceived as whole and intact, but displaced from the real arm. When the real arm is viewed, the phantom appears to jump into the arm and may emerge and reenter intermittently while the anesthesia wears off. These sensory phantoms demonstrate that the central machinery for processing somatic sensory information is not idle in the absence of peripheral stimuli; apparently, central sensory maps and processing systems continue to operate independently of the periphery.

Phantoms might simply be a curiosity—or a provocative clue about higher-order somatic sensory processing—were it not for the fact that a substantial number of amputees also develop phantom pain. This common problem is usually described as a tingling or burning sensation in the missing part. Sometimes, however, the sensation becomes a much more serious pain that patients find increasingly debilitating. Phantom pain is, in fact, one of the more common causes of chronic pain syndromes, a condition that is extraordinarily difficult to treat. Because of the relative independence of the central processing of pain, ablation of the spinothalamic tract, portions of the thalamus, or even primary sensory cortex does not generally relieve the discomfort felt by these patients.

References

MELZACK, R. (1989) Phantom limbs, the self and the brain. The D.O. Hebb Memorial Lecture. Canad. Psychol. 30: 1–14.

MELZACK, R. (1990) Phantom limbs and the concept of a neuromatrix. Trends Neurosci. 13: 88–92.

NASHOLD, B. S., JR. (1991) Paraplegia and pain. In *Deafferentation Pain Syndromes: Pathophysiology and Treatment*. B. S. Nashold, Jr. and J. Ovelmen-Levitt (eds.). New York: Raven Press, pp. 301–319.

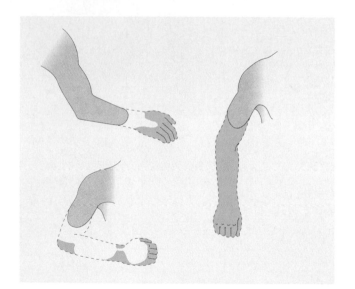

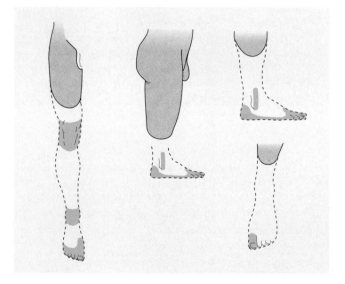

Drawings of phantom arms and legs based on patients' reports. The phantom is indicated by a dashed line, with the colored regions showing the most vividly experienced parts. Note that some of the phantoms are telescoped into the stump. (After Solonen, 1962.)

nucleus receive nociceptive information from the rest of the body. The similar arrangement for mechanosensory and noxious stimuli is evidently responsible for discriminative aspects of pain (the ability to locate a pain and judge its intensity). A parallel projection to the reticular formation of the medulla, pons, and midbrain is probably responsible for the general arousal that pain causes, and for the autonomic activation that follows a noxious stimulus (the classic fight-or-flight reaction) (see Figure 9.5). Other thalamic nuclei, such as the central lateral nucleus and the intralaminar complex, receive projections from the reticular formation and also participate in the arousal response evoked by a noxious stimulus. Despite the location of nociceptive neurons in the same general regions of the thalamus as the mechanosensory neurons, at a more detailed level they nonetheless appear to be separate systems.

The cortical representation of pain is the least documented aspect of the central pathways for nociception. Although the thalamic neurons that relay noxious sensations via the VP complex project to the primary somatic sensory cortex, ablations of the relevant regions of the parietal cortex do not generally alleviate chronic pain (although they impair contralateral mechanosensory perception, as expected). Perhaps this is because widespread cortical activation, mediated by projections from the central lateral nucleus and the intralaminar complex is observed following a potent noxious stimulus. Whatever the explanation, the cortical processing of pain remains something of a mystery.

■ CENTRAL REGULATION OF PAIN PERCEPTION

Observers have long commented on the difference between the objective reality of a painful stimulus and the subjective response to it. Modern studies of this discrepancy have provided considerable insight into how the emotions affect pain perception and, ultimately, into the anatomy and pharmacology of the pain system.

During World War II, Henry Beecher and his colleagues at Harvard Medical School made a fundamental observation. In the first systematic study of its kind, they found that soldiers suffering from severe battle wounds often suffered little or no pain. Indeed, many of the wounded expressed surprise at this odd dissociation. Beecher, an anesthesiologist, concluded that the perception of pain depends very much on its context. The pain of an injured soldier on the battlefield would presumably be mitigated by the imagined benefits of being removed from danger, whereas a similar injury in a domestic setting might raise quite a different set of thoughts that could exacerbate the pain (loss of work, financial liability, and so on). Such observations, together with the well-known placebo effect (discussed in the next section), made clear that, much more so than other sensations, the perception of pain is subject to central modulation. This statement should not be taken as a vague acknowledgment about the importance of psychological factors. On the contrary, there has been a gradual realization among neuroscientists and neurologists that such "psychological" effects are as real and important as any other neural phenomenon. This appreciation has provided a much more rational view of psychosomatic problems in general.

■ THE PLACEBO EFFECT

The placebo effect is defined as a physiological response following the administration of a pharmacologically inert remedy. The word *placebo* means "I will please," and the effect has a long history of use (and abuse) in medicine. The reality of the placebo effect is undisputed. In one classic study, medical students were given one of two different pills, one said to be a sedative and the other a stimulant. In fact, both pills contained only an inert substance. Of the students who received the "sedative," more than two-thirds reported that they felt drowsy. Moreover, the students who took two such pills felt sleepier than those who had taken only one. Conversely, a large fraction of the students who took the "stimulant" reported that they felt less tired. Moreover, about a third of the entire group reported side effects ranging from headaches and dizziness to tingling extremities and a staggering gait! Only 3 of the 56 students studied reported that the pills they took had no effect.

Typically, 3 out of 4 patients suffering from postoperative wound pain report satisfactory pain relief after an injection of sterile saline. The researchers who carried out this study noted that the responders were indistinguishable from the nonresponders, both in the apparent severity of their pain and in the nature of their characters. Most tellingly, the placebo effect in postoperative patients can be blocked by naloxone (a competitive antagonist of opiate receptors; see the next section). A common misunderstanding about the placebo effect is the view that patients who respond to a therapeutically meaningless reagent are not suffering real pain, but only "imagining" it.

Among other things, the placebo effect probably explains such phenomena as acupuncture anesthesia and analgesia achieved by hypnosis. In China, surgery has often been carried out under the effect of a needle being twirled in an earlobe or some other part of the anatomy dictated by ancient acupuncture charts. Indeed, before the era of modern anesthesia, surgery without apparent pain was often observed. In Europe before the twentieth century, operations such as thyroidectomies for goiter were commonly done without anesthesia and without great discomfort, particularly among populations where stoicism was the cultural norm. In modern American medicine, hypnosis is often used successfully to ameliorate chronic pain. The mechanisms of pain amelioration on the battlefield, in the use of acupuncture, and in hypnosis are presumably related.

In short, the placebo effect is quite real. Although the mechanisms by which the brain affects the perception of pain are only beginning to be understood, the effect is neither magical nor a sign of a suggestible intellect.

■ THE PHYSIOLOGICAL BASIS OF PAIN MODULATION

In fact, the placebo effect and related phenomena have a very concrete physiological basis. Understanding the central modulation of pain perception was greatly advanced by the finding that electrical or pharmacological stimulation of certain regions of the midbrain produces relief of pain (Figure 9.5). This analgesic effect arises from activation of descending pain-modulating pathways that project, via the medulla, to neurons in the dorsal horn—particularly in Rexed's lamina II—that control the ascending information in the nociceptive system. The major brainstem regions that produce this effect are located in the **periaqueductal gray matter** and the **rostral ventral medulla**. Electrical stimulation at each of these sites in experimental animals not only produces analgesia by behavioral criteria, but demonstrably inhibits the activity of nociceptive projection neurons in the dorsal horn of the spinal cord.

An everyday example of the modulation of painful stimuli is the ability to reduce the sensation of sharp pain by activating low-threshold mechanoreceptors: if you crack your shin or stub a toe, a natural (and effective) reaction is to vigorously rub the site of injury for a minute or two. Such observa-

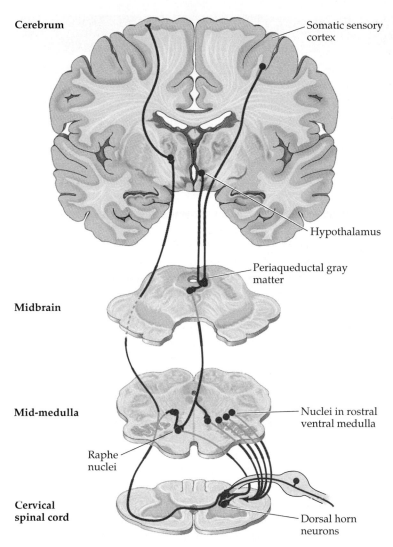

Cerebrum

Somatic sensory cortex

Hypothalamus

Periaqueductal gray matter

Midbrain

Mid-medulla

Nuclei in rostral ventral medulla

Raphe nuclei

Cervical spinal cord

Dorsal horn neurons

Figure 9.5 The descending systems that modulate the transmission of ascending pain signals. These modulatory systems originate in the somatic sensory cortex, the hypothalamus, the periaqueductal gray matter of the midbrain, the raphe nuclei, and other nuclei of the rostral ventral medulla. Complex modulatory effects occur at each of these sites, as well as in the dorsal horn.

tions led Ronald Melzack and Patrick Wall to propose that the flow of nociceptive information through the spinal cord is modulated by concomitant activation of the large myelinated fibers associated with low-threshold mechanoreceptors. Even though further investigation led to the modification of some of the original assumptions of Melzack and Wall's **gate theory of pain**, the proposition triggered a great deal of work on pain modulation.

Perhaps the most exciting advance in this effort was the discovery of endogenous opioids. For centuries it had been apparent that opium derivatives such as morphine are powerful analgesics (indeed, they remain a mainstay of analgesic therapy today). Modern animal studies have shown that a variety of brain regions are susceptible to the action of opiate drugs, particularly—and significantly—the periaqueductal gray matter and the rostral ventral medulla. There are, in addition, opiate-sensitive regions at the level of the spinal cord. In other words, the areas that produce analgesia when stimulated are also responsive to exogenous opiates. It seems likely, then, that opiate drugs act at all of the sites shown in Figure 9.5 in producing their dramatic pain-relieving effects.

The analgesic action of opiates implied the existence of specific brain receptors for these drugs long before the receptors were actually found dur-

TABLE 9.2 Endogenous Opioids[a]	
Name	*Amino acid sequence*[b]
Leucine-enkephalin	*Tyr-Gly-Gly-Phe*-Leu-OH
Methionine-enkephalin	*Tyr-Gly-Gly-Phe*-Met-OH
β-Endorphin	*Tyr-Gly-Gly-Phe*-Met-Thr-Ser-Glu-Lys-Ser-Gln-Thr-Pro-Leu-Val-Thr-Leu-Phe-Lys-Asn-Ala-Ile-Val-Lys-Asn-Ala-His-Lys-Gly-Gln-OH
α-Neoendorphin	*Tyr-Gly-Gly-Phe*-Leu-Arg-Lys-Tyr-Pro-Lys
Dynorphin	*Tyr-Gly-Gly-Phe*-Leu-Arg-Arg-Ile-Arg-Pro-Lys-Leu-Lys-Trp-Asp-Asn-Gln-OH

[a]The role of these agents as neurotransmitters is discussed in Chapter 6.
[b]Note the initial homology, indicated by italics.

ing the 1960s and 1970s. Since such receptors are unlikely to exist for the purpose of responding to the administration of opium and its derivatives, the conviction grew that there must be *endogenous* compounds for which these receptors had evolved. Several categories of endogenous opioids have now been isolated from the brain and intensively studied (Table 9.2). These agents are found in the same regions that are involved in the modulation of nociceptive afferents, although each of the families of endogenous opioid peptides has a somewhat different distribution. All three of the major groups (**enkephalins**, **endorphins**, and **dynorphins**; see also Chapter 7) are present in the periaqueductal gray matter. The enkephalins and dynorphins have also been found in the rostral ventral medulla and in the spinal cord regions involved in the modulation of pain.

An impressive feature of this story is the wedding of physiology, pharmacology, and clinical research to yield a much richer understanding of the intrinsic modulation of pain; this information has finally begun to explain the subjective variability of painful stimuli and the striking dependence of pain perception on the context of the experience. Precisely how pain is modulated is being explored in many laboratories at present, motivated by the tremendous clinical (and economic) benefits that would accrue from still deeper knowledge of this phenomenon.

■ SUMMARY

Whether from a structural or functional perspective, pain is an extraordinarily complex sensory modality. Because of the importance of warning an animal about dangerous stimuli, the mechanisms and pathways that subserve nociception are widespread and redundant. The major nociceptive pathway, like other somatic sensory modalities, comprises a three-neuron relay from periphery to cortex. The arrangement differs from the mechanosensory pathway primarily in that the central axons of dorsal root ganglion cells synapse on second-order neurons in the spinal cord, which then cross the midline and project to brainstem and thalamic nuclei; the thalamic neurons, in turn, project to the same cortical areas as other somatic sensory modalities. The molecular basis of pain modulation is particularly intricate and is only beginning to be deciphered. The major features are the modulation of pain peripherally by the release of a variety of agents at the injury site, and the central modulation of afferent pain pathways by endogenous opioids at the level of both the spinal cord and the brainstem. Tremendous progress in understanding pain has been made in the last 20 years, and more seems likely, given the importance of the problem. No patients are more distressed or more difficult to treat than those with chronic pain. Indeed, some aspects of pain seem much more destructive to the sufferer than required by any physiological purposes; consider, for example, the pain of a chronic illness, such as invasive cancer. Perhaps such effects are a necessary but unfortunate by-product of the protective benefits of this all-important sensory modality.

Additional Reading

Reviews

FIELDS, H. L. AND A. I. BASBAUM (1978) Brain stem control of spinal pain transmission neurons. Ann. Rev. Physiol. 40: 217–248.

NASHOLD, B. S., JR. AND J. OVELMEN-LEVITT (1993) Chronic pain. In Neuroscience Year, B. Smith and G. Edelman (eds.). Boston: Birkhäuser, pp. 35–39.

Important Original Papers

BASBAUM, A. I. AND H. L. FIELDS (1979) The origin of descending pathways in the dorsolateral funiculus of the spinal cord of the cat and rat: further studies on the anatomy of pain modulation. J. Comp. Neurol. 187: 513–522.

BEECHER, H. K. (1946) Pain in men wounded in battle. Ann. Surg. 123: 96.

BLACKWELL, B., S. S. BLOOMFIELD AND C. R. BUNCHER (1972) Demonstration to medical students of placebo response and non-drug factors. Lancet 1: 1279–1282.

CRAIG, A. D., M. C. BUSHNELL, E.-T. ZHANG AND A. BLOMQVIST (1994) A thalamic nucleus specific for pain and temperature sensation. Nature 372: 770–773.

Books

FIELDS, H. L. (1987) Pain. New York: McGraw-Hill.

FIELDS, H. L. (ed.). (1990) Pain Syndromes in Neurology. London: Butterworths.

KOLB, L. C. (1954) The Painful Phantom. Springfield, IL: Charles C. Thomas.

SKRABANEK, P. AND J. MCCORMICK (1990) Follies and Fallacies in Medicine. New York: Prometheus Books.

WALL, P. D. AND R. MELZACK (1989) Textbook of Pain. New York: Churchill Livingstone.

CHAPTER 10

VISION: THE EYE

■ OVERVIEW

The human visual system is extraordinary in the quantity and quality of information it supplies about the world. A quick glance is sufficient to describe the location, size, shape, color, and texture of objects and, if the objects are moving, their direction and relative speed. Equally remarkable is the fact that much of this information can be discerned over a wide range of stimulus intensities, from the faint light of stars at night to bright sunlight. The next two chapters describe the molecular, cellular, and circuit mechanisms that allow us to see. The initial stages of the process are determined by the optics of the eye, the molecular mechanisms by which light is transduced into electrical signals in the retina, and the retinal circuitry that determines the information relayed from eye to brain.

■ THE FORMATION OF IMAGES ON THE RETINA

The formation of focused images on the photoreceptors of the retina depends in large part on the refraction (bending) of light by the **cornea** and the **lens** (Figure 10.1). The cornea is responsible for most of the necessary refraction, a contribution that is easily appreciated by considering the hazy out-of-focus images experienced when swimming underwater. Water, unlike air, has a refractive index close to that of the cornea; as a result, immersion in water virtually eliminates the refraction that normally occurs at the air/cornea interface. The lens has considerably less refractive power than the cornea; however, the refraction supplied by the lens is adjustable, allowing objects that lie at various distances from the observer to be brought into sharp focus on the retinal surface.

These dynamic changes in the refractive power of the lens are referred to as **accommodation**. When viewing distant objects, the lens is made relatively thin and flat and has the least refractive power. For near vision, the lens becomes thicker and rounder, and has the most refractive power (Figure 10.2). The lens is held in place by radially arranged connective tissue bands (called zonule fibers) that are attached to the **ciliary muscle** that runs circumferentially near the inner surface of the eye. The shape of the lens is determined by two opposing forces: the elasticity of the lens, which tends to keep it rounded up (removed from the eye, the lens becomes spheroidal), and the force exerted by zonule fibers, which tends to flatten it. Under normal conditions, the force of the zonule fibers is greater than the elasticity of the lens and the lens assumes the flatter shape that allows focusing on distant objects. Focusing on closer objects requires relaxing the tension in the zonule fibers, allowing the inherent elasticity of the lens to increase its curvature. This relaxation is accomplished by the contraction of the ciliary muscle. Because the ciliary muscle forms a ring, the attachment points of the zonule fibers move toward the center of the eye when the muscle contracts, thus reducing the tension on the lens. Unfortunately, changes in the shape of the lens are not always able to produce a focused image on the retina, in which case a sharp image can be achieved only with the help of additional corrective lenses (Box A).

Figure 10.1   Anatomy of the human eye.

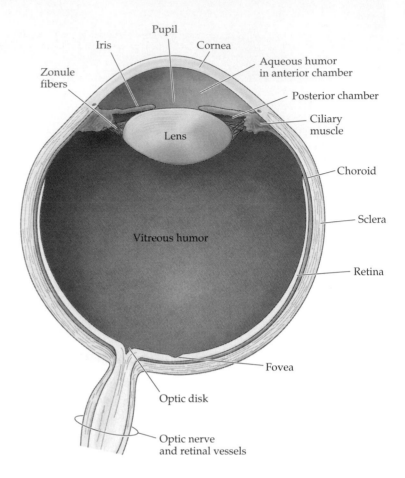

Adjustments in the size of the **pupil** (defined by the circular opening in the iris) also contribute to the clarity of images formed on the retina. Like the images formed by other optical instruments, those generated by the eye are not perfect; spherical and chromatic aberrations can cause blurring of the retinal image, a problem that is greatest for the light rays that pass farthest from the center of the lens. Narrowing the pupil therefore reduces both spherical and chromatic aberrations, just as closing the iris diaphragm on a camera lens improves the sharpness of a photographic image. Reducing the

Figure 10.2   Diagram showing the anterior part of the human eye in the unaccommodated (left) and accommodated (right) state. Accommodation for focusing on near objects involves the contraction of the ciliary muscle, which reduces the tension in the zonule fibers and allows the elasticity of the lens to increase its curvature.

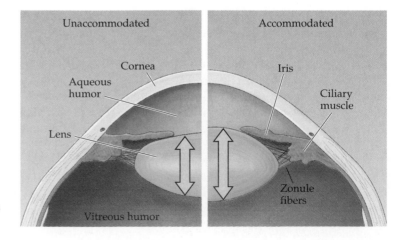

Box A
REFRACTIVE ERRORS

Optical discrepancies between the various components of the eye cause a majority of the human population to have some form of refractive error, or ametropia. People who are unable to bring distant objects into clear focus are said to be nearsighted, or myopic. Myopia can be caused by the front of the cornea being too curved or the eyeball being too long. In either case, with the lens as flat as it can be, the image of distant objects focuses in front of, rather than on, the retina. People who are unable to focus on near objects are said to be farsighted, or hyperopic. Hyperopia can be caused by the eyeball being too short or the refracting system being too weak. Even with the lens in its most rounded up state, the image is out of focus on the retinal surface (focusing at some point behind it). Both hyperopia and myopia are correctable by appropriate lenses, convex (plus) and concave (minus) respectively.

Even people with normal (emmetropic) vision as young adults eventually experience difficulty focusing on near objects. One of the consequences of aging is that the lens loses its elasticity; as a result, the maximum curvature the lens can achieve when the ciliary muscle contracts is reduced. The near point (the closest point that can be brought into clear focus) thus recedes, and it becomes necessary to hold objects (such as a book) farther and farther away in order to focus them on the retina. At some point, usually during early middle age, the accommodative ability of the eye is so reduced that near vision tasks like reading become difficult or impossible. This condition is referred to as presbyopia, and can be corrected by convex lenses for near-vision tasks, or by bifocal lenses if myopia is also present. Bifocal correction presents a particular problem for devotees of contact lenses. Because contact lenses float on the surface of the eye, having the distance correction above and the near correction below (as in conventional bifocal glasses) doesn't work. A solution to this problem for some contact lens wearers is to put a near correcting lens in one eye and a distance correcting lens in the other! The success of this approach is a testament to the remarkable ability of the visual system to adjust to a wide variety of unusual circumstances (see also Chapters 18 and 19).

References

COSTER, D. J. (1994) *Physics for Opthalmologists*. Edinburgh: Churchill Livingston.

HART JR., W. M. (ed.) (1992) *Adler's Physiology of the Eye: Clinical Application*, 9th Edition. St. Louis: Mosby Year Book.

(A) Emmetropia (normal)

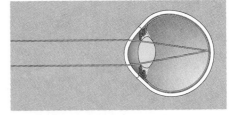

(B) Myopia (nearsighted)

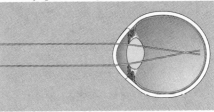

(C) Hyperopia (farsighted)

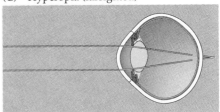

(D)

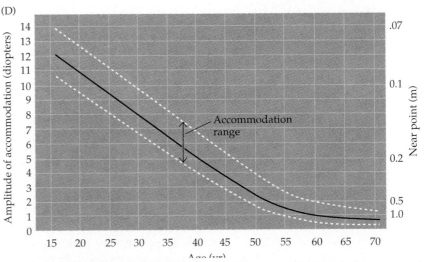

Refractive errors. (A) In the normal eye, with ciliary muscles relaxed, an image of a distant object is focused on the retina. (B) In myopia, light rays are focused in front of the retina. (C) In hyperopia images are focused at a point beyond the retina. (D) Changes in the ability of the lens to round up (accommodate) with age. The graph shows how the near point (the closest point to the eye that can be brought into focus) changes as a function of age. Accommodation, which is an optical measurement of the refractive power of the lens, is given in diopters.

size of the pupil also increases the depth of field—that is, the range of distances over which objects can be moved toward and away from the eye without appearing blurred. However, a small pupil also limits the amount of light that reaches the retina; under conditions of dim illumination, visual acuity becomes limited by the number of available photons rather than by optical aberrations. An adjustable pupil thus provides an effective compromise: it limits optical aberrations and maximizes depth of field as different levels of illumination permit.

The spaces in front of and behind the lens account for most of the eye's volume, and are filled with two different types of fluid. **Aqueous humor**, a clear, watery liquid, fills the space between the lens and the cornea (the anterior chamber) and supplies nutrients to both of these structures. Aqueous humor is produced in the posterior chamber (the region between the lens and the iris) and flows into the anterior chamber through the pupil. A specialized meshwork of cells that lies at the junction of the iris and the cornea is responsible for its uptake. Under normal conditions, the rates of aqueous humor production and uptake are in equilibrium, ensuring a constant intraocular pressure. Glaucoma, a disease in which intraocular pressure increases as a result of abnormal fluid regulation, reduces the vascular supply to the eye and eventually damages retinal neurons.

The space between the back of the lens and the surface of the retina is filled with a thick, gelatinous substance called **vitreous humor**. In addition to maintaining the shape of the eye, the vitreous humor contains cells with phagocytic activity that remove blood and other debris that might interfere with light transmission. The housekeeping powers of the vitreous are limited, however, as the large number of middle aged individuals with vitreal "floaters" will attest. Floaters are collections of debris too large for phagocytic consumption that remain to cast annoying shadows on the retina; they typically arise when the aging vitreous membrane pulls away from the overly long eyeball of myopics (Box B).

■ THE RETINA

Despite its peripheral location, the **retina** or neural portion of the eye, is actually part of the central nervous system. During development, the retina forms as an outpocketing of the diencephalon, called the optic vesicle, which then undergoes invagination to form the optic cup (Figure 10.3). The inner wall of the optic cup gives rise to the retina, while the outer wall gives rise to the **pigment epithelium**, a melanin-containing structure that reduces back scattering of light that enters the eye and plays an important role in the maintenance of photoreceptors.

Consistent with its central nervous system status, the retina comprises complex neural circuitry that converts the graded electrical activity of photoreceptors into action potentials that travel to the brain via the optic nerve. Because of its accessibility and relative simplicity, retinal circuitry provides a good opportunity to study synaptic organization in the brain. Although it has the same types of functional elements and neurotransmitters found in other parts of the central nervous system, there are only a few classes of neurons in the retina, and these are arranged in a manner that has been relatively easy to unravel.

There are five types of neurons within the retina: **photoreceptors, bipolar cells, ganglion cells, horizontal cells**, and **amacrine cells**. The cell bodies and processes of these retinal neurons are stacked in five alternating layers (Figure 10.4); the cell bodies are located in the inner nuclear, outer nuclear,

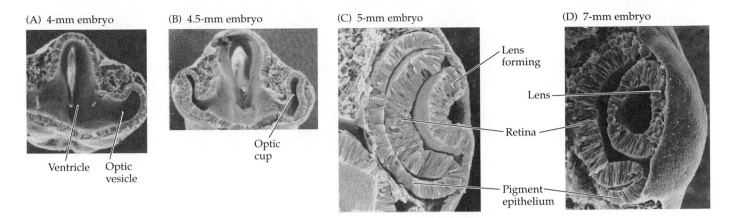

(A) 4-mm embryo (B) 4.5-mm embryo (C) 5-mm embryo (D) 7-mm embryo

Ventricle Optic vesicle

Optic cup

Lens forming

Lens

Retina

Pigment epithelium

Figure 10.3
Development of the human eye. (A) The retina develops as an outpocketing from the neural tube, called the optic vesicle. (B) The optic vesicle invaginates to form the optic cup. (C, D) The inner wall of the optic cup becomes the neural retina, while the outer wall becomes the pigment epithelium. (A–C from Hilfer and Yang, 1980; D, micrograph by K. Tosney, University of Michigan.)

and ganglion cell layers, while the processes and synaptic contacts are located in the inner plexiform and outer plexiform layers. The terms *inner* and *outer* designate relative distances from the center of the eye: inner, near the center of the eye; outer, away from the center, or toward the pigment epithelium. A three-neuron chain—photoreceptor cell, bipolar cell, ganglion cell—is the most direct route for information flow from photoreceptors to the optic nerve.

There are two types of photoreceptors, **rods** and **cones**, and these are the only elements of the retina that are sensitive to light. Both types of photoreceptors have an outer segment that contains photopigment and an inner segment that contains the cell nucleus and gives rise to the synaptic terminals that contact bipolar or horizontal cells. Absorption of light by the photopigment in the outer segment of the photoreceptors initiates a cascade of events that changes the membrane potential of the receptor and therefore the amount of neurotransmitter released by the photoreceptor synapses onto the cells they contact. The synapses between photoreceptor terminals and bipolar cell (and horizontal cell) processes occur in the outer plexiform layer; the cell bodies of photoreceptors make up the outer nuclear layer, whereas the cell bodies of bipolar cells lie in the inner nuclear layer. The axonal processes of bipolar cells make synaptic contacts in turn on the dendritic processes of ganglion cells in the inner plexiform layer. The axons of the ganglion cells form the **optic nerve**, which carries information about retinal stimulation to the rest of the central nervous system.

The two other types of neurons in the retina, horizontal cells and amacrine cells, have their cell bodies in the inner nuclear layer and are primarily responsible for lateral interactions within the retina. For example, lateral interactions between receptors, horizontal cells, and bipolar cells in the outer plexiform layer are largely responsible for the visual system's sensitivity to luminance contrast (see below). The processes of amacrine cells, which extend laterally in the inner plexiform layer, are postsynaptic to bipolar cell terminals and presynaptic to the dendrites of ganglion cells (see Figure 10.4). There are numerous subclasses of amacrine cells that can be distinguished by

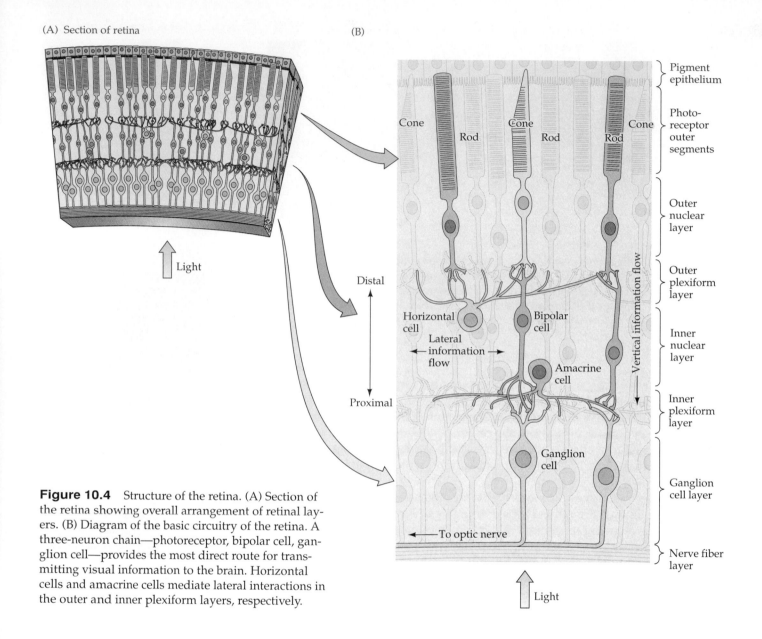

(A) Section of retina

(B)

Light

Distal

Proximal

Pigment epithelium

Photo-receptor outer segments

Cone

Rod

Cone

Rod

Cone

Outer nuclear layer

Outer plexiform layer

Horizontal cell

Lateral information flow

Bipolar cell

Amacrine cell

Vertical information flow

Inner nuclear layer

Inner plexiform layer

Ganglion cell

Ganglion cell layer

To optic nerve

Nerve fiber layer

Light

Figure 10.4 Structure of the retina. (A) Section of the retina showing overall arrangement of retinal layers. (B) Diagram of the basic circuitry of the retina. A three-neuron chain—photoreceptor, bipolar cell, ganglion cell—provides the most direct route for transmitting visual information to the brain. Horizontal cells and amacrine cells mediate lateral interactions in the outer and inner plexiform layers, respectively.

the particular neuropeptide transmitter they contain. Amacrine cells remain the least understood of the retinal neurons.

At first glance, the organization of the cellular layers in the retina seems counterintuitive, since light rays must pass through all of the neural circuitry of the retina (not to mention the retinal vasculature) before striking the outer segments of the photoreceptors (see Figure 10.4). This is certainly not the way an engineer would have designed things. However, this peculiar arrangement allows the tips of the outer segments of the photoreceptors to contact the pigment epithelium. The outer segments contain membranous disks that house the photopigment and other proteins involved in the transduction process. The disks are continuously formed near the inner segment and pushed toward the tip of the outer segment, where they are shed. The pigment epithelium plays an essential role in removing the expended receptor disks; this is no small task, since all the disks in the outer segments are replaced every 12 days! It is presumably the life cycle of the photoreceptor

Box B
MYOPIA AND THE GROWTH OF THE EYE

As noted in Box A, myopia, or nearsightedness—a condition in which light rays are brought to focus in front of the retina—is extraordinarily common (an estimated 50% of the population in the United States). Given the large number of people who need glasses or contact lenses to correct this refractive error, one naturally wonders how nearsighted people coped in the days (indeed, eons) before spectacles were invented. From what is now known about myopia, most people's vision may have been considerably better in ancient times. The basis for this assertion is the surprising finding that the growth of the eyeball is strongly influenced by focused light falling on the retina. This phenomenon was first described in 1977 by Torsten Wiesel and Elio Raviola, who studied monkeys reared with their lids sutured closed (the same approach used to demonstrate the effects of visual deprivation on cortical connections in the visual system; see Chapter 22). This procedure deprives the eye of focused retinal images. Animals growing to maturity under these conditions show a remarkable axial growth (elongation) of the eyeball. The effect of focused light deprivation appears to be a local one, since the abnormal growth of the eye occurs in experimental animals even if the optic nerve is cut. Indeed, if only a portion of the retinal surface is deprived of focused light, then only that region of the eyeball grows abnormally.

Although the mechanism of light-mediated control of eye growth is not fully understood, many workers in this field believe that some aspect of modern civilization—perhaps learning to read and write—interferes with the normal feedback control of vision on eye development, leading to abnormal elongation of the eyeball. A corollary of this hypothesis is that if children (or, more likely,

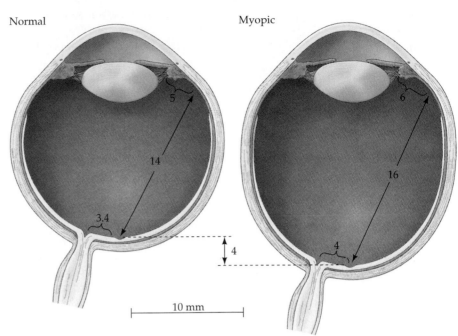

Normal   Myopic

Abnormal growth of the eye in rhesus monkeys reared with one eyelid sutured closed. The normal eye is shown on the left and the deprived eye on the right. Numbers indicate distances in millimeters. (After Wiesel and Raviola, 1977)

their parents) wanted to improve their vision, they might be able to do so by practicing far vision to counterbalance the near work "overload." Practically, of course, most people would probably choose wearing glasses or contacts rather than indulging in the onerous daily practice that would presumably be required. A few individuals, however, who especially value high-acuity vision—for example, those who anticipate careers as pilots—might be sufficiently motivated to undertake such training. (There is evidence that the susceptibility to abnormal eye growth continues into the teenage years.) Not everyone agrees, however, that such a remedy would be effective, and a number of investigators (and drug companies) are exploring the possibility of pharmacological intervention during the

period of childhood when abnormal eye growth is presumed to occur. In any event, it is a remarkable fact that deprivation of focused light on the retina causes a compensatory growth of the eye and that this feedback loop is so easily perturbed.

References

BOCK, G. AND K. WIDDOWS (1990) *Myopia and the Control of Eye Growth*. Ciba Foundation Symposium 155. Chichester: Wiley.

SHERMAN, S. M., T. T. NORTON AND V. A. CASAGRANDE (1977) Myopia in the lid sutured tree shrew. Brain Res. 124: 154–157.

WALLMAN, J., J. TURKEL AND J. TRACTMAN (1978) Extreme myopia produced by modest changes in early visual experience. Science 201: 1249–1251.

WIESEL, T. N. AND E. RAVIOLA (1977) Myopia and eye enlargement after neonatal lid fusion in monkeys. Nature 266: 66–68.

disk that explains why photoreceptors are found in the outermost rather than the innermost layer of the retina.

■ PHOTOTRANSDUCTION

In most sensory systems, activation of a receptor by the appropriate stimulus causes the cell membrane to depolarize, stimulating transmitter release and ultimately a postsynaptic potential in the neurons it contacts. Thus, it may come as a surprise to learn that shining light on a photoreceptor, either a rod or a cone, leads to membrane *hyperpolarization* rather than depolarization (Figure 10.5). In the dark, the receptor is in a depolarized state, with a membrane potential of roughly −40 mV. Progressive increases in the intensity of illumination cause the potential across the receptor membrane to become more negative, a response that saturates when the membrane potential reaches about −65 mV. Transmitter release from the synaptic terminals of the photoreceptor, like that from any other nerve cell, is dependent on the potential difference across the terminal membrane. Thus, the depolarized photoreceptors continually release transmitter in the dark; when they are hyperpolarized by light, the level of transmitter release is *reduced*. Although this arrangement may seem odd, the only logical requirement for subsequent visual processing is a consistent relationship between luminance changes and the activity of photoreceptors. In any event, the reason for this apparent "sign reversal" in the activation of photoreceptor cells is not known.

The depolarized state of photoreceptors in the dark depends on the presence of ion channels in the outer segment membrane that permit sodium, calcium, and magnesium ions to flow into the cell, thus reducing the degree of inside negativity (Figure 10.6). The probability of these channels in the outer segment being open or closed is regulated by the levels of the nucleotide **cyclic guanosine monophosphate (cGMP)**. In darkness, high levels of cGMP in the outer segment keep the channels open. In light, cGMP levels drop and some of the channels close, leading to hyperpolarization of the outer segment membrane.

The absorption of a photon by a molecule of photopigment in the disks of the photoreceptor outer segment initiates a biochemical cascade that ultimately decreases intracellular levels of cGMP. The photopigment in the receptor discs contains a light absorbing component (**11-*cis* retinal**) coupled to one of a variety of proteins (**opsins**) that tunes the molecule's absorption

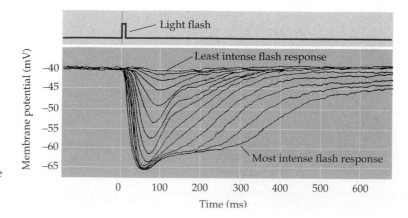

Figure 10.5 An intracellular recording from a single cone stimulated with different amounts of light. Each trace represents the response to a brief flash that was varied in intensity. At the highest light levels, the response amplitude saturates (at about −65 mV). The hyperpolarizing response is characteristic of vertebrate photoreceptors; interestingly, some invertebrate photoreceptors depolarize in response to light.

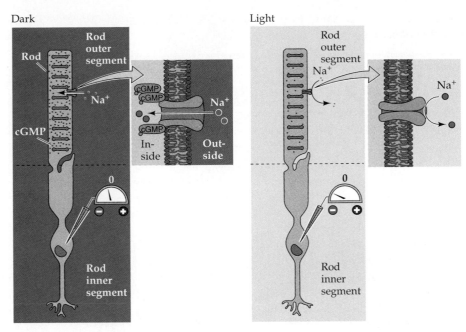

Figure 10.6 Cyclic GMP-gated channels in the outer segment membrane are responsible for the light-induced changes in the electrical activity of photoreceptors (a rod is shown here, but the same scheme applies to cones). In the dark, cGMP levels in the outer segment are high; this molecule binds to the Na^+-permeable channels in the membrane, keeping them open and allowing sodium (and other cations) to enter, thus depolarizing the cell. Exposure to light leads to a decrease in cGMP levels, a closing of the channels, and receptor hyperpolarization.

of light to a particular region of the spectrum. The different protein component of the photopigment in rods and cones contributes critically to the functional specialization of the two receptor types. Most of what is known about the molecular events of phototransduction has been gleaned from experiments in rods, in which the photopigment is **rhodopsin**. When the retinal moiety in rhodopsin absorbs a photon of light, its configuration changes and triggers a series of alterations in the protein component of the molecule (Figure 10.7). These changes lead, in turn, to the activation of an intracellular messenger called **transducin**, which activates a **phosphodiesterase** that hydrolyzes cGMP. Thus, absorption of light results in structural changes in rhodopsin, the activation of transducin, the activation of a cGMP phosphodiesterase, and finally the breakdown of cGMP.

This complex cascade provides enormous amplification. A single light-activated rhodopsin molecule can activate hundreds of transducin molecules, which in turn can lead to the hydrolysis of hundreds of cGMP molecules. It has been estimated that the absorption of a single photon by a rhodopsin molecule results in the closure of 300 ion channels, or about 3% of the number of channels in each rod that are open in the dark.

■ SPECIALIZATIONS OF THE ROD AND CONE SYSTEMS

The two types of photoreceptors, rods and cones (Figure 10.8), are distinguished by their shape (from which they derive their names), the type of photopigment they contain (see above), their distribution across the retina,

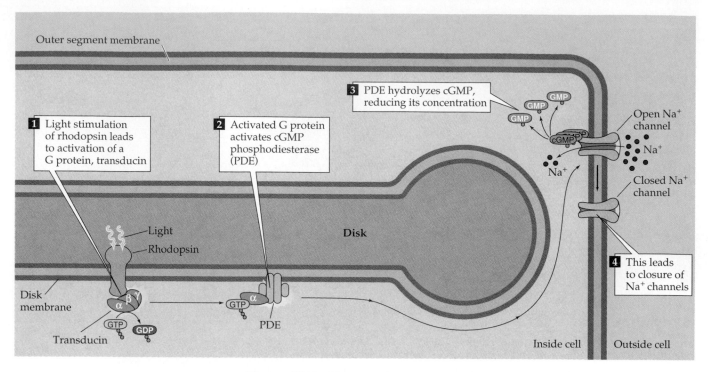

Figure 10.7 The second-messenger cascade of phototransduction. Light stimulation of rhodopsin in the receptor disks leads to the activation of a G-protein (transducin), which in turn activates a phosphodiesterase (PDE). The phosphodiesterase hydrolyzes cGMP, reducing its concentration in the outer segment and leading to the closure of sodium channels in the outer segment membrane.

and their pattern of synaptic connections. These properties reflect the fact that the rod and cone systems (the receptors and their connections within the retina) are specialized for different aspects of vision. The rod system has very low spatial resolution, but is extremely sensitive to light; it is therefore specialized for sensitivity at the expense of resolution. In contrast, the cone system has very high spatial resolution but is relatively insensitive to light; it is therefore specialized for acuity at the expense of sensitivity. The cone system also allows us to see color.

The contributions of the rod and cone systems to the range of illumination over which the visual system can operate are shown in Figure 10.9. At the lowest levels of illumination (below the level of starlight), rods are the only receptors activated; such rod-mediated perception is called **scotopic vision**. Everyone is familiar with the difficulty of making visual discriminations under very low light conditions, where only the rod system is active. (The problem is the poor acuity of this system, and the fact that all perception of color is lost.) Cones begin to contribute to visual perception at about the level of starlight, and they are the only receptors that function under relatively bright conditions such as normal indoor lighting or sunlight. Cone-mediated vision, called **photopic vision**, occurs at high levels of illumination because the response of rod photoreceptors to light saturates—that is, the membrane potential of individual rods no longer varies as a function of illumination (see Figure 10.5). Finally, **mesopic vision** occurs at light levels in which both rods and cones contribute. From these considerations it should be clear that most of what we think of as seeing is mediated by the cone sys-

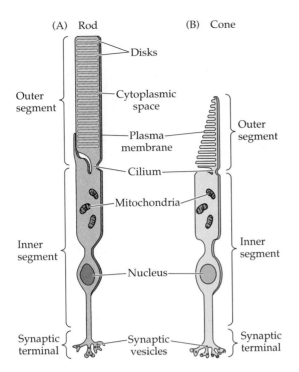

(A) Rod   (B) Cone

Disks
Cytoplasmic space
Plasma membrane
Cilium
Mitochondria
Nucleus
Outer segment
Inner segment
Synaptic terminal
Synaptic vesicles
Outer segment
Inner segment
Synaptic terminal

Figure 10.8 Structural differences between rods and cones. Although generally similar in structure, rods (A) and cones (B) differ in their size and shape, and in the arrangement of the membranous disks in their outer segments.

tem, and that loss of cone function is devastating. Individuals who have lost cone function are legally blind, whereas those who have lost rod function only experience difficulty seeing at low levels of illumination (called night blindness).

The factors that contribute to the functional differences in the rod and cone systems are several. First, differences in the structure of rods and cones, including the amount of photopigment and the shape of the outer segment, make rod receptors more sensitive (see Figure 10.8). Rods are longer and contain more photopigment than cones, enabling them to capture more

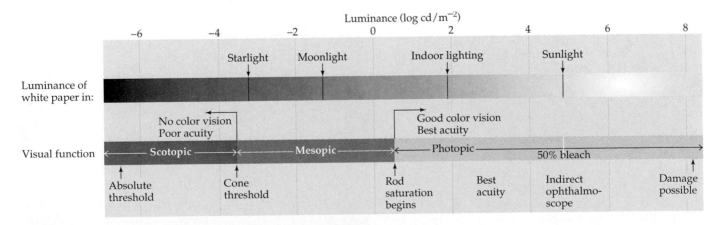

Figure 10.9 The range of luminance values over which the visual system operates. At the lowest levels of illumination, only the rods are activated. Cones begin to contribute to perception at about the level of starlight, and they are the only receptors that function under relatively bright conditions.

light. Even more important, the transduction mechanism in rods is capable of greater amplification than that of cones. A rod can respond to a single photon, whereas more than 100 photons are required to activate a cone.

Another critical distinction between the two receptor systems concerns color vision. Both rods and cones transmit information about the wavelength of light as a function of the types of photopigments they contain. All rods contain the same photopigment—rhodopsin—whereas individual cones contain one of three different photopigments, collectively called **cone opsins**, that have different but overlapping absorption spectra (Figure 10.10). The relative activity of these three sets of cones (referred to as short, middle, and long wavelength, or blue, green, and red) generates the retinal signals that ultimately give rise to the sensation of color. Besides its aesthetic appeal, color vision makes it possible to distinguish objects that might be difficult to identify on the basis of luminance differences alone. In a sense, color vision enhances our ability to detect objects by increasing the contrast with their surroundings. Seeing color is not essential, however; many mammals lack this ability, and color blindness in humans is a relatively minor problem. Defective color vision can result either from the loss of one or more of the cone mechanisms, from a change in the absorption spectra of individual cone pigments (Box C), or from lesions in the central stations that process color information (see Chapter 12).

The distribution of rods and cones across the surface of the retina varies markedly (Figure 10.11). Cones are the only photoreceptors located in the **fovea**. In this specialized region of the retina, the layers of cell bodies and processes that generally overlie the photoreceptors are displaced, so light rays are subjected to a minimum of scattering before they strike the receptors (Figure 10.12). Another potential source of distortion, blood vessels, are also absent from a small region in the center of the fovea (called the **foveola**); because this region lacks a capillary bed, it is dependent on the underlying choroid and pigment epithelium for sustenance.

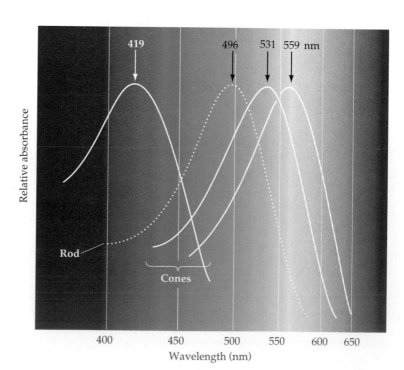

Figure 10.10   The absorption spectra of the four photopigments in the normal human retina. The solid curves are for the three kinds of cone opsins, the dashed curve for rod rhodopsin. Absorbance is defined as the log value of the intensity of incident light divided by intensity of transmitted light.

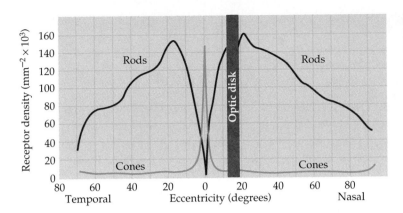

Figure 10.11 Distribution of rods and cones in the human retina. Cones are present throughout the retina but are most prevalent in the fovea. Conversely, there are no rods in the center of the fovea.

Although cones are not restricted to the fovea, their lower density outside the fovea, as well as the lower density of the ganglion cells that they supply, explains why visual acuity declines so markedly as a function of eccentricity. Indeed, the greater acuity of foveal vision is the main reason humans spend so much time moving their eyes (and heads) around—in effect directing the foveas of the two eyes to objects of interest (see Chapter 19). Acuity is reduced by 75% just 6° eccentric to the line of sight, a fact that can readily be appreciated by trying to read the words on any line of this page away from the word being fixated on. Conversely, the exclusion of rods from the fovea, and their presence in high density away from the fovea, explain why the threshold for detecting a light stimulus is much lower outside the region of central vision. It is easier to see a dim object (such as a faint star) by looking away from it, so that the starlight stimulates the region of the retina that is rich in rods.

Another distinction between the rod and cone systems is the degree of receptor convergence onto other cell types in the retina. The rod system is highly convergent: many rods synapse on a single bipolar cell, and many bipolar cells that receive rod input converge on the same ganglion cell. In contrast, the cone system is much less convergent. In the center of the fovea, a single cone may contact only one bipolar cell, which in turn contacts a single

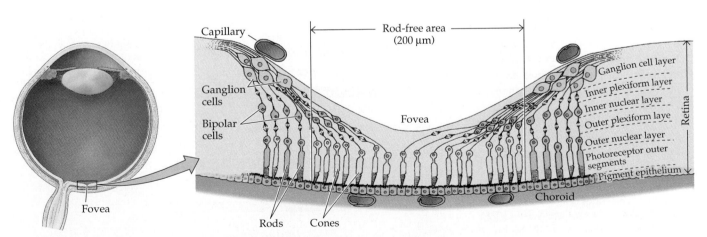

Figure 10.12 Diagrammatic cross-section through the human fovea. The overlying cellular layers and blood vessels are displaced so that light rays are subject to a minimum of scattering before they strike the outer segments of the cones in the center of the fovea.

Box C
DEFICIENCIES IN COLOR VISION

Under normal conditions, most people can match any color in a test stimulus by adjusting the intensity of three superimposed colored light sources (blue, green, and red). The fact that only three colors are necessary to match all the colors perceived is a reflection of the fact that our sense of color is based on the relative levels of activity in three sets of cones with different absorption spectra (see Figure 10.10). That color vision is *trichromatic* was first recognized by Thomas Young at the beginning of the 19th century. For about 2% of the male population and 0.03% of the female population, color vision is more limited. Only two colors of light are needed to match all the colors that these individuals can perceive; the third color is not seen. Such dichromacy, or color blindness as it is commonly called, is inherited as a recessive, sex-linked characteristic and exists in two forms: protanopia, in which all color matches can be achieved by using only green and blue light, and deuteranopia, in which all matches can be achieved by using only blue and red wavelengths. In another major class of color deficiencies all three wavelengths of light are needed to make all possible color matches, but the matches are made using values that are significantly different from those used by most individuals. Some of these anomalous trichromats require more red than normal to match other colors (protanomalous trichromats); others require more green than normal (deuteranomalous trichromats).

Jeremy Nathans and his colleagues have provided a deeper understanding of the basis for such color vision deficiencies by identifying and sequencing the genes that encode the three human cone pigments. The genes that encode the red and green pigments lie adjacent to each other on the X chromosome and show a high degree of sequence homology. (This sex linkage explains the prevalence of color blindness in males.) In contrast, the blue-sensitive pigment gene is found on chromosome 7 and is considerably different in its amino acid sequence. These facts suggest that the red and green pigment genes have evolved relatively recently, perhaps as the result of duplication of a single ancestral gene. This genetic knowledge also explains why most color vision abnormalities involve the red and green cone pigments, while the blue cone pigment remains relatively stable. Because they are located adjacent to each other on the X chromosome, crossing over during meiosis can result in an unequal distribution of the genes such that one chromosome contains multiple copies, while the other contains none. Crossing over can also result in hybrid genes that code for pigments with different absorption spectra.

Human dichromats lack one of the three cone pigments, either because the corresponding gene is missing or because it exists as a hybrid of the red and green pigment genes. For example, some deuteronopes lack the green pigment gene altogether; others have a hybrid gene that is thought to produce a redlike pigment in the "green" cones. Anomalous trichromats also possess hybrid genes, but these are thought to elaborate pigments whose spectral properties lie between those of the normal red and green pigments. Thus, although most anomalous trichromats have two distinct sets of long-wavelength cones (one normal, one hybrid), there is more overlap in their absorption spectra than in normal trichromats, and less of a difference in how the two sets of cones respond to a given wavelength.

References

NATHANS, J., D. THOMAS AND D. S. HOGNESS (1986) Molecular genetics of human color vision: The genes encoding blue, green and red pigments. Science 232: 193–202

NATHANS, J., T. P. PIANTANIDA, R. EDDY, T. B. SHOWS AND D. S. HOGNESS (1986) Molecular genetics of inherited variation in human color vision. Science 232: 203–210.

NATHANS, J. (1987) Molecular biology of visual pigments. Annu. Rev. Neurosci. 10: 163–194.

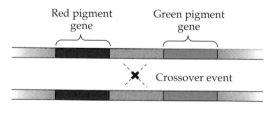

Red pigment gene   Green pigment gene

Crossover event

Different crossover events can lead to:

(1) Hybrid gene

(2) Loss of gene

Patterns in color-blind men

(3) Duplication of gene (does not affect color vision)

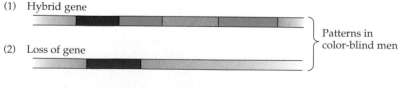

Many deficiencies of color vision are the result of genetic alterations in the red or green cone pigments due to the crossing over of chromosomes during meiosis. This recombination can lead to the loss of a gene, the duplication of a gene, or the formation of a hybrid with characteristics distinct from those of normal genes.

ganglion cell. Convergence makes the rod system a good detector of light because small signals from many rods can combine to generate a larger response in the bipolar cell. At the same time, convergence reduces the spatial resolution of the rod system, since the source of a signal in a rod bipolar cell or retinal ganglion cell could have come from anywhere within a relatively large area of the retinal surface. The one-to-one relationship of foveal cones to bipolar and ganglion cells is, of course, just what is required to maximize acuity.

Finally, the routes by which rod and cone information reaches ganglion cells for transmission to central visual targets are different. The rod pathway involves a class of bipolar cells (rod bipolars) that does not contact ganglion cells directly; instead, these bipolar cells synapse on the processes of a specialized class of amacrine cells that in turn synapse with the processes of ganglion cells. Although the routes through the retina are distinct, extrafoveal information from the rod and cone systems ultimately converges on the same ganglion cells. Thus, individual ganglion cells can display both rod- and cone-driven characteristics, depending on the level of illumination. Within the fovea, however, the ganglion cells are driven entirely by cones.

■ RETINAL GANGLION CELL RECEPTIVE FIELDS

Some understanding of the purpose of the complex synaptic interactions in the retina has come from physiological studies in which small spots of light are used to examine the responses of individual retinal neurons. Stephen Kuffler pioneered this approach in the 1950s by characterizing the responses of single ganglion cells in the cat retina. He found that each ganglion cell responds to stimulation of a small, restricted, circular patch of the retina, which defines the cell's receptive field (see Chapter 8 for a discussion of receptive fields in the somatic sensory system). Based on the these responses, Kuffler distinguished two classes of ganglion cells, "on"-center and "off"-center. Turning on a spot of light in the center of an **"on"-center ganglion cell** receptive field produces a burst of electrical activity (an "on"response) (Figure 10.13). Turning the light on in the center of an **"off"-center ganglion cell** receptive field has the opposite effect: the spontaneous rate of firing decreases, and when the spot of light is turned off, the cell responds with a burst of action potentials (an "off" response). "On"- and "off"-center ganglion cells are present in roughly equal numbers. The receptive fields have overlapping distributions, so that every point on the retinal surface (that is, every part of visual space) is analyzed by several "on"-center and "off"-center ganglion cells. The significance of these two distinct types of retinal ganglion cells was further demonstrated by Peter Schiller and his colleagues, who examined the effects of pharmacologically inactivating "on"-center ganglion cells on a monkey's ability to detect a variety of visual stimuli. After silencing "on"-center ganglion cells, the animals showed a dramatic and specific deficit in their ability to detect stimuli that were brighter than the background; however, they could still see objects that were darker than the background.

These several observations suggest that information about increases or decreases in luminance (perceived as brightness and darkness, respectively) is carried separately to the brain by these two types of retinal ganglion cells. Having two separate luminance channels means that changes in light intensity are always conveyed to the brain by an excitatory process, rather than relying on decreases in activity below some set resting level to signal diminished luminance. For example, a rapid increase in the firing rate of "on"-center cells, which have a low rate of firing in dim illumination, unambiguously

Figure 10.13 The responses of "on"-center and "off"-center retinal ganglion cells to stimulation of different regions of their receptive fields. See text for explanation.

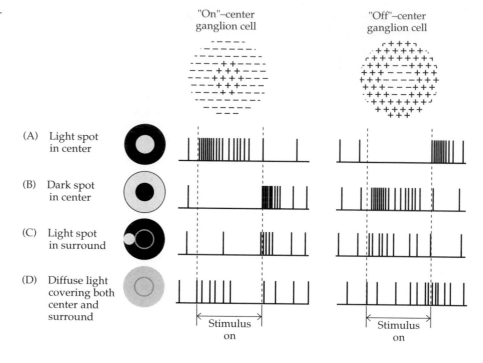

Figure 10.14 Rate of discharge of an "on"-center ganglion cell to a spot of light as a function of the distance of the spot from the receptive field center. Zero on the *x* axis corresponds to the center; at a distance of 5 degrees, the spot falls outside the receptive field.

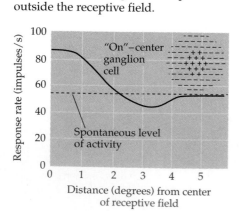

signals a rapid increase in luminance (Figure 10.13A). However, these cells could not reliably signal a rapid decrease in luminance from the original level (Figure 10.13B). The identification of two distinct classes of retinal ganglion cells, and the demonstration that their activity conveys different types of information to central visual structures, illustrates one strategy for coping with the wealth of information in the visual scene: the emergence of **parallel systems** for analyzing different features of the visual stimulus. Other examples of the parallel processing of different categories of visual information (such as color and motion) are discussed in Chapter 11.

Kuffler's work called attention to another important feature of visual processing. Retinal ganglion cells do not act as simple photodetectors; in fact, most ganglion cells are relatively poor at signaling differences in the level of diffuse illumination. Instead, they are sensitive to differences between the level of illumination that falls on the receptive field center and the level of illumination that falls on the surround—that is, **luminance contrast**. Kuffler noticed that the center of a ganglion cell receptive field is surrounded by a concentric region that, when stimulated, antagonizes the response to stimulation of the receptive field center. For example, as a spot of light is moved from the center of the receptive field of an "on"-center cell toward its periphery, the response of the cell to the spot of light decreases (Figure 10.14). When the spot falls completely outside the center (that is, in the surround), the response of the cell falls below its resting level; the cell is effectively inhibited until the distance from the center is so great that the spot no longer falls on the receptive field at all, in which case the cell returns to its resting level of firing. "Off"-center cells also show an antagonistic surround. Light stimulation of the surround of an "off"-center cell increases the firing rate of the cell, a response that opposes the decrease in firing rate that occurs when the center is stimulated (Figure 10.13C). Because of their antagonistic surrounds, ganglion cells respond much more vigorously to small spots of light confined to their receptive field centers than to large spots or uniform illumination (Figure 10.13D).

To appreciate how center-surround antagonism helps to detect luminance contrast, consider the activity levels in a hypothetical population of "on"-center ganglion cells whose receptive fields are distributed across a retinal image of a light-dark edge (Figure 10.15). The neurons whose firing rates are most affected by this stimulus—either increased (neuron D) or decreased (neuron B)—are those with receptive fields that lie along the light/dark border; those with receptive fields completely illuminated (or completely darkened) remain relatively unaffected (neurons A and E). Thus, the signal supplied by the retina to central visual structures does not give equal weight to all regions of the visual scene; rather, it emphasizes the regions that contain the most information—namely, the regions where there are differences in luminance.

This property of retinal ganglion cells explains why our perception of the brightness or darkness of a given region in the visual scene is influenced so strongly by the luminance of adjacent regions. For example, the middle panel in Figure 10.16 reflects the same amount of light on the left side of the figure as on the right; yet the part on the right appears significantly brighter than the part on the left. (You can convince yourself that the middle panel is really equiluminant by using two pieces of paper to block out the regions above and below it.) The *perceived* brightness of the panel is computed on the basis of the activity of those retinal ganglion cells whose receptive fields intersect its borders. Their activity in turn depends on the **contrast**—the difference in the amount of light that falls on their receptive field centers and surrounds.

The special sensitivity of retinal ganglion cells to contrast rather than absolute levels of luminance also explains why the perceived brightness of objects remains constant over a wide range of lighting conditions. Increases or decreases in the overall level of illumination have equal effects on the cen-

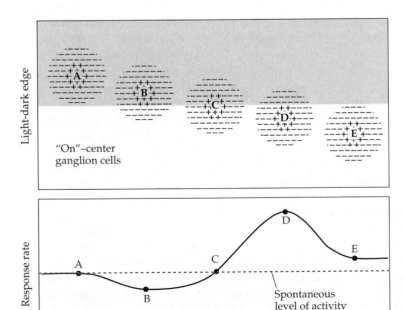

Figure 10.15 Responses of a hypothetical population of "on"-center ganglion cells whose receptive fields (A–E) are distributed across a light-dark edge. Those cells whose activity is most affected have receptive fields that lie along the light-dark edge.

Figure 10.16 The effect of background contrast on the perception of brightness. Although the gray panel in the middle reflects exactly the same amount of light on the left-hand side of the figure as on the right, the right-hand part of the stripe appears brighter because of the greater contrast with the surrounding area.

ter and surround of each ganglion cell's receptive field, and thus do little to affect the cell's level of activity. In bright sunlight, for example, the print on this page reflects considerably more light to the eye than it does in room light. In fact, the *print* reflects more light in sunlight than the *paper* reflects in room light; yet the print looks just as black (and the page just as white) indoors or out. The signal sent to the brain from the retina therefore downplays the background level of illumination while enhancing the salient features of a visual stimulus—in particular, its contrast with the surroundings.

■ RETINAL CIRCUITRY UNDERLYING GANGLION CELL RECEPTIVE FIELD PROPERTIES

The responses of retinal ganglion cells are clearly quite different from the responses of the photoreceptors. How this transformation comes about has been explored by recording the responses of the several other retinal cell types. The basic features of ganglion cell receptive fields described by Kuffler are actually generated in the outer plexiform layer as a result of synaptic interactions among photoreceptor terminals, bipolar cells, and horizontal cells. These interactions define two distinct populations of bipolar cells with properties almost identical to those of ganglion cells: that is, they have receptive fields with "on"- or "off"- centers and antagonistic surrounds. Not surprisingly, the "on"-center bipolar cells synapse with "on"-center ganglion cells, and "off"-center bipolar cells synapse with "off"-center ganglion cells. The principal difference between ganglion cells and bipolar cells lies in the nature of their electrical response. Like most other cells in the retina, bipolar cells have graded potentials rather than action potentials; because the distances involved are so small, action potentials are unnecessary. Graded depolarization of bipolar cells leads to an increase in transmitter release at their synapses, while graded hyperpolarization leads to a decrease.

The two classes of bipolar cells, "on"- and "off"-center, differ in the types of glutamate receptors they express and it is this that explains why they respond so differently to changes in light intensity. Glutamate released from photoreceptor terminals causes "on"-center bipolar cells to hyperpolarize and "off"-center cells to depolarize. The sequence of events that follows light onset in the center of an "on"-center ganglion cell's receptive field is shown in Figure 10.17A. Photoreceptors that contribute to the center of the cell's receptive field hyperpolarize, decreasing their release of neurotransmitter. "On"-center bipolar cells contacted by the photoreceptors are then freed from the hyperpolarizing influence of the photoreceptor's transmitter,

(A)

(B)

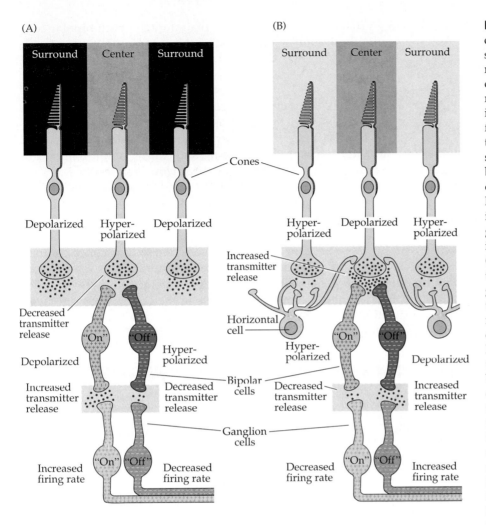

Cones

Surround | Center | Surround

Surround | Center | Surround

Depolarized | Hyper-polarized | Depolarized

Hyper-polarized | Depolarized | Hyper-polarized

Decreased transmitter release

Increased transmitter release

Horizontal cell

"On" "Off"

"On" "Off"

Depolarized | Hyper-polarized

Hyper-polarized | Depolarized

Increased transmitter release | Decreased transmitter release | Bipolar cells

Decreased transmitter release | Increased transmitter release

Ganglion cells

"On" "Off"

"On" "Off"

Increased firing rate | Decreased firing rate

Decreased firing rate | Increased firing rate

Figure 10.17 Functional interactions of retinal neurons. (A) Circuitry responsible for generating center responses of retinal ganglion cells. In this case, the centers of the on- and off-ganglion cell receptive fields are illuminated without illumination of the surrounds. Light falling on the receptor leads to a reduction in transmitter release that has opposite effects on the two populations of bipolar cells: "on"-center bipolar cells depolarize, and "off"-center bipolar cells hyperpolarize. This, in turn, leads to an increase in the firing rate of "on"-center ganglion cells and a decrease in the firing rate of "off"-center ganglion cells. (B) Circuitry responsible for generating the receptive field surrounds of retinal ganglion cells. In this case, the centers of the ganglion cell receptive fields are illuminated as in A; but with the addition of light in the surrounds. The effect of light on the photoreceptors that supply the on and off bipolar cells is opposed by the action of horizontal cells that are connected to photoreceptors that lie in the surround. Changes in the release of transmitter from horizontal cells onto the receptor terminals supplying "on"- and "off"-center bipolar cells lead to receptor depolarization and an increase in glutamate release. The effects of glutamate on the "on"- and "off"-center bipolar cells cause "on"-center bipolars to hyperpolarize and "off"-center bipolars to depolarize. A comparison of (A) and (B) illustrates why diffuse illumination of a ganglion cell receptive field is an ineffective stimulus.

and they consequently depolarize. The depolarization of the "on"-center bipolar cells causes them to increase their release of transmitter, which leads, in turn, to a depolarization of the ganglion cells they contact. Light onset in the center of an "off"-center ganglion cell's receptive field has exactly the opposite effect.

The antagonistic surround of the bipolar cell's receptive field is largely the result of lateral connections established by horizontal cells (Figure 10.17B). These neurons are well suited to mediate surround effects in that they receive inputs from photoreceptor terminals as well as from a vast network of other horizontal cells distributed over a wide area of the retinal surface. By synapsing directly on photoreceptor terminals, horizontal cells can regulate the amount of transmitter that the receptors release onto bipolar cell dendrites. The general effect of horizontal cell input is to antagonize the photoreceptor's response to light. When the receptive field surround of a bipolar cell is illuminated, activity conveyed through the lateral connections of horizontal cells causes a depolarization of the photoreceptor terminals that contact the bipolar cell. These effects are exactly the opposite of what occurs when light strikes the receptors. Simultaneous stimulation of both the center and the surround with diffuse light therefore produces light-induced changes in photoreceptor membrane potential and horizontal circuitry-induced changes that tend to cancel out, explaining why both bipolar cells and retinal ganglion cells are largely unresponsive to diffuse illumination.

■ SUMMARY

The light that falls on the retina is transformed by the retinal circuitry into a pattern of action potentials that ganglion cell axons convey to the visual centers in the brain. This process begins with phototransduction, a biochemical cascade that ultimately regulates the opening and closing of ion channels in the membrane of the photoreceptor's outer segment. Two systems of photoreceptors—rods and cones—and their associated connections allow the visual system to meet the conflicting demands of high sensitivity and high acuity, respectively. Retinal ganglion cells operate quite differently from the photoreceptor cells. The center-surround arrangement of ganglion cell receptive fields makes these neurons particularly sensitive to luminance contrast and relatively insensitive to the overall level of illumination. Center-surround organization is generated via the synaptic interactions between photoreceptors, horizontal cells, and bipolar cells in the outer plexiform layer. Thus, the signal sent to central visual targets is already highly processed when it leaves the retina.

Additional Reading

Reviews

SCHNAPF, J. L. AND D. A. BAYLOR (1987) How photoreceptor cells respond to light. Sci. Amer. 256: 40–47.

STERLING, P. (1990) Retina. In *The Synaptic Organization of the Brain*, G. M. Shepherd (ed). New York: Oxford University Press, pp. 170–213.

STRYER, L. (1986) Cyclic GMP cascade of vision. Annu. Rev. Neurosci. 9: 87–119.

Important Original Papers

BAYLOR, D. A., M. G. F. FUORTES AND P. M. O'BRYAN (1971) Receptive fields of cones in the retina of the turtle. J. Physiol. (Lond.) 214: 265–294.

DOWLING, J. E. AND F. S. WERBLIN (1969) Organization of the retina of the mud puppy, *Necturus maculosus*. I. Synaptic structure. J. Neurophysiol. 32: 315–338.

FASENKO, E. E., S. S. KOLESNIKOV AND A. L. LYUBARSKY (1985) Induction by cyclic GMP of cationic conductance in plasma membrane of retinal rod outer segment. Nature 313: 310–313.

KUFFLER, S. W. (1953) Discharge patterns and functional organization of mammalian retina. J. Neurophysiol. 16: 37–68.

SCHILLER, P. H., J. H. SANDELL AND J. H. R. MAUNSELL (1986) Functions of the "on" and "off" channels of the visual system. Nature 322: 824–825.

WERBLIN, F. S. AND J. E. DOWLING (1969) Organization of the retina of the mud puppy, *Necturus maculosus*. II. Intracellular recording. J. Neurophysiol. 32: 339–354.

Books

BARLOW, H. B. AND J. D. MOLLON (1982) *The Senses*. London: Cambridge University Press.

DOWLING, J. E. (1987) *The Retina: An Approachable Part of the Brain*. Cambridge, MA: Belknap Press.

HART, W. M. J. (ed.) (1992) *Adler's Physiology of the Eye: Clinical Application*, 9th Ed. St. Louis: Mosby Year Book.

HOGAN, M. J., J. A. ALVARADO AND J. E. WEDDELL (1971) *Histology of the Human Eye: An Atlas and Textbook*. Philadelphia: Saunders.

RODIECK, R. W. (1973) *The Vertebrate Retina*. San Francisco: W. H. Freeman.

WANDELL, B. A. (1995) *Foundations of Vision*. Sunderland, MA: Sinauer Associates.

CHAPTER 11

CENTRAL VISUAL PATHWAYS

■ OVERVIEW

Visual perception is the product of complex interactions between multiple subdivisions of the brain. Separate areas in the occipital lobe and parts of the parietal and temporal lobes function together to provide a unified picture of objects in the environment. Other centers use the information supplied by the retina to perform reflexive tasks such as adjusting the size of the pupil and directing the eyes to targets of interest. Still other visual subsystems use retinal information to govern behaviors that are tied to the day/night cycle. The pathways and structures that mediate these various functions are necessarily diverse. Of these, the primary visual pathway—from the retina to the dorsal lateral geniculate nucleus in the thalamus and to the primary visual cortex—is the most important and the best understood. Individual neurons at each station of the primary visual pathway are specifically tuned to extract and encode different types of visual information, such as contrast, color, form, and movement. Normal vision therefore depends on the integrity of this projection.

■ CENTRAL PROJECTIONS OF RETINAL GANGLION CELLS

All the visual pathways to the brain arise from ganglion cell axons that exit the retina through a circular region in its nasal part called the **optic disk** (or optic papilla), where they bundle together to form the **optic nerve**. Because this region of the retina contains no photoreceptors, it is insensitive to light and produces the perceptual phenomenon known as the **blind spot** (Box A). The optic disk is easily identified when the retina is examined with an ophthalmoscope as a whitish circular area; it also is recognized as the site from which the ophthalmic artery and veins enter (or leave) the eye (Figure 11.1). In addition to being a conspicuous retinal landmark, the appearance of the optic disk can be a useful gauge of intracranial pressure. The subarachnoid space surrounding the optic nerve is continuous with that of the brain; as a result, increases in intracranial pressure—a sign of serious neurological problems—can be detected as a swelling of the optic disk (called papilledema).

Ganglion cell axons in the optic nerve run a straight course to the optic chiasm at the base of the diencephalon. In humans, about 60% of the fibers cross in the chiasm, while the other 40% continue toward the brain on the same side. Once past the chiasm, the ganglion cell axons on each side form the **optic tract**. Thus, the optic tract, unlike the optic nerve, contains fibers from both eyes. The partial crossing (decussation) of ganglion cell axons at the optic chiasm allows information from corresponding points on the two retinas to be processed by approximately the same cortical site in each hemisphere (this important issue is considered further in the next section).

The ganglion cell axons in the optic tract reach a number of structures in the diencephalon and midbrain (Figure 11.2). The major target in the diencephalon is the **lateral geniculate nucleus** of the thalamus. Neurons in the lateral geniculate nucleus, like their counterparts in the thalamic relays of other sensory systems, send their axons to the cerebral cortex via the internal

Figure 11.1 The retinal surface of the right eye, viewed with an ophthalmoscope. The optic disk is the region where the ganglion cell axons leave the retina to form the optic nerve; it is also characterized by the entrance and exit, respectively, of the ophthalmic arteries and veins that supply the retina. The macula lutea can be seen as a pale yellowish area at the center of the optical axis (the optic disk lies nasally); the macula is the region of the retina that has the highest visual acuity. The fovea is a small pit (about 1.5 mm in diameter) that lies at the center of the macula (see Chapter 10).

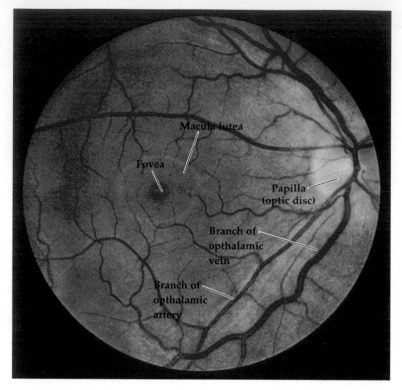

capsule. These axons pass through a portion of the internal capsule called the **optic radiation** and terminate in the **primary visual (or striate) cortex (Brodmann's area 17)**, which lies largely along the calcarine fissure in the occipital lobe. The striate cortex is often referred to as V1 by analogy with other primary sensory cortices such as S1 (primary somatic sensory) and A1 (primary

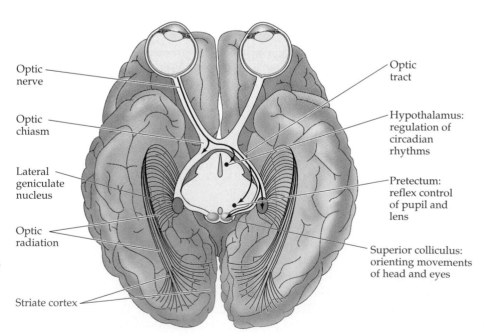

Figure 11.2 Central projections of retinal ganglion cells. Ganglion cell axons terminate in the lateral geniculate nucleus of the thalamus, the superior colliculus, the pretectum, and the hypothalamus. For clarity, only the crossing axons of the right eye are shown.

Box A
THE BLIND SPOT

One would imagine a visual field defect (called a scotoma) that results from damage to the retina or central visual pathways would be obvious to the individual who has one. When the deficit involves a peripheral region of the visual field, however, a scotoma often goes unnoticed until an accident occurs that may all too dramatically reveal the sensory loss. In fact, we all have a physiological scotoma that we are quite unaware of: the substantial gap in each of our monocular visual fields that corresponds to the location of the optic disk (the receptor-free region where the optic nerve leaves the eye; see Figure 11.1).

To find the "blind spot" of the right eye, close the left eye and fixate on the **X** shown below, holding the book about 30–40 centimeters away. Now take a pencil in your right hand and, without breaking fixation, move it slowly toward the **X** from the right side of the page. At some point, the tip of the pencil will disappear; mark this point and continue to move the pencil to the left until the point reappears; make another mark. The borders of the blind spot along the vertical axis can be determined in the same way, by moving the pencil up and down so that its path falls between the two horizontal marks. To

prove that the information from the region of visual space bounded by the marks is really not perceived, put a penny inside the demarcated area. When you fixate the **X** with both eyes and then close the left eye, the penny will disappear.

How can we be unaware of such a large defect in the visual field (typically about 5° × 8°)? The scotoma created by the optic disk arises from the nasal retina of each eye. With both eyes open, information about the corresponding region of visual space is, of course, available from the temporal retina of the other eye. But this fact does not explain why the blind spot remains undetected with one eye closed. When the world is viewed monocularly, the visual system appears to "fill in" the missing part of the scene based on the information supplied by the regions surrounding the optic disk. To observe this phenomenon, notice what happens when a pencil or some other object lies across the optic disk representation that you have mapped. Remarkably, the pencil looks complete! Electrophysiological recordings have shown that neurons in the visual cortex whose receptive fields lie in the optic disk representation can be activated by stimulating the regions that surround the optic disk

of the contralateral eye. Perhaps, "filling in" the blind spot is based on cortical mechanisms that integrate information from different points in the visual field. However, as Herman Von Helmholtz pointed out in the nineteenth century, it may just be that this part of the visual world is ignored. In this conception, the pencil is coupled across the blind spot because the rest of the scene "collapses" around it.

References

FIORANI, M., M. G. P. ROSA, R. GATTASS AND C. E. ROCHA-MIRANDA (1992) Dynamic surrounds of receptive fields in striate cortex: A physiological basis for perceptual completion? Proc. Natl. Acad. Sci. USA 89: 8547–8551.

GILBERT, C. D. (1992) Horizontal integration and cortical dynamics. Neuron 9: 1–13.

HELMHOLTZ, H. VON (1968). *Helmholtz's Treatise on Physiological Optics*, Vols. I–III (Translated from the 3rd German Edition published in 1910). J. P. C. Southall (ed.). New York: Dover Publications. see pp. 204ff in Vol. III.

RAMACHANDRAN, V. S. AND T. L. GREGORY (1991) Perceptual filling in of artificially induced scotomas in human vision. Nature 350: 699–702.

X

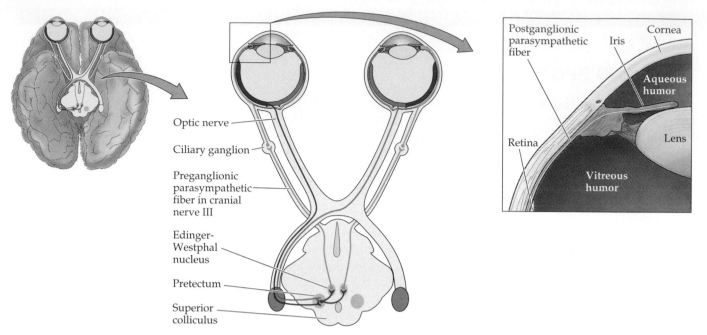

Figure 11.3 The circuitry responsible for the pupillary light reflex. This pathway includes bilateral projections from the retina to the pretectum and projections from the pretectum to the Edinger-Westphal nucleus. Neurons in the Edinger-Westphal nucleus terminate in the ciliary ganglion, and neurons in the ciliary ganglion innervate the pupillary constrictor muscles. Notice that this afferent axons activate both Edinger-Westphal nuclei via the neurons in the pretectum.

auditory). The **retinogeniculostriate pathway**, also called the **primary visual pathway**, is responsible for most of what is thought of as seeing. Thus, damage anywhere along this pathway results in serious visual impairment.

A second major target of the ganglion cell axons is a collection of neurons that lies between the thalamus and the midbrain in a region known as the **pretectum**. Although small in size compared to the lateral geniculate nucleus, the pretectum is particularly important as the coordinating center for the **pupillary light reflex**: the reduction in the diameter of the pupil following stimulation of the retina with light (Figure 11.3). The first part of the pathway responsible for the pupillary light reflex involves a bilateral projection from the retina to the pretectum. Pretectal neurons, in turn, project to the **Edinger-Westphal nucleus**, a group of nerve cells that lies close to the nucleus of the third cranial nerve in the midbrain. The Edinger-Westphal nucleus contains the preganglionic parasympathetic neurons that send their axons via the third cranial nerve to terminate on neurons in the ciliary ganglion. Neurons in the ciliary ganglion innervate the constrictor muscle in the iris, which decreases the diameter of the pupil when it is activated. Shining light in the eye leads to an increase in the activity of pretectal neurons, the Edinger-Westphal neurons, and the ciliary ganglion neurons, thus constricting the pupil.

In addition to its normal role in regulating the amount of light that enters the eye, the pupillary reflex provides an important diagnostic tool that allows the physician to test the intactness of the visual sensory apparatus, the motor outflow to the pupillary muscles, and the central pathways that mediate the reflex. Under normal conditions, the pupils of both eyes respond identically, regardless of which eye is stimulated; that is, light in one eye produces constriction of both the stimulated eye (the direct response) and the nonstimulated eye (the consensual response; see Figure 11.3). Comparing the response in the two eyes is often helpful in localizing a lesion. For example, a direct response in the left eye without a consensual response in the right eye suggests a problem with the motor outflow to the right eye, possibly damage to the third nerve. Failure to elicit response (either direct or indirect) to stimulation of the left eye if both eyes respond normally to stimulation of the right eye suggests damage to the sensory input from the left eye, possibly to the left optic nerve.

There are two other important targets of retinal ganglion cell axons. One is the **suprachiasmatic nucleus** of the hypothalamus, a small group of cell bodies at the base of the diencephalon (see Figure 26.7). The **retinohypothalamic pathway** is the route by which variation in light levels influences the broad spectrum of visceral functions that are entrained to the day/night cycle (see Chapter 26). The other target is the **superior colliculus**, a prominent structure visible on the dorsal surface of the midbrain (see Figure 1.12). The superior colliculus coordinates head and eye movements (see Chapter 19).

■ THE RETINOTOPIC REPRESENTATION OF THE VISUAL FIELD

The spatial relationships among the ganglion cells in the retina are maintained in their central targets. Central visual structures therefore exhibit an orderly representation, or map, of visual space. Importantly, information from the left half of the visual world is represented in the right half of the brain, and vice versa. Understanding the neural basis for this arrangement requires considering how images are projected onto the two retinas, and which parts of the two retinas cross at the optic chiasm.

Each eye sees a part of visual space that defines its **visual field** (Figure 11.4A). For descriptive purposes, each retina and its corresponding visual field are divided into quadrants. In this scheme, the surface of the retina is subdivided by vertical and horizontal lines that intersect at the center of the fovea (Figure 11.4B). The vertical line divides the retina into **nasal** and **temporal divisions** and the horizontal line divides the retina into **superior** and **inferior divisions**. Corresponding vertical and horizontal lines in visual space intersect at the **point of fixation** (the point in visual space that the fovea is aligned with) and define the quadrants of the visual field. The passage of light through the optical elements of the eye causes the images of objects in the visual field to be inverted and left-right reversed on the retinal surface. As a result, objects in the temporal part of the visual field are seen by the nasal part of the retina, and objects in the superior part of the visual field are seen by the inferior part of the retina. (It may help in understanding Figure 11.4B to imagine that you are looking at the back surfaces of the retinas, with the corresponding visual fields projected onto them.)

With both eyes open, the two foveas are normally aligned with a single target in visual space, causing the visual fields of both eyes to overlap extensively (see Figure 11.4B and Figure 11.5). This **binocular field** of view consists of two symmetrical visual hemifields (left and right). The left binocular hemifield includes the nasal visual field of the right eye and the temporal visual field of the left eye; the right hemifield includes the temporal visual field of the right eye and the nasal visual field of the left eye. Except for the extreme periphery of the field of view, which is seen by one eye or the other, all points in each visual hemifield are seen by both eyes; thus, most points in visual space lie in the nasal visual field of one eye and the temporal visual field of the other. The monocular portions of the visual field are seen by the most medial portion of the nasal retina of each eye, which has no temporal equivalent in the other eye (the shape of the face and nose prevent the lateral most temporal retina from seeing this portion of the field).

Figure 11.4 Projection of the visual fields onto the left and right retinas. (A) Projection of an image onto the surface of the retina. The passage of light rays through the optical elements of the eye results in images that are inverted and left/right reversed on the retinal surface. (B) Retinal quadrants and their relation to the organization of monocular and binocular visual fields, as viewed from the back surface of the eyes. Vertical and horizontal lines drawn through the center of the fovea define retinal quadrants (bottom). Comparable lines drawn through the point of fixation define visual field quadrants (center). Color coding illustrates corresponding retinal and visual field quadrants. The overlap of the two monocular visual fields is shown at the top.

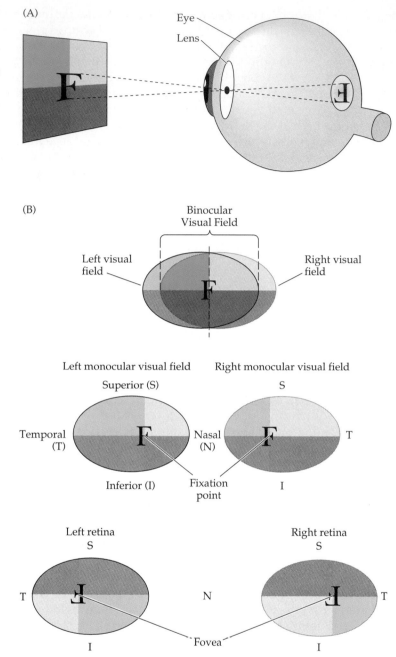

Ganglion cells that lie in the nasal division of each retina give rise to axons that cross in the chiasm, while those that lie in the temporal retina give rise to axons that remain on the same side (Figure 11.5). The boundary between contralaterally and ipsilaterally projecting ganglion cells (the line of decussation) is a line that runs through the center of the fovea and defines the border between the nasal and temporal hemiretinas. Images of objects in the left visual hemifield (such as point B in Figure 11.5) fall on the nasal retina of the left eye and the temporal retina of the right eye, and the axons from ganglion cells in these regions of the two retinas project through the right optic tract. Objects in the right visual hemifield (such as point C in Figure 11.5) fall on the nasal retina of the right eye and the temporal retina of

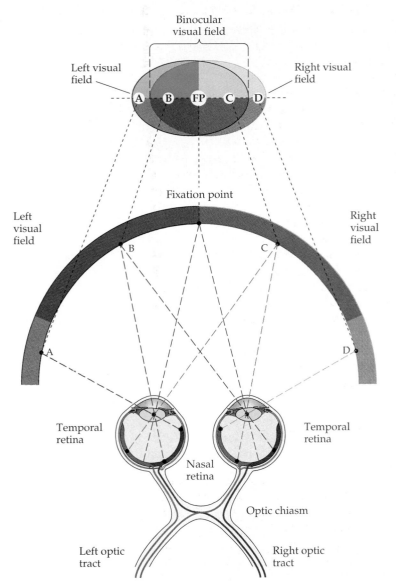

Figure 11.5 Projection of the binocular field of view onto the two retinas and its relation to the crossing of fibers in the optic chiasm. Points in the binocular portion of the left visual field (B) fall on the nasal retina of the left eye and the temporal retina of the right eye. Points in the binocular portion of the right visual field (C) fall on the nasal retina of the right eye and the temporal retina of the left eye. Points that lie in the monocular portions of the left and right visual fields (A and D) fall on the left and right nasal retinas, respectively. The axons of ganglion cells in the nasal retina cross in the optic chiasm, whereas those from the temporal retina do not. As a result, information from the left visual field is carried in the right optic tract, and information from the right visual field is carried in the left optic tract.

the left eye; the axons from ganglion cells in these regions project through the left optic tract. As mentioned previously, objects in the monocular portions of the visual hemifields (points A and D in Figure 11.5) are seen only by the extreme nasal retina of each eye; the axons of ganglion cells in these regions (like the rest of the nasal retina) project through the contralateral optic tract.

When the axons in the optic tract reach the lateral geniculate nucleus, they terminate in an orderly fashion creating a map of the contralateral hemifield (albeit in separate right and left eye layers; see Figure 11.14). The lateral geniculate neurons, in turn, maintain this topography in their projection to the striate cortex (Figure 11.6). The fovea is represented in the posterior part of the striate cortex, whereas the more peripheral regions of the retina are represented in progressively more anterior parts of the striate cortex. The upper visual field is represented below the calcarine sulcus, and the lower visual field above it. As in the case of the somatosensory system, the amount of cortical area devoted to a unit area of the sensory surface is not

(A)

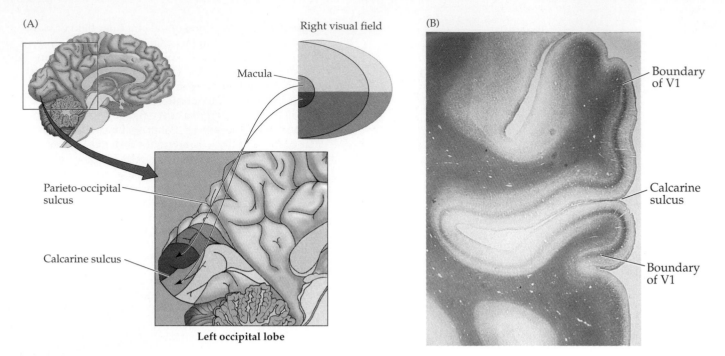

Right visual field

Macula

Parieto-occipital sulcus

Calcarine sulcus

Left occipital lobe

(B)

Boundary of V1

Calcarine sulcus

Boundary of V1

Figure 11.6 Visuotopic organization of the striate cortex in the left occipital lobe. (A) The primary visual cortex occupies a large part of the occipital lobe. The area of central vision (the fovea) is represented over a disproportionately large part of the caudal portion of the lobe, whereas peripheral vision is represented more anteriorly. The upper visual field is represented below the calcarine fissure, the lower field above the calcarine fissure. (B) Photomicrograph of a coronal section of the human striate cortex, showing the characteristic myelinated band, or stria, that gives this region of the cortex its name. The calcarine sulcus on the medial surface of the occipital lobe is indicated. (B courtesy of Tim Andrews and Dale Purves.)

uniform, but reflects the density of receptors and sensory fibers that supply the peripheral region. Thus, like the representation of the hand region in the somatosensory cortex, the representation of the fovea is disproportionately large, occupying most of the caudal pole of the occipital lobe.

A rough topography is also maintained in the projections from the lateral geniculate nucleus to the striate cortex. Some of these optic radiation axons loop out into the temporal lobe on their route to the striate cortex, an anomaly called **Meyer's loop** (Figure 11.7). Meyer's loop carries information from the superior portion of the contralateral visual field. More medial parts of the optic radiation, which pass under the parietal lobe, carry information from the inferior portion of the contralateral visual field.

Figure 11.7 Course of the optic radiation to the striate cortex. Axons carrying information about the superior portion of the visual field sweep around the lateral horn of the ventricle in the temporal lobe (Meyer's loop) before reaching the occipital lobe. Those carrying information about the inferior portion of the visual field travel in the parietal lobe.

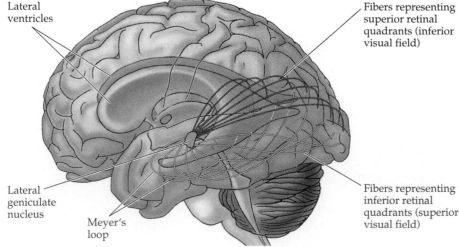

Lateral ventricles

Lateral geniculate nucleus

Meyer's loop

Fibers representing superior retinal quadrants (inferior visual field)

Fibers representing inferior retinal quadrants (superior visual field)

■ VISUAL FIELD DEFICITS

A wide variety of retinal or more central pathologies can cause visual field deficits that are limited to particular regions of visual space. Because the spatial relationships in the retinas are maintained in central visual structures, a careful mapping of the visual fields can often indicate the site of neurological damage. Relatively large visual field deficits are called **anopsias** (smaller ones are called scotomas; see Box A), a term that is combined with various prefixes to indicate the specific region of the visual field from which sight has been lost (Figure 11.8).

Damage to the retina or one of the optic nerves before it reaches the chiasm results in a loss of vision that is limited to the eye of origin. In contrast, damage in the region of the optic chiasm—or more centrally—results in specific types of deficits that involve the visual fields of both eyes. Damage to structures that are central to the optic chiasm, including the optic tract, lateral geniculate nucleus, optic radiation, and visual cortex, results in deficits that are limited to the contralateral visual hemifield. For example, interruption of the optic tract on the right results in a loss of sight in the left visual field (that is, blindness in the temporal visual field of the left eye and

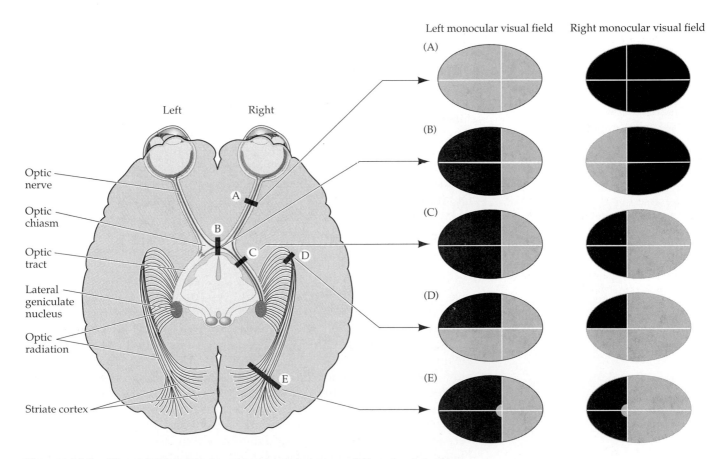

Figure 11.8   Visual field deficits resulting from damage at different points along the primary visual pathway. The panel on the left illustrates the basic organization of the primary visual pathway and indicates the location of various lesions. The right panels illustrate the visual field deficits associated with each lesion. (A) Loss of vision in right eye. (B) Bitemporal (heteronomous) hemianopsia. (C) Left homonomous hemianopsia. (D) Left superior quadrantanopsia. (E) Left homonomous hemianopsia with macular sparing.

the nasal visual field of the right eye). Because such damage affects corresponding parts of the visual field in each eye, there is a complete loss of vision in the affected region of the binocular visual field, and the deficit is referred to as a homonomous hemianopsia (in this case, a left homonomous hemianopsia).

In contrast, damage to the optic chiasm results in visual field deficits that involve noncorresponding parts of the visual field of each eye. For example, damage to the middle portion of the optic chiasm, commonly associated with pituitary tumors, can affect the fibers that are crossing from the nasal retina of each eye, leaving the uncrossed fibers from the temporal retinas intact. The resulting loss of vision is confined to the temporal visual field of each eye and is known as bitemporal hemianopsia. It is also called heteronomous hemianopsia to emphasize that the parts of the visual field that are lost in each eye do not overlap. Individuals with this condition are able to see in both left and right visual fields, provided both eyes are open. However, all information from the most peripheral parts of visual fields (which are seen only by the nasal retinas) is lost.

Damage to central visual structures is rarely complete; as a result, the deficits associated with damage to the chiasm, optic tract, optic radiation, or visual cortex will be more limited than those shown in Figure 11.8. This is especially true for damage along the optic radiation, which fans out under the temporal and parietal lobes in its course from the lateral geniculate nucleus to the striate cortex (see Figure 11.7). Damage to parts of the temporal lobe with involvement of Meyer's loop often result in a superior homonomous quadrantanopsia; damage to the optic radiation underlying the parietal lobe results in an inferior homonomous quadrantanopsia.

Damage to central visual structures can also lead to a phenomenon called macular sparing: the loss of vision throughout wide areas of the visual field, with the exception of foveal vision. Macular sparing is commonly found with damage to the cortex, but can be a feature of damage anywhere along the length of the visual pathway. Although several explanations for macular sparing have been offered, the basis for this selective preservation is not known.

■ THE FUNCTIONAL ORGANIZATION OF THE STRIATE CORTEX

The discovery by Stephen Kuffler that retinal ganglion cell receptive fields have a center-surround structure that conveys luminance contrast information to the brain (see Chapter 10) led David Hubel and Torsten Wiesel to consider the sorts of information extracted by more central visual structures. An analysis of the receptive field properties of neurons in the lateral geniculate nucleus showed surprisingly little difference from what had been found in the retina. In the striate cortex, however, the small spots of light that were so effective at stimulating neurons in the retina and lateral geniculate nucleus were largely ineffective. Instead, cortical neurons in cats and monkeys responded vigorously to light-dark bars or edges, and only if these bars were presented at a particular orientation within the cell's receptive field (Figure 11.9). Moreover, each cortical cell responded maximally to a narrow range of edge orientations (the cell's preferred orientation). Thus, all the orientations present in visual scenes appear to be encoded in the activity of distinct populations of **orientation-selective neurons**.

Hubel and Wiesel also found that within a class of neurons that preferred the same orientation, there were subtly different subtypes. For exam-

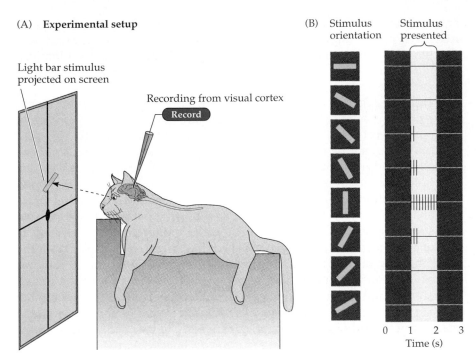

(A) **Experimental setup**

Light bar stimulus projected on screen

Recording from visual cortex

Record

(B) Stimulus orientation   Stimulus presented

0   1   2   3
Time (s)

Figure 11.9   Neurons in the visual cortex respond selectively to oriented edges. (A) An anesthetized animal is fitted with contact lenses to focus the eyes on a screen, where images can be projected; an extracellualr electrode records the responses of neurons in the visual cortex. (B) Neurons in visual cortex typically respond vigorously to a bar of light oriented at a particular angle and weakly—or not at all—to other orientations.

ple, the receptive fields of some cells, which they called **simple cells**, were composed of spatially separate "on" and "off" response zones, as if the "on" and "off" centers of the retinal ganglion cells that supplied these neurons were arrayed in separate parallel bands. Other neurons, referred to as **complex cells**, exhibited mixed "on" and "off" responses throughout the receptive field, as if they received their inputs from a number of simple cells. Further analysis uncovered cortical neurons sensitive to the length of the bar of light that was moved across their receptive field, decreasing their rate of response when the bar exceeded a certain length. Hubel and Wiesel called such neurons **hypercomplex** (or end-stopped) **cells**. Still other cells responded selectively to the direction in which an edge moved across their receptive field. Although the mechanisms responsible for generating these selective responses are still not fully understood, there is little doubt that the specificity of the receptive field properties of neurons in the striate cortex (and beyond) is essential for perceiving different aspects of a visual scene.

Another feature that distinguishes the responses of neurons in the striate cortex from those at earlier stages in the pathway is **binocularity**. Although the lateral geniculate nucleus receives axons from both eyes, these axons terminate in separate layers, so that individual neurons are monocular, driven by either the left or right eye but not by both (Figure 11.10; see also Figure 11.14). In some species, including most (but not all) primates, inputs from the left and right eyes remain segregated to some degree even beyond the geniculate because the axons of geniculate neurons terminate in alternating eye-specific columns within cortical layer IV—the so-called **ocu-**

lar dominance columns (see the next section). Beyond this point, the signals from the two eyes are combined at the cellular level. Thus, most cortical neurons have binocular receptive fields, and these fields are almost identical, having the same size, shape, preferred orientation, and roughly the same position in the visual field of each eye.

Bringing together the inputs from the two eyes at the level of the striate cortex provides a basis for one of the important cues for depth perception: **stereopsis**. Because the two eyes look at the world from slightly different angles, objects that lie in front of or behind the plane of fixation project to noncorresponding points on the two retinas. To convince yourself of this fact, hold your hand at arm's length and fixate on the tip of one finger. Maintain fixation on the finger as you hold a pencil in your other hand about half as far away. At this distance, the image of the pencil falls on noncorresponding points on the two retinas and will therefore be perceived as two separate pencils (a phenomenon called double vision, or diplopia). If the pencil is now moved toward the finger (the point of fixation), the two images of the pencil fuse and a single pencil is seen in front of the finger. Thus, for a small distance on either side of the plane of fixation, where the disparity between the two views of the world remains modest, a single image is perceived, and the disparity between the two eye views is interpreted as depth (Figure 11.11).

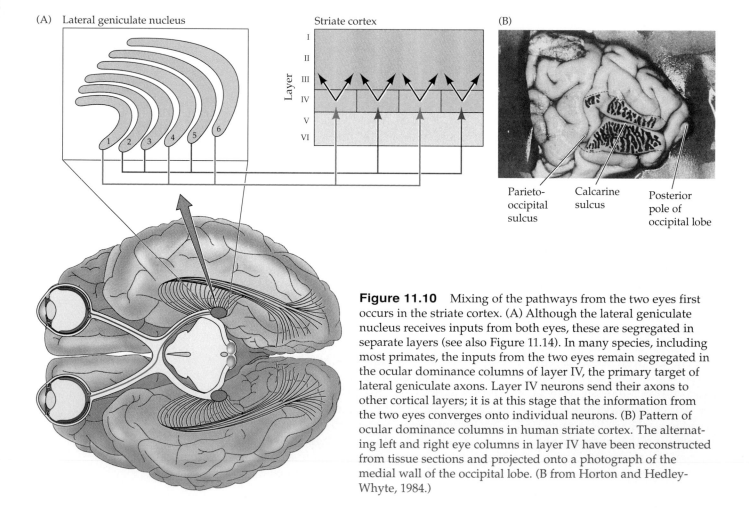

Figure 11.10 Mixing of the pathways from the two eyes first occurs in the striate cortex. (A) Although the lateral geniculate nucleus receives inputs from both eyes, these are segregated in separate layers (see also Figure 11.14). In many species, including most primates, the inputs from the two eyes remain segregated in the ocular dominance columns of layer IV, the primary target of lateral geniculate axons. Layer IV neurons send their axons to other cortical layers; it is at this stage that the information from the two eyes converges onto individual neurons. (B) Pattern of ocular dominance columns in human striate cortex. The alternating left and right eye columns in layer IV have been reconstructed from tissue sections and projected onto a photograph of the medial wall of the occipital lobe. (B from Horton and Hedley-Whyte, 1984.)

Figure 11.11 Binocular disparities are the basis of stereopsis. When the eyes are fixated on b, points that lie beyond the plane of fixation (point c) or in front of the point of fixation (point a) project to noncorresponding points on the two retinas. When these disparities are small, the images are fused and the disparity is interpreted by the brain as differences in depth. When the disparities are greater, we experience double vision.

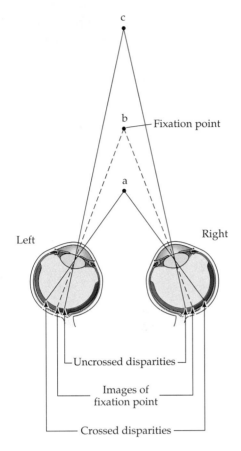

Although the neurophysiological basis of stereopsis is not completely understood, some neurons in the striate cortex have receptive field properties that make them good candidates for extracting information about depth. Unlike many binocular cells whose monocular receptive fields sample from the same region of visual space, these neurons have monocular fields that are slightly displaced, so that the cell is maximally activated by stimuli that fall on noncorresponding parts of the retinas. Some neurons (**far cells**) discharge to disparities beyond the plane of fixation, while others (**near cells**) respond to disparities in front of the plane of fixation. The pattern of activity in these different classes of neurons is thought to give rise to the sensation of stereoscopic depth (Box B).

The preservation of the binocular responses of cortical neurons is contingent on the normal activity from the two eyes during early postnatal life (see Chapter 21). Anything that creates an imbalance in the activity of the two eyes—for example, the clouding of one lens or the abnormal alignment of the eyes during infancy (strabismus)—can permanently reduce the effectiveness of one eye in driving cortical neurons and thus impair the ability to use binocular disparity information as a cue for depth. Early detection and correction of visual problems is therefore essential for normal visual function in maturity.

■ THE COLUMNAR ORGANIZATION OF THE STRIATE CORTEX

A special feature of the functional organization of the striate cortex is the grouping together of neurons that have similar response properties into radial arrays that span the thickness of the cortex. For example, all of the neurons encountered in a vertical penetration into the striate cortex (that is, perpendicular to its surface) have similar preferred orientations. Furthermore, electrode penetrations made tangential to the cortical surface show a systematic shift in preferred orientations, such that movements across the surface of about a millimeter are sufficient to encounter a series of neurons that cover the full range of orientations (Figure 11.12).

The columnar organization of the striate cortex is equally apparent in the binocular responses of cortical neurons. Although most neurons in the striate cortex respond to both eyes, the relative strength of the inputs from the two eyes varies from neuron to neuron. At the extremes of this continuum are neurons that respond almost exclusively to the left or right eye; in the middle are those that respond equally well to both eyes. As with orientation preference, vertical electrode penetrations encounter neurons with similar ocular preference (or ocular dominance, as it is usually called), and tangential penetrations show gradual shifts in ocular dominance as the electrode moves across the plane of the cortical surface (Figure 11.13). These shifts reflect the columnar segregation of the inputs from the two eyes in layer IV in ocular dominance columns (see Figure 11.10).

It may come as a surprise to learn that within the retinotopic map of visual space, response properties other than location are represented in an

Figure 11.12 Columnar organization of orientation selectivity in the monkey striate cortex. Vertical electrode penetrations encounter neurons with the same preferred orientations, whereas oblique penetrations show a systematic change in orientation across the cortical surface. Notice that there are no orientation-selective cells in layer IVc.

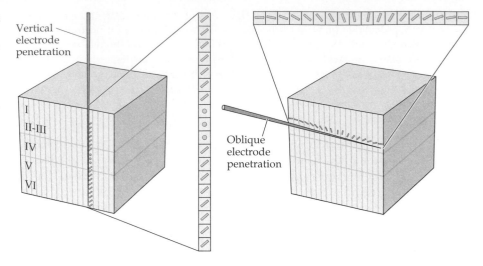

orderly fashion. The map of visual space is relatively gross; on a finer scale, each small region in visual space is represented within the receptive fields of neurons that are distributed over several millimeters of the cortical surface. This is more than enough cortical area to accommodate the columns of cells that are required to cover the complete range of orientation preferences and ocular dominance values (Box C).

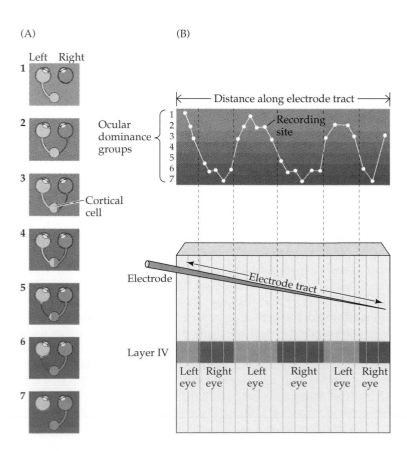

Figure 11.13 Columnar organization of ocular dominance. (A) Cortical neurons in all layers vary in the strength of their response to the inputs from the two eyes, from complete domination by one eye to equal influence of the two eyes. (B) Tangential electrode penetration across the superficial cortical layers reveals a gradual shift in the ocular dominance of the recorded neurons from one eye to the other. In contrast, all neurons encountered in a vertical electrode penetration (other than those neurons that lie in layer IV) tend to have the same ocular dominance.

Although the modular arrangement of the visual cortex was first recognized on the basis of orientation and ocular dominance columns, further work has shown that other stimulus features, such as color, direction of motion, and spatial frequency are also distributed in iterated patterns that are systematically related to each other (for example, orientation columns tend to intersect ocular dominance columns at right angles). Thus, the striate cortex is composed of repeating units, or modules, that contain all the neuronal machinery necessary to analyze a small region of visual space for a variety of different stimulus attributes. As described in Box B in Chapter 8, a number of other cortical regions show a similar columnar arrangement of their processing circuitry.

■ PARALLEL STREAMS OF INFORMATION FROM THE RETINA

The information passed on by the retina to central visual structures has been treated so far as if it were derived from a relatively uniform population of ganglion cells that differ only in sign ("on"-center or "off"-center). In fact, there are several functionally distinct populations of retinal ganglion cells, each of which has "on"- and "off"-center subtypes distributed across the surface of the retina. In primates, two that are of particular interest are called P and M ganglion cells (because of their relationship to the parvocellular and magnocellular layers of the geniculate, respectively) (Figure 11.14). M ganglion cells have larger cell bodies, larger dendritic fields, and larger diameter axons than P cells. These differences are expressed in their response properties; M ganglion cells have larger receptive fields than P cells, and their axons have faster conduction velocities.

M and P cells also differ in ways that are not so obviously related to their morphology. M cells respond transiently to the presentation of visual stimuli, while P cells respond in a sustained fashion. Moreover, P ganglion

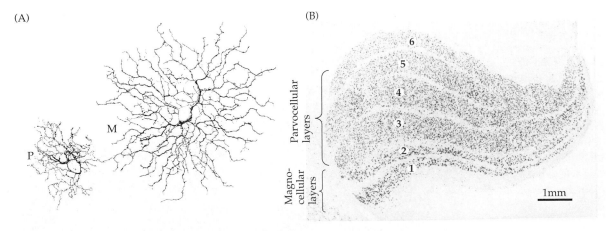

(A)

M

P

(B)

6
5
4
3
2
1

Parvocellular layers

Magno-cellular layers

1mm

Figure 11.14 Magno- and parvocellular streams. (A) Tracings of M and P ganglion cells as seen in flat mounts of the retina after staining by the Golgi method. M cells have large-diameter cell bodies and large dendritic fields. They supply the magnocellular layers of the lateral geniculate nucleus. P cells have smaller bodies and dendritic fields. They supply the parvocellular layers of the lateral geniculate nucleus. (B) Photomicrograph of the human lateral geniculate nucleus showing the magnocellular and parvocellular layers. (Courtesy of Tim Andrews and Dale Purves.)

Box B
RANDOM DOT STEREOGRAMS AND RELATED AMUSEMENTS

An important advance in understanding stereopsis was made in 1959 when Bela Julesz, then working at the Bell Laboratories in Murray Hill, New Jersey, discovered an ingenious way of showing that stereoscopy depends on matching information seen by the two eyes without any prior recognition of what object(s) such matching might generate. Julesz, a Hungarian whose background was in engineering and physics, was working on the problem of how to "break" camouflage. He surmised that the brain's ability to fuse the slightly different views of the two eyes to bring out new information would be an aid in overcoming military camouflage. Julesz also realized that, if his hypothesis was correct, a hidden figure in a random pattern presented to the two eyes should emerge when a portion of the otherwise identical pattern was shifted horizontally in the view of one eye or the other. A horizontal shift in one direction would cause the hidden object to appear in front of the plane of the background, whereas a shift in the other direction would cause the hidden object to appear in back of the plane. Such a figure, called a random dot stereogram, and the method of its creation are shown in the accompanying figure. The two images can be easily fused in a stereoscope (like the familiar Viewmaster® toy) but can also be fused simply by allowing the eyes to diverge. Most people find it easiest to do this by imagining that they are looking "through" the figure; after some seconds, during which the brain tries to make sense of what it is presented with, the two images merge and the hidden figure appears (in this case, a square that occupies the middle portion of the figure). The random dot stereogram has been widely used in stereoscopic research for about 30 years.

An impressive—and extraordinarily popular—derivative of the random dot stereogram is the autostereogram. The possibility of autostereograms was first discerned by the nineteenth-century British physicist Sir David Brewster. While staring at a Victorian wallpaper with an iterated but offset pattern, he noticed that when the patterns were fused, he perceived two different planes. The plethora of autostereograms that can be seen today in posters, books, and even newspapers are close cousins of the random dot stereogram in that computers are used to shift patterns of iterated information with respect to each other. The result is that different planes emerge from what appears to be a meaningless array of visual information (or, depending on the taste of the creator, an apparently "normal" scene in which the iterated and displaced information is hidden). Some autostereograms are designed to reveal the hidden figure

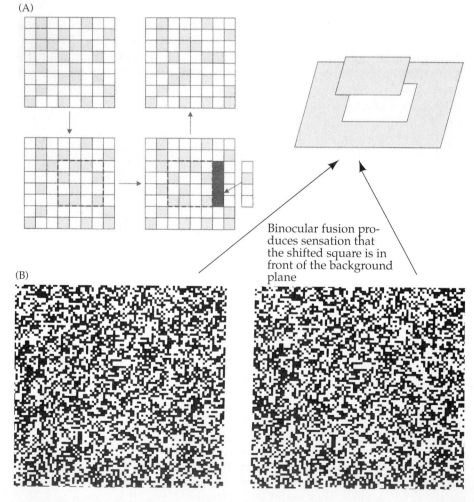

(A)

(B)

Binocular fusion produces sensation that the shifted square is in front of the background plane

when the eyes diverge, and others when they converge. (Looking at a plane more distant than the plane of the surface causes divergence; looking at a plane in front of the picture causes the eyes to converge.)

The elevation of the autostereogram to a popular art form should probably be attributed to Chris W. Tyler, a student of Julesz's and a visual psychophysicist, who created the first commercial autostereograms. Numerous graphic artists, preeminently in Japan, where the popularity of the autostereogram is enormous, are now involved in the elaboration of such images. As with the random dot stereogram, the task in viewing the autostereogram is not clear to the observer. Nonetheless, the hidden figure emerges, often after minutes of effort in which the brain automatically tries to make sense of the occult information. Still further entertainment possibilities are probably forthcoming. For example, images of moving objects can also be fused. Despite the failure of three-dimensional (3-D) movies in the 1950s (the need for polaroid glasses made the fad too cumbersome), the entertainment possibilities for viewing fused motion remain obvious.

References

JULESZ, B. (1971) *Foundations of Cyclopean Perception.* Chicago: The University of Chicago Press.

JULESZ, B. (1995) *Dialogues on Perception.* Cambridge, MA: MIT Press.

N. E. THING ENTERPRISES (1993) *Magic Eye: A New Way of Looking at the World.* Kansas City: Andrews and McMeel.

(C)

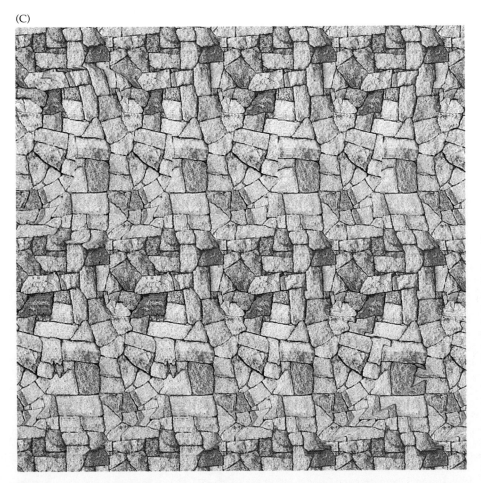

Random dot stereograms and autostereograms. (A) To construct a random dot stereogram, a random dot pattern is created to be observed by one eye. The stimulus for the other eye is created by copying the first image, displacing a particular region horizontally and then filling in the gap with a random sample of dots. (B) When the right and left images are viewed simultaneously, but independently by the two eyes (by using a stereoscope or fusing the images by converging or diverging the eyes) the shifted region (a square) appears to be in a different plane from the other dots. (C) An autostereogram. The hidden figure (3 geometrical forms) emerges by diverging the eyes in this case. (A from Wandell, 1995; C courtesy of Jun Oi.)

Box C
OPTICAL IMAGING OF FUNCTIONAL DOMAINS IN THE VISUAL CORTEX

The recent availability of optical imaging techniques has made it possible to visualize how response properties, such as the selectivity for edge orientation, direction of motion or ocular dominance are mapped across the cortical surface. These methods generally rely on intrinsic signals (changes in the amount of light reflected from the cortical surface) that are correlated with levels of neural activity. Such signals are thought to arise at least in part from local changes in the ratio of oxyhemoglobin and deoxyhemoglobin that accompany

neural activity; more active areas have a higher deoxyhemoglobin/oxyhemoglobin ratio (see also Box E in Chapter 1). Because the absorption spectrum for deoxyhemoglobin is shifted toward shorter wavelengths of light, when the cortical surface is illuminated with red light (605–700 nm), active cortical regions absorb more light than less active ones. With the use of a sensitive video camera, and averaging over a number of trials (the changes are small, 1 or 2 parts per thousand) it is possible to visualize these differences and use

them to map cortical patterns of activity (Figure A).

This approach has now been successfully applied to both striate and extrastriate areas in both experimental animals and human patients undergoing neurosurgery. The results emphasize that maps of stimulus features are a general principle of cortical organization. For example, orientation preference is mapped in a continuous fashion such that adjacent positions on the cortical surface tend to have only slightly shifted orientation preferences. However, there

(A)

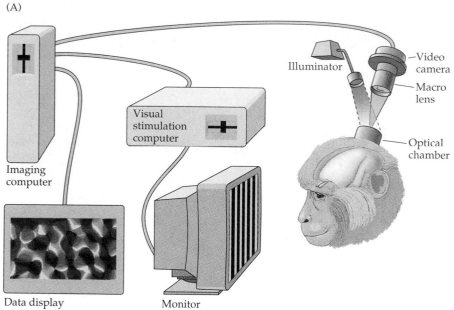

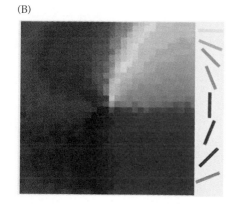

(B)

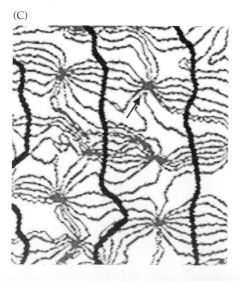

(C)

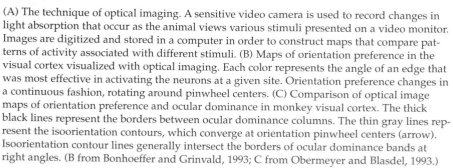

(A) The technique of optical imaging. A sensitive video camera is used to record changes in light absorption that occur as the animal views various stimuli presented on a video monitor. Images are digitized and stored in a computer in order to construct maps that compare patterns of activity associated with different stimuli. (B) Maps of orientation preference in the visual cortex visualized with optical imaging. Each color represents the angle of an edge that was most effective in activating the neurons at a given site. Orientation preference changes in a continuous fashion, rotating around pinwheel centers. (C) Comparison of optical image maps of orientation preference and ocular dominance in monkey visual cortex. The thick black lines represent the borders between ocular dominance columns. The thin gray lines represent the isoorientation contours, which converge at orientation pinwheel centers (arrow). Isoorientation contour lines generally intersect the borders of ocular dominance bands at right angles. (B from Bonhoeffer and Grinvald, 1993; C from Obermeyer and Blasdel, 1993.)

are points where continuity breaks down. Around these points, orientation preference is represented in a radial pattern resembling a pinwheel, covering the whole 180° of possible orientation values (Figure B).

This powerful technique can also be used to determine how maps for different stimulus properties are arranged relative to one another. A comparison of ocular dominance bands and orientation preference maps, for example, shows that pinwheel centers are generally located in the center of ocular dominance bands, and that the isoorientation contours that emanate from the pinwheel centers run orthogonal to the borders of ocular dominance bands (Figure C). An orderly relationship between maps of orientation selectivity and direction selectivity has also been demonstrated. These systematic relationships between the functional maps that coexist within primary visual cortex are thought to ensure that all combinations of stimulus features (orientation, direction, ocular dominance, and spatial frequency) are analyzed for all regions of visual space.

References

BLASDEL, G. G. AND G. SALAMA (1986) Voltage sensitive dyes reveal a modular organization in monkey striate cortex. Nature 321: 579–585.

BONHOEFFER, T. AND A. GRINVALD (1993) The layout of iso-orientation domains in area 18 of the cat visual cortex: Optical imaging reveals a pinwheel-like organization. J Neurosci 13: 4157–4180.

BONHOEFFER, T. AND A. GRINVALD (1996) Optical imaging based on intrinsic signals: The methodology. In Brain Mapping: The Methods, A. Toge, ed. New York: Academic Press

OBERMAYER, K. AND G. G. BLASDEL (1993) Geometry of orientation and ocular dominance columns in monkey striate cortex. J. Neurosci. 13: 4114–4129.

WELIKY, M., W. H. BOSKING AND D. FITZPATRICK (1996) A systematic map of direction preference in primary visual cortex. Nature 379: 725–728.

cells can transmit information about color, while M cells cannot. P cells convey color information because their receptive field centers and surrounds are driven by different classes of cones (either red, green, or blue). For example, some P ganglion cells have centers that receive inputs from long-wavelength (red) cones and surrounds that receive inputs from medium-wavelength (green) cones. Others have centers that receive inputs from green cones and surrounds from red cones. As a result, P cells are sensitive to differences in the wavelengths of light striking their receptive field center and surround. Although M ganglion cells also receive inputs from cones, there is no difference in the type of cone input to the receptive field center and surround; the center and surround of each M cell receptive field are driven by all cone types. Thus, M cells are largely insensitive to differences in the wavelengths of light striking their receptive field centers and surrounds and are therefore unable to transmit such information to their central targets.

M and P ganglion cells terminate in different layers of the lateral geniculate nucleus (Figure 11.14B). In addition to being specific for input from one eye or the other, the geniculate layers are also distinguished on the basis of cell size: M cells terminate selectively in the **magnocellular layers** of the lateral geniculate nucleus, while P cells terminate in the **parvocellular layers** (see Figure 11.14B). Because these two pathways remain segregated in the lateral geniculate nucleus and at least in the initial stages of cortical processing, the terms **magnocellular stream** and **parvocellular stream** are often used to describe the pathways that convey information derived from the two types of ganglion cells.

These differences in the response properties of M and P ganglion cells suggest that the magno- and parvocellular streams make different contributions to visual perception. This idea has been tested experimentally by examining the visual capabilities of monkeys after selectively damaging either the magno- or parvocellular layers of the lateral geniculate nucleus.

Damage to the magnocellular layers has little effect on visual acuity or color vision, but sharply reduces the ability to perceive quickly moving stimuli. In contrast, damage to the parvocellular layers has no effect on motion perception, but severely impairs visual acuity, and virtually eliminates color perception. These observations suggest that the visual information conveyed by the parvocellular stream is critical for high-resolution vision—the detailed analysis of the shape, size, and color of objects; the magnocellular stream conveys information that is critical for analyzing the movement of objects in space.

■ THE FUNCTIONAL ORGANIZATION OF EXTRASTRIATE VISUAL AREAS

Anatomical and electrophysiological studies in monkeys have led to the discovery of a multitude of areas in the occipital, parietal, and temporal lobes that are involved in processing visual information (Figure 11.15). Each of these areas contains a map of visual space, and each is largely dependent on the primary visual cortex for its activation. The response properties of the neurons in some of these areas suggest that they are specialized to deal with different aspects of the visual scene. For example, the **middle temporal area** (MT) contains neurons that respond selectively to the direction of a moving edge without regard to its color. In contrast, neurons in another cortical area called **V4**, respond selectively to the color of a visual stimulus without regard to its direction of movement. These physiological findings are supported by behavioral evidence; thus, damage to area MT leads to a specific impairment in a monkey's ability to perceive the direction of motion in a stimulus pattern, while other aspects of visual perception remain intact.

Is there, as one would expect, a comparable parceling out of function in the extrastriate areas of the human brain? Evidence on this point is still fragmentary, but an organization similar to that in the monkey is consistent with the clinical description of selective visual deficits after localized damage to the occipital lobe. For example, a patient who suffered a stroke that damaged an extrastriate region of the occipital cortex comparable to area MT in the monkey was unable to see objects in motion. The neurologist who treated her noted that she had difficulty in pouring tea or coffee into a cup because the fluid seemed to be "frozen." In addition, she could not stop pouring at the right time because she was unable to perceive when the fluid level had moved to the brim. The patient also had trouble following a dialogue because she could not see the movements of the speaker's mouth. Crossing the street was terrifying because she couldn't judge the movement of approaching cars. As the patient related, "when I'm looking at the car first, it seems far away. But then, when I want to cross the road, suddenly the car is very near." Her ability to perceive other features of the visual scene, such as color and form, was quite intact.

Another intriguing example of a specific visual deficit as a result of damage to extrastriate cortex is **cerebral achromatopsia**. These patients lose the ability to see the world in color, whereas other aspects of vision remain in good working order. The normal colors of a visual scene are described as being replaced by "dirty" shades of gray, much like looking at a black-and-white movie. These individuals know the normal colors of objects—that a school bus is yellow, an apple red—but can no longer see them. Thus, when asked to draw objects from memory, they have no difficulty with shapes but are unable to appropriately color the objects they have represented. It is important to distinguish this condition from the color blindness that arises

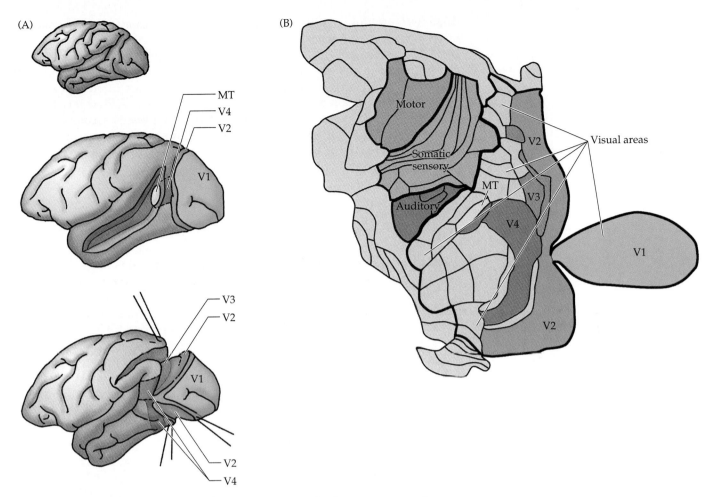

Figure 11.15 Subdivisions of the extrastriate cortex in the macaque monkey. (A) Each of these subdivisions contains neurons that respond to visual stimulation. Many are buried in sulci so that the overlying cortex must be removed in order to expose them. Some of the more extensively studied extrastriate areas are identified (V2, V3, V4, and MT). V1 is the primary visual cortex; MT is the middle temporal area. (B) The appearance of extrastriate and other areas of neocortex in a flattened view of the monkey neocortex. There are at least 25 areas that are predominantly or exclusively visual in function, plus 7 other areas suspected to play a role in visual processing. (A after Maunsell and Newsome, 1987; B after Felleman and Van Essen, 1991.)

from the congenital absence of one or more cone pigments in the retina (see Box C in Chapter 10). In achromatopsia, the three types of cones are functioning normally; it is damage to specific extrastriate cortical areas that renders the patient unable to use the information supplied by the retina.

Recent functional imaging studies in humans (see Boxes C and E in Chapter 1) have further supported the idea that separate areas of the extrastriate cortex process color and motion information. For example, normal subjects were asked to examine a color collage and one in which the colors had been replaced with shades of gray. When the patterns of cortical activity under these two conditions were compared, a particular area in the extrastriate cortex (the fusiform gyrus) was activated by the color pattern but not by the gray pattern. Neurons in the striate cortex were equally active under these two stimulus conditions. In another experiment, subjects were asked to view a stationary pattern of black-and-white squares and another pattern in which the squares moved in a random fashion. The distribution of neuronal activity in the extrastriate cortex under these conditions was quite different from that in the color experiments; the focus was more lateral, in an area located at the junction of the occipital and temporal lobes (Figure 11.16).

Based on the anatomical connections between visual areas, as well as differences in electrophysiological response properties, a consensus has emerged that extrastriate cortical areas are organized into two separate systems that feed information into cortical association areas in the temporal and parietal lobes (see Chapter 24). One system, which leads from the striate cortex into the temporal lobe (including area V4), is evidently responsible for high-resolution form vision and object recognition. The other, which leads from striate cortex into the parietal lobe (including the middle temporal area), is responsible for spatial aspects of vision, such as the analysis of motion and understanding the positional relationships between objects in the visual scene (Figure 11.17). This general idea has been supported by the observations of Mortimer Mishkin and his colleagues, who compared the effects of lesions in the parietal and inferotemporal cortices on the ability of monkeys to perform tasks that required spatial vision and object recognition. Lesions of the parietal cortex severely impaired the animals' ability to distinguish objects on the basis of their position, while having little effect on their ability to perform object recognition tasks. In contrast, lesions of the inferotemporal cortex produced profound impairments in the ability to perform

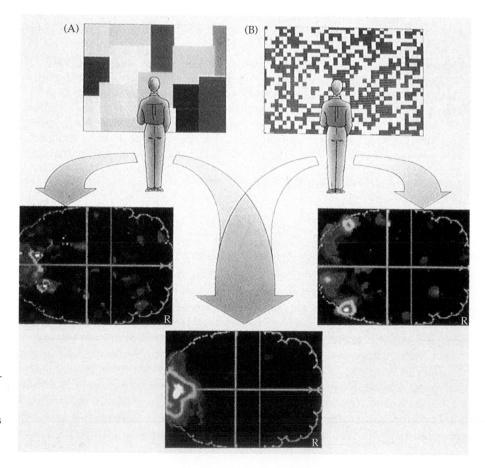

Figure 11.16 Comparison of functional maps of the extrastriate cortex in human subjects viewing color patterns (A) and moving patterns (B). These horizontal PET scan images (see Box E in Chapter 1) show that colored and moving visual stimuli activate different parts of the extrastriate cortex. (From Zeki, 1990.)

recognition tasks but no impairment in spatial tasks. These deficits are remarkably similar to the signs and symptoms in human patients after damage to the parietal and temporal lobes (see Chapter 24).

What, then, is the relationship between these "higher order" extrastriate visual pathways and the magno- and parvocellular streams that supply the primary visual cortex? Not long ago, it seemed that these intra-cortical pathways were simply a continuation of the geniculostriate pathways—that is, the magnocellular stream provided input to the parietal pathway and the parvocellular stream provided inputs to the temporal pathway. However, recent studies have indicated that the situation is more complicated. The temporal pathway clearly has access to the information conveyed by both the magno- and parvocellular streams, and the parietal pathway, while dominated by inputs from the magnocellular stream, may also receive inputs from the parvocellular stream. Thus, interaction and cooperation between the magno- and parvocellular streams appear to occur in most complex visual perceptions. Nevertheless, fundamental aspects of visual perception, such as the detection of rapidly moving objects, and the discrimination of color, represent distinct contributions of the magnocellular and parvocellular streams.

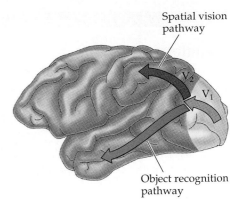

Figure 11.17 The visual areas beyond the striate cortex are broadly organized into two pathways: a ventral pathway that leads to the temporal lobe, and a dorsal pathway that leads to the parietal lobe. The ventral pathway plays an important role in object recognition, the dorsal pathway in spatial vision.

■ SUMMARY

Retinal ganglion cells send their axons to a number of central visual structures that serve different functions. The most important projections are to the pretectum for mediating the pupillary light reflex, to the hypothalamus for the regulation of circadian rhythms, to the superior colliculus for the regulation of eye and head movements, and—most important of all—to the lateral geniculate nucleus, for mediating visual perception. The retinogeniculate-striate projection (the primary visual pathway) is arranged topographically such that central visual structures contain an organized map of the contralateral visual field. Damage anywhere along the primary visual pathway, which includes the optic nerve, optic tract, lateral geniculate nucleus, optic radiation and striate cortex, results in a loss of vision confined to a predictable region of visual space. Compared to retinal ganglion cells, neurons at higher levels of the visual pathway become increasingly selective in their stimulus requirements. Thus, most neurons in the striate cortex respond to light-dark edges only if they are presented at a certain orientation; some are selective for the length of the edge, and others to movement of the edge in a specific direction. Indeed, a point in visual space is analyzed by a set of cortical neurons, each of which is specialized for detecting a limited set of the attributes present in the visual stimulus. The neural circuitry in the striate cortex also brings together information from the two eyes; most cortical neurons (other than those in layer IV, which are segregated into eye-specific columns) have binocular responses. Binocular convergence is essential for the detection of binocular disparity, an important component of depth perception. Finally, the visual system shows a remarkable degree of parallel function that begins at the retina. Separate classes of retinal ganglion cells supply the magno- and parvocellular layers of the lateral geniculate nucleus; these functional streams are specialized for the detection of rapidly moving stimuli (magnocellular stream) and color and acuity (parvocellular stream). Parceling of function continues in the pathways that lead from the striate cortex to the extrastriate and association areas in the temporal and parietal lobes. Areas in the inferotemporal cortex are especially important in object recognition, whereas areas in the parietal lobe are critical for understanding the spatial relations between objects in the visual field.

Additional Reading

Reviews

CHAPMAN, B. AND M. P. STRYKER (1992) Origin of orientation tuning in the visual cortex. Current Biology 2: 498–501.

DEYOE, E. A. AND D. C. VAN ESSEN (1988) Concurrent processing streams in monkey visual cortex. Trends Neurosci. 11: 219–226.

FELLEMAN, D. J. AND D. C. VAN ESSEN (1991) Distributed hierarchical processing in primate cerebral cortex. Cerebral Cortex 1: 1–47.

HORTON, J. C. (1992) The central visual pathways. In *Alder's Physiology of the Eye*. W. M. Hart (ed.). St. Louis: Mosby Yearbook.

HUBEL, D. H. AND T. N. WIESEL (1977) Functional architecture of macaque monkey visual cortex. Proc. R. Soc. (Lond.) 198: 1–59.

LIVINGSTON, E. M. AND D. H. HUBEL (1988) Segregation of form, color, movement, and depth: anatomy, physiology, and perception. Science 240: 740–749.

MAUNSELL, J. H. R. (1992) Functional visual streams. Curr. Opin. Neurobiol. 2: 506–510.

SCHILLER, P. H. AND N. K. LOGOTHETIS (1990) The color-opponent and broad-band channels of the primate visual system. Trends Neurosci. 13: 392–398.

UNGERLEIDER, J. G. AND M. MISHKIN (1982) Two cortical visual systems. In *Analysis of Visual Behavior*, D. J. Ingle, M. A. Goodale and R. J. W. Mansfield (eds.). Cambridge, MA: MIT Press, pp. 549–586.

Important Original Papers

HUBEL, D. H. AND T. N. WIESEL (1962) Receptive fields, binocular interaction and functional architecture in the cat's visual cortex. J. Physiol. (Lond.) 160: 106–154.

HUBEL, D. H. AND T. N. WIESEL (1968) Receptive fields and functional architecture of monkey striale cortex. J. Physiol. (Lond.) 195: 215–243.

ZEKI, S. (1974) Functional organization of a visual area in the posterior bank of the superior temporal sulcus of the rhesus monkey. J. Physiol. (Lond.) 236: 549–573.

ZIHL, J., D. CRAMON AND D. VON N MAI (1983) Selective disturbance of movement vision after bilateral brain damage. Brain 106: 313–340.

Books

HUBEL, D. H. (1988) *Eye, Brain, and Vision*. New York: Scientific American Library.

ZEKI, S. (1993) *A Vision of the Brain*. Oxford: Blackwell Scientific Publications.

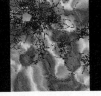

CHAPTER 12

THE AUDITORY SYSTEM

■ OVERVIEW

The auditory system is one of the engineering masterpieces of the human body. At the heart of the system is an array of miniature acoustical detectors packed into a space no larger than a pea. These detectors can faithfully transduce vibrations as small as the diameter of an atom, and respond a thousand times faster than visual photoreceptors. Although humans are highly visual creatures, much of human communication is mediated by the auditory system; indeed, deafness can be more socially debilitating than blindness. From a cultural perspective, the auditory system is essential not only to language, but also to music, one of the most aesthetically sophisticated forms of human expression. For these and other reasons, audition represents a fascinating and especially important dimension of brain function.

■ SOUND

Before considering the auditory system as such, it is useful to review some basic properties of sound waves and their propagation in elastic media such as air. When people speak of sound, they are usually referring to pressure waves generated by vibrating air molecules. Sound waves are much like the ripples that radiate outward when a rock is thrown in a pool of water. However, instead of occurring across a two-dimensional surface, sound waves propagate in three dimensions, creating spherical shells of alternating compression and rarefaction. Like all wave phenomena, sound waves have four major features: **waveform**, **phase**, **amplitude** (usually expressed in decibels, abbreviated dB) and **frequency** (expressed in cycles per second or Hertz, abbreviated Hz). For the human listener, the amplitude and frequency of a sound roughly correspond to **loudness** and **pitch**, respectively.

The waveform of a sound is its amplitude plotted against time. It helps to begin by visualizing an acoustical waveform as a sine wave. At the same time, it must be kept in mind that sounds composed of single sine waves are extremely rare in nature; most sounds in speech, for example, consist of acoustically complex waveforms. Interestingly, such complex waveforms can often be modeled as the sum of sinusoidal waves of varying amplitudes, frequencies, and phases. In engineering applications, an algorithm called the Fourier transform decomposes a complex signal into its sinusoidal components. In the auditory system, as will be apparent later in the chapter, the inner ear acts as a sort of acoustical prism, decomposing complex sounds into a myriad of constituent tones.

Figure 12.1 diagrams the behavior of air molecules near a tuning fork that vibrates sinusoidally when struck. The vibrating tines of the tuning fork produce local displacements of the surrounding molecules, such that when the tine moves in one direction, there is molecular condensation; when it moves in the other direction, there is rarefaction. These changes in density of the air molecules are equivalent to local changes in air pressure.

Such regular, sinusoidal cycles of compression and rarefaction can be thought of as a form of circular motion, with one complete cycle equivalent to one full revolution (360°). This point can be illustrated with two sinusoids

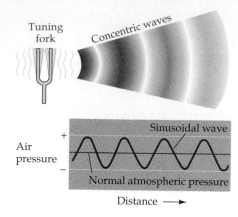

Figure 12.1 Diagram of the periodic condensation and rarefaction of air molecules produced by the vibrating tines of a tuning fork. The molecules are pictured as if frozen at the instant they responded to the resultant pressure wave. Shown below is a plot of the air pressure versus distance from the fork. Note its sinusoidal quality.

of the same frequency projected onto a circle, a strategy that also makes it easier to understand the concept of phase (Figure 12.2). Imagine that two tuning forks, both of which resonate at the same frequency, are struck at slightly different times. At a given time, $t = 0$, one wave is at position P and the other at position Q. By projecting P and Q onto the circle, their respective phase angles, θ_1 and θ_2, are apparent. The sine wave that starts at P reaches a particular point on the circle, say 180°, at time t_1, whereas the wave that starts at Q reaches 180° at time t_2. Thus, phase differences have corresponding time differences, a concept that is important in appreciating how the auditory system locates sounds in space (see below).

The human ear is extraordinarily sensitive to sound. At the threshold of hearing, air molecules are displaced an average of only 10 picometers (10^{-11}m), a distance 10,000 times smaller than the wavelength of visible light. The intensity of such a sound is about one-trillionth of a watt per square meter! This means a listener on an otherwise noiseless planet could hear a 1-watt, 3-kHz sound source located over 300 miles away (consider that very dim light bulbs consume more than 1 watt of power). Even dangerously high sound pressure levels exert power at the ear only in the milliwatt range (Box A).

■ THE AUDIBLE SPECTRUM

Humans can detect sounds in a frequency range from about 20 Hz to 20 kHz. (Human infants can actually hear frequencies slightly higher than 20 kHz, but lose some high-frequency sensitivity as they mature; the upper limit in average adults is often closer to 15–17 kHz.) Not all mammalian species are sensitive to the same range of frequencies. Most small mammals are sensitive to very high frequencies, but not to low frequencies. For instance, some species of bats are sensitive to tones as high as 200 kHz, but their lower limit is around 20 kHz—the upper limit for young people with normal hearing. One reason for these differences is that small objects are better resonators for high frequencies, whereas large objects are better for low frequencies (which is why the violin has a higher pitch than the cello).

■ A SYNOPSIS OF AUDITORY FUNCTION

Figure 12.2 A sine wave and its projection as circular motion. The two sinusoids shown are at different phases, such that point P corresponds to phase angle θ_1 and point Q corresponds to phase angle θ_2.

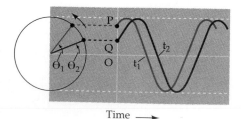

The auditory system transforms sound waves into distinct patterns of neural activity, which are then integrated with information from other sensory systems to guide behavior. The first stage of this transformation occurs at the external and middle ears, which collect sound waves and amplify their pressure, so that the sound energy in the air can be successfully transmitted to the fluid-filled cochlea of the inner ear. In the inner ear, a series of biomechanical processes occur that break up the signal into simpler, sinusoidal components, with the result that the frequency, amplitude, and phase of the original signal are all faithfully transduced by the sensory **hair cells** and encoded by the electrical activity of the **auditory nerve fibers**. The systematic representation of sound frequency along the length of the cochlea is referred to as **tonotopy**, an important feature that is preserved throughout the central auditory pathways. The earliest stage of central processing occurs at the cochlear nucleus, where the peripheral auditory information diverges into a number of parallel central pathways. Accordingly, the output of the cochlear nucleus has several targets. One of these is the superior olivary complex, the first place that information from the two ears interacts and the site of the initial processing of the cues that allow us to localize sound in

BOX A
THE WORLD'S LOUDEST ROCK AND ROLL BAND

Although noise-induced hearing loss is most often associated with industrial settings such as sawmills, or high-impulse noise such as gunfire, another cause is extremely loud rock and roll music. The advent of heavy-metal music has sometimes resulted in exposure of the performers, as well as the audience, to sound pressure levels of greater than 135 dB! (The decibel scale is a log scale that provides a standardized way to express the relative sound pressure level.) These sound levels eclipse previous marks set by "The Who," one of the loudest bands of the 1960s and 70s. To produce such high sound pressure levels, large banks of amplifiers and high-output public address systems are used to generate tens of thousands of watts of acoustical power.

The consequences of exposure to such loud sounds are considerable. The members of what has been billed as the world's loudest rock and roll band, "Man o' War," kindly agreed to undergo audiometric analyses immediately before and after a typical 90-minute concert. Distinct increases in the sound pressure levels required to reach hearing threshold were observed following the performance. These differences were as large as 16 dB (that is, 40-fold) for some frequency ranges. Only the lead singer's right ear was largely unaffected below 2 kHz, apparently because of his inclination to wear a plug in that ear. Some of this loss of sensitivity appears to be transient and typifies a phenomenon known as temporary threshold shift, which is thought to involve a transient softening

of the hair cells' stereocilia in response to loud noise. However, even in the preconcert audiogram, all the band members and the chief roadie had a detectable loss of sensitivity centered at 6 kHz. This deficiency probably reflects the early stages of a permanent hearing loss due to repeated acoustical trauma. The hearing of both the performers and the audience is evidently at risk in the outer realms of musical experience.

References

DRAKE-LEE, A. B. (1992) Beyond music: Auditory temporary threshold shift in rock musicians after a heavy metal concert. J. Roy. Soc. Med. 85(10): 617–619.

space. The cochlear nucleus also projects to the inferior colliculus of the midbrain, a major integrative center and the first place where auditory information can interact with the motor system. The inferior colliculus is an obligatory relay for information traveling to the thalamus and cortex, where more complex aspects of sound, especially those germane to speech, are processed.

■ THE EXTERNAL EAR

The external ear, which consists of the **pinna**, **concha**, and **auditory meatus**, gathers sound energy and focuses it on the eardrum, or **tympanic membrane** (Figure 12.3). One consequence of the configuration of the external ear is to selectively boost of the sound pressure 30- to 100-fold for frequencies around 3000 Hz. This amplification makes humans most sensitive to frequencies in this range—and also explains why they are particularly prone to acoustical injury and hearing loss near this frequency. Not surprisingly, most human speech sounds are distributed in the bandwidth around 3 kHz. Most vocal communication occurs in the low-kilohertz range because transmission of airborne sound is less efficient at higher frequencies, and the detection of lower frequencies is difficult for animals our size.

A second important function of the pinna and concha is to filter different sound frequencies in order to provide cues about the elevation of the sound source. The convolutions of the pinna are shaped so that the external ear transmits more high-frequency components from an elevated source than from the same source at ear level. This effect can be demonstrated by

Figure 12.3 The human ear. Note the large surface area of the tympanic membrane (eardrum) relative to the oval window.

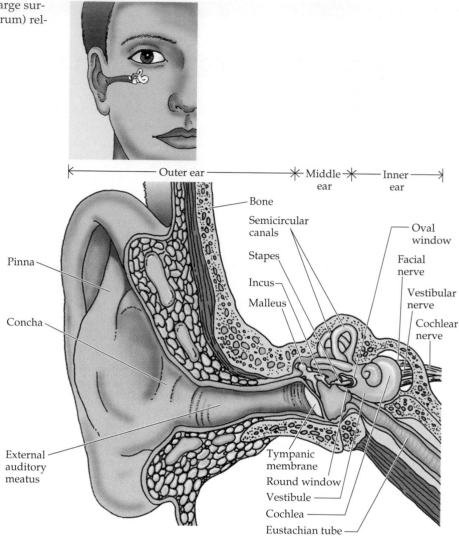

recording sounds from different elevations after they have passed through an artificial external ear; when the recorded sounds are played back via earphones, so that the whole series is at the same elevation relative to the listener, the recordings from higher elevations seem to come from positions higher in space than the recordings from lower elevations.

■ THE MIDDLE EAR

Sounds impinging on the external ear are airborne; however, the environment within the inner ear, where the sound-induced vibrations are converted to neural impulses, is aqueous. The major function of the middle ear is to match relatively low impedance airborne sounds to the higher impedance fluid of the inner ear. The term impedance in this context describes a medium's resistance to movement. Normally, when sound travels from a low-impedance medium, like air, to a much higher impedance medium like water, almost all (more than 99.9%) of the acoustical energy is reflected. The middle ear (see Figure 12.3) overcomes this problem and ensures transmission of the sound energy across the air-fluid boundary by boosting the pressure measured at the tympanic membrane almost 200-fold by the time it reaches the inner ear.

Two mechanical processes occur within the middle ear to achieve this large pressure gain. The first and major boost is achieved by focusing the force impinging on the relatively large diameter tympanic membrane onto the much smaller diameter **oval window**, the site where the bones of the middle ear contact the inner ear. A second and related process relies on the mechanical advantage gained by the lever action of the three small interconnected middle ear bones, or **ossicles** (i.e., the malleus, incus, and stapes; see Figure 12.3) which connect the tympanic membrane to the oval window.

Bone and soft tissue have impedances close to that of water. Therefore, even in the absence of an intact tympanic membrane or middle ear ossicles, acoustical vibrations can still be transferred directly through the bones and tissues of the head to the oval window and hence to the inner ear. By applying a vibrating source, such as a tuning fork, directly to a patient's skull, the physician can therefore determine if a hearing loss is due to mechanical damage in the middle ear or a problem in the inner ear or central auditory pathways.

■ THE INNER EAR

The **cochlea** of the inner ear is the most critical structure in the auditory pathway, for it is here that the energy from sonically generated pressure waves is transformed into neural impulses. The cochlea not only amplifies sound waves and converts them into neural impulses, but also acts as a mechanical frequency analyzer, decomposing complex acoustical waveforms into simpler elements. Many features of auditory perception derive directly from the physical properties of the cochlea; hence, it is worth examining this structure in some detail.

The cochlea (from the Latin for "snail") is a small (about 10 mm wide) coiled structure, which when uncoiled forms a tube about 35 mm long (Figure 12.4). Both the oval window and the **round window** are at the basal end of this tube. The cochlea is bisected by the cochlear partition, a flexible structure that supports the **basilar membrane** and the **tectorial membrane**. There are fluid-filled spaces on each side of the cochlear partition, named the **scala vestibuli** and the **scala tympani**; the **scala media** is a distinct channel that runs within the cochlear partition. The cochlear partition does not extend all the way to the apical end of the cochlea; instead there is an opening, known as the **helicotrema**, that joins the scala vestibuli to the scala tympani. As a result, a brief application of inward pressure at the oval window causes the round window to bulge out slightly and deforms the basilar membrane.

The manner in which the basilar membrane vibrates in response to sound is the key to understanding cochlear function. Measurements of the vibration of different parts of the basilar membrane, as well as the discharge rates of individual auditory nerve fibers, show that both these features are tuned; that is, they are greatest in response to a sound of a specific frequency. Frequency tuning within the inner ear is attributable in part to the geometry of the basilar membrane, which is wider and more flexible at the apical end and narrower and stiffer at the basal end. Georg von Békésy showed that a membrane of varying width and flexibility generates vibrations at different positions in response to different frequencies (Figure 12.5). Using tubular models and human cochleae taken from cadavers, he found that an auditory stimulus initiates a traveling wave of the same frequency in the cochlea, which propagates from the base to the apex of the basilar membrane, growing in amplitude and slowing in velocity until a point of maxi-

Figure 12.4 The cochlea, viewed face-on (upper left) and in cross section (subsequent panels). The stapes transfers force from the tympanic membrane to the oval window. The hair cells are named for their tufts of stereocilia; inner hair cells receive afferent inputs from the VIII nerve, whereas outer hair cells receive mostly efferent input. As a result, the round window bulges outward when the stapes compresses the oval window, thus deforming the basilar membrane, which in turn deflects the stereocilia of the hair cells. The only point of fluid continuity between the scala vestibuli and scala tympani is at the helicotrema. The cross sections show the scala media between the scalae vestibuli and tympani; the hair cells are located between the basilar and tectorial membranes.

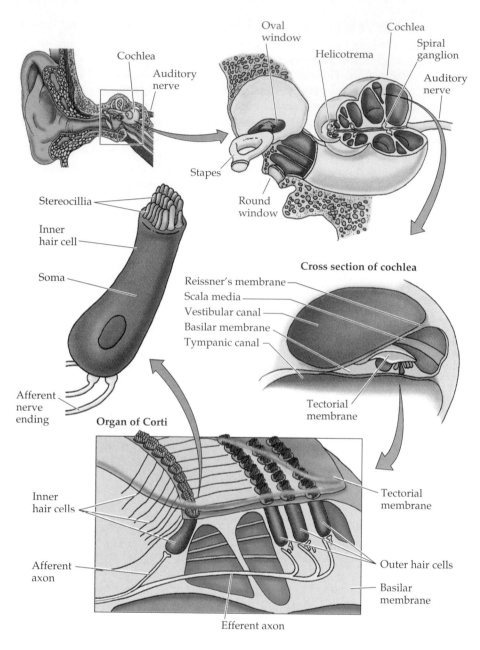

mum displacement is reached. This point is determined by the frequency of the sound. The points responding to high frequencies are at the base of the basilar membrane, and the points responding to low frequencies are at the apex, giving rise to a topographical mapping of frequency (that is, to tonotopy). An important feature of the tonotopically organized basilar membrane is that complex sounds cause a pattern of vibration equivalent to the superposition of the vibrations generated by the individual tones making up that complex sound, thus accounting for the decompositional aspects of cochlear function mentioned earlier.

Von Békésy's model of cochlear mechanics was a passive one, resting on the premise that the basilar membrane acts like a series of linked resonators, much like a concatenated set of tuning forks. Each point on the basilar membrane was postulated to have a characteristic frequency at which it

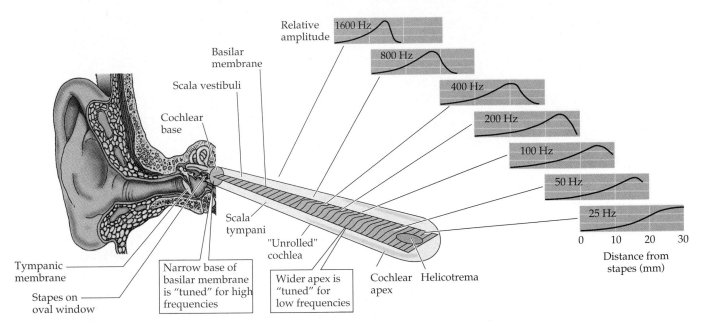

Figure 12.5 Traveling waves along the cochlea. A traveling wave is shown at a given instant along the cochlea, which has been uncoiled for clarity. The graphs profile the amplitude of the traveling wave along the basilar membrane for different frequencies, and show that the position where the traveling wave reaches its maximum amplitude varies directly with the frequency of stimulation. (Drawing after Dallos, 1992; graphs after von Békésy, 1960.)

vibrated most efficiently, but because it was physically linked to adjacent areas of the membrane, it also vibrated (somewhat less readily) at other frequencies, thus permitting propagation of the traveling wave. It is now clear that the tuning of the auditory periphery, whether measured at the basilar membrane or recorded in the electrical activity of auditory nerve fibers, is too sharp to be explained by solely passive mechanics. At very low sound intensities, the basilar membrane vibrates much more than would be predicted by linear extrapolation from the motion measured at high intensities. Therefore, the ear's sensitivity arises from an active biomechanical process, as well as from its passive resonant properties. The outer hair cells, which together with the inner hair cells comprise the sensory cells of the inner ear, are the most likely candidates for driving this active process. The details of this process are only poorly understood.

The motion of the traveling wave initiates sensory transduction by displacing the hair cells that sit atop the basilar membrane. A shearing motion occurs between the basilar membrane and the overlying tectorial membrane in response to the wave, because these structures are anchored at different positions (Figure 12.6). This motion bends tiny processes, called **stereocilia**, that protrude from the apical ends of the hair cells, leading to voltage changes across the hair cell membrane. How the bending of stereocilia leads to receptor potentials in hair cells is considered in the following section.

■ HAIR CELLS AND THE MECHANOELECTRICAL TRANSDUCTION OF SOUND WAVES

The hair cell is an evolutionary triumph that solves the problem of transforming vibrational energy into an electrical signal. The scale at which the

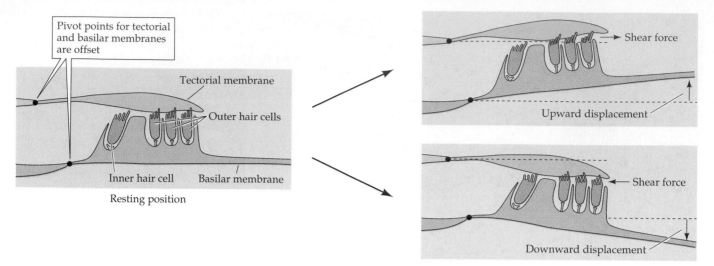

Figure 12.6 Movement of the basilar membrane creates a shearing force that bends the stereocilia of the hair cells. The pivot point of the basilar membrane is offset from the pivot point of the tectorial membrane, so that when the basilar membrane is displaced, the tectorial membrane moves across the tops of the hair cells, bending the stereocilia.

hair cell operates is truly amazing. At the limits of human hearing, hair cells can faithfully detect movements of atomic dimensions and respond in the tens of microseconds. Furthermore, hair cells can adapt rapidly to constant stimuli, thus allowing the listener to extract signals from a noisy background.

The hair cell is a flask-shaped epithelial cell named for the bundle of hairlike processes that protrude from its apical end into the scala media. Each hair bundle contains anywhere from 30 to a few hundred hexagonally arranged stereocilia, with one taller **kinocilium** (Figure 12.7A). Despite their names, only the kinocilium is a true ciliary structure, with the characteristic two central tubules surrounded by nine doublet tubules (Figure 12.7B). The function of the kinocilium is unclear, and in the cochlea of humans and other mammals it actually disappears shortly after birth. The stereocilia are simpler, containing only an actin cytoskeleton. Each stereocilium tapers where it inserts into the apical membrane, forming a hinge about which each stereocilium can pivot (Figure 12.7C). The stereocilia are graded in height, with the taller ones being closest to the kinocilium. This arrangement creates a plane of bilateral symmetry running through the kinocilium. Displacement of the hair bundle in this plane towards the tallest stereocilia depolarizes the hair cell, while movements toward the shortest stereocilia cause hyperpolarization. In contrast, displacements perpendicular to the plane of symmetry do not alter the hair cell's membrane potential. The hair bundle movements at the threshold of hearing are approximately 0.3 nm, about the diameter of a single atom of gold!

Hair cells can convert the displacement of the stereociliary bundle into an electrical potential in as little as 10 μs; such speed is required to faithfully transduce high-frequency signals and enable the accurate localization of the source of the sound. This need for microsecond resolution places certain constraints on the transduction mechanism, ruling out the relatively slow second messenger pathways used in visual and olfactory transduction; a

(A)

(B)

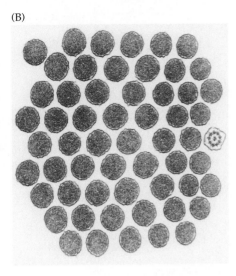

(C)

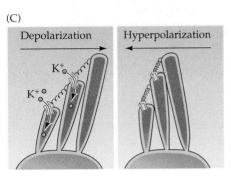

Figure 12.7 The structure and function of the hair bundle. The vestibular hair bundles shown here resemble those of cochlear hair cells, except for the presence of the kinocilium, which disappears in the mammalian cochlea shortly after birth. (A) The hair bundle of a guinea pig vestibular hair cell. This view shows the increasing height leading to the kinocilium (arrow). (B) A cross section through the hair bundle shows the 9+2 array of microtubules in the kinocilium (on right), which contrasts with the simpler actin filament structure of the stereocilia. (C) Diagram of the stereocilia and tip links, which when stretched by movement toward the kinocilium open channels that generate a depolarizing current. When compressed, these same structures lead to hyperpolarization of the hair cell (A from Lindeman, 1973; B from Hudspeth, 1983; C after Pickles et al., 1984.)

direct, mechanically gated transduction channel is needed to operate this quickly. Evidently the filamentous structures that link the tips of adjacent stereocilia, known as **tip links**, directly open cation-selective transduction channels when stretched, allowing potassium ions to flow into the cell (see Figure 12.7C). As the linked stereocilia pivot from side to side, the tension on the tip link varies, modulating the ionic flow and resulting in a graded receptor potential that follows the movements of the stereocilia (Figure 12.8). The tip link model also explains why only deflections along the axis of the hair bundle activate transduction channels since tip links join adjacent stereocilia along the axis directed toward the tallest stereocilia (see also Box A in Chapter 13).

Understanding the ionic basis of hair cell transduction has been revolutionized by intracellular recordings made from these tiny structures. The hair cell has a resting potential between -45 and -60 mV relative to the fluid that bathes the basal end of the cell. At the resting potential, only a small fraction of the potassium-selective transduction channels are open. When the hair bundle is displaced in the direction of the tallest stereocilium, more transduction channels open, causing depolarization as K^+ enters the cell. Depolarization in turn opens voltage-gated calcium channels in the hair cell membrane, and the resultant Ca^{2+} influx causes more transmitter release from the basal end of the cell onto the auditory nerve endings (Figure 12.8A). Such calcium-dependent exocytosis is similar to chemical neurotransmission elsewhere in the brain. Because some of the transduction channels are open at rest, the

(A)

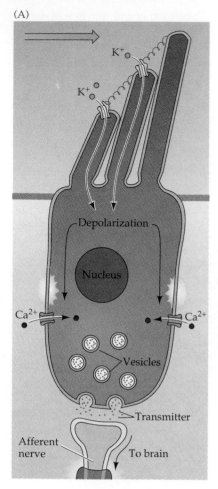

(B)

(C)

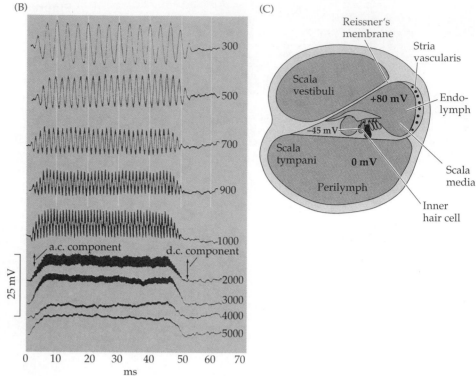

Figure 12.8 Mechanoelectrical transduction mediated by hair cells. (A) When the hair bundle is deflected toward the tallest stereocilium, cation-selective channels open near the tips of the stereocilia, allowing K⁺ ions to flow into the hair cell down their electrochemical gradient (see text for the explanation of this peculiar situation). The resulting depolarization of the hair cell opens voltage-gated Ca^{2+} channels in the cell soma, allowing calcium entry and release of neurotransmitter onto the nerve endings of the auditory nerve. (B) Potential changes within an individual hair cell in the cochlea in response to pure tones (indicated in Hz at the right of the tracings). Note that the hair cell potential faithfully follows the waveform of the stimulating sinusoids for low frequencies. (C) The stereocilia of the hair cells protrude into the endolymph, which is high in K⁺ and has an electrical potential of +80 mV relative to the perilymph. (A after Hudspeth, 1983; B after Palmer and Russell, 1986.)

receptor potential is biphasic: movement toward the tallest stereocilia depolarizes the cell, while movement in the opposite direction leads to hyperpolarization. This situation allows the hair cell to generate a sinusoidal receptor potential in response to a sinusoidal stimulus, thus preserving the temporal information present in the original signal (Figure 12.8B).

An unusual adaptation of the hair cell is that K⁺ serves both to depolarize and repolarize the cell. As with other epithelial cells, the basal and apical surfaces of the hair cell are separated by tight junctions. The apical end is exposed to the potassium-rich, sodium-poor **endolymph**, which is produced by the **stria vascularis** (Figure 12.8C). The basal end is bathed in the same fluid that fills the scala tympani, known as **perilymph**, which resembles other extracellular fluids in that it is potassium-poor and sodium-rich. The compartment containing endolymph is about 80 mV more positive than the perilymph compartment, while the inside of the hair cell is about 45 mV more negative than the perilymph (and 125 mV more negative than the endolymph). The resulting electrical gradient across the membrane of the stereocilia (about 125 mV) drives K^+ through open transduction channels into the hair cell, even though these cells already have a high internal K^+ concentration. K^+ entry via the transduction channels leads to depolarization of the hair cell, which in turn opens voltage-gated Ca^{2+} and K^+ channels located on the hair cell soma (see Box B in Chapter 13). The opening of somatic K^+ channels favors K^+ efflux, and thus repolarization; the efflux occurs because the perilymph surrounding the basal end is low in K^+ relative to the cytosol, and because the equilibrium potential for K^+ is more negative than the hair cell's resting potential.

Repolarization of the hair cell via K^+ efflux is also facilitated by Ca^{2+} entry. In addition to modulating the release of neurotransmitter, Ca^{2+} entry opens Ca^{2+}-dependent K^+ channels, which provide another avenue for K^+ to enter the perilymph. Indeed, the interaction of Ca^{2+} influx and Ca^{2+}-dependent K^+ efflux can lead to electrical resonances that enhance the tuning of response properties within the inner ear (see Box B in Chapter 13). In short, the hair cell exploits the different ionic milieus of its apical and basal surfaces to provide extremely fast and energy-efficient repolarization.

■ TWO KINDS OF HAIR CELLS IN THE COCHLEA

The cochlear hair cells in humans consist of one row of **inner hair cells** and three rows of **outer hair cells**. The inner hair cells are the actual sensory receptors, and 95% of the fibers of the auditory nerve that project to the brain arise from this subpopulation. The terminations on the outer hair cells are almost all from *descending* axons that arise from cells elsewhere in the brain. Why do the outer hair cells, which make up 75% of the total population, give rise to only 5% of the afferent fibers? For years this anatomical fact remained a puzzle. A clue to the mystery was provided by the discovery that basilar membrane motion is influenced by an active process within the cochlea, as noted above. First, it was found that the cochlea actually emits sound under certain conditions. These otoacoustical emissions can be detected by placing a sensitive microphone at the eardrum and monitoring the response after briefly presenting a tone. Such emissions can also occur spontaneously. These observations indicate that a process within the cochlea is capable of producing sound. Second, the outer hair cells, unlike the inner ones, contain many actin filaments, raising the possibility that they contract. Finally, isolated outer hair cells move in response to small electrical currents, apparently due to the transduction process being driven in reverse. Thus, it seems likely that the outer hair cells sharpen the frequency-resolving power of the cochlea by actively contracting and relaxing, thus changing the stiffness of the tectorial membrane at particular locations. An active process of this sort is necessary in any event to explain the nonlinear vibration of the basilar membrane at low sound intensities (Box B).

■ TUNING AND TIMING IN THE AUDITORY NERVE

The rapid response time of the transduction apparatus allows the membrane potential of the hair cell to follow deflections of the hair bundle up to quite high frequencies of oscillation. In humans, the receptor potentials of certain hair cells and the action potentials of their associated auditory nerve fiber can follow stimuli of up to 3 kHz in a one-to-one fashion. Such real-time encoding of stimulus frequency by the pattern of action potentials in the auditory nerve is known as the "volley theory of auditory information transfer." Even these extraordinarily rapid processes, however, fail to follow

Box B
THE SWEET SOUND OF DISTORTION

As early as the first half of the eighteenth century, musical composers such as Giuseppe Tartini and W. A. Sorge discovered that upon playing pairs of tones, other tones not present in the original stimulus are also heard. These combination tones, fc, are mathematically related to the played tones, f_1 and f_2 ($f_2 > f_1$), by the formula

$$fc = mf_1 \pm nf_2$$

where m and n are positive integers. Combination tones have been used for a variety of compositional effects, as they can strengthen the harmonic texture of a chord. Furthermore, organ builders sometimes use the difference tone ($f_2 - f_1$) created by two smaller organ pipes to produce the extremely low tones that would otherwise require building one especially large pipe.

Modern experiments indicate that this distortion product is actually due to the nonlinear properties of the inner ear. M. Ruggero and his colleagues placed small glass beads (10–30 μm in diameter) on the basilar membrane of an anesthetized animal and then determined the velocity of the basilar membrane in response to different combinations of tones by measuring the Doppler shift of laser light reflected from the beads. When two tones were played into the ear, the basilar membrane vibrated not only at those two frequencies, but also at other frequencies predicted by the above formula.

Related experiments on hair cells studied in vitro suggest that these nonlinearities result from the properties of the mechanical linkage of the transduction apparatus. By moving the hair bundle sinusoidally with a metal-coated glass fiber, A. J. Hudspeth and his coworkers found that the hair bundle exerts a force at the same frequency. However, when two sinusoids were applied simultaneously, the forces exerted by the hair bundle occurred not only at the primary frequencies, but at several combination frequencies as well. These distortion products are due to the transduction apparatus, since blocking the transduction channels causes the forces exerted at the combination frequencies to disappear, even though the forces at the primary frequencies remain unaffected. Apparently the tip links add a certain extra springiness to the hair bundle in the small range of motions over which the transduction channels are changing between closed and open states. If nonlinear distortions of basilar membrane vibrations arise from the properties of the hair bundle, then it is likely that hair cells can indeed influence basilar membrane motion, thereby accounting for the cochlea's extreme sensitivity. Apparently, when we hear difference tones, we are paying the price in distortion for an exquisitely fast and sensitive transduction mechanism.

References

PLANCHART, A. E. (1960) A study of the theories of Giuseppe Tartini. J. Music Theory, 4(1): 32–61.

ROBLES, L., M. A. RUGGERO AND N. C. RICH (1991) Two-tone distortion in the basilar membrane of the cochlea. Nature 439: 413–414.

JARAMILLO, F., V. S. MARKIN AND A. J. HUDSPETH (1993) Auditory illusions and the single hair cell. Nature 364: 527–529.

frequencies above 3 kHz (see Figure 12.8B). Accordingly, some other mechanism must be used to transmit auditory information at higher frequencies. The tonotopically organized basilar membrane provides an alternative to temporal coding, namely a "labeled-line" coding mechanism. In this case, frequency information is specified by preserving the tonotopy of the cochlea at higher levels in the auditory pathway. Because the auditory nerve fibers innervate the inner hair cells in approximately a one-to-one ratio, each auditory nerve fiber transmits information about only a small part of the audible frequency spectrum. As a result, auditory nerve fibers related to the apical end of the cochlea respond to low frequencies, and fibers that are related to the basal end respond to high frequencies (see Figure 12.5). The limitations of specific fibers can be seen in electrophysiological recordings of responses to sound (Figure 12.9). These threshold functions are called **tuning curves**, and the peak (i.e., the lowest threshold) of the tuning curve is called the

characteristic frequency. Since the topographical order of the characteristic frequency of neurons is retained throughout the system, information about frequency is also preserved.

The other prominent feature of hair cell function—their ability to follow the waveform of low-frequency sounds—is also important in more subtle aspects of auditory coding. As mentioned earlier, hair cells have biphasic response properties. Because hair cells release transmitter only when depolarized, auditory nerve fibers fire only during the positive phases of low-frequency sounds (Figure 12.10). The "phase locking" that results provides temporal information from the two ears to neural centers that compare interaural time differences. The evaluation of interaural time differences provides a critical cue for sound localization, by means of which a person is able to perceive auditory "space." That auditory space can be perceived in this way is especially remarkable, given that the cochlea, unlike the retina, cannot represent space directly.

■ HOW INFORMATION FROM THE COCHLEA REACHES TARGETS IN THE BRAINSTEM

A hallmark of the ascending auditory system is its parallel organization. This arrangement becomes evident as soon as the auditory nerve enters the brainstem, where it branches to innervate the three divisions of the cochlear nucleus. The auditory nerve (the major component of cranial nerve VIII) comprises the central processes of the bipolar spiral ganglion cells in the cochlea (see Figure 12.4); each of these cells sends a peripheral process to contact one or more hair cells and a central process to innervate the cochlear nucleus. Within the cochlear nucleus, each auditory nerve fiber branches, sending an ascending branch to the anteroventral cochlear nucleus and a descending branch to the posteroventral cochlear nucleus and the dorsal cochlear nucleus (see Figure 12.11). The tonotopic organization of the cochlea is maintained in the three parts of the cochlear nucleus, each of which contains different populations of cells with quite different properties. In addition, the patterns of termination of the auditory nerve axons differ in density and type; thus, there are several opportunities at this level for transformation of the information from the hair cells.

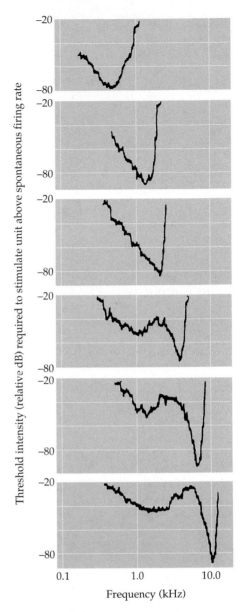

Figure 12.9 Frequency tuning curves of six different fibers in the auditory nerve. Each graph plots, across all frequencies to which the fiber responds, the minimum sound level required to increase the fiber's firing rate above its spontaneous firing level. The lowest point in the plot is the weakest sound intensity to which the neuron will respond. The frequency at this point is called the neuron's characteristic frequency. (After Kiang and Moxon, 1972.)

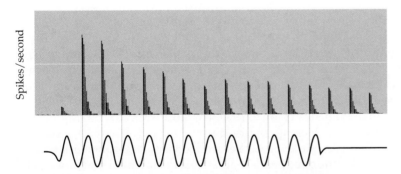

Figure 12.10 Temporal response patterns of a low-frequency axon in the auditory nerve. The stimulus waveform is indicated beneath the histograms, which show the phase-locked responses to a 50-ms tone pulse of 260 Hz. Note that the spikes are all timed to the same phase of the sinusoid. (After Kiang, 1984.)

■ INTEGRATING INFORMATION FROM THE TWO EARS

Just as the auditory nerve branches to innervate several different targets in the cochlear nuclei, the neurons in these nuclei give rise to several different pathways (Figure 12.11). Given the relatively byzantine organization already present at the level of the auditory brainstem, it is helpful to consider these pathways in the context of their functions. The best-understood function mediated by the auditory brainstem nuclei, and certainly the most intensively studied, is sound localization.

Humans use at least two different strategies to localize the horizontal position of sound sources, depending on the frequencies in the stimulus. For frequencies below 3 kHz (which can be followed in a phase-locked manner—see above), interaural time differences are used to localize the source; above these frequencies, interaural intensity differences are used as cues. Parallel pathways originating from the cochlear nucleus serve each of these strategies for sound localization.

The human ability to detect interaural time differences is remarkable. The longest interaural time differences, which are produced by sounds arising directly lateral to one ear, are on the order of only 700 μs (a value given by the width of the head divided by the speed of sound in air, about 340 m/s). Psychophysical experiments show that humans can actually detect interaural time differences as small as 10 μs; two sounds presented through earphones separated by such small interaural time differences are perceived as being localized towards the side of the leading ear. This sensitivity translates into an accuracy for sound localization of about 1°.

How is timing in the 10 μs range accomplished by neural components that operate in the millisecond range? The neural circuitry that computes such tiny interaural time differences consists of binaural inputs to the **medial superior olive (MSO)** that arise from the right and left anteroventral cochlear nuclei (Figures 12.11 and 12.12). The medial superior olive contains cells with bipolar dendrites that extend both medially and laterally. The lateral dendrites receive input from the ipsilateral anteroventral cochlear nucleus, and the medial dendrites receive input from the contralateral anteroventral cochlear nucleus (both inputs are excitatory). The MSO cells work as coincidence detectors, responding when both excitatory signals arrive at the same time. For a coincidence mechanism to be useful in localizing sound, different neurons must be maximally sensitive to different interaural time delays. The axons that project from the anteroventral cochlear nucleus evidently vary systematically in length to create delay lines. (Remember that the length of an axon multiplied by its conduction velocity equals the conduction time.). These anatomical differences compensate for sounds arriving at slightly different times at the two ears, so that the resultant neural impulses arrive at a particular MSO neuron simultaneously, making each cell especially sensitive to sound sources in a particular place.

Sound localization perceived on the basis of interaural time differences requires phase-locked information from the periphery, which is available to humans only for frequencies below 3 kHz. (In barn owls, the champions of sound localization, phase locking occurs at up to 9 kHz.) Therefore, a second mechanism must come into play at higher frequencies. At frequencies higher than about 2 kHz, the human head begins to act as an acoustical obstacle because the wavelengths of the sounds are too short to bend around it. As a result, when high-frequency sounds are directed toward one side of the head, an acoustical "shadow" of lower intensity is created at the far ear. These intensity differences provide a second cue about the location of a sound. The circuits that compute the position of a sound source on this basis

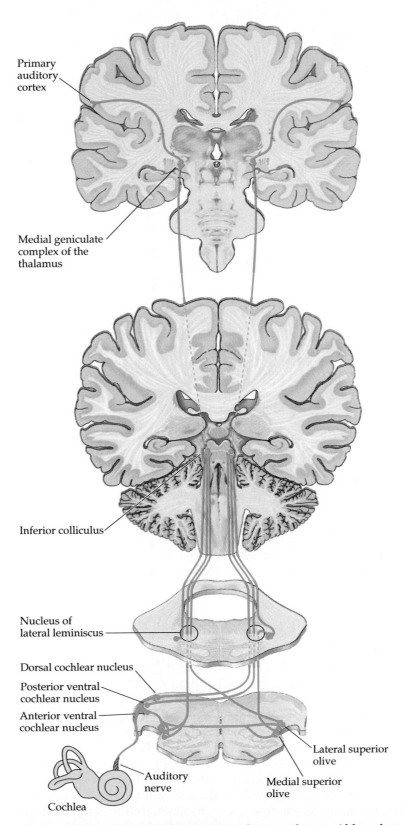

Primary
auditory
cortex

Medial geniculate
complex of the
thalamus

Inferior colliculus

Nucleus of
lateral leminiscus

Dorsal cochlear nucleus

Posterior ventral
cochlear nucleus

Anterior ventral
cochlear nucleus

Lateral superior
olive

Auditory
nerve

Medial superior
olive

Cochlea

Figure 12.11 Diagram of the major auditory pathways. Although many details
are missing from this diagram, two important points are evident: (1) the auditory
system entails several parallel pathways, and (2) information from each ear reaches
both sides of the system.

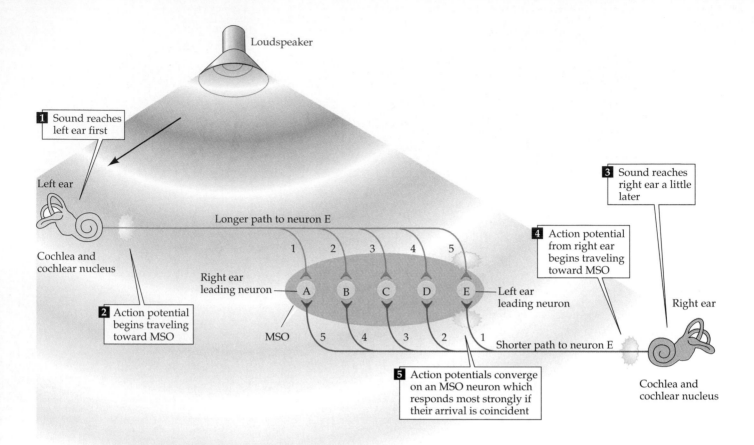

Figure 12.12 Diagram illustrating how the MSO computes the location of a sound by interaural time differences. A given MSO neuron responds most strongly when the two inputs arrive simultaneously, as occurs when the contralateral and ipsilateral inputs precisely compensate (via their different lengths) for differences in the time of arrival of a sound at the two ears. The systematic (and inverse) variation in the delay lengths of the two inputs creates a map of sound location: in this model, E would be most sensitive to sounds located to the left, and A to sounds from the right; C would respond best to sounds coming from directly in front of the listener. (After Jeffress, 1948.)

are found in the **lateral superior olive (LSO)** and the **medial nucleus of the trapezoid body (MNTB)**. Excitatory axons project directly from the ipsilateral anteroventral cochlear nucleus to the LSO (as well as to the MSO; see Figure 12.11). Note that the LSO also receives inhibitory input from the contralateral ear, via an inhibitory neuron in the MNTB (Figure 12.13). This excitatory/inhibitory interaction results in a net excitation of the LSO on the same side of the body as the sound source. For sounds arising directly lateral to the listener, firing rates will be highest in the LSO on that side; in this circumstance, the excitation via the ipsilateral anteroventral cochlear nucleus will be maximal, and inhibition from the contralateral MNTB minimal. In contrast, sounds arising closer to the listener's midline will elicit lower firing rates in the ipsilateral LSO because of increased inhibition arising from the contralateral MNTB. For sounds arising at the midline, or from the other side, the increased inhibition arising from the MNTB is powerful enough to completely silence LSO activity. Note that each LSO only encodes sounds

(A)

(B)

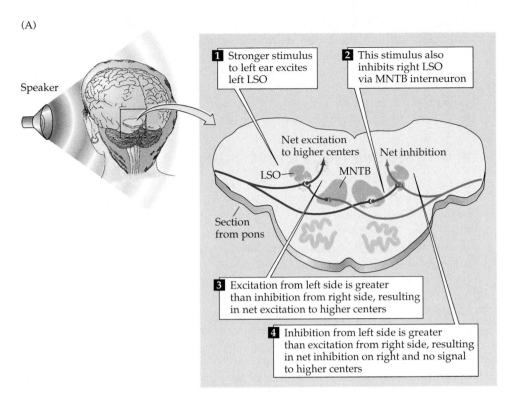

Figure 12.13 Lateral superior olive neurons encode sound location through interaural intensity differences. (A) LSO neurons receive direct excitation from the ipsilateral cochlear nucleus; input from the contralateral cochlear nucleus is relayed via inhibitory interneurons in the MNTB. (B) This arrangement of excitation–inhibition makes LSO neurons fire most strongly in response to sounds arising directly lateral to the listener on the same side as the LSO, because excitation from the ipsilateral input will be great and inhibition from the contralateral input will be small. In contrast, sounds arising from in front of the listener, or from the opposite side, will silence the LSO output, because excitation from the ipsilateral input will be minimal, but inhibition driven by the contralateral input will be great. Note that LSOs are paired and bilaterally symmetrical; each LSO only encodes the location of sounds arising on the same side of the body as its location.

arising in the ipsilateral hemifield; it takes both LSOs to represent the full range of horizontal positions.

In summary, there are two separate pathways—and two separate mechanisms—for localizing sound. Interaural time differences are processed in the medial superior olive, and interaural intensity differences are processed in the lateral superior olive. These two pathways are eventually merged in the midbrain auditory centers.

■ MONAURAL PATHWAYS FROM THE COCHLEAR NUCLEUS TO THE LATERAL LEMNISCUS

The binaural pathways for sound localization are only part of the output of the cochlear nucleus. This fact is hardly surprising, given that auditory perception involves much more than locating the position of the sound source. A second major set of pathways from the cochlear nucleus bypasses the superior olive and terminates in the **nuclei of the lateral lemniscus** on the contralateral side of the brainstem (see Figure 12.11). These pathways convey information about sound at one ear only, and are thus referred to as monaural. Some cells in the lateral lemniscus nuclei signal the onset of sound, regardless of its intensity or frequency. Other cells in the lateral lemniscus nuclei process other temporal aspects of sound, such as duration. The precise role of these pathways in processing temporal features of sound is not yet known. As with the outputs of the superior olivary nuclei, the pathways from the nuclei of the lateral lemniscus converge at the midbrain.

■ INTEGRATION IN THE INFERIOR COLLICULUS

Auditory pathways ascending via the olivary and lemniscal complexes, as well as other projections that arise directly from the cochlear nucleus, project to the midbrain auditory center, the **inferior colliculus** (see Figure 12.11). In examining how integration occurs in the inferior colliculus, it is again instructive to turn to the most completely analyzed auditory mechanism, the binaural system for localizing sound. As already noted, space is not mapped on the auditory receptor surface; thus the perception of auditory space must somehow be synthesized by circuitry in the lower brainstem and midbrain. Experiments in the barn owl, perhaps the most proficient animal at localizing sounds, show that the convergence of binaural inputs in the midbrain produces something entirely new relative to the periphery, namely, a computed topographical representation of auditory space. Neurons within this **auditory space map** of the colliculus respond best to sounds originating in a specific region of space and thus have both a preferred elevation and a preferred horizontal location, or azimuth. Although comparable maps of auditory space have not yet been found in mammals, humans have a clear perception of both the elevational and azimuthal components of a sound's location, suggesting that we have a similar auditory space map.

Another important property of the inferior colliculus is its ability to process sounds with complex temporal patterns. Many neurons in the inferior colliculus respond only to frequency-modulated sounds, while others respond only to sounds of specific durations. Such sounds are typical components of biologically relevant sounds, such as those made by predators, or intraspecific communication sounds, which in humans include speech. The inferior colliculus is evidently the first stage in a system, continued in the auditory thalamus and cortex, that analyzes sounds like these that have particular significance.

■ THE AUDITORY THALAMUS

Despite the parallel pathways in the auditory stations of the brainstem and midbrain, the **medial geniculate complex (MGC)** in the thalamus is an obligatory relay for all ascending auditory information destined for the cortex (see Figure 12.11). Most input to the MGC arises from the inferior colliculus, although a few auditory fibers from the lower brainstem bypass the inferior colliculus to reach the auditory thalamus directly. The MGC has several divisions, including the ventral division, which functions as the major thalamocortical relay, and the dorsal and medial divisions, which are organized like a belt around the ventral division.

In some mammals, the strictly maintained tonotopy of the lower brainstem areas is exploited by convergence onto MGC neurons, generating specific responses to certain spectral combinations. The original evidence for this statement came from research on the response properties of cells in the MGC of echolocating bats. Some cells in the belt areas of the bat MGC respond only to combinations of widely spaced frequencies that are specific components of the bat's echolocation signal and of the echoes that are reflected from objects in the bat's environment. In the mustached bat, where this phenomenon has been most thoroughly studied, the echolocation pulse has a changing frequency (frequency-modulated, or FM) component that includes a fundamental frequency and one or more harmonics. The fundamental frequency (FM_1) has low intensity and sweeps from 30 kHz to 20 kHz. The second harmonic (FM_2) is the most intense component and sweeps from 60 kHz to 40 kHz. Note that these frequencies do not overlap. Most of the echoes are from the intense FM_2 sound, and virtually none arise from the weak FM_1, even though the emitted FM_1 is loud enough for the bat to hear. Apparently, the bat measures the distance to an object by measuring the delay between the FM_1 emission and the FM_2 echo. Certain MGC neurons respond when FM_2 follows FM_1 by a specific delay, providing a mechanism for sensing such frequency combinations. Because each neuron responds best to a particular delay, a range of distances is encoded by the population of MGC neurons.

Bat sonar illustrates two important points about the function of the auditory thalamus. First, the MGC is the first station in the auditory pathway where selectivity for combinations of frequencies is found. The mechanism responsible for this selectivity is presumably the ultimate convergence of inputs from cochlear areas with different spectral sensitivities. Second, cells in the MGC are selective not only for frequency combinations, but also for specific time intervals between the two frequencies. The principle is the same as that described for binaural neurons in the medial superior olive, but here, two monaural signals with different frequency sensitivity coincide, and the time difference is in the millisecond rather than the microsecond range.

In summary, neurons in the medial geniculate complex receive convergent inputs from spectrally and temporally separate pathways. This complex, by virtue of its convergent inputs, mediates the integration of spectral and temporal aspects of sounds. It is not known whether cells in the human medial geniculate are selective to combinations of sounds, but the processing of speech certainly requires both spectral and temporal combination sensitivity.

■ THE AUDITORY CORTEX

The ultimate target of ascending auditory information is the auditory cortex. Although the auditory cortex has a number of subdivisions, a broad distinction can be made between a primary area and peripheral, or belt, areas. The **primary auditory cortex (A1)** receives point-to-point input from the ventral division of the medial geniculate complex and thus contains a precise tonotopic map. The **belt areas** of the auditory cortex receive more diffuse input from the belt areas of the medial geniculate complex and therefore are less precise in their tonotopic organization.

The primary auditory cortex (A1) has a topographical map of the cochlea (Figure 12.14), just as the primary visual cortex (V1) and the primary somatosensory cortex (S1) have topographical maps of their respective sensory epithelia. Unlike the visual and somatosensory systems, however, the cochlea has already decomposed the acoustical stimulus so that it is arrayed tonotopically along the length of the basilar membrane. Thus, A1 is said to represent a tonotopic map. Orthogonal to the tonotopic map is a striped arrangement of binaural properties. The cells in one stripe are excited by both ears (and are therefore called EE), while the cells in the next stripe are excited by one ear and inhibited by the other ear (EI). The EE and EI stripes alternate, an arrangement that is reminiscent of the ocular dominance

Figure 12.14 The human auditory cortex. (A) Diagram showing the brain in left lateral view, including the depths of the lateral sulcus, where part of the auditory cortex normally lies hidden. The primary auditory cortex (A1) is shown in blue; the surrounding belt areas of the auditory cortex are in red. (B) The primary auditory cortex has a tonotopic organization, as shown in this diagram of a segment of A1.

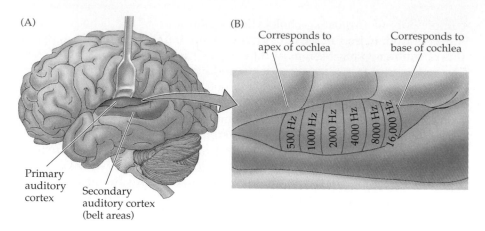

columns in V1 (see Chapter 11). The sorts of sensory processing that occur in the other divisions of the auditory cortex are not understood. It appears that some areas are specialized for processing combinations of frequencies, while others are specialized for processing modulations of amplitude or frequency.

Sounds that are especially important often have a highly ordered temporal structure. In humans, the best example is speech. Behavioral studies in cats and monkeys show that the auditory cortex is especially important for processing temporal sequences of sound. If the auditory cortex is ablated in these animals, they lose the ability to discriminate between two complex sounds that have the same frequency components but differ in the temporal sequence of the components. Thus, without the auditory cortex monkeys cannot discriminate one conspecific communication sound from another. Studies of human patients with bilateral damage to the auditory cortex also reveal severe problems in processing the temporal order of sounds. It seems likely, therefore, that specific regions of the human auditory cortex are specialized for processing elementary speech sounds, as well as other temporally complex acoustical signals, such as music. Indeed, Wernicke's area, which is critical to the comprehension of human language, lies within the secondary auditory area (Figure 12.15; see also Chapter 25).

Figure 12.15 The human auditory cortical areas related to processing speech sounds. (A) Diagram of the brain in left lateral view, showing locations in the intact hemisphere. (B) An oblique section (plane of dashed line in A) shows the cortical areas on the superior surface of the temporal lobe. Note that Wernicke's area, a region important in comprehending speech, is just posterior to the primary auditory cortex.

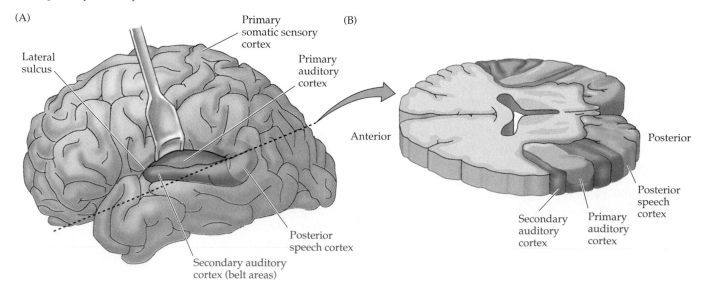

■ SUMMARY

Sound waves are transmitted via the external and middle ear to the cochlea of the inner ear, which produces a traveling wave when stimulated. For high-frequency sounds, the amplitude of the traveling wave reaches a maximum at the base of the cochlea; for low-frequency sounds, the traveling wave reaches a maximum at the apical end. The associated motions of the basilar membrane are transduced by the inner and outer hair cells. The tonotopic organization of the cochlea is retained at all levels of the central auditory system. Projections from the cochlea innervate the three main divisions of the cochlear nucleus. The targets of the cochlear nucleus neurons include the superior olivary complex and nuclei of the lateral lemniscus, where the binaural cues for sound localization are processed. The inferior colliculus is the target of nearly all of the auditory pathways in the lower brainstem and carries out important integrative functions, such as processing sound frequencies and integrating the cues for localizing sound in space. The primary auditory cortex, which is also organized tonotopically, is essential for basic auditory functions, such as frequency discrimination and sound localization. The belt areas of the auditory cortex have a less strict tonotopic organization and probably analyze complex sounds, such as those that mediate communication. In the human brain, the major speech comprehension areas are located in the zone immediately adjacent to the auditory cortex.

Additional Reading

Reviews

COREY, D. P. AND A. J. HUDSPETH (1979) Ionic basis of the receptor potential in a vertebrate hair cell. Nature 281: 675–677.

DALLOS, P. (1992) The active cochlea. J. Neurosci. 12: 4575–4585.

HEFFNER, H. E. AND R. S. HEFFNER (1990) Role of primate auditory cortex in hearing. In *Comparative Perception Volume II: Complex Signals.* W. C. Stebbins and M. A. Berkley (eds.). New York: John Wiley.

KIANG, N. Y. S. (1984) Peripheral neural processing of auditory information. In *Handbook of Physiology*, Section 1: *The Nervous System*, Volume III. *Sensory Processes*, Part 2. J. M. Brookhart, V. B. Mountcastle, I. Darian-Smith and S. R. Geiger (eds.). Bethesda, MD: American Physiological Society.

LEWIS, R. S. AND A. J. HUDSPETH (1983) Voltage- and ion-dependent conductances in solitary vertebrate hair cells. Nature 304: 538–541.

MIDDLEBROOKS, J. C., A. E. CLOCK, L. XU AND D. M. GREEN (1994) A panoramic code for sound location by cortical neurons. Science 264: 842–844.

NEFF, W. D., DIAMOND, I. T. AND J. H. CASSEDAY (1975) Behavioral studies of auditory discrimination. In *Handbook of Sensory Physiology*, Volumes V–II. W. D. Keidel and W. D. Neff (eds.). Berlin: Springer-Verlag.

SUGA, N. (1990) Biosonar and neural computation in bats. Sci. Amer. 262: 60–68.

Important Original Papers

CRAWFORD, A. C. AND R. FETTIPLACE (1981) An electrical tuning mechanism in turtle cochlear hair cells. J. Physiol. 312: 377–412.

FITZPATRICK, D. C., J. S. KANWAL, J. A. BUTMAN AND N. SUGA (1993) Combination-sensitive neurons in the primary auditory cortex of the mustached bat. J. Neurosci. 13: 931–940.

KNUDSEN, E. I. AND M. KONISHI (1978) A neural map of auditory space in the owl. Science 200: 795–797.

JEFFRESS, L. A. (1948) A place theory of sound localization. J. Comp. Physiol. Psychol. 41: 35–39.

SUGA, N., W. E. O'NEILL AND T. MANABE (1978) Cortical neurons sensitive to combinations of information-bearing elements of biosonar signals in the mustache bat. Science 200: 778–781.

VON BÉKÉSY, G. (1960) *Experiments in Hearing*. New York: McGraw-Hill. (A collection of von Békésy's original papers.)

Books

PICKLES, J. O. (1988) *An Introduction to the Physiology of Hearing*. London: Academic Press.

YOST, W. A. AND G. GOUREVITCH (eds.) (1987) *Directional Hearing,*. Berlin: Springer Verlag.

YOST, W. A. AND D. W. NIELSEN (1985) *Fundamentals of Hearing*. Fort Worth: Holt, Rinehart and Winston.

THE VESTIBULAR SYSTEM

■ OVERVIEW

The peripheral portion of the vestibular system is a part of the inner ear that acts as a miniaturized accelerometer and inertial guidance device, continually reporting information about the motions and position of the head and body to integrative centers located in the brainstem and cerebellum. Although we are normally unaware of its function, the vestibular system is a key component in both postural reflexes and eye movements. If the system is damaged, balance, the control of eye movements when the head is moving, and the sense of orientation in space are all affected. These manifestations of vestibular damage are especially important in the evaluation of brainstem injury; the vestibular circuitry extends through a large part of the brainstem, and simple clinical tests of vestibular function can be performed to determine brainstem involvement, even on comatose patients.

■ THE VESTIBULAR LABYRINTH

The main peripheral component of the vestibular system is an elaborate set of interconnected canals—the **labyrinth**—that has much in common with the cochlea. Like the cochlea (see Chapter 12), the vestibular system is derived from the otic placode of the embryo, and it uses the same specialized set of sensory cells—hair cells—to transduce motion into neural impulses. In the cochlea, the motion is due to airborne sounds; in the vestibular system, the motions that are transduced arise from head movements, inertial effects due to gravity, and ground-borne vibrations (Box A).

The labyrinth is buried deep in the temporal bone, and consists of three parts: the two **otolith organs** (the **utricle** and the **sacculus**) and the **semicircular canals** (Figure 13.1). The elaborate and tortuous architecture of these components explains why this part of the vestibular system is called the labyrinth. The utricle and sacculus are specialized to respond to *linear accelerations* of the head and *static head position*, whereas the semicircular canals are specialized for responding to *rotational accelerations* of the head.

The intimate relationship between the cochlea and the labyrinth goes beyond their common embryonic origin; the cochlear and vestibular spaces are actually joined (see Figure 13.1). The membranous sacs within the bone are filled with fluid (endolymph) and are collectively called the membranous labyrinth. The endolymph (like the cochlear endolymph) is similar to intracellular solutions in that it is high in K^+ and low in Na^+. Between the bony walls (the osseous labyrinth) and the membranous labyrinth is another fluid, the perilymph, which is similar in composition to cerebrospinal fluid (that is, low in K^+ and high in Na^+). The vestibular hair cells are located in the utricle and sacculus and in three juglike swellings, called **ampullae**, located at the base of the semicircular canals next to the utricle.

■ VESTIBULAR HAIR CELLS

The vestibular hair cells, which transduce minute displacements into behaviorally relevant receptor potentials, provide the basis for vestibular function.

Box A
A PRIMER ON (VESTIBULAR) NAVIGATION

The function of the vestibular system can be simplified by remembering some basic terminology of classical mechanics. All bodies moving in a three-dimensional framework have six degrees of freedom: three of these are translational and three are rotational. The translational elements refer to linear movements in the x, y, and z axes (the horizontal and vertical planes). Translational motion in these planes (linear acceleration and static displacement of the head) is the primary concern of the otolith organs. The three degrees of rotational freedom refer to a body's rotation relative to the x, y, and z axes, and are commonly referred to as roll, pitch and yaw. The semicircular canals are primarily responsible for sensing rotational accelerations around these three axes.

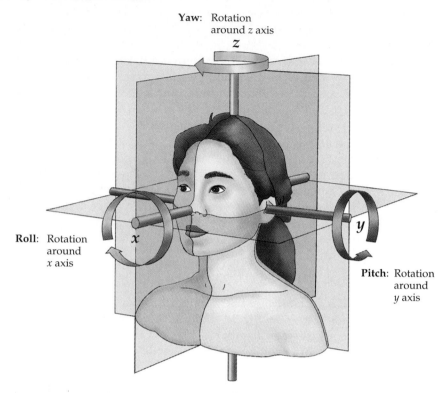

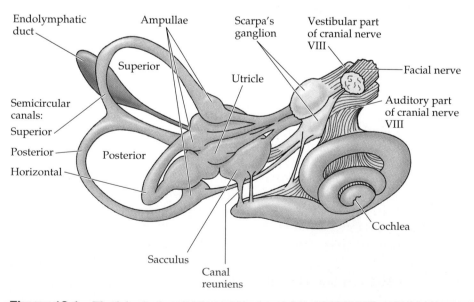

Figure 13.1 The labyrinth and its innervation. The vestibular and auditory portions of the eighth nerve are shown; the small connection from the vestibular nerve to the cochlea contains auditory efferent fibers.

Vestibular and auditory hair cells are quite similar; a detailed description of hair cell structure and function has already been given in Chapter 12. As in the case of auditory hair cells, movement of the stereocilia toward the kinocilium in the vestibular end organs opens mechanically gated transduction channels located at the tips of the stereocilia, depolarizing the hair cell and causing neurotransmitter release onto (and excitation of) the vestibular nerve fibers. Movement of the stereocilia in the direction away from the kinocilium closes the channels, hyperpolarizing the hair cell and thus reducing vestibular nerve activity. The biphasic nature of the receptor potential indicates that some transduction channels are open in the absence of stimulation, with the result that hair cells tonically release transmitter, thereby generating considerable spontaneous activity in vestibular fibers (Box B).

The hair cell bundles have specific orientations in each vestibular organ (Figure 13.2). As a result, the organ as a whole is responsive to displacements in specific directions. The directional polarization of the receptor surfaces is a basic principle of organization in the vestibular system, as will become apparent in the following descriptions of the individual organs. In a given semicircular canal, the hair cells are all polarized in the same direction (Figure 13.2C). In the utricle and sacculus, a specialized area called the **striola** divides the hair cells into two populations with opposing polarities (Figures 13.2C; see also Figure 13.4A).

Figure 13.2 The morphological polarization of vestibular hair cells and the polarization maps of the vestibular organs. (A) A cross section of hair cells shows that the kinocilia of a group of hair cells are all located on the same side of the hair cell. The arrow indicates the direction of deflection that depolarizes the hair cell. (B) View looking down on the hair bundles. (C) In the ampullae, located at the base of each semicircular canal, the hair bundles are oriented in the same direction. In the sacculus and utricle, the striola divides the hair cells into populations with opposing hair bundle polarities.

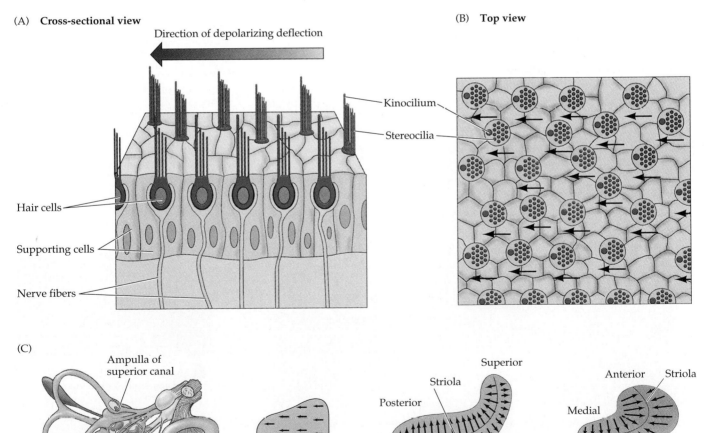

(A) **Cross-sectional view**

Direction of depolarizing deflection

Kinocilium

Stereocilia

Hair cells

Supporting cells

Nerve fibers

(B) **Top view**

(C)

Ampulla of superior canal

Utricular macula

Sacullar macula

Ampulla

Striola

Posterior

Superior

Anterior

Inferior

Sacculus

Anterior   Striola

Medial

Posterior   Lateral

Utricle

Box B
ADAPTATION AND TUNING OF VESTIBULAR HAIR CELLS

Hair Cell Adaptation

The minuscule movement of the hair bundle at sensory threshold has been compared to the displacement of the top of the Eiffel Tower by a thumb's breadth! Despite its great sensitivity, the hair cell can still adapt quickly and continuously to static displacements of the hair bundle caused by large movements. Such adjustments are especially useful in the otolith organs, where adaptation permits hair cells to maintain sensitivity to small linear and angular accelerations of the head despite the constant input from gravitational forces that are over a million times greater. In other receptor cells, such as photoreceptors, adaptation is accomplished by regulating the second-messenger cascade induced by the initial transduction event. The hair cell has to depend on a different strategy, however, because there is no second-messenger system between the initial transduction event and the subsequent

receptor potential (as might be expected for receptors that respond so rapidly).

Adaptation occurs in both directions in which the hair bundle displacement generates a receptor potential, albeit at different rates for each direction. When the hair bundle is pushed toward the kinocilium, tension is initially increased in the gating spring. During adaptation, tension decreases back to the resting level, perhaps because one end of the gating spring repositions itself along the shank of the stereocilium. When the hair bundle is displaced in the opposite direction, away from the kinocilium, tension in the spring initially decreases; adaptation then involves an increase in spring tension. One theory is that a calcium-regulated motor such as a myosin ATPase climbs along actin filaments in the stereocilium and actively resets the tension in the transduction spring. During sus-

tained depolarization, some Ca^{2+} enters through the transduction channel, along with K^+. Ca^{2+} then causes the motor to spend a greater fraction of its time unbound from the actin, resulting in slippage of the spring down the side of the stereocilium. During sustained hyperpolarization (Figure A), Ca^{2+} levels drop below normal resting levels and the motor spends more of its time bound to the actin, thus climbing up the actin filaments and increasing the spring tension. As tension increases, though, some of the previously closed transduction channels open, admitting Ca^{2+} and thus slowing the motor's progress until a balance is struck between the climbing and slipping of the motor. In support of this model, when internal Ca^{2+} is reduced artificially, spring tension increases. This model of hair cell adaptation presents an elegant molecular solution to the regulation of a mechanical process.

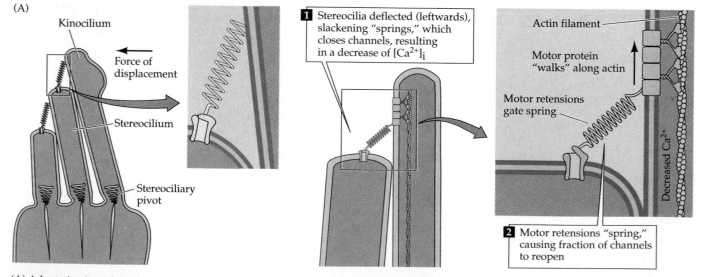

(A) Adaptation is explained in the gating spring model by adjustment of the insertion point of tips links. Movement of the insertion point up or down the shank of the stereocilium, perhaps driven by a Ca^{2+}-dependent protein motor, can continually adjust the resting tension of the tip link. (After Hudspeth and Gillespie, 1994.)

Electrical Tuning

Although mechanical tuning plays an important role in generating frequency selectivity in the cochlea, there are other mechanisms that contribute to this process in vestibular and auditory nerve cells. These other tuning mechanisms are especially important in the otolith organs, where, unlike the cochlea, there are no obvious macromechanical resonances to selectively filter and/or enhance biologically relevant movements. One such mechanism is an electrical resonance displayed by hair cells in response to depolarization: the membrane potential of a hair cell undergoes damped sinusoidal oscillations at a specific frequency in response to the injection of depolarizing current pulses (Figure B).

The ionic mechanism of this process involves two major types of ion channels located in the membrane of the hair cell soma. The first of these is a voltage-activated Ca^{2+} conductance, which lets Ca^{2+} into the cell soma in response to depolarization, such as that generated by the transduction current. The second is a Ca^{2+}-activated K^+ conductance, which is triggered by the rise in internal Ca^{2+} concentration. These two currents produce an interplay of depolarization and repolarization that results in electrical resonance (Figure C). Activation of the hair cell's calcium-activated K^+ conductance occurs 10 to 100 times faster than that of similar currents in other cells. Such rapid kinetics allow this conductance to generate an electrical response that usually requires the fast properties of a voltage-gated channel.

Although a hair cell responds to hair bundle movement over a wide range of frequencies, the resultant receptor potential is largest at the frequency of electrical resonance. The resonance frequency represents the characteristic frequency of the hair cell, and motion at that frequency will be most efficient. This electrical resonance has important implications for structures like the utricle and sacculus, which may encode a range of characteristic frequencies based on the different resonance frequencies of their constituent hair cells. Thus, electrical tuning in the otolith organs can generate enhanced tuning to biologically-relevant frequencies of stimulation, even in the absence of macromechanical resonances within these structures.

References

ASSAD, J. A. AND D. P. COREY (1992) An active motor model for adaptation by vertebrate hair cells. J. Neurosci. 12: 3291–3309.

CRAWFORD, A. C. AND R. FETTIPLACE (1981) An electrical tuning mechanism in turtle cochlear hair cells. J. Physiol. 312: 377–412.

HUDSPETH, A. J. (1985) The cellular basis of hearing: the biophysics of hair cells. Science 230: 745–752.

HUDSPETH, A. J. AND P. G. GILLESPIE. (1994) Pulling strings to tune transduction: adaptation by hair cells. Neuron 12: 1–9.

LEWIS, R. S. AND A. J. HUDSPETH (1988) A model for electrical resonance and frequency tuning in saccular hair cells of the bull-frog, *Rana catesbeiana*. J. Physiol. 400: 275–297.

LEWIS, R. S. AND A. J. HUDSPETH (1983) Voltage- and ion-dependent conductances in solitary vertebrate hair cells. Nature 304: 538–541.

SHEPHERD, G. M. G. AND D. P. COREY (1994) The extent of adaptation in bullfrog saccular hair cells. J. Neurosci. 14: 6217–6229.

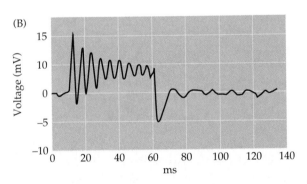

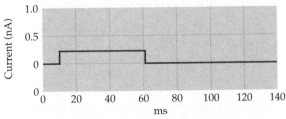

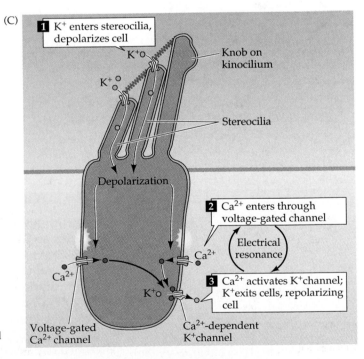

(B) Voltage oscillations (upper trace) in an isolated hair cell in response to a depolarizing current injection (lower trace). (C) Proposed ionic basis for electrical resonance in hair cells. (B after Lewis and Hudspeth, 1983; C after Hudspeth, 1985.)

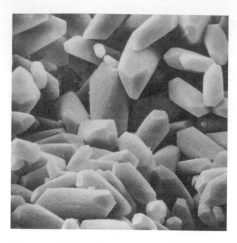

Figure 13.3 Scanning electron micrograph of calcium carbonate crystals (otoconia) in the utricular macula of the cat. Each crystal is about 50 μm long. (From Lindeman, 1973.)

■ THE OTOLITH ORGANS

Displacements and linear accelerations of the head, such as those induced by tilting or translational movements of the head, are detected by the two otolith organs: the sacculus and the utricle. Both of these organs contain a sensory epithelium, the **macula**, which consists of hair cells and associated supporting cells. Overlying the hair cells and their hair bundles is a gelatinous layer, and above this is a fibrous structure, the **otolithic membrane**, in which are embedded crystals of calcium carbonate called **otoconia** (Figures 13.3 and 13.4). The crystals give the otolith organs their name (*otolith* is Greek for "ear stones"). The otoconia make the otolithic membrane considerably heavier than the structures and fluids surrounding it, so when the head tilts, gravity causes the membrane to shift relative to the sensory epithelium (Figure 13.4B). The resulting shearing motion between the otolithic membrane and the macula displaces the hair bundles, which are embedded in the lower, gelatinous surface of the membrane. This displacement of the hair bundle generates a receptor potential in the hair cell. A shearing motion between the macula and the otolithic membrane also occurs when the head undergoes linear accelerations (see Figure 13.5); the greater relative mass of the otolithic membrane causes it to lag behind the macula temporarily, leading to transient displacement of the hair bundle.

As already mentioned, the orientation of the hair cell bundles is organized relative to the striola, which demarcates the overlying layer of otoconia (see Figure 13.4A). The striola forms an axis of mirror symmetry such that hair cells on opposite sides of the striola have opposing morphological polarizations. Thus, a tilt along the axis of the striola will excite the hair cells on one side while inhibiting the hair cells on the other side. The saccular macula is oriented vertically and the utricular macula is oriented horizontally, with a continuous variation in the morphological polarization of the hair cells located in each macula (in Figure 13.4C, the arrows indicate the direction of movement that produces excitation). Inspection of the excitatory orientations in the maculae indicates that the utricle responds to movements of the head in the horizontal plane, such as sideways head tilts and rapid lateral displacements, whereas the sacculus responds to movements in the vertical plane (up-down and forward-backward movements in the sagittal plane). An interesting feature of this system is that the saccular and utricular maculae on one side of the head are mirror images of those on the other side. Thus, a tilt of the head to one side has opposite effects on corresponding hair cells of the two utricular maculae. This concept will be important later in understanding how the central connections of the vestibular periphery mediate the interaction of inputs from the two sides of the head.

■ HOW OTOLITH NEURONS SENSE LINEAR FORCES

The sacculus and utricle are primarily involved in discerning translational movements and tilting of the head. In lower vertebrates, the sacculus also serves to detect ground-borne vibrations; it is less clearly specialized for this function in humans. The structure of the otolith organs enables them to sense both static displacements, as would be caused by tilting the head, and linear accelerations caused by translational movements of the head. The mass of the otolithic membrane relative to the surrounding endolymph, as well as the otolithic membrane's physical uncoupling from the underlying macula, means that hair bundle displacement will occur transiently in response to linear accelerations and tonically in response to tilting of the head. Therefore, at least two types of information can be conveyed by these sense organs.

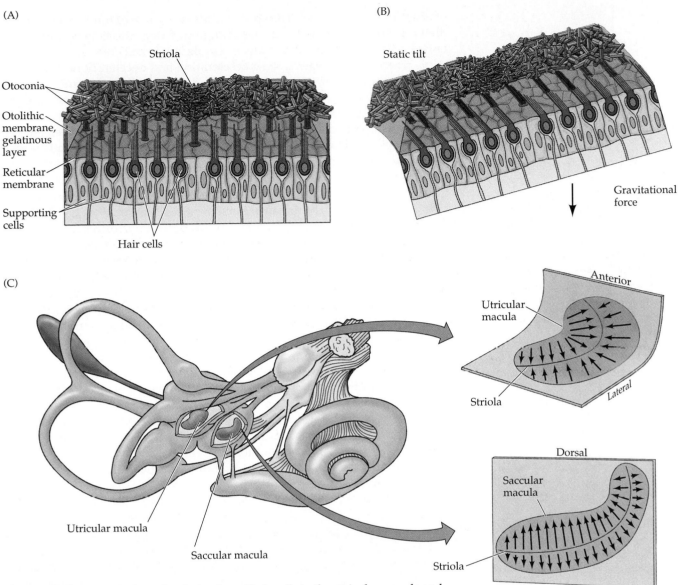

Figure 13.4 Morphological polarization of hair cells in the utricular macula and saccular macula. (A) Cross section of the utricular macula showing hair bundles projecting into the gelatinous layer when the head is level. (B) Cross section of the utricular macula when the head is tilted. (C) Orientation of the utricular and saccular maculae in the head; arrows show orientation of the kinocilia, as in Figure 13.2. The sacculae are oriented more or less vertically, and the utricules more or less horizontally. The striola is a structural landmark consisting of small otoconia arranged in a narrow trench that divides each otolith organ. In the utricular macula, the kinocilia are directed toward the striola. In the saccular macula, the kinocilia point away from the striola. Note that, taken with the utricle and sacculus on the other side of the body, there is a continuous representation of all directions of body movement.

The vestibular neurons that innervate the otolith organs are characterized by a steady and relatively high discharge rate when the head is upright. The change in firing rate in response to a given movement can either be tonic (constant and sustained), or phasic-tonic (a transient change that decays to a tonic level), implying that the neurons can encode either linear acceleration or absolute head position. The spontaneous firing rate of tonic axons is modulated by the tilt of the head; the change in firing rate is generally maintained as long as the head remains tilted. Because the firing rate of the tonic nerve fibers only changes during acceleration and deceleration of the head, such alterations serve to encode angular velocity and head position. Figure 13.5 illustrates some of the forces produced by head tilt and linear accelerations on the utricular macula.

Tonically responding axons that innervate the utricle have been studied in monkeys seated in a chair that could be tilted for several seconds to produce a steady force. As shown in Figure 13.6, tilting the monkey increases the firing rate because of the force generated along the depolarizing axis of the relevant hair cells. Notice that the response remains at a high level as long as the tilting force remains constant; thus, this neuron faithfully encodes the linear force being applied to the head (Figure 13.6A). When the force abates, the firing level of the neuron returns to baseline value. Conversely, when the force is in the opposite direction, the neuron responds by decreasing its firing rate below the resting level (Figure 13.6B) and remains depressed as long as the static force continues. In summary, the otolith organs detect linear forces acting on the head, whether by static displacement of hair bundles due to gravity or by transient displacement of hair bundles due to linear accelerations.

The range of orientations of hair bundles within the otolith organs shows that they are capable of transmitting information about linear force in every direction the body moves (see Figure 13.4C). The utricle, which is primarily concerned with motion in the horizontal plane, and the sacculus, which is concerned with vertical motion, combine to effectively gauge in three dimensions the linear forces acting on the head at any instant. Tilts of the head off the horizontal plane and translational movements of the head in any direction stimulate a distinct subset of hair cells in the saccular macula

Head tilt

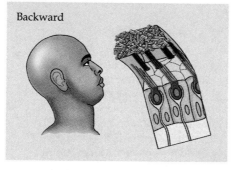

Backward

Forward

No head tilt

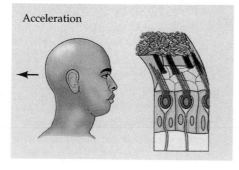

Acceleration

Deceleration

Upright

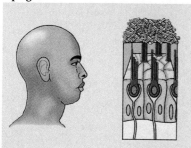

Figure 13.5 Forces acting on the head and the resulting displacement of the otolithic membrane of the utricular macula. For each of the positions and accelerations due to translational movements, some set of hair cells will be maximally excited, whereas another set will be maximally inhibited. Note that head tilts produce displacements similar to certain accelerations.

(A)

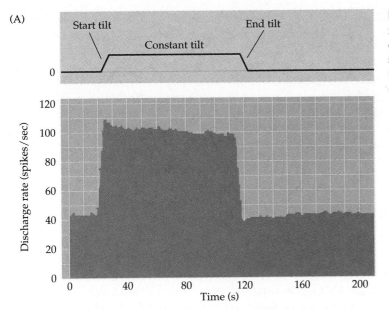

Figure 13.6 Response of a vestibular nerve axon from an otolith organ. (A) The stimulus (top) is a change that causes the head to tilt. The histogram shows the neuron's response to tilting in one direction. (B) A response of the same fiber to tilting in the opposite direction. (After Fernandez and Goldberg, 1976.)

(B)

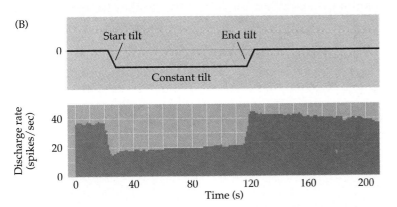

and utricular macula, while simultaneously suppressing the responses of other hair cells in these organs.

■ THE SEMICIRCULAR CANALS

Each of the three semicircular canals has at its base a bulbous expansion called the ampulla (Figure 13.7), which houses the sensory epithelium, or **crista**, that contains the hair cells. The structure of the canals suggests how they detect the angular accelerations that arise through rotation of the head. The hair bundles extend out of the crista into a gelatinous mass, the **cupula**, that bridges the width of the ampulla, forming a fluid barrier through which endolymph cannot circulate. As a result, the compliant cupula is distorted by movements of the endolymphatic fluid. Thus, when the head turns in the plane of one of the semicircular canals, the inertia of the endolymph produces a force across the cupula, distending it away from the direction of

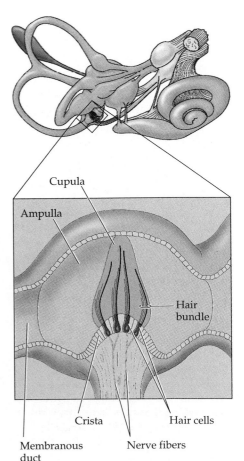

Figure 13.7 The ampulla of the posterior semicircular canal showing the crista, hair bundles, and cupula. The cupula is distorted by the fluid in the membranous canal when the head rotates.

head movement and causing a displacement of the hair bundles within the crista (Figure 13.8A,B). In contrast, linear accelerations of the head produce equal forces on the two sides of the cupula, so the hair bundles are not displaced.

Unlike the saccular and utricular maculae, all of the hair cells in the crista within each semicircular canal are organized with their kinocilia pointing in the same direction (see Figure 13.2C). Thus, when the cupula moves in the appropriate direction, the entire population of hair cells is depolarized, and activity in the innervating axons increases; when the cupula moves in the opposite direction, the population is hyperpolarized, and neuronal activity decreases. Deflections orthogonal to the excitatory-inhibitory direction produce little or no response.

Each semicircular canal works in concert with a partner located on the other side of the head, which has its hair cells aligned oppositely. There are three such pairs: the two horizontal canals, and the anterior canal on each side working with the posterior canal on the other side (Figure 13.8C). Head rotation deforms the cupula in opposing directions for the two partners, resulting in opposite changes in their firing rates (see Box C). For example, the orientation of the horizontal canals makes them selectively sensitive to rotation in the horizontal plane. When the head turns to the left, the cupula is pushed toward the kinocilium in the left horizontal canal, and the firing rate of the relevant axons in the left vestibular nerve increases. In contrast, the cupula in the right horizontal canal is pushed away from the kinocilium, with a concomitant decrease in the firing rate of the related neurons. If the head

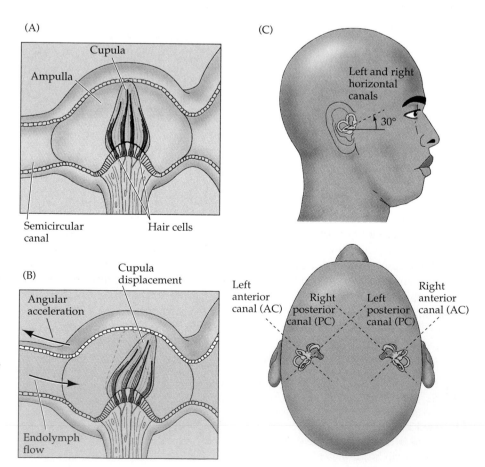

Figure 13.8 Functional organization of the semicircular canals. (A) The position of the cupula without angular acceleration. (B) Distortion of the cupula during angular acceleration. When the head is rotated in the plane of the canal (arrow outside canal), the inertia of the endolymph creates a force (arrow inside the canal) that displaces the cupula. (C) Arrangement of the canals in pairs. The two horizontal canals form a pair; the right anterior canal (AC) and the left posterior canal (PC) form a pair; the left AC and the right PC form a pair.

movement is to the right, the result is just the opposite. This push-pull arrangement operates for all three pairs of canals; the pair whose activity is modulated is in the rotational plane, and the member of the pair whose activity is increased is on the side toward which the head is turning. The net result is a system that provides information about the rotation of the head in any direction.

■ HOW SEMICIRCULAR CANAL NEURONS SENSE ANGULAR ACCELERATIONS

An important feature of the neurons in the vestibular ganglion (the location of the nerve cell bodies that innervate the end organs of the labyrinth) is their high rate of spontaneous activity. As a result, they can transmit information by either increasing or decreasing their firing rate. Responses of fibers innervating the hair cells of the semicircular canal have been studied by recording the axonal firing rates in a monkey's vestibular nerve. Seated in a chair that could be rotated, the monkey was first rotated at an accelerated rate, then at constant velocity for several seconds, and finally the chair was decelerated to a stop (Figure 13.9). The maximum firing rates observed correspond to the period of acceleration; the maximum inhibition, which falls below the resting level, corresponds to the period of deceleration. During the constant-velocity phase, the response adapts so that the firing rate subsides to resting level; after the movement stops, the neuronal activity decreases transiently before returning to the to resting level. Neurons innervating paired canals have an opposite response pattern. Note that the rate of adaptation (on the order of tens of seconds) corresponds to the time it takes the cupula to return to its undistorted state (and for the hair bundles to return to their undeflected position); adaption therefore can occur even while the head is still turning, as long as a constant angular velocity is maintained. Such constant forces are rare in nature, although they are sometimes encountered on ships, airplanes, and space vehicles.

Although angular accelerations are the effective stimulus for the semicircular canals, the relevant axons in the vestibular nerve actually encode the angular velocity of the head, at least for the physiological range of motions. Remember that the force that acts to deform the cupula is generated by the mass of the endolymph multiplied by the angular acceleration of the head (that is, $f = ma$). A constant force should therefore generate a constant acceleration, and thus an ever-increasing velocity; however, the endolymph is a viscous fluid that interacts with the narrow walls of the canal. When viscosity is significant, as in this case, a constant force will produce a constant velocity (which explains why the constant force of gravity does not accelerate a skydiver past the terminal velocity of 50 meters per second). Although angular acceleration displaces the cupula, its deformation is actually proportional to the angular velocity of the head. As a result, any angular acceleration will be encoded by three component vectors of angular velocity (one from each semicircular canal). By integrating the responses of the various

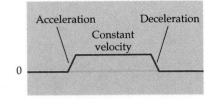

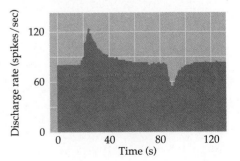

Figure 13.9 Response of a vestibular nerve axon from the semicircular canal to angular acceleration. The stimulus (top) is a rotation that first accelerates, then maintains constant velocity, and then decelerates the head. The axon increases its firing above resting level in response to the acceleration, returns to resting level during constant velocity, then decreases its firing rate below resting level during deceleration. (After Goldberg and Fernandez, 1971.)

vestibular neurons over time, the actual movement of the head in three dimensions can be computed by the nervous system.

■ CENTRAL VESTIBULAR PATHWAYS: EYE, HEAD, AND BODY REFLEXES

The targets of the vestibular system are centers in the brainstem and the cerebellum that perform much of the processing necessary to compute head position and motion. The cell bodies of the bipolar neurons that innervate the semicircular canals and the otolith organs are in the **vestibular nerve ganglion** (called **Scarpa's ganglion**; see Figure 13.1) and project via the vestibular portion of the eighth cranial nerve to the vestibular nuclei and also directly to the cerebellum. The **vestibular nuclei** are important centers of integration, receiving input from the vestibular nuclei of the opposite side, as well as from the cerebellum and the visual and somatic sensory systems.

One of the main functions of the vestibular system is to coordinate head and eye movements. The **vestibulo-ocular reflex (VOR)** is a mechanism for producing eye movements that counter head movements, thus permitting the gaze to remain fixed on a particular point (Box C; see also Chapter 19). For example, activity in the left horizontal canal, induced by leftward rotation of the head, results in reflexive eye movements to the right. Horizontal movement of the two eyes toward the right requires contraction of the left medial and right lateral rectus muscles. Vestibular nerve fibers originating in the left horizontal semicircular canal project to the medial and lateral vestibular nuclei (Figure 13.10). Excitatory fibers from the medial vestibular nucleus cross to the contralateral abducens nucleus, which has two outputs. One of these is a motor pathway that causes the lateral rectus of the right eye to contract; the other is an excitatory projection that crosses the midline and ascends via the **medial longitudinal fasciculus** to the left oculomotor nucleus, where it activates neurons that cause the medial rectus of the left eye to contract. The medial rectus of the left eye also receives input from the lateral vestibular nucleus, which sends excitatory axons to the ipsilateral oculomotor nucleus. Finally, inhibitory neurons project from the medial vestibular nucleus to the left abducens nucleus, directly causing the motor drive on the lateral rectus of the left eye to decrease and also indirectly causing the right medial rectus to relax. The consequence of these several connections is that excitatory input from the horizontal canal on one side produces eye movements toward the opposite side. Therefore, turning the head to the left causes eye movements to the right. In a similar fashion, head turns in other planes activate other semicircular canals, causing other appropriate compensatory eye movements.

Eye movements can also be activated by direct electrical stimulation of the ampullae, each of which exerts specific actions on individual extraocular muscles. For example, stimulation of the right anterior canal moves the left eye upward and outward by activating the inferior oblique and superior rectus muscles while simultaneously inhibiting the antagonistic muscles. Simultaneous stimulation of the two vertical canals is required for **conjugate** (paired) eye movements; stimulation of only one canal results in movement of just the contralateral eye. Thus, the push-pull interaction of the paired semicircular canals is necessary for coordinated reflex movements of the two eyes.

The loss of the VOR can have severe consequences. A patient with vestibular damage finds it difficult or impossible to fixate on visual targets while the head is moving. This condition is called **oscillopsia**. If the damage is unilateral, the patient usually recovers the ability to fixate objects during

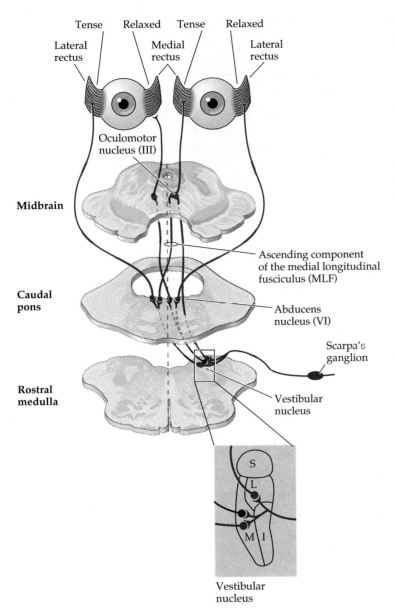

Tense   Relaxed   Tense   Relaxed

Lateral rectus   Medial rectus   Lateral rectus

Oculomotor nucleus (III)

Midbrain

Ascending component of the medial longitudinal fusciculus (MLF)

Caudal pons

Abducens nucleus (VI)

Scarpa's ganglion

Rostral medulla

Vestibular nucleus

S
L
M I

Vestibular nucleus

Figure 13.10 Connections underlying the vestibulo-ocular reflex. Projections of the vestibular nucleus to the nuclei of cranial nerves III (oculomotor) and VI (abducens). The connections to the oculomotor nucleus and to the contralateral abducens nucleus are excitatory (red), whereas the connections to ipsilateral abducens nucleus are inhibitory (black). Connections exist from the oculomotor nucleus to the medial rectus of the left eye and from the adbucens nucleus to the lateral rectus of the right eye. This circuit pulls the eyes to the right, that is, in the direction away from the left horizontal canal, when the head rotates to the left. Turning to the right, which causes increased activity in the right horizontal canal, would have the opposite effect on eye movements. The projections from the right vestibular nucleus are omitted for clarity. The letters in the blowup of the vestibular nucleus indicate its superior (S), lateral (L), medial (M), and inferior (I) subdivisions.

head movements. However, a patient with bilateral loss of vestibular function has the persistent and disturbing sense that the world is moving when the head moves. The problem in such cases is that information about head and body movements, normally generated by the vestibular organs, is not available to the oculomotor centers, so that corrective eye movements cannot be made.

Descending projections from the vestibular nuclei are essential for postural adjustments of the head and body. Axons from the medial vestibular nucleus descend in the medial longitudinal fasciculus to reach the upper cervical levels of the spinal cord (Figure 13.11). This pathway regulates head position by reflex activity of neck muscles in response to stimulation of the semicircular canals from rotational accelerations of the head. For example, during a downward pitch of the body, such as tripping forward, the superior canals are activated and the head muscles reflexively pull the head up. The dorsal flexion of the head initiates other reflexes, such as forelimb extension and hindlimb flexion, to stabilize the body (see Chapter 11).

Figure 13.11 Descending projections from the medial and lateral vestibular nuclei to the spinal cord (and input to the lateral vestibular nucleus from cerebellum). The medial vestibular nuclei project bilaterally in the medial longitudinal fasciculus to reach the medial part of the ventral horns. The lateral vestibular nucleus sends axons via the lateral vestibular tract to contact anterior horn cells innervating the axial and proximal limb muscles. Neurons in the lateral vestibular nucleus receive input from the cerebellum, allowing the cerebellum to influence posture and equilibrium.

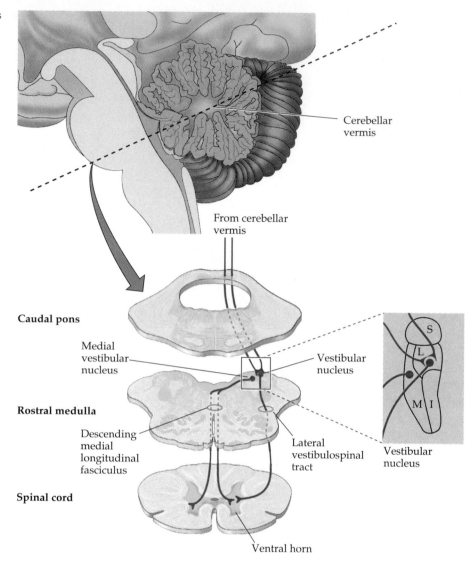

The inputs from the otolith organs project mainly to the lateral vestibular nucleus, which in turn sends axons in the lateral vestibulospinal tract to the spinal cord. The input from this tract exerts a powerful excitatory influence on the extensor (antigravity) muscles (see Chapter 16). When hair cells in the otolith organs are activated, signals reach the medial part of the ventral horn. By activating the ipsilateral pool of motor neurons innervating extensor muscles in the trunk and limbs, this pathway mediates balance and the maintenance of upright posture.

■ VESTIBULAR PATHWAYS TO THE THALAMUS AND CORTEX

In addition to these several descending projections, the superior and vestibular nuclei send axons to the ventral posterior nuclear complex of the thalamus, which projects to two cortical areas for vestibular sensations (Figure 13.12). One cortical target is just posterior to the primary somatosensory cortex, near the representation of the face, and the other is at the transition between the somatosensory cortex and the motor cortex (Brodmann's area

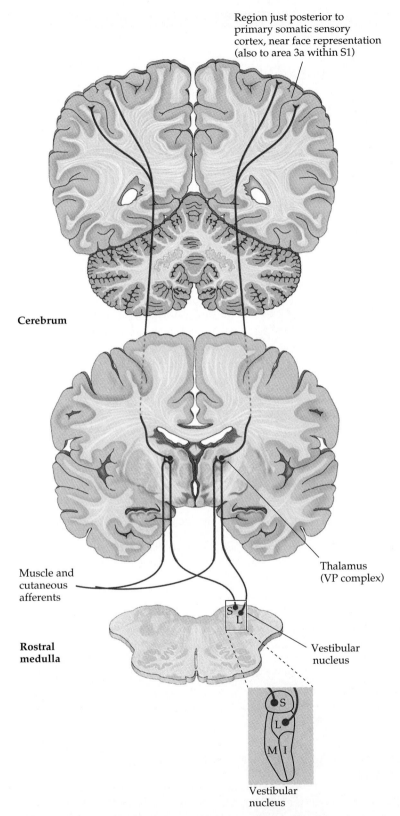

Region just posterior to
primary somatic sensory
cortex, near face representation
(also to area 3a within S1)

Cerebrum

Muscle and
cutaneous
afferents

Thalamus
(VP complex)

Rostral
medulla

Vestibular
nucleus

S
L
M I

Vestibular
nucleus

Figure 13.12 Thalamocortical pathways carrying vestibular information. The lateral and superior vestibular nuclei project to the thalamus. From the thalamus, the vestibular neurons project to the vicinity of the central sulcus near the face representation. Sensory inputs from the muscles and skin also converge on thalamic neurons receiving vestibular input (see Chapter 8).

Box C
THROWING COLD WATER ON THE VESTIBULAR SYSTEM

Testing the integrity of the vestibular system can indicate much about the condition of the brainstem, particularly in comatose patients.

Normally, when the head is not being rotated, the output of the nerves from the right and left sides are equal; thus, no eye movements occur. When the head is rotated in the horizontal plane, the vestibular afferent fibers on the side toward the turning motion will increase in their firing rate, while the afferents on the opposite side will decrease their firing rate (Figures A and B). The net difference in firing rates then

leads to slow movements of the eyes counter to the turning motion. This reflex response generates the slow component of a normal eye movement pattern called physiological nystagmus

(B)
1. Physiological nystagmus

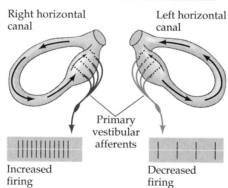

Head rotation

Slow eye movement

Fast eye movement

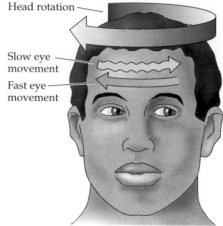

Right horizontal canal

Left horizontal canal

Primary vestibular afferents

Increased firing

Decreased firing

2. Spontaneous nystagmus

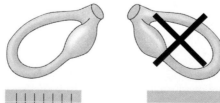

Baseline firing

No firing

(Figure B1). (The fast component is a saccade that resets the eye position; see Chapter 19.)

Unwanted and deleterious nystagmus (which means "nodding" or oscillatory movements of the eyes) can occur if there is unilateral damage to the vestibular system. In this case, the silencing of the spontaneous output from the damaged side will result in a pathological difference in firing rate because the spontaneous discharge from the intact side remains (Figure B2). The difference in firing rates (spontaneous versus none at all) will cause nystagmus, even though no head movements are being made.

Such responses can also be used to assess the integrity of the brainstem in patients when damage is suspected. If a patient is placed on his back and the head is elevated to about 30° above horizontal, the horizontal semicircular canals lie in an almost vertical orientation. Irri-

(A)

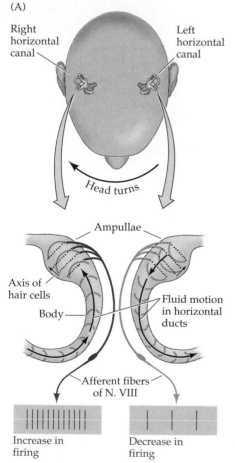

Right horizontal canal

Left horizontal canal

Ampullae

Axis of hair cells

Body

Fluid motion in horizontal ducts

Head turns

Afferent fibers of N. VIII

Increase in firing

Decrease in firing

(A) View looking down on the top of a person's head illustrates the fluid motion generated in the left and right horizontal canals, and the changes in vestibular nerve firing rates when the head turns to the right. (B) In normal individuals, rotating the head elicits physiological nystagmus (1), which consists of a slow eye movement counter to the direction of head turning. The slow component of the eye movements is due to the net differences in left and right vestibular nerve firing rates acting via the central circuit diagrammed in Figure 13.10. When the vestibular system is damaged, nystagmus can occur spontaneously. Spontaneous nystagmus (2), where the eyes move rhythmically from side to side in the absence of any head movements, occurs when one of the canals is damaged. In this situation, net differences in vestibular nerve firing rates exist even when the head is stationary because the vestibular nerve innervating the intact canal fires steadily when at rest, in contrast to a lack of activity on the damaged side.

(C)

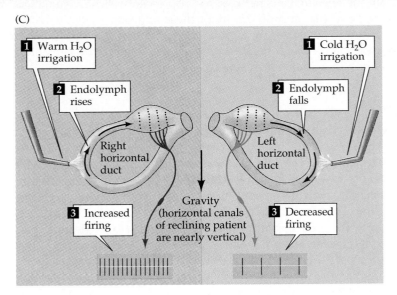

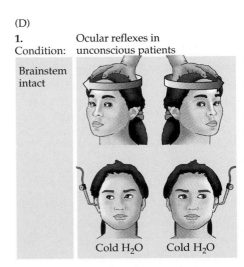

1. Warm H₂O irrigation

2. Endolymph rises

Right horizontal duct

3. Increased firing

Gravity (horizontal canals of reclining patient are nearly vertical)

1. Cold H₂O irrigation

2. Endolymph falls

Left horizontal duct

3. Decreased firing

(C) Caloric testing of vestibular function is possible because irrigating an ear with water slightly warmer than body temperature generates convection currents in the canal that mimic the endolymph movement induced by turning the head to the irrigated side. Irrigation with cold water induces the opposite effect. These currents ultimately result in changes in the firing rate of the associated vestibular nerve, with an increased rate on the warmed side and a decreased rate on the chilled side. As in head rotation and spontaneous nystagmus, net differences in firing rates generate eye movements.

gating one ear with cold water will then lead to spontaneous eye movements because convection currents in the canal mimic rotatory head movements away from the irrigated ear (Figure C). In alert patients, these eye movements consist of a slow movement toward the irrigated ear and a fast movement away from it.

The fast movement is most readily detected by the observer, and the significance of its direction can be kept in mind by using the mnemonic COWS (cold opposite, warm same). This test can also be used in comatose patients, but because saccadic movements are no longer made, the response consists of only the slow

movement component. When cold water is used, the slow movement will be toward the irrigated ear unless damage to the brainstem has occurred (Figure D). Lesions in the caudal pons or rostral medulla (the site of the vestibular nuclei) can abolish or alter these responses.

(D)

1. Condition: Brainstem intact — Ocular reflexes in unconscious patients

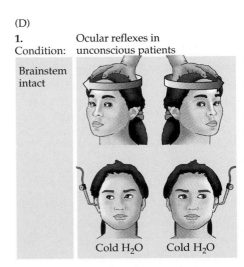

Cold H₂O   Cold H₂O

2. Condition: MLF lesion (bilateral) — Ocular reflexes in unconscious patients

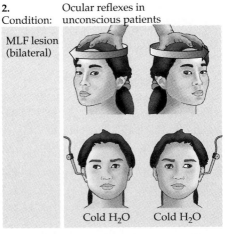

Cold H₂O   Cold H₂O

3. Condition: Low brainstem lesion — Ocular reflexes in unconscious patients

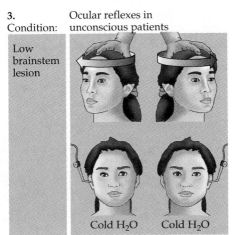

Cold H₂O   Cold H₂O

(D) Manual turning of the head and caloric testing can be used together to test the function of the brainstem in an unconscious patient. The figures show the slow eye movements, resulting from either manual rotation of the head or from cold water irrigation in one ear, for three different conditions: (1) with the brainstem intact; (2) with a lesion of the medial longitudinal fasciculus (MLF; note that unlike head rotation, irrigation in this case results in movement of the eye only on the irrigated side); and (3) with a low brainstem lesion.

3a; Chapter 8). Electrophysiological studies of individual neurons in these areas show that the relevant cells respond to proprioceptive and visual stimuli as well as to vestibular stimuli. Many of these neurons are activated by moving visual stimuli as well as by rotation of the body (even with the eyes closed), suggesting that these cortical regions are involved in the perception of body orientation in extrapersonal space.

■ SUMMARY

The receptors of the vestibular system are located in the otolith organs and the semicircular canals of the inner ear. These structures are responsible for the two major functions of the vestibular system. The otolith organs provide information necessary for postural adjustments of the somatic musculature, particularly the axial musculature, when the head tilts in various directions or undergoes linear accelerations. This information represents linear forces acting on the head that arise through static effects of gravity or from translational movements. The semicircular canals provide information about rotational accelerations of the head. This latter information generates reflex movements that adjust the eyes, head, and body during motor activities. The most important of these reflexes are eye movements that compensate for head movements, thereby stabilizing the visual scene when the head turns. Input from all the vestibular organs is integrated with input from the visual and somatic sensory systems to provide our perceptions of body position and orientation in extrapersonal space, enabling appropriate reflex movements.

Additional Reading

Reviews

BENSON, A. (1982) The vestibular sensory system. In *The Senses*, H. B. Barlow and J. D. Mollon (eds.). New York: Cambridge University Press.

BRANDT, T. (1991) Man in motion: Historical and clinical aspects of vestibular function. A review. Brain 114: 2159–2174.

FURMAN, J. M. AND R. W. BALOH (1992) Otolith-ocular testing in human subjects. Ann. New York Acad. Sci. 656: 431–451.

GOLDBERG, J. M. (1991) The vestibular end organs: Morphological and physiological diversity of afferents. Curr. Opin. Neurobiol. 1: 229–235.

GOLDBERG, J. M. AND C. FERNANDEZ (1984) The vestibular system. In *Handbook of Physiology, Section 1: The Nervous System*, Volume III: *Sensory Processes*, Part II, J. M. Brookhart, V. B. Mountcastle, I. Darian-Smith and S. R. Geiger (eds.). Bethesda, MD: American Physiological Society.

Important Original Papers

GOLDBERG, J. M. AND C. FERNANDEZ (1971) Physiology of peripheral neurons innervating semicircular canals of the squirrel monkey, Parts 1, 2, 3. J. Neurophysiol. 34: 635–684.

GOLDBERG, J. M. AND C. FERNANDEZ (1976) Physiology of peripheral neurons innervating otolith organs of the squirrel monkey, Parts 1, 2, 3. J. Neurophysiol. 39: 970–1008.

LINDEMAN, H. H. (1973) Anatomy of the otolith organs. Adv. Oto.-Rhino.-Laryng. 20: 405–433.

Book

BALOH, R. W. AND V. HONRUBIA (1990) *Clinical Neurophysiology of the Vestibular System*, 2nd Ed. Philadelphia: F. A. Davis Co.

CHAPTER 14

THE CHEMICAL SENSES

■ OVERVIEW

Three sensory systems are dedicated to the detection of chemicals in the environment: olfaction, taste, and the trigeminal chemosensory system. The olfactory system detects airborne molecules; the taste system detects ingested, water-soluble molecules; and the trigeminal chemosensory system detects noxious molecules that can actually damage the affected surfaces. In humans, olfaction provides information about chemicals from food, one's self, other people, and a variety of animals, plants, and other aspects of the environment. Olfactory information can influence feeding behavior, social interactions, and reproduction. Taste, or gustation, provides information about the quality, quantity, and safety of ingested food. Trigeminal chemosensation provides information about irritating or noxious chemicals that come into contact with skin or mucous membranes. The chemosensory systems rely on receptors in the nasal cavity, in the mouth, or on the face to transduce and transmit the effects of these molecular stimuli to the central nervous system.

■ THE ORGANIZATION OF THE OLFACTORY SYSTEM

Although from an evolutionary perspective olfaction is considered the "oldest" sense, in many ways it remains the most mysterious. The olfactory system encodes information about the molecular identity and concentration of a wide range of chemical stimuli. These stimuli, called **odorants**, interact with olfactory receptor neurons in an epithelial sheet—the **olfactory epithelium**—that lines the interior of the nose (Figure 14.1A). Olfactory receptor neurons send their axons directly to the **olfactory bulb**. Neurons in the olfactory bulb project in turn to a number of targets in the forebrain, including the hypothalamus, the amygdala, and several regions of the cerebral cortex (Figure 14.1B). These several brain regions mediate odorant identification, as well as the emotional reactions and motor responses to chemical stimuli.

   The olfactory system abides by many of the principles that govern other sensory modalities. Thus, the effects of stimuli are transduced and transmitted from the receptors to a major relay station that projects to higher-order centers. There are, however, some important differences between the olfactory system and other sensory pathways. In both the somatic sensory and visual systems, distinct submodalities (form and motion, for example) are transmitted centrally along parallel pathways (see Chapter 11). It is not yet clear whether such submodalities and parallel pathways exist for olfaction. Moreover, the central representation of olfaction remains enigmatic. The topographical maps in the somatic sensory or visual cortices that reflect the geometry and functional organization of the receptor surface do not exist in any obvious way in the olfactory areas of the central nervous system.

■ OLFACTORY PERCEPTION IN HUMANS

In the other sensory systems, the perceptual effect of a stimulus is directly related to its physical properties. Thus, for vision, the shorter the wavelength, the bluer the color; and for audition, the higher the frequency of sound

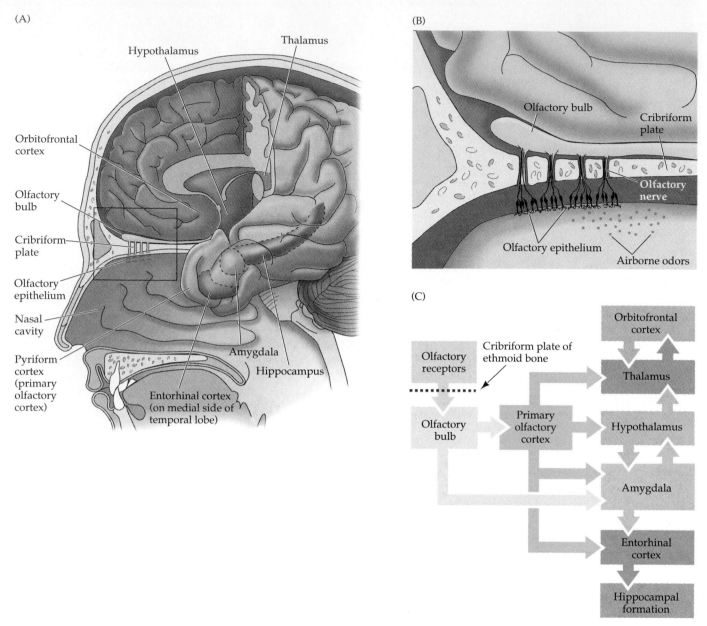

Figure 14.1 Organization of the human olfactory system. (A) Peripheral and central components of the olfactory pathway. (B) Enlargement of region boxed in (A) showing the relationship between the olfactory epithelium, containing the olfactory receptor neurons, and the olfactory bulb, the central target of olfactory receptor neurons. (C) Diagram of the basic pathways for processing olfactory information.

pressure waves, the higher the perceived pitch. No such simplifying relationships for olfactory stimuli have been found. A potentially useful classification was developed in the 1950s by John Amoore, who divided odors into a number of classes based on their perceived quality and molecular structure: pungent, floral, musky, camphor, peppermint, ether, and putrid. These categories are still used routinely to describe odors, to assess the cellular process of olfactory transduction, and to discuss the central representation of

OH
O
O
H
O
O
O
Ethanol
alcoholic
2 mM
Ethyl acetate
ethereal
0.06 mM
Benzaldehyde
bitter almond
0.3 μM
4-Hydroxyoctanoic acid lactone
coconut
0.05 μM
O
O
S
O
O
Cl
Cl
Cl
Pentadecalactone
musky
7 nM
Dimethylsulfide
putrid
5 nM
5α-Androst-16-en-3-one
urinous
0.6 nM
2,3,6-Trichloroanisole
moldy
0.1 nM
OH
H
O
O
N
O
N
Geosmin
earthy
0.1 nM
2-*trans*-6-*cis*-Nonadienal
cucumber
0.07 nM
β-Ionone
violet
0.03 nM
2-Isobutyl-3-methoxypyrazine
bell pepper
0.01 nM

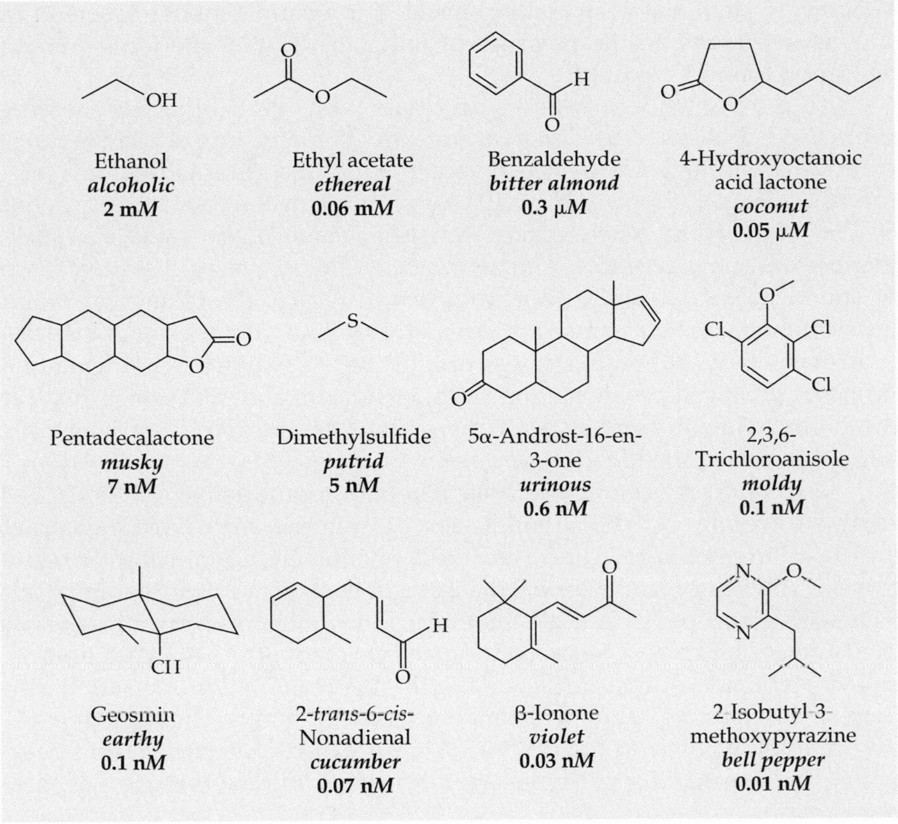

Figure 14.2 Chemical structure and olfactory thresholds for 12 common odorants. Molecules perceived at low concentration thresholds are more lipid-soluble, whereas those with higher thresholds are more water-soluble. (After Pelosi, 1994.)

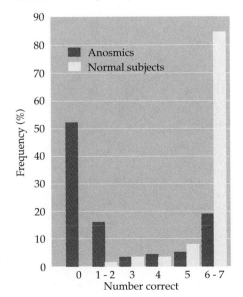

olfactory information. Nonetheless, this classification remains entirely empirical. Despite its limitations, the utility of this sort of schema makes clear that the olfactory system can identify individual odorants that have distinct chemical identities. Thus, coconuts, violets, cucumbers, and bell peppers all have a unique odorant molecule that we readily recognize (Figure 14.2). Most naturally occurring odors are actually blends of several odorant molecules. Nonetheless, these odors—like a favorite perfume or wine—are experienced as a single perception. The unitary perception of complex odors is a remarkable feature of olfaction.

Although a number of other animals are superior in their olfactory abilities, humans are remarkably good at detecting and identifying particular molecules in the environment. The threshold concentrations for odorant identification vary over a range of about 10,000,000. The major aromatic constituent of bell pepper, 2-isobutyl-3-methoxypyrazine, can be detected at a concentration of 0.01 n*M*, whereas ethanol, for example, cannot be detected until its concentration reaches approximately 2 m*M*. This range of sensitivity to different odorants depends on physical properties like lipid solubility and vapor pressure. Thresholds for odorant identification generally decrease as the lipid solubility of the odorant molecule increases. In other words, the more hydrophobic the odorant, the lower the threshold for its identification.

Psychologists and neurologists have developed a battery of tests that measure ability to name common odors. Most people are able to identify such odors consistently. Some, however, fail to identify one or more common odorants (Figure 14.3). Such chemosensory deficits, called **anosmias,** are often restricted to a single odorant, suggesting that a specific element in the

Figure 14.3 Anosmia is the inability to identify common odors. When normal subjects or subjects identified as anosmic are presented with a battery of seven common odors (a test frequently used by neurologists), 85% of the normals can identify all seven odors correctly. Surprisingly, a few "normal" subjects have difficulty identifying even these common odors (in this case, baby powder, chocolate, cinnamon, coffee, mothballs, peanut butter, and soap). When individuals previously identified as anosmics are presented with the same battery of common odors, only a few can identify all of the odors (less than 15%) and more than half cannot identify any of the odors. (After Cain and Gent, in Meiselman and Rivlin, 1986.)

90
80
70
60
50
40
30
20
10
0
Frequency (%)
Anosmics
Normal subjects
0
1 - 2
3
4
5
6 - 7
Number correct

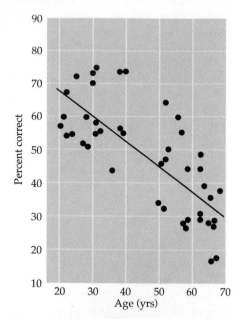

Figure 14.4 Normal decline in olfactory sensitivity with age. The ability to identify 80 common odorants was scored for individuals between the ages of 20 and 70. There is an average decline of 50% in the number of odors identified correctly over this period. (After Murphy, 1986.)

olfactory system has been compromised. For example, about 1 person in 1000 is insensitive to the presence of butyl mercaptan, the foul-smelling odorant released by skunks.

The ability to identify odors decreases with age. If otherwise healthy subjects are challenged to identify a battery of common odors, there is a dramatic difference in performance of young adults and older individuals (Figure 14.4). People between 20 and 40 years of age can correctly identify about 50–75% of the odors, whereas those between 50 and 70 years of age correctly identify only about 30–45%. A more radically diminished or distorted sense of smell can accompany eating disorders, diabetes, the taking of certain medications, and Alzheimer's disease; the reasons for these deficiencies remain obscure. Although the loss of olfactory sensitivity is not usually a source of great concern, it can diminish enjoyment, and, if severe, can affect the ability to identify and respond appropriately to unsanitary or dangerous odors (such as spoiled food, or smoke).

Olfactory cues can also influence human reproductive functions. For example, women housed in single-sex dormitories may have menstrual cycles at the same time. The basis of this phenomenon is apparently olfactory since synchronization of the menses can be induced by presenting subjects with gauze pads from the underarms of women at different stages of their menstrual cycles. The olfactory system evidently detects chemicals, called pheromones, that influence the timing of ovulation. Olfaction also helps mediate recognition between mothers and infants. Changes in suckling rate in response to their mother's odor can be detected in infants as young as 6 weeks of age. Mothers can also discriminate the smell of their own infant.

■ THE OLFACTORY EPITHELIUM AND OLFACTORY RECEPTOR NEURONS

The transduction of olfactory information occurs in the olfactory epithelium, the sheet of neurons and supporting cells that lines the nasal cavities. The olfactory epithelium includes several distinct cell types (Figure 14.5A). The most important of these, the **olfactory receptor neuron**, is a bipolar neuron that gives rise on its basal surface to a small-diameter, unmyelinated axon that carries the olfactory information to the brain. At its apical surface, the receptor neuron gives rise to a single process that expands into a knoblike protrusion from which several microvilli, called **olfactory cilia**, extend into the thick layer of mucus that lines the nasal cavity and controls the ionic milieu of the olfactory cilia. The mucus is produced by secretory specializations called Bowman's glands that are distributed throughout the epithelium. Two other cell classes, basal cells and sustentacular (supporting) cells, are also present in the olfactory epithelium. This entire apparatus—mucus layer and epithelium with neural and supporting cells—is called the **nasal mucosa**.

The arrangement of the nasal mucosa allows the olfactory receptor neurons direct access to odorant molecules; but it also means that these neurons are exceptionally exposed. Airborne pollutants, allergens, microorganisms, and other potentially harmful substances subject the olfactory receptor neurons to daily damage. Several mechanisms help maintain the integrity of the olfactory epithelium in the face of this problem. Thus, immunoglobulins are secreted into the mucus, providing an initial defense against harmful antigens. In addition, the sustentacular cells catabolize a variety of organic chemicals that enter the nasal cavity.

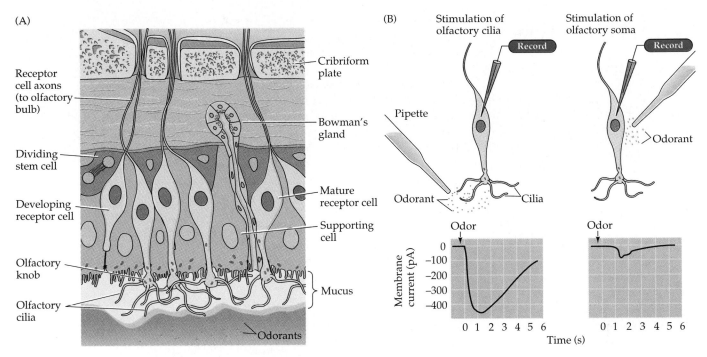

Figure 14.5 Structure and function of the olfactory epithelium. (A) Diagram of the olfactory epithelium showing the major cell types and the projection of the olfactory receptor neurons to the olfactory bulb (see below). Bowman's glands produce mucus, and supporting (or sustantacular) cells help to detoxify chemicals that come in contact with the epithelium. New olfactory receptor neurons are generated from stem cells. (B) Generation of receptor potentials in response to odors takes place in the cilia of receptor neurons. Odorants evoke a large inward (depolarizing) current when applied to the cilia (left), but only a small inward current when applied to the cell body (right). (B after Firestein et al., 1991.)

Despite these defenses, olfactory receptor neurons normally degenerate and must be replaced. Estimates in rodents suggest that the entire population of olfactory neurons is renewed every 6 to 8 weeks. This feat is accomplished by maintaining among the basal cells a population of precursors (stem cells) that divide to give rise to new receptor neurons (see Figure 14.5A). How the new olfactory receptor neurons extend axons to the brain and reestablish appropriate functional connections is not fully known. The naturally occurring degeneration and regeneration of olfactory receptor cells provides an opportunity to investigate how neural precursor cells can successfully generate new neurons and reconstitute function in the mature central nervous system, a topic of considerable clinical interest (see Unit 4).

■ THE TRANSDUCTION OF OLFACTORY SIGNALS

The cellular and molecular machinery for olfactory transduction is located in the olfactory cilia (Figure 14.5B). Odorants bind to specific receptors on the external surface of cilia (see below); this binding may occur directly, or by way of proteins in the mucus that sequester the odorant and shuttle it to the receptor. Several additional steps then generate a receptor potential by opening ion channels. At least two second-messenger pathways mediate this process. As in photoreceptors, cyclic nucleotide-gated channels play an

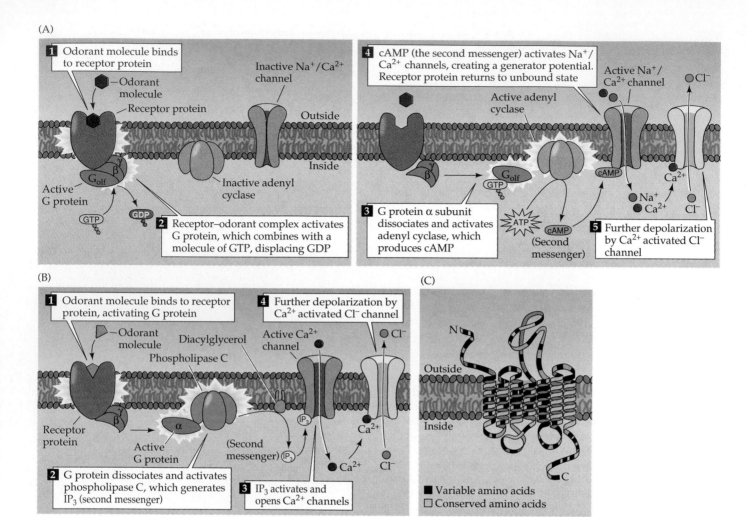

Figure 14.6 Olfactory transduction. (A) Diagram of the events leading to the generation of cyclic AMP (cAMP) in response to odorant stimulation of olfactory receptor neurons. The ultimate target of cAMP is a cation-selective channel. The influx of Na$^+$ and Ca^{2+} through this channel depolarizes the cell. The increase in intracellular Ca^{2+} opens a Cl$^-$ channel that further depolarizes the olfactory receptor neuron. (B) Generation of IP$_3$-mediated intracellular signaling in response to odorant binding on olfactory receptor neurons. Increased IP$_3$ leads to the activation of Ca^{2+} channels. (C) Structure of the putative olfactory odorant receptor; these proteins (there may be as many as 2000 varieties) have seven transmembrane domains plus a variable cell surface region and a cytoplasmic domain that interacts with G-proteins.

important role in olfactory transduction. The receptor neurons contain an olfactory-specific G-protein (G$_{olf}$), which activates an olfactory-specific adenylate cyclase (Figure 14.6A). The resulting increase in cyclic AMP (cAMP) opens a channel that permits Na$^+$ and Ca^{2+} entry, thus depolarizing the neuron. This depolarization, amplified by a Ca^{2+}-activated Cl$^-$ current, is conducted passively from the cilia to the axon hillock region of the olfactory receptor neuron, where action potentials are generated and transmitted to the olfactory bulb. In another second-messenger pathway, odorant binding activates a different G-protein, which in turn activates phospholipase C, leading to an increase in inositol triphosphate (IP$_3$; Figure 14.6B). Increased IP$_3$ leads to an increase in intracellular Ca^{2+} by opening Ca^{2+} channels. Odorants with distinct qualities differentially activate these two second-messenger pathways. Fruity odors such as citralva—the essence of lemon—activate primarily the cAMP pathway, whereas putrid odorants such as isovaleric acid—a molecule found in sweat—activate the IP$_3$ pathway (Table 14.1). Whether the IP$_3$ and cAMP pathways coexist in the same mammalian olfactory receptor neuron is not yet established.

Olfactory receptor neurons are very good at extracting a signal from chemosensory noise. For example, fluctuations in the cAMP concentration in an olfactory receptor neuron could, in theory, cause the receptor cell to be activated in the absence of odorants. Such nonspecific responses do not occur, however, because the cAMP-gated channels are blocked at the resting potential by the high Ca^{2+} and Mg^{2+} concentrations in mucus. To overcome this voltage-dependent block and generate an action potential, several channels must be opened at once. This requirement ensures that olfactory receptor neurons fire only in response to stimulation by odorants.

TABLE 14.1
Generation of Second Messengers in Response to Different Odorants

Odorant	Cyclic AMP[a]	IP$_3$[a]
Citralva (fruity)	100	4
Eugenol (herbaceous)	70	4
Geraniol (floral)	62	6
Pyrroline (putrid)	0	70
Isovaleric acid (putrid)	0	68
Lyral (floral)	0	100

Adapted from Breer and Boekhoff, 1991.
[a]Values are expressed as percentages of the effect produced by citralva and lyral.

■ SPECIFICITY OF ODORANT DETECTION: ODORANT RECEPTORS

Odorant receptor molecules presumably generate most of the specificity of olfactory signal transduction. Between 1000 and 2000 genes identified from an olfactory epithelium cDNA library have defined a family of putative odorant receptors (Figure 14.6C). This number of odorant receptors is probably large enough to account for the number of distinct odors that can be discriminated by the olfactory system (estimated to be about 2000). The olfactory receptor molecules are homologous to a large family of G-protein-linked receptors that include β-adrenergic receptors and rod opsin. The odorant receptor proteins all have seven membrane-spanning hydrophobic domains, potential odorant binding sites in the extracellular domain of the protein, and the ability to interact with G-proteins at the carboxyl terminal region of their cytoplasmic domain. The family of receptor genes displays substantial variability, particularly in regions that code for the membrane-spanning domains. Messenger RNAs for these genes are expressed in subsets of olfactory neurons that occur in bilaterally symmetric patches of olfactory epithelium. This patchiness suggests that different odors activate spatially distinct subsets of olfactory receptor neurons. Moreover, genetic analysis shows that each olfactory receptor neuron expresses only one or a few of the 1000 to 2000 odorant receptor genes. On the other hand, physiological experiments show that individual olfactory receptor neurons are activated by more than one odorant (see below). In any event, the odorant sensitivity of each receptor neuron is determined by the expression of a small number of odorant receptors.

■ NEURAL CODING IN THE OLFACTORY SYSTEM

Like other sensory receptor cells, olfactory receptor neurons are sensitive to a subset of stimuli that define a "tuning curve." Different odorants (Figure 14.7A) or changing concentrations of the same odorant (Figure 14.7B), change the latency of response, the duration of the response, and/or the firing frequency of individual neurons. Subsets of olfactory receptor neurons respond best to a general category of stimuli—for example, fruity or putrid odorants. This observation implies that the molecular machinery of each

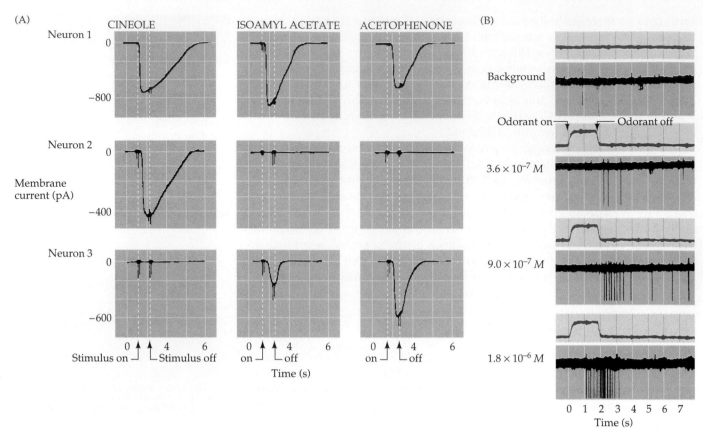

Figure 14.7 Responses of olfactory receptor neurons to selected odorants. (A) Neuron 1 responds similarly to three different odorants. In contrast, neuron 2 responds to only one of these odorants. Neuron 3 responds to two of the three stimuli. The responses of these receptor neurons were recorded by whole-cell patch clamp recording; downward deflections represent inward currents measured at a holding potential of −55mV. (B) Responses of a single olfactory receptor neuron to changes in the concentration of a single odorant, isoamyl acetate. The upper trace in each panel (red) indicates the duration of the odorant stimulus; the lower trace the neuronal response. The frequency and number in each panel of action potentials increases as the odorant concentration increases. (A after Firestein et al., 1992; B after Getchell, 1986.)

olfactory receptor neuron for odorant binding and transduction is selective for particular odorant molecules, but not exclusively so. However achieved, it is clear that the individual receptor neurons can be activated by several odorants.

Olfactory receptor neurons respond to the continued presence of an odorant by reducing their frequency of firing. Thus, when stimulated for long periods, these neurons initially respond with a burst of action potential activity that soon returns to a lower but constant level. Such adaptation probably arises from an increase in intracellular Ca^{2+} generated by the initial odorant-dependent depolarization. The rise in intracellular Ca^{2+} increases the probability of closing the second messenger-activated channels, thus decreasing the frequency of action potential firing. Perceptually, adaptation is reflected in a decreased ability to discriminate odors during repeated inhalations. Recordings of the overall electrical activity of the human olfactory epithelium (called the electroolfactogram) show that identical responses to an odorant can be obtained over several presentations only if the exposures are spaced 1 to 3 minutes apart. This recovery interval is evidently necessary for the signal transduction machinery to reset itself. The reality of olfactory adaptation can be appreciated by recalling how one eventually becomes less aware of the smell of cigarette smoke after entering a "smoking" motel room.

Two general proposals have been put forward to explain how olfactory receptor responses are assembled into a code that identifies the type and concentration of a given odorant. The first is the **labeled line hypothesis**, which postulates a specific set of receptor neurons for each odor that relays the information to the brain along a separate pathway. The second proposal is the **computational model** of olfaction, which suggests that odor identity is computed centrally by comparing and combining patterns of activity across populations of olfactory receptor neurons. The response properties of individual receptor neurons are not in complete accord with either of these models. In fact, the organization of the projections from the receptors to the brain is consistent with a combination of labeled line and computational strategies for the representation of olfactory information (see Box A below).

■ CENTRAL PROCESSING OF OLFACTORY SIGNALS

Odorant transduction and coding by olfactory receptor neurons are only the first steps in processing olfactory signals. The events that follow rely on the synaptic organization of a number of regions of the brain that receive and compute chemosensory information (Figure 14.8A). As the olfactory receptor axons leave the olfactory epithelium, they coalesce to form a large number of bundles that together make up the **olfactory nerve** (cranial nerve I). The olfactory nerve runs directly to the **olfactory bulb**, which lies on the ventral anterior aspect of the ipsilateral forebrain (see Figure 14.1).

The most distinctive feature of the olfactory bulb is an array of spherical accumulations of neuropil called **glomeruli** that lie just beneath its surface and receive the primary olfactory axons (Figure 14.8A,B). In addition to these units, the bulb comprises several cell and neuropil layers that are specialized for receiving, processing, and relaying olfactory information (Figure 14.8B,C). In humans, the olfactory bulb is relatively small; in other mammals, it is comparatively larger, sometimes dwarfing other forebrain regions. These differences in relative size correspond, not surprisingly, to differences in the olfactory abilities of various species. The surface area of the olfactory epithelium, and therefore the number of olfactory receptor neurons, also differs greatly among species. The olfactory epithelium is approximately 10 cm^2 in a 70-kg human, for example, and 20 cm^2 in a 3-kg cat.

Within the glomeruli, the axons of the receptor neurons contact the apical dendrites of **mitral cells**. The cell bodies of the mitral cells, the principal projection neurons of the olfactory bulb, are located in a distinct layer deep in the olfactory glomeruli (Figure 14.8B,C). Mitral cells extend a primary dendrite into a glomerulus, giving rise to an elaborate glomerular tuft onto which the primary olfactory axons synapse. Each glomerulus in the mouse (where this matter has been studied quantitatively) includes the apical dendrites of approximately 25 mitral cells, and approximately 25,000 olfactory receptor axons innervate each mouse glomerulus. Accordingly, individual mitral cells are innervated by approximately 1000 receptor axons, a degree of convergence that is presumably representative of other species. The glomerulus also includes dendritic processes from two other

(A)

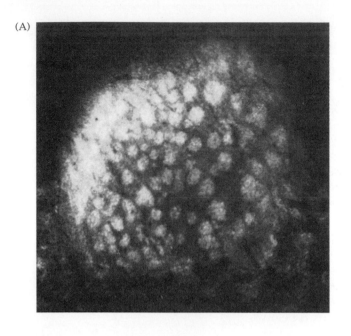

(B)

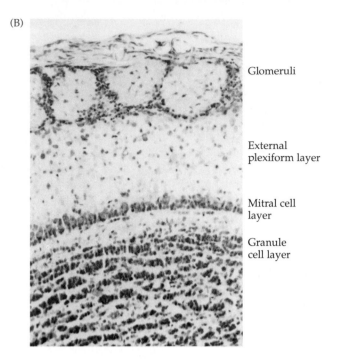

Glomeruli

External
plexiform layer

Mitral cell
layer

Granule
cell layer

(C)

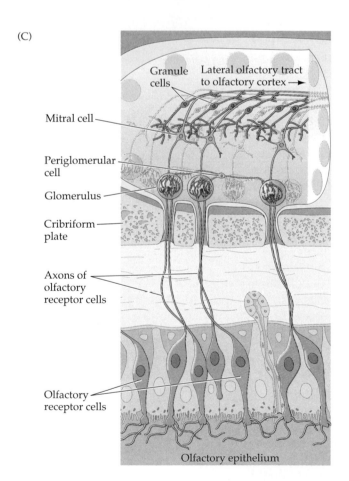

Granule
cells

Lateral olfactory tract
to olfactory cortex →

Mitral cell

Periglomerular
cell

Glomerulus

Cribriform
plate

Axons of
olfactory
receptor cells

Olfactory
receptor cells

Olfactory epithelium

(D)

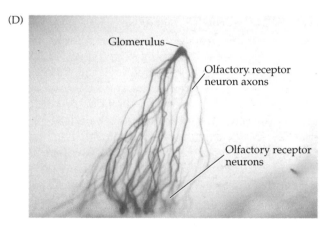

Glomerulus

Olfactory receptor
neuron axons

Olfactory receptor
neurons

(E)

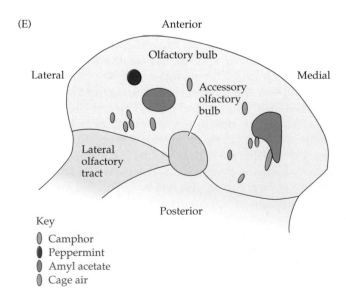

Anterior

Olfactory bulb

Lateral

Accessory
olfactory
bulb

Medial

Lateral
olfactory
tract

Posterior

Key
Camphor
Peppermint
Amyl acetate
Cage air

◀ **Figure 14.8**   Olfactory glomeruli, the initial processing circuits of the olfactory pathway. (A) Glomeruli near the surface of the olfactory bulb, visualized in a living mouse after removing the overlying bone. These dense accumulations of dendrites and synapses are stained here with a vital fluorescent dye that recognizes neuronal processes. (B) Coronal section through a mouse olfactory bulb stained to show the distribution of cell bodies rather than processes. The glomeruli occur in a distinct layer among several others, including the external plexiform layer, where dendrites and axons from the mitral cell layer are found, and the granule cell layer. (C) Diagram of the laminar and synaptic organization of the olfactory bulb. The axons of olfactory receptor neurons terminate on mitral cell apical dendrites within glomeruli. In the external plexiform layer, the mitral cells also receive intrinsic connections from granule cells and other mitral cells. Only the mitral cell projects out of the olfactory bulb. (D) Individual glomeruli receive input from receptor neurons that express particular odorant receptor molecules. This suggests that particular glomeruli are specialized to process distinct odorants or classes of odorants. The receptor neurons are labeled here by a reporter gene coupled to the promoter of a specific receptor type. (E) Patterns of focal neuronal activity (determined by 2-deoxy-glucose uptake) in a diagram of the rat olfactory bulb evoked by cage air, peppermint, amyl acetate, and camphor. Particular regions of the bulb respond specifically to different odors. (A from LaMantia et al., 1992; B from Pomeroy et al., 1990; D from Axel, 1995; E modified from Shepherd, 1994.)

classes of local circuit neurons: tufted cells and periglomerular cells (approximately 50 tufted cells and 25 periglomerular cells contribute to each glomerulus). Other local circuit neurons, called granule cells, also participate in the synaptic interactions within the olfactory bulb. Granule cells synapse on mitral cell dendrites in different glomeruli and sharpen the chemical selectivity of the mitral cells.

Only the mitral cells transmit olfactory information to targets in the rest of the brain. Their axons form a bundle—the **lateral olfactory tract**—that projects to the accessory olfactory nuclei, the olfactory tubercle, the entorhinal cortex, and portions of the amygdala (see Figure 14.1B). The major target of the olfactory tract is the three-layered **pyriform cortex** in the ventromedial aspect of the cerebral hemispheres, near the optic chiasm. The axons of pyramidal cells in the pyriform cortex project in turn to several thalamic nuclei, to neocortical regions, and to the hippocampus and amygdala. Some of these pathways also converge on cortical regions in the inferotemporal lobe. Consequently, information about odors reaches a wide variety of forebrain regions to influence cognitive, emotional and homeostatic behaviors (see Chapter 27).

Receptor neurons that express particular odorant receptor genes send their axons to distinct subsets of olfactory glomeruli (Figure 14.8D), an arrangement that is apparently invariant between individual animals of the same species. These observations raise the possibility of a systematic representation of receptors across the surface of the olfactory bulb based on the odorant receptor molecules they express (Box A), and perhaps the specific odors to which they respond (Figure 14.8E).

■ THE ORGANIZATION OF THE TASTE SYSTEM

The taste system, together with the olfactory and visual systems, tells us if food should be ingested, whether the experience is pleasurable, and when we are satiated. When placed in the mouth, chemical constituents of food interact with receptors on taste cells that transduce stimuli and convey information regarding identity and concentration. In parallel, information regard-

Box A
MAPPING THE SENSE OF SMELL

The peripheral receptor surfaces of the visual, somatic sensory, and auditory systems are mapped to the brain in an orderly fashion. These maps arise from a combination of specific molecular cues that match ingrowing axons to their targets, and activity-dependent mechanisms that promote the construction of functionally appropriate connections between presynaptic and postsynaptic cells (see Chapters 21 and 22). It is still not clear whether such maps exist for the primary olfactory pathway, although recent evidence suggests that they may.

There are at least two possibilities for representing odor-specific information in the brain. Distinct odors might be mapped as "labeled lines" between neurons containing a specific odorant receptor in the olfactory epithelium and corresponding central areas. Alternately, odor information might be represented by computational circuitry that does not require an anatomically discernible map.

In the past few years, several investigators, including Richard Axel, Robert Vassar, and Peter Mombert at Columbia and Linda Buck and her colleagues at Harvard, have shown that the olfactory epithelium is in fact mapped onto the olfactory bulb. This mapping evidently establishes a correspondence between contiguous patches of cells in the olfactory epithelium that express specific odorant receptor molecules and subsets of glomeruli (the primary processing circuits) in the olfactory bulb. Because individual olfactory receptor neurons apparently express only one or a few odorant receptor genes, matching molecularly distinct populations of olfactory receptor neurons with small numbers of glomeruli in the bulb generates an odor-specific map in the brain. Such a map presumably arises from molecular cues that match each population of olfactory receptor neurons with their appropriate glomerular targets.

Humans can identify approximately 2000 distinct odors, there are approximately 2000 glomeruli in the human olfactory bulb, and we possess approximately 2000 putative olfactory G-protein receptor genes. These observations, together with the mapping studies described here, suggest that there may be labeled lines in the olfactory pathway. Nonetheless, it is important to note that we also perceive a continuum of odorant combinations as single odors. This aspect of odor perception (and memory) presumably reflects patterns of activity across ensembles of neurons throughout the olfactory system, and supports a computational scheme. In all likelihood, both means of representing olfactory information are important.

References

AXEL, R. (1995) The molecular logic of smell. Sci. Am. 273(10): 154–159.

CHESS, A., I. SIMON, H. CEDAR AND R. AXEL (1994) Allelic inactivation regulates olfactory receptor gene expression. Cell 78: 823–834.

RESSLER, K. J., S. L. SULLIVAN AND L. B. BUCK (1994) Information coding in the olfactory system: Evidence for a stereotyped and highly organized epitope map in the olfactory bulb. Cell 79: 1245–1255.

SINGER, M. S., G. M. SHEPHERD AND C. A. GREER (1995) Olfactory receptors guide axons. Nature 377: 19–20.

VASSAR, R., S. K. CHAO, R. SITCHERAN AND J. M. NUTEM (1994) Topographic organization of sensory projections to the olfactory bulb. Cell 79: 981–992.

ing the temperature and texture of food is transduced and relayed via somatic sensory receptors from the trigeminal and other sensory cranial nerves (see Chapter 8). Of course, food is not simply eaten for nutritional value; in this sense, "taste" also depends on cultural and psychological factors. How else could one rationalize why some people enjoy consuming "hot" peppers, Stilton cheese, and bitter-tasting liquids such as beer?

Like all sensory systems, the taste system includes peripheral receptors and a number of central pathways (Figure 14.9). The peripheral receptors, called **taste cells**, are found in **taste buds** distributed throughout the oral cavity, pharynx, and the upper part of the esophagus (Figures 14.9A). Primary sensory axons in the chorda tympani branch of cranial nerve VII (facial), the lingual branch of cranial nerve IX (glossopharyngeal), and the superior laryngeal branch of cranial nerve X (vagus) innervate the taste buds and carry taste information from the tongue, palate, epiglottis and esophagus, respectively (Figure 14.9B). These axons project to neurons in the rostral and lateral regions of the **nucleus of the solitary tract** in the medulla, known as the **gustatory nucleus** of the nucleus of the solitary tract. The pattern of

(A) Labeled line scheme

(B) Computational scheme

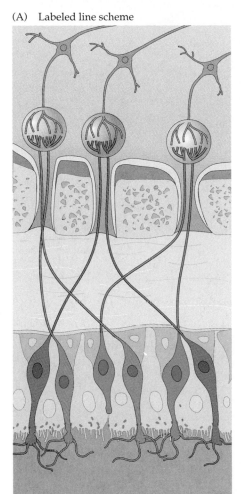

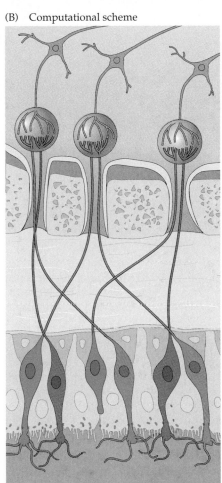

Two schemes for the representation of olfactory information in the olfactory bulb. (A) Contiguous patches of olfactory receptor neurons that express distinct odorant receptor molecules send their axons to corresponding glomeruli in the olfactory bulb; an odor-specific map would emerge from this labeled line system. (B) Olfactory receptor neurons and glomeruli are not specified by receptor type; this sort of arrangement would give rise to perceptions based on overall patterns of activity rather than labeled lines. (After Axel, 1995.)

innervation of cranial nerve branches in the oral cavity is topographically represented along the rostral-caudal axis of the gustatory nucleus; the terminations from the seventh nerve are most rostral, those from the tenth nerve are most caudal. Axons from the gustatory nucleus project to the ventral posterior complex of the thalamus, where they terminate in the medial half of the parvocellular portion of the **ventral posterior medial nucleus** (VPMpc). In the rhesus monkey, neurons from the VPMpc project to several regions of the cortex, including the anterior insula in the temporal lobe and the operculum of the frontal lobe. There is also a secondary cortical taste area in the caudolateral orbitofrontal cortex, where neurons respond to combinations of visual, gustatory, and olfactory stimuli. This region is thought to represent the "flavor" center of the cerebral cortex. **Flavor** is defined psychophysically as the perceptual combination of taste and olfactory information from a particular stimulus. Finally, there are indirect projections from the nucleus of the solitary tract via the pons to the hypothalamus and amygdala. These projections probably influence satiety and other affective states associated with taste.

(A)

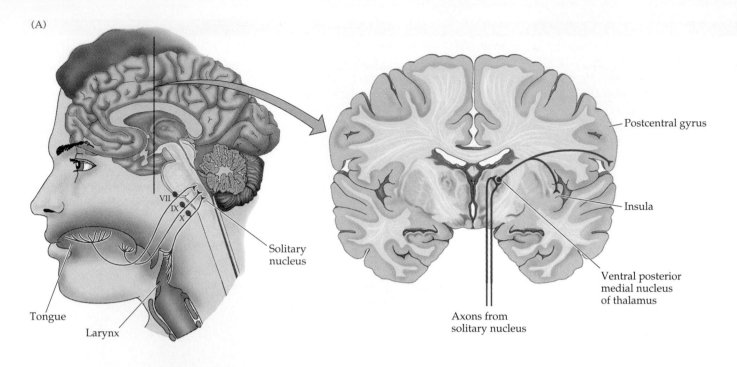

(B)

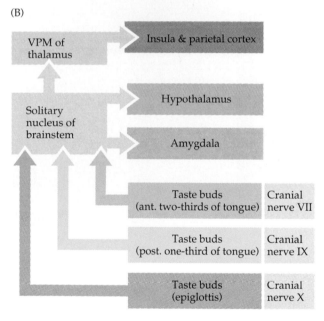

Figure 14.9 Organization of the human taste system. (A) Drawing on the left shows the relationship between receptors in the oral cavity and upper alimentary canal, the nucleus of the solitary tract in the medulla. The coronal section on the right shows the VPM nucleus of the thalamus and its connection with gustatory regions of the cerebral cortex. (B) Diagram of the basic pathways for processing taste information.

■ TASTE PERCEPTION IN HUMANS

Most taste stimuli are nonvolatile, hydrophilic molecules that are soluble in saliva. Examples include salts such as NaCl that are required for electrolyte balance; essential amino acids such as glutamate that are needed for protein synthesis; sugars such as glucose that are required for energy; and acids such as citric acid that indicate the palatability of various foods (oranges, in the case of citrate). Bitter-tasting molecules consist mainly of plant alkaloids, such as atropine and strychnine, that may be poisonous. Placing bitter compounds in the mouth usually deters ingestion unless one "acquires a taste" for them, as in drinking tonic water that contains quinine. Psychophysical experiments have shown that the higher the stimulus concentration, the greater the perceived intensity of taste. As with olfaction, sensitivity to tastants declines with age. This fact may have unhealthy consequences for elderly people who are also hypertensive, since they tend to put more salt on their food. Unfortunately, a safe and effective substitute for NaCl has not yet been developed. For those who wish to limit caloric intake, however, many synthetic sweeteners are available.

Threshold concentrations for most ingested tastants are quite high. For example, the threshold concentration for NaCl is 10 mM; for sucrose, 20 mM; and for citric acid, about 2 mM. The body requires substantial concentrations of salts and carbohydrates, and taste receptor cells may respond to relatively high concentrations of these essential substances to insure an adequate intake. However, the taste system must also be able to detect potentially dangerous bitter-tasting plant compounds at much lower concentrations. Thus, the threshold concentration for quinine is 0.008 mM, and for strychnine it is 0.0001 mM.

There are two common misconceptions about taste. The first is that sweet is perceived at the tip of the tongue, salt along its posterolateral edges, sour along the mediolateral edges, and bitter in the posterior region. This arrangement was initially proposed in 1901 by D. Hanig, who measured taste thresholds for NaCl, sucrose, quinine, and hydrochloric acid (HCl). Hanig never said that other regions of the tongue were insensitive to these chemicals; he only indicated which regions were the *most* sensitive. People missing the anterior part of their tongue can still taste sweet and salty stimuli. In fact, all of these tastes can be detected throughout the tongue. However, different regions of the tongue do have different thresholds. Because the tip of the tongue is most responsive to sweet-tasting compounds, and since these compounds produce pleasurable perceptions, information from this region activates feeding behaviors such as mouth movements, salivary secretion, insulin release, and swallowing. In contrast, responses to bitter compounds are greatest in the posterior region (see Figure 14.10A). Activation of this part of the tongue by bitter-tasting substances elicits gagging and other protective reactions that prevent ingestion.

A second misconception is that there are only four "primary" tastes: salt (NaCl), sweet (glucose or sucrose), sour (acid), and bitter (quinine). If this were true, then all tastes could be represented as a combination of these primary tastes. While these four tastes do indeed represent distinct categories, such a classification is obviously limited. We all experience additional taste sensations, such as astringency (cranberries and tea), pungency (chili pepper and ginger), and various metallic tastes (to name but a few). None of these fits into the four "primary" categories. Moreover, other cultures consider other tastes to be "primary." For example, the Japanese consider the taste of monosodium glutamate to be distinct from that of salt, and give it a different name ("umami," which means delicious). Finally, mixtures of various chemicals may elicit an entirely new taste sensation. While it is possible to estimate the number of perceived odors (approximately 2000), these uncertainties have made it difficult to estimate the number of tastes.

■ THE ORGANIZATION OF THE PERIPHERAL TASTE SYSTEM

In humans, approximately 4000 taste buds are distributed throughout the oral cavity and upper alimentary canal (Figure 14.10A). Taste buds are about 50 μm wide at their base and approximately 80 μm long, each containing 30

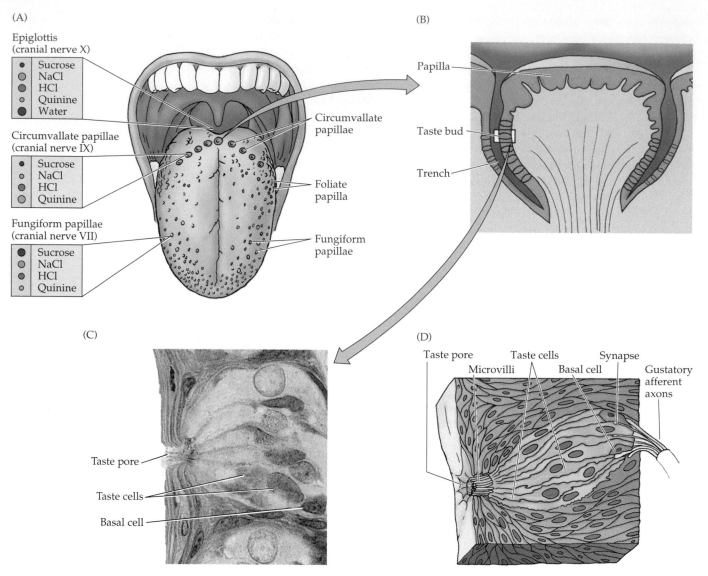

Figure 14.10 Taste buds and the peripheral innervation of the tongue. (A) Distribution of taste papillae on the dorsal surface of the tongue. Different responses to sweet, salty, sour, and bitter tastants recorded in the three cranial nerves that innervate the tongue and epiglottis. The size of the circles representing sucrose, NaCl, HCl, quinine, and water corresponds to the relative response of the papillae to these stimuli. (B) Diagram of a circumvallate papilla showing location of individual taste buds. (C) Light micrograph of a taste bud from the trench of circumvallate papillae. (D) Diagram of a taste bud, showing various types of taste cells and the associated gustatory nerves. The apical surface of the receptor cells have microvilli that are oriented toward the taste pore. (C from Ross, Rommell, and Kaye, 1995.)

to 100 taste cells (the primary sensory receptor cells) plus a few basal cells. About 75% percent of all taste buds are found on the dorsal surface of the tongue in small elevations called **papillae**. There are three types of papillae: **fungiform** (which contain 24% of the total number of taste buds), **circumvallate** (which contain 48% of the taste buds), and **foliate** (which contain 28%). Fungiform papillae are found only on the anterior two-thirds of the tongue; the highest density (about 30 per cm^2) is at the tip. They have a mushroomlike structure (hence their name) and contain about three taste buds at their apical surface. There are nine circumvallate papillae arranged as a chevron at the rear of the tongue, each consisting of a circular trench containing about 250 taste buds along the trench walls (Figure 14.10B). Two foliate papillae are present on the posterolateral margin at the back of the tongue, each having about 20 parallel ridges with about 600 taste buds in their walls. Thus, chemical stimuli on the tongue initially stimulate fungiform papillae and then foliate and circumvallate papillae. Tastants subsequently stimulate scattered taste buds in the palate, pharynx, larynx, and upper esophagus.

Taste cells in individual taste buds (Figure 14.10C) synapse with primary afferent axons from branches of the facial (cranial VII), glossopharyngeal (IX) and vagus (X) nerves (see Figures 14.9 and 14.10D). The taste cells in fungiform papillae are innervated exclusively by the chorda tympani branch of the facial nerve; on the palate, the taste cells are innervated by the greater superior petrosal branch of the facial nerve, and in circumvallate papillae they are innervated exclusively by the lingual branch of the glossopharyngeal nerve. Foliate papillae on the posterior two-thirds of the tongue are innervated by the glossopharyngeal nerve, whereas those on the anterior one-third are innervated by the chorda tympani branch of the facial nerve. Taste buds of the epiglottis and esophagus are innervated by the superior laryngeal branch of the vagus nerve.

The primary events of chemosensory transduction occur in the taste cells, which are specialized epithelial cells within the taste buds. The receptors for tastants are located on microvilli that emerge from the apical surface of the taste cell; synapses onto the afferent axons of the various cranial nerves are made at the basal surface (Figure 14.10D). The apical surfaces of individual taste cells in taste buds are clustered in a small opening near the surface of the tongue called a **taste pore**. Like olfactory receptor neurons (and presumably for the same reasons), taste cells have a lifetime of only about 2 weeks; they regenerate from basal cells recruited from the surrounding epithelium.

■ RESPONSES TO TASTANTS

Taste stimuli evidently work through a variety of cellular and molecular mechanisms. For example, the bitter tasting substances phenylthiocarbamide (PTC) and quinine do not activate the same pathway. Many people (about 30–40%) cannot taste PTC but can taste quinine. Indeed, humans can be divided into two groups with quite different thresholds for bitter compounds containing the N—C=S group found in PTC. The difference between these groups is the presence of a single autosomal gene (*Ptc*) with one dominant (tasters) and one recessive (nontasters) allele.

Similarly, a number of vastly different compounds taste sweet to humans. These include saccharides (glucose, sucrose, and fructose), amino acids (D-amino acids), peptides (aspartame, or Nutrasweet®: L-aspartyl, L-phenylalanine methyl ester), organic anions (saccharin), and proteins (monellin and thaumatin). Most people can distinguish among the different sweeteners, and some people find that saccharin has a bitter-tasting component. Again, these compounds activate separate receptors. For example, saccharides activate cAMP pathways, whereas nonsaccharide sweeteners activate IP_3 pathways. Thus, the perceptual experience of sweet can be elicited by various sensory transduction mechanisms.

Taste sensitivity for salt also relies on a number of distinct mechanisms. Not all salts, or even all monovalent chloride salts, activate the same pathway. Psychophysical studies have shown that amiloride, a potassium-

Figure 14.11 Selective inhibition of salty taste by amiloride. (A) A psychophysical experiment with normal human subjects showed that amiloride (0.20M) inhibits the perceived taste intensity of NaCl and LiCl by approximately 50%. In contrast, the taste intensity of KCl was statistically unchanged. (B) Amiloride inhibition of inward (Na^+) currents in a taste cell. Note that removal of amiloride results in an inward current (downward direction). (A after Schiffman et al., 1983; B after Avenet and Lindemann, 1988.)

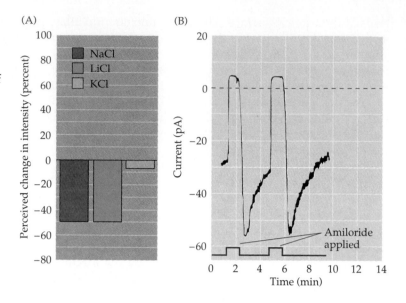

sparing diuretic that blocks Na^+ entry through epithelial Na^+ channels, decreases the taste intensity of NaCl and LiCl but not KCl (Figure 14.11). Although LiCl, like NaCl, tastes salty, it would not make a good salt substitute because of its profound effects on the central nervous system (it is used clinically to treat manic depression). Na succinate, NH_4Cl, and CsCl do not taste exclusively salty. Indeed, CsCl has a bitter or salty-bitter taste. Additional evidence for a distinct receptor for NaCl comes from developmental studies. Infants up to 4 months old can distinguish between water and sucrose (and lactose), water and acid, and water and bitter tastants, but they cannot distinguish between water and 0.2 M NaCl. Thus, either the receptor for Na^+ has not yet been expressed, or, if expressed, it is not yet functional. Infants between the ages of 4 and 6 months, however, can discriminate between NaCl and water, and children can detect the full salty taste of NaCl at about 4 years of age.

Sour taste is produced by relatively high concentrations of acid. At the same H^+ activity, weak organic acids such as lactic acid and citric acid exhibit distinctly different tastes than HCl, whereas strong inorganic acids such as HCl, HNO_3, and H_2SO_4 have similar tastes. Clearly, the anion influences the taste of acids.

In short, the taste system uses a wide variety of mechanisms to distinguish among the various chemicals placed in the mouth.

■ TASTE RECEPTORS AND THE TRANSDUCTION OF TASTE SIGNALS

The effects of taste stimuli are mediated by cell surface receptors coupled to intracellular signalling pathways or by direct actions on ion channels. In some cases tastants can directly activate ion channels. Receptor molecules that bind various tastants are found primarily on the apical microvilli of the taste cells (see Figure 14.10C). The channels and pumps typically found in axonal membranes are located on the basolateral aspect of the cell. These include voltage-gated Na^+, K^+, and Ca^{2+} channels as well as Na^+,K^+-ATPase (Figure 14.12). As already noted, taste cells form synapses with the terminals of peripheral sensory neurons. The synaptic vesicles in taste cells contain neurotransmitters and modulators, and require an increase in intracellular

(A)

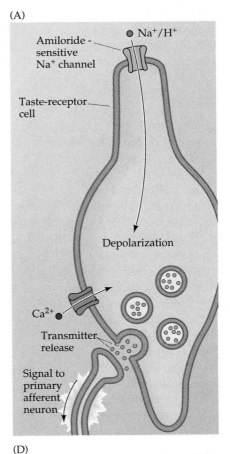

(B)

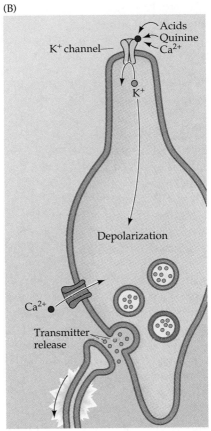

(C)

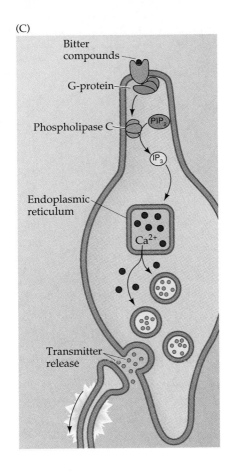

(D)

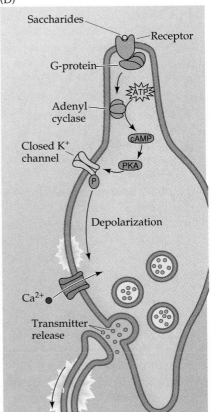

(E)

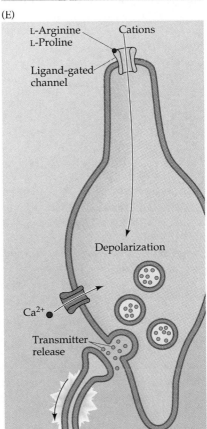

Figure 14.12 Some of the transduction mechanisms in taste buds. (A) Na$^+$ and/or H$^+$ enter the taste cell through amiloride-sensitive channels. Voltage-dependent Ca^{2+} channels are present on the basolateral surface of the receptor cells. Opening of the Na$^+$ channels depolarizes the cell, which in turn activates voltage-dependent Ca^{2+} channels, resulting in transmitter release onto primary afferent neurons. (B) Acids, quinine, and Ca^{2+} depolarize taste cells by inhibiting K$^+$ channels. (C) Bitter-tasting stimuli bind to receptors on the apical membrane via G-proteins coupled to the IP$_3$ pathway. Activation of this cascade causes an increase in intracellular Ca^{2+}, resulting in transmitter release. (D) Receptors for saccharides are coupled to G-proteins that activate cAMP, inhibiting K$^+$ channels on the basolateral membrane, thereby leading to cell depolarization. (E) Some amino acids activate ligand-gated channels on the apical membranes of taste cells. Others (not shown) activate G-protein-coupled receptors. In short, a variety of mechanisms generate receptor responses.

Ca^{2+} to fuse with the presynaptic membrane and release their contents. Increases in intracellular Ca^{2+} can occur through the opening of voltage-dependent Ca^{2+} channels (via a depolarizing receptor potential), through the release of Ca^{2+} from intracellular stores (via IP_3), or by other means. The greater the depolarization of the taste cell—that is, the steeper the rise in intracellular Ca^{2+}—the greater the frequency of action potentials produced in the associated axons of the peripheral gustatory neurons.

The cellular mechanisms of taste transduction have been examined in several sorts of experimental animals, including nonhuman primates. For salt (NaCl) taste, Na^+ enters a subset of taste cells through an amiloride-inhibitable Na^+ channel (Figures 14.11B and 14.12A). The larger the NaCl concentration, the larger the depolarizing current carried by ingested Na^+. Thus, the "receptor" for Na^+ is an epithelial-type Na^+ channel on the apical membrane of some taste cells. Protons (H^+) can also diffuse through this channel, albeit more slowly than Na^+, which may explain why the addition of acids like lemon juice to salty foods reduces their salty taste.

Protons, which are primarily responsible for sour tastes, also interact with a distinct channel on the apical membranes of a subset of taste cells. In such cells, protons have been found to block K^+ channels (Figure 14.12B). One bitter-tasting compound, quinine, also inhibits K^+ channels (albeit of a different type than the ones inhibited by H^+). In addition, the bitter taste associated with Cs^+ or Mg^{2+} salts may be a consequence of these cations blocking K^+ channels associated with bitter taste. Other bitter-tasting molecules, such as quinine and caffeine, bind to G-protein-coupled receptors (these G-proteins include gusducin and transducin) that increase the concentration of IP_3, resulting in increased intracellular Ca^{2+} (Figure 14.12C). Sweet-tasting molecules such as sucrose bind to receptors on the apical surface of taste cells. These receptors activate adenylate cyclase via G-proteins, leading to the production of cAMP, which in turn activates a protein kinase that phosphorylates (and closes) K^+ channels on the basolateral surface, resulting in a depolarization of the taste cell (Figure 14.12D). Recent studies with saccharin indicate that it binds to receptors and activates an IP_3-mediated pathway. Amino acids such as L-proline depolarize taste cells by activating cation-selective ligand-gated channels (Figure 14.12E). Glutamate receptors have also been identified on taste cells and may be involved in the taste of foods containing the amino acid glutamate.

As in olfaction, these taste receptor mechanisms adapt to the ongoing presence of a stimulus. Adaptation of the taste system is a common experience. If a chemical is left on the tongue for a sufficient time, it ceases to be perceived (consider saliva, for example). Thus, to obtain the full taste of foods, one must either frequently change the types of foods placed in the mouth or wait a sufficient time between helpings, facts that have long since been taken to heart by restauranteurs.

■ NEURAL CODING IN THE TASTE SYSTEM

Taste coding refers to the way that information about the identity and concentration of tastants is represented in the patterns of action potentials relayed to the brain. As noted, physiological measurements indicate that different regions of the tongue and oral cavity have different sensitivities to various tastants (see Figure 14.10A). This same selectivity is found in recordings from primary afferent sensory axons in cranial nerves VII, IX, and X. Specifically, the chorda tympani branch of the facial nerve responds best to NaCl and sucrose (that is, a greater number of axons are activated); the glossopharyngeal nerve responds best to acid and quinine; and the superior laryngeal branch of the vagus nerve responds best to acid and water. Individual taste cells, like olfactory receptors, are broadly tuned, usually responding to several chemical stimuli (Figure 14.13). For example, a single peripheral neuron

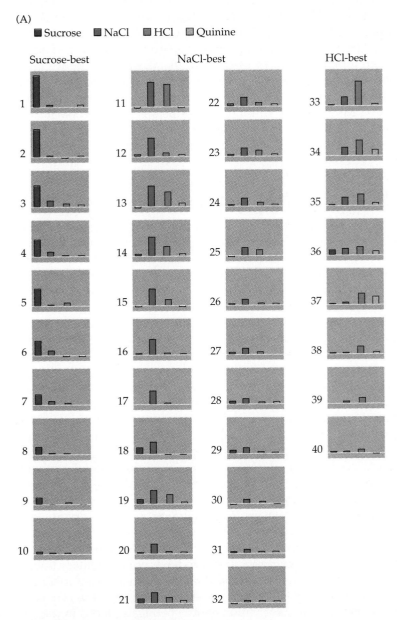

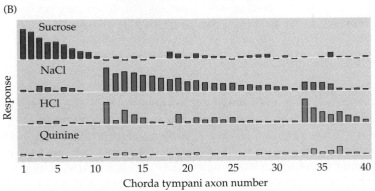

Figure 14.13 Encoding taste. (A) Response profiles of individual chorda tympani axons to four different stimuli. The numbers indicate individual axons. The responses reflect the net activity for 5 seconds after application of each tastant. The response patterns suggest a labeled line mechanism; axons 1–10 are sucrose-best, axons 11–32 are NaCl-best, and axons 33–40 are HCl-best. (B) When plotted another way, however, the responses of the same fibers are consistent with a computational scheme of taste coding. Each row depicts the pattern elicited by a single tastant in the full population of 40 axons. Each taste has its own distinct pattern. (After Smith and Frank, 1993.)

might respond best to sucrose, but also be activated by fructose, NaCl, and acetic acid.

Given these observations, how is sensory information for taste encoded? As with olfaction, two competing ideas have been proposed: the labeled line hypothesis and the computational hypothesis. The labeled line paradigm classifies individual taste cells into types such as NaCl-best, sugar-best, and so on, meaning that of all the stimuli tested for a particular cell, one of the four "primary" stimuli will evoke the greatest number of action potentials per unit time (Figure 14.13A). A corollary is that gustatory information is transmitted to the brain via pathways that preserve the "best" fiber classes. In contrast, the computational paradigm proposes that the pattern of the response to a particular stimulus across *all* fibers is the central feature of coding (Figure 14.13B). In this model, the taste of sucrose is computed from the response evoked from all 40 taste cells illustrated in Figure 14.13, whether active or silent.

■ CENTRAL PROCESSING OF TASTE SIGNALS

If the labeled line theory is correct, one might expect the physiological properties of distinct taste receptor cells in the periphery to be faithfully projected to the central relay stations of the taste system. There is, in fact, evidence for "best" fiber responses in the nucleus of the solitary tract, thalamus, and cortex. For instance, amiloride has been shown to inhibit the responses to NaCl only from NaCl-best neurons in the central nervous system, as it does for the NaCl-best primary afferents. It is important to note, however, that "best" types are determined by evaluating a limited number of taste stimuli at a limited number of concentrations. Under these conditions, it is very difficult to prove the validity of the labeled line theory.

If the computational model is correct, one should be able to identify patterns of activation across subsets of neurons that are independent of stimulus concentration, and these patterns should be associated with particular tastes. There is also evidence that the pattern of activity elicited by a single stimulus in a set of taste-responsive neurons remains stable, regardless of the stimulus concentration. In keeping with the computational model, mixtures of stimuli should have unique patterns of activity; in addition, there should be (and in fact there are) unique tastes that cannot easily be described as a combination of four "primary" tastes. Some neurons in the solitary nucleus, thalamus, and cortex respond to several tastants. Indeed, axons or cells in the solitary nucleus appear even more broadly tuned than in the periphery, a fact more consistent with the computational model. From the solitary tract to the cortex, the chemical selectivity of individual gustatory neurons increases, suggesting that central processing sharpens response profiles, thus enabling the system to extract specific information about chemicals placed in the oral cavity. In summary, neural coding for taste, as for olfaction, probably involves ensembles of neurons across the entire gustatory system.

■ TRIGEMINAL CHEMORECEPTION

The third of the major chemosensory systems, the trigeminal chemosensory system, consists primarily of polymodal nociceptive axons in the trigeminal nerve (cranial nerve V), although it includes nociceptive axons from the glossopharyngeal and vagus nerves (IX and X). These fibers are typically activated by chemicals classified as irritants, including air pollutants (sulfur dioxide), ammonia (smelling salts), ethanol (liquor), acetic acid (vinegar), menthol, which elicits a cold sensation, and capsaicin, the compound in chili

Box B
CAPSAICIN

Capsaicin, the ingredient in chili pepper responsible for its pungent taste, is eaten daily by over a third of the world's population. Capsaicin activates responses in a subset of nociceptive C fibers (polymodal nociceptors; see Chapter 9) by opening ligand-gated ion channels that permit the entry of Na^+ and Ca^{2+}. Receptors for capsaicin have been found in polymodal nociceptors of all mammals so far examined. When applied to mucosal membranes of the oral cavity, capsaicin acts as an irritant, producing protective reactions. When injected into skin, it produces a burning pain and elicits hyperalgesia to thermal and mechanical stimuli. Capsaicin also desensitizes pain fibers and prevents neuromodulators (such as substance P, VIP, and somatostatin) from being released from peripheral and central nerve terminals. Consequently, capsaicin is used clinically as an analgesic and anti-inflammatory agent; it is usually applied topically in a cream (0.075%) to relieve the pain associated with arthritis, postherpetic neuralgia, mastectomy, and trigeminal neuralgia. It is remarkable that this chemical irritant not only gives gustatory pleasure on an enormous scale, but is a useful pain reliever!

Reference

WOOD, J. (1993) *Capsaicin and the Study of Pain*. Academic Press: New York.

(A)

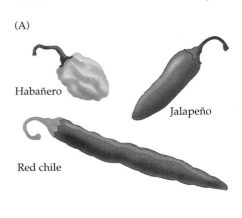

Habañero

Jalapeño

Red chile

(B)   Capsaicin

(C)

(A) Some popular peppers that contain capsaicin. (B) The chemical structure of capsaicin. (C) The shape of capsaicin.

peppers that elicits a burning sensation (Box B). The irritant-sensitive polymodal nociceptors alert the organism to potentially harmful chemical stimuli that have been ingested, respired, or come in contact with the face, and is closely tied to the trigeminal pain system discussed in Chapter 9.

In the trigeminal system, chemosensory information from the face, scalp, cornea, and mucous membranes of the oral and nasal cavities is relayed via the three major sensory branches of the trigeminal nerve: the ophthalmic, maxillary, and mandibular (Figure 14.14). The central target of these components is the spinal trigeminal nucleus, which relays this information to the ventral posterior medial nucleus of the thalamus and thence to the somatic sensory cortex and to other cortical areas that process facial pain and irritation (see Chapter 9).

Many compounds classified as irritants can also be recognized as odors or tastes; however, the threshold concentrations for trigeminal chemoreception are usually much higher than those for olfaction or taste. When potentially irritating compounds are given to people who have lost their sense of smell, their perceptual thresholds for these compounds are approximately 100 times higher than those of normal subjects who perceive the compounds

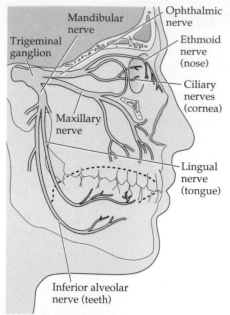

Figure 14.14 Diagram of the branches of the trigeminal nerve that innervate the oral and nasal cavities. The chemosensitive structures innervated by each trigeminal branch are indicated in parentheses.

as odors (Figure 14.15). Similar differences are found for identifying chemicals as tastes rather than irritants. Thus, 0.1 M NaCl has a salty taste, but 1.0 M NaCl is perceived as an irritant. Another common irritant is ethanol. When placed on the tongue at moderate temperatures and high concentrations—as in drinking vodka "neat"—ethanol produces a burning sensation.

Several physiological responses mediated by the trigeminal chemosensory system follow exposure to irritants. Increased salivation, vasodilation, tearing, nasal secretion, sweating, decreased respiratory rate, and bronchoconstriction are commonly experienced after the ingestion of a chemical irritant such as capsaicin (see Box B). These reactions are generally protective. They dilute the stimulus (tearing, salivation, sweating) and prevent inhaling or ingesting more of the irritant. In addition, irritants activate gross motor responses that help remove the noxious stimulus.

The receptors for irritants are primarily on the terminals of polymodal nociceptive neurons (see Chapter 9). Although these receptors respond to the same stimuli as olfactory receptor neurons, they are probably not activated by the same mechanism; for instance, the G-protein-coupled receptors for odorants are found only in olfactory receptor neurons. With the exception of capsaicin and acidic stimuli, both of which activate cation-selective ion channels, little is known about the transduction mechanisms for irritants, or their central processing.

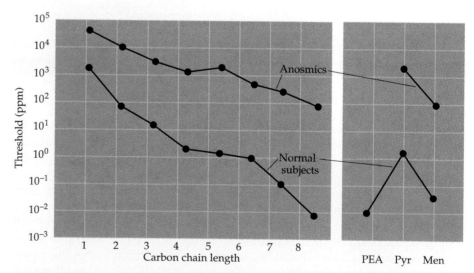

Figure 14.15 Perceptual thresholds in anosmic and normal subjects for 11 related organic chemicals. In anosmics, these chemicals are only detected as irritants at relatively high concentrations (indicated here in parts per million—ppm); in normal subjects, they are first detected at much lower concentrations as odors. The numbers 1–8 stand for the aliphatic alcohols from methanol to 1-octanol. Perceptual thresholds for three additional common irritants—phenylethyl alcohol (PEA), pyridine (Pyr), and menthol (Men)—are shown at the far right. (After Commetto-Muniz and Cain, 1990.)

■ SUMMARY

The chemical senses—olfaction, taste, and trigeminal chemosensation—all contribute to our awareness of airborne or soluble molecules from a variety of sources. Humans rely on this information for behaviors as diverse as attraction, avoidance, reproduction, and feeding. Receptor neurons in the olfactory epithelium transduce chemical stimuli into neuronal activity via the stimulation of G-protein-linked receptors. Receptor activation leads to elevated levels of second messengers such as cAMP which open cation-selective channels and thereby generate potentials in the olfactory receptor neuron. Taste receptor cells use a variety of mechanisms for transducing chemical stimuli. These include ion channels that are directly activated by salts and amino acids, as well as G-protein linked receptors that activate second messengers. For both smell and taste, the frequency and pattern of action potential activity encodes information about the identity and intensity of chemical stimuli. Each of the approximately 2000 odors (and an undetermined number of tastes) is probably encoded by distinct receptor cells, although smells and tastes are recognized computationally by the overall patterns of activity among receptor neurons in the nose, tongue, and oral cavity. Olfaction, taste, and trigeminal chemosensation all are relayed and processed via specific pathways in the central nervous system. In the olfactory system, receptor neurons project directly to the olfactory bulb; in the taste system, information is relayed centrally to the solitary nucleus in the brainstem; in the trigeminal chemosensory system the peripheral receptors project to the spinal trigeminal nucleus. Subsequently, these structures send chemical information to many sites in the brain, which give rise to some of the most sublime pleasures that humans can experience.

Additional Reading

Reviews

ERICKSON, R. P. (1985) Definitions: A matter of taste. In *Taste, Olfaction, and the Central Nervous System*. D. W. Pfaff (ed.). New York: Rockefeller University Press, p. 129.

GILBERTSON, T. A. (1993) The physiology of vertebrate taste reception. Curr. Opin. Neurobiol. 3: 532–539.

KRUGER, L. AND P. W. MANTYH (1989) Gustatory and related chemosensory systems. In *Handbook of Chemical Neuroanatomy*, Vol. 7, *Integrated Systems of the CNS*, Part II. A. Björkland, T. Hökfelt and L. W. Swanson (eds.). Elsevier Science: New York, pp. 323–410.

LAURENT, G. (1996) Odor, images and tunes. Neuron 16: 473–476.

REED, R. R. (1992) Signaling pathways in odorant detection. Neuron 8: 205–209.

ROPER, S. D. (1992) The microphysiology of peripheral taste organs. J. Neurosci. 12: 1127–1134.

SHEPHERD, G. M. (1994) Discrimination of molecular signals by the olfactory receptor neuron. Neuron 13: 771–790.

Important Original Papers

AVANET, P. AND B. LINDEMANN (1988) Amiloride-blockable sodium currents in isolated taste receptor cells. J. Membrane Biol. 105: 245–255.

BUCK, L. AND R. AXEL (1991) A novel multigene family may encode odorant receptors: a molecular basis for odor recognition. Cell 65: 175–187.

GRAZIADEI, P. P. C. AND G. A. MONTI-GRAZIADEI (1980) Neurogenesis and neuron regeneration in the olfactory system of mammals. III. Deafferentation and reinnervation of the olfactory bulb following section of the fila olfactoria in rat. J. Neurocytol. 9: 145–162.

ROLLS, E. T. AND L. L. BAYLIS (1994) Gustatory, olfactory and visual convergence within primate orbitofrontal cortex. J. Neurosci. 14: 5437–5452.

SCHIFFMAN, S. S., LOCKHEAD, E. AND MAES, F. W. (1983) Amiloride reduces taste intensity of salts and sweeteners. Proc. Natl. Acad. Sci. USA 80: 6136–6140.

SCHOENBAUM, G. AND EICHENBAUM, H. (1995) Information coding in the rodent prefrontal cortex. I. Single-neuron activity in orbital frontal cortex compared with that of pyriform cortex. J. Neurophysiol. 74: 733–750.

WONG, G. T., GANNON, K. S. AND R. F. MARGOLSKEE (1996) Transduction of bitter and sweet taste by gustducin. Nature 381: 796–800.

Books

BARLOW, H. B. AND J. D. MOLLON (1989) *The Senses*. Cambridge: Cambridge University Press, Chapters 17, 18, and 19.

DOTY, R. L. (ED.) (1995) *Handbook of Olfaction and Gustation*. New York: Marcel Dekker.

FARBMAN, A. I. (1992) *Cell Biology of Olfaction*. New York: Cambridge University Press.

GETCHELL, T. V., L. M. BARTOSHUK, R. L. DOTY AND J. B. SNOW, JR. (1991) *Smell and Taste in Health and Disease*. New York: Raven Press.

MEISELMAN, H. L. AND R. S. RIVLIN (1986) *Clinical Measurement of Taste and Smell*. New York: Macmillan.

SIMON, S. A. AND S. D. ROPER (1993) *Mechanisms of Taste Transduction*. Boca Raton: CRC Press, Chapters 2, 6, 9, 10, 12, 13, and 14.

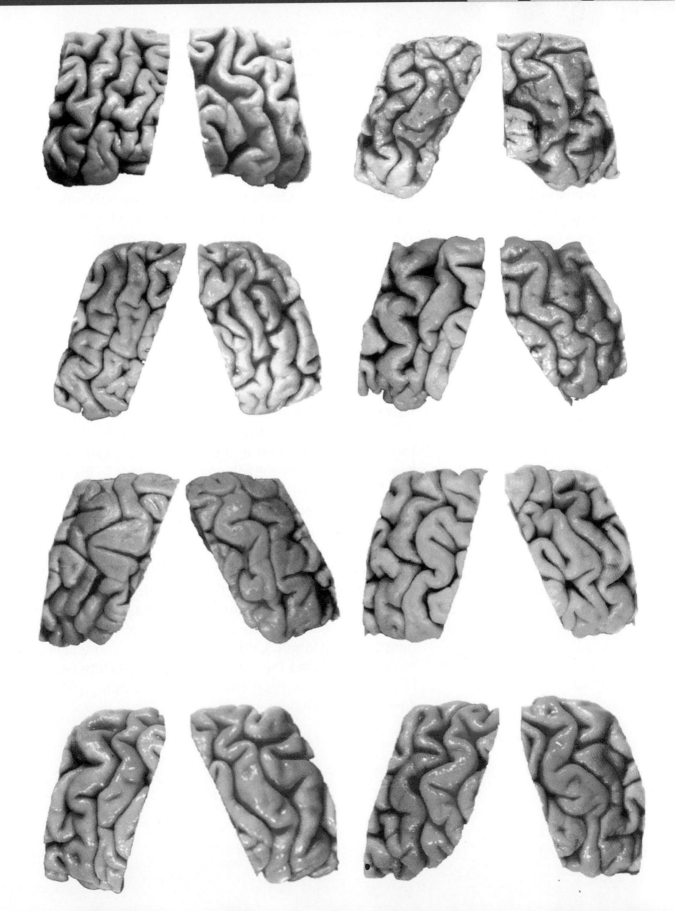

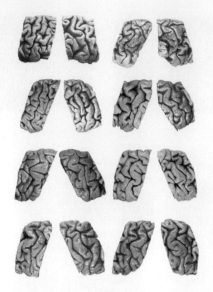

UNIT **III**

MOVEMENT AND ITS CENTRAL CONTROL

The human primary motor cortex, shown here in red, lies just anterior to the central sulcus in the precentral gyrus. This figure illustrates the primary motor cortex in the right and left hemispheres of eight different subjects. (Courtesy of Len White and Dale Purves.)

Every behavior, whether conscious or unconscious, is based on a set of muscular contractions orchestrated by the brain. Analyzing how the brain conducts this motor symphony is fundamental to understanding both normal behavior and the etiology of a variety of neurological disorders. This unit considers the spinal circuitry that makes elementary reflex movements possible, as well as brain systems that govern the successful performance of more complex motor acts. All movements are elicited by motor neurons in the spinal cord that directly innervate the muscle fibers whose contraction changes the position of skeletal elements. In addition to their modulation by local reflex circuitry, these "lower" motor neurons are controlled and coordinated by the brain—principally by the "upper" motor neurons located in the motor cortex and brainstem. The basal ganglia and cerebellar systems provide the motor cortex and brainstem with the sensory, perceptual, and cognitive information that enable complex movements appropriate to a given situation.

Disorders of movement are often a cardinal sign that a particular region of the brain is damaged. For example, three of the most intensively studied neurodegenerative disorders— Parkinson's disease, Huntington's disease, and amyotrophic lateral sclerosis—all have a component of the motor system as their primary target. Understanding the various levels of motor control and their interactions is therefore essential in identifying the location and cause of the many pathologies that affect motor behavior.

CHAPTER 15

SPINAL CORD CIRCUITS AND MOTOR CONTROL

■ OVERVIEW

The proximate control of movement is provided by neurons in the spinal cord and brainstem. The primary motor neurons located in the ventral horn of the spinal cord gray matter (and the corresponding motor neurons in brainstem motor nuclei) send axons directly to skeletal muscles via the ventral roots and peripheral nerves (or cranial nerves, in the case of the brainstem nuclei). The activity of these "lower" motor neurons is determined by local circuitry within the spinal cord and brainstem and by descending pathways from the "upper" motor neurons in the cortex and other brainstem centers, such as the vestibular nucleus and the reticular formation. Circuitry in the spinal cord also mediates a variety of important sensorimotor reflexes. Lower motor neurons, therefore, provide a common pathway for transmitting neural impulses to the skeletal muscles.

■ NEURAL STRUCTURES RESPONSIBLE FOR MOVEMENT

The neuronal assemblies responsible for the control of movement are most easily understood as four distinct but highly interactive subsystems, each of which makes a unique contribution to motor control (Figure 15.1). The first of these subsystems is the circuitry within the gray matter of the spinal cord. The relevant cells include the primary or **alpha motor neurons**, which send their axons out of the spinal cord to innervate skeletal muscle fibers; and spinal cord **interneurons**, which are a major source of the synaptic input to motor neurons. All commands for movement, whether reflexive or voluntary, are ultimately conveyed to muscles by the activity of alpha motor neurons (also referred to as **lower motor neurons**); thus they are, in the words of Charles Sherrington, the final common path for motor behavior. Spinal cord interneurons receive sensory inputs as well as descending projections from higher centers and provide much of the reflexive coordination between muscle groups that is essential for movement. The contribution of this subsystem to motor control is substantial; even after the spinal cord is disconnected from higher motor centers of the brain, appropriate stimulation can elicit highly coordinated motor reflexes.

The second motor subsystem consists of neurons whose cell bodies lie in the brainstem and cerebral cortex. The axons of these higher-order or **upper motor neurons** descend to synapse with interneurons and/or with alpha motor neurons in the spinal cord gray matter. The descending pathways are essential for the control of voluntary movements and are, in a very real sense, the link between thoughts and actions. Descending systems originating in the brainstem are responsible for integrating vestibular, somatic, and visual sensory information to adjust the reflex activity of the spinal cord. Their contributions are critical for basic steering movements of the body, and in the control of posture. Descending projections from cortical areas in the frontal lobe, including Brodmann's area 4 (the **primary motor cortex**), area 6 (the **premotor cortex**), and the **supplementary motor cortex**

Figure 15.1 Overall organization of neural structures involved in the control of movement. Four distinct systems—spinal cord circuits, descending systems, the basal ganglia, and the cerebellum—make essential and distinct contributions to motor control.

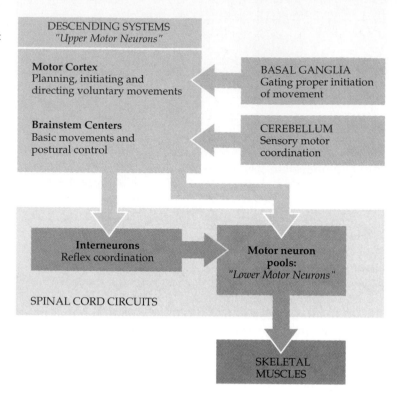

are essential for planning, initiating, and directing voluntary movements. The influence of these cortical areas is conveyed to spinal cord circuits directly via the **corticospinal pathway** and indirectly via projections to the brainstem centers, which in turn project to the spinal cord.

The third and fourth subsystems are structures (or groups of structures) that have no direct access to alpha motor neurons or spinal cord interneurons; rather, they exert control over movement by regulating the activity of the upper motor neurons that give rise to the descending pathways. One of these subsystems, the **cerebellum**, is located on the dorsal surface of the pons (see Chapters 1 and 17). Its principal function is to correct errors of movement by comparing the movement commands issued by the cortex and brainstem with sensory feedback about the movements that have actually occurred. Thus, the cerebellum coordinates the components of complex movements. Changes in the output of cerebellar circuits also underlie certain aspects of motor learning. The other subsystem, embedded in the depths of the forebrain, is the **basal ganglia** (see Chapters 1 and 17). It is more difficult to characterize the contribution of the basal ganglia to motor control, but disorders of basal ganglia function, such as Parkinson's disease and Huntington's disease, attest to their importance in the initiation of voluntary movements (see Chapters 17 and 18).

Despite many years of effort, there is still no complete understanding of the sequence of events that leads from thought to movement, and it is fair to say that the picture becomes increasingly blurred the farther one moves from the muscles themselves. It is appropriate then to begin a more detailed account of motor behavior by considering the anatomical and physiological relationships between alpha motor neurons and the muscle fibers they innervate.

■ THE TOPOGRAPHY OF MOTOR NEURON–MUSCLE RELATIONSHIPS

The majority of the neurons that innervate the body's skeletal muscles are located in the ventral horn of the spinal cord. Each motor neuron innervates muscle fibers within a single muscle, and all the motor neurons innervating a single muscle (called the **motor neuron pool** for that muscle) are grouped together into rod-shaped clusters that run parallel to the long axis of the cord for one or more spinal cord segments (Figure 15.2).

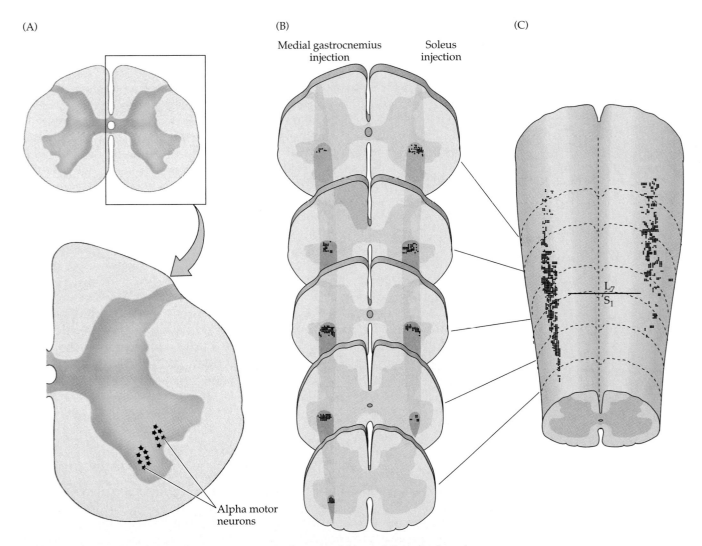

Figure 15.2 Organization of motor neurons in the ventral horn of the spinal cord demonstrated by retrograde labeling from individual muscles. Neurons were labeled by placing a retrograde tracer into the medial gastrocnemius or soleus muscle of the cat. (A) Section through the lumbar level of the spinal cord showing the distribution of labeled cell bodies. Alpha motor neurons form two distinct clusters (motor pools) in the ventral horn. Spinal cord cross sections (B) and a reconstruction seen from the dorsal surface (C) illustrate the distribution of motor neurons innervating individual skeletal muscles in the long axis of the cord. The cylindrical shape and distinct distribution of different pools are especially evident in the dorsal view of the reconstructed cord. The dashed lines in (C) represent individual lumbar and sacral spinal cord segments. (After Burke et al., 1977.)

Figure 15.3 Somatotopic organization of motor neurons in a cross section of the ventral horn at the cervical level of the spinal cord. Alpha motor neurons innervating axial musculature are located medially, and those innervating the distal musculature more laterally.

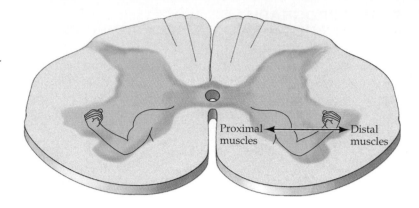

An orderly relationship between the location of motor neuron pools and the muscles they innervate is apparent both along the length of the spinal cord and across the mediolateral dimension of the cord. Thus, the motor neuron pools that innervate the upper extremity are located in the cervical enlargement of the cord, those that innervate the leg in the lumbar enlargement, and so on (see Figure 1.15). The topography of motor neuron pools in the mediolateral dimension can best be appreciated in a cross section through the cervical enlargement (the level illustrated in Figure 15.3). Neurons that innervate the axial musculature (the muscles of the trunk) are located most medially in the cord. Lateral to these cell groups are motor neuron pools innervating muscles located progressively more laterally in the body. Neurons that innervate the muscles of the shoulders (or pelvis), for example, are next, whereas those that innervate the proximal muscles of the arm (or leg) are located more laterally. The motor neuron pools that innervate the distal parts of the extremities lie farthest from the midline. This organizational feature is worth remembering, since it provides a clue about the function of some of the descending pathways described in Chapter 16.

Although the following discussion focuses on the motor neurons in the spinal cord, comparable sets of motor neurons responsible for the control of muscles in the head and neck are located in the brainstem. These neurons are in the motor nuclei of the cranial nerves distributed in the medulla, pons, and midbrain.

■ THE MOTOR UNIT

Most mature skeletal muscle fibers in mammals are innervated by only a single motor neuron; however, individual motor axons branch within muscles to synapse on many different fibers. The muscle fibers innervated by a single motor neuron are distributed over a wide area within a muscle, presumably to insure that the contractile force of the motor unit is spread more evenly (Figure 15.4). In addition, this arrangement reduces the chance that damage to one or a few spinal motor neurons will significantly alter a muscle's action. Because an action potential generated by a motor neuron normally brings all of the muscle fibers it contacts to threshold, a single motor neuron and its associated muscle fibers together constitute the smallest unit that can be activated to produce movement. Sherrington was again the first to recognize this fundamental relationship between a motor neuron and the muscle fibers it innervates, for which he coined the term **motor unit.**

For most muscles, three types of motor units can be identified based on their speed of contraction, the maximum amount of tension they generate,

(A)   (B)

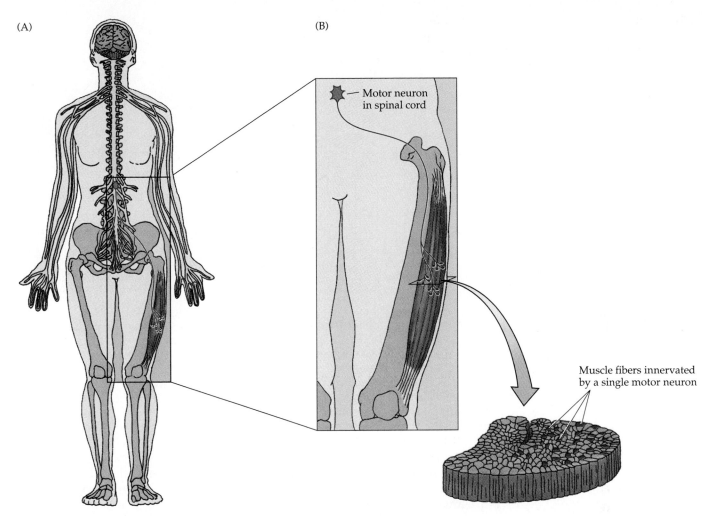

Motor neuron
in spinal cord

Muscle fibers innervated
by a single motor neuron

Figure 15.4   The motor unit. (A) Diagram showing a motor neuron in the spinal cord and the course of its axon to the muscle. (B) Each alpha motor neuron synapses with multiple muscle fibers. The motor neuron and the fibers it contacts defines the motor unit. Cross section through the muscle shows the distribution of muscle fibers (red dots) contacted by the motor neuron.

and the degree to which they fatigue (Figure 15.5). **Fast fatigable (FF) motor units** contract and relax rapidly and are capable of generating the largest force. However, as the name suggests, they fatigue after several minutes of repeated stimulation. At the other extreme are **slow (S) motor units**. These are highly resistant to fatigue (they can maintain a constant force for over an hour of repeated stimulation), but contract more slowly and are capable of generating only a fraction of the force generated by FF units. The third group has properties that are intermediate between the other two. These **fast fatigue-resistant (FR) motor units** are not quite as fast as FF units but are substantially more resistant to fatigue. An individual FR motor unit generates about twice the force of a slow unit.

The differences in the functional properties of the various motor unit types are based largely on differences in the physiological and biochemical characteristics of their constituent muscle fibers. Most muscles contain a mixture of three types of muscle fibers, and all of the muscle fibers belonging to a given motor unit are of the same type. In addition to the type of

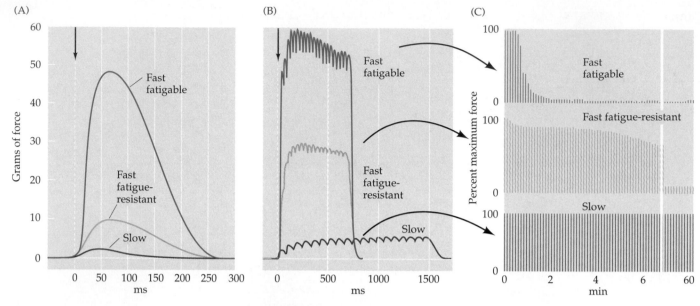

Figure 15.5 Comparison of the force and fatigability of the three different types of motor units. In each case, the response reflects stimulation of a single motor neuron. (A) Change in tension in response to single motor neuron action potentials. (B) Tension in response to repetitive stimulation of the motor neurons. (C) Response to repeated stimulation at a level that evokes maximum tension. The y axis represents the force generated by each stimulus. Note the strikingly different rates of fatigue. (After Burke et al., 1974.)

myosin in the constituent fibers, both the number and the cross-sectional area of a motor unit's muscle fibers determine its contractile properties. For any particular muscle, individual FF units comprise a greater number of muscle fibers than individual S units. Moreover, the muscle fibers associated with FF units are larger in cross-sectional area than those in S units. (FR units are intermediate in these respects.) Thus, compared to an S unit, activation of an FF unit entails the contraction of a greater number of muscle fibers, each of which is capable of generating a greater amount of force.

Individual muscles differ in the proportions of different motor unit types they contain. Muscles involved in postural support, which must supply steady tension for long periods, tend to have a high proportion of slow units. Conversely, muscles that generate extraordinarily rapid movements contain mostly fast motor units (a good example is the extraocular muscles that generate saccades; see Chapter 19). The majority of muscles, however, contain a more even balance of the different motor unit types. The rationale for motor unit diversity within a single muscle lies in how the different types of motor units are recruited to generate force.

■ THE REGULATION OF MUSCLE FORCE

Increasing or decreasing the number of active motor units regulates the amount of force produced by a muscle. In the 1960s, Elwood Henneman and his colleagues at Harvard Medical School found that steady increases in muscle tension could be produced by progressively increasing the activity of the sensory axons that synapsed with the relevant pool of spinal motor neurons. The gradual increase in tension results from the recruitment of motor units in a fixed order according to the conduction velocity of the motor axons. Since conduction velocity is a function of axon diameter (which is in turn correlated with cell size), Henneman recognized that the smallest motor neurons in a motor pool must have the lowest threshold for activation, and that they must be the only units activated by weak synaptic stimulation. As the synaptic drive increases, progressively larger motor neurons are recruited. Within the motor pool that innervates a given muscle, S units tend to have small cell bodies and slow conduction velocities and FF units have comparatively large cell bodies and fast conduction velocities. (FR units again have intermediate characteristics.) Thus, as synaptic activity driving a motor neuron pool increases, low threshold S units are recruited first, then FR units, and finally, at the highest levels of activity, the FF units. Since these original experiments, evidence for the orderly recruitment of motor units has been found in a variety of voluntary and reflexive movements. This relationship has come to be known as the **size principle**.

An illustration of how the size principle operates for the motor units of the medial gastrocnemius muscle in the cat is shown in Figure 15.6. When the animal is standing quietly, the force measured directly from the muscle tendon is a small fraction (about 5%) of the total force that the muscle can generate. The force is provided by the S motor units, which make up about 25% of the motor units in this muscle. When the cat begins to walk, larger forces are necessary: locomotor activities that range from slow walking to fast running require up to 25% of the muscle's total force capacity. This additional need is met by the recruitment of FR units. Only movements such as galloping and jumping, which are performed infrequently and for short periods, require the full power of the muscle; such demands are met by the recruitment of the FF units. Thus, the size principle provides a simple solution to the problem of grading muscle force. The combination of motor units activated by such orderly recruitment optimally matches the physiological properties of different motor unit types with the range of forces required to perform different motor tasks.

The firing rate of motor neurons also contributes to the regulation of muscle tension. The increase in force that occurs with increased firing rate

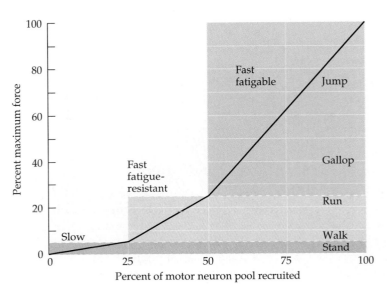

Figure 15.6 The recruitment of motor neurons in the cat medial gastrocnemius muscle under different behavioral conditions. Slow motor units provide the tension required for standing. Fast fatigue-resistant units provide the additional force needed for walking and running. Fast fatigable units are recruited for the most strenuous activities. (After Walmsley et al., 1978.)

(A)

(B)

(C)

(D)

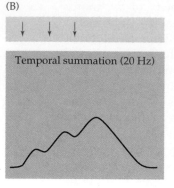

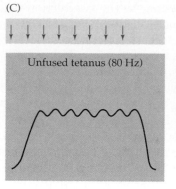

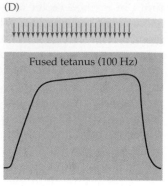

Single muscle twitches (5 Hz)

Temporal summation (20 Hz)

Unfused tetanus (80 Hz)

Fused tetanus (100 Hz)

Figure 15.7 The effect of stimulation rate on muscle tension. (A) At low frequencies of stimulation, each action potential in the motor neuron results in a single twitch of the related muscle fibers. (B) At higher frequencies, the twitches sum to produce a force greater than that produced by single twitches. (C) At a still higher frequency of stimulation, the force produced is greater, but individual twitches are still apparent. This response is referred to as unfused tetanus. (D) At the highest rates of motor neuron activation, individual twitches are no longer apparent (fused tetanus).

reflects the summation of successive muscle contractions: the muscle fibers are activated by the next action potential before they have completely relaxed (Figure 15.7). The lowest firing rates during a voluntary movement are on the order of 8 per second (Figure 15.8). As the firing rate of individual units rises to a maximum of about 20–25 per second, the amount of force produced goes up. At the highest firing rates, individual muscle fibers are in a state of **fused tetanus**—that is, the tension produced in individual motor units no longer has peaks and troughs that correspond to the individual twitches evoked by the motor neuron's action potentials. Under normal conditions, the maximum firing rate of motor neurons is less than 30 Hz (see Figure 15.8), resulting in unfused tetanus. However, asynchronous firing of motor units averages out the variations among individual units, allowing the movements to be executed smoothly.

Figure 15.8 Motor units recorded in the extensor *digitorum communis* muscle of the human hand as the amount of voluntary force produced is progressively increased. Motor units (represented by the lines between the dots) are initially recruited at a low frequency of firing (8 Hz); the rate of firing for each unit increases as the subject generates more and more force. (After Monster and Chan, 1977.)

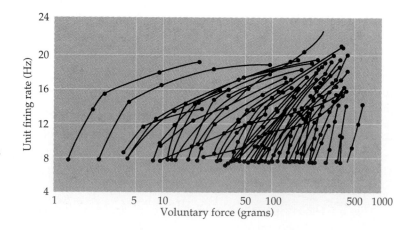

■ THE LOWER MOTOR NEURON SYNDROME

Clinically, a distinction is made between lower motor neurons (alpha motor neurons in the spinal cord and brainstem) and upper motor neurons (neurons in the brainstem and cortex that give rise to descending motor pathways that modulate the activity of the lower motor neurons). These terms are somewhat misleading in that upper motor neurons are not really motor neurons at all, since they do not synapse with muscle fibers. Nevertheless, the clinical signs and symptoms associated with damage to alpha motor neurons differ markedly from those that occur after damage to descending systems; thus, recognizing whether a neurological problem reflects damage to lower or upper motor neurons is an important first step in localizing the lesion (Box A; see Chapter 16 for a full discussion of the symptoms associated with damage to descending pathways).

Damage to spinal motor neurons or their peripheral processes results in **paralysis** or **paresis** (muscle weakness) of the affected muscle, the severity

Box A
AMYOTROPHIC LATERAL SCLEROSIS

Amyotrophic lateral sclerosis (ALS) is a neurodegenerative disease that affects an estimated 0.05% of the population. It is also called Lou Gehrig's disease, after the New York Yankees baseball player who died of the disorder in 1936. ALS is characterized by the slow but inexorable degeneration of alpha motor neurons in the ventral horn of the spinal cord and brainstem, and eventually neurons in the motor cortex. Affected individuals show progressive weakness and wasting of skeletal muscles and usually die within 5 years of onset. Sadly, these patients are condemned to watch their own demise, since the intellect remains intact. There is no effective therapy.

Approximately 10% of ALS cases are familial. Familial ALS (FALS) is inherited as an autosomal dominant trait, but is otherwise indistinguishable from the sporadic form of the disease. The familial form has, however, provided an opportunity for geneticists and neurologists to decipher the genetic basis of the disease in this subset of patients: a mutation in the long arm of

chromosome 21. Interestingly, this same region contains the gene that encodes the cytosolic antioxidant enzyme copper/zinc superoxide dismutase (SOD). Evidence that superoxide radicals can destroy nerve cells suggests that a mutation of the *SOD1* gene may cause FALS. This implication has been strengthened by finding mutations of *SOD1* in roughly 40% of families with FALS. Transgenic mice that overexpress mutant SOD protein develop a neurological disease that mimics ALS both behaviorally and pathologically. Overexpression of normal SOD, however, is not associated with motor neuron disease in mice. Together these data suggest that the mutant SOD molecule is in some way cytotoxic.

These discoveries raise a host of questions. For instance, how does mutant SOD1 damage motor neurons, and what accounts for the curious predilection of the disease for motor neurons? It is not yet clear that insights arising from the molecular pathogenesis of the familial form will shed light on the mechanisms of the sporadic forms of

ALS, or whether any of these insights will lead to more effective therapy. Nonetheless many laboratories are presently pursuing these clues in an attempt to ameliorate or cure this devastating disease.

References

BROWN, R. H., JR. (1995) Amyotrophic lateral sclerosis: Recent insights from genetics and transgenic mice. Cell 80: 687–692.

DENG, H. X. AND 19 OTHERS (1993) Amyotrophic lateral sclerosis and structural defects in Cu,Zn superoxide dismutase. Science 261: 1047–1051.

MULDER, D W., L. T. KURLAND, K. P. OFFORD AND C. M. BEARD (1986) Familial adult motor neuron disease: amyotrophic lateral sclerosis. Neurol. 36: 511–517.

ROSEN, D. R. AND 32 OTHERS (1993) Mutations in Cu/Zn superoxide dismutase gene are associated with familial amyotrophic lateral sclerosis. Nature 362: 59–62.

SIDDIQUE, T. AND 14 OTHERS (1991) Linkage of a gene causing familial amyotrophic lateral sclerosis to chromosome 21 and evidence of a genetic-locus heterogeneity. N. Engl. J. Med. 324: 1381–1384.

depending on the extent of the damage. In addition to paralysis and paresis, the lower motor neuron syndrome includes **areflexia** (loss of reflexes) and loss of muscle tone (evidenced by a decreased resistance to passive stretch). Reflexes and muscle tone are discussed in more detail in the next section, but it should be obvious that damage to motor neurons prevents, or at least reduces, the transfer of all types of neural activity to muscle fibers, whether voluntary or reflexive. Damage to alpha motor neurons also results in **atrophy** of the affected muscles, and **fibrillations** and **fasciculations**—spontaneous twitches of single denervated muscle fibers or motor units, respectively. The spontaneous contraction of denervated fibers can be recognized in an electromyogram, an especially helpful clinical tool in diagnosing lower motor neuron disorders.

■ THE SPINAL CORD CIRCUITRY UNDERLYING SENSORIMOTOR REFLEXES

The synaptic circuitry within the spinal cord mediates a number of sensorimotor reflex actions. The simplest of these reflex arcs entails the response to muscle stretch, which provides direct feedback to the motor neurons innervating the muscle that has been stretched (Figure 15.9). The sensory signal for the **stretch reflex** originates in specialized structures called **muscle spindles** that are embedded within most muscles (see Chapter 8). Spindles are composed of 8–10 modified muscle fibers called intrafusal fibers arranged in parallel with the ordinary (extrafusal) fibers that make up the bulk of the muscle (Figure 15.9A). Sensory fibers (Ia afferents, the largest class of myelinated sensory nerve fibers) are coiled around the central part of the spindle. Stretching the muscle deforms the intrafusal muscle fibers, which leads to increased activity of the sensory fibers that innervate each spindle. The sensory fibers synapse directly on the alpha motor neurons in the ventral horn of the spinal cord (as well as transmitting sensory information to higher centers; see Chapter 8). Thus, activation of muscle spindles produces a rapid increase in muscle tension that opposes the stretch (Figure 15.9B).

Borrowing a concept from engineering, the stretch reflex arc can be viewed as a negative feedback loop that tends to maintain muscle length at a constant value (Figure 15.9C). The desired muscle length is specified by the activity of descending pathways that influence the motor neuron pool. Deviations from the desired length are detected by the muscle spindles; thus, increases or decreases in the stretch of the intrafusal fibers change the level of activity in the sensory fibers that innervate the spindles. These changes, in turn, lead to appropriate adjustments in the activity of the alpha motor neurons, returning the muscle to the desired length.

One of the most important determinants of any feedback system is its ability to adjust the regulated variable, a property called **gain**. The larger the gain of the stretch reflex, the greater the change in muscle force that results

Figure 15.9 Stretch reflex circuitry. (A) Diagram of muscle spindle, the sensory receptor that initiates the stretch reflex. (B) Stretching a muscle spindle leads to increased activity in Ia afferents and an increase in the activity of alpha motor neurons that innervate the same muscle. Ia afferents also excite the motor neurons that innervate synergists and inhibit the motor neurons that innervate antagonists. (C) The stretch reflex operates as a negative feedback loop to regulate muscle length.

(A) Muscle spindle

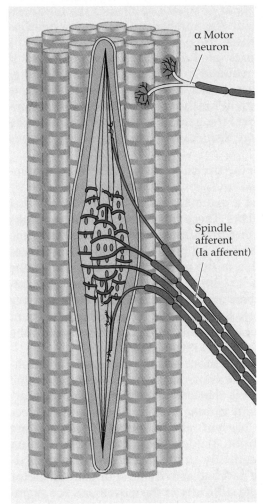

α Motor neuron

Spindle afferent (Ia afferent)

(B)

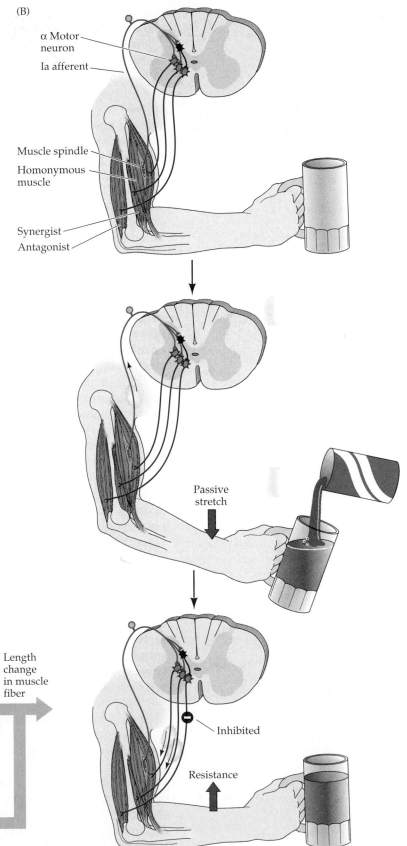

α Motor neuron

Ia afferent

Muscle spindle

Homonymous muscle

Synergist

Antagonist

Passive stretch

Inhibited

Resistance

(C)

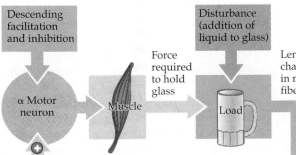

Descending facilitation and inhibition

Disturbance (addition of liquid to glass)

Force required to hold glass

Length change in muscle fiber

α Motor neuron

Muscle

Load

Increase spindle afferent discharge

Spindle receptor

from a given amount of stretch applied to the intrafusal fibers. If the gain of the reflex is high, then a small amount of stretch applied to the intrafusal fibers will produce a large increase in the number of alpha motor neurons recruited and a large increase in their firing rates; this, in turn, leads to a large increase in the amount of tension produced by the extrafusal fibers. If the gain is low, a greater stretch is required to generate the same amount of tension in the extrafusal muscle fibers. In fact, the gain of the stretch reflex is continuously adjusted to meet different functional requirements. For example, when standing in a moving bus, the gain of the stretch reflex can be increased to compensate for the large disturbances that ensue when the bus stops or starts abruptly.

The need to adjust the gain of the stretch reflex explains why the spindle sensors are themselves modified muscle fibers. The gain is adjusted by changing the level of activation of a distinct class of motor neurons that innervate the spindle (intrafusal) fibers. These small **gamma motor neurons** are interspersed among the alpha motor neurons in the ventral horn of the spinal cord. An increase in the activity of gamma motor neurons produces an increase in the amount of tension in the intrafusal fibers. Although the intrafusal fibers are much too sparse to generate a net increase in muscle tension, contraction of the intrafusal fibers increases the sensitivity of Ia sensory fibers to muscle stretch. The same stretch can then produce a larger amount of Ia afferent activity, which causes an increase in the activity of the alpha motor neurons that innervate the extrafusal muscle fibers.

The activation of gamma motor neurons and the subsequent shortening of intrafusal muscle fibers can also ensure that spindle afferents continue to transmit information when a muscle shortens following contraction. During voluntary movements, alpha and gamma motor neurons are often co-activated by higher centers to prevent muscle spindles from being "unloaded" (Figure 15.10). In addition, the level of gamma motor neuron activity can be modulated independently of alpha activity, in a context-dependent fashion. In general, the baseline activity level of gamma motor neurons increases with the speed and difficulty of the movement. For example, experiments on cat hindlimb muscles show that gamma activity is high when the animal has to perform a difficult movement, such as walking across a narrow beam. Unpredictable conditions, as when the animal is picked up or handled, also lead to marked increases in gamma activity and greatly increased spindle responsiveness. Gamma motor neuron activity, however, is not the only factor setting the gain of the stretch reflex. The gain also depends on the level of excitability of the alpha motor neurons that serve as the effector side of this reflex loop. Thus, other local circuits in the spinal cord, as well as descending projections, can influence the gain of the stretch reflex via excitation or inhibition of either alpha or gamma motor neurons (Box B).

The action of the stretch reflex is not limited to the muscle that has been stretched. Spindle afferents also make direct (although fewer) connections to motor neurons innervating muscles that have similar actions (called synergists) and to interneurons in the gray matter of the spinal cord (referred to as Ia inhibitory interneurons) that synapse on motor neurons innervating antagonist muscles. Thus, increasing the stretch on a particular muscle leads to a coordinated contraction of the synergists and relaxation of the antagonists that move the same joint. This arrangement, whereby the activity of a muscle spindle excites its own muscle and its synergists while inhibiting the action of its antagonists, is an example of **reciprocal innervation**. Reciprocal innervation is of value in controlling voluntary movements

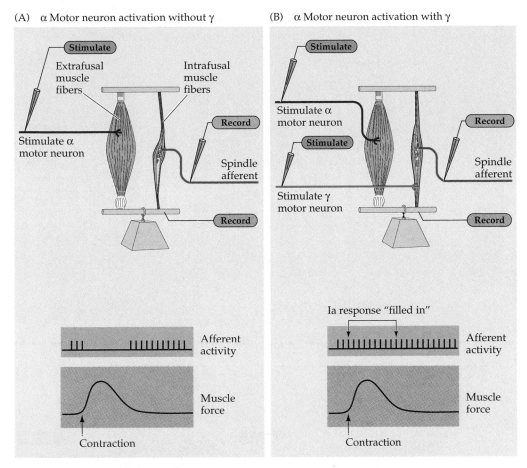

(A) α Motor neuron activation without γ

Stimulate

Extrafusal muscle fibers

Intrafusal muscle fibers

Record

Stimulate α motor neuron

Spindle afferent

Record

Afferent activity

Muscle force

Contraction

(B) α Motor neuron activation with γ

Stimulate

Stimulate α motor neuron

Stimulate

Record

Spindle afferent

Stimulate γ motor neuron

Record

Ia response "filled in"

Afferent activity

Muscle force

Contraction

Figure 15.10 The role of gamma motor neuron activity in regulating the responses of muscle spindles. (A) When alpha motor neurons are stimulated without activation of gamma motor neurons, the response of the Ia fiber decreases as the muscle contracts. (B) When both alpha and gamma motor neurons are activated, there is no decrease in Ia firing during muscle shortening. Thus, the gamma motor neurons can regulate the gain of muscle spindles so that they can operate efficiently at any length of the parent muscle. (After Hunt and Kuffler, 1951.)

as well as reflexes; thus, descending axons from the motor cortex that make direct excitatory connections with motor neurons also send collaterals to Ia inhibitory neurons.

Another sensory structure that is important in the reflex regulation of motor unit activity is the Golgi tendon organ. **Golgi tendon organs** are encapsulated endings located at the junction of the muscle and tendon (Figure 15.11A; see also Table 8.1). Each tendon organ is related to a single group Ib sensory axon (the Ib axons are slightly smaller than the Ia axons that innervate the muscle spindles). In contrast to the parallel arrangement of extrafusal muscle fibers and spindles, Golgi tendon organs are in series with the muscle fibers. When a muscle is passively stretched, most of the change in length occurs in the muscle fibers, since they are more elastic than the fibrils of the tendon. When a muscle actively contracts, however, the force acts directly on the tendon, leading to an increase in the tension of the collagen fibrils in the tendon organ and compression of the intertwined sensory receptors. As a result, Golgi tendon organs are sensitive to increases in muscle

(A)

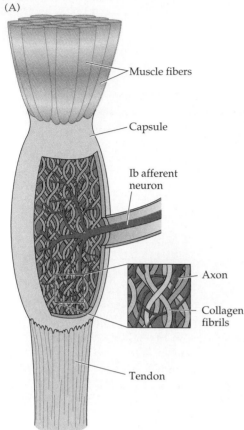

Muscle fibers

Capsule

Ib afferent neuron

Axon

Collagen fibrils

Tendon

Figure 15.11 Comparison of the function of muscle spindles and Golgi tendon organs. (A) Golgi tendon organs are arranged in series with extrafusal muscle fibers because of their location at the junction of muscle and tendon. (B) The two types of muscle receptors, the muscle spindles and the Golgi tendon organs, have different responses to passive muscle stretch (*top*) and active muscle contraction (*bottom*). Both afferents discharge in response to passively stretching the muscle, although the Golgi tendon organ discharge is much less than that of the spindle. When the extrafusal muscle fibers are made to contract by stimulation of their motor neurons, however, the spindle is unloaded and therefore falls silent, whereas the rate of Golgi tendon organ firing increases. (B after Patton, 1965.)

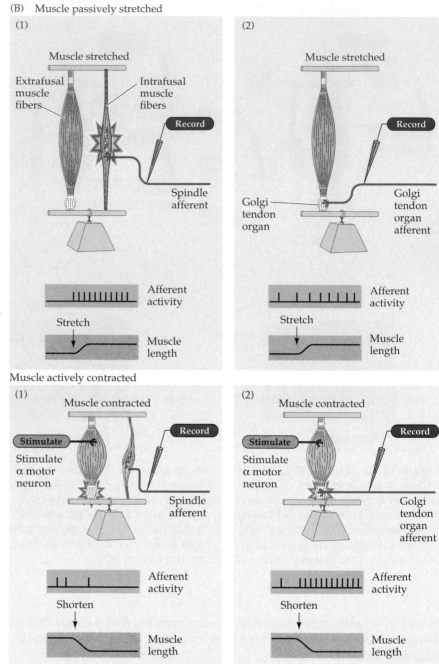

(B) Muscle passively stretched

(1) Muscle stretched

Extrafusal muscle fibers

Intrafusal muscle fibers

Record

Spindle afferent

Afferent activity

Stretch

Muscle length

(2) Muscle stretched

Record

Golgi tendon organ

Golgi tendon organ afferent

Afferent activity

Stretch

Muscle length

Muscle actively contracted

(1) Muscle contracted

Stimulate

Stimulate α motor neuron

Record

Spindle afferent

Afferent activity

Shorten

Muscle length

(2) Muscle contracted

Stimulate

Stimulate α motor neuron

Record

Golgi tendon organ afferent

Afferent activity

Shorten

Muscle length

tension that arise from muscle contraction and, unlike spindles, are much less sensitive to passive stretch (Figure 15.11B).

The Ib axons from Golgi tendon organs contact inhibitory interneurons in the spinal cord (called Ib inhibitory interneurons) that synapse, in turn, with the alpha motor neurons that innervate the same muscle. The Golgi tendon circuit is thus a negative feedback system that regulates muscle tension, decreasing the activation of muscles when exceptionally large forces are generated. This reflex circuit also operates at reduced levels of muscle

Box B
MUSCLE TONE

Muscle tone is the resting level of tension in a muscle; in general, it prepares the muscle for a rapid and reliable response to voluntary or reflexive commands. Tone in the extensor muscles of the legs, for example, helps maintain posture while standing. By keeping the muscles in a state of readiness to resist stretch, tone in the leg muscles prevents the amount of sway that normally occurs while standing from becoming too large. During activities such as walking or running, the "background" level of tension in leg muscles also helps to store mechanical energy, in effect enhancing the muscle tissue's springlike qualities.

Muscle tone depends on the resting level of discharge of alpha motor neurons. Activity in the Ia spindle afferents—the neurons responsible for the stretch reflex—is the major contributor to this tonic level of firing. The gamma efferent system (by its action on intrafusal muscle fibers) regulates the resting level of activity in the Ia afferents and establishes the baseline level of alpha motor neuron activity in the absence of muscle stretch.

Clinically, muscle tone is assessed by judging the resistance of a patient's limb to passive stretch. Damage to either the alpha motor neurons or the Ia afferents carrying sensory information to the alpha motor neurons results in a decrease in muscle tone, called *hypotonia*. In general, damage to descending pathways that terminate in the spinal cord has the opposite effect, leading to an increase in muscle tone, or *hypertonia*. The neural changes responsible for hypertonia following damage to higher centers are not well understood; however, at least part of this change is due to an increase in the responsiveness of alpha motor neurons to Ia sensory inputs. Thus, in experimental animals in which descending inputs have been severed, the resulting hypertonia can be eliminated by sectioning the dorsal roots.

Increased resistance to passive movement following damage to higher centers is called *spasticity*, and is associated with two other characteristic signs: the clasp-knife phenomenon and clonus. When first stretched, a spastic muscle provides a high level of resistance to the stretch and then suddenly yields, much like the blade of a pocket knife (or clasp knife, in old-fashioned terminology).

Hyperactivity of the stretch reflex loop is the reason for the increased resistance to stretch in the clasp knife phenomenon. The physiological basis for the inhibition that causes the sudden collapse of the stretch reflex (and loss of muscle tone) is thought to involve the activation of the Golgi tendon organs (see text).

Clonus refers to a rhythmic pattern of contractions (3–7 contractions per second) due to alternate stretching and unloading of the muscle spindles in a spastic muscle. Clonus can be demonstrated in the extensor muscles of the leg by pushing up on the sole of patient's foot to dorsiflex the ankle. If there is damage to descending pathways, holding the ankle loosely in this position will reveal rhythmic contractions of both the gastrocnemius and soleus muscles. Both the increase in muscle tone and the pathological oscillations seen after damage to descending pathways are very different from the tremor at rest and cogwheel rigidity present in basal ganglia disorders such as Parkinson's disease, phenomena that are discussed in Chapters 16 and 17.

force, counteracting small changes in muscle tension by increasing or decreasing the inhibition of alpha motor neurons. Under these conditions, the Golgi tendon system tends to maintain a steady level of muscle force, counteracting effects such as fatigue, which diminish muscle force. If the muscle spindle system is viewed as a feedback system that monitors and maintains muscle *length*, then the Golgi tendon system is a feedback system that monitors and maintains muscle *force*. Like the muscle spindle system, the Golgi tendon organ system is not a closed loop. Ib inhibitory interneurons also receive synaptic inputs from a variety of other sources, including cutaneous receptors, joint receptors, muscle spindles, and descending pathways (Figure 15.12). Together these inputs regulate the responsiveness of Ib interneurons to activity arising in Golgi tendon organs.

Figure 15.12 Negative feedback regulation of muscle tension by Golgi tendon organs. Ib afferents from tendon organs contact inhibitory interneurons that decrease the activity of alpha motor neurons innervating the same muscle. Ib inhibitory interneurons also receive input from other sensory fibers, as well as from descending pathways. This arrangement prevents muscles from generating excessive tension.

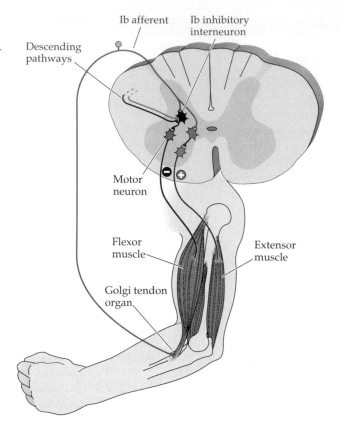

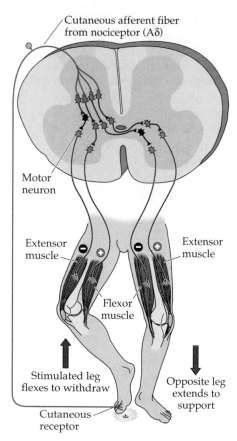

Figure 15.13 Spinal cord circuitry responsible for the flexion reflex. Stimulation of cutaneous receptors in the foot leads to activation of spinal cord circuits that withdraw (flex) the stimulated extremity and extend the other extremity to provide compensatory support.

■ FLEXION REFLEX PATHWAYS

So far, the discussion has focused on reflexes whose sensory receptors are located within muscles or tendons. Other reflex circuitry, however, mediates the withdrawal of a limb from a sudden painful stimulus, such as a pinprick or the heat of a flame. Contrary to what might be imagined, given the speed with which we are able to withdraw from such stimuli, this **flexion reflex** involves several synaptic links (Figure 15.13). As a result of this circuitry, stimulation of nociceptive sensory fibers leads to excitation of ipsilateral flexor muscles and inhibition of ipsilateral extensor muscles. Flexion of the stimulated limb is accompanied by an opposite reaction in the contralateral limb; extensor muscles are excited while flexor muscles are inhibited. This **crossed extension reflex** serves to enhance postural support during withdrawal from the painful stimulus.

Like the other reflex pathways, interneurons in the flexion reflex pathway receive converging inputs from several different sources, including cutaneous receptors, other spinal cord interneurons, and descending pathways. Although the functional significance of this complex pattern of connectivity is uncertain, changes in the character of the reflex following damage to descending pathways provide a clue. Under normal conditions, a noxious stimulus is required to evoke the flexion reflex; following damage to descending pathways, however, other types of stimulation, such as moderate squeezing of a limb, can produce the same response. Thus, the descending projections to the cord may function, at least in part, to gate the responsiveness of interneurons in the flexion reflex pathway to a variety of other sensory inputs.

■ SPINAL CORD CIRCUITRY AND LOCOMOTION

The contribution of spinal cord circuitry to motor control is not limited to reflexive responses to sensory inputs. Studies of rhythmic movements such as locomotion and swimming in animal models have demonstrated that spinal cord circuits are fully capable of controlling the timing and coordination of such complex patterns of movement, and of adjusting them in response to altered circumstances.

    The movement of a single limb during locomotion can be thought of as a cycle consisting of two phases: a stance phase, during which the limb is extended and placed in contact with the ground to propel the animal forward; and a swing phase, during which the limb is flexed to leave the ground and then brought forward to begin the next stance phase (Figure 15.14A). Increases in the speed of locomotion reduce the amount of time taken to complete a cycle, and most of the change in cycle time is due to a shortening of the stance phase; the swing phase remains relatively constant over a wide range of locomotor speeds.

Figure 15.14 The cycle of locomotion is organized by central pattern generators. (A) The step cycle, showing leg flexion (F) and extension (E) and their relation to the swing and stance phases of locomotion. EMG indicates electromyographic recordings. (B) Comparison of the stepping movements of the cat for different gaits. Brown bars, foot lifted (swing phase); gray bars, foot planted (stance phase). (C) Transection of the spinal cord of a cat at the thoracic level isolates the hindlimb segments of the cord. The hindlimbs are still able to walk on a treadmill after recovery from surgery. Reciprocal bursts of electrical activity can be recorded from flexors during the swing phase and from extensors during the stance phase of walking. (After Pearson, 1976.)

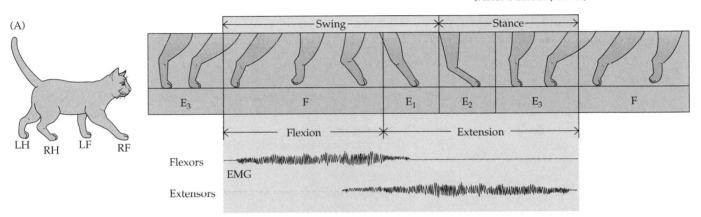

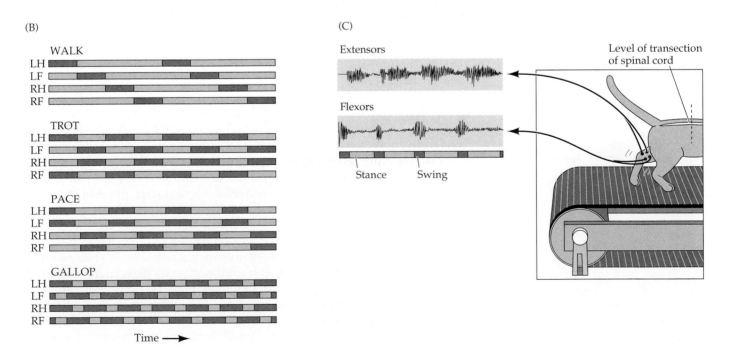

In quadrapeds, changes in locomotor speed are also accompanied by changes in the sequence of limb movements. At low speeds, for example, there is a back-to-front progression of leg movements, first on one side and then on the other. As the speed increases to a trot, the movements of the right forelimb and left hindlimb are synchronized (as are the movements of the left forelimb and right hindlimb). At the highest speeds (gallop) the movements of the two front legs are synchronized, as are the movements of the two hindlimbs (Figure 15.14B).

Given the precise timing of the movement of individual limbs and the coordination between limbs, it is natural to assume that locomotion is accomplished by higher centers that organize the action of the individual limbs. However, following transection of the spinal cord at the thoracic level, a cat's hindlimbs will still make coordinated locomotor movements if the cat is supported and placed on a moving treadmill (Figure 15.14C). Under these conditions, the speed of locomotor movements is determined by the speed of the treadmill; one could therefore argue that the movement is nothing more than a reflexive response to stretching the limb muscles. This possibility has been ruled out, however, by experiments in which the dorsal roots are also sectioned. Although the speed of walking is slowed and the movements are less coordinated than under normal conditions, appropriate locomotor movements are still observed. These and other experiments show that the basic rhythmic patterns of limb movement during locomotion are not dependent on sensory input; nor are they dependent on descending projections from higher centers. Each limb appears to have its own **central pattern generator**—an oscillatory spinal cord circuit responsible for the alternating flexion and extension of the limb during locomotion. Under normal conditions, the central pattern generators for the limbs are variably coupled to achieve the different sequences of movements that occur at different speeds of locomotion.

Although some locomotive movements can also be elicited in humans following damage to descending pathways, these are considerably less effective than the movements seen in the cat. The reduced ability of the transected spinal cord to mediate rhythmic stepping movements in humans presumably reflects an increased dependence of spinal centers on descending pathways. Perhaps bipedal locomotion brings with it requirements for postural control greater than can be accommodated by spinal cord circuitry alone. Whatever the explanation, there is no doubt that the basic oscillatory circuits that control such rhythmic behaviors as flying, walking, and swimming in animals also play an important part in human locomotion.

■ SUMMARY

Four distinct but highly interactive motor subsystems—spinal cord circuits, descending pathways, the basal ganglia, and the cerebellum—make essential contributions to motor control. Alpha motor neurons within the spinal cord, as well as neurons in the brainstem cranial nerve nuclei, provide the direct link between the nervous system and muscles. Each alpha motor neuron and its associated muscle fibers constitute a functional entity, the motor unit. Motor units vary in size, amount of tension produced, and degree of fatigability. Graded increases in muscle tension are mediated by both the orderly recruitment of different types of motor units and an increase in motor neuron firing frequency. Damage to alpha motor neurons or their axons leads to paralysis of the associated muscle and to other changes, including loss of reflex activity, loss of muscle tone, the onset of muscle atrophy, and spontaneous muscle fiber contractions. Spinal cord circuits involving sensory inputs, interneurons, and alpha and gamma motor neurons are especially important in the reflexive control of muscle activity. The stretch reflex is mediated by monosynaptic connections between sensory fibers that innervate muscle spindles and alpha motor neurons in the ventral horn of the spinal cord. Gamma motor neurons regulate the gain of the stretch reflex by adjusting the level of tension in the intrafusal muscle fibers in the muscle spindle. These elements of the stretch reflex circuit are responsible for setting the baseline level of activity in motor neurons that regulates muscle length and produces muscle tone. Other reflex circuits provide feedback control of muscle tension and mediate the rapid withdrawal of limbs from painful stimuli. Finally, much of the timing of muscle activation required for rhythmic movements such as locomotion is provided by specialized circuits called central pattern generators that also reside in the gray matter of the spinal cord.

Additional Reading

Reviews

BURKE, R. E. (1981) Motor units: Anatomy, physiology and functional organization. In *Handbook of Physiology*, V. B. Brooks (ed.). Section 1: *The Nervous System*. Volume 1, Part 1. Bethesda, MD: American Physiological Society, pp. 345–422.

BURKE, R. E. (1990) Spinal cord: ventral horn. In *The Synaptic Organization of the Brain*, 3rd Ed. G. M. Shepherd (ed.). New York: Oxford University Press, pp. 88–132.

GRILLNER, S. AND P. WALLEN (1985) Central pattern generators for locomotion, with special reference to vertebrates. Annu. Rev. Neurosci. 8: 233–261.

HENNEMAN, E. (1990) Comments on the logical basis of muscle control. In *The Segmental Motor System*, M. C. Binder and L. M. Mendell (eds). New York: Oxford University Press, pp. 7–10.

HENNEMAN, E. AND L. M. MENDELL (1981) Functional organization of the motoneuron pool and its inputs. In *Handbook of Physiology*, V. B. Brooks (ed). Section 1: *The Nervous System*. Volume 1, Part 1. Bethesda, MD: American Physiological Society, pp. 423–507.

LUNDBERG, A. (1975) Control of spinal mechanisms from the brain. In *The Nervous System*, Volume 1: *The Basic Neurosciences*. D. B. Tower (ed.). New York: Raven Press, pp. 253–265.

PATTON, H. D. (1965) Reflex regulation of movement and posture. In *Physiology and Biophysics*, 19th Ed., T. C. Rugh and H. D. Patton (eds.). Philadelphia: Saunders, pp. 181–206.

PEARSON, K. (1976) The control of walking. Sci. Amer. 235: 72–86.

PROCHAZKA, A., M. HULLIGER, P. TREND AND N. DURMULLER (1988) Dynamic and static fusimotor set in various behavioral contexts. In *Mechanoreceptors: Development, Structure, and Function*. P. Hnik, T. Soulup, R. Vejsada and J. Zelena (eds.). New York: Plenum, pp. 417–430.

SCHMIDT, R. F. (1983) Motor Systems. In *Human Physiology*. R. F. Schmidt and G. Thews (eds.). Berlin: Springer Verlag, pp. 81–110.

Important Original Papers

BURKE, R. E., D. N. LEVINE, M. SALCMAN AND P. TSAIRES (1974) Motor units in cat soleus muscle: Physiological, histochemical, and morphological characteristics. J. Physiol. (Lond.) 238: 503–514.

BURKE, R. E., P. L. STRICK, K. KANDA, C. C. KIM AND B. WALMSLEY (1977) Anatomy of medial gastrocnemius and soleus motor nuclei in cat spinal cord. J. Neurophysiol. 40: 667–680.

HENNEMAN, E., E. SOMJEN, AND D. O. CARPENTER (1965) Excitability and inhibitability of motoneurons of different sizes. J. Neurophysiol. 28: 599–620.

HUNT, C. C. AND S. W. KUFFLER (1951) Stretch receptor discharges during muscle contraction. J. Physiol. (Lond.) 113: 298–315.

LIDDELL, E. G. T. AND C. S. SHERRINGTON (1925) Recruitment and some other factors of reflex inhibition. Proc. R. Soc. London 97: 488–518.

LLOYD, D. P. C. (1946) Integrative pattern of excitation and inhibition in two-neuron reflex arcs. J. Neurophysiol. 9: 439–444.

MONSTER, A. W. AND H. CHAN (1977) Isometric force production by motor units of extensor digitorum communis muscle in man. J. Neurophysiol. 40: 1432–1443.

WALMSLEY, B., J. A. HODGSON AND R. E. BURKE (1978) Forces produced by medial gastrocnemius and soleus muscles during locomotion in freely moving cats. J. Neurophysiol. 41: 1203–1216.

Books

BRODAL, A. (1981) *Neurological Anatomy in Relation to Clinical Medicine*, 3rd Ed. New York: Oxford University Press.

SHERRINGTON, C. (1947) *The Integrative Action of the Nervous System*, 2nd Ed. New Haven: Yale University Press.

CHAPTER 16

DESCENDING CONTROL OF SPINAL CORD CIRCUITRY

■ OVERVIEW

Descending projections from higher centers control the primary brainstem and spinal cord circuitry and are essential for producing voluntary, goal-oriented movements. These projections originate in several structures within the brainstem that participate in the motor system, and in the motor areas of the cerebral cortex. Two particularly important components in the brainstem—the vestibular nucleus and the reticular formation—are responsible for much of the postural stability that is a prerequisite for any movement. The motor and premotor areas of the cerebral cortex are responsible for planning movements and executing them efficiently. The cortex exercises this influence by projections onto both the brainstem centers and the interneurons and motor neurons of the spinal cord and cranial nerve motor nuclei.

■ THE ORGANIZATION AND DESCENDING CONTROL OF MEDIAL AND LATERAL SPINAL CORD CIRCUITRY

The general arrangement of descending projections is most easily understood by considering the way motor neurons and interneurons—the targets of descending systems—are organized within the spinal cord. As described in Chapter 15, motor neurons in the ventral horn are organized in a somatotopic fashion: the most medial part of the ventral horn contains motor neuron pools that innervate the axial muscles and proximal muscles of the limbs, whereas the lateral part contains motor neurons that innervate the distal muscles of the limbs. The interneurons that lie in the intermediate zone of the spinal cord, which supply much of the input to the ventral horn motor neurons, are also topographically arranged. Thus, the medial region of the intermediate zone contains the interneurons that synapse with motor neurons in the medial part of the ventral horn, whereas the lateral parts of the intermediate zone contain interneurons that synapse primarily with neurons in the lateral part of the ventral horn.

The pattern of connections made by interneurons in the medial part of the intermediate zone are strikingly different from the patterns made by those in the lateral part (Figure 16.1). The medially located interneurons that supply the medial ventral horn have long axons that project to many spinal cord segments; indeed, some project the entire length of the cord. Many of these neurons also have axonal branches that cross the midline in the ventral commissure of the spinal cord to innervate the medial ventral horn in the other hemicord. In contrast, interneurons in the lateral part of the intermediate zone have shorter axons that course less than five segments and are predominantly ipsilateral. The widespread connections that characterize the interneurons that innervate the medial part of the ventral horn provide the coordinated control of multiple groups of axial and proximal limb muscles that is required for postural support. The more restricted patterns of connectivity characteristic of the interneurons that innervate the lateral regions of the ventral horn reflects the finer and more independent control that must be exerted over the muscles of the distal extremities.

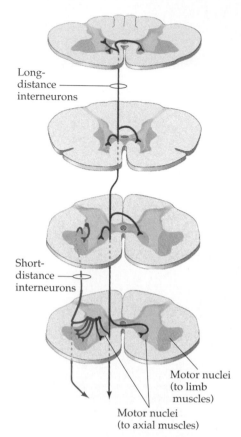

Long-distance interneurons

Short-distance interneurons

Motor nuclei
(to limb muscles)

Motor nuclei
(to axial muscles)

Figure 16.1 Interneurons that supply the medial region of the ventral horn are situated medially in the intermediate zone of the spinal cord gray matter and have axons that extend over a number of spinal cord segments and terminate bilaterally. Those that supply the lateral parts of the ventral horn are located more laterally, have axons that extend over a few spinal cord segments, and terminate only on the same side of the cord. Descending pathways that contact the medial parts of the spinal cord gray matter are involved primarily in the control of posture; those that contact the lateral parts are involved in the fine control of the distal extremities.

In general, the extent and pattern of terminations from the descending brainstem and cortical pathways conform to the arrangement of spinal cord circuits for the control of axial and distal muscle groups. Most of the descending axons that project to the medial part of the ventral horn also project to the medial part of the intermediate zone, giving rise to collaterals that terminate over a number of spinal cord segments and innervating medial cell groups on both sides of the cord. In contrast, descending axons that target the more lateral parts of the spinal cord gray matter have much more focused terminal fields in both the ventral horn and intermediate zone, and these are limited to only a few spinal cord segments. Furthermore, some of the structures that give rise to descending pathways restrict their terminal fields to either the medial or lateral aspects of the spinal cord gray matter. Thus, spinal cord circuits as well as the descending pathways that govern them are specialized for controlling different types of motor functions.

■ SOME GENERAL POINTS ABOUT BRAINSTEM PROJECTIONS TO THE SPINAL CORD

The projections to the spinal cord from the brainstem originate in four distinct structures (Figure 16.2). By and large, the most significant projections arise from the vestibular nucleus and the reticular formation. The projections of both these structures target the medial part of the intermediate zone and ventral horn, and are primarily involved in the control of the axial and proximal limb muscle groups. The organization of these two systems is considered in more detail in the next section. Two additional brainstem structures, the superior colliculus and the red nucleus, also contribute projections to the spinal cord, but these are more limited. The superior colliculus projects to medial cell groups in the cervical cord; this pathway is important for generating orienting movements of the head. Much more is known about the role of the superior colliculus in the generation of eye movements, a topic that is covered separately in Chapter 19. The red nucleus projections are also limited to the cervical level of the cord, but these terminate in the lateral aspect of the ventral horn and intermediate zone. In humans, this system controls primarily the muscles of the arm. The limited distribution of rubrospinal ("rubro-" means red) projections may seem surprising given the large size of the red nucleus in humans. In fact, most of the red nucleus is taken up by a subdivision that does not project to the spinal cord at all, but forms a pathway from the cortex to the cerebellum (see Chapter 18).

■ THE VESTIBULAR NUCLEUS AND RETICULAR FORMATION: MAINTAINING BALANCE AND POSTURE

As described in Chapter 13, the **vestibular nucleus** is a target of the axons of the eighth nerve, receiving sensory information from the semicircular canals and the otolith organs that signal the position of the head in space. Many neurons in the vestibular nucleus have descending axons that terminate in the medial region of the spinal cord gray matter, but which also extend laterally to contact the neurons that control the limbs. The projections from the vestibular nucleus that control neck and back muscles, as well as those that influence proximal limb muscles, actually originate in different parts of the vestibular nucleus and take different routes—the medial and lateral vestibulospinal tracts.

The **reticular formation** is a complicated network of neuronal circuits located in the core of the brainstem that extends from the rostral midbrain to

(A) MEDIAL BRAINSTEM PATHWAYS   (B) LATERAL BRAINSTEM PATHWAYS

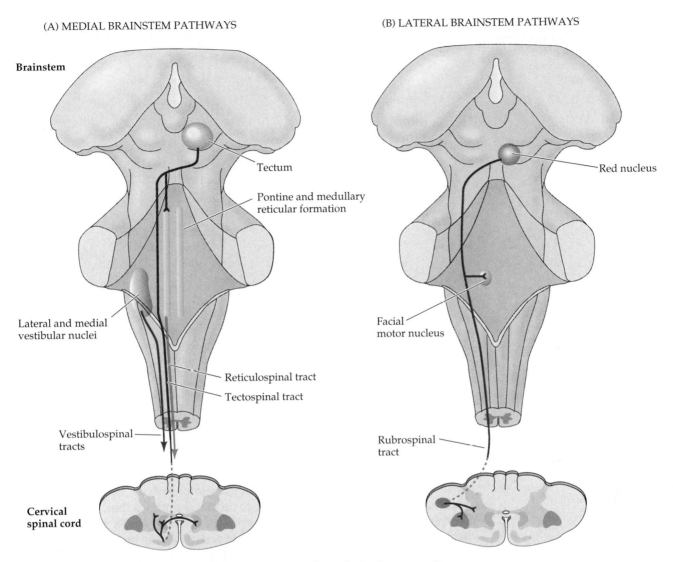

Brainstem

Tectum

Pontine and medullary reticular formation

Lateral and medial vestibular nuclei

Reticulospinal tract

Tectospinal tract

Vestibulospinal tracts

Cervical spinal cord

Red nucleus

Facial motor nucleus

Rubrospinal tract

Figure 16.2   Diagram of the descending projections from the brainstem to the spinal cord. Pathways that influence motor neurons in the medial part of the ventral horn originate in the superior colliculus, reticular formation, and vestibular nucleus. Those that influence motor neurons in the lateral part of the ventral horn originate in the red nucleus.

the caudal medulla (see Figure 16.2). Unlike the well-defined sensory and motor nuclei of the cranial nerves, the reticular formation comprises small clusters of neurons scattered among a welter of interdigitating axon bundles (Figure 16.3). The reticular formation has a variety of functions, including cardiovascular and respiratory control, regulation of sleep and wakefulness, and, most important for present purposes, various aspects of motor control. The descending projections from the reticular formation are similar to those of the vestibular nucleus, terminating primarily in the medial parts of the spinal cord gray matter to exert their influence on axial and proximal limb muscles. Together, the vestibular nuclei and the reticular formation provide the signals to the spinal cord that maintain posture in the face of environmental (or self-induced) disturbances of body stability.

Figure 16.3 The location of the reticular formation in relation to some other major landmarks at different levels of the brainstem. Neurons in the reticular formation are scattered among the axon bundles that course through the medial portion of the midbrain, pons and medulla.

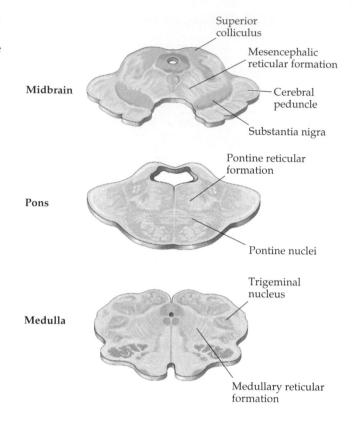

Midbrain

Superior colliculus

Mesencephalic reticular formation

Cerebral peduncle

Substantia nigra

Pons

Pontine reticular formation

Pontine nuclei

Medulla

Trigeminal nucleus

Medullary reticular formation

How these brainstem signals maintain posture can be appreciated by analyzing what happens during voluntary motor activity. Even the simplest motor acts are accompanied by the activation of muscles that, on the face of it, seem to have little to do with the primary purpose of the movement. For example, Figure 16.4 shows the pattern of activity in the biceps muscle when a subject uses his arm to pull on a handle immediately after an auditory tone is presented. Activity in the biceps muscle begins roughly 200 ms after the tone. However, as the records show, the contraction of the biceps is accompanied by a significant increase in the activity of the gastrocnemius muscle. In fact, contraction in the gastrocnemius muscle begins well before contraction of the biceps.

These results illustrate that postural control is mediated in part by an anticipatory, or feedforward, mechanism (Figure 16.5). As part of the motor plan for moving the arm, the effect of the upcoming movement on body stability is evaluated and used to generate a change in the activity of the gastrocnemius muscle that actually precedes the movement of the arm. (In this case, contraction of the biceps tends to pull the body forward, an action that is opposed by the contraction of the gastrocnemius muscle.) In short, a feedforward mechanism predicts a disturbance in body stability and formulates an appropriate stabilizing response.

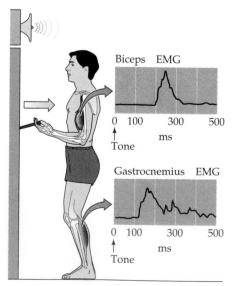

Figure 16.4 Anticipatory maintenance of body posture. At the onset of a tone, the subject pulls on a handle, contracting the biceps muscle. Contraction of the gastrocnemius muscle precedes that of the biceps to ensure postural stability. (After Nashner, 1979.)

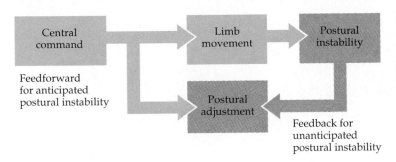

Figure 16.5 Feedforward and feedback mechanisms of postural control. Feedforward postural responses are "pre-programmed" and often precede the onset of limb movement. Feedback responses are initiated by sensory inputs that detect postural instability.

The importance of the reticular formation for feedforward mechanisms of postural control has been explored in more detail in cats trained to use a forepaw to strike an object. The movement of the forelimb is accompanied by feedforward postural adjustments in the other legs that maintain the animal upright. These adjustments shift the animal's weight from an even distribution over all four feet to a diagonal pattern, in which the weight is carried mostly by the nonreaching forelimb and the ipsilateral hindlimb. Lifting of the forepaw and the postural adjustments in the other limbs can also be induced in an alert cat by electrical stimulation of the motor cortex. After pharmacological inactivation of the reticular formation, however, electrical stimulation of the cortex evokes only the forelimb movements, without the feedforward postural adjustments that normally accompany them. As described in the following sections, the motor cortex influences the spinal cord motor neurons by two routes: direct projections to the spinal cord and projections onto brainstem centers that in turn project to the spinal cord. The reticular formation is one of the major targets of these latter projections from the motor cortex; thus, cortical neurons initiate both the movement of the forelimb and the postural adjustments necessary to maintain body stability. The forelimb movements are mediated via the direct connections between the cortex and the spinal cord (and possibly via the red nucleus), whereas the postural adjustments are mediated via connections from the motor cortex to the reticular formation (the corticoreticulospinal pathway).

Of course, feedforward mechanisms are not always sufficient to achieve postural stability; unexpected disturbances must be corrected by relying on sensory signals. Because it is so sensitive to motion of the head, the vestibular system is a particularly important source of such sensory information. Moreover, the direct projections from the vestibular nucleus to the spinal cord (see Chapter 13) ensure a rapid compensatory response to any indication of postural instability. Other sources of sensory signals that generate feedback postural responses are the muscle spindles and Golgi tendon organs, as well as the components of the visual system that detect motion.

■ THE MOTOR CORTEX

In addition to the brainstem, the other major source of descending motor control is the cerebral cortex. The part of the cortex from which movements can most easily be elicited by electrical stimulation lies in the precentral gyrus and is referred to as the **primary motor cortex** or **Brodmann's area 4**. Systematic stimulation studies by Wilder Penfield in neurosurgical patients (and by Clinton Woolsey in monkeys) showed many years ago that the primary motor cortex contains a topographical representation of the body's musculature (Figure 16.6 and Box A). Regions controlling facial muscles are

Figure 16.6 Topographic map of the body musculature in the primary motor cortex. (A) Location of primary motor cortex in the precentral gyrus. (B) Section along the precentral gyrus, illustrating the somatotopic organization of the motor cortex. The most medial parts of the motor cortex are responsible for controlling muscles in the legs; the most lateral portions are responsible for controlling muscles in the face. (C) Disproportionate representation of various portions of the body musculature in motor cortex. Representations of parts of the body that exhibit fine motor control capabilities (such as the hands and face) occupy a greater amount of space than those that exhibit less precise motor control (such as the trunk).

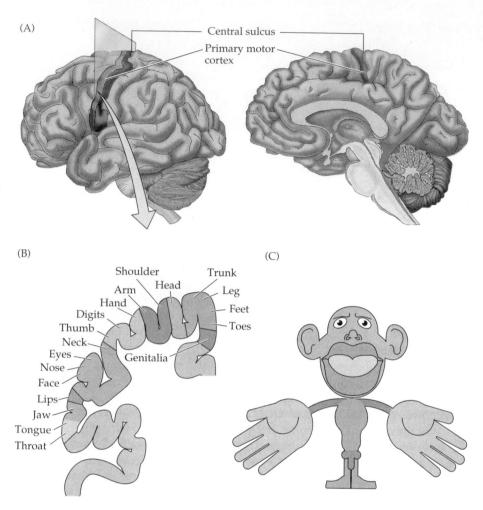

located on the lateral surface of the hemisphere, those controlling the leg are located on the medial surface, and areas controlling neck, arm, and trunk muscles are distributed in between. The motor map shows the same disproportions of somatic representation as does the somatosensory map in the postcentral gyrus (see Figure 8.8); somatic musculature used in tasks requiring extraordinarily fine motor control (such as that of the face and hands) occupies a greater amount of space in the map than the representation of musculature requiring relatively less precise motor control (such as that of the trunk). (Some behavioral aspects of the disproportions in cortical motor maps are considered in Box B.)

Electrophysiological stimulation of regions rostral to area 4, called the **premotor area**, also elicits motor responses. The premotor area includes the **premotor cortex (area 6)** on the lateral surface of the hemisphere, and the **supplementary motor cortex** located more medially (Figure 16.7). However, movements evoked by stimulation of these additional motor areas differ from those elicited by stimulation of the primary motor cortex in two respects. First, larger stimulating currents are required to evoke a response, suggesting that the connections of the premotor area with spinal cord motor neurons are less direct. Second, the movements elicited are generally more complex than those evoked by stimulation of area 4, often involving actions at multiple joints and on both sides of the body.

Box A
WHAT DO MOTOR MAPS REPRESENT?

Electrical stimulation studies by Wilder Penfield in human patients and Clinton Woolsey in animals clearly demonstrated a systematic map of the body's musculature in the primary motor cortex. The fine structure of this map, however, has been a continuing source of controversy. Is the map in the motor cortex a "piano keyboard" for the control of individual muscles, or is it a map of movements, in which specific sites control multiple muscle groups that contribute to the generation of particular actions?

Initial experiments implied that the map in the motor cortex is a fine-scale representation of individual muscles. Thus, stimulation of small regions of the map with very low levels of current (cortical microstimulation) activated single muscles, suggesting that vertical columns of cells in the motor cortex were responsible for controlling the actions of particular muscles, much as columns in the somatic sensory map are thought to analyze particular types of stimulus information (see Chapter 8).

More recent studies using anatomical and physiological techniques, however, indicate that the map in the motor cortex is far more complex than a columnar representation of individual muscles. For instance, individual pyramidal tract axons are now known to terminate on sets of spinal motor neurons that innervate different muscles. This relationship is evident even for neurons in the hand representation of the motor cortex, the region that controls the most discrete, fractionated movements. Furthermore, additional microstimulation experiments have shown that a single muscle is represented multiple times over a wide region of the motor cortex (about 2–3 mm in primates) in a complex, mosaic fashion. It seems likely that horizontal connections within the motor cortex create ensembles of neurons that coordinate the pattern of firing in the population of ventral horn cells that contribute to a given movement.

Thus, while the somatotopic maps in the motor cortex generated by early studies are correct in their overall topography, the fine structure of the map appears to be far more intricate. Unraveling the details of motor maps still holds the key to understanding how the patterns of activity in the motor cortex generate a given movement.

References

BARINAGA, M. (1995) Remapping the motor cortex. Science 268: 1696–1698.

LEMON, R. (1988) The output map of the primate motor cortex. Trends Neurosci. 11: 501–506.

PENFIELD, W. AND E. BOLDREY (1937) Somatic motor and sensory representation in the cerebral cortex of man studied by electrical stimulation. Brain 60: 389–443.

SCHIEBER, M. H. AND L. S. HIBBARD (1993) How somatotopic is the motor cortex hand area? Science 261: 489–491.

WOOLSEY, C. N. (1958) Organization of somatic sensory and motor areas of the cerebral cortex. In Biological and Biochemical bases of Behavior, H. F. Harlow and C. N. Woolsey (eds.). Madison, WI: University of Wisconsin Press, pp. 63-81.

(A) Lateral view

Premotor cortex

Supplementary motor cortex

Primary motor cortex

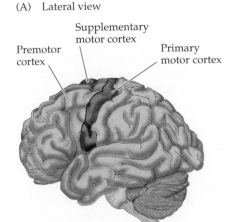

(B) Medial view

Supplementary motor cortex

Primary motor cortex

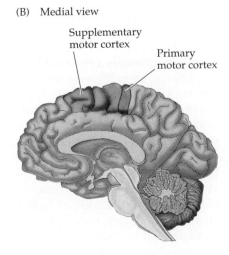

Figure 16.7 The primary motor cortex and the premotor area in the human cerebral cortex as seen in lateral (A) and medial (B) views. The primary motor cortex is located in the precentral gyrus. The premotor area (premotor cortex and supplementary motor cortex) is more rostral.

Box B
SENSORIMOTOR TALENTS AND CORTICAL SPACE

Are special sensorimotor talents, such as the exceptional speed and coordination displayed by premier atheletes, ballet dancers, or concert musicians, visible in the structure of the nervous system? The widespread use of noninvasive brain imaging techniques (see Boxes C and E in Chapter 1) has generated a spate of studies that have tried to answer this and related questions. Most of these studies have sought to relate particular sensorimotor skills to the amount of brain space devoted to such talents. For example, a study of professional violinists, cellists, and classical guitarists purported to show that representations of the "fingering" digits of the left hand in the right primary somatic sensory cortex are larger than the corresponding representations in nonmusicians.

In general, the idea that greater motor talents (or any other ability) will be reflected in a greater amount of brain space devoted to that task makes good sense. Comparisons across species show that special talents are invariably based on commensurately sophisticated brain circuitry, which means more neurons, more synaptic contacts between neurons, and more supporting glial cells—all of which occupies more space within the brain. The size and proportion of bodily representations in the primary somatic sensory and motor cortices of various animals reflects species-specific nuances of mechanosensory discrimination and motor control. Thus, the representations of the paws are disproportionately large in the sensorimotor cortex of racoons; rats and mice devote a great deal of cortical space to representations of their prominent facial whiskers; and a large fraction of the sensorimotor cortex of the star-nosed mole is given over to representing the elaborate nasal appendages that provide critical mechanosensory information for this burrowing species. The link between behavioral competence and the allocation of space is equally apparent in animals in which a particular ability has diminished, or never developed fully, during the course of evolution.

Nevertheless, it remains uncertain how—or if—this principle applies to variations in behavior among members of the same species, including humans. For example, there does not appear to be any average hemisphere asymmetry in the allocation of space in either the primary sensory or motor area, as measured cytoarchitectonically. Some asymmetry might be expected simply because 90% of humans prefer to use the right hand when they perform challenging manual tasks. It seems likely that individual sensorimotor talents among humans will be reflected in the allocation of an appreciably different amount of space to those behaviors, but this issue is just beginning to be explored.

References

CATANIA, K. C. AND J. H. KAAS (1995) Organization of the somatosensory cortex of the star-nosed mole. J. Comp. Neurol. 351: 549–567.

ELBERT, T., C. PANTEV, C. WIENBRUCH, B. ROCKSTROH, AND E. TAUB. (1995) Increased cortical representation of the fingers of the left hand in string players. Science 270: 305–307.

WELKER, W. I. AND S. SEIDENSTEIN (1959) Somatic sensory representation in the cerebral cortex of the racoon (Procyon lotos). J. Comp. Neurol. 111: 469–501.

WHITE, L. E., T. J. ANDREWS, C. HULETTE, A. RICHARDS, M. GROELLE, J. PAYDARFAR AND D. PURVES (in press) Structure of the human sensorimotor system. II. Lateral symmetry. Cereb. Cortex.

WOOLSEY, T. A. AND H. VAN DER LOOS (1970) The structural organization of layer IV in the somatosensory region (SI) of mouse cerebral cortex. The description of a cortical field composed of discrete cytoarchitectonic units. Brain Res. 17: 205–242.

It is generally believed that the premotor area operates mainly to plan or program movements, whereas area 4 governs the execution of movements. An experiment using PET imaging to study local cerebral blood flow illustrates this relationship (Figure 16.8). When the individuals in the study performed a simple motor act, in this case pressing a finger against a spring, regional blood flow increased in the hand representation of area 4. (Blood flow increased in the somatosensory cortex as well, presumably as a result of the activation of somatosensory receptors in the hand.) When subjects were asked to perform a more complex movement, such as a particular sequence of finger movements, blood flow also increased in the supplementary motor area. Finally, when the subjects were told to rehearse the finger sequence mentally but not actually perform it, blood flow increased *only* in the sup-

(A) Simple finger flexion
(performance)

Somatic sensory
cortex

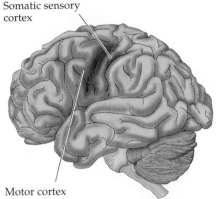

Motor cortex

(B) Finger movement sequence
(performance)

Supplementary
motor cortex

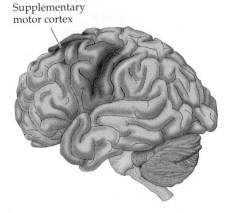

(C) Finger movement sequence
(mental rehearsal)

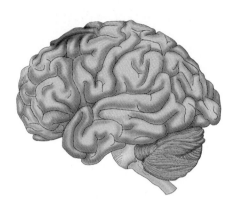

Figure 16.8 Regional blood flow to the motor cortices under three different behavioral conditions. (A) Blood flow during simple movements of the fingers increases in the primary motor and somatosensory cortices. (B) During more complex movements, blood flow also increases in the supplementary motor cortex. (C) During a mental rehearsal of the movement without execution, the increase in blood flow is largely confined to the supplementary motor cortex. (After Roland et al., 1980.)

plementary motor area. Since blood flow is a good index of neuronal activity (see Box E in Chapter 1), these results support the idea that the supplementary motor cortex operates at a level of abstraction "above" the execution of movement, whereas the actual execution is closely tied to the activity of neurons in area 4.

■ DESCENDING PROJECTIONS FROM THE MOTOR CORTEX

The projections from the motor cortex to the spinal cord and brainstem originate from pyramidal neurons in cortical layer V. As described in Chapter 1, the axons of these neurons travel in the internal capsule of the forebrain and in the cerebral peduncle at the base of the midbrain. They then pass through the base of the pons, where they become scattered among the transverse pontine fibers and the nuclei of the pontine gray matter. The axons destined to reach the spinal cord coalesce again on the ventral surface of the medulla as the right and left **pyramidal tracts**, or simply the **pyramids** (Figure 16.9). (The name of the tract refers to the pyramidal shape of the axon bundles on the ventral surface of the medulla, not to the pyramidal shape of the cells in the motor cortex that give rise to these bundles.) Axons that innervate cranial nerve nuclei, as well as the reticular formation and the red nucleus, leave the pathway at various levels of the brainstem (Box C).

The contribution of the primary motor cortex (area 4) actually accounts for only about half the fibers in the pyramids. The other axons come from neurons in the premotor area, as well as from the somatosensory cortex. The projections from motor and premotor areas terminate in the ventral horn and the intermediate zone of the cord. In contrast, those from the somatosensory cortex terminate in the dorsal horn. Given their targets, the projections from the somatosensory cortex are unlikely to be involved in motor control; it has been suggested that they regulate the transmission of sensory information from the dorsal horn to higher centers. The following discussion therefore focuses on those fibers that originate in the motor cortex.

The descending motor axons are arrayed somatotopically in both the internal capsule and the cerebral peduncle (Figure 16.10). Thus, axons that

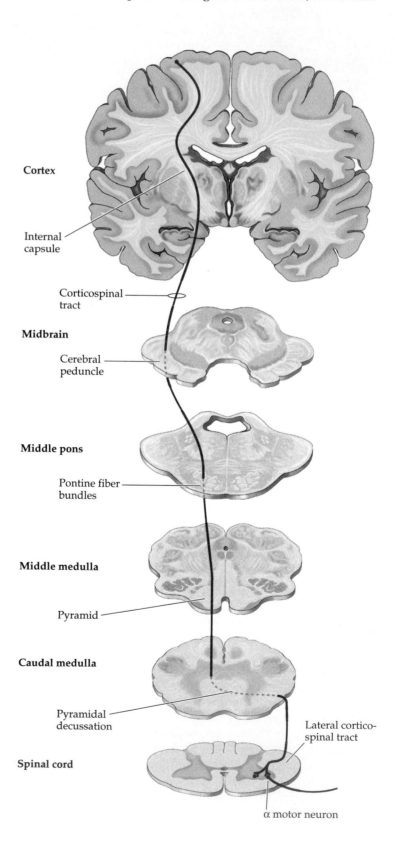

Cortex

Internal capsule

Corticospinal tract

Midbrain

Cerebral peduncle

Middle pons

Pontine fiber bundles

Middle medulla

Pyramid

Caudal medulla

Pyramidal decussation

Lateral cortico-spinal tract

Spinal cord

α motor neuron

Figure 16.9 The pyramidal system. Neurons in the motor cortex give rise to axons that travel through the internal capsule and coalesce on the ventral surface of the midbrain, within the cerebral peduncle. These axons continue through the pons and come to lie on the ventral surface of the medulla, forming the pyramids. Most of these fibers cross in the caudal part of the medulla to form the lateral corticospinal tract in the spinal cord. Those axons that do not cross (not illustrated here) descend on the same side and form the ventral cortico-spinal tract (see Figure 16.12).

control arm, trunk, and leg muscles are distributed from rostral to caudal in the posterior limb of the internal capsule; those that control muscles of the face course more rostrally in the genu (the bend, or "knee") of the internal capsule. In the cerebral peduncle, the fibers that control facial muscles are located medially; those that control the leg are located more laterally.

At the caudal end of the medulla, about three-quarters of the axons in the pyramidal tract cross (or **decussate**) and enter the lateral columns of the spinal cord, where they form the **lateral corticospinal tract**. These axons originate largely from the parts of the motor cortex that represent the limbs,

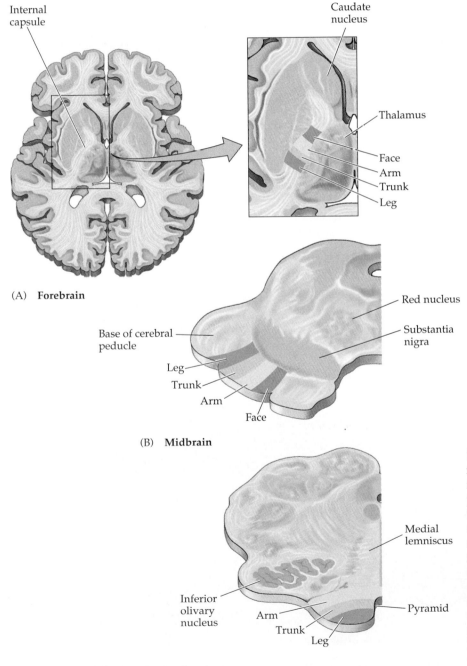

(A) **Forebrain**

(B) **Midbrain**

(C) **Medulla**

Figure 16.10 Somatotopic arrangement of cortical motor axons within the internal capsule, cerebral peduncle, and pyramidal tract. (A) Horizontal section through the forebrain showing the location of the internal capsule. Fibers that originate in motor cortex are arrayed in a topographic fashion. Those controlling the muscles of the face are located rostrally; those controlling the legs are located caudally. (B) Fibers controlling muscles of the face are located in the medial part of the cerebral peduncle; those controlling the leg are located more laterally. (C) A somatotopic arrangement is also present in the pyramid.

Box C
DESCENDING PROJECTIONS TO CRANIAL NERVE MOTOR NUCLEI

Axons descending from neurons in the face representation of the motor cortex leave the corticospinal pathway at various levels of the brainstem to innervate the somatic sensory and motor nuclei of the cranial nerves, thus forming the corticobulbar pathways. These axons terminate either contralaterally or bilaterally, depending on the nucleus, a pattern of termination that has considerable significance for understanding the neurologic deficits seen following cortical damage. The projections to the motor nuclei (trigeminal nucleus, facial nucleus, nucleus ambiguus, and spinal accessory nucleus) are mostly bilateral. Those to the hypoglossal nucleus and a part of the facial nucleus are contralateral. The control of muscles innervated by motor nuclei that receive *bilateral* projections from the cortex is relatively unaffected by *unilateral* damage to the cortex or the descending fiber pathways; the remaining intact projection is sufficient for normal (or near normal) muscle control. This sparing of function is not apparent for those nuclei that receive descending projections only from the contralateral side of the cortex.

The descending projections to the facial nucleus deserve special comment. Motor neurons in the part of the nucleus that innervates the upper face receive bilateral projections from the motor cortex, while those that innervate the lower face receive projections only from the contralateral cortex. Thus, damage of the descending projections to the facial nucleus will have different effects on muscles of the upper and lower parts of the face; upper facial muscles (frontalis, orbicularis oculi) will continue to operate normally, while lower facial muscles (orbicularis oris, buccinator) will be weakened. Accordingly, patients with damage to the face representation in the motor cortex on one side can wrinkle the forehead and close the eyes voluntarily, but lose voluntary control of the contralateral muscles of the mouth. The most obvious sign is a sagging of the corner of the mouth and a stilted smile. This arrangement can be especially informative in distinguishing damage to the brainstem (or to the facial nerve itself) from damage to higher centers. Damage to the brainstem or facial nerve affects all the muscles of facial expression (both upper and lower face) on the side of the lesion. Damage to the motor cortex affects control of only the lower part of the face on the side contralateral to the lesion.

Reference

Brodal, A. (1981) *Neurological Anatomy in Relation to Clinical Medicine*, 3rd Edition. New York: Oxford University Press.

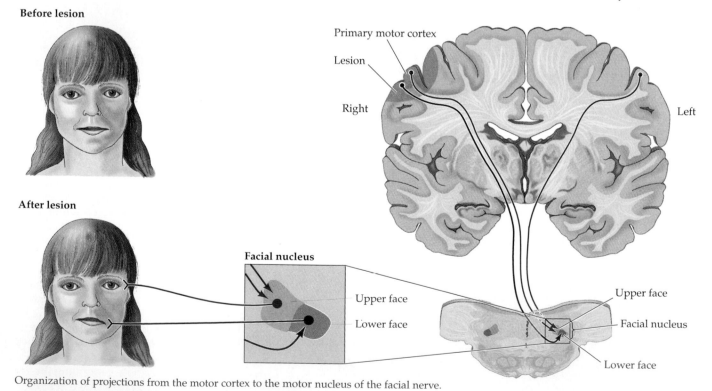

Organization of projections from the motor cortex to the motor nucleus of the facial nerve.

and they terminate preferentially on lateral cell groups in the ventral and intermediate portions of the spinal gray matter (Figures 16.11 and 16.12). The remaining axons in the pyramids pass directly into the spinal cord without crossing to form the **ventral corticospinal tract**. These axons originate from the parts of the motor cortex that represent the neck, shoulder and trunk regions; as described earlier, they terminate on neurons in the medial parts of the ventral horn and intermediate zone. Many ventral corticospinal axons also have branches that cross the midline of the spinal cord to innervate the medial cell groups in the other hemicord.

Figure 16.12 illustrates the projections from the motor cortex to the two major brainstem sources of descending projections, the red nucleus and the reticular formation. The axons that innervate the red nucleus originate from the parts of motor cortex that also project to the lateral part of the spinal cord gray matter; the axons to the reticular formation originate from the parts of the motor cortex that project to the medial part of the spinal cord gray matter. Thus, the motor cortex has two routes by which it can independently influence spinal cord motor neurons: a **direct route** via the pyramids to motor neurons and interneurons in the medial and lateral parts of the spinal cord gray matter, and an **indirect route** via projections onto the reticular formation and the red nucleus.

The functions of the direct and indirect pathways from the motor cortex to the spinal cord were first suggested by Dutch neurobiologist Hans Kuypers when he examined the behavior of monkeys that had their pyramidal tracts transected (without interruption of the descending projections from brainstem centers). Immediately after the surgery, the animals were able to stand, walk, run, and climb, but they had great difficulty using their extremities, especially their hands, independently of other body movements. For example, they could cling to the cage but were unable to pick up food. After several weeks, the animals recovered the use of their hands and were again able to pick up objects of interest, but this action now involved the concerted closure of all of the fingers. The ability to make independent, fractionated movements of the fingers never returned. In addition, all move-

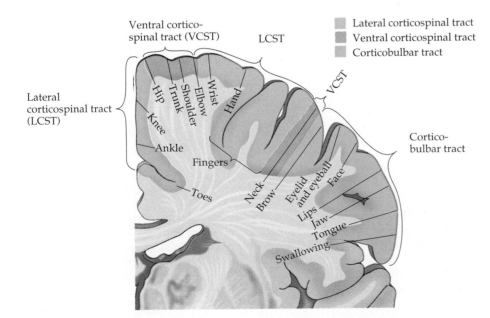

Figure 16.11 The origin within the motor cortex of projections to the lateral and ventral corticospinal tracts, as well as to the corticobulbar tracts. The lateral corticospinal tract is largely concerned with control of the limbs, while the ventral corticospinal tract is concerned with axial and proximal limb muscle groups. The corticobulbar tract is responsible for the control of cranial nerve nuclei that innervate the muscles of the head and neck.

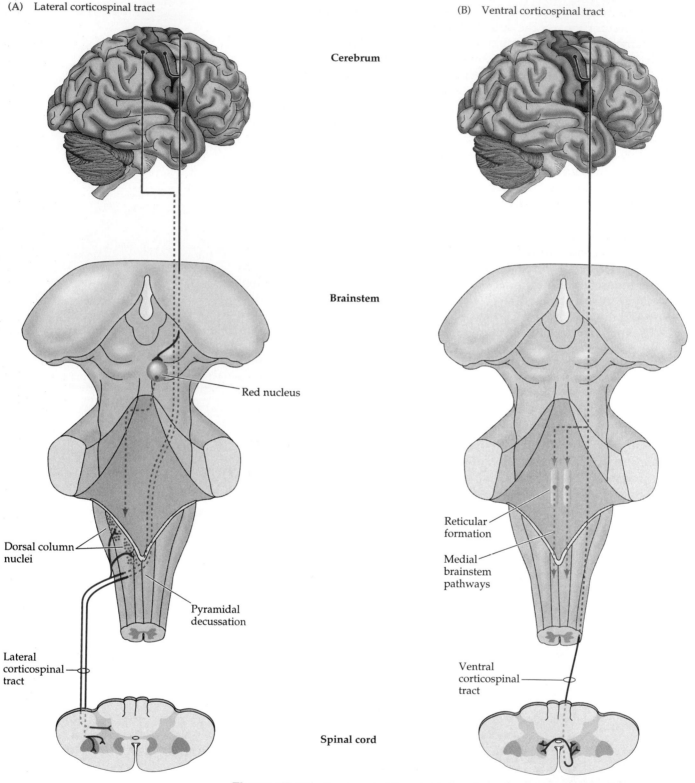

(A) Lateral corticospinal tract

(B) Ventral corticospinal tract

Cerebrum

Brainstem

Red nucleus

Dorsal column nuclei

Reticular formation

Medial brainstem pathways

Pyramidal decussation

Lateral corticospinal tract

Ventral corticospinal tract

Spinal cord

Figure 16.12 Summary of the projections from the motor cortex to the lateral (A) and medial (B) gray matter of the spinal cord. Neurons in the motor cortex that supply the lateral part of the ventral horn also terminate on neurons in the red nucleus. Neurons in the motor cortex that supply the medial part of the ventral horn also terminate on neurons in the reticular formation. Thus, the motor cortex has both direct and indirect routes by which it can influence the activity of spinal cord neurons.

ments were slower and fatigued more rapidly than those in normal animals. These observations show that following damage to the pyramidal tract, the projections from the motor cortex to brainstem centers (or from brainstem centers alone) are capable of sustaining much of the motor behavior of monkeys. By the same token, the direct projections from the motor cortex enable greater speed and agility of movements and a higher degree of precision in fractionated movements than is provided by the indirect pathways.

Fractionated movements appear to depend on direct connections between neurons in the motor cortex and motor neurons in the ventral horn. A comparison of the distribution of corticospinal fibers across different species shows that descending projections from the motor cortex to the ventral horn of the spinal cord are a relatively recent phylogenetic addition. Thus, most of the descending projections to the spinal cord in rodents and carnivores end in the dorsal horn and intermediate zone, rather than in the ventral horn. Exceptions to this rule are found only in animals that have unusual manual dexterity, such as the raccoon. Other evidence that supports the conclusion that fractionated movements depend on direct cortical projections to the ventral horn comes from developmental studies. The ability to generate individual movements of the fingers in humans and monkeys is not present at birth. In newborns, the corticospinal pathway terminates primarily on neurons in the dorsal horn and intermediate gray matter and has not yet innervated the ventral horn. In monkeys, fractionated movements emerge gradually and reach an adult level of proficiency at 7 or 8 months—about the same time that the adult density of cortical axon terminals appears in the ventral horn.

Selective damage to the pyramidal tract in humans is rarely seen in the clinic. Nonetheless, the evidence that direct projections from the cortex to the spinal cord are essential for the performance of discrete finger movements helps explain the limited recovery that follows damage to the motor cortex or to the internal capsule. Immediately after such an injury, the patient is typically paralyzed (see the next section). With time, however, some ability to perform voluntary movements reappears; these movements are crude for the most part, and the ability to perform discrete finger movements such as those required for writing, typing, or buttoning typically remains impaired.

■ DAMAGE TO DESCENDING MOTOR PATHWAYS: THE UPPER MOTOR NEURON SYNDROME

Damage to descending motor pathways gives rise to a set of signs and symptoms that is called the **upper motor neuron syndrome**. Injuries of this sort are common because of the large amount of cortex occupied by motor areas, and because motor pathways extend all the way from the cerebral cortex to the spinal cord. Damage anywhere along the length of these pathways produces characteristic deficits in the ability to control movements.

Injury to the motor cortex or to the descending fibers in the internal capsule leads to an initial flaccid paralysis (loss of tone) of the muscles on the contralateral side of the body and face. Because of the topographical arrangement of the motor system, the parts of the body or face that are affected help to localize the lesion. The manifestations of damage tend to be most severe in the arms and legs; if an arm or leg is elevated and released, it drops passively and all reflex activity on the affected side is abolished. Control of trunk muscles is typically preserved, either by the remaining brainstem pathways or the bilateral projections of the ventral corticospinal pathway. This initial period of hypotonia is called **spinal shock**. After several days, a consistent pattern of additional motor signs emerges, including:

1. *The Babinski sign.* The normal response to stroking the sole of the foot from the heel to the toes is a flexion of the big toe, often accompanied by flexion of the other toes. Following damage to descending pathways, the same stimulus elicits extension of the big toe and a fanning of the other toes (Figure 16.13). A similar response occurs in human infants before the maturation of the corticospinal pathway.

2. *Spasticity.* Spasticity includes increased muscle tone, hyperactive stretch reflexes, and clonus (see Chapter 15). Severe lesions are often accompanied by rigidity of the extensor muscles of the leg and the flexor muscles of the arm (decerebrate rigidity; see below).

3. *Hyporeflexia of superficial reflexes.* These signs refer to decreased vigor (and increased threshold) of superficial reflexes such as the corneal reflex, superficial abdominal reflex (tensing of abdominal muscles in response to stroking of the overlying skin), and the cremasteric reflex in males (elevation of the scrotum in response to stroking the inner aspect of the thigh). The diminishment of superficial reflexes is thought to result from a loss of facilitating cortical input.

Although upper motor neuron signs and symptoms may arise from damage anywhere along the descending pathways, the alterations in muscle tone vary according to the location of the lesion. For example, the increase in tone that follows damage to descending pathways in the spinal cord is somewhat less than that seen following damage to the cortex or internal capsule. The extensor muscles in the legs of an individual with spinal cord damage above the lumbosacral enlargement cannot support the individual's body weight, whereas those of a patient with damage to the cortex often can. Lesions that interrupt the descending pathways in the brainstem above the level of the vestibular nuclei but below the level of the red nucleus produce an even greater amount of extensor tone than that seen following damage to higher regions. Charles Sherrington was the first to describe such **decerebrate rigidity** in cats, where the extensor tone in all four limbs is so great that the animal is able to stand without any other support. An unconscious patient may exhibit similar signs of decerebration—arms and legs stiffly extended, jaw clenched, and neck retracted—after injury to the pons.

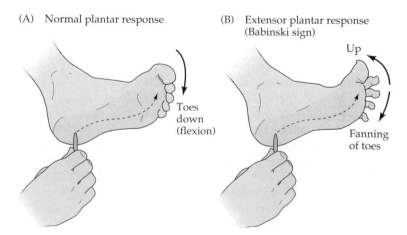

(A) Normal plantar response

(B) Extensor plantar response (Babinski sign)

Up

Toes down (flexion)

Fanning of toes

Figure 16.13 The Babinski sign. Following damage to descending motor pathways, stroking the sole of the foot causes an abnormal fanning of the toes and the extension of the big toe.

The increased muscle tone that follows damage at any level along the descending pathways is the result of an increase in the gain of spinal cord stretch reflexes. The evidence for this statement comes from experiments showing that the enhanced muscle tone in animals with decerebrate rigidity disappears following section of the dorsal roots. The relatively greater hypertonia following damage to the nervous system above the level of the spinal cord is explained by the activity of the intact descending pathways. The elevated levels of tone in decerebrate posture, for example, are thought to be due to an abnormally high level of activity in reticulospinal and vestibulospinal pathways, which increase the gain of spinal cord stretch reflexes.

■ SUMMARY

There are two major sets of descending motor projections, each making particular contributions to motor control. One set originates from neurons in the brainstem—primarily in the reticular formation and the vestibular nucleus—and is responsible for postural regulation. The reticular formation is especially important for feedforward control of posture (that is, movements that occur in anticipation of a change in body stability). The neurons in the vestibular nucleus that project to the spinal cord play an important role in feedback postural mechanisms (movements generated in response to sensory signals that indicate an existing postural disturbance). The other set of descending motor projections originates from the frontal lobe and includes projections from the primary motor cortex (area 4) and the premotor area. The premotor area appears to be involved in the planning of movements, whereas the primary motor cortex is involved in the execution of movements. The motor cortex exerts its influence over movements directly by contacting motor neurons and interneurons in the spinal cord and cranial nerve motor nuclei, and indirectly by innervating neurons in brainstem centers (the reticular formation and red nucleus). Whereas gross motor control can be mediated via brainstem pathways, direct projections from the motor cortex to alpha motor neurons in the spinal cord are essential for fine, fractionated movements of the extremities.

Additional Reading

Reviews

GAHERY, Y. AND J. MASSION (1981) Co-ordination between posture and movement. Trends Neurosci. 4: 199–202.

KUYPERS, H. G. J. M. (1981) Anatomy of the descending pathways. In *Handbook of Physiology*, Section 1: *The Nervous System*, Volume II, *Motor Control*, Part 1, V. B. Brooks (ed.). Bethesda, MD: American Physiological Society.

NASHNER, L. M. (1979) Organization and programming of motor activity during posture control. In *Reflex Control of Posture and Movement*, R. Granit and O. Pompeiano (eds.). Prog. Brain Res. 50: 177–184.

NASHNER, L. M. (1982) Adaptation of human movement to altered environments. Trends Neurosci. 5: 358–361.

Important Original Papers

LAWRENCE, D. G. AND H. G. J. M. KUYPERS (1968) The functional organization of the motor system in the monkey. I. The effects of bilateral pyramidal lesions. Brain 91: 1–14.

ROLAND, P. E., B. LARSEN, N. A. LASSEN AND E. SKINHOF (1980) Supplementary motor area and other cortical areas in organization of voluntary movements in man. J. Neurophysiol. 43: 118–136.

Books

ASANUMA, H. (1989) *The Motor Cortex*. New York: Raven Press.

BRODAL, A. (1981) *Neurological Anatomy in Relation to Clinical Medicine*, 3rd Ed. New York: Oxford University Press.

BROOKS, V. B. (1986) *The Neural Basis of Motor Control*. New York: Oxford University Press.

PENFIELD, W. AND T. RASMUSSEN (1950) *The Cerebral Cortex of Man: A Clinical Study of Localization of Function*. New York: MacMillan.

PHILLIPS, C. G. AND R. PORTER (1977) *Corticospinal Neurones. Their Role in Movement*. Academic Press.

SHERRINGTON, C. (1947) *The Integrative Action of the Nervous System*, 2nd Ed. New Haven: Yale University Press.

SJÖLUND, B. AND A. BJÖRKLUND (1982) *Brainstem Control of Spinal Mechanisms*. Amsterdam: Elsevier.

CHAPTER 17

MODULATION OF MOVEMENT BY THE BASAL GANGLIA AND CEREBELLUM

■ OVERVIEW

The initiation and execution of any particular movement is based on information about the intention of the movement, the sensory context in which it will be made, and the status of the body parts making the movement. All this requires neural circuitry capable of providing the motor cortex with highly integrated signals about the state of the mover and the state of the world. The basal ganglia and the cerebellum are major sources of such information. Each of these structures includes a central processing region—the caudate and putamen in the basal ganglia, the cerebellar cortex in the cerebellum—and several associated nuclei. The central processing region receives massive inputs from secondary sensory and higher-order association cortices. These processing regions exert their control primarily by projecting back to the thalamus via intermediate relay nuclei, thereby modulating the output of the primary motor cortex and related cortical areas in the frontal lobe. Although there are similarities in the organization of the basal ganglia and cerebellum, there are also essential differences between these two systems for motor modulation. The basal ganglia participate primarily in the planning of complex movements, whereas the cerebellum is concerned primarily with the coordinated execution of ongoing movements. Although movement can occur without either the basal ganglia or cerebellum (as several clinical syndromes make plain), these systems ensure that movement is appropriately planned and smoothly executed.

■ SENSORY INFORMATION AND MOTOR COMMANDS

The basal ganglia and the cerebellum form two distinct modulatory systems that integrate sensory information for the purpose of motor control (Figure 17.1). The terminology used to describe these structures can be confusing (Table 17.1). The **basal ganglia** refers to a collection of nuclei found primarily in the basal forebrain. These include the **caudate nucleus** and the **putamen** (collectively called the **striatum** or **neostriatum**), the **globus pallidus**, and the **subthalamic nucleus** (which is actually part of the diencephalon). In addition, a mesencephalic structure, the **substantia nigra**, is included as a member of the basal ganglia. The cerebellar system includes the **cerebellar cortex**, the **deep cerebellar nuclei**, the **pontine relay nuclei** and the associated fiber tracts known as the **cerebellar peduncles**.

The purpose of the modulation provided by both the basal ganglia and the cerebellum is to assist in the appropriate planning, initiation, coordination, guidance, and termination of voluntary movements. Not surprisingly, then, these two systems share some general anatomical and functional features: each receives massive cortical input, has one main processing station, has an intermediate relay, and ultimately sends information to the thalamus; the thalamus then relays this information back to the motor and premotor areas of the cortex to influence the cortical control of voluntary movement. Despite these general similarities, the functions of the basal ganglia and cerebellum in motor modulation are quite different. The basal ganglia play a featured part in the planning, initiation, and termination of movements, par-

Figure 17.1 Summary diagram of motor modulation by the basal ganglia and cerebellum. The central processing component in each of these structures receives massive input from the cerebral cortex; each system then generates feedback signals about the state of the mover and the state of the world. Note that various modulatory inputs also influence the processing of information within the caudate and putamen and the cerebellar cortex. The output signals from the caudate and putamen and from the cerebellar cortex are relayed indirectly to the thalamus and then back to the motor cortex, where they modulate motor commands.

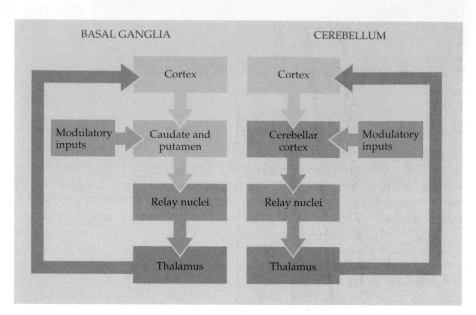

ticularly those with a complex cognitive dimension. The cerebellum is more important for the smooth execution and appropriate completion of ongoing movements, particularly those guided by vision.

The distinctive contribution to the modulation of motor function provided by the basal ganglia and cerebellum can be appreciated by analyzing the execution of an everyday motor act such as signing one's name (Figure 17.2). To appreciate the complexity of this seemingly simple task, consider the information relevant to completing a signature. The data needed to initi-

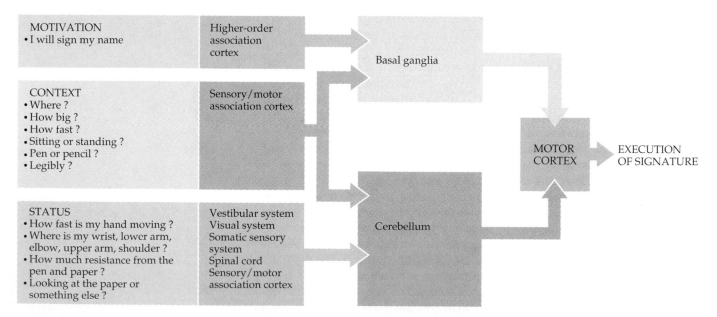

Figure 17.2 Diagram showing how the basal ganglia and cerebellum contribute to the initiation and execution of voluntary movements, in this case the execution of a signature.

ate a signature includes where the name will be written, the size of the writing surface, the script size appropriate for the space allocated, the position of the body when signing (standing or sitting), the type of writing surface (paper, chalkboard), the writing implement (ballpoint pen, pencil, magic marker), the size of the writing implement, and the amount of pressure needed to ensure the signature is legible. All this information needs to be acquired, processed, and made available to the motor cortex whenever this apparently straightforward motor act is called for!

Once the movement is begun, additional information is necessary to insure its proper execution. Signing one's name—or performing virtually any complex motor task—represents a coordinated series of movements made over time in a definite sequence. To get from the first letter to the last, information is needed about where the hand is in respect to the paper. How must the wrist, elbow, and shoulder change position as signing proceeds? Is the pen or pencil moving too slowly or too quickly across the surface as each letter is executed? Is the signature slipping above or below the line? To successfully complete the signature, the motor cortex must be continually updated about the ongoing movement and its context by several sensory systems, most notably vision and somatic sensation.

For the most part, the basal ganglia contribute to the preparatory phase of such motor control, while the cerebellum governs the coordination of ongoing movements. This assertion is based largely on the sort of sensory information received and processed by each of the two systems, and on the deficits observed after damage to one system or the other. This division of labor is not absolute, however. The basal ganglia make some contribution to the coordination of movements, and the cerebellum can be involved in some aspects of the planning and initiation of movements.

The importance of motor modulation by the basal ganglia and cerebellum is dramatically illustrated by the deficits in motor function that occur when either of these two regions is compromised, as described in the following sections. The major effect of both basal ganglia and cerebellar damage is difficulty in the initiation, execution, and coordination of complex sequences of movements; the ability to contract the muscles, the production of involuntary movements, and autonomic (visceral) motor functions are all left intact.

■ BASAL GANGLIA LESIONS: DEFICITS IN THE INITIATION OF MOVEMENT

Deficits that result from basal ganglia lesions are characterized by difficulty in planning and performing entire motor acts. **Parkinson's disease** (Box A) and **Huntington's disease** (Box B) provide two examples of pathological processes that have such consequences. Each disorder involves the degeneration of a specific processing or relay station in the basal ganglia. In Parkinson's disease, the substantia nigra degenerates bilaterally (Figure 17.3A); in Huntington's disease, the caudate nucleus degenerates bilaterally (Figure 17.3B).

Patients with Parkinson's disease, the most common basal ganglia disorder, exhibit a constant tremor at rest, muscle and limb rigidity, limited initiation of movements (sometimes resulting in a vacant facial expression and general passivity), diminished spontaneous movements (less scratching, fidgeting, tapping fingers, smiling, and grimacing), and slowness in performing complex voluntary movements. This latter problem is referred to as **bradykinesia**; characteristic signs include a shuffling gait, slow speech, and difficulty manipulating objects. The deficits of Parkinson's disease are particularly pronounced in the preparatory and initial phases of movement.

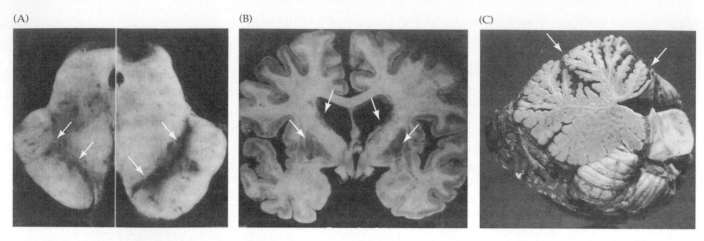

Figure 17.3 The pathological changes in certain neurological diseases provide insights about the function of the basal ganglia and cerebellum. (A) Left: the midbrain from a patient with Parkinson's disease. The substantia nigra (pigmented area) is largely absent in the region above the cerebral peduncles (arrows). Right: the mesencephalon from a normal subject, showing intact substantia nigra (arrows). (B) The size of the caudate and putamen (the striatum) (arrows) is dramatically reduced in patients with Huntington's disease. (C) Chronic alcohol abuse causes degeneration of the anterior cerebellum (arrows) while leaving other cerebellar regions intact. (A and B from Bradley et al., 1991; C from Victor et al., 1959).

Parkinson's patients sometimes describe this aspect of their problem by comparing it to lifting a heavy weight: once the initial "resistance" is overcome, the movement can proceed more smoothly.

The manifestations of Huntington's disease provide some insight into the contribution of the basal ganglia to the cognitive aspects of motor control. This disorder results in the indiscriminate "release" of entire motor behaviors. In its early stages, the disease is noticed as an inability to match complex motor acts to the social context in which they are performed; initially, this can be as subtle as an occasional facial twitch or inappropriate fidgeting. Inappropriate motor behaviors are normally kept in check by an "awareness" of one's situation. Accordingly, the early signs and symptoms of Huntington's disease imply that motor programs with cognitive significance are normally initiated at appropriate times by information processed by the basal ganglia. Eventually, Huntington's is made evident by jerky, random movements that engage various body parts in an apparently random sequence (referred to as **chorea**), as well as involuntary, repetitive writhing movements and abnormal postures (referred to as **dystonia**, or **athetosis** when the writhing movement is limited to the limbs). The full range of motor signs in Huntington's disease is thus called **choreoathetosis**. Evidently, choreoathetosis reflects the release of motor behaviors that the basal ganglia would normally hold in check.

■ CEREBELLAR LESIONS: DEFICITS IN COORDINATING AND TERMINATING MOVEMENTS

Cerebellar lesions lead first and foremost to an incoordination of ongoing movements. This deficit, called **cerebellar ataxia**, represents the inappropriate operation of groups of muscles that normally rely on sensory feedback to produce smooth, concerted actions. Patients with cerebellar lesions exhibit

BOX A
CAN GENE THERAPY HELP PARKINSON'S PATIENTS?

Gene therapy—the correction of a disease by the introduction of genetic information into the affected individual—holds considerable promise in the treatment of nervous system diseases. The neurological disease in which the greatest progress has been made is Parkinson's disease. Described by James Parkinson in 1817, this neurodegenerative disorder is characterized by tremor at rest, slowness of movement (bradykinesia), rigidity of the extremities and neck, and an expressionless face. The gait of Parkinsonian patients is characterized by short steps, stooped posture, and a paucity of normal limb movements. To make matters worse, these abnormalities of motor function are often associated with dementia. Following a gradual onset between the ages of 50 and 70, the disease progresses slowly toward death 10 to 20 years later. The defects in motor function are due to a loss of dopaminergic neurons in the substantia nigra, a population of neurons in the midbrain that projects to targets in the caudate and putamen. The cause of their insidious deterioration is not known.

The degeneration of a chemically defined neuronal population in Parkinson's disease presents an attractive opportunity for gene therapy, since it has long been clear that enhancing the release of dopamine in the caudate and putamen by conventional pharmacological strategies ameliorates the symptoms of the disease (most successful has been oral supplementation with L-DOPA).

Gene therapy could be accomplished in principle by increasing the expression of tyrosine hydroxylase, the enzyme that converts tyrosine to L-DOPA, which in turn is converted by a ubiquitous decarboxylase to the neurotransmitter dopamine (see Chapter 6). A controversial surgical approach has been to transplant tissue derived from the midbrain of a human fetus into the deteriorating caudate and putamen; the fetal midbrain is enriched in developing neurons that express tyrosine hyroxylase, and that therefore synthesize dopamine. Implants of fetal tissue into rats with an induced disorder similar to Parkinson's disease indeed show some anatomical and functional evidence of repair. Implantation of fetal midbrain tissue into the caudate and putamen of humans with Parkinson's disease has also produced significant improvement of motor performance in some patients for as long as 2 years. However, concerns about the use of fetal tissues, as well as the difficulty in obtaining pathogen-free tissue of the proper age, have stimulated the search for other approaches.

One attractive strategy is to implant genetically engineered cells that have been carried in tissue culture. For example, transfection of muscle cells in primary culture with a tyrosine hydroxylase expression plasmid produces a stable cell line that expresses this enzyme in vitro. Intracerebral transplantation of these cells reduces the motor defects in the rat model of Parkinson's disease, and this approach will soon be tested in humans with the disease. Transplantation of genetically modified cells expressing a protein that ameliorates the progression of a disease like Parkinson's could of course be useful for a wide variety of neurological disorders. The problems that must be overcome in gene therapy, however, are still daunting, even in the case of Parkinson's disease where circumstances favor this strategy.

References

BJÖRKLUND, A. AND U. STENEVI (1979) Reconstruction of the nigrostriatal dopamine pathway by intracerebral nigral transplants. Brain Res. 177: 555–560.

FREED, C. R. AND 18 OTHERS (1992) Survival of implanted fetal dopamine cells and neurologic improvement 12 to 46 months after transplantation for Parkinson's disease. N. Engl. J. Med. 327: 1549–1555.

JIAO, S., V. GUREVICH AND J. A. WOLFF (1993) Long-term correction of rat model of Parkinson's disease by gene therapy. Nature 362: 450–453.

ROEMER, K. AND T. FRIEDMANN (1992) Concepts and strategies for human gene therapy. Eur. J. Biochem. 208: 211–225.

SPENCER, D. D. AND 15 OTHERS (1992) Unilateral transplantation of human fetal mesencephalic tissue into the caudate nucleus of patients with Parkinson's disease. N. Engl. J. Med. 327: 1541–1548.

WIDNER, H., J. TETRUD, S. REHNCRONA, B. SNOW, P. BRUNDIN, B. GUSTAVII, A. BJÖRKLUND, O. LINDVALL AND J. W. LANGSTON (1992) Bilateral fetal mesencephalic grafting in two patients with Parkinsonism induced by 1-methyl-4-phenyl-1,2,3,6-tetrahydropyridine (MPTP). N. Engl. J. Med. 327: 1556–1563.

tremor during movement (called **intention tremor**, as opposed to the tremor at rest seen in Parkinsonian patients). They may also show a more pronounced instability as the limb approaches the target, a clinical sign referred to as **dysmetria**. Cerebellar patients tend to overestimate or underestimate the amount of force necessary to complete a movement. As a result, they are less able to stop movements, often overshooting targets and rebounding after

BOX B
HUNTINGTON'S DISEASE

In 1872, a physician named George Huntington described a group of patients seen by his father and grandfather in their practice in East Hampton, Long Island. The disease he defined, which became known as Huntington's disease (HD), is characterized by the gradual onset of defects in behavior, cognition, and movement beginning in the fourth and fifth decades of life. The disorder is inexorably progressive, resulting in death within 10 to 20 years. HD is inherited in an autosomal dominant pattern, a feature that has led to a much better understanding of its cause in molecular terms.

One of the more common of the neurodegenerative diseases, HD usually presents as an alteration in mood, especially depression, or a change in character, such as increased irritability, suspiciousness, or impulsive or eccentric behavior. Defects of memory and attention may also occur. The hallmark of the disease, however, is a movement disorder consisting of rapid, jerky motions with no clear purpose; these choreiform movements may be confined to a finger or may involve a whole extremity, the facial musculature, or even the vocal apparatus. The movements themselves are involuntary, but the patient often incorporates them into an apparently deliberate action, presumably in an effort to obscure the problem. There is no paralysis, ataxia, or deficit of sensory function. Occasionally, the disease begins in childhood or adolescence. The clinical manifestations in juveniles include rigidity, seizures, more marked dementia, and a rapidly progressive course. A distinctive neuropathology is associated with these clinical manifestations, namely, a profound but selective atrophy of the caudate and putamen, with some associated degeneration of the frontal and temporal cortices. This pattern of destruction is thought to explain the disorders of movement, cognition, and behavior, as well as the sparing of other neurological functions.

The availability of extensive HD pedigrees has recently allowed molecular geneticists to decipher some aspects of the mechanism of this disease. HD was one of the first human diseases in which DNA polymorphisms were used to localize the mutant gene, which in 1983 was mapped to the short arm of chromosome 4. This discovery led to an intensive effort to identify the HD gene within this region by positional cloning. Ten years later, these efforts culminated in identification of the gene (*IT15* in segment 4p16.3) that is almost certainly responsible for the disease. The HD mutation consists of an unstable DNA segment (a CAG repeat) that codes for glutamine. The *IT15* gene in normal individuals contains between 15 and 34 repeats, whereas the gene in HD patients contains from 42 to over 66 repeats.

HD is one of a growing number of diseases that can be attributed to unstable DNA segments; other examples are fragile X syndrome, myotonic dystrophy, spinal and bulbar muscular atrophy, and spinocerebellar ataxia type 1. Nonetheless, some puzzles remain. The HD gene is expressed predominantly in the expected neurons, but is also expressed in regions of the brain that are not affected in HD. Indeed, the gene is expressed in many organs outside the nervous system.

Identification of the HD gene promises to advance both the diagnosis and management of these patients. Improved methods of detecting affected individuals and establishing diagnoses prenatally are already available, and an intensive effort is underway to elucidate how the defective gene leads to the selective degeneration of neurons in the basal ganglia and particular regions of the cerebral cortex.

References

GUSELLA, J. F. AND 13 OTHERS (1983) A polymorphic DNA marker genetically linked to Huntington's disease. Nature 306: 234–238.

HUNTINGTON, G. (1872) On chorea. Med. Surg. Reporter. 26: 317.

HUNTINGTON'S DISEASE COLLABORATIVE RESEARCH GROUP (1993) A novel gene containing a trinucleotide repeat that is expanded and unstable on Huntington's disease chromosomes. Cell 72: 971–983.

LI, S. H. AND 12 OTHERS (1993) Huntington's disease gene (IT15) is widely expressed in human and rat tissues. Neuron 11: 985–993.

SOTREL, A., P. A. PASKEVICH, D. K. KIELY, E. D. BIRD, R. S. WILLIAMS AND R. H. MYERS (1991) Morphometric analysis of the prefrontal cortex in Huntington's disease. Neurology 41: 1117–1123.

WEXLER, A. (1995) *Mapping Fate: A Memoir of Family, Risk and Genetic Research.* New York: Times Books.

the intended movement is complete. Cerebellar lesions thus result in the inability to perform smooth, directed movements. Rapid alternating movements, fine repetitive movements, and some types of eye movements are also compromised (see Chapter 19).

The functional deficits attending cerebellar lesions are always on the same side of the body as the damage to the cerebellum. This fact reflects the cerebellum's unique status as a brain structure in which sensory and motor information is represented ipsilaterally rather than contralaterally. Furthermore, somatic sensation and vision are represented topographically within the cerebellum; as a result, cerebellar deficits may be quite specific. For example, one of the most common cerebellar syndromes is caused by degeneration in the anterior portion of the cerebellar cortex in patients with a long history of alcohol abuse (Figure 17.3C). Such damage specifically affects movement in the lower limbs, which are represented in the anterior cerebellum. The consequences include a wide and staggering gait, with little impairment of arm or hand movements; speech also tends to remain normal. Thus the topographical organization of the cerebellum allows cerebellar damage to disrupt the coordination of movements performed by specific groups of muscles. The implication of these pathologies is that the cerebellum is normally capable of integrating the moment-to-moment actions of muscles and joints throughout the body to ensure the smooth execution of a full range of motor behaviors.

■ THE STRUCTURAL ELEMENTS OF THE BASAL GANGLIA AND CEREBELLUM

The basal ganglia and cerebellum have several well-defined components that constitute the core circuitry for each system (Figure 17.4; Table 17.1). As noted, the basal ganglia and cerebellum share some key organizational features (see Figure 17.1). Thus, each has a central processing region that sends information to intermediate relay nuclei, which in turn project to thalamic nuclei. The ultimate targets of the thalamic nuclei are the motor and premotor cortices. As a result, both systems modulate and control motor activity that is initiated by the cerebral cortex.

For the basal ganglia, the central processing region is a set of nuclei called the caudate and putamen, located on the floor of the forebrain. The caudate and putamen are large, homogeneous groups of neurons with little or no apparent substructure when examined grossly. The central processing region of the cerebellum is the cerebellar cortex, a continuous sheet of cells folded into the compact structure that one recognizes grossly as the cerebellum.

A series of intermediate relay nuclei process and send information from the caudate and putamen or the cerebellar cortex to the thalamus. For the basal ganglia, the relay nuclei include the internal and external segments of the globus pallidus and the subthalamic nucleus, as well as the substantia nigra (Figure 17.4A). For the cerebellum, the relay nuclei are the various cell groups in the deep cerebellar nuclei (Figure 17.4B). The ultimate target of these pathways is a cluster of thalamic nuclei collectively referred to as the ventral anterior and ventral lateral thalamic complex (VA/VL complex). These nuclei project in turn to primary motor cortex and premotor area of the frontal lobe (see Chapter 15). It is via these intermediate relay nuclei and their thalamic projections that the basal ganglia and cerebellum gain access to the cortical centers that control movement.

Both the basal ganglia and cerebellum receive their principal input from the cerebral cortex, most heavily from association regions of the cortex (Table 17.2). This relationship suggests that much of the information reaching the basal ganglia and cerebellum is already highly processed. Both the basal ganglia and the cerebellum also receive modulatory inputs from nuclei in the brainstem and spinal cord.

TABLE 17.1
Major Components of the Basal Ganglia and Cerebellum

Basal ganglia	Cerebellum
Striatum:	Cerebellar cortex:
Caudate nucleus	Cerebrocerebellum
Putamen	Spinocerebellum
	Vestibulocerebellum
Globus pallidus:	Deep cerebellar nuclei:
External segment	Dentate nucleus
Internal segment	Interposed nuclei
Subthalamic nucleus	Pontine relay nuclei
Substantia nigra	Cerebellar peduncles
Pars compacta	
Pars reticulata	

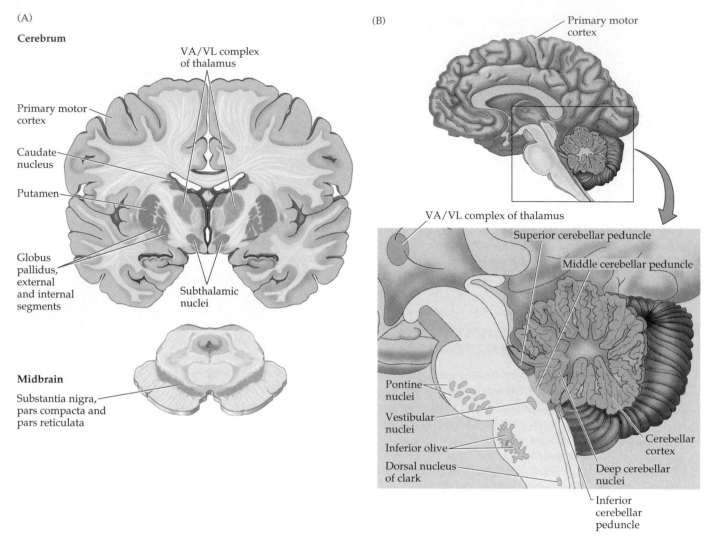

Figure 17.4 Components of the human basal ganglia and cerebellum. (A) Coronal section showing the components of the basal ganglia. Most of these structures are in the telencephalon, although the substantia nigra is in the midbrain and the subthalamic nucleus is in the diencephalon. The ventral anterior and ventral lateral thalamic nuclei (VA/VL complex) are also indicated. These thalamic nuclei are the targets of the basal ganglia, relaying the modulatory effects of the basal ganglia to the cortex. (B) Sagittal section showing the major structures of the cerebellar system (including the cerebellar cortex, the deep cerebellar relay nuclei, and their thalamic target, the VA/VL complex).

◼ PROJECTIONS TO THE BASAL GANGLIA

Most regions of the neocortex project directly to the caudate and putamen; the only exceptions are the primary visual and primary auditory cortices (Figures 17.5 and 17.6; Table 17.2). The heaviest projections are from association areas in the frontal and parietal lobes; temporal, insular, and cingulate cortices also project to the caudate and putamen, but to a lesser extent. The cortical projections run in the internal capsule to reach the caudate and putamen directly.

(A) Cortical areas projecting to caudate/putamen

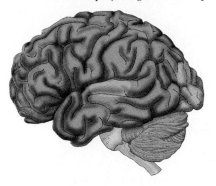
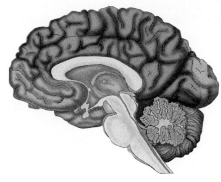

Figure 17.5 Regions of the cerebral cortex that project to the caudate and putamen (shown in blue) and to the cerebellar cortex (shown in green), in both lateral and medial views. (A) The caudate and putamen receive cortical projections from the association areas of the frontal, parietal, and temporal lobes, whereas (B) the cortical projections to the cerebellum are mainly from the sensory association cortex of the parietal lobe and motor association areas of the frontal lobe.

(B) Cortical areas projecting to cerebellum

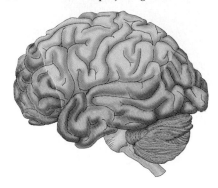
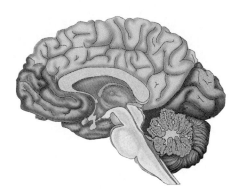

The sources of the cortical inputs to the caudate and putamen indicate significant functional differences between these two nuclei. The caudate receives cortical input primarily from these multimodal association cortical areas (see Chapter 24). As the name implies, these multimodal areas do not process any one type of sensory information; rather, they receive inputs from a number of primary and secondary sensory cortices and associated thalamic nuclei. Based on the origin of the cortical inputs to the caudate nucleus, it seems likely that this component of the striatum processes cognitive information that contributes to the initiation of complex motor acts. In contrast, the putamen receives input from the primary and secondary somatic sensory cortices in the parietal lobe, from the secondary visual cortices in the occipital and temporal lobes, from the premotor and motor cortices in the frontal lobe, and from the auditory association areas in the temporal lobe. The visual and somatic sensory cortical projections are topographically mapped within the putamen. This arrangement suggests that the putamen processes information about the sensory context in which an intended movement must be performed.

Both the caudate and putamen also receive subcortical modulatory inputs (Figure 17.6), mainly from the substantia nigra. Axons from dopaminergic neurons in the **pars compacta** of the substantia nigra (the ventral half of the nucleus, where cell bodies are packed more densely) are distributed widely throughout both the caudate and putamen. Even though much is known about the pharmacology and physiology of these connections (see Chapter 18), the nature of the information provided by the substantia nigra

(A)

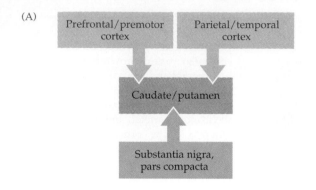

(B) Cerebrum

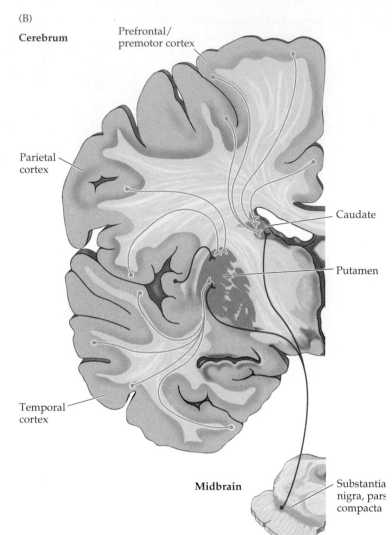

Figure 17.6 Functional organization of the inputs to the basal ganglia. (A) Diagram of the major inputs to the caudate and putamen. (B) An idealized coronal section through the human brain, showing the projections from the cerebral cortex and the substantia nigra to the caudate and putamen.

TABLE 17.2
Major Inputs to the Basal Ganglia and Cerebellum

Basal ganglia inputs[a]	*Cerebellar inputs*[b]
From cerebral cortex:	**From cerebral cortex:**
Parietal cortex (secondary visual, primary and secondary somatic sensory)	Parietal cortex (secondary visual, primary and secondary somatic sensory)
Temporal cortex (secondary visual, primary and secondary auditory)	Cingulate cortex (limbic)
Cingulate cortex (limbic)	Frontal cortex (primary and secondary motor)
Frontal cortex (primary and secondary motor)	
Prefrontal cortex	
Other sources:	**Other sources:**
Substantia nigra, pars compacta	Spinal cord (Clarke's column)
	Vestibular labyrinth and nuclei
	Inferior olivary nucleus
	Locus ceruleus

[a]Via internal capsule
[b]Via inferior and middle cerebellar peduncles

to the caudate and putamen is not clear. The substantia nigra also receives substantial input from both the cerebral cortex and the caudate and putamen. As a result, this structure can provide both feedforward and feedback modulation to the basal ganglia.

■ PROJECTIONS TO THE CEREBELLUM

The cerebellum is divided into three parts, each of which receives a different input: the **cerebrocerebellum** receives input from the cerebral cortex; the **vestibulocerebellum** receives input from the vestibular nuclei in the brainstem; and the **spinocerebellum** receives input from the spinal cord. In humans, the cerebrocerebellum is by far the largest subdivision.

The cortical projections to the cerebrocerebellum are from a somewhat more circumscribed area than those to the basal ganglia (Figures 17.6 and 17.7; see also Table 17.2). The majority originate in the motor and premotor cortices of the frontal lobe, the primary and association somatic sensory cortices of the anterior parietal lobe, and the visual association regions of the

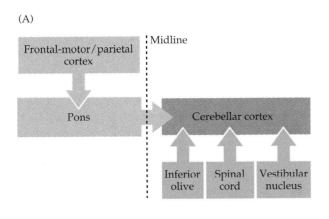

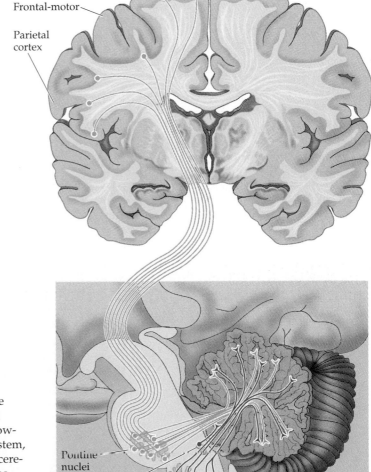

Figure 17.7 Functional organization of the inputs to the cerebellum. (A) Diagram of the major inputs. (B) Idealized coronal and sagittal sections through the human brain, showing inputs to the cerebellum from the cortex, vestibular system, spinal cord, and brainstem. The cortical projections to the cerebellum are made via relay neurons in the pons. These axons then cross the midline within the pons and run to the cerebellum via the middle cerebellar peduncle. Axons from the inferior olive, spinal cord, and vestibular nuclei enter via the inferior cerebellar peduncle.

posterior parietal lobe. The somatic sensory input remains topographically mapped such that there is an orderly representation of the body surface within the cerebellum. The visual input originates mostly in association areas concerned with processing moving visual stimuli—that is, from the cortical targets of the magnocellular stream of the central visual pathway (see Chapter 11). These visual association projections to the cerebellum imply that visually guided coordination of ongoing movement is one of the major tasks carried out by the cerebellum.

In contrast to the basal ganglia, the cortical input to the cerebellum is indirect: cortical axons relay information to the cerebellum via neurons in the pons. These neurons are grouped into clusters called **pontine relay nuclei**, whose axons cross in the pons before projecting to the contralateral cerebellum via the **middle cerebellar peduncle**. The large size of the pons and middle cerebral peduncles in humans indicates the predominance of the cortical input to the cerebellum.

There are also direct sensory inputs to the cerebellum from the brainstem (see Figure 17.7). Vestibular axons from the eighth cranial nerve and axons from the vestibular nuclei in the medulla innervate the vestibular portion of the cerebellar cortex (the vestibulocerebellum). In addition, relay neurons in the dorsal nucleus of Clarke in the spinal cord (a group of spinal cord relay neurons innervated by proprioceptive axons from the periphery) send their axons to the spinal portion of the cerebellar cortex (the spinocerebellum). These vestibular and spinal cord inputs provide the cerebellum with information from the labyrinth in the ear, from muscle spindles, and other mechanoreceptors that monitor the position of the body. The vestibular and spinal cord inputs to the cerebellum remain ipsilateral from their point of entry in the brainstem, running in the **inferior cerebellar peduncle**. Finally, the cerebellum receives modulatory inputs from the inferior olive and the locus ceruleus in the brainstem. These nuclei are thought to participate in the learning and memory functions served by cerebellar circuitry (see Chapters 18 and 19).

■ PROJECTIONS FROM THE BASAL GANGLIA

The output of the basal ganglia must reach the motor cortex to influence motor commands. With few exceptions, the only way to the cortex is via the thalamus. Accordingly, information from the basal ganglia must be relayed through the relevant thalamic nuclei (in this case, the VA/VL complex). The caudate and putamen, however, do not project directly to the thalamus. Rather, it is the intermediate nuclei in the globus pallidus that provide the major output of the basal ganglia (Figure 17.8; Table 17.3).

The globus pallidus is divided into internal and external divisions, each with a distinct pattern of connections. The internal segment of the globus pallidus receives input from the caudate and putamen and sends a projection directly to the thalamus (specifically, the oral and medial subdivisions of the VA/VL complex). Accordingly, this circuit is known as the **direct pathway**. The **indirect pathway** from the basal ganglia involves the external segment of the globus pallidus and an additional structure, the subthalamic nucleus. The external segment also receives input from the caudate and putamen, but projects to the subthalamic nucleus. The subthalamic nucleus then sends axons back to the internal segment of the globus pallidus, which in turn projects to the thalamus. There is one other pathway from the caudate and putamen, namely, a sizable output to the substantia nigra. This pathway projects to the **pars reticulata** of the substantia nigra (where cells are spaced farther apart in

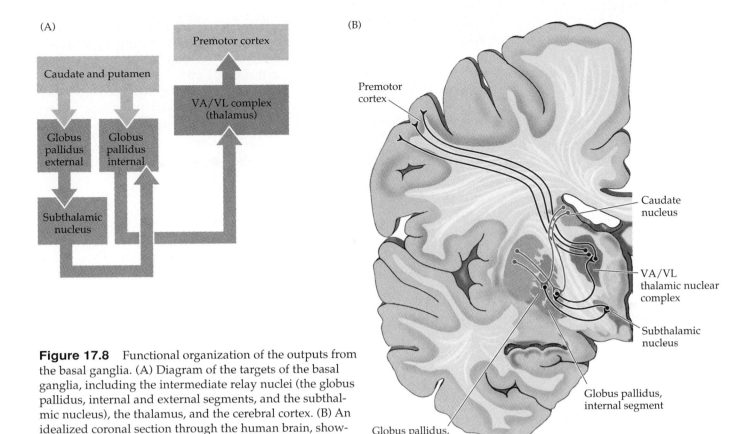

Figure 17.8 Functional organization of the outputs from the basal ganglia. (A) Diagram of the targets of the basal ganglia, including the intermediate relay nuclei (the globus pallidus, internal and external segments, and the subthalamic nucleus), the thalamus, and the cerebral cortex. (B) An idealized coronal section through the human brain, showing the structures and pathways diagrammed in (A).

a reticulum, or meshwork), which in turn sends axons to either the thalamus or to the superior colliculus (see Chapters 18 and 19).

The neurons of the internal segment of the globus pallidus send their axons primarily to the portion of the ventral lateral (VL) nuclear complex that projects to the premotor area and other association regions of the frontal lobe

TABLE 17.3 Outputs of the Basal Ganglia and Cerebellum		
	Outputs of the caudate and putamen	*Outputs of the cerebellar cortex*

Direct targets	Globus pallidus:
	    Internal division
	    External division
	Subthalamic nucleus
	Substantia nigra, pars reticulata
		Deep cerebellar nuclei:
		    Dentate nucleus
		    Interposed nuclei
		Red nucleus
		Vestibular nuclei
		Spinal cord

Thalamic targets[a]	Ventral anterior nucleus/
	    ventral lateral nuclear
	    complex: oral and medial
	    subdivisions
	Mediodorsal nucleus
	Centromedian nucleus
		Ventral lateral nuclear complex:
		    oral posterolateral subdivision
		    and area X

[a]Via intermediate relay nuclei

anterior to the primary motor cortex, completing a modulatory loop. Thus, the basal ganglia influence the primary motor cortex indirectly, by modulating activity in motor association areas. The globus pallidus also sends axons to the mediodorsal and centromedian thalamic nuclei. These nuclei innervate the prefrontal and association cortices, allowing the basal ganglia to further influence motor commands via cortico-cortical connections between these association cortices, the premotor, and the primary motor cortex.

■ PROJECTIONS FROM THE CEREBELLUM

The cerebellar cortex, like the caudate and putamen, projects to an intermediate relay station—the deep cerebellar nuclei—that in turn projects to the thalamus (Figure 17.9; Table 17.3). There are four major deep nuclei: the den-

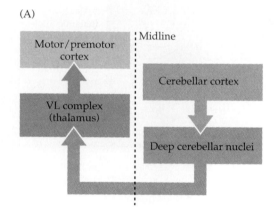

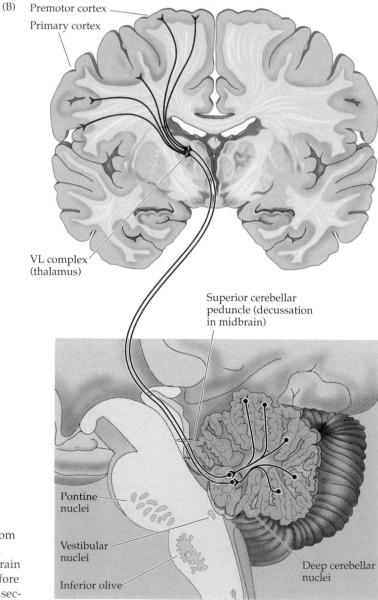

Figure 17.9 Functional organization of the outputs from the cerebellum. (A) Diagram of targets of the cerebellum. The axons of the deep cerebellar nuclei cross in the midbrain in the decussation of the superior cerebellar peduncle before reaching the thalamus. (B) Idealized coronal and sagittal sections through the human brainstem, showing the location of the structures and pathways diagrammed in (A).

tate nucleus (by far the largest), and a group called the interposed nuclei—the emboliform nucleus, the globose nucleus, and the fastigial nucleus. Each sends axons to the posterior portion of the ventral lateral complex in the thalamus. These outputs from the deep nuclei must cross the midline if the motor cortex in each hemisphere is to receive information from the cerebellum about the appropriate side of the body. Thus, the deep nuclear axons exit the cerebellum via the **superior cerebellar peduncle**, cross at the decussation of the superior cerebellar peduncle in the midbrain, and then ascend to the thalamus. The thalamic projections from the deep cerebellar nuclei are segregated from those of the basal ganglia and terminate in two distinct subdivisions of the ventral lateral nuclear complex: the oral, or anterior, part of the posterolateral segment, and a region simply called area X. Both of these thalamic centers project directly to primary motor and premotor association cortices. Thus, the cerebellum has access to the cortical-spinal projections that organize the sequence of muscular contractions in complex movements.

■ SUMMARY

The contribution of the basal ganglia and cerebellum to motor control is apparent from the deficits that result from damage to either system, as well as from their structural and functional organization. Lesions of the basal ganglia compromise the initiation and performance of programs of voluntary movement, as exemplified by the paucity of movement in Parkinson's disease and the inappropriate release of entire motor programs in Huntington's disease. In contrast, cerebellar lesions tend to disrupt the modulation and coordination of ongoing movements. Patients with cerebellar damage have difficulty producing well-coordinated movements. Instead, movements tend to be jerky and imprecise. The organization of the basal ganglia and cerebellum indicates how these systems modulate movement. Complex sensory information about the context and status of movements reaches both the basal ganglia and cerebellum from a number of sensory, motor, and association cortical areas. Not surprisingly, then, a wide variety of sensory and cognitive information can influence motor performance. The basal ganglia and cerebellum each comprise several interconnected structures that integrate, process, and relay such information. Both systems ultimately send a major projection to the thalamus, thereby influencing the motor commands generated by the cortex.

Additional Reading

Reviews

ALEXANDER, G. E. AND M. D. CRUTCHER (1990) Functional architecture of basal ganglia circuits: neural substrates of parallel processing. Trends Neurosci. 13: 266–271.

ALLEN, G. AND N. TSUKAHARA (1974) Cerebrocerebellar communication systems. Physiol. Rev. 54: 957–1006.

GRAYBIEL, A. M. AND C. W. RAGSDALE (1983) Biochemical anatomy of the striatum, in *Chemical Neuroanatomy*. P. C. Emson (ed.). New York: Raven Press, 427–504.

MINK, J. W. AND W. T. THACH (1993) Basal ganglia intrinsic circuits and their role in behavior. Curr. Opin. Neurobiol. 3: 950–957.

THACH, W. T., H. P. GOODKIN AND J. G. KEATING (1992) The cerebellum and adaptive coordination of movement. Ann. Rev. Neurosci. 15: 403–442.

Important Original Papers

ANDEN, N.-E., A. DAHLSTROM, K. FUXE, K. LARSSON, K. OLSON AND U. UNGERSTEDT (1966) Ascending monoamine neurons to the telencephalon and diencephalon. Acta Physiol. Scand. 67: 313–326.

ASANUMA, C., W. T. THACH AND E. G. JONES (1983) Distribution of cerebellar terminals and their relation to other afferent terminations in the ventral lateral thalamic region of the monkey. Brain Res. Rev. 5: 237–265.

BRODAL, P. (1978) The corticopontine projection in the rhesus monkey: origin and principles of organization. Brain 101: 251–283.

KEMP, J. M. AND T. P. S. POWELL (1970) The cortico-striate projection in the monkey. Brain 93: 525–546.

KIM, R., K. NAKANO, A. JAYARAMAN AND M. B. CARPENTER (1976) Projections of the globus pallidus and adjacent structures: an autoradiographic study in the monkey. J. Comp. Neurol. 169: 217–228.

VICTOR, M., R. D. ADAMS, AND E. L. MANCALL (1959) A restricted form of cerebellar cortical degeneration occurring in alcoholic patients. Arch. Neurol. 1: 579–688.

Books

BRADLEY, W. G., R. B. DAROFF, G. M. FENICHEL AND C. D. MARSDEN (eds.) (1991) *Neurology in Clinical Practice*. Boston: Butterworth-Heinemann, Chapters 29 and 77.

KLAWANS, H. L. (1989) *Toscanini's Fumble and Other Tales of Clinical Neurology*. New York: Bantam, Chapters 7 and 10.

■ OVERVIEW

The inputs and outputs of the basal ganglia and cerebellum, as well as the clinical syndromes that result from damage to these structures, imply that these two systems coordinate behavior by integrating sensory and cognitive information about the intention, context, and status of movements. The processing capabilities of both systems are defined by the types of neurons they contain, the neurotransmitters used by their constituent cells, and their synaptic relationships. Both the basal ganglia and the cerebellum are notable for the great convergence of inputs in their central processing regions. Thus, individual projection neurons in the caudate and putamen and in the cerebellar cortex are innervated by multiple cortical sources, as well as several classes of neurons from a variety of subcortical centers. The organization of these connections produces characteristic patterns of neuronal activity in each structure that modulate specific aspects of movement. In general, signals generated in the basal ganglia precede the onset of motor programs, whereas those generated in the cerebellum coincide with the execution of particular movements. The basal ganglia and cerebellum are also sites for activity-dependent synaptic modification in response to experience. Evidently these structures are important not only for the planning and successful execution of movement, but for learning and remembering complex motor tasks.

■ NEURONS AND CIRCUITS IN THE BASAL GANGLIA AND CEREBELLUM

The caudate and putamen and the cerebellar cortex have principal projection neurons that send their axons to relatively distant targets (Figures 18.1 and 18.2). In the caudate and putamen, the principal neurons are called **medium spiny neurons**; in the cerebellar cortex they are the **Purkinje cells**. Both types of neurons process information derived primarily from the cerebral cortex. Cortical axons synapse directly on medium spiny neurons, whereas cortical input to Purkinje cells is relayed via the pontine nuclei and then through local circuit neurons called **granule cells**. In addition, several other classes of local circuit neurons (Table 18.1) directly influence the activity of medium spiny neurons or Purkinje cells, as do several cell groups in the brainstem. The targets of the medium spiny neurons are neurons in the globus pallidus and substantia nigra; the targets of the Purkinje cells are the neurons of the deep cerebellar nuclei. The medium spiny neuron and the Purkinje cell therefore act as the final integrators and conveyors of the complex information processed in the caudate and putamen and the cerebellar cortex.

Given the degree of convergence upon them, it is not surprising that medium spiny neurons and Purkinje cells are specialized to accommodate a large number of diverse inputs. Both have highly branched dendritic arborizations studded with thousands of spines that receive synaptic contacts. The cortical input to medium spiny neurons is made directly, via excitatory synapses onto dendritic spines (see Figure 18.1B). The modulatory input from interneurons and most of the brainstem aminergic nuclei is made primarily on dendritic shafts. An important exception to this arrangement is the loca-

(A)

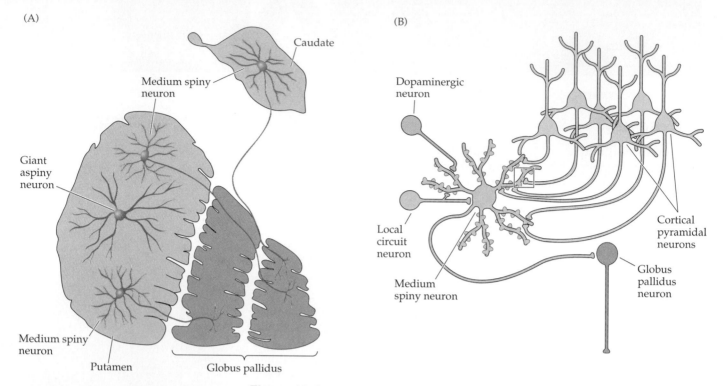

(C)

Figure 18.1 Neurons and circuits of the basal ganglia. (A) Neuronal types in the caudate and putamen. (B) Diagram showing convergent inputs onto the medium spiny neuron from cortical neurons, dopaminergic cells of the substantia nigra, and local circuit neurons. The primary output of the medium spiny cells is to the globus pallidus [boxed region shown at higher magnification in (C)]. (C) Electron micrograph of an immunocytochemically labeled medium spiny neuron dendrite and dendritic spine receiving a synaptic input. (C courtesy of A.-S. LaMantia and C. Ouimet.)

tion of synapses from the substantia nigra on the medium spiny neurons. The contacts from these dopaminergic neurons are also made onto spines, usually in close proximity to the cortical synapses. This distinctive geometry of inputs on the medium spiny neuron places cortical and dopaminergic synapses relatively far from the cell body, where individual inputs are less likely to fire the cell; other modulatory inputs occur closer to the cell body, where they can influence the effectiveness of cortical or dopaminergic synaptic activation.

The organization of inputs on the cerebellar projection neurons is similar to that on the medium spiny neurons (Figure 18.2). Thus, there is a massive convergence of information from the cerebral cortex onto the dendritic spines of Purkinje cells. The activity of these inputs is modulated by other inputs onto dendritic shafts and the cell body. In the cerebellar circuit, however, the cortical input is indirect. Neurons in the pontine nuclei receive the cortical projection directly, and then send the information to the cerebellar cortex (see Chapter 17). The axons from the pontine nuclei synapse upon granule cells in the granule cell layer of the cerebellar cortex (Figure 18.2B) The synaptic endings of the pontine axons (as well as other inputs to granule cells) are called **mossy fibers**. Cerebellar granule cells—the most abundant class of neurons in the human brain—give rise to specialized axons called **parallel fibers** that run in the **molecular layer** of the cerebellar cortex. These axons bifurcate to form T-shaped branches that relay cortical information via excitatory synapses made on the dendritic spines of Purkinje cells. In con-

(A)

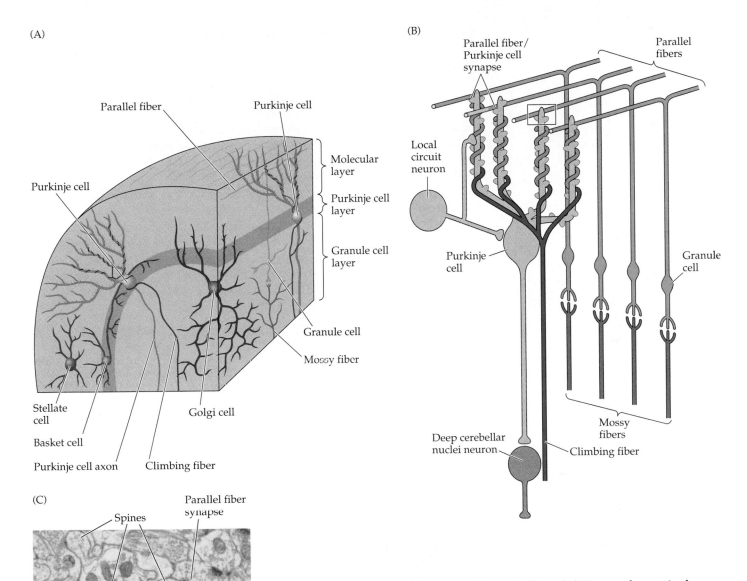

Parallel fiber

Purkinje cell

Molecular
layer

Purkinje cell
layer

Granule cell
layer

Purkinje cell

Granule cell

Mossy fiber

Stellate
cell

Golgi cell

Basket cell

Purkinje cell axon

Climbing fiber

(B)

Parallel fiber/
Purkinje cell
synapse

Parallel
fibers

Local
circuit
neuron

Purkinje
cell

Granule
cell

Deep cerebellar
nuclei neuron

Mossy
fibers

Climbing fiber

(C)

Spines

Parallel fiber
synapse

Purkinje cell dendrite

Figure 18.2 Neurons and circuits of the cerebellum. (A) Neuronal types in the cerebellar cortex. Note that the various neuron classes are found in distinct layers. (B) Diagram showing convergent inputs onto the Purkinje cell from parallel fibers and local circuit neurons [boxed region shown at higher magnification in (C)]. The output of the Purkinje cells is to the deep cerebellar nuclei. (C) Electron micrograph showing Purkinje cell dendritic shaft with three spines contacted by synapses from a trio of parallel fibers. (C courtesy of A.-S. La Mantia and P. Rakic.)

Table 18.1		
Neurons of the Caudate and Putamen and the Cerebellar Cortex		
	Caudate and putamen	*Cerebellar cortex*
Principal neurons	Medium spiny neurons	Purkinje cells
Local circuit neurons	Medium aspiny cells Giant aspiny cells	Granule cells Basket cells Golgi cells Stellate cells

trast, modulatory inputs to Purkinje cells are found on the dendritic shaft or the cell body. Particularly striking is the inhibitory complex of synapses made around the Purkinje cell body by the **basket cells** and the wealth of excitatory synapses made by **climbing fibers** from the inferior olive on the shafts of the highly branched Purkinje cell dendrites.

The functional significance of the convergence of cortical information onto medium spiny neurons in the basal ganglia and Purkinje cells in the cerebellum can be appreciated by thinking of these principal projection neurons as detectors of temporal and spatial coincidence. Both cell classes integrate a continual barrage of excitatory and inhibitory impulses, firing only when an appropriate pattern of signals is present among their inputs. Since movements occur in time and space, any information that influences movement must be integrated along both of these dimensions. The integration of signals in medium spiny neurons or Purkinje cells must provide information about when to move, where to move, and in what sequence. Evidently, this information is derived from coincident patterns of activity among the inputs that impinge on the principal neurons of the basal ganglia and cerebellum.

■ DISINHIBITION: DIRECT AND INDIRECT PATHWAYS THROUGH THE BASAL GANGLIA

An important concept in understanding how basal ganglia circuitry influences the motor cortex is **disinhibition**. In a simple disinhibitory circuit, a transiently active inhibitory neuron synapses on a tonically active inhibitory

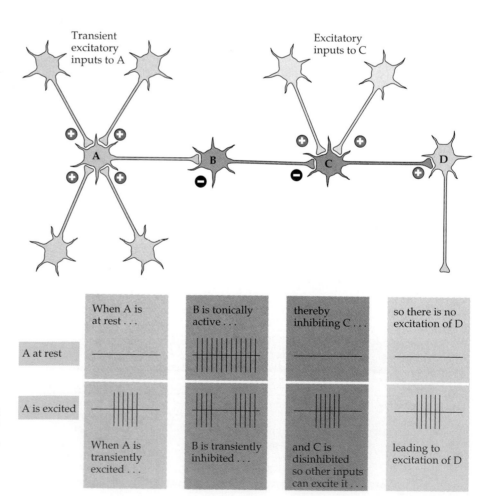

Figure 18.3   A chain of nerve cells arranged in a disinhibitory circuit. Top: Diagram of the connections between two inhibitory neurons, A and B, and an excitatory neuron, C. Bottom: Pattern of the action potential activity of cells A, B, and C when A is at rest, and when neuron A fires transiently as a result of its excitatory inputs.

neuron that then synapses on an excitatory neuron (Figure 18.3). When the tonic inhibition generated by the circuit is interrupted by the transient inhibitory activity of the first neuron in the chain, other inputs can excite the target cell and cause it to fire. In this way, two inhibitory neurons can actually facilitate the transient excitation of a third target neuron.

Such disinhibition occurs in one of the major functional circuits in the basal ganglia, the so-called **direct pathway** (Figure 18.4A). The transiently active inhibitory neuron is the medium spiny neuron in the caudate and putamen; the tonically active inhibitory neuron is in the globus pallidus; and the transiently excited downstream neuron is in the VA/VL complex of the thalamus. The neurotransmitters used by each of these cells, as well as the pattern of electrical activity recorded from them, support this interpretation. The transmitter used by medium spiny neurons in the caudate and putamen

(A) Direct pathway

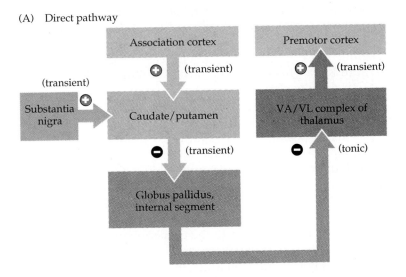

(B) Indirect pathway

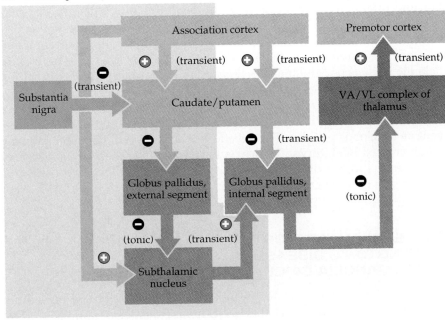

Figure 18.4 Disinhibition in the direct and indirect pathways through the basal ganglia. (A) In the direct pathway, transiently inhibitory projections from the caudate and putamen project to tonically inhibitory neurons in the *internal* segment of the globus pallidus, which project in turn to the VA/VL complex of the thalamus. Transiently excitatory inputs to the caudate and putamen from the cortex and substantia nigra are also shown, as is the transiently excitatory input from the thalamus back to the cortex. (B) In the indirect pathway (tan box), transiently inhibitory neurons from the caudate and putamen project to tonically inhibitory neurons of the *external* segment of the globus pallidus. Note that the influence of nigral dopaminergic input to neurons in the indirect pathway is inhibitory. This is due to a particular class of dopamine receptors—the D_1 receptors— found on these neurons. The globus pallidus (external segment) neurons project to the subthalamic nucleus, which also receives a strong excitatory input from the cortex. The subthalamic nucleus in turn projects to the globus pallidus (internal segment), where its transiently excitatory drive acts to oppose the disinhibitory action of the direct pathway.

Figure 18.5 Example of activity in a medium spiny neuron in the putamen before, during, and after a movement. In this case, a monkey was trained to move a lever forward (A) or backward (B). The upper traces show the position of the lever (the lever is moving when this line slopes upward or downward); the lower traces show the activity of the putamen cell (the duration of movement is indicated by the color block). Rasters below indicate multiple trials. Just before the movement begins, the cell fires transiently, consistent with its role in the disinhibitory circuitry of the basal ganglia. The cell stops firing before the movement is over, suggesting that sustained activity of the neuron is not necessary to complete the movement. (After DeLong and Strick, 1974.)

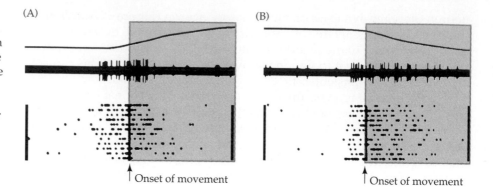

(A)

(B)

↑ Onset of movement

↑ Onset of movement

is GABA, the major inhibitory neurotransmitter of the central nervous system (see Chapter 6). The medium spiny neurons have little or no spontaneous activity, but are transiently activated by cortical innervation. The cortical activity, and consequently activity in the medium spiny neurons, usually begins before the onset of movement and ceases once the movement is underway (Figure 18.5). This pattern is consistent with the idea that the circuitry of the basal ganglia contributes to the initiation of movement, but not to its ongoing coordination. Neurons in the globus pallidus also use GABA as their neurotransmitter; they are, however, tonically active—consistent with their postulated position in the disinhibitory circuit. Accordingly, they continually inhibit the target neurons in the VA/VL complex of the thalamus. When pallidal neurons are transiently inhibited by the activity of the medium spiny neurons, the VA/VL thalamic neurons can be excited by other inputs (primarily from various cortical regions), stimulating the generation of a signal back to the premotor cortices. As a result of this circuitry, activity ceases in pallidal neurons prior to the execution of complex movements and resumes once the movement is underway (see also Figure 18.10).

The **indirect pathway** through the basal ganglia antagonizes the direct pathway. The globus pallidus is divided into an **internal segment**, whose neurons project directly to the thalamus (the direct pathway just described) and an **external segment,** whose tonically inhibitory neurons receive caudate and putamen input and project to the **subthalamic nucleus**. The subthalamic nucleus in turn projects to the internal segment of the globus pallidus, which allows the indirect pathway from the caudate and putamen to influence the activity of the thalamus (Figure 18.4B). The neurons of the subthalamic nucleus use glutamate as their primary neurotransmitter, providing additional excitation to the tonically inhibitory neurons in the internal segment of the globus pallidus. Accordingly, the subthalamic nucleus can increase the tonic inhibition of the thalamus arising from the internal segment of the globus pallidus. The net effect of activity in the indirect pathway is an increased inhibitory influence on the thalamus (rather than a release of inhibition, as in the direct pathway). Consequently, these two pathways oppose each other in controlling the excitatory output of the VA/VL complex to the motor and premotor cortices.

■ AN EXPLANATION OF PARKINSON'S AND HUNTINGTON'S DISEASES IN TERMS OF BASAL GANGLIA CIRCUITRY

Knowledge about the direct and indirect pathways is helpful in explaining the deficits seen in disorders that compromise the basal ganglia, such as

Parkinson's and Huntington's diseases (see Boxes A and B in Chapter 17). Parkinson's disease is caused by the loss of dopaminergic neurons in the substantia nigra. The major effect of the nigral input on the caudate and putamen is the excitation of the medium spiny neurons, which is mediated by the D_2 class of dopaminergic receptors on these cells. The dopaminergic input from the substantia nigra presumably reinforces or amplifies cortical signals that activate the direct pathway. Consequently, when the influence of the dopaminergic input on the direct pathway is diminished (Figure 18.6A), excitation of medium spiny neurons in the direct pathway is diminished, and thalamic activation of the motor cortex is less likely to occur. In this model, the paucity of movement seen in Parkinson's disease (and in other hypokinetic movement disorders) reflects a failure of disinhibition in the basal ganglia.

(A) Parkinson's disease (hypokinetic)

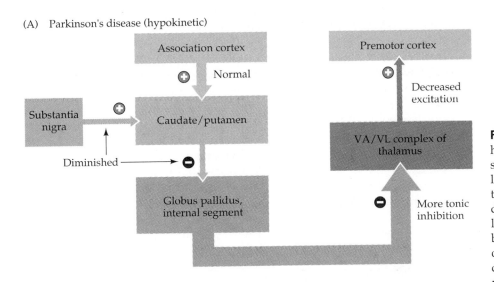

(B) Huntington's disease (hyperkinetic)

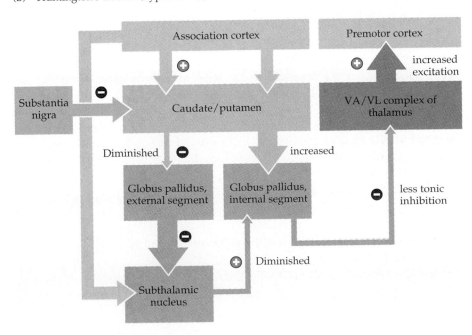

Figure 18.6 Cellular explanation of hypokinetic disorders such as Parkinson's disease and hyperkinetic disorders like Huntington's disease. In both cases, the balance of inhibitory signals in the direct and indirect pathways is altered, leading to a diminished ability of the basal ganglia to control the thalamic output to the cortex. (A) In Parkinson's disease, the excitatory input provided by the substantia nigra is diminished (thinner arrow), making it more difficult to generate the transient inhibition from the caudate and putamen. The result of this change in the direct pathway is to sustain the tonic inhibition from the globus pallidus (internal segment) to the thalamus, making thalamic excitation of the motor cortex less likely (thinner arrow from thalamus to cortex). (B) In hyperkinetic diseases such as Huntington's, the projection from the caudate and putamen to the globus pallidus (external segment) is diminished (thinner arrow). This effect increases the tonic inhibition from the globus pallidus to the subthalamic nucleus (larger arrow), making the excitatory subthalamic nucleus less effective in opposing the action of the direct pathway (thinner arrow). Thus, thalamic excitation of the cortex is increased (larger arrow), leading to greater and often inappropriate motor activity. (After DeLong, 1991.)

Similarly, knowledge of the indirect pathway in the basal ganglia helps explain the motor deficits seen in Huntington's disease. In patients with Huntington's, the medium spiny neurons that project to the external segment of the globus pallidus degenerate. The net effect of this neuronal loss is a decrease in the excitatory output of the subthalamic nucleus to the internal segment of the globus pallidus (Figure 18.6B). Consequently, the direct pathway is abnormally efficient in disinhibiting the thalamus. The unopposed disinhibition increases the probability that inappropriate signals will reach the motor cortex, resulting in the ballistic and choreic movements that characterize Huntington's disease.

■ CEREBELLAR CIRCUITRY AND THE COORDINATION OF ONGOING MOVEMENT

In contrast to the basal ganglia, neuronal activity in the cerebellum changes continually during the course of a movement. For instance, the execution of a relatively simple task like flipping the wrist back and forth elicits a dynamic pattern of activity in both the Purkinje cells and the deep cerebellar nuclear cells that closely follows the ongoing movement (Figure 18.7; see also Figure 18.5). Both types of cells are tonically active at rest and change their frequency of firing as movements occur. Neurons in the cerebellum respond selectively to various aspects of movement, including extension or contraction of specific muscles, the position of the joints, and the direction of the next movement that will occur. All this information is evidently encoded by changes in the firing frequency of Purkinje cells and deep cerebellar nuclear cells.

Another important function of the cerebellum is to correct errors in ongoing movements. Error correction ensures that movements are modified

(A) PURKINJE CELL

At rest

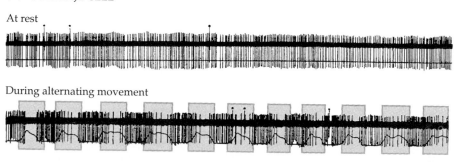

During alternating movement

Figure 18.7 Activity of Purkinje cells (A) and deep cerebellar nuclear cells (B) at rest (upper traces) and during movement of the wrist (lower traces). The lines below the action potential records show changes in muscle tension recorded by electromyography, which therefore reflects movement of the wrist. The duration of the wrist movements is indicated by the colored blocks. Both classes of cells are tonically active at rest. Rapid alternating movements result in the transient inhibition of the tonic activity of both cell types. (After Thach, 1968.)

(B) DEEP NUCLEAR CELL

At rest

During alternating movement

to cope with unanticipated circumstances. The Purkinje cells and the deep cerebellar nuclear cells recognize potential errors by comparing patterns of convergent activity that are concurrently available to both cell types; the deep nuclear cells then generate corrective signals to maintain the accuracy of the movement. A key feature of this circuitry is that both cell groups receive basically the same excitatory input from mossy fibers (a collective term for axons from pontine relay neurons, vestibular neurons, and spinal cord neurons) and climbing fibers (Figure 18.8). These excitatory inputs modulate the ongoing pattern of tonic activity in the cerebellar output. Each Purkinje cell samples a much broader array of mossy fiber inputs than the corresponding deep nuclear cell. The "comparisons" that take place are evidently mediated by the connections between the Purkinje cells and the deep nuclear cells. The local circuit neurons of the cerebellar cortex (the granule cells, basket cells, stellate cells, and Golgi cells) also contribute to the information converging on the Purkinje cells. The patterns of connectivity among these cells are essential for normal cellular function (Box A). Changes in the

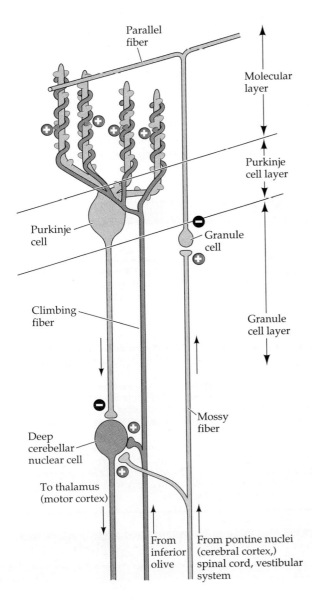

Figure 18.8 Excitatory and inhibitory connections in the cerebellar cortex and deep cerebellar nuclei. The excitatory input from mossy fibers and climbing fibers to Purkinje cells and deep nuclear cells is basically the same. Additional convergent input onto the Purkinje cell from local circuit neurons and other Purkinje cells establishes a basis for the comparison of ongoing movement and sensory feedback derived from it. The Purkinje cell output to the deep cerebellar nuclear cell thus generates an error correction signal that can modify movements already begun. (After Stein, 1986.)

Box A
GENETIC ANALYSIS OF CEREBELLAR FUNCTION

Since the early 1950s, investigators interested in motor behavior have identified and studied strains of mutant mice in which movement is compromised. These mutant mice were easy to spot: the "screen" following induced or spontaneous mutagenesis was simply to look for animals that had difficulty moving. Genetic analysis suggested that some of these abnormal behaviors could be explained by single autosomal recessive or semidominant mutations, in which homozygotes are most severely affected. The strains were given names like *reeler*, *weaver*, *lurcher*, *staggerer*, and *leaner* that reflected the nature of the motor dysfunction they exhibited (see table). The relatively large number of mutations that compromise movement suggested it might be possible to understand motor circuits and function at the genetic level.

A common feature of the mutants was ataxia resembling that associated with cerebellar dysfunction in humans. Indeed, all the mutations were associated with some form of cerebellar pathology. The pathologies associated with the *reeler* and *weaver* mutations were particularly striking. In the *reeler* cerebellum, Purkinje cells, granule cells, and interneurons are all displaced from their usual laminar positions, and there are fewer granule cells than normally. In *weaver*, most of the granule cells are lost prior to their migration from the external granule layer (a proliferative region where cerebellar granule cells are generated during development), leaving only Purkinje cells and interneurons to carry on the work of the cerebellum. Thus, these mutations causing deficits in motor behavior impair the development and final disposition of the neurons that comprise the major processing circuits of the cerebellum.

Efforts to characterize the cellular mechanisms underlying these motor deficits were unsuccessful and the molecular identity of the affected genes remained obscure until recently. In the past few years, however, both the *reeler* and *weaver* genes have been identified and cloned.

The *reeler* gene was cloned through a combination of good luck and careful observation. In the course of making transgenic mice by inserting DNA fragments in the mouse genome, investigators in Tom Curran's laboratory created a new strain of mice that behaved much like *reeler* mice and had similar cerebellar pathology. This "synthetic" *reeler* mutation was identified by finding the position of the novel DNA fragment—which turned out to be on

		Motor Mutations in Mice	
Mutation	*Inheritance*	*Chromosome affected*	*Behavioral and morphological characteristics*
reeler (rl)	Autosomal recessive	5	Reeling ataxia of gait, dystonic postures, and tremors. Systematic malposition of neuron classes in the forebrain and cerebellum. Small cerebellum, reduced number of granule cells.
weaver (wv)	Autosomal recessive	?	Ataxia, hypotonia, and tremor. Cerebellar cortex reduced in volume. Most cells of external granular layer degenerate prior to migration.
leaner (tg^{1a})	Autosomal recessive	8	Ataxia and hypotonia. Degeneration of granule cells, particularly in the anterior and nodular lobes of the cerebellum. Degeneration of a few Purkinje cells.
lurcher (lr)	Autosomal semidominant	6	Homozygote dies. Heterozygote is ataxic with hesitant, lurching gait and has seizures. Cerebellum half normal size; Purkinje cells degenerate; granule cells reduced in number.
nervous (nr)	Autosomal recessive	8	Hyperactivity and ataxia. Ninety percent of Purkinje cells die between 3 and 6 weeks of age.
Purkinje cell degeneration (pcd)	Autosomal recessive	13	Moderate ataxia. All Purkinje cells degenerate between the fifteenth embryonic day and third month of age.
staggerer (sg)	Autosomal recessive	9	Ataxia with tremors. Dendritic arbors of Purkinje cells are simple (few spines). No synapses of Purkinje cells with parallel fibers. Granule cells eventually degenerate.

(Adapted from Caviness and Rakic, 1978.)

the same chromosome as the original *reeler* mutation. Further analysis showed that the same gene had indeed been mutated, and the *reeler* gene was subsequently identified. Remarkably, the protein encoded by this gene is homologous to known extracellular matrix proteins such as tenascin, laminin, and fibronectin (see Chapter 21). This finding makes good sense, since the pathophysiology of the *reeler* mutation entails altered cell migration, resulting in misplaced neurons in the cerebellar cortex as well as the cerebral cortex and hippocampus.

Molecular genetic techniques have also led to cloning the *weaver* gene. Using linkage analysis and the ability to clone and sequence large pieces of mammalian chromosomes, Andy Peterson and his colleagues "walked" (i.e., sequentially cloned) over several kilobases of DNA in the chromosomal region to find where the *weaver* gene mapped. By comparing normal and mutant sequences within this region, they determined *weaver* to be a mutation in a K$^+$ channel that resembles the Ca^{2+}-activated K$^+$ channels found in cardiac muscle. How this particular molecule influences the development of granule cells or causes their death in the mutants is not yet clear.

The story of the proteins encoded by the *reeler* and *weaver* genes indicates both the promise and the challenges of a genetic approach to understanding cerebellar function. Identifying motor mutants and their pathology is reasonably straightforward, but understanding their molecular genetic basis depends on hard work and good luck.

References

CAVINESS, V. S. JR. AND P. RAKIC (1978) Mechanisms of cortical development: A view from mutations in mice. Annu. Rev. Neurosci. 1: 297–326.

D'ARCANGELO, G., G. G. MIAO, S. C. CHEN, H. D. SOARES, J. I. MORGAN AND T. CURRAN (1995) A protein related to extracellular matrix proteins deleted in the mouse mutation *reeler*. Nature 374: 719–723.

PATIL, N., D. R. COX, D. BHAT, M. FAHAM, R. M. MEYERS AND A. PETERSON (1995) A potassium channel mutation in *weaver* mice implicates membrane excitability in granule cell differentiation. Nature Genetics 11: 126–129.

RAKIC, P. AND V. S. CAVINESS JR. (1995) Cortical development: A view from neurological mutants two decades later. Neuron 14: 1101–1104.

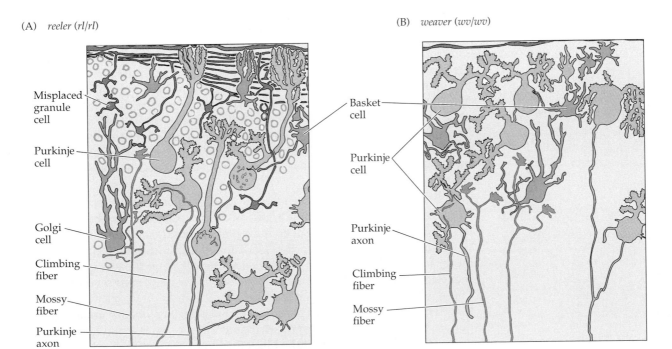

(A) *reeler (rl/rl)* (B) *weaver (wv/wv)*

Misplaced granule cell — Purkinje cell — Golgi cell — Climbing fiber — Mossy fiber — Purkinje axon

Basket cell — Purkinje cell — Purkinje axon — Climbing fiber — Mossy fiber

The cerebellar cortex is disrupted in both the *reeler* and *weaver* mutations. (A) The *reeler* mutation causes the major cell types of the cerebellar cortex to be displaced from their normal laminar positions. Despite the disorganization of the cerebellar cortex in reeler mutants, the major inputs—mossy fibers and climbing fibers—find appropriate targets. (B) The cerebellar cortex in homozygous *weaver* mice. The granule cells are missing, and the major cerebellar inputs synapse inappropriately on the remaining neurons. (After Rakic, 1977.)

ongoing pattern of tonic inhibition from the Purkinje cells evidently modify the excitatory output of the deep cerebellar nuclei to reflect this integrated information about the broader context of any particular movement. Such modulation provides corrective signals to the motor cortex that promote coordinated motor behavior.

■ MOTOR LEARNING AND MEMORY IN THE BASAL GANGLIA AND CEREBELLUM

The circuitry of the basal ganglia and cerebellum was long thought of as feedback loops that contributed little, if anything, to cognitive function. Such "higher order" tasks were considered the exclusive province of the cerebral cortex and hippocampus (see Chapters 24 and 29). Recent work in rats, rabbits, and monkeys, however, together with a better understanding of the wide range of cognitive deficits that results from basal ganglia and cerebellar disorders in humans, has led to a revision of this view. The basal ganglia and cerebellum are now regarded as major players in the acquisition and storage of learned movements and complex behaviors.

The role of the cerebellum in acquiring and storing motor skills has been examined primarily in the context of the modification of two reflexes: the **blink reflex** in rabbits and the **vestibulo-ocular reflex** (**VOR**) in monkeys and humans. (The VOR keeps the eyes trained on a visual target during head movements; see Chapter 19). The relative simplicity of these reflexes has made it possible to analyze some of the mechanisms that enable motor learning.

Conditioning is a basic form of learning in which a reflex elicited by a stimulus becomes associated with another stimulus that would not normally call forth the response. Rabbits blink in reaction to a puff of air directed at the eye (actually, the rabbit moves a specialized lid structure called the nictitating membrane). If a tone (the conditioned, or normally ineffective, stimulus) precedes the puff of air (the unconditioned, or normally effective, stimulus), rabbits eventually begin to blink in response to the tone alone—they have learned to associate the tone, which has nothing to do with blinking, with the puff of air. Once conditioning is complete, the association—that is, the ability to respond to the conditioned stimulus—is stored somewhere in the brain for long periods.

Several lines of evidence support the conclusion that the site of such storage is the cerebellum. First, ablation of the ipsilateral cerebellum (both the cerebellar cortex and the deep nuclei) abolishes the ability of fully conditioned animals to respond to the conditioned stimulus (the tone). Furthermore, such animals cannot be newly conditioned using the eye ipsilateral to the lesion (whereas the contralateral eye continues to support conditioning). These observations have been reinforced by electrophysiological recordings that demonstrate changes in the activity of the deep cerebellar nuclei accompanying the conditioned response. Finally, lesions of the cerebellar cortex alone abolish conditioning of the blink response, but not the performance of already-conditioned responses. Thus, the circuitry of the cerebellar cortex and the deep nuclei evidently work in concert to encode and store learned information of this sort.

The question remains, however, whether the cerebellum participates in the acquisition and storage of more complex motor behaviors. This issue has been explored using the vestibulo-ocular reflex. When the visual image on the retina shifts its position as a result of head movement, the eyes must move in a complementary way to maintain a stable percept (see Chapter 19).

The adaptability of the VOR to changes in the nature of incoming sensory information can be challenged by fitting subjects (either humans or monkeys) with magnifying or minifying spectacles (Figure 18.9). Because the glasses alter the size of the visual image on the retina, compensatory eye movements that would normally have maintained a stable image of an object are now either too large or too small. Over time, subjects (whether monkeys or humans) learn to adjust the distance the eyes must move in response to head movements to accord with the artificially altered size of the visual field. Moreover, this change is retained for significant periods after the spectacles are removed. Information that reflects this change in the sensory context of the VOR must therefore be learned and remembered. Once again, if the cerebellum is damaged or removed, the ability of the VOR to adapt to the new conditions is lost. Electrophysiological recordings in monkeys during the period of VOR adjustment confirm that the cerebellum is instrumental in modifying the reflex; the responses of both Purkinje cells and deep cerebellar nuclear cells are altered. These observations support the conclusion that the cerebellum is critically important in motor learning.

A compelling argument for the participation of the basal ganglia in learning and memory also comes from studies of eye movements. Monkeys can be trained to make a saccade to a target that suddenly appears away from the point of fixation, and the activity of cells in the substantia nigra, pars reticulata (the part of the substantia nigra that lacks dopaminergic neurons) can be monitored during this task. These GABA-ergic neurons function much like globus pallidus (internal segment) cells in the direct pathway: they receive transient inhibitory input from the caudate and putamen and are tonically active, thus inhibiting the VA/VL complex. Evidently these cells are specifically concerned with the modulation of eye movements by the basal ganglia. Thus, their activity decreases or ceases before the onset of

Normal vestibulo-ocular reflex (VOR)

Head and eyes move in a coordinated manner to keep image on retina

VOR out of register

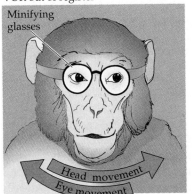

Eyes move too far in relationship to image movement on the retina when the head moves

After several hours

VOR gain reset

Eyes move smaller distances in relationship to head movement to compensate

Figure 18.9 Learned changes in the vestibulo-ocular reflex in monkeys. Normally, the reflex operates to move the eyes as the head moves, so that the retinal image remains stable. When the animal observes the world through minifying spectacles, the eyes initially move too far with respect to the "slippage" of the visual image on the retina. After some practice, however, the VOR is reset and the eyes move an appropriate distance in relation to head movement, thus compensating for the altered size of the visual image.

Figure 18.10 Eye movements to a remembered target. When a monkey makes random saccades, the activity of the nigral neuron being monitored here remains steady. When, however, the monkey makes a saccade to a significant target for which he receives a reward, the activity of the neuron is diminished before the onset of the movement, and returns to normal levels only after the movement is complete. When, the monkey is trained to make a saccade to the place where the target *had* been (a remembered target), the nigral neuron is again active prior to the initiation of the movement, as if the target were still present. These observations indicate that circuits in the basal ganglia can use internally generated or remembered information to guide motor behaviors. (After Hikosaka and Wurtz, 1989.)

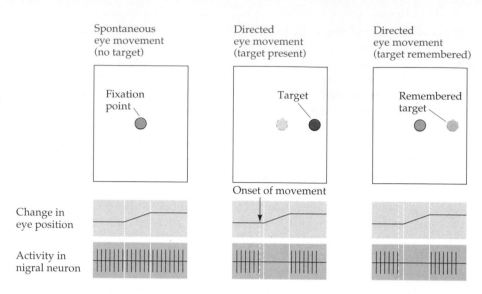

a saccade to a novel target, remains suppressed during the movement, and then returns to normal levels.

Monkeys can also learn to make saccades to targets whose locations are remembered rather than perceived (Figure 18.10). For example, a monkey can be trained to fixate on a point and watch for a novel stimulus to appear elsewhere without moving his eyes. In order to get a reward on the next trial, the monkey must make a saccade to the location where the novel stimulus appeared during the previous trial. Thus, the monkey has to remember where the stimulus was and move his eyes to that spot. The cells of the substantia nigra appear to participate in remembering the task: they are silent before the saccade to the remembered location and become active again once the movement is complete. Therefore, the output of the caudate and putamen to the substantia nigra must fire transiently in response to a target that is not there. In short, activity of the basal ganglia neurons (and movement) in this case is initiated by remembered rather than immediately perceived information.

This brief description of the behavioral physiology of motor learning and memory underscores a basic point about the central modulation of motor control: the basal ganglia and cerebellum use sensory and cognitive information to learn, remember, and plan movements. Thus, these two modulatory circuits are capable of contributing to motor control using concurrent sensory information or internally generated representations; these properties indicate circuitry far more sophisticated than the simple feedback loops once thought to characterize both the basal ganglia and the cerebellum. The significance (and continuing mystery) of these two systems lies in their capacity to guide a seemingly endless and dynamic repertoire of movements—from the graceful gestures of Mikhail Baryshnikov, to the explosive precision of Michael Jordan, to the remarkable ability of each of us to move more or less efficiently through the world each day.

■ SUMMARY

The function of the basal ganglia and cerebellar cortex is predicated on massive convergence of sensory information—mostly from the cerebral cortex—onto the principal neurons (the medium spiny neurons and the Purkinje cells) of these two structures. The principal neurons sum the concurrent synaptic activity of their cortical inputs, while being modulated by local circuit neurons and other subcortical innervation. In the basal ganglia, transiently inhibitory synapses made by the medium spiny neurons modulate the activity of tonically inhibitory neurons to allow the momentary release of motor programs. This circuitry influences the excitatory signals from the thalamus to the motor cortex through both a direct pathway and an opposing indirect pathway. In contrast, cerebellar circuitry is arranged to provide error correction of ongoing movements. This function is accomplished by changes in the tonically inhibitory activity of Purkinje cells that influence the tonically excitatory deep cerebellar nuclear cells. The resulting effects on the ongoing activity of the deep cerebellar nuclear cells adjust the cerebellar output signal to the thalamus. These intricate circuits in the cerebellum and basal ganglia are not simply feedback loops that modulate activity the motor cortex. They also participate in learning and remembering motor tasks.

Additional Reading

Reviews

DELONG, M. R. (1990) Primate models of movement disorders of basal ganglia origin. Trends Neurosci. 13: 281–285.

GLICKSTEIN, M. AND C. YEO (1990) The cerebellum and motor learning. J. Cog. Neurosci. 2: 69–80.

GOLDMAN-RAKIC, P. S. AND L. D. SELEMON (1990) New frontiers in basal ganglia research. Trends Neurosci. 13: 241–244.

HIKOSAKA, O. AND R. H. WURTZ (1989) The basal ganglia. In *The Neurobiology of Saccadic Eye Movements*. R. H. Wurtz et al.. (eds.). London: Elsevier.

LISBERGER, S. G. (1988) The neural basis for learning of simple motor skills. Science 242: 728–735.

STEIN, J. F. (1986) Role of the cerebellum in the visual guidance of movement. Nature 323: 217–221.

WILSON, C. J. (1990) Basal Ganglia. In *Synaptic Organization of the Brain*. G. M. Shepherd (ed.). Oxford: Oxford University Press, Chapter 9.

Important Original Papers

CRUTCHER, M. D. AND M. R. DELONG (1984) Single cell studies of the primate putamen. Exp. Brain Res. 53: 233–243.

DELONG, M. R. AND P. L. STRICK (1974) Relation of basal ganglia, cerebellum, and motor cortex units to ramp and ballistic movements. Brain Res. 71: 327–335.

DIFIGLIA, M., P. PASIK AND T. PASIK (1976) A Golgi study of neuronal types in the neostriatum of monkeys. Brain Res. 114: 245–256.

ECCLES, J. C. (1967) Circuits in the cerebellar control of movement. Proc. Natl. Acad. Sci. 58: 336–343.

KOCSIS, J. D., M. SUGIMORI AND S. T. KITAI (1977) Convergence of excitatory synaptic inputs to caudate spiny neurons. Brain Res. 124: 403–413.

McCORMICK, D. A., G. A. CLARK, D. G. LAVOND AND R. F. THOMPSON (1982) Initial localization of the memory trace for a basic form of learning. Proc. Natl. Acad. Sci. USA 79: 2731–2735.

THACH, W. T. (1968) Discharge of Purkinje and cerebellar nuclear neurons during rapidly alternating arm movements in the monkey. J. Neurophysiol. 31: 785–797.

THACH, W. T. (1978) Correlation of neural discharge with pattern and force of muscular activity, joint position, and direction of intended next movement in motor cortex and cerebellum. J. Neurophysiol. 41: 654–676.

Book

ITO, M. (1984) *The Cerebellum and Neural Control*. New York: Raven Press.

CHAPTER 19

EYE MOVEMENTS AND SENSORY-MOTOR INTEGRATION

■ OVERVIEW

Eye movements are, in many ways, simpler than movements of other parts of the body. There are only six extraocular muscles, each of which has a specific role in adjusting eye position. Moreover, there are only five stereotyped kinds of eye movements, each with its own control circuitry. The relative simplicity of eye movements has made them a useful model for understanding the basic mechanisms of motor control. Indeed, much of what is known about the regulation of movements by the cerebellum, basal ganglia, and vestibular system has come from the study of eye movements. Here the major features of eye movement control are used to illustrate the principles of sensory-motor integration for more complex motor behaviors.

■ WHAT EYE MOVEMENTS ACCOMPLISH

Eye movements are especially important in humans because high visual acuity is restricted to the fovea, the small circular region (about 1.5 mm in diameter) in the center of the retina that is densely packed with cone photoreceptors (see Chapter 10). Eye movements can direct the fovea to new objects of interest or compensate for disturbances that cause the fovea to be displaced from a target already being attended to.

As demonstrated several decades ago by the Russian physiologist Alfred Yarbus, eye movements reveal a good deal about the strategies used to inspect a scene. Yarbus used contact lenses with small mirrors on them (see Box A) to document (by the position of a reflected beam) the pattern of eye movements made while subjects examined a variety of objects and scenes. Figure 19.1 shows the direction of a subject's gaze as he viewed a picture of Queen Nefertiti. The thin, straight lines represent the quick, ballistic eye movements (saccades) that are used to align the foveas with particular parts of the scene; the denser spots along these lines represent points of fixation where the individual paused for a variable period to take in visual information (little or no visual perception occurs during a saccade). The results obtained by Yarbus, and subsequently many others, showed that vision is an active process in which eye movements shift the view to selected parts of the scene to examine especially interesting features. The spatial distribution of the fixation spots indicates that much more time is spent scrutinizing Nefertiti's eye, nose, mouth, and ear than examining the middle of her cheek or neck. Thus, eye movements allow us to focus attention on the portions of an image that convey the most significant information. (In consequence, the tracking of eye movements can be used to determine what aspects of a scene are particularly arresting. Advertisers now use modern versions of this method to determine which pictures and scene arrangements will best market their product.)

The importance of eye movements for visual perception has also been demonstrated by experiments in which a visual image is stabilized on the retina, either by paralyzing the extraocular eye muscles or by moving a scene in exact register with eye movements so that the different features of the image always fall on exactly the same parts of the retina (Box A). Such stabilized visual images rapidly disappear, for reasons that remain poorly

Figure 19.1 The eye movements of a subject viewing a picture of Queen Nefertiti. The bust on the left is what the subject saw; the diagram on the right shows his eye movements over a 2-minute viewing period. (From Yarbus, 1967.)

understood. Nonetheless, these observations on stabilized images make it plain that eye movements are also essential for normal visual perception.

■ THE ACTIONS AND INNERVATION OF EXTRAOCULAR MUSCLES

Eye movements are controlled by three antagonistic pairs of muscles: the **lateral** and **medial rectus muscles**, the **superior** and **inferior rectus muscles**, and the **superior** and **inferior oblique muscles**. These muscles are responsible for movements of the eye along three different axes: horizontal, either toward the nose (adduction) or away from the nose (abduction); vertical, either elevation or depression; and torsional, movements that bring the top of the eye toward the nose (intorsion) or away from the nose (extorsion). Horizontal movements are controlled entirely by the medial and lateral rectus muscles; the medial rectus muscle is responsible for adduction, the lateral rectus muscle for abduction. Vertical movements involve the coordinated action of the superior and inferior rectus muscles, as well as the oblique muscles. The relative contribution of the rectus and oblique groups depends on the horizontal position of the eye (Figure 19.2). In the primary

Figure 19.2 The contributions of the six extraocular muscles to vertical and horizontal eye movements. Horizontal movements are mediated by the medial and lateral rectus muscles, while vertical movements are mediated by the superior and inferior rectus and the superior and inferior oblique muscle groups.

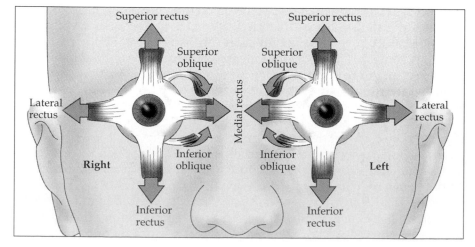

Box A
THE PERCEPTION OF STABILIZED RETINAL IMAGES

Visual perception depends critically on frequent changes of scene. Normally, our view of the world is changed by saccades (see text); tiny saccades that continue to move the eyes abruptly over a fraction of a degree of visual arc occur even when we stare intently at an object of interest; and continual drift of the eyes during fixation progressively shifts the image onto a nearby but different set of photoreceptors (figure below). As a consequence of these several sorts of eye movements, our point of view changes more or less continually.

The importance of a continually changing scene for normal vision is dramatically revealed when the retinal image is stabilized. If a small mirror is

Diagram illustrating one means of producing stabilized retinal images. By attaching a small mirror to the eye, the scene projected onto the screen will always fall on the same set of retinal points, no matter how the eye is moved.

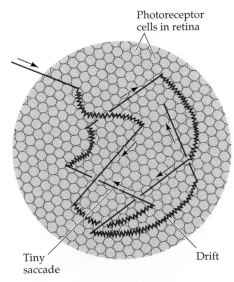

Diagram of the types of eye movements that continually change the retinal stimulus during fixation. The straight lines indicate microsaccades and the curved lines drift; the structures in the background are photoreceptors drawn approximately to scale. The normal scanning movements of the eyes (saccades) are much too large to be shown here, but obviously contribute to the changes of view that we continually experience, as do slow tracking eye movements (although the fovea tracks a particular object, the scene nonetheless changes). (After Pritchard, 1961.)

attached to the eye by means of a contact lens and an image reflected off the mirror onto a screen, then the subject necessarily sees the same thing, whatever the position of the eye: every time the eye moves, the projected image moves exactly the same amount (figure above). Under these circumstances, the stabilized image "disappears" from perception within a few seconds!

A much simpler way to demonstrate the rapid disappearance of a stabilized retinal image is to visualize one's own retinal blood vessels. The blood vessels, which lie in front of the photoreceptor layer, cast a shadow on the underlying receptors. Although normally invisible, the vascular shadows can be seen by moving a source of light across the eye, a phenomenon first noted by J. E. Purkinje more than 150 years ago. This perception can be elicited with an ordinary pen light pressed gently against the lateral side of the closed eyelid. When the light is wiggled vigorously, a rich network of black blood vessel shadows appears against an orange background. (The vessels appear black because they are shadows.) By starting and stopping the movement, it is readily

apparent that the image of the blood vessel shadows disappears within a few seconds after the light source is stilled.

The conventional interpretation of the rapid disappearance of stabilized images is retinal adaptation. In fact, the phenomenon is at least partly of central origin. Thus, stabilizing the retinal image in one eye diminishes perception through the other eye, an effect known as interocular transfer. Although the explanation of these remarkable effects is not entirely clear, they emphasize the point that the visual system is designed to deal with novelty.

References

BARLOW, H. B. (1963) Slippage of contact lenses and other artifacts in relation to fading and regeneration of supposedly stable retinal images. Q. J. Exp. Psychol. 15: 36–51.

COPPOLA, D. AND D. PURVES (1996) The extraordinarily rapid disappearance of entopic images. Proc. Natl. Acad. Sci. 96: 8001-8003.

HECKENMUELLER, E. G. (1965) Stabilization of the retinal image: a review of method, effects and theory. Psychol. Bull. 63: 157–169.

KRAUSKOPF, J. AND L. A. RIGGS (1959) Interocular transfer in the disappearance of stabilized images. Amer. J. Psychol. 72: 248–252.

position (eyes straight ahead), both groups contribute to vertical movements. Elevation is due to the action of the superior rectus and inferior oblique muscles, while depression is due to the action of the inferior rectus and superior oblique muscles. When the eye is abducted, the rectus muscles are the prime vertical movers. Elevation is due to the action of the superior rectus, and depression is due to the action of the inferior rectus. When the eye is adducted, the oblique muscles are the prime vertical movers. Elevation is due to the action of the inferior oblique muscle, while depression is due to the action of the superior oblique muscle. The oblique muscles are also primarily responsible for torsional movements.

The extraocular muscles are innervated by three cranial nerves: the abducens, the trochlear, and the oculomotor (Figure 19.3). The **abducens nerve** (cranial nerve VI) exits the brainstem from the pons-medullary junction and innervates the lateral rectus muscle. The **trochlear nerve** (IV) exits from the caudal portion of the midbrain and supplies the superior oblique muscle. In contrast to other cranial nerves, the trochlear nerve exits from the dorsal surface of the brainstem and crosses the midline to innervate the superior oblique muscle on the contralateral side. The **oculomotor nerve** (III), which exits from the rostral midbrain near the cerebral peduncle, sup-

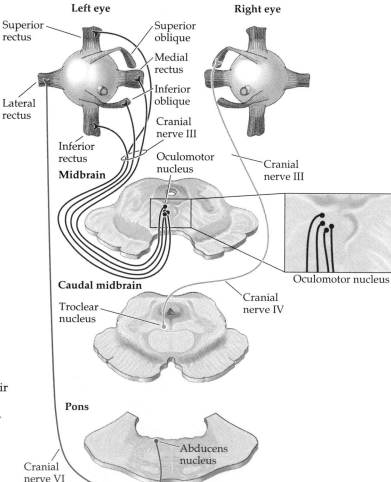

Figure 19.3 Organization of the several cranial nerve nuclei that govern eye movements, showing their innervation of the extraocular muscles. The abducens nucleus innervates the lateral rectus muscle; the trochlear nucleus innervates the superior oblique muscle; and the oculomotor nucleus innervates all the rest of the extraocular muscles. These include the medial rectus, inferior rectus, superior rectus, and inferior oblique.

plies all the rest of the extraocular muscles. Although the oculomotor nerve governs several different muscles, each receives its innervation from a separate group of neurons within the third nerve nucleus. In addition to supplying the extraocular muscles, a distinct cell group within the oculomotor nucleus innervates the levator muscles of the eyelid; the axons from these neurons also travel in the third nerve. Finally, the third nerve carries axons from the nearby Edinger-Westphal nucleus that are responsible for pupillary constriction (see Chapter 11). Thus, damage to the third nerve results in three characteristic deficits: impairment of eye movements, drooping of the eyelid (ptosis), and pupillary dilation.

■ TYPES OF EYE MOVEMENTS AND THEIR FUNCTIONS

There are five basic classes of eye movements: saccades, smooth pursuit movements, vergence movements, vestibulo-ocular movements, and optokinetic movements. The functions of each type of eye movement are introduced here; in subsequent sections, the neural circuitry responsible for some of these movements is presented in more detail (see Chapters 13 and 18 for further discussion of vestibulo-ocular movements).

Saccades are rapid, ballistic movements of the eyes that abruptly change the point of fixation. They range in amplitude from the small movements made while reading to the much larger movements made while gazing around a room. Saccades can be elicited voluntarily, but occur reflexively whenever the eyes are open, even when fixated on a target (see Box A). The rapid eye movements that occur during a particular phase of sleep (see Chapter 26) are also saccades. The time course of a saccadic eye movement is shown in Figure 19.4. After a decision is made to move the eyes (in this example, the stimulus was the movement of an already fixated target), it takes about 200 ms for eye movement to begin. During this delay, the position of the target with respect to the fovea is computed (that is, how far the eye has to move). The difference between the initial and intended position is converted into a motor command that activates the extraocular muscles to move the eyes the correct distance in the appropriate direction. Saccadic eye movements are said to be ballistic because the saccade-generating system cannot respond to subsequent changes in the position of the target during the delay period. If the target moves again during this time, the saccade will miss the target, and a second saccade must be made to correct the error.

Smooth pursuit movements are much slower tracking movements of the eyes designed to keep a moving stimulus on the fovea. Such movements are under voluntary control in the sense that you "decide" whether to track a moving stimulus (Figure 19.5). Surprisingly, however, only highly trained observers can make a smooth pursuit movement in the absence of a moving target. Most people who try to move their eyes in a smooth fashion without a moving target simply make a saccade.

Vergence movements align the fovea of each eye with targets located at different distances from the observer. Unlike other types of eye movements in which the two eyes move in the same direction (**conjugate eye movements**), vergence movements are **disconjugate** (or **disjunctive**); they involve either a convergence or divergence of the lines of sight of each eye to see an object that is nearer or farther away. Convergence is one of the three reflexive visual responses elicited by interest in a near object. The other components of the so-called **near reflex triad** are accommodation of the lens, which brings the object into focus, and pupillary constriction, which increases the depth of field and sharpens the image on the retina (see Chapter 10).

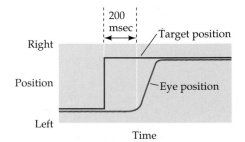

Figure 19.4 The metrics of a saccadic eye movement. The red line indicates the position of a fixation target and the blue line the position of the fovea. When the target moves suddenly to the right, there is a delay of about 200 ms before the eye begins to move to the new target position. (After Fuchs, 1967.)

Figure 19.5 The metrics of smooth pursuit eye movements. These traces show eye movements (blue lines) tracking a stimulus moving at different velocities (red lines). After a quick saccade to capture the target, the eye movement attains a velocity that matches the velocity of the target. (After Fuchs, 1967.)

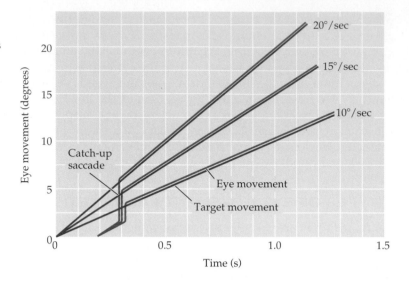

Vestibular eye movements and **optokinetic movements** operate together to stabilize the eyes relative to the external world, thus compensating for head movements. These reflex responses prevent visual images from "slipping" on the surface of the retina as head position varies. The action of vestibular and optokinetic eye movements can be appreciated by fixating an object and moving the head from side to side; the eyes automatically compensate for the head movement, thus keeping the image of the object at more or less the same place on the retina.

The **vestibular system** detects brief, transient changes in head position and produces rapid corrective eye movements (see Chapter 13). Sensory information from the semicircular canals directs the eyes to move in a direction opposite to the head movement. While the vestibular system operates effectively to counteract rapid movements of the head, it is relatively insensitive to slow movements or to persistent rotation of the head. For example, if the vestibulo-ocular reflex is tested with continuous rotation and without visual cues about movement of the image (with eyes closed or in the dark), the compensatory eye movements cease after only about 30 seconds of rotation. However, if the same test is performed with visual cues, eye movements persist. The compensatory eye movements in this case are due to the activation of another system that relies not on vestibular information, but on visual cues indicating motion of the visual field. This **optokinetic system** is especially sensitive to slow movements of large areas of the visual field, and it responds slowly, complementing the properties of the vestibular system.

The optokinetic system can be tested by placing a subject inside a rotating cylinder with vertical stripes. (In practice, this is usually done by simply seating the subject in front of a screen on which a series of horizontally moving vertical bars is presented.) The eyes automatically track the stripes until they reach the end of their excursion. There is then a quick saccade in the direction opposite to the movement, followed once again by smooth pursuit of the stripes. This alternating slow and fast movement of the eyes in response to such stimuli is called **optokinetic nystagmus**. Optokinetic nystagmus is a normal reflexive response of the eyes in response to large-scale movements of the visual scene, and should not be confused with the pathological nystagmus that can result from certain kinds of brain injury (for example, damage to the vestibular system or the cerebellum; see Chapter 17).

■ THE NEURAL CONTROL OF SACCADIC EYE MOVEMENTS

The problem of moving the eyes to fixate a new target in space (or indeed any other movement) entails two separate issues: controlling the *amplitude* of movement (how far), and controlling the *direction* of the movement (which way). The amplitude of a saccadic eye movement is encoded by the duration of neuronal activity in the oculomotor nuclei. As shown in Figure 19.6, for instance, neurons in the abducens nucleus fire a burst of action potentials prior to abducting the eye (by causing the lateral rectus muscle to contract), and are silent when the eye is adducted. The amplitude of the movement is correlated with the duration of the burst of action potentials in the abducens neuron. With each saccade, the abducens neurons reach a new baseline level of discharge that is correlated with the position of the eye in the orbit. The steady baseline level of firing holds the eye in its new position.

The direction of the movement is determined by which eye muscles are activated. Although in principle any given direction of movement could be specified by independently adjusting the activity of individual eye muscles, the complexity of the task would be overwhelming. Instead, the direction of eye movement is controlled by the activation of two **gaze centers** in the reticular formation, each of which is responsible for generating movements along a particular axis. The **paramedian pontine reticular formation (PPRF)** or **horizontal gaze center** is a collection of neurons near the midline in the pons responsible for generating horizontal eye movements. The **rostral interstitial nucleus** or **vertical gaze center** is located in the rostral part of the midbrain reticular formation and is responsible for vertical movements. Activation of each gaze center separately results in movements of the eyes along a single axis, either horizontal or vertical. Activation of the gaze centers in concert, results in oblique movements whose trajectories are specified by the relative contribution of each center.

An example of how the PPRF works with the abducens and oculomotor nuclei to generate a horizontal saccade to the right is shown in Figure 19.7. Neurons in the PPRF innervate cells in the abducens nucleus on the

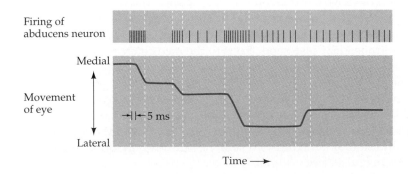

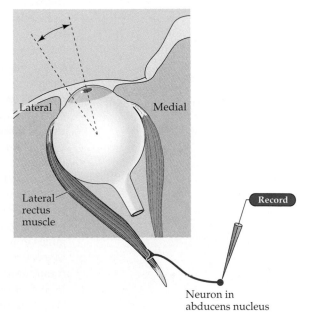

Figure 19.6   Motor neuron activity in relation to saccadic eye movements. In this example, an abducens motor neuron fires a burst of activity (upper trace) that precedes and extends throughout the movement (solid line). An increase in the tonic level of firing is associated with more lateral displacement of the eye. Note also the decline in firing rate during a saccade in the opposite direction. (After Fuchs and Luschei, 1970.)

Figure 19.7 Simplified diagram of synaptic circuitry responsible for horizontal movements of the eyes to the right. Activation of neurons in the right horizontal gaze center (the PPRF) leads to increased activity of motor neurons (red) and internuclear neurons (purple) in the right abducens nucleus. The motor neurons innervate the lateral rectus muscle of the right eye. The internuclear neurons innervate motor neurons in the contralateral oculomotor nucleus, which in turn innervate the medial rectus muscle of the left eye.

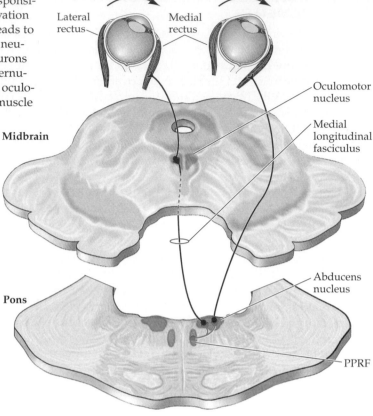

same side of the brain. There are two types of neurons in the abducens nucleus. One is a motor neuron that innervates the lateral rectus muscle on the same side. The other, called an internuclear neuron, sends its axon across the midline and ascends in a fiber tract called the medial longitudinal fasciculus (MLF), terminating in the portion of the oculomotor nucleus that contains neurons innervating the medial rectus muscle. As a result of this arrangement, activation of PPRF neurons on the right side of the brainstem causes horizontal movements of both eyes to the right; the converse is of course true for the PPRF neurons in the left half of the brainstem.

Neurons in the PPRF also send axons to the medullary reticular formation, where they contact inhibitory interneurons. The interneurons, in turn, project to the contralateral abducens nucleus, where they terminate on motor neurons and internuclear neurons. In consequence, activation of neurons in the PPRF on the right results in a reduction in the activity of the motor neurons whose muscles would oppose movements of the eyes to the right. This inhibition of antagonists resembles the strategy used to control limb muscle antagonists (see Chapter 16).

Although they continue to occur in complete darkness, saccades are often elicited when something attracts our attention and we direct our foveas to it. How is sensory information about the location of a target in space transformed into an appropriate pattern of activity in the horizontal and vertical gaze centers? Two structures that project to the gaze centers are demonstrably important for the initiation and accurate targeting of saccadic eye movements: the **superior colliculus** of the midbrain, and a region of the frontal lobe that lies just rostral to premotor cortex, known as the **frontal eye field** (**Brodmann's area 8**). Neurons in both of these structures discharge immediately prior to saccades, and both structures contain a topographical motor map. Thus, activation of a particular site in the superior colliculus or in the frontal eye field produces saccadic eye movements in a specified direction and for a specified distance that is independent of the position of the eyes in the orbit. The direction and distance are always the same for a given stimulation site, changing systematically when different sites are activated.

Both the superior colliculus and the frontal eye field contain cells that respond to visual stimuli; however, the relation between the sensory and motor responses of individual cells is better understood for the superior colliculus. There is an orderly map of visual space within the superior colliculus, and this sensory map is in register with the motor map that generates eye movements. Thus, neurons in a particular region of the superior colliculus are activated by the presentation of visual stimuli in a limited region of visual space; such activation leads in turn to the generation of a saccade that moves the eye by an amount just sufficient to align the fovea with the region of visual space that provided the stimulation (Figure 19.8).

Superior colliculus neurons also respond to auditory and somatic stimuli, and the location in space for these other modalities is also mapped in register with the motor map in the colliculus. Topographically organized maps of auditory space and of the body surface in the superior colliculus can therefore orient the eyes (and the head) to a variety of different sensory stimuli. This registered organization of sensory and motor maps illustrates an important function of topographical maps in other components of the motor system: to provide an efficient mechanism for sensory-motor transformations.

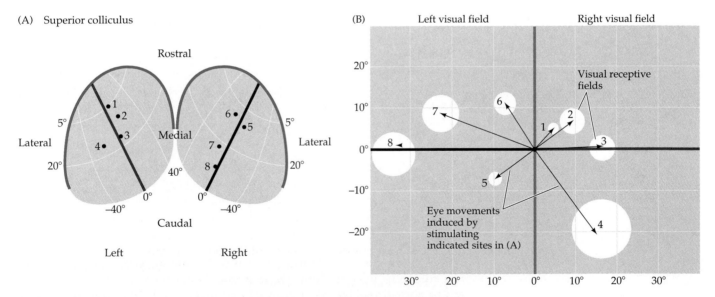

Figure 19.8 Evidence for sensory-motor transformation obtained from electrical recording and stimulation in the superior colliculus. (A) Surface views of the superior colliculus illustrating the location of eight separate electrode recording and stimulation sites. (B) Map of visual space showing the receptive field location of the sites in (A) (white circles), and the amplitude and direction of the eye movements elicited by stimulating these sites electrically (arrows). In each case, electrical stimulation results in eye movements that align the fovea with a region of visual space that corresponds to the visual receptive field of the site. (After Schiller and Stryker, 1972.)

Figure 19.9 The relationship of the frontal eye field (Brodmann's area 8) to the superior colliculus and the horizontal gaze center (PPRF). There are two routes by which the frontal eye field can influence eye movements: by projections to the superior colliculus, which in turn project to the contralateral PPRF, and by projections directly to the contralateral PPRF.

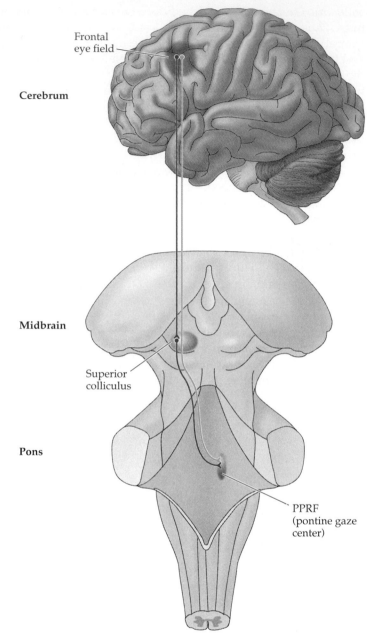

The functional relationship between the frontal eye field and the superior colliculus in controlling eye movements is similar to that between the motor cortex and the red nucleus in controlling of limb movements (see Chapter 16). The frontal eye field projects to the superior colliculus, and the superior colliculus projects to the PPRF on the contralateral side (Figure 19.9). (It also projects to the vertical gaze center, but for simplicity the discussion here is limited to PPRF.) The frontal eye field can thus control eye movements by activating selected populations of superior colliculus neurons. This cortical area also projects directly to the contralateral PPRF; as a result, the frontal eye field can also control eye movements independently of the superior colliculus. The parallel inputs to the PPRF from the frontal eye field and superior colliculus are reflected in the deficits that result from damage to these structures. Injury to the frontal eye field results is an inability to make saccades to the contralateral side and a deviation of the eyes to the side of the lesion. These effects are transient, however; in monkeys with experimentally induced lesions of this cortical region, recovery is virtually complete in 2 to 4 weeks. Lesions of the superior colliculus change the accuracy, frequency, and velocity of saccades; yet saccades still occur, and the deficits also improve with time. These results suggest that the frontal eye fields and the superior colliculus represent complementary pathways for the control of saccades. One of the structures appears to be able to compensate (at least partially) for the loss of the other. In support of this interpretation, combined lesions of the frontal eye field and the superior colliculus produce a dramatic and permanent loss in the ability to make saccadic eye movements.

■ NEURAL CONTROL OF SMOOTH PURSUIT MOVEMENTS

Smooth pursuit movements are also mediated by neurons in the PPRF, but are under the influence of centers other than the superior colliculus and frontal eye field. (The superior colliculus and frontal eye field are exclusively involved in the generation of saccades.) The exact route by which visual information reaches the PPRF to generate smooth pursuit movements is not known (a pathway through the cerebellum has been suggested). It is clear, however, that neurons in the striate and extrastriate visual areas provide sensory information that is essential for the initiation and accurate guidance of smooth pursuit movements. In monkeys, neurons in the middle temporal area (which is largely concerned with the perception of moving stimuli and a target of the magnocellular stream; see Chapter 11) respond selectively to targets moving in a specific direction. Moreover, damage to this area disrupts smooth pursuit movements. In humans, damage of comparable areas in the parietal and occipital lobes also results in deficits in smooth pursuit movements. Unlike the effects of lesions to the frontal eye field and the superior colliculus, the deficits are in eye movements made toward the side of the lesion. For example, a lesion of the left parieto-occipital region is likely to result in an inability to track an object moving from right to left.

Although a detailed description of the pathways responsible for the optokinetic reflex is beyond the scope of this chapter, it is worth noting that the same areas of the cortex that provide sensory information necessary for smooth pursuit also provide the sensory information that stimulates the slow phase of optokinetic nystagmus. Thus, patients with damage to the visual areas in the occipital and parietal lobes have deficits in optokinetic nystagmus for stimuli that are moving toward the side of the lesion.

■ THE ROLE OF THE CEREBELLUM AND BASAL GANGLIA IN EYE MOVEMENTS

Damage to the cerebellum results in deficits in eye movements that are similar to the cerebellar signs manifested in other body movements. Thus, saccadic eye movements become inaccurate and uncoordinated following damage to the cerebellum; they overshoot or undershoot the target, and multiple saccades are necessary to fixate a novel target. Patients with damage to the cerebellum also have difficulty holding fixation—their eyes tend to wander from the target. As might be expected, damage to the vestibular parts of the cerebellum results in nystagmus and an inability to generate smooth pursuit movements. Moving targets, no matter how slow, are tracked by a series of saccades rather than by smooth pursuit.

Just as the cerebellum sets the gain of the vestibulo-ocular reflex (see Chapter 18), it also affects experience-dependent modification of saccadic eye movements. In an experiment carried out on a monkey (Figure 19.10), the lateral rectus muscle of one eye was weakened by partially cutting its tendon. A patch was then placed over the eye and the animal trained to make a saccade to a target with the normal eye. Under these conditions, the normal eye makes a saccade that accurately fixates the target, while the saccade made by the weakened eye falls far short of the target (the position of the eyes is precisely monitored by a magnetic search coil technique that works equally well with or without the patch). If the patch is then removed from the weakened eye and placed over the normal eye so that the monkey has to use the weakened eye, the saccades initially continue to fall short, and

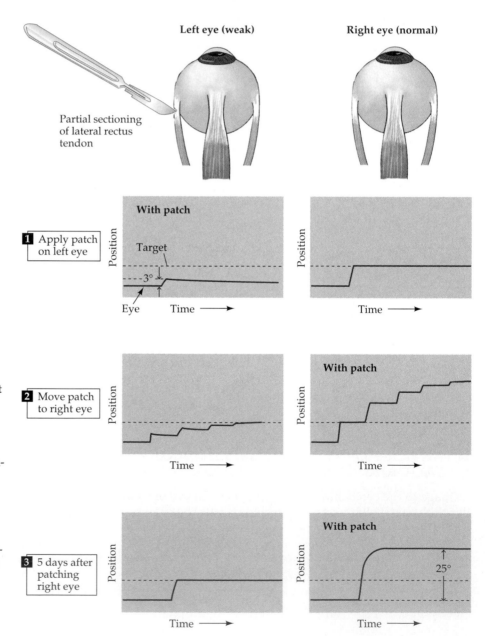

Figure 19.10 Contribution of the cerebellum to the experience-dependent modification of saccadic eye movements. Weakening of the lateral rectus muscle of the left eye causes the eye to undershoot the target (1). When the experimental subject (in this case a monkey) is forced to use this eye by patching the right eye, multiple saccades must be generated to acquire the target (2). After 5 days of experience with the weak eye, the gain of the saccadic system has been increased and a single saccade is used to fixate the target (3). This adjustment in the gain of the saccadic eye movement system depends on an intact cerebellum. (After Optican and Robinson, 1980.)

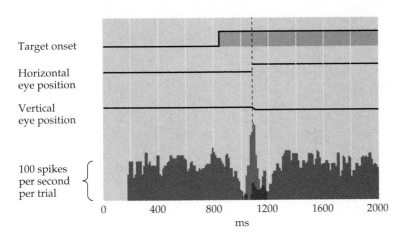

Target onset

Horizontal eye position

Vertical eye position

100 spikes per second per trial

0   400   800   1200   1600   2000
ms

Figure 19.11   Activity of a substantia nigra neuron and one of its presumed target cells in the superior colliculus during a gaze-orienting task. In this experiment, a monkey performs a saccade to a visual target after a fixation point has been extinguished. The upper trace indicates the onset of this cue. The two traces below the target trace indicate horizontal and vertical eye positions. Substantia nigra (blue) and collicular (red) activities are shown in spike histograms. The histograms are aligned to the onset of the saccadic movement (dashed line). Note that the tonic activity of the substantia nigra cell is reduced at the same time that the collicular cell discharges a vigorous burst of action potentials that triggers the saccadic eye movement. (After Hikosaka and Wurtz, 1989.)

several saccades have to be made to reach the target. However, after as little as 5 days of use, the oculomotor system recalibrates itself to the characteristics of the weakened muscle, and the distance from the initial point of fixation to the target is covered by a single saccade of the appropriate amplitude. In compensating, the saccadic system adjusts the movements of both eyes to the strength of the weakened eye, so that the normal eye now makes saccades that have too great an amplitude and overshoot the target. This result reflects the brainstem circuitry that normally mediates conjugate movements. In some people, a weak eye may be the eye that has better vision, and adjustments of this sort can significantly improve visual performance. This ability to modify the saccadic system to compensate for changes in muscle strength is dependent on an intact cerebellum. Following cerebellar damage, the amplitude of the saccades fails to adjust.

The role of the basal ganglia in the control of eye movements alluded to in the previous chapter can now be appreciated more fully. One of the main output structures of the basal ganglia, the substantia nigra pars reticulata (which is comparable to the globus pallidus), projects directly to the superior colliculus. The nigral axons terminate on neurons that are in turn the source of projections to the vertical and horizontal gaze centers in the reticular formation. The outflow from the basal ganglia is inhibitory (these particular neurons in the substantia nigra, like those in the globus pallidus, are GABAergic). The relevant neurons in the substantia nigra have a high resting level of discharge and thus exert a tonic inhibitory influence on the neurons in the superior colliculus (Figure 19.11). Immediately preceding an eye movement, the level of activity in the substantia nigra is reduced, and it is during this period of reduced inhibition that the neurons in the superior colliculus are activated to produce the movement. Local injection of substances that reduce the activity of nigral neurons results in an exuberance of saccadic eye movements in monkeys: the animals are unable to inhibit inappropriate saccades and have difficulty maintaining fixation. In contrast, substances that increase the activity of nigral neurons lead to a paucity of eye movements and an increase in their latency. The latter effect is similar to the oculomotor signs that occur in patients suffering from Parkinson's disease.

■ SUMMARY

Despite their high degree of specialization, the systems that control eye movements have much in common with the motor systems that govern movements of other parts of the body. Just as the spinal cord provides the basic circuitry for coordinating the actions of muscles around a joint, the reticular formation of the pons and midbrain provides the basic circuitry that mediates movements of the eyes. Descending projections from higher-order centers in the superior colliculus and the frontal eye field innervate the brainstem gaze centers, providing a basis for integrating eye movements with a variety of sensory information that indicates the location of objects in space. The superior colliculus and the frontal eye field are organized in a parallel as well as a hierarchical fashion, enabling one of these structures to compensate for the loss of the other. Eye movements, like other movements, are also under the control of the basal ganglia and cerebellum, which ensure the proper initiation and successful execution of these relatively simple motor behaviors that allow us to interact efficiently with the universe of things that can be seen.

Additional Reading

Reviews

FUCHS, A. F., C. R. S. KANEKO AND C. A. SCUDDER (1985) Brainstem control of eye movements. Ann. Rev. Neurosci. 8: 307–337.

HIKOSAKA, O AND R. H. WURTZ (1989) The basal ganglia. In *The Neurobiology of Saccadic Eye Movements: Reviews of Oculomotor Research*, Volume 3. R. H. Wurtz and M. E. Goldberg (eds.). Amsterdam: Elsevier, pp. 257–281.

ROBINSON, D. A. (1981) Control of eye movements. In *Handbook of Physiology*, Section 1: *The Nervous System*, Volume II: *Motor Control*, Part 2. V. B. Brooks (ed.). Bethesda, MD: American Physiological Society. pp. 1275–1320.

SPARKS, D. L. AND L. E. MAYS (1990) Signal transformations required for the generation of saccadic eye movements. Annu. Rev. Neurosci. 13: 309–336.

ZEE, D. S. AND L. M. OPTICAN (1985) Studies of adaption in human oculomotor disorders. In *Adaptive Mechanisms in Gaze Control: Facts and Theories*. A Berthoz, G Melvill Jones (Eds). Amsterdam: Elsevier, pp. 165–176.

Important Original Papers

FUCHS, A. F. AND E. S. LUSCHEI (1970) Firing patterns of abducens neurons of alert monkeys in relationship to horizontal eye movements. J. Neurophysiol. 33: 382–392.

OPTICAN, L. M. AND D. A. ROBINSON (1980) Cerebellar-dependent adaptive control of primate saccadic system. J. Neurophysiol. 44: 1058–1076.

SCHILLER, P. H. AND M. STRYKER (1972) Single unit recording and stimulation in superior colliculus of the alert rhesus monkey. J. Neurophysiol. 35: 915–924.

SCHILLER, P. H., S. D. TRUE AND J. L. CONWAY (1980) Deficits in eye movements following frontal eye-field and superior colliculus ablations. J. Neurophysiol. 44: 1175–1189.

Books

LEIGH, R. J. AND D. S. ZEE (1983) *The Neurology of Eye Movements*. Contemporary Neurology Series. Philadelphia: Davis

YARBUS, A. L. (1967) *Eye Movements and Vision*. Basil Haigh (trans.). New York: Plenum Press.

SCHOR, C. M. AND K. J. CIUFFREDA (EDS.) (1983) *Vergence Eye Movements: Basic and Clinical Aspects*. Boston: Butterworth.

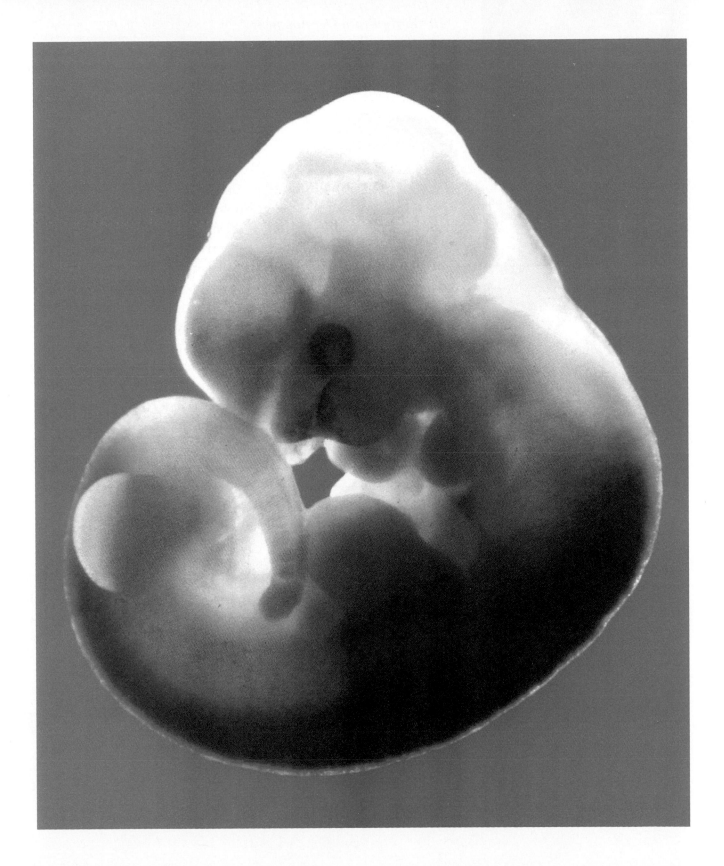

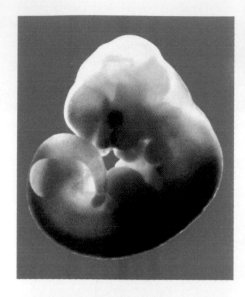

UNIT IV
THE CHANGING BRAIN

A mammalian embryo in which cells in the developing nervous system responding to the signaling molecule retinoic acid have been labeled by means of a reporter gene. (Courtesy of Anthony LaMantia and Elwood Linney.)

Although we think of ourselves as the same person throughout life, the functional and structural state of the brain changes dramatically over the human lifespan. The initial development of the nervous system entails the birth of neurons, the formation of specific axonal pathways, the elaboration of vast numbers of synapses, and the emergence of increasingly complex behaviors. After birth, experiences during postnatal life shape behavioral repertories and cognitive abilities in highly specific ways, typically during restricted temporal windows. Even in maturity, synaptic connections continue to be modified as new memories are laid down and older ones forgotten. And, like any other organ system, the brain is subject to diseases and traumatic insults that call repair mechanisms into play.

Understanding these changes is a goal shared by several subfields in contemporary neuroscience. Developmental neurobiologists are examining the mechanisms involved in the growth and differentiation of neurons and their connections. Neuroethologists explore how such changes affect behavior. Molecular neurobiologists, synaptic physiologists, and cognitive neuroscientists are unraveling the mechanisms by which brain circuits are modified and memories encoded, stored, and retrieved. Finally, clinical investigators are beginning to apply this knowledge to ameliorate or reverse the devastating consequences of neurological injury and disease.

CHAPTER 20

EARLY BRAIN DEVELOPMENT

■ OVERVIEW

The elaborate architecture of the adult brain is the final product of genetic instructions, cellular interactions, and the interplay between the newborn and the external world. The early development of the nervous system is dominated by events that occur prior to the formation of synapses and are therefore activity-independent. These events include the establishment of the primordial nervous system in the early embryo, the initial generation of neurons from undifferentiated precursor cells, the formation of the major brain regions, and the migration of neurons from the sites of generation to their final positions. (The subsequent formation of axon pathways and synaptic connections is considered in the following chapter.) When any of these processes goes awry—because of genetic mutation, disease, or exposure to drugs or chemicals—the consequences can be disastrous. Indeed, most congenital brain defects result from interference in the normal programs of activity-independent neuronal development. With the advent of powerful new techniques, the cellular and molecular machinery underlying these extraordinarily complex events is beginning to be understood.

■ THE INITIAL FORMATION OF THE NERVOUS SYSTEM: GASTRULATION AND NEURULATION

Well before the patch of cells that will become the nervous system appears, embryonic polarity and the primitive cell layers required for the subsequent formation of the nervous system are established. Critical to this early framework in all vertebrate embryos is the process of **gastrulation,** the invagination of the covering of the developing embryo that produces the three **germ layers** of the embryo: the outer layer, or **ectoderm**; the middle layer, or **mesoderm**; and the inner layer, or **endoderm** (see Figure 20.1). Gastrulation also defines the midline and the anterior-posterior axes of all vertebrate embryos, including the human embryo.

One key consequence of gastrulation is the formation of the **notochord,** a distinct cylinder of mesodermal cells that extends along the midline of the embryo from anterior to posterior. The notochord forms from an aggregation of mesoderm that invaginates and extends forward from a surface indentation called the **primitive pit**, which subsequently extends to form the **primitive streak**. As a result of these cell movements during gastrulation, the notochord comes to define the embryonic midline. Because of its proximity, the ectoderm that lies immediately above the notochord becomes the nervous system. In addition to specifying the basic topography of the embryo and determining the position of the nascent nervous system, the notochord is required for the subsequent neural differentiation (Figure 20.1). The notochord (along with the primitive pit) sends **inductive signals** to the overlying ectoderm that cause a subset of ectodermal cells to differentiate into neural precursor cells. During this process, called **neurulation**, the midline ectoderm that contains these cells thickens into a distinct columnar epithelium called the **neural plate**. The lateral margins of the neural plate then fold inward, eventually transforming the neural plate into a tube. This structure, the **neural tube**, subsequently gives rise to the entire brain and spinal cord.

Figure 20.1 Neurulation in the mammalian embryo. On the left are dorsal views of the embryo at several different stages of early development; each boxed view on the right is a mid-line cross section through the embryo at the same stage. (A) During late gastrulation and early neurulation, the notochord forms by invagination of the ectoderm in the region of the primitive streak. (B) As neurulation proceeds, the neural plate begins to fold upon itself, forming the neural groove. The neural plate immediately above the notochord differentiates into the floorplate. The neural crest becomes distinct at the margins of the neural plate. (C) Once the edges of the neural plate meet in the midline, the neural tube is complete. Adjacent to the tube, the mesoderm thickens and subdivides into structures called somites—the precursors of the axial musculature and skeleton. (D) As development continues, the neural tube adjacent to the somites becomes the rudimentary spinal cord, and the neural crest gives rise to sensory ganglia. Finally the anterior ends of the neural plate (anterior neural folds) grow together at the midline and continue to expand, eventually giving rise to the brain.

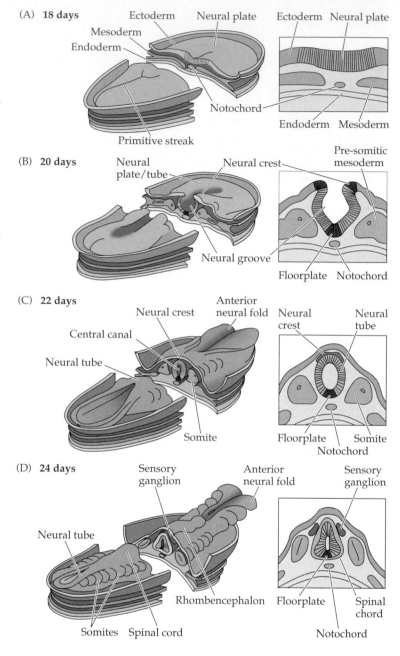

The progenitor cells of the neural tube are known as **neural precursor cells**. The precursors are dividing stem cells that produce more precursors and, eventually, nondividing **neuroblasts** that differentiate into neurons. Due to their proximity to the notochord, cells at the ventral midline of the neural tube differentiate into a strip of epithelial-like cells called the **floorplate**. The position of the floorplate at the ventral midline determines the polarity of the neural tube and further influences the differentiation of neural precursor cells. Inductive signals from the floorplate lead to the differentiation of cells in the ventral portion of the neural tube that eventually give rise to motor neurons (which are thus closest to the ventral midline). Precursor cells that are farther away from the ventral midline eventually give rise to sensory neurons. At the most dorsal limit of the neural tube, a

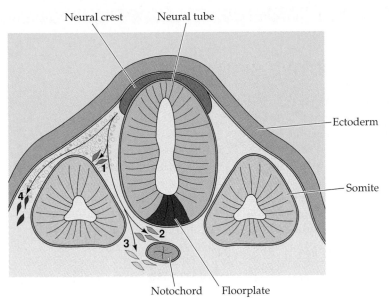

Neural crest   Neural tube

Ectoderm

Somite

Notochord   Floorplate

Figure 20.2   The neural crest. Diagram of a cross section through a developing mammalian embryo at a stage similar to that in Figure 20.1C. The neural crest cells follow four distinct migratory paths that lead to differentiation of distinct cell types and structures. Cells that follow the pathways labeled (1) and (2) give rise to sensory and sympathetic ganglia, respectively. The precursors of adrenal neurosecretory cells migrate along pathway (3), while cells destined to become non-neural tissues migrate along pathway (4). Each pathway permits cells to interact with different kinds of cellular environments, from which they receive inductive signals (see Figure 20.11). (After Sanes, 1988.)

third distinct population of cells emerges in the region where the edges of the folded neural plate join together. Because of their location, these precursors are called **neural crest cells** (Figure 20.2). The neural crest cells migrate away from the neural tube over specific pathways that expose them to another set of inductive signals (discussed in the next section). As a result, these cells subsequently form the neurons and glia of the sensory and sympathetic ganglia, the neurosecretory cells of the adrenal gland, and the enteric nervous system (as well as a variety of non-neural elements such as cartilage and pigment cells).

■ THE MOLECULAR BASIS OF NEURAL INDUCTION

The essential consequence of gastrulation and neurulation for the development of the nervous system is the emergence of a population of neural precursors from a subset of ectodermal cells. Through a variety of experimental manipulations, primarily involving transplantation of different portions of developing embryos, classical embryologists recognized early on that this process depends on signals arising from cells in the primitive pit and notochord. Because a wide variety of chemical agents and physical manipulations were able to mimic the effects of these endogenous signals, their nature remained a mystery for decades. It is now clear that the generation of cell identity—of which neural induction is but one example—results from the spatial and temporal control of different sets of genes. The inducing signals from the primitive pit and notochord are, not surprisingly, molecules that modulate gene expression.

The increasingly sophisticated effort to understand exactly how these inducers work has therefore focused on molecules that can modify patterns of gene expression. An instructive example is **retinoic acid**, a derivative of vitamin A (Box A) and a member of the steroid/thyroid superfamily of hormones. Retinoic acid activates a unique class of **transcription factors**—the retinoid receptors—that modulate the expression of particular genes. Another class of inducers is the peptide hormones, including those that belong to the **fibroblast growth factor** (FGF) and **transforming growth factor**

Box A
RETINOIC ACID: TERATOGEN AND INDUCTIVE SIGNAL

In the early 1930s, investigators noticed that vitamin A deficiency during pregnancy led to fetal malformations. At about the same time, experimental studies in animals yielded the surprising finding that *excess* vitamin A could cause similar defects. These studies of vitamin A suggested that an entire family of compounds—the retinoids—might have similar teratogenic effects. (*Teratogenesis* is the term for birth defects induced by exogenous agents.) The retinoids include the alcohol form of vitamin A, called retinol, the aldehyde form, retinal, and the acid form, retinoic acid. Subsequent studies in animals confirmed that other retinoids produce birth defects similar to those generated by too much or too little vitamin A. The disastrous consequences of retinoid exposure during pregnancy were underscored in the early 1980s when the drug Accutane® was introduced as a treatment for intractable acne. Accutane® is the trademark name for isoretinoin, or 13-*cis*-retinoic acid. Pregnant women who took this drug had an increased number of spontaneous abortions and children born with a range of birth defects. The most severe abnormalities affected the developing brain, which was often grossly malformed. Despite these dramatic observations, the reason for the adverse effects of retinoids on fetal development remained obscure.

An important insight into teratogenic potential of retinoids came when embryologists working on limb development in chicks found that retinoic acid mimics the inductive ability of a portion of the limb bud. Still the mystery remained as to just what retinoic acid (or its absence) was doing to influence or compromise development. The answer came in the mid-1980s, when the receptors for retinoic acid were discovered. These receptors are members of the steroid/thyroid hormone receptor superfamily; when they bind retinoic acid or similar ligands, the receptors act as transcription factors to activate specific genes. Subsequent studies have shown that retinoic acid activates gene expression in the developing brain (see the illustration on page 376). Either an excess or a deficiency of such factors can therefore affect normal development, presumably by eliciting inappropriate patterns of retinoid-induced gene expression.

The saga of retinoic acid—from teratogen to endogenous signaling molecule—shows that the retinoids cause birth defects by mimicking the normal signals that influence gene expression. The story provides a good example of how clinical, cellular, and molecular observations can be combined to explain seemingly bizarre developmental pathology.

References

EVANS, R. M. (1988) The steroid and thyroid hormone receptor superfamily. Science 240: 889-895.

LAMMER, E. J., AND 11 OTHERS (1985) Retinoic acid embryopathy. N. Engl. J. Med. 313: 837–841.

SCHARDEIN, J. L. (1993) *Chemically Induced Birth Defects*, 2nd Ed. New York: Marcel Dekker.

TICKLE, C., B. ALBERTS, L. WOLPERT AND J. LEE (1982) Local application of retinoic acid to the limb bud mimics the action of the polarizing region. Nature 296: 564–565.

WARKANY, J. AND E. SCHRAFFENBERGER (1946) Congenital malformations induced in rats by maternal vitamin A deficiency. Arch. Ophthalmol. 35: 150–169.

(**TGF**) families. These molecules, like retinoic acid, are produced by a variety of embryonic tissues including the notochord, the floorplate, and the neural ectoderm itself, and bind to cell surface receptors, many of which are protein kinases. These peptide hormones activate a cascade of specific genes in ectodermal cells, which are evidently the first steps leading to formation of the nervous system. Both retinoids and the peptide hormones can elicit aspects of neural induction. If signaling via these molecules is disrupted, the formation of the nervous system is compromised.

Knowledge about some of the molecules involved in neural induction has provided a much more informed way of thinking about the etiology and prevention of a number of congenital disorders. Embryonic exposure to a variety of substances—alcohol and thalidomide come readily to mind—can elicit pathological differentiation of the embryonic nervous system by providing inductive signals at inappropriate times or places. Anomalies like **spina bifida** (failure of the posterior neural tube to close completely), **anencephaly** (failure of the anterior neural tube to close at all), and other brain malformations (often accompanied by mental retardation) probably result from defects in inductive signaling or the genes that mediate this process. In animal models, exposure to exogenous retinoic acid and its derivatives gives rise to spina bifida. At least two rare genetic disorders—Waardenburg's syndrome (the characteristics of which include deafness and neural tube anomalies) and aniridia (malformations of the iris and anosmia)—result from mutations of single transcription factor genes. Because the consequences of disordered neural induction are so severe, pregnant women are now advised to avoid virtually all drugs, especially during the first trimester.

■ THE FORMATION OF THE MAJOR BRAIN SUBDIVISIONS

Soon after neural tube formation, the forerunners of the major brain regions become apparent as a result of morphogenetic movements that bend, fold, and constrict the neural tube. Initially, the anterior end of the tube forms a crook, giving it the shape of a candy cane (Figure 20.3). The end of the candy cane nearest the sharper bend, or cephalic flexure, balloons out to form the **forebrain**, or **prosencephalon**. The **midbrain**, or **mesencephalon**, forms as a bulge on the cephalic flexure. The **hindbrain**, or **rhombencephalon**, forms in the long, relatively straight stretch between the cephalic flexure and the more caudal cervical flexure. Caudal to the cervical flexure, the neural tube forms the precursor of the spinal cord. The lumen enclosed by the developing neural tube is also altered by these bends and folds, forming what will eventually become the ventricles of the mature brain (Figure 20.3C; see also Chapter 1).

Figure 20.3 Regional specification of the developing brain. (A) The neural tube becomes subdivided into the prosencephalon (at the anterior end of the embryo), mesencephalon, and rhombencephalon. The spinal cord differentiates from the posterior end of the neural tube. View is looking down on a longitudinal section. (B) The initial bending of the neural tube at its anterior end leads to a candy cane shape, as seen in a lateral view of the same stage shown in (A). (C) Further development distinguishes the telencephalon and diencephalon from the prosencephalon; two other subdivisions—the metencephalon and myelencephalon—derive from the rhombencephalon. These subregions give rise to the rudiments of the major functional subdivisions of the brain, while the spaces they enclose eventually form the ventricles of the mature brain. (D) Lateral view of the embryo at the developmental stage shown in (C).

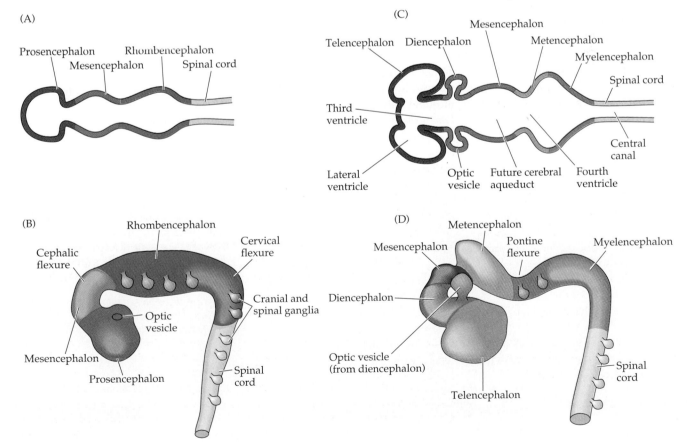

Once these primitive brain regions are established, they undergo at least two more rounds of partitioning, each of which produces additional brain regions in the adult (Figure 20.3C,D). Thus, the rostral prosencephalon forms the **telencephalon**, which contains the rudiments of the cerebral cortex, hippocampus, basal ganglia, basal forebrain nuclei, and olfactory bulb. The more caudal portion of the prosencephalon forms the **diencephalon** which contains the rudiments of the thalamus and hypothalamus, as well as a pair of lateral outpocketings (the **optic cups**) from which the neural portion of the retina will form. Finally, the rostral part of the rhombencephalon becomes the **metencephalon** (which gives rise to the adult cerebellum and pons) while the caudal part of the rhombencephalon becomes the **myelencephalon** (which gives rise to the adult medulla).

How can a simple tube of neuronal precursor cells produce such a variety of brain structures? At least part of the answer comes from the observation made early in the twentieth century that the neural tube is organized into repeating units called **neuromeres**. This discovery led to the idea that

Box B
HOMEOTIC GENES AND HUMAN BRAIN DEVELOPMENT

The notion that particular genes can influence the establishment of distinct regions in an embryo arose from efforts to catalog single-gene mutations that affect development of the fruit fly *Drosophila*. In the 1960s and 70s, E. B. Lewis at the California Institute of Technology reported a number of mutations that resulted in either the duplication of a distinct body segment or the appearance of an inappropriate structure at an ectopic location in the fly. These genes were called homeotic genes because they were able to convert segments of one sort to those of another (*homeo* is Greek for "similar"). Subsequently, studies by C. Nusslein-Volhard and E. Wieschaus demonstrated the existence of numerous such "master control" genes, each forming part of a cascade of gene expression leading to segmentation of the developing embryo. (In 1995, Lewis, Nusslein-Volhard, and Wieschaus shared a Nobel Prize for these discoveries.)

Homeotic genes code for DNA-binding proteins—that is, transcription factors—that bind to a distinctive sequence of genomic DNA called the homeobox. Similar genes have been found in many species. Using an approach known as cloning by homology, at least four "clusters" of homeobox genes have been identified in frogs, mice, and humans, among others. The genes of each cluster are closely, but not consecutively, spaced on a single chromosome. Other motifs identified in *Drosophila* have led to the discovery of additional families of DNA-binding proteins, which have again been found in a variety of species.

A number of developmental anomalies in mice and humans have turned out to be mutations in homeotic or other developmental control genes initially identified in the fly. These include genes for several human congenital abnormalities of the central and peripheral nervous system, including DiGeorge syndrome and Waardenburg's syndrome. DiGeorge syndrome is caused by a mutation in the *Hox-1.3* gene; the disease affects peripheral and central nervous system morphogenesis and neural crest migration. Waardenburg's syndrome is caused by a mutation in the *Pax-3* gene. The disease affects the inner ear and several regions of the brain. The longstanding pursuit of developmental control genes in fruit flies has thus led to a much deeper understanding of genetic diseases that affect the developing human nervous system.

References

GEHRING, W. J. (1993) Exploring the homeobox. Gene 135: 215–221.

GRUSS, P. AND C. WALTHER (1992) *Pax* in development. Cell 69: 719–722.

LEWIS, E. B. (1978) A gene complex controlling segmentation in *Drosophila*. Nature 276: 565–570.

NUSSLEIN-VOLHARD, C. AND E. WIESCHAUS. (1980) Mutations affecting segment number and polarity in *Drosophila*. Nature 287: 795–801.

REDLINE, R., A. NEISH, L. B. HOLMES AND T. COLLINS (1992) Biology of disease: Homeobox genes and congenital malformations. Lab. Invest. 66: 659–670.

the process of segmentation—used by all animal embryos at the earliest stages of development to establish regional identity in the body—might also establish regional identity in the developing brain. Support for this hypothesis has emerged from the development of the body plan of the fruit fly *Drosophila*. In the fly, early expression of a class of genes called **homeobox genes** (Box B) guides the differentiation of the embryo into distinct segments that give rise to the head, thorax, and abdomen (Figure 20.4). Similar homeotic genes in mammals (referred to as *Hox* genes) have been identified, and in some cases their patterns of expression coincide with, or even precede, the formation of morphological features such as the various bends, folds, and constrictions that signify the progressive regionalization of the developing neural tube (Box C). The patterned expression of homeobox genes, as well as other developmentally regulated transcription factors and signaling molecules, does not by itself determine the fate of a group of embryonic neural

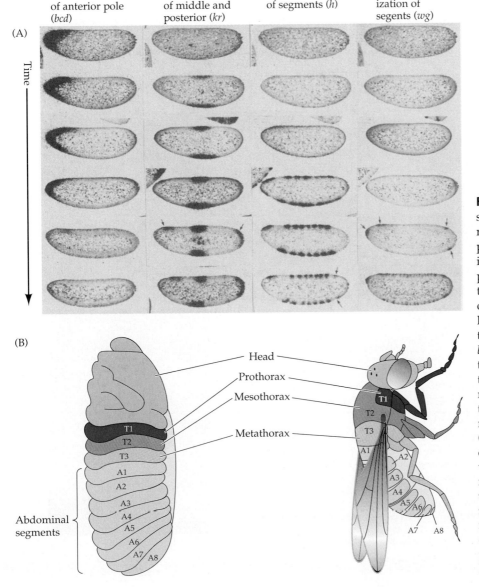

Figure 20.4 Sequential gene expression divides the *Drosophila* embryo into regions and segments. (A) Temporal pattern of expression of four genes that influence the establishment of the body plan in *Drosophila*. A series of sections through the anterior-posterior midline of the embryo are shown from early to later stages of development (top to bottom in each row). Initially, the gene *bicoid* (*bcd*) is expressed and helps define the anterior pole of the embryo. Next, the gene *krüppel* (*kr*) is expressed in the middle and then at the posterior end of the embryo, to further define the anterior-posterior axis. Then, the gene *hairy* (*h*) is expressed, which helps to delineate the position of the domains that will eventually form the mature segmented body of the fly. Finally, the gene *wingless* (*wg*) is expressed, further defining the organization of individual segments. (B) The relationship of embryonic segments in the *Drosophila* larva to the body plan of the mature fly. (A from Ingham, 1988; B after Gilbert, 1994.)

Box C
RHOMBOMERES

An interesting parallel between early embryonic segmentation and early brain development was noticed around the turn of the century. Several embryologists reported repeating units in the early neural plate and neural tube, which they called neuromeres. In the late 1980s Andrew Lumsden, Roger Keynes, and their colleagues, as well as R. Krumlauf, R. Wilkinson, and colleagues, noticed further that combinations of homeobox (*Hox*) genes are expressed in banded patterns in the developing chick nervous system, especially in the hindbrain (the common name for the rhombencephalon and its derivatives). These *Hox* expression domains corresponded to the rhombomeres, which in the chick are a series of seven transient bulges in the developing rhombencephalon. Rhombomeres are sites of differential cell proliferation (cells at rhombomere boundaries divide faster than cells in the rest of the rhombomere), differential cell mobility (cells from any one rhombomere cannot easily cross into adjacent rhombomeres), and differential cell adhesion (cells prefer to stick to those of their own rhombomere).

Later in development, the pattern of axon outgrowth from the cranial motor nerves also correlates with the earlier rhombomeric pattern. Cranial motor nerves originate either from a single rhombomere or from specific pairs of neighboring rhombomeres (transplantation experiments indicate that rhombomeres are specified in pairs). *Hox* gene expression probably represents an early step in the formation of cranial nerves in the developing brain. Thus, mutation or ectopic activation of *Hox* genes in mice alters the position of specific cranial nerves, or prevents their formation. Mutation of the *Hox-1.6* gene by homol-

(A)

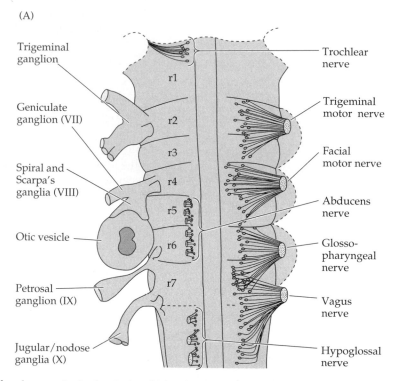

(B)

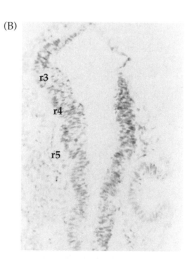

(C)

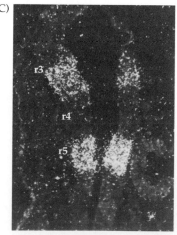

Rhombomeres in the developing chicken hindbrain and their relationship to the differentiation of the cranial nerves. (A) Diagram of the chick hindbrain, indicating the position of the cranial ganglia and nerves and their rhombomeric origin (rhombomeres denoted as r1 to r8). (B) Section through early chicken hindbrain, showing bulges that will eventually become rhombomeres (r3 to r5). (C) Differential patterns of transcription factor expression (in this case, *krx* 20, a *Hox*-like gene) define rhombomeres at early stages of development, well before the cranial nerves that will eventually emerge from them are apparent. (A courtesy of Andrew Lumsden; B and C from Wilkinson and Krumlauf, 1990.)

ogous recombination—the so-called "knockout" strategy for targeting mutations to specific genes—prevents formation of rhombomeres. In these animals, development of the external, middle and inner ear is also compromised, and cranial nerve ganglia are fused and located incorrectly. It is likely that problems in rhombomere formation are the underlying cause of congenital nervous system defects involving cranial nerves, ganglia, and peripheral structures derived from the neural crest.

While the exact relationship between early patterns of rhombomere-specific gene transcription and subsequent cranial nerve development remains a puzzle, the correspondence between these repeating units in the embryonic brain and similar iterated units in the development of the insect body (see Figure 20.4) suggests that differential expression of transcription factors in distinct regions is essential for the normal development of many species. In each case, spatially and temporally distinct patterns of transcription factor expression coincide with spatially and temporally distinct patterns of differentiation. The idea that the bulges and folds in the neural tube are segments defined by patterns of gene expression provides an attractive framework for understanding the molecular basis of pattern formation in the developing vertebrate brain.

References

CARPENTER, E. M., J. M. GODDARD, O.

CHISAKA, N. R. MANLEY AND M. CAPECCHI (1993) Loss of *HoxA-1* (*Hox-1.6*) function results in the reorganization of the murine hindbrain. Development 118: 1063–1075.

LUMSDEN, A. AND R. KEYNES (1989) Segmental patterns of neuronal development in the chick hindbrain. Nature 337: 424–428.

WILKINSON, D. G. AND R. KRUMLAUF (1990) Molecular approaches to the segmentation of the hindbrain. Trends in Neurosci. 13: 335–339.

VON KUPFFER, K. (1906) Die morphogenie des central nerven systems, in Handbuch der vergleichende und experiementelle Entwicklungslehreder Wirbeltiere, Vol. 2, 3: 1–272. Fischer Verlag, Jena.

ZHANG, M., H. J. KIM, H. MARSHALL, M. GENDRON-MAGUIRE, D. A. LUCAS, A. BARON, L. J. GUDAS, T. GRIDLEY, R. KRUMLAUF AND J. F. GRIPPO (1993) Ectopic *HoxA-1* induces rhombomere transformation in mouse hindbrain. Development 120: 2431–2442.

precursors. But as in the case of neural induction, regionally distinct transcription factor expression contributes to a series of genetic and cellular processes that eventually produce a fully differentiated brain.

■ THE INITIAL DIFFERENTIATION OF NEURONS AND GLIA

The mature human brain contains about 100 billion neurons and many more glial cells, all generated over the course of only a few months from a small population of precursor cells. Except for a few specialized cases, the entire neuronal complement of the adult brain is produced during a time window that closes before birth; thereafter, precursor cells disappear and no new neurons can be added to replace those lost by age or injury. The precursor cells are located in the **ventricular zone**, the innermost cell layer surrounding the lumen of the neural tube. The ventricular zone is a region of extraordinary mitotic activity: in humans, it has been estimated that about 250,000 new neurons are generated each minute during the peak of cell proliferation during gestation.

The dividing precursor cells in the ventricular zone undergo a stereotyped pattern of cell movements as they progress through the mitotic cycle (Figure 20.5), leading to the formation of either new stem cells or postmitotic neuroblasts that differentiate into neurons. As cells become postmitotic, they leave the ventricular zone and migrate to their final positions in the developing brain. Knowing when the neurons destined to populate a given brain region are "born"—that is, when they become postmitotic—has provided considerable insight into how different regions of the brain are constructed. For example, in the cerebral cortex, birthdating studies (Box D) have shown that the six layers of the cortex form in an inside-out manner: the firstborn

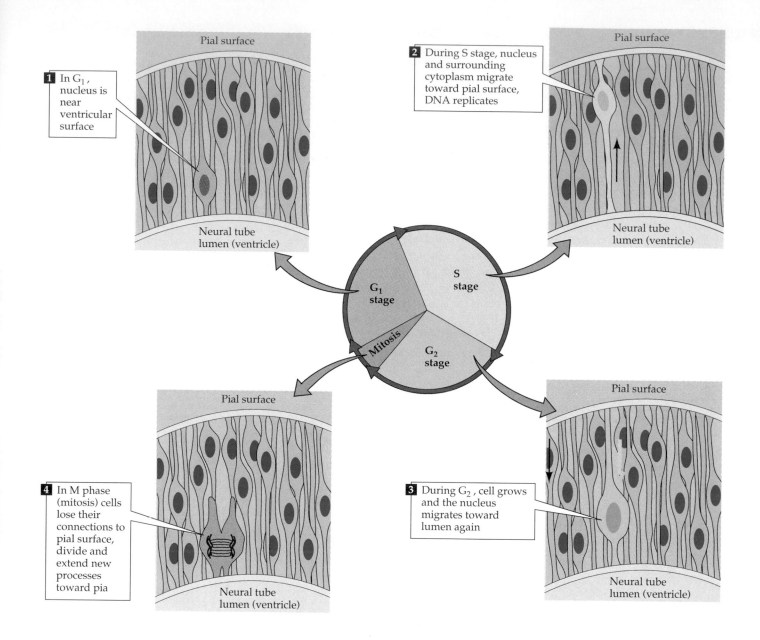

1 In G₁, nucleus is near ventricular surface

2 During S stage, nucleus and surrounding cytoplasm migrate toward pial surface, DNA replicates

3 During G₂, cell grows and the nucleus migrates toward lumen again

4 In M phase (mitosis) cells lose their connections to pial surface, divide and extend new processes toward pia

Pial surface

Neural tube lumen (ventricle)

G₁ stage

S stage

G₂ stage

Mitosis

cells are eventually located in the deepest layers, whereas later generations of neurons migrate through the older cells and come to lie superficial to them (Figure 20.6). This fact implies that each layer of the cortex—which has distinct patterns of connections and cell types—is populated by a cohort of cells generated during a specific developmental period. The nuclei of the brain are also distinguished by specific times of generation.

■ THE GENERATION OF NEURONAL DIVERSITY

All the neuronal precursor cells in the ventricular zone of the embryonic brain look and act about the same. Yet the postmitotic cells that these precursors ultimately give rise to are enormously diverse in form and function. The cerebral cortex, for example, contains at least several dozen neuronal cell

◀ **Figure 20.5** Dividing precursor cells in the vertebrate neuroepithelium (neural plate and neural tube stages) are attached both to the pial (outside) surface of the neural tube and to its ventricular (lumenal) surface. The nucleus of the cell translocates between these two limits within a cylinder of cytoplasm. The position of the nucleus corresponds to the stage of the mitotic cycle that the cell is in. When cells are closest to the outer surface of the neural tube, they enter a phase of DNA synthesis (the S stage); after the nucleus moves back to the ventricular surface (the G_2 stage) the precursor cells lose their connection to the outer surface and enter mitosis (the M stage). When mitosis is complete, the two daughter cells extend processes back to the outer surface of the neural tube, and the new precursor cells enter a resting (G_1) phase of the cell cycle. At some point a precursor cell generates both a precursor cell and a daughter cell that will not divide further (that is, a neuroblast), instead of another pair of precursor cells. The entire process takes several hours.

types distinguished by morphology, neurotransmitter content, cell surface molecules, and the types of synapses they make and receive. On an even more basic level, the stem cells of the ventricular zone produce both neurons and glia, cells with markedly different properties and functions. How and when are these different cell types determined?

At one extreme, it might be that different populations of neurons (or neurons and glia) are established very early in development, perhaps at the formation of the neural plate (Figure 20.7A). Separate types of precursor cells would then exist in the ventricular zone, each giving rise to a particular type of cell in the adult. According to this school of thought, a cell's fate is a func-

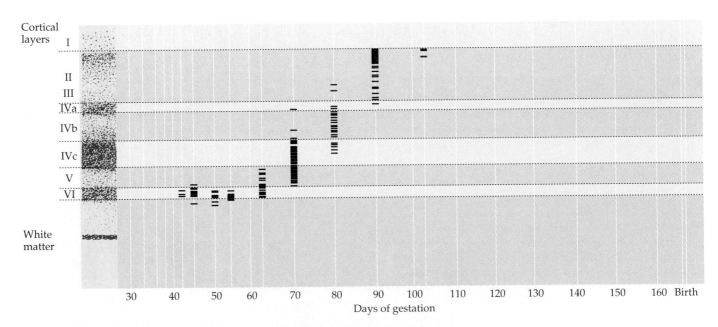

Figure 20.6 Generation of cortical neurons during the gestation of a rhesus monkey (a span of about 165 days). The final cell divisions of the neuronal precursors, determined by maximal incorporation of radioactive thymidine administered to the pregnant mother (see Box D), occur primarily during the first half of pregnancy and are complete by about embryonic day 105. Each short horizontal line represents the position of a neuron heavily labeled by maternal injection of radiolabeled thymidine at the time indicated by the corresponding vertical line. The numerals on the left designate the cortical layers. (After Rakic, 1974.)

(A) Cell lineage model

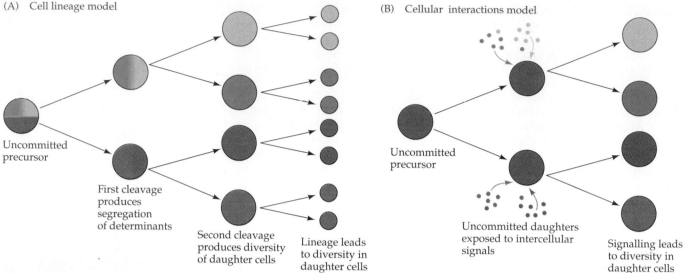

Uncommitted precursor

First cleavage produces segregation of determinants

Second cleavage produces diversity of daughter cells

Lineage leads to diversity in daughter cells

(B) Cellular interactions model

Uncommitted precursor

Uncommitted daughters exposed to intercellular signals

Signalling leads to diversity in daughter cells

Figure 20.7 Two hypotheses about the generation of cell diversity during embryonic development. (A) Cells acquire diverse fates while still at the precursor stage, relying primarily on information intrinsic to each cell. Subsequent divisions result in the proliferation of these cells, which differentiate according to their lineage. (B) Cells are descended from a pleuripotential precursor. Diversity is generated among daughter cells by distinct signals from other cells. Experimental evidence in vertebrates favors this second model.

tion of its lineage: neurons of different types would have distinct "ancestors," as would neurons and glia. At the other extreme, precursor cells might be providing essentially no information about eventual phenotype; in this scenario all such determinations would be derived from interactions with other cells in particular brain microenvironments (Figure 20.7B).

Despite much effort, there is little evidence that precursor cells are committed early on to produce particular types of daughter cells; in fact precursor cells continue to generate postmitotic daughter cells that assume a number of different phenotypes (Figure 20.8). In the retina, for example,

Figure 20.8 Cells derived from the last divisions of a single precursor can assume different fates, implying that lineage has little influence on cellular phenotype. In the chick optic tectum, injection of a replication-incompetent retrovirus at early stages in development inserts a reporter gene into the genome of a single progenitor cell; the progeny of the cell can then be detected using a simple histochemical stain for expression of the reporter gene (in this case for the enzyme β-galactosidase). As shown in (A), the offspring of a single progenitor cell form a narrow column spanning almost the entire thickness of the optic tectum. Both neurons recognized by the shape of their cell body (B and C) and glia recognized by the halo of fine, hair-like processes (D and E) can be derived from the same progenitor. (From Galileo et al., 1990.)

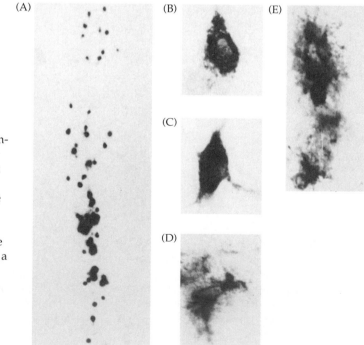

Box D
NEURONAL BIRTHDATING

At some point in development, stem cells—the dividing cells that populate the proliferative zones of the developing brain—undergo asymmetrical divisions that produce both another stem cell and a neuronal precursor that will never again undergo cell division. Because neurons are generally unable to reenter the cell cycle once they have left it, the point at which a neuronal precursor leaves the cycle—often called the cell's birthdate—is an important developmental milestone in the series of events leading to a neuron's adult phenotype. Knowing exactly when during the protracted course of development specific cell types or brain areas are generated is important for understanding both the etiology of congenital brain defects and the basic mechanisms of brain development.

In animals with extraordinarily simple nervous systems, such as the worm *Caenorhabditis elegans*, it is possible to directly monitor in a microscope each embryonic stem cell as it undergoes its characteristic series of cell divisions and thereby determine when a specific neuron is born. In the vastly more complex vertebrate brain, this approach is not feasible. Instead, neurobiologists rely on the characteristics of the cell cycle itself to label cells according to their date of birth. When cells are actively replicating DNA, they take up nucleotides—the building blocks of DNA—from the

extracellular space. Cell birthdating studies introduce a labeled nucleotide that can only be incorporated into newly synthesized DNA—usually tritium-labeled thymidine or a chemically distinctive analog of thymidine such as bromodeoxyuridine—at a known time in the organism's developmental history. All stem cells that are actively synthesizing DNA incorporate the radiolabeled thymidine and pass it on to their descendants. Because the labeled probe is only available for a few hours after being injected into an animal, if a cell continues to divide, the levels of the labeled probe in the cell's DNA are quickly diluted. However, if a cell undergoes only a single division after incorporating the label and produces a postmitotic neuron, that neuron retains high levels of the labeled thymidine indefinitely. Once the animal has matured, histological sections prepared from the brain and coated with a sensitive photographic emulsion show the radioactive decay from the labeled neurons. The most heavily labeled cells are those that incorporated label shortly before their final division; they are therefore said to have been "born" at the time of injection. In more recent variants of this approach, nucleotide analogues like bromodeoxyuridine that can be detected with antibodies are employed, but the principle is the same.

One of the earliest insights obtained from this approach was that the layers of the cerebral cortex develop in an "inside-out" fashion (see Figure 20.6). In certain mutant mice, such as *reeler* (see Box A, Chapter 18), birthdating studies show that the oldest cells end up in the most superficial layers and the most recently generated cells in the deepest as a result of defective migration. Although neuronal birthdates do not, in themselves, tell when cells acquire specific phenotypic features, they mark a major transition in the genetic programs dictating cell behavior.

References

ANGEVINE, J.B., JR. AND R. L. SIDMAN (1961) Autoradiographic study of cell migration during histogenesis of the cerebral cortex in the mouse. Nature 192: 766–768.

CAVINESS, V. S. JR. AND R. L. SIDMAN (1973) Time of origin of corresponding cell classes in the cerebral cortex of normal and *reeler* mutant mice: an autoradiographic analysis. J. Comp. Neurol. 148: 141–151.

GRATZNER, H. G. (1982) Monoclonal antibody to 5-bromo and 5-iododeoxyuridine. A new reagent for the detection of DNA replication. Science 218: 474–475.

MILLER, M. W. AND R. S. NOWAKOWSKI (1988) Use of bromodeoxyuridine immunohistochemistry to examine the proliferation, migration, and time of origin of cells in the central nervous system. Brain Res. 457: 44–52.

experiments using lineage-marking techniques have shown that even as late as the last cell division, a precursor cell can generate any combination of cell types found in the retina, including bipolar cells, ganglion cells, amacrine cells, and glial cells. Lineage, then, appears to play only a minor role in specifying cell fate. Similar experiments in the chicken optic tectum (superior colliculus) and the mouse cerebral cortex have led to the same conclusion.

The bulk of the evidence favors the view that neuronal differentiation is based on cell-cell interactions. Historically, most experimental approaches to this issue have relied on transplantation strategies: moving bits of a particular brain region to a host animal to determine whether the transplanted cells acquire the host phenotype or retain their original fate during subsequent development. When very young precursor cells are transplanted, they usually acquire the host phenotype; transplants at increasingly older ages, however, usually retain the original phenotype. The progressive restriction of possible phenotypes that a given cell can assume almost certainly results from local cellular and molecular cues that progressively limit the complement of genes expressed in that cell.

One of the best-understood examples of how such cellular and molecular interactions generate diverse neuronal phenotypes is the development of the eye in *Drosophila*. The fly eye is a stereotyped structure consisting of hundreds of repeating cellular units called ommatidia, each comprising eight distinct receptor cells (R1–R8; Figure 20.9). The assignment of phenotypes to individual cells is independent of the lineage of precursor cells, depending instead on a series of cell-cell interactions involving cell-specific ligands and receptors. These interactions have been analyzed in a strain of mutant flies called *sevenless* (because they are missing one receptor, R7). A similar phenotype has been observed in another mutation, called *boss* (for *bride of sevenless*). These studies have shown that the *boss* gene encodes a tyrosine kinase-linked membrane receptor, while *sevenless* encodes a membrane-bound ligand for this receptor. Each gene is expressed in separate populations of cells in the developing eye. The interaction between these two proteins leads to a series of intracellular events resulting in the differentiation of the R7 neuroblast. Based on this and much other work, alterations in receptor kinases are thought to be broadly related to cellular differentiation (and its pathologies, such as the uncontrolled cell proliferation that takes place in some types of cancer, including retinoblastoma).

In short, the emergence of diverse cell types in the mammalian nervous system does not result from the unfolding of a rigid program based on lineage, but from elaborately orchestrated spatial and temporal interactions between neuronal precursors and molecular signals derived from other cells.

(A)

(B)

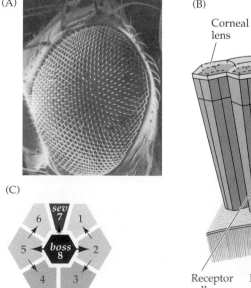

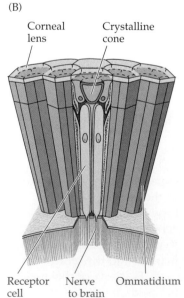

Figure 20.9   Development of the compound eye of *Drosophila* illustrates how cell-cell interactions can determine cell fate. (A) Scanning electron micrograph of the eye in the fruit fly *Drosophila*. (B) Diagram of the structure of the fly eye. The eye consists of an array of identical ommatidia, each comprising an array of eight photoreceptors. (C) Arrangement of photoreceptors within each ommatidium and the cell-cell signalling that determines their fate. A membrane-bound ligand on R8 (the *boss* gene product) binds to a receptor (encoded by the *sevenless* gene, *sev*) on the R7 cell. These interactions eventually lead to the changes in gene expression that determine the fate of an R7 cell. The arrows between R8 and the remaining receptor cells indicate interactions necessary for determining the fates of R1–6. (A courtesy of T. Venkatesh; B, C after Rubin, 1989.)

■ NEURONAL MIGRATION

Migration is a ubiquitous feature of development that brings groups of cells into specific spatial relationships. The final location of postmitotic nerve cells is particularly important because neural function depends on precise connections between neurons and their targets: the developing presynaptic and postsynaptic elements must be in the right place at the right time. Most developing neurons arising from the ventricular zone of the neural tube or the neural crest migrate substantial distances. The distance traversed is especially obvious in large animals like primates. To form the cerebral cortex, for example, neurons must sometimes traverse several millimeters from the ventricular zone to the pial surface. Defects in neuronal migration lead to major neurological problems in both humans and in experimental animals (see Box A in Chapter 18).

Quite a bit is known about the mechanics of how neurons move from their birthplace to their final destination. Depending on the area of the developing nervous system in which they originate, migrating neurons follow one of two strategies. Neural crest cells are largely guided along distinct migratory pathways by specialized adhesive molecules in the extracellular matrix (see Figure 20.2); at different developmental stages similar molecules are probably used to guide axonal outgrowth (see Chapter 21). In contrast, neurons in many regions, including the cortex, cerebellum, hippocampus, and spinal cord, are guided to their final destinations by crawling along a particular type of glial cell, called **radial glia**, which acts as a cellular guide (Figure 20.10A).

That some neurons migrate along glial guides was inferred from histological observations of embryonic brains and further supported by an analysis of electron microscopic images of fixed tissue (Figure 20.10B, C). Subsequently, innovations in cell culture systems and microscopy made it possible to observe the process of migration directly. If radial glial cells and immature neurons are isolated from the cerebellum and mixed together in vitro, the neurons attach to the glial cells, assume the characteristic shape of migrating cells seen in vivo, and begin moving along the glial processes. Indeed, the membranes of glial cells, when coated onto thin glass fibers, support normal migration. Several cell surface adhesion molecules and extracellular matrix adhesion molecules apparently mediate this process. In many regions of the brain, however, neurons migrate without the benefit of glial guides; thus, glial scaffolds are not required for neuronal migration. Nonetheless, migration along radial glia figures prominently in regions where cells are organized into layers, like the cerebral cortex, hippocampus, and cerebellum.

Relatively little is known about the specific messages that neurons receive as they migrate, although moving through a changing cellular environment clearly has important effects on the differentiation of neurons. Such effects are most evident in the migration of neural crest cells, where the migratory paths of precursor cells determine both their ultimate identity and position in the body (see Figure 20.2). These pathways are characterized by differential distributions of diverse cell adhesion molecules on the surfaces of somites, extracellular matrix, and the crest cells themselves. Of particular significance is the fact that specific peptide hormone growth factors, probably secreted by target cells, cause neural crest cells to differentiate into distinct phenotypes (Figure 20.11). These effects depend on the location of the precursor cell along a migratory pathway as it receives a signal, different signals being available at different points. Such position-dependent cues are probably not restricted to the peripheral nervous system; in the cerebellum, for example, different patterns of genes are expressed in migrating granule

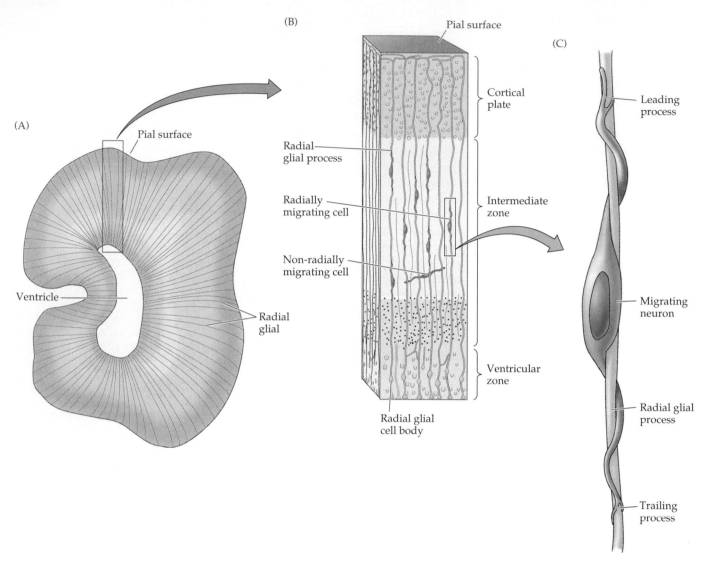

Figure 20.10 Migration of neurons along radial glia in the developing central nervous system. (A) A section through the developing forebrain showing radial glial processes that stretch from the ventricular to the pial surfaces of the neuroepithelium. (B) Enlargement of boxed area in (A). The migrating neurons are intimately apposed to radial glial cells, which form a pathway that guides some postmitotic neurons from their birthplace in the ventricular zone to their final position in the developing cerebral cortex. Some cells also take a non-radial migratory route, which can lead to a wide dispersion of post-mitotic cortical neurons derived from the same precursor. (C) A further enlargement showing a single migrating neuroblast (based on serial reconstruction of electron microscope sections). (After Rakic, 1974.)

neurons at different locations, implying the existence of different signals (as yet unknown) along the migratory path.

Thus, neuronal migration involves much more than the mechanics of moving cells from one place to another. As is the case for inductive events during the initial formation of the nervous system, stereotyped movements bring different classes of cells into contact with one another, thereby providing a means of constraining cell-cell signaling to specific times and places.

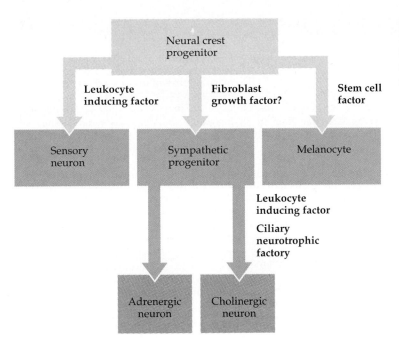

Figure 20.11   Cell signaling during the migration of neural crest cells. The establishment of each precursor type relies upon signals provided by one of several specific peptide hormones. The availability of each signal depends on the migratory pathway.

■ SUMMARY

The initial development of the nervous system depends on an intricate interplay of cellular movements and inductive signals. In addition to an early reordering of cellular positions as a result of morphogenesis, substantial migrations of neuronal precursors are necessary for the subsequent differentiation of distinct classes of neurons, and for the eventual formation of specialized patterns of neural connections. The fate of individual precursor cells is not determined by their mitotic history; rather, the information required for differentiation arises largely from the interactions between the developing cells and their local environment. All of these events are dependent on the same categories of molecular and cellular phenomena: cell-cell signaling, transcriptional regulation, and, ultimately, gene expression. The molecules that participate in signaling and regulation in the developing brain are similar to the agents that carry out these tasks in the rest of the body: hormones, transcription factors, second messengers, and cell adhesion molecules. As might be expected, understanding the functions of these molecules has begun to explain a variety of congenital neurological defects. These key signaling events early in neural development are special points of vulnerability for the expression of genetic mutations, and for the action of the many drugs and toxins that can compromise the generation of a normal nervous system.

Additional Reading

Reviews

ANDERSON, D. J. (1993) Molecular control of cell fate in the neural crest: the sympathoadrenal lineage. Annu. Rev. Neurosci. 16: 129–158.

CAVINESS, V. S. JR. AND P. RAKIC (1978) Mechanisms of cortical development: a view from mutations in mice. Annu. Rev. Neurosci. 1: 297–326.

HATTEN, M. E. (1993) The role of migration in central nervous system neuronal development. Curr. Opin. Neurobiol. 3: 38–44.

INGHAM, P. (1988) The molecular genetics of embryonic pattern formation in *Drosophila*. Nature 335: 25–34.

JESSELL, T. M. AND D. A. MELTON (1992) Diffusible factors in vertebrate embryonic induction. Cell 68: 257–270.

KESSLER, D. S. AND D. A. MELTON (1994) Vertebrate embryonic induction: mesodermal and neural patterning. Science 266: 596–604.

KEYNES, R. AND R. KRUMLAUF (1994) *Hox* genes and regionalization of the nervous system. Annu. Rev. Neurosci. 17: 109–132.

LEWIS, E. M. (1992) The 1991 Albert Lasker Medical Awards. Clusters of master control genes regulate the development of higher organisms. JAMA 267: 1524–1531.

LINNEY, E. AND A. S. LaMANTIA (1994) Retinoid signaling in mouse embryos. Adv. Dev. Biol. 3: 73–114.

SANES, J. R. (1989) Extracellular matrix molecules that influence neural development. Annu. Rev. Neurosci. 12: 491–516.

SELLECK, M. A., T. Y. SCHERSON AND M. BRONNER-FRASER (1993) Origins of neural crest cell diversity. Dev. Biol. 159: 1–11.

ZIPURSKY, S. L. AND G. M. RUBIN (1994) Determination of neuronal cell fate: lessons from the R7 neuron of *Drosophila*. Annu. Rev. Neurosci. 17: 373–397.

Important Original Papers

ANGEVINE, J. B. AND R. L. SIDMAN (1961) Autoradiographic study of cell migration during histogenesis of cerebral cortex in the mouse. Nature 192: 766–768.

GALILEO, D. S., G. E. GRAY, G. C. OWENS, J. MAJORS AND J. R. SANES (1990) Neurons and glia arise from a common progenitor in chicken optic tectum: Demonstration with two retroviruses and cell type-specific antibodies. Proc. Natl. Acad. Sci. USA 87: 458–462.

GRAY, G. E. AND J. R. SANES (1991) Migratory paths and phenotypic choices of clonally related cells in the avian optic tectum. Neuron 6: 211–225.

HAFEN, E., K. BASLER, J. E. EDSTROEM AND G. M. RUBIN (1987) *Sevenless*, a cell-specific homeotic gene of *Drosophila*, encodes a putative transmembrane receptor with a tyrosine kinase domain. Science 236: 55–63.

HEMMATI-BRIVANLOU, A. AND D. A. MELTON (1994) Inhibition of activin receptor signaling promotes neuralization in *Xenopus*. Cell 77: 273–281.

KRAMER, H., R. L. CAGAN, AND S. L. ZIPURSKY (1991) Interaction of bride of sevenless membrane-bound ligand and the sevenless tyrosine-kinase receptor. Nature 352: 207–212.

LANDIS, S. C. AND D. L. KEEFE (1983) Evidence for transmitter plasticity in vivo: Developmental changes in properties of cholinergic sympathetic neruons. Dev. Biol. 98: 349–372.

McMAHON, A. P. AND A. BRADLEY (1990) The *wnt-1* (*int-1*) protooncogene is required for the development of a large region of the mouse brain. Cell 62: 1073–1085.

NODEN, D. M. (1975) Analysis of migratory behavior of avian cephalic neural crest cells. Dev. Biol. 42: 106–130.

PATTERSON, P. H. AND L. L. Y. CHUN (1977) The induction of acetylcholine synthesis in primary cultures of dissociated rat sympathetic neurons. Dev. Biol. 56: 263–280.

RAKIC, P. (1971) Neuron-glia relationship during granule cell migration in developing cerebral cortex. A Golgi and electronmicroscopic study in *Macacus rhesus*. J. Comp. Neurol. 141: 283–312.

RAKIC, P. (1974) Neurons in rhesus monkey visual cortex: Systematic relation between time of origin and eventual disposition. Science 183: 425–427.

SAUER, F. C. (1935) Mitosis in the neural tube. J. Comp. Neurol. 62: 377–405.

SPEMANN, H. AND H. MANGOLD (1924) Induction of embryonic primordia by implantation of organizers from a different species. Translated into English by V. Hamburger and reprinted in *Foundations of Experimental Embryology*, B. H. Willier and J. M. Oppenheimer (eds.) (1974). New York: Hafner Press.

STEMPLE, D. L. AND D. J. ANDERSON (1992) Isolation of a stem cell for neurons and glia from the mammalian neural crest. Cell 71: 973–985.

WALSH, C. AND C. L. CEPKO (1992) Widespread dispersion of neuronal clones across functional regions of the cerebral cortex. Science 255: 434–440.

YAMADA, T., M. PLACZEK, H. TANAKA, J. DODD AND T. M. JESSELL (1991) Control of cell pattern in the developing nervous system. Polarizing activity of the floor plate and notochord. Cell 64: 635–647.

Books

LAWRENCE, P. A. (1992) *The Making of a Fly: The Genetics of Animal Design*. Oxford: Blackwell Scientific.

MOORE, K. L. (1988) *The Developing Human: Clinically Oriented Embryology*, 4th Ed. Philadelphia: W.B. Saunders Company.

CHAPTER 21

CONSTRUCTION OF NEURAL CIRCUITS

■ OVERVIEW

After neurons have differentiated and migrated to their intended destinations (see Chapter 20), they must extend axons, select targets from a myriad of possibilities, and initiate the formation of synapses with appropriate cells in the target. These events rely on a wealth of cellular and molecular information that guides axons and facilitates correct synaptic partnerships. Cues include cell adhesion molecules that regulate the interactions between axons and the surfaces upon which they grow, diffusible molecules that attract growing axons, and an important family of molecules known as neurotrophins that promotes and maintains stable synapses between axons and their targets. The circuitry of the developing nervous system is gradually constructed by means of these intricate interactions.

■ AXON OUTGROWTH AND PATHFINDING

Among the many extraordinary features of nervous system development, perhaps the most fascinating is the ability of growing axons to navigate through a complex cellular terrain to find appropriate synaptic partners that may be millimeters or even centimeters away. In 1910, R. G. Harrison, who first observed axons extending in vitro, noted:

> The growing fibers are clearly endowed with considerable energy and have the power to make their way through the solid or semi-solid protoplasm of the cells of the neural tube. But we are at present in the dark with regard to the conditions which guide them to specific points.

The "energy and power" of growing axons are now known to arise from the properties of the **growth cone**, a specialized structure at the tip of the extending axon. Growth cones are highly motile structures, designed to explore the extracellular environment and respond to local cues by changing the speed or direction of growth. They are typically endowed with numerous fine processes, called **filopodia**, that rapidly form and disappear, like fingers reaching out to touch or sense the environment (Figure 21.1).

Growing axons make many choices in their journey, and these are especially significant at the biological equivalent of forks in the road. For example, at the optic chiasm (see Chapter 11), axons arising from retinal ganglion cells in the temporal retina of humans and other mammals remain on the same (ipsilateral) side of the brain, whereas those from the nasal retina cross to the contralateral side. Depending on their location in the retina, therefore, developing retinal ganglion cells must decide whether or not to cross the midline (Figure 21.2A). Observations of living axons in animal models show that growth cones slow down and change shape at such decision points, becoming less streamlined and more complex (Figure 21.2B). These functional and structural changes presumably reflect an assessment of especially important cues in the local environment.

Figure 21.1 Photomicrograph of a growth cone at the tip of a sensory ganglion cell axon that is extending in tissue culture. Lamellapodia (flat, sheetlike protrusions) and filopodia (long, fingerlike processes) can be seen arising from the growth cone. These highly motile extensions evidently sample the local environment in order to regulate the speed and direction of axonal growth. (Courtesy of P. Forscher.)

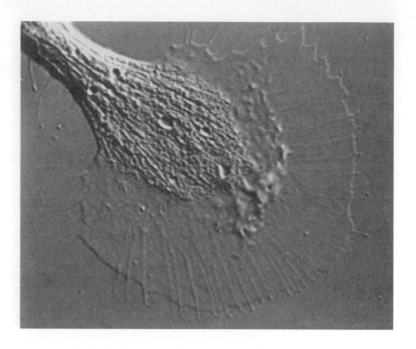

■ FIXED SIGNALS FOR AXON GUIDANCE: EXTRACELLULAR MATRIX AND CELL SURFACE MOLECULES

What is the nature of the cues that guide growing axons and elicit these changes at decision points? Based on a large body of work over the past 25 years, two broad categories of informational molecules have been described: cell surface and extracellular markers, which are primarily adhesive molecules that form the equivalent of molecular highways; and diffusible molecules that set up attractive or repellent gradients for growing axons.

As they navigate through the tissues of the developing embryo, growth cones encounter not only the surfaces of other neuronal and non-neuronal cells, but a variety of molecules located in the **extracellular matrix**, a complex of substances produced by cells but not directly attached to cell membranes.

Figure 21.2 Growth cone behavior at a decision point (the optic chiasm). (A) In ▶ the embryonic mouse visual system, the growing axons of retinal ganglion cells reach optic chiasm at about embryonic day 13 (E13); some temporal axons from each retina remain on the same side of the brain, while most axons cross to the opposite side (the diagram does not reflect the actual proportions of each). (B) As growth cones approach the chiasm, they change their speed and shape, as seen in silhouettes of a living, dye-labeled growth cone of a retinal ganglion cell at various times (shown in hours). As the growth cone advances toward the chiasm, it has a tapered, streamlined appearance. As it crosses the midline, however, the growth cone slows down and becomes spread out and more complex; after crossing (at about 3 hours in this time-lapse recording), it regains a streamlined shape and advances more rapidly. (C) Cells at the optic chiasm express a number of extracellular cues that may elicit these changes. One candidate is a glycoprotein, known as CD44, found on cell surfaces at the chiasm. Eliminating either the protein or the cells expressing it prevents the normal decussation of the ganglion cell axons and the formation of the chiasm. (A and B from Godemont, 1994; C from Sretavan et al., 1995.)

(A)

Embryonic day 11

Nasal

Nasal

Anterior

Temporal

Temporal

Area represented
in photo (C)

Embryonic day 12

Retina

Optic nerve

Growth
cones

Embryonic day 15

Optic chiasm

(B)

0

Midline

1

Hours

2

3

4

0   100   200   300   400
μm

(C)

Normal development

CD44 not expressed

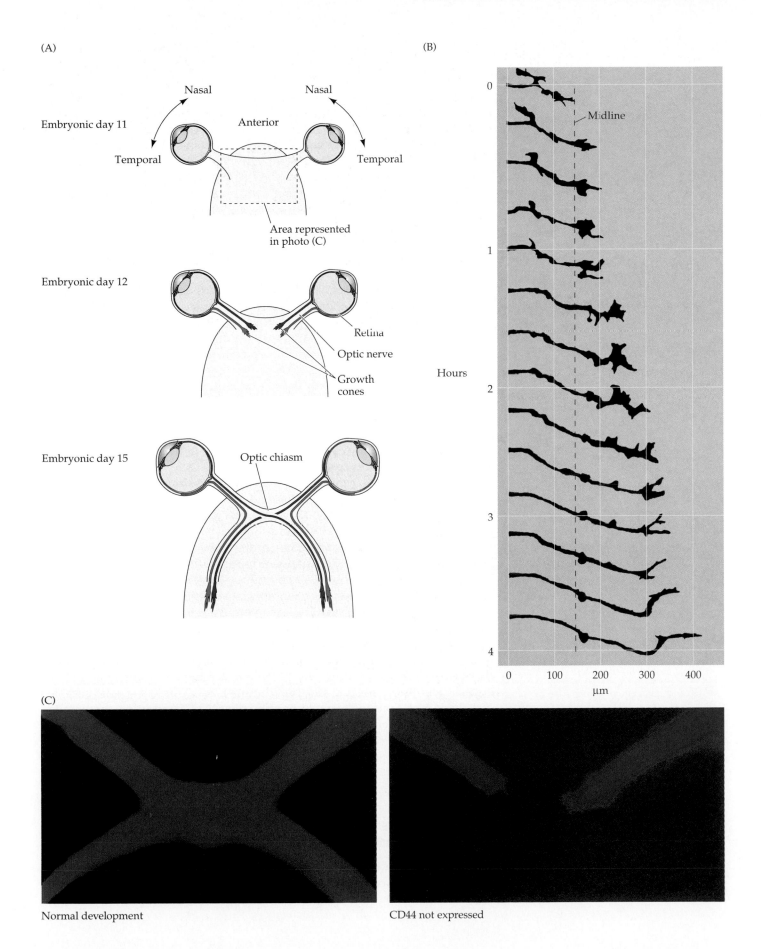

The extracellular matrix is particularly rich in several classes of large molecules that growth cones stick to, the most prominent being **laminin** and **fibronectin**. The growth cone membrane contains specific receptors, collectively known as **integrins**, that bind to these molecules; the binding of laminin and fibronectin to integrins triggers a cascade of events within the growth cone (involving changes in levels of intracellular messengers such as calcium and inositol trisphosphate) that promotes and guides axon extension.

The cell surfaces encountered by growth cones also express molecules that influence axon extension. One class of such proteins, the **cadherins** (of which there are at least ten different forms, each encoded by a different gene), is found on both the growth cone and the surfaces of cells with which the growth cone interacts. Cadherin molecules on growth cones recognize and bind to identical molecules on cell surfaces in a *calcium-d*ependent manner (hence their name); this homophilic binding is common to many cell-cell interactions.

Although the homophilic interactions of cadherins depend on the presence of extracellular calcium, interactions among another major group of cell surface molecules, the calcium-independent **cell adhesion molecules** (**CAMs**) do not. Cell adhesion molecules, which are structurally similar to immunoglobulin molecules, come in two principal flavors. The neuron-glial cell adhesion molecule, or **Ng-CAM**, promotes elongation of axons along astrocytes and Schwann cells in the central and peripheral nervous systems, respectively. The neural cell adhesion molecules, or **N-CAMs**, are found on many growing axons; homophilic interactions between N-CAMs allow adjacent axons to form bundles as they extend, with later growing axons following pathways formed by earlier "pioneers." Some of the ways that these various molecules interact are diagrammed in Figure 21.3.

Some extracellular matrix and cell surface molecules, such as laminin, N-cadherin, and N-CAM, are common to many developing pathways in the nervous system and probably act as generic growth-promoting molecules for many classes of axons. Variants of these molecules, which may be encoded by individual members of a multigene family through differential splicing of mRNA or posttranslational modifications of the proteins, have more restricted patterns of expression during development and presumably guide specific subsets of axons to appropriate targets.

The importance of these adhesive interactions for axon guidance is highlighted by several inherited human conditions that lead to mental retardation and hydrocephalus (an abnormal enlargement of the cerebral ventricles that often causes atrophy of the cerebral cortex). Individuals with these syndromes—X-linked hydrocephalus, MASA (an acronym for *m*ental retardation, *a*phasia, *s*huffling gait, and *a*dducted thumbs), and X-linked spastic paraplegia—have mutations in the gene encoding Ng-CAM. In addition, these mutations can lead to the absence of the corpus callosum, which connects the two cerebral hemispheres, and of the corticospinal tract, which carries cortical information to the spinal cord. Congenital anomalies such as these (which are fortunately quite rare) are now understood as errors in the signaling mechanisms responsible for axon navigation.

■ DIFFUSIBLE SIGNALS FOR AXON GUIDANCE: CHEMOTROPIC FACTORS

The great Spanish neuroanatomist Santiago Ramón y Cajal proposed long ago that growing axons are guided to their targets by secretion of "attractive factors" emanating from target tissues. With remarkable prescience, Cajal

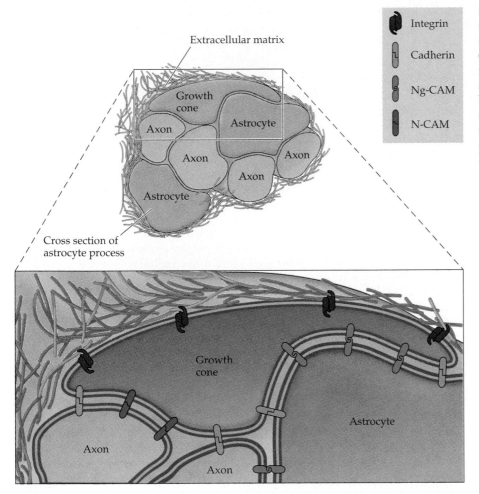

Figure 21.3 Molecular interactions that enable growth cone navigation through the developing nervous system. Growth cones bind to extracellular matrix molecules such as laminin and fibronectin by means of integrins; they bind to other neuronal elements (such as other axons) and to non-neuronal cells (such as glia) via cadherins and the cell adhesion molecules Ng-CAM and N-CAM. (After Reichart et al., 1991.)

deduced that signals originating from different targets could selectively influence the movement of axonal growth cones and thereby attract them to appropriate destinations. A variety of experiments carried out both in vivo and in vitro have confirmed this basic idea. The identity of the signals, however, has proven remarkably difficult to establish. One problem is the vanishingly small amounts of such factors in the developing embryo. Another is that of distinguishing **tropic** molecules—which guide growing axons toward a source—from **trophic** molecules—which support the survival and growth of neurons and their processes (see below).

The first family of molecules to meet all the criteria for a genuine chemotropic attractant is the **netrins** (from the Sanskrit "to guide"). These agents were isolated and their genes cloned a little more than a century after their existence was proposed by Cajal. In the developing spinal cord, a specific group of axons, called the commissural fibers, grow from the dorsal region of the cord to its ventral part (Figure 21.4A). The floor plate—a small, well-defined area at the ventral margin of the developing spinal cord—attracts growing commissural axons (Figure 21.4B). When bits of floor plate are placed next to pieces of spinal cord developing in vitro, the same attractive effects are observed. The netrins reproduce all of these attractive behaviors. Moreover, inserting netrin genes into non-neuronal cells causes the cells to secrete netrins, and to act as chemotropic sources for the commissural

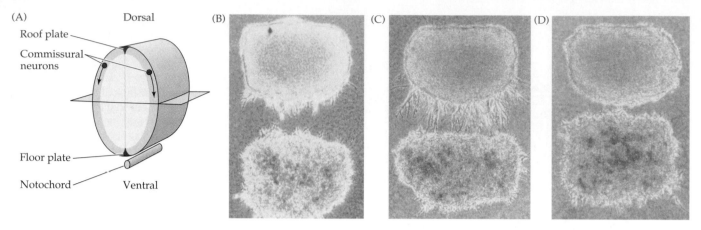

Figure 21.4 Chemotropic molecules (netrins) in the developing spinal cord. (A) The embryonic spinal cord and the cuts used to prepare explants from chick embryos. The commissural neurons send axons to the ventral region of the spinal cord, including a specific region called the floor plate. (B) Axons from commissural neurons in an explant culture (upper piece of tissue) grow directly toward a floor plate explant (lower piece of tissue) about 100 μm away. The axon bundles from the spinal cord explant grow only in the direction of the floor plate, indicating the presence of a diffusible chemotropic factor originating from the floor plate tissue. (C) If the floor plate is replaced by an aggregate of cells from a cell line transfected with the netrin gene, the identical behavior is observed. (D) Control cells, which do not contain the netrin gene, elicit no outgrowth. (A after Serafini et al., 1994; B, C, and D from Kennedy et al., 1994.)

axons (Figure 21.4C,D). In addition to their positive chemotropic actions on the commissural axons, netrins also exert negative, growth-inhibiting effects on other groups of growing axons. Thus, the same signal can tell one set of growing axons, the commissural fibers in this case, to "come hither," while simultaneously sending a message to other axons to "stay away" (see below).

A further unexpected finding was that the netrin genes are remarkably similar to a previously described gene from the nematode *C. elegans* (called *unc-6*) that helps guide neuronal outgrowth in the worm. This cross-species evidence suggests that at least some netrins were present early in evolution; permutations of a simpler signaling system have evidently produced a variety of netrin-like molecules in vertebrates to guide different types of neurons.

■ NEGATIVE REGULATION OF AXON GROWTH: INHIBITORS AND CHEMOREPELLENTS

Most research on axon guidance has focused on molecules that enhance axon outgrowth or attract growing neurons. But constructing the nervous system also requires telling axons where *not* to grow (as suggested by the netrin story). Recently, molecules that inhibit or repel growing axons have received increasing attention. Some of the thorniest problems in neurology—such as the failure of central nervous system fiber tracts damaged by trauma or disease to regenerate—may be explained by the presence of such inhibitors rather than by the absence of growth-promoting factors (Box A).

Two broad classes of inhibitors have been described. One class of these molecules is bound to cell surfaces or to the extracellular matrix, where it acts to prevent the extension of nearby axons. An example is the protein IN-1 (inhibitor-1), which is secreted by central nervous system oligodendrocytes (the myelin-producing glial cells) and incorporated into the myelin sheaths surrounding central axons. Experiments both in vivo and in vitro indicate that this protein inhibits axonal growth. A second class of molecules acts as a diffusible inhibitory factor, or chemorepellent. Evidence from two separate lines of research—one in the developing chick brain and the other in the developing nervous systems of grasshoppers and fruit flies—led to the discovery of a family of such diffusible growth-inhibiting molecules, the **semaphorins** (after the word *semaphore*, which means "signal"). The discovery of one member of this family resulted from experiments carried out by Jonathan Raper that showed that brain extracts from chicks cause collapse and retraction of axon growth cones in tissue culture. The responsible molecule—**collapsin**—was eventually purified and its gene cloned. At about the same time, Corey Goodman and co-workers discovered that a monoclonal antibody to a cell surface glycoprotein in grasshoppers (fasciclin IV) bound to one member of a family of proteins that guide axons in *Drosophila*. Using the polymerase chain reaction to find similar molecules in a variety of species, including humans, they identified and sequenced the genes for four structurally related proteins (the semaphorins). One of these, semaphorin III, is closely related to the chick protein collapsin. The semaphorins can exist as either secreted or cell surface molecules, and their genes are expressed in specific patterns in both developing vertebrate and invertebrate embryos. Studies in cell culture have shown that the secreted semaphorins repel specific groups of growing axons while having no effects on others (Figure 21.5).

It is likely that the pattern of expression of these and other chemorepellent agents, coupled with the patterns of expression of the chemoattractants, provide the molecular scaffolding that guides the formation of the major axon tracts in the developing nervous system.

■ THE FORMATION OF TOPOGRAPHICAL MAPS: RECOGNITION MOLECULES

As described in earlier chapters, the central representations of the sensory periphery are highly ordered in the visual and somatic sensory systems,

(A)

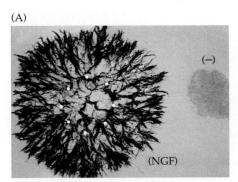

(B)

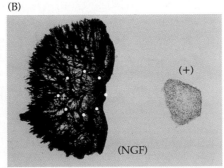

Figure 21.5 Semaphorins act as chemorepulsive cues. (A) In the presence of nerve growth factor (NGF; see below) explant cultures of chick dorsal root ganglia extend halos of neurites that originate from different neuronal subpopulations. (B) Co-culture of a ganglion with non-neuronal cells (+) transfected with the gene for semaphorin III (collapsin) results in asymmetrical growth of the ganglion cell neurites as a result of chemorepulsion. Control cells not transfected with the gene [(−) in panel A] have no effect on the pattern of outgrowth. (From Messersmith et al., 1995.)

Box A
WHY AREN'T WE MORE LIKE FISH AND FROGS?

The central nervous system of adult mammals, including humans, recovers only poorly from injury. Once severed, major axon tracts, such as those in the spinal cord, never regenerate. The devastating consequences of these injuries—loss of movement and the inability to control basic bodily functions—has led many neurobiologists to seek ways of restoring the connections of severed axons. There is no a priori reason for this biological failure, since "lower" vertebrates—such as lampreys, fish, and frogs—can regenerate a severed spinal cord or optic nerve. Even in mammals, the inability to regenerate axonal tracts is a special failing of the central nervous system; peripheral nerves can and do regenerate in adult animals, including humans. Why, then, not the central nervous system?

The answer to this puzzle probably lies in the molecular cues that promote and inhibit axon outgrowth. In mammalian peripheral nerves, axons are surrounded by a basement membrane—a proteinaceous extracellular layer composed of collagens, glycoproteins, and proteoglycans—that is secreted in part by Schwann cells, the glial cells associated with peripheral axons. After a peripheral nerve is crushed, the axons within it degenerate; the basement membrane around each axon, however, persists for months. One of the major components of the basement membrane is laminin, which (along with other growth-promoting molecules in the basement membrane) forms a hospitable environment for regenerating growth cones. The surrounding Schwann cells also react by

releasing neurotrophic factors, which further promote axon elongation.

This peripheral environment is so favorable to regrowth that even neurons from the central nervous system can be induced to extend into transplanted segments of peripheral nerve. Albert Aguayo and his colleagues at the Montreal General Hospital found that grafts derived from peripheral nerves can act as "bridges" for central neurons (in this case, retinal ganglion cells), allowing them to grow for over a centimeter; they even form a few functional synapses in their target tissues.

These several observations suggest that the failure of central neurons to regenerate is not due to an intrinsic inability to sprout new axons, but rather to something in the local environment that prevents growth cones from extend-

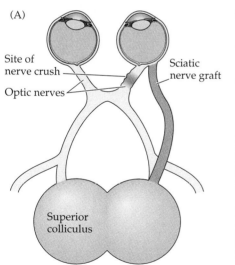

(A)

Site of nerve crush

Optic nerves

Sciatic nerve graft

Superior colliculus

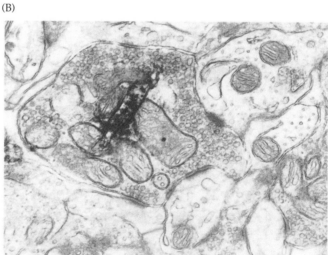

(B)

Implantation of a section of peripheral nerve into the central nervous system facilitates the extension of central axons. (A) Mammalian retinal ganglion neurons, which do not normally regenerate following a crush injury, will grow for many millimeters into a graft derived from the sciatic nerve. (B) If the distal end of the graft is inserted into a normal target of retinal ganglion cells, such as the superior colliculus, a few regenerating axons invade the target and form functional synapses, as shown in this electron micrograph. The dark material is an intracellularly transported label that identifies particular synaptic terminals as originating from a regenerated retinal axon. (A after So and Aguayo, 1985; B from Bray et al., 1991.)

ing. This impediment could result either from the absence of growth-promoting factors—such as the neurotrophins—or from the presence of molecules that actively prevent axon outgrowth. Studies by Martin Schwab and his colleagues point to the latter possibility. Schwab found that central nervous system myelin contains an inhibitory component that causes growth cone collapse in vitro and prevents axon growth in vivo (see discussion of collapsins and related chemorepellent molecules in the text). This component, called IN-1, is found in the myelinated portions of the central nervous system, but is absent from peripheral nerves. It is also found in the optic nerve and spinal cord of mammals, but is absent from the same sites in fish, which do regenerate these central tracts. IN-1 is secreted by oligodendrocytes, but not by Schwann cells in the peripheral nervous system. Most dramatically, antibodies to IN-1 substantially increase the extent of spinal cord regeneration in rats. All this implies that the human central nervous system differs from that of many "lower" vertebrates in that humans and other mammals present an unfavorable molecular environment for regrowth after injury. Why this state of affairs occurs is not known. One speculation is that brains that store extraordinary amounts of information may put a premium on stabilizing adult connectivity.

At present there is only one modestly helpful treatment for spinal cord injury. High doses of a steroid, methylprednisolone, immediately after the injury prevents some of the secondary damage to neurons resulting from the initial trauma. Although it may never be possible to fully restore function after such injuries, enhancing axon regeneration—by blocking inhibitory molecules and providing additional trophic support to surviving neurons—could allow sufficient recovery of motor control to give afflicted individuals a much better quality of life than they now enjoy.

References

BRAY, G. M., M. P. VILLEGAS-PEREZ, M. VIDAL-SANZ AND A. J. AGUAYO (1987) The use of peripheral nerve grafts to enhance neuronal survival, promote growth and permit terminal reconnections in the central nervous system of adult rats. J. Exp. Biol. 132: 5–19.

SCHNELL, L. AND M. E. SCHWAB (1990) Axonal regeneration in the rat spinal cord produced by an antibody against myelin-associated neurite growth inhibitors. Nature 343: 269–272.

VIDAL-SANZ, M., G. M. BRAY, M. P. VILLEGAS-PEREZ, S. THANOS AND A. J. AGUAYO (1987) Axonal regeneration and synapse formation in the superior colliculus by retinal ganglion cells in the adult rat. J. Neurosci. 7: 2894–2909.

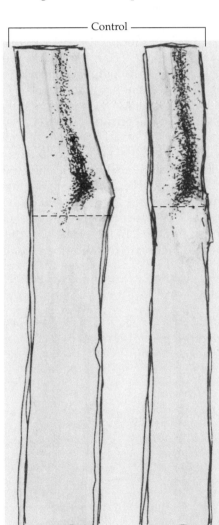

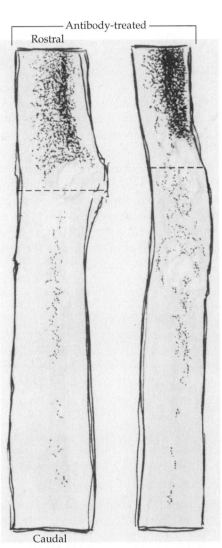

Blockade of a specific component of myelin facilitates spinal cord regeneration. Rat spinal cords were cut at the sites indicated by the dashed lines. Labeled corticospinal axons appear as black dots. Axons in control spinal cords stop at the site of transection. In animals in which antibodies to IN-1, an axon growth-inhibiting molecule, were infused at the site of transection, a small but significant proportion of axons grow past the cut, as indicated by the presence of the label in the more caudal regions of the cord. (From Schnell and Schwab, 1990.)

such that neighboring points in the periphery are represented by adjacent locations in the central nervous system (see Chapters 8 and 11). How do growing axons distribute themselves with such topographical orderliness across their target regions in the brain?

In the early 1960s, Roger Sperry, who later did pioneering work on the functional specialization of the cerebral hemispheres (Chapter 25), articulated the **chemoaffinity hypothesis**, based primarily on his work in the visual system of frogs and goldfish. In these animals, the terminals of retinal ganglion cells form a precise topographical map in the optic tectum (homologous to the mammalian superior colliculus). When Sperry crushed the optic nerve and allowed it to regrow (fish and amphibians, unlike mammals, can regenerate axonal tracts in their central nervous system; see Box A), he found that retinal axons reestablished a normal pattern of connections in the tectum. Even if the eye was rotated 180°, the regenerating axons grew back to their usual tectal destinations (Figure 21.6). Accordingly, Sperry proposed that each tectal cell carries a unique "identification tag"; he further supposed that the growing terminals of retinal ganglion cells have complementary tags, such that they seek out a specific location in the tectum. In modern parlance, these "chemical" tags are **recognition molecules**, and the "affinity" they engender is a selective binding of molecules on the growth cone to corresponding molecules on the tectal cells.

Further experiments in the lower vertebrate visual system have made the strictest form of the chemoaffinity hypothesis—labeling of each tectal location by a different recognition molecule—untenable. However, gradients of cell surface molecules to which growing axons respond *are* present in the tectum, a fact shown convincingly in a series of in vitro experiments carried out by Friedrich Bonhoeffer and his group (Figure 21.7). Normally, axons from the temporal region of the chick retina innervate the anterior pole of

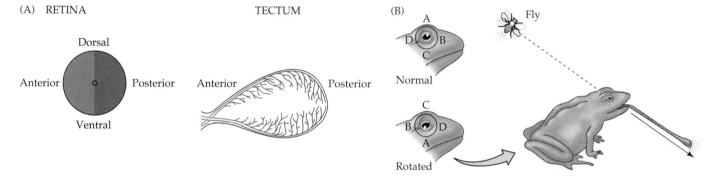

Figure 21.6 The axons of the retinal ganglion cells project to appropriate positions in the optic tectum during both development and regeneration in frogs and other "lower" vertebrates. (A) Posterior retinal axons project to the anterior tectum and anterior retinal axons to the posterior tectum. When the optic nerve of a frog is surgically interrupted, the axons regenerate with the appropriate specificity. (B) Even if the eye is rotated after severing the optic nerve, the axons regenerate to their original position in the tectum. That the topographical visual map in the tectum remains unchanged is evident from the frog's behavior: when a fly is presented above, the frog consistently strikes downward, and vice versa. This outcome indicates a specific matching of retinal neurons to their target cells in the tectum, a phenomenon taken to explain topographical mapping in the mammalian brain as well. (After Sperry, 1963.)

(A)

A   P   A   P   A   P   A

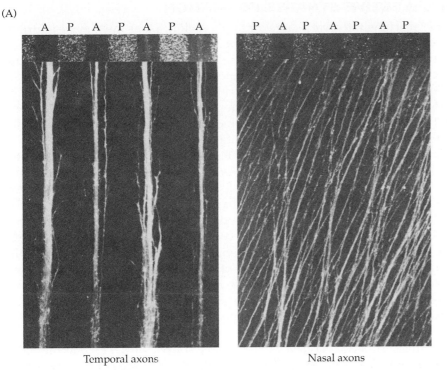

P   A   P   A   P   A   P

Temporal axons   Nasal axons

(B)

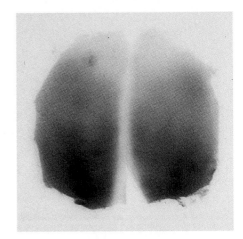

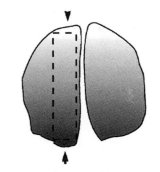

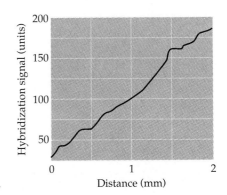

Figure 21.7 A possible molecular basis of retinotectal chemoaffinity. (A) An in vitro assay for cell surface molecules that may contribute to topographical specificity in the optic tectum. A set of alternating stripes (90 μm wide) of membranes from anterior (A) and posterior (P) optic tectum of chicks were formed on a glass coverslip. The posterior membranes have had fluorescent green particles added to make this set of stripes visible (top of panels). Subsequently, explants of retina were taken from either nasal or temporal retina and placed on the stripes. Temporal axons prefer to grow on anterior membranes and are repulsed by posterior membranes. In contrast, nasal retinal axons grow equally well on both stripes. (B) The messenger RNA for a mammalian homologue of the repellent protein from chick tectum, termed ELF-1, is distributed in an linear, anterior-to-posterior gradient in the developing mouse tectum, as indicated both by in situ hybridization (top) and measurements of the hybridization signal along the region indicated by the dashed rectangle. (A from Walter et al., 1987; B from Cheng et al., 1995.)

the tectum and avoid the posterior pole. In an in vitro assay using cell membranes derived from these two tectal regions, temporal retinal axons, when presented with a choice, grew exclusively on anterior membranes and avoided growing on membranes derived from the "wrong" region of the tectum.

The positive interactions probably are due to increased adhesion of the growth cones to the substrate; the failure to grow into inappropriate regions may result from interactions that collapse the growth cones (see above). A likely candidate for this negative guidance signal has been purified from the tectum and the gene encoding it cloned. The protein—called RAGS (*r*epulsive *a*xon *g*uidance *s*ignal)—is a membrane-anchored ligand for a specific type of tyrosine kinase-coupled receptor; this system of cellular interactions is analogous to the *boss-sevenless* system described in Chapter 20. Both the chick protein and a homologous mammalian protein are distributed in a well-defined gradient in the tectum (Figure 21.7B) and could thus help explain the topographical organization of the visual system.

■ SELECTIVE SYNAPSE FORMATION

After reaching the correct target or target region, axons must make a further local determination about which particular target cells to innervate among a variety of potential synaptic partners. Because of the complexity of central circuits, this issue has been studied most thoroughly in the peripheral nervous system, particularly in the innervation of autonomic ganglion cells. The problem was first explored by British physiologist John Langley at the end of the nineteenth century. Preganglionic sympathetic neurons located at different levels of the spinal cord innervate cells in sympathetic chain ganglia in a stereotyped and selective manner (Figure 21.8; see also Box A in Chapter 1). In the superior cervical ganglion, for example, cells from the highest thoracic level (T1) innervate ganglion cells that project in turn to targets in the eye, whereas neurons from a somewhat lower level (T4) innervate ganglion cells that cause constriction of the blood vessels of the ear. Since the axons of all these neurons run together in the cervical sympathetic trunk to arrive at the ganglion, the mechanisms underlying the differential innervation of the ganglion cells must occur at the level of synapse formation rather than axon guidance. Anticipating Sperry by more than 50 years in a quite different context, Langley concluded that selective synapse formation is based on differential chemoaffinities of the pre-and postsynaptic elements.

Modern studies based on intracellular recordings from individual neurons in the superior cervical ganglion have shown, however, that the selective affinities between pre- and postsynaptic neurons are not terribly restrictive. Thus, synaptic connections to ganglion cells made by preganglionic neurons of a particular spinal level are preferred, but synaptic contacts from neurons at other levels are not excluded. Furthermore, if the innervation to the superior cervical ganglion from a particular spinal level is surgically interrupted, recordings made some weeks later indicate that new connections are established by residual axons arising from what would normally be inappropriate spinal segments. The novel connections also establish a pattern of segmental preferences, as if the system attempts to achieve the best match it can under the altered circumstances. Despite this relative promiscuity during synapse formation, a quite different line of work has shown that *where* a synapse forms on the target cell is tightly controlled by a set of molecules that are now beginning to be understood (Box B).

From a broader vantage, of course, there are some absolute restrictions to synaptic associations. Thus, neurons do not innervate nearby glial or connective tissue cells, and many instances have been described in which various nerve and target cell types show little or no inclination to establish connections. When synaptogenesis does proceed, however, neurons and their targets in both the central and peripheral nervous systems appear to associate according to a continuously variable system of preferences. Such biases guide the pattern of innervation that arises in development or reinnervation without limiting it in any absolute way. The target cells residing in the optic

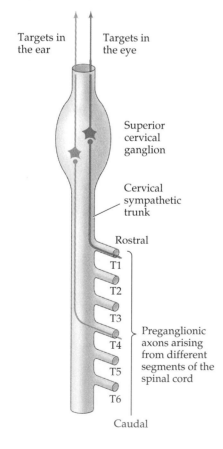

Targets in the ear

Targets in the eye

Superior cervical ganglion

Cervical sympathetic trunk

Rostral

T1
T2
T3
T4
T5
T6

Preganglionic axons arising from different segments of the spinal cord

Caudal

Figure 21.8 Evidence that synaptic connections between mammalian neurons form according to specific affinities between different classes of pre- and postsynaptic cells. In the superior cervical ganglion, preganglionic neurons located in particular spinal cord segments (T1, for example) innervate ganglion cells that project to particular peripheral targets (the eye, for example). The establishment of these preferential synaptic relationships indicates that selective neuronal affinities are a major determinant of neural connectivity.

tectum or in autonomic ganglia are certainly not equivalent, but neither are they unique with respect to the innervation they can receive.

■ TROPHIC INTERACTIONS AND THE ULTIMATE SIZE OF NEURONAL POPULATIONS

The formation of synaptic contacts between growing axons and their synaptic partners signals the beginning of a new stage of development. Once synaptic contacts are made, neurons become dependent in some degree on the presence of their targets for continued survival and differentiation; in the absence of synaptic targets, the axons and dendrites of developing neurons atrophy and the nerve cells may eventually die. This long-term dependency between neurons and their targets is referred to as **trophic interaction**. The word *trophic* is taken from the Greek *trophē*, meaning, roughly, "nourishment." Despite this name, the sustenance provided to neurons by trophic interactions is not the sort derived from metabolites such as glucose or ATP. Rather the dependence is based on specific signaling molecules called **neurotrophic factors**. Neurotrophic factors, like some other intercellular signaling molecules (mitogens and cytokines, for example), originate from target tissues and regulate neuronal survival and subsequent growth and differentiation.

Why do neurons depend so strongly on their targets, and what specific cellular and molecular interactions mediate this dependence? In part, the answer lies in the changing scale of the developing nervous system and the body it serves, and the related need to precisely match the number of neurons in particular populations with the size of their targets. The mechanisms by which neurons are initially generated have already been considered in Chapter 20. A general—and surprising—strategy in the development of many vertebrates is the production of an initial surplus of nerve cells (on the order of two- or threefold); the final population is subsequently established by the degeneration of those neurons that fail to interact successfully with their intended targets, a process now known to be mediated by neurotrophic factors.

Evidence that targets play a major role in determining the size of the neuronal populations that innervate them has come from a long series of studies dating from the start of the twentieth century. The seminal observation was that the removal of a limb bud from a chick embryo results, at later embryonic stages, in a striking reduction in the number of nerve cells in the corresponding portions of the spinal cord (Figure 21.9A,B). The interpretation of these experiments is that neurons, in the spinal cord in this case, compete with one another for a resource present in the target (the developing limb) that is available in limited supply. In support of this idea, many neurons that would normally have died can be rescued by augmenting the amount of target available, thereby providing extra trophic support (Figure 21.9C,D). Thus, the size of nerve cell populations in the adult is not fully determined in advance, but is governed in part by idiosyncratic neuron–target interactions in each developing individual.

The death of neurons deprived of trophic support from their targets is different from the cell death resulting from injury or disease. Trophically deprived neurons degenerate and die through a process called **apoptosis**, which depends on the active transcription of a host of specific genes that when "turned on" cause neurons or other cells to degenerate. The histopathology of neuronal death on this basis contrasts sharply with the demise of nerve cells as a result of trauma, stroke, or neurodegenerative disease. The

Box B
MOLECULAR SIGNALS THAT PROMOTE SYNAPSE FORMATION

Synapses require a precise organization of presynaptic and postsynaptic elements in order to function properly (see Chapters 5–7). At the neuromuscular junction, for example, synaptic vesicles and the related release machinery are located at sites in the nerve terminal called active zones; and, in the postsynaptic muscle cell, acetylcholine receptors are localized in high density exactly subjacent to the presynaptic active zones. During the past 20 years, a number of investigators have tried to identify the molecular cues that guide the formation of these carefully apposed elements. Their efforts have met with the greatest success at the neuromuscular junction, where a molecule called agrin is now known to be responsible for initiating at least some of the events that lead to the formation of a fully functional synapse.

Agrin was originally identified as a result of its influence in the reinnervation of frog neuromuscular junctions following damage to the motor nerve. In mature skeletal muscle, each fiber typically receives a single synaptic contact at a highly specialized region called the end plate (see Chapter 5). Jack McMahan and his colleagues found that regenerating axons precisely reinnervate the original end plate site. In seeking to determine the molecular signals underlying this phenomenon, they took advantage of the fact that each muscle fiber is surrounded by a sheath of extracellular matrix called the basal lamina (see text). When muscle fibers degenerate, they leave the basal lamina behind (as do degenerating axons); moreover, a specific infolding of the basal lamina at the former end plate site allows its continued identification. Remarkably, presynaptic nerve terminals differentiate at these original sites even when the associated muscle fibers are absent. Equally

remarkable is that regenerating muscle fibers form postsynaptic specializations—such as densely packed acetylcholine receptors—at precisely these same basal lamina locations in the absence of nerve fibers! These findings show that the signal(s) guiding synapse formation remain in the extracellular environment after removal of either

nerve or muscle, presumably in the basal lamina "ghost" that surrounds each muscle fiber.

Using a bioassay based on the aggregation of acetylcholine receptors to analyze the constituents of the basal lamina, one of the relevant molecules—agrin—was isolated and purified. Agrin is a protein found in both mammalian

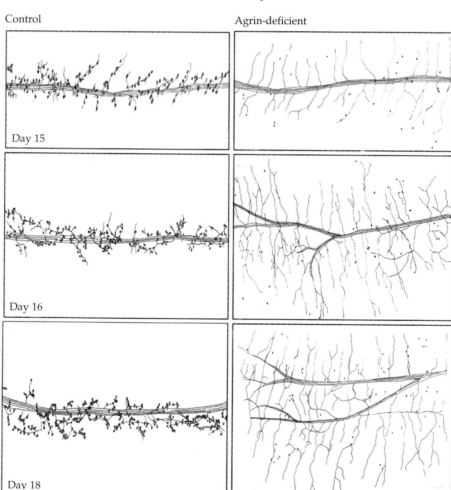

Control   Agrin-deficient

Day 15

Day 16

Day 18

Development of neuromuscular junctions in agrin-deficient mice. Diaphragm muscles from control (left) and agrin-deficient (right) mice at embryonic day 15, 16, and 18 were double-stained for acetylcholine receptors and axons, then drawn with a camera lucida. The developing muscle fibers run vertically. In both control and mutant muscles, an intramuscular nerve (black) and aggregates of AChRs (red) are present by embryonic day 15. In controls, axonal branches and AChR clusters are confined to a band at the central end plate at all stages. Mutant AChR aggregates are smaller, less dense, and less numerous; axons form fewer branches and their synaptic relationships are disorganized. (From Gautam et al., 1996.)

motor neurons and muscle fibers; it is also abundant in brain tissue. The neuronal form of agrin is synthesized by motor neurons, transported down their axons, and released from growing nerve fibers. Agrin binds to a postsynaptic receptor whose activation leads to a clustering of acetylcholine receptors and, evidently, to subsequent events in synaptogenesis. Strong support for the role of agrin as an organizer of synaptic differentiation is the recent finding by Josh Sanes and his collaborators that genetically engineered mice that lack the gene for agrin develop in utero with few neuromuscular junctions. Animals missing the agrin receptor also fail to develop neuromuscular junctions and die at birth. Agrin is therefore one of the first examples of a molecule that promotes the formation of synapses at specific sites on target cells.

Because synapse formation requires an ongoing dialogue between pre- and postsynaptic partners, it is likely that a number of signaling systems are involved at various stages in the process. Another such molecule may be s-laminin, a synapse-specific form of this extracellular matrix molecule. So far, however, agrin and its receptor are the best example of molecules whose primary function is to guide synapse formation. The effects of agrin at developing synapses in the central nervous system remain to be explored.

References

DeChiara, T. M. and 14 others (1996) The receptor tyrosine kinase MuSK is required for neuromuscular junction formation in vivo. Cell 85: 501–512.

Gautam, M., P. G. Noakes, L. Moscoso, F. Rupp, R. H. Scheller, J. P. Merlie and J. R. Sanes (1996) Defective neuromuscular synaptogenesis in agrin-deficient mutant mice. Cell 85: 525–535.

McMahan, U. J. (1990) The agrin hypothesis. Cold Spring Harbor Symp. Quant. Biol. 50: 407–418.

Noakes, P. G., M. Gautam, J. Mudd, J. R. Sanes and J. P. Merlie (1995) Aberrant differentiation of neuromuscular junctions in mice lacking s-laminin/laminin β2. Nature 374: 258–262.

Reist, N. E., M. J. Werle and U. J. McMahan (1992) Agrin released by motor neurons induces the aggregation of AChRs at neuromuscular junctions. Neuron 8: 677–689.

Sanes, J. R., L. M. Marshall and U. J. McMahan (1978) Reinnervation of muscle fiber basal lamina after removal of myofibers. J. Cell Biol. 78: 176–198.

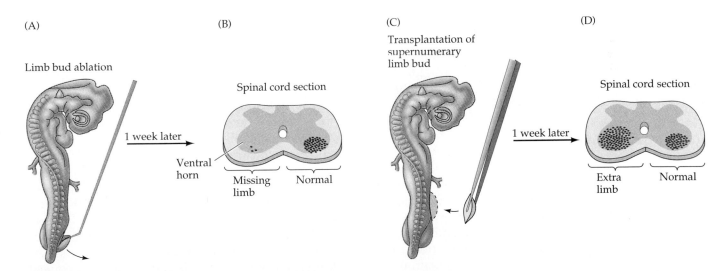

Figure 21.9 Effect of removing or augmenting neural targets on the survival of related neurons. (A) Limb bud amputation in a chick embryo at the appropriate stage of development (about 2.5 days of incubation) depletes the pool of motor neurons that would have innervated the missing extremity. (B) A cross section of the lumbar spinal cord in an embryo that underwent this surgery about a week earlier. The motor neurons (dots) in the ventral horn that would have innervated the hindlimb degenerate almost completely after embryonic amputation; a normal complement of motor neurons is present on the other side. (C) Adding an extra limb bud before the normal period of cell death rescues neurons that normally would have died. (D) Such augmentation leads to an abnormally large number of limb motor neurons (dots) on the side related to the extra limb. (After Hamburger, 1958, 1977; Hollyday and Hamburger, 1976.)

cellular and molecular processes underlying apoptosis appear to involve many of the same mechanisms that govern cell differentiation and control of the cell cycle. Thus, cell death by apoptosis is an actively determined state of cell differentiation.

■ TROPHIC INTERACTIONS AND THE FORMATION OF NEURONAL CONNECTIONS

Once neuronal populations are established by this competitive winnowing, trophic interactions continue to modulate the formation of synaptic connections, a process that begins in embryonic life, but extends far beyond birth. Among the problems that must be solved during the establishment of innervation is insuring that each target cell is innervated by the right number of axons, and that each axon innervates the right number of target cells. Getting these numbers right is evidently another major achievement of trophic interactions between developing nerve and target cells.

This aspect of trophic interactions has been studied most thoroughly in the peripheral nervous system. For example, most skeletal muscle fibers in adult mammals are innervated at a single end plate site by one motor axon (see Chapter 15). Neonatal muscle fibers, however, are innervated by terminals arising from several different axons, a relationship generally referred to as polyneuronal innervation (see Chapter 22). The transition from polyneuronal innervation to the mature one-on-one relationship of motor axons and muscle fibers occurs gradually in postnatal life. Moreover, the elimination of a portion of the initial innervation to a muscle occurs without the loss of any of the spinal motor neurons involved; thus, this transition is not simply a result of ongoing motor neuron death. A similar reorganization is evident in a variety of other peripheral and central nervous system regions. In the peripheral nervous system, the number of presynaptic axons innervating each neuron can also fall markedly, as demonstrated by studies of certain autonomic ganglia. In the cerebellum, each adult Purkinje cell is innervated by a single climbing fiber (see Chapter 18); early in development, however, each Purkinje cell receives multiple climbing fiber inputs.

Clearly, the pattern of synaptic connections that emerges in the adult is not simply a consequence of the biochemical identities of synaptic partners or of other determinate developmental rules. Rather, the wiring plan in maturity is also the result of a much more flexible process in which neuronal connections are formed or removed according to local circumstances. These interactions guarantee that every target cell is innervated—and continues to be innervated—by the right number of inputs and synapses, and that every innervating axon contacts the right number of target cells with an appropriate number of synaptic endings. The regulation of convergence (the number of inputs to a target cell) and divergence (the number of connections made by a neuron) in the developing nervous system is another key consequence of trophic interactions among neurons and their targets.

■ A MOLECULAR PARADIGM FOR TROPHIC INTERACTIONS: THE NEUROTROPHINS

The two major functions of neurotrophic signaling—the survival of a subset of neurons from a considerably larger population, and the formation of appropriate numbers of connections—can be understood in terms of the same basic rules about the supply and availability of trophic factors. These rules, which define the **neurotrophic hypothesis**, entail several assumptions about neurons and their targets (which may be other neurons, muscles, or other peripheral structures). First, neurons depend on the availability of some minimum amount of trophic factor for survival, and subsequently for the persistence of their target connections. Second, target tissues synthesize and make available to developing neurons appropriate trophic factors. Third, targets produce trophic factors in limited amounts; in consequence, the survival of developing neurons (and later, the persistence of neuronal connections) depends on neuronal competition for the available factor. One much-studied molecule, the peptide called **nerve growth factor** (**NGF**), has provided strong support for these several assumptions. Although the story of nerve growth factor certainly does not explain all aspects of trophic interactions, it has proved to be a remarkably useful paradigm for understanding the manner in which neural targets influence the survival and connections of the nerve cells that innervate them.

Nerve growth factor was discovered in the early 1950s by Rita Levi-Montalcini and Viktor Hamburger at Washington University. On the basis of experiments involving the survival of motor neurons after removal of developing limb buds (see Figure 21.9), they made an informed guess that targets provided some sort of signal to the relevant neurons, and that limited amounts of this agent explained the apparently competitive nature of nerve cell death. Accordingly, Levi-Montalcini and Hamburger undertook a series of experiments to explore the source and nature of the postulated signal, focusing on dorsal root and sympathetic ganglion neurons rather than the spinal cord cells. A former student of Hamburger's had earlier removed a limb from a chick embryos and replaced it with a piece of mouse tumor. The remarkable outcome of this experiment was that the tumor apparently furnished an even more potent stimulus than the limb, causing an enlargement of the sensory and sympathetic ganglia that normally innervate the appendage. In further experiments, Levi-Montalcini and Hamburger provided evidence that the tumor (a mouse sarcoma) secreted a soluble factor that stimulated the survival and growth of both sensory and sympathetic ganglion cells. Levi-Montalcini then devised a bioassay for the presumed agent, and, in collaboration with Stanley Cohen, isolated and characterized the molecule—which had by then been named nerve growth factor for its ability to induce the massive outgrowth of neurites from explanted ganglia (Figure 21.10). (The term *neurite* is used to describe neuronal branches when it is not known whether they are axons or dendrites.) NGF was identified as a protein and was substantially purified from a rich biological source, the salivary glands of the male mouse. Subsequently, its amino acid sequence was determined and the cDNAs encoding NGF cloned in several species; in 1991, the crystal structure of the active moiety of NGF was finally solved (Figure 21.11).

Support for the idea that NGF is important for neuronal survival in more physiological circumstances emerged from a number of further observations. Depriving developing mice of NGF by the chronic administration of an NGF antiserum resulted in adult mice lacking most sympathetic neurons (Figure 21.12). Conversely, injection of exogenous NGF into newborn rodents caused enlargement of sympathetic ganglia, an effect opposite that of NGF deprivation. Neurons in these treated ganglia were both more numerous and larger; there was also more neuropil between cell bodies, suggesting an overgrowth of axons, dendrites, and other cellular elements. The dramatic influence of NGF on cell survival, together with what was known about the significance of neuronal death in development, suggested that NGF was indeed a target-derived signal that served to match the number of nerve cells to the number of target cells.

Figure 21.10 Effect of NGF on the outgrowth of neurites. (A) A chick sensory ganglion taken from an 8-day-old embryo and grown in organ culture for 24 hours in the absence of NGF. Few, if any, neuronal branches grow out into the plasma clot in which the explant is embedded. (B) A similar ganglion in identical culture conditions 24 hours after the addition of NGF to the medium. NGF stimulates a halo of neurite outgrowth from the ganglion cells. (From Purves and Lichtman, 1985; courtesy of R. Levi-Montalcini.)

(A)　　　　　　　　　(B)

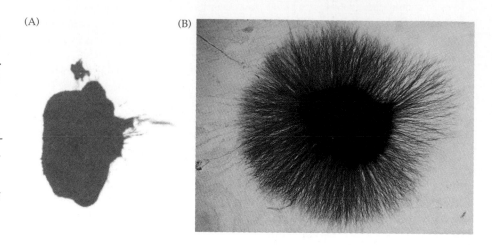

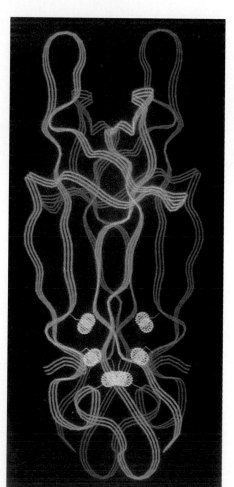

The ability of NGF to support neuronal survival (and of NGF antiserum to enhance cell death) is not in itself unassailable proof of a physiological role for this factor in development. In particular, these observations provided no direct evidence for NGF synthesis by (and uptake from) neuronal targets. This gap was filled by another series of ingenious experiments in several laboratories that showed NGF to be present in sympathetic targets, and to be quantitatively correlated with the density of sympathetic innervation. Furthermore, messenger RNA for NGF was demonstrated in targets innervated by sympathetic and sensory ganglia, but not in the ganglia themselves or in targets innervated by other types of nerve cells. As might be expected from such specificity, the NGF-sensitive neurons were also shown to have receptor molecules for the trophic factor. Importantly, the NGF message appears only after ingrowing axons have reached their targets; this fact makes it unlikely that secreted NGF acts as a chemotropic guidance molecule (like the netrins discussed earlier). Finally, the great majority of sympathetic neurons are lacking in genetic "knockout" mice in which the gene encoding NGF has been deleted.

In sum, several decades of work in a number of laboratories have shown that NGF mediates cell survival among two specific neuronal populations in birds and mammals (sympathetic and a subpopulation of sensory ganglion cells). These observations include the death of the relevant neurons in the absence of NGF; the survival of a surplus of neurons in the presence of augmented levels of the factor; the presence and production of NGF in neuronal targets; and the existence of receptors for NGF in innervating nerve terminals. Indeed, these observations define the criteria that must be satisfied in order to conclude that a given molecule is indeed a trophic factor. Although NGF remains the prototypical (and certainly most thoroughly studied) neurotrophic factor, it was apparent from the outset that only certain classes of nerve cells respond to NGF. A flurry of work in the last few years has shown that NGF is only one member of a family of related trophic molecules, the **neurotrophins**. At present, there are four characterized members of the neurotrophin family in addition to NGF: **brain derived neurotrophic factor (BDNF)**, **neurotrophin-3 (NT-3)**, and **NT-4/5**.

Figure 21.11 The crystal structure of the active component of the nerve growth factor complex. The active moiety of NGF consists of a dimer of identical β subunits. (From McDonald et al., 1991.)

(A)

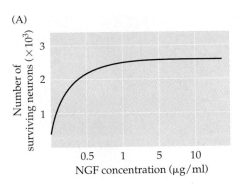

(B)

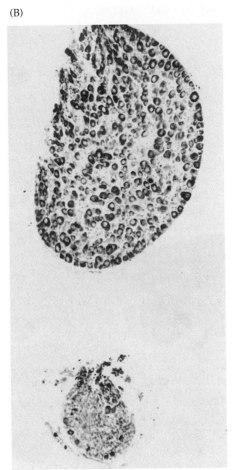

Figure 21.12 Effect of NGF on the survival of sympathetic ganglion cells. (A) The survival of newborn rat sympathetic ganglion cells grown in culture for 30 days evaluated quantitatively as a function of NGF concentration. Dose-response curves such as this one confirm the strict dependence of these neurons on the availability of NGF. (B) Cross section of a superior cervical ganglion from a normal 9-day-old mouse (*top*) compared to a similar section from a littermate injected daily since birth with NGF antiserum (*below*). The ganglion of the treated mouse shows marked atrophy, with obvious loss of nerve cells. (A after Chun and Patterson, 1977; B from Levi-Montalcini, 1972.)

■ NEUROTROPHIN RECEPTORS

Although all the neurotrophins are highly homologous in amino acid sequence and structure, they are very different in their specificity (Figure 21.13). For example, whereas NGF supports the survival of (and neurite outgrowth from) sympathetic neurons, another family member—BDNF—cannot. Conversely, BDNF, but not NGF, can support the survival of nodose ganglion neurons, which have a different embryonic origin. NT-3 supports both of these normal populations, indicating that the specificity of neurotrophins is both distinct and overlapping (see Figure 21.13). These specificities of action are exactly what one might expect, given the diverse systems that must be coordinated during neural development.

The selective actions of the neurotrophins arise from a family of receptor proteins, the **Trk receptors**, that are activated by different neurotrophins. (These proteins were originally identified as receptors associated with *tyrosine kinase* activity and were thus abbreviated as Trk, pronounced "track".) **TrkA** is primarily a receptor for NGF, **TrkB** a receptor for BDNF, and **TrkC** a receptor for NT-3 (Figure 21.14). The expression of a particular Trk receptor subtype confers on that neuron the capacity to respond to the corresponding neurotrophin. Since neurotrophins and Trk receptors are expressed only in certain cell types in the nervous system, the binding between ligand and receptor accounts for the specificity of neurotrophic interactions.

The structure and activation of Trk receptors closely resemble those of non-neuronal growth factor receptors, such as the epidermal growth factor

Figure 21.13 The effects of the neurotrophins NGF, BDNF, and NT-3 on the
outgrowth of neurites from explanted,
dorsal root ganglia (left column), nodose ganglia (middle column), and sympathetic ganglia (right column). The
specificities of these several neurotrophins are evident in the ability of NGF to
induce neurite outgrowth from sympathetic and dorsal root ganglia, but not
from nodose ganglia; of BDNF to induce
neurite outgrowth from dorsal root and
nodose ganglia, but not from sympathetic ganglia; and of NT-3 to induce
neurite outgrowth from all three types
of ganglia. (From Maisonpierre et al.,
1990.)

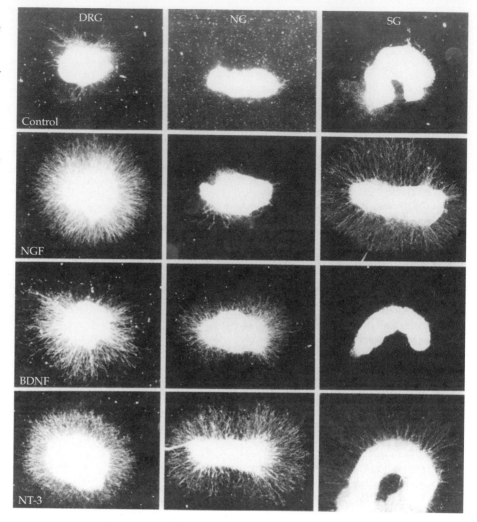

Figure 21.14 The Trk family of receptor tyrosine
kinases for the neurotrophins. TrkA is primarily a receptor for NGF, TrkB a receptor for BDNF and NT-4,
and TrkC a receptor for NT-3. Because of the high
degree of structural homology among both the neurotrophins and the Trk receptors, there is some degree
of cross-activation between factors and receptors. For
example, NT-3 can bind to and activate TrkB under
some conditions, as indicated by the dashed arrow.
Moreover, all neurotrophins bind to a low-affinity neurotrophin receptor called p75 (not shown). Although
the function of p75 remains unclear, it is probably
involved in the formation of high-affinity receptor complexes (along with Trk receptors) for neurotrophins.
These distinct receptors allow various neurons to
respond selectively to the different neurotrophins.

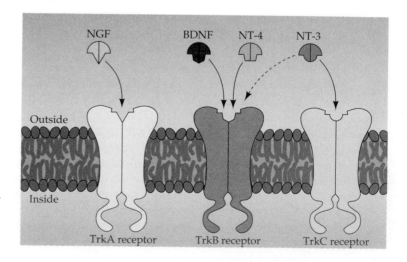

(EGF) receptor. Interestingly, activation of Trk receptors transfected into non-neuronal cells induces cell proliferation, which is the normal response of cells like fibroblasts to EGF. These similarities imply that the intracellular actions of the tyrosine kinase moiety of the Trk receptors are similar to those of other growth factor receptors. Activation of Trk receptors by neurotrophins presumably leads to a cascade of intracellular signaling events that eventually elicits changes in the pattern of gene expression in target neurons, thereby leading to all the manifestations of trophic interactions.

■ THE EFFECT OF NGF ON THE DIFFERENTIATION OF NEURONAL FORM

As already noted, a second major role of nerve growth factor and other trophic molecules (in addition to their influence on neuron survival) is to modulate the growth of neuronal branches. In the case of NGF, a compelling indication of this further action is that explanted sensory or sympathetic ganglia exposed to a culture medium containing NGF show a marked outgrowth of neurites within 24 hours (see Figure 21.10). More specific evidence has been obtained using a culture system that distinguishes the local effects of NGF on neurites from effects mediated through the neuronal cell body. In this system, dissociated ganglion cells are placed in the central well of a chamber with three compartments whose NGF concentration can be varied independently (Figure 21.15). If the three compartments contain adequate concentrations of NGF, then neurites from the ganglion cells in the central well extend into both peripheral compartments. However, if NGF is removed from one of the peripheral wells, then the neurites that have grown into that compartment gradually retract. Conversely, if NGF is removed from the central compartment but retained in the peripheral wells, the neurites remain in place. These results indicate that neurites extend or retract as a function of the *local* concentration of NGF. If the effects of NGF on neurites depended only on sufficient trophic factor being supplied to the nerve cell as a whole, then the neurites in the different compartments should grow regardless of which compartments contained NGF.

In short, neurite outgrowth can be controlled locally by trophic stimuli and does not simply depend on the overall effects of trophic agents on the parent cell. As a result, some branches of a neuron may extend while others

Figure 21.15 Evidence that NGF can influence neurite growth by local action. Three compartments of a culture dish (A, B, C) are separated from one another by a plastic divider sealed to the bottom of the dish with grease. Isolated rat sympathetic ganglion cells plated in compartment A can grow through the grease seal and into compartments B and C. (A magnified view looking down on the compartments is shown below.) Growth into a lateral chamber occurs as long as the compartment contains an adequate concentration of NGF. Subsequent removal of NGF from a compartment causes a local regression of neurites without affecting the survival of cells or neurites in the other compartments. These observations show that neuritic growth can be locally controlled by neurotrophins. (After Campenot, 1981.)

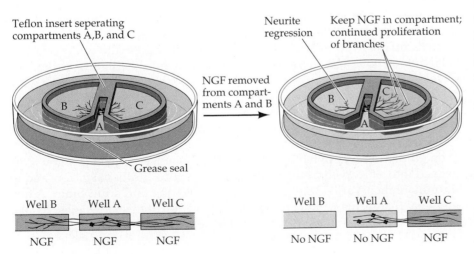

retract, which is what actually happens during the establishment of synaptic connections in normal development (Chapter 22).

■ A GENERAL SCHEME FOR THE ACTION OF TROPHIC MOLECULES

Taken together, the evidence about NGF and the other neurotrophins suggests a general scheme for the regulation of neuronal connections in the nervous system. Neuronal targets, whether non-neural cells or other neurons, produce trophic molecules in limited amounts. In embryonic and early postnatal life—and to a more limited extent in maturity—the survival of innervating neurons depends on exposure to a critical amount of these agents. In consequence, neurons sensitive to a particular trophic molecule initially compete with one another, and those that fail in this competition die. Following the establishment of definitive neuronal populations in this way, ongoing trophic dependency is apparent in the growth and retraction of neuronal processes, again as a function of target-derived support. In early postnatal life, this dependence is evident in the continuing growth and rearrangement of the initial connections whereby each neuron comes to innervate an appropriate number of target cells and each target cell comes to be contacted by an appropriate number of axons. Later in life, neural connections made by a fixed number of nerve cells continue to adjust by sprouting and retraction as targets change in size, form, and function during a prolonged period of maturation. In addition to mediating the compensatory adjustments required by growth, competition for trophic molecules allows neuronal branches and their connections to change in response to a variety of other circumstances, including injury and altered patterns of neural activity associated with experience, as described in the next chapter.

■ SUMMARY

Neurons in the developing brain must integrate a variety of molecular signals in order to determine whether to live or die, where to form synapses, how many synapses to make, and whether to retain them. Fixed and/or diffusible, chemotropic, chemorepulsive, and trophic molecules all regulate the trajectory of growing axons and the synaptic connections they make with target cells. Moreover, these interactions transpire over weeks, months, and indeed the lifetime of the animal. In the early stages of development, the most salient effects of trophic agents are on cell survival and differentiation. Once the adult population of neurons is established, trophic signals continue to govern the establishment of neural connections, particularly the extent of axonal and dendritic arborizations. Dysfunction of these molecular and cellular interactions has the potential for producing devastating pathology. Defects in the early guidance of axons are probably responsible for a wide range of congenital neurological syndromes; and conditions thought to reflect trophic dysfunction range from rare diseases such as familial dysautonomia to degenerative diseases such as amyotrophic lateral sclerosis and Parkinson's disease (see Chapter 18). Understanding the molecular basis of axon guidance and trophic signaling began a century ago and has now burgeoned into a broad effort that continues to identify additional factors and to illuminate their varied roles in both the developing and adult brain. A further goal that seems within reach is the application of this knowledge to alleviate a spectrum of previously intractable neurological diseases.

Additional Reading

Reviews

CHAO, M. V. (1992) Neurotrophin receptors: a window into neuronal differentiation. Neuron 9: 583–593.

KORSCHING, S. (1993) The neurotrophic factor concept: a reexamination. J. Neurosci. 13: 2739–2748.

LEVI-MONTALCINI, R. (1987) The nerve growth factor 35 years later. Science 237: 1154–1162.

PURVES, D. AND J. W. LICHTMAN (1978) Formation and maintenance of synaptic connections in autonomic ganglia. Physiol. Rev. 58: 821–862.

REICHARDT, L. F. AND K. J. TOMASELLI (1991) Extracellular matrix molecules and their receptors: functions in neural development. Annu. Rev. Neurosci. 14: 531–570.

RUTISHAUSER, U. (1993) Adhesion molecules of the nervous system. Curr. Opin. Neurobiol. 3: 709–715.

SCHWAB, M. E., J. P. KAPFHAMMER AND C. E. BANDTLOW (1993) Inhibitors of neurite growth. Annu. Rev. Neurosci. 16: 565–595.

SILOS-SANTIAGO, I., L. J. GREENLUND, E. M. JOHNSON JR. AND W. D. SNIDER (1995) Molecular genetics of neuronal survival. Curr. Opin. Neurobiol. 5: 42–49.

SNIDER, W. D. (1994) Functions of the neurotrophins during nervous system development: what the knockouts are teaching us. Cell 77: 627–638.

TAKEICHI, M. (1991) Cadherin cell adhesion receptors as a morphogenetic regulator. Science 251: 1451–1455.

Original Papers

BAIER, H. AND F. BONHOEFFER (1992) Axon guidance by gradients of a target-derived component. Science 255: 472–475.

CAMPENOT, R. B. (1977) Local control of neurite development by nerve growth factor. Proc. Natl Acad. Sci. USA 74: 4516–4519.

DRESCHER, U., C. KREMOSER, C. HANDWERKER, J. LOSCHINGER, M. NODA AND F. BONHOEFFER (1995) In vitro guidance of retinal ganglion cell axons by RAGS, a 25 kDa tectal protein related to ligands for Eph receptor tyrosine kinases. Cell 82: 359–370.

FREDETTE, B. J. AND B. RANSCHT (1994) T-cadherin expression delineates specific regions of the developing motor axon-hindlimb projection pathway. J. Neurosci. 14: 7331–7346.

FARINAS, I., K. R. JONES, C. BACKUS, X. Y. WANG AND L. F. REICHARDT (1994) Severe sensory and sympathetic deficits in mice lacking neurotrophin-3. Nature 369: 658–661.

HOHN, A., J. LEIBROCK, K. BAILEY AND Y. A. BARDE (1990) Identification and characterization of a novel member of the nerve growth factor/brain-derived neurotrophic factor family. Nature 344: 339–341.

KAPLAN, D. R., D. MARTIN-ZANCA AND L. F. PARADA (1991) Tyrosine phosphorylation and tyrosine kinase activity of the *trk* proto-oncogene product induced by NGF. Nature 350: 158–160.

KENNEDY, T. E., T. SERAFINI, J. R. DE LA TORRE AND M. TESSIER-LAVIGNE (1994) Netrins are diffusible chemotropic factors for commissural axons in the embryonic spinal cord. Cell 78: 425–435.

KOLODKIN, A. L., D. J. MATTHES AND C. S. GOODMAN (1993) The semaphorin genes encode a family of transmembrane and secreted growth cone guidance molecules. Cell 75: 1389–1399.

LANGLEY, J. N. (1895) Note on regeneration of pre-ganglionic fibres of the sympathetic. J. Physiol. (Lond.) 18: 280–284.

LEVI-MONTALCINI, R. AND S. COHEN (1956) In vitro and in vivo effects of a nerve growth-stimulating agent isolated from snake venom. Proc. Natl. Acad. Sci. USA 42: 695–699.

LUO, Y., D. RAIBLE AND J. A. RAPER (1993) Collapsin: a protein in brain that induces the collapse and paralysis of neuronal growth cones. Cell 75: 217–227.

MESSERSMITH, E. K., E. D. LEONARDO, C. J. SHATZ, M. TESSIER-LAVIGNE, C. S. GOODMAN AND A. L. KOLODKIN (1995) Semaphorin III can function as a selective chemorepellent to pattern sensory projections in the spinal cord. Neuron 14: 949–959.

OPPENHEIM, R. W., D. PREVETTE AND S. HOMMA (1990) Naturally occurring and induced neuronal death in the chick embryo in vivo requires protein and RNA synthesis: Evidence for the role of cell death genes. Dev. Biol. 138: 104–113.

SPERRY, R. W. (1963) Chemoaffinity in the orderly growth of nerve fiber patterns and connections. Proc. Natl. Acad. Sci. 50: 703–710.

WALTER, J., S. HENKE-FAHLE AND F. BONHOEFFER (1987) Avoidance of posterior tectal membranes by temporal retinal axons. Development 101: 909–913.

Books

LETOURNEAU, P. C., S. B. KATER AND E. R. MACAGNO (EDS.) (1991) *The Nerve Growth Cone.* New York: Raven Press.

LOUGHLIN, S. E. AND J. H. FALLON (EDS.) (1993) *Neurotrophic Factors.* San Diego, CA: Academic Press.

PURVES, D. (1988) *Body and Brain: A Trophic Theory of Neural Connections.* Cambridge, MA: Harvard University Press.

RAMÓN Y CAJAL, S. (1928) *Degeneration and Regeneration of the Nervous System.* R. M. May (ed.). New York: Hafner Publishing.

CHAPTER 22

MODIFICATION OF
DEVELOPING
BRAIN CIRCUITS
BY NEURAL
ACTIVITY

■ OVERVIEW

The rich diversity of human personalities, abilities, and behavior is presumably generated by the uniqueness of individual human brains. These fascinating neurobiological differences among humans derive from both heritable and environmental influences. The first steps in the construction of the brain's circuitry—the formation of major axon tracts, the guidance of growing axons to appropriate targets, and the initiation of synaptogenesis—rely largely on intrinsic cellular and molecular processes, some of which were examined in the previous chapter. Once the outlines of brain wiring are established, however, patterns of neural activity gradually increase the precision of synaptic connections by the selective addition (or, in some cases, removal) of connections throughout the developing brain. Although some patterns of early brain activity arise from intrinsic interactions in the newly formed circuitry, the neuronal activity generated by interactions with the outside world eventually provides a mechanism by which the early environment can also influence brain structure and function. Modification of the developing brain by the environment is typically restricted to temporal windows called critical periods: as the organism matures, the brain becomes increasingly refractory to the lessons of experience.

■ INSIGHTS FROM THE DEVELOPMENT OF PERIPHERAL SYNAPSES

Studying synaptic refinement in the complex circuitry of the cerebral cortex or other regions of the central nervous system is a formidable challenge. As a result, many basic ideas about the ongoing modification of developing brain circuitry have come from simpler, more accessible systems, most notably the vertebrate neuromuscular junction and the innervation of autonomic ganglion cells (Figure 22.1 and Box A). As briefly noted in Chapter 21, adult skeletal muscle fibers and neurons in some classes of autonomic ganglia are each innervated by a single axon. Initially, however, each of these target cells receives innervation from several neurons, a condition termed **polyneuronal innervation**. In such cases, inputs are gradually lost during early postnatal development until only one remains. This process of loss is generally referred to as synapse elimination, although the elimination actually refers to a reduction in the number of different inputs to the target cells, not to a reduction in the overall number of synapses made on the postsynaptic cells. In fact, the overall number of synapses in the peripheral nervous system increases steadily during the course of development, as is the case throughout the brain.

A variety of experiments have shown that the elimination of some initial inputs to muscle and ganglion cells is a process in which synapses originating from different neurons compete with one another for "ownership" of an individual target cell (Box B). As described in the preceding chapter, the object of such competition is evidently molecular feedback from the target cells. Trophic factors are good candidates for such feedback, although the putative factors have not been conclusively identified in either muscles or autonomic

419

Figure 22.1 Major features of synaptic rearrangement during the first few weeks of post-natal life in the mammalian peripheral nervous system. In ganglia comprising neurons without dendrites (A) and in muscles (B), each axon innervates more target cells at birth than in maturity. In both muscles and ganglia, however, the size and complexity of the terminal arbor on each target cell increases. Thus, each axon elaborates more and more terminal branches and synaptic endings on the target cells it will innervate in maturity. The common denominator of this process is not a net loss of synapses, but the focusing by each axon of a progressively increasing amount of synaptic machinery on fewer target cells. (After Purves and Lichtman, 1980.)

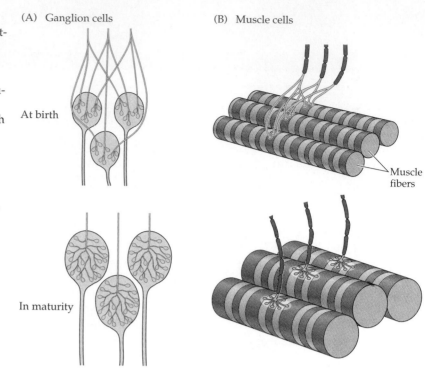

ganglia (the neurotrophin molecules fulfill some, but not all, of the requirements for a muscle-derived trophic factor). Moreover, other molecules, such as receptors for neurotransmitters, also appear to be involved (see Box B).

Importantly, such competition is regulated by patterns of electrical activity in the pre- and postsynaptic partners. For example, if acetylcholine receptors at the neuromuscular junction are blocked by α–bungarotoxin (a potent and essentially irreversible antagonist of the acetylcholine receptor; see Chapter 6), polyneuronal innervation persists. Blocking presynaptic action potentials in the motor neuron axons (by silencing the nerve with tetrodotoxin, a sodium channel blocker) also prevents the reduction of polyneuronal innervation. Blocking neural activity, therefore, reduces or prevents competitive interactions and the associated synaptic rearrangements.

Many of the phenomena of activity-dependent competition in muscles and ganglia (as well as in more complex central nervous system structures) can be accounted for by postulating that (1) synapses require a certain minimal level of trophic support to persist, (2) the relevant factors are secreted in limited amounts by the postsynaptic (target) cells in response to synaptic activation, and (3) synapses can only avail themselves of trophic support if their activity and that of the target cell coincide. Although the first two points have been best established in the peripheral nervous system (see Chapter 21), the role of activity in competitive interactions has been much further advanced by studies of the visual system, as described in the following sections.

■ DEVELOPMENT OF OCULAR DOMINANCE COLUMNS IN THE VISUAL CORTEX

In a profoundly influential series of experiments, David Hubel and Torsten Wiesel, working at Harvard Medical School, found that depriving an experimental animal of normal visual experience during a restricted period of

Box A
WHY NEURONS HAVE DENDRITES

Perhaps the most striking feature of neurons is their diverse morphology (see Figure 1.2). Some classes of neurons have no dendrites at all; others have a modest dendritic arborization; still others have an arborization that rivals the complex branching of a fully mature tree. Why should this be? Although there are probably many reasons for this diversity, neuronal geometry influences the number of different inputs that a target neuron receives by modulating competitive interactions among the innervating axons.

Evidence that the number of inputs a neuron receives depends on its geometry has come from studies of the peripheral autonomic system, where it is possible to stimulate the full complement of axons innervating an autonomic ganglion and its constituent neurons. This approach is not usually feasible in the central nervous system because of the anatomical complexity of most central circuits. Since individual postsynaptic neurons can also be labeled via an intracellular recording electrode, electrophysiological measurements of the number of different axons innervating a neuron can routinely be correlated with target cell shape. In both parasympathetic and sympathetic ganglia, the degree of preganglionic convergence onto a neuron is proportional to its dendritic complexity. Thus, neurons that lack dendrites altogether are generally innervated by a single input, whereas neurons with increasingly complex dendritic arbors are innervated by a proportionally greater number of different axons. This correlation of neuronal geometry and input number holds within a single ganglion, among different ganglia in a single species, and among homologous ganglia across a range of species. Since ganglion cells that have few or no dendrites are initially innervated by several different inputs, confining inputs to the limited arena of the developing cell soma evidently enhances competition between them, whereas the addition of dendrites to a neuron allows multiple inputs to persist in peaceful coexistence.

A neuron innervated by a single axon will clearly be more limited in the scope of its responses than a neuron innervated by 100,000 inputs (1 to 100,000 is the approximate range of convergence in the mammalian brain). By regulating the number of inputs that neurons receive, dendritic form greatly influences function.

References

HUME, R. I. AND D. PURVES (1981) Geometry of neonatal neurons and the regulation of synapse elimination. Nature 293: 469–471.

PURVES, D. AND R. I. HUME (1981) The relation of postsynaptic geometry to the number of presynaptic axons that innervate autonomic ganglion cells. J Neurosci 1: 441–452.

PURVES, D. AND J. W. LICHTMAN (1985) Geometrical differences among homologous neurons in mammals. Science 228: 298–302.

PURVES, D., E. RUBIN, W. D. SNIDER AND J. W. LICHTMAN (1986) Relation of animal size to convergence, divergence and neuronal number in peripheral sympathetic pathways. J Neurosci 6: 158–163.

(A)

0.2 mm

Number of innervating axons = 1   2   3   4   5   7

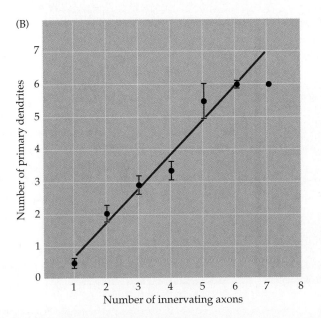

(B)

Number of primary dendrites (y-axis, 0–7)
Number of innervating axons (x-axis, 1–8)

The number of axons innervating ciliary ganglion cells in adult rabbits. (A) Neurons studied electrophysiologically and then labeled by intracellular injection of a marker enzyme have been arranged in order of increasing dendritic complexity. The number of axons innervating each neuron is indicated. (B) This graph summarizes observations on a large number of cells. There is a strong correlation between dendritic geometry and input number. (After Purves and Hume, 1981.)

Box B
DIRECT OBSERVATION OF SYNAPTIC REARRANGEMENT IN DEVELOPING MUSCLE

Using different-colored fluorescent dyes that stain either the presynaptic terminal or the postsynaptic receptors, Jeff Lichtman and his colleagues at Washington University have directly observed the process of synaptic rearrangement in living animals. With this technique they followed the same neuromuscular junction over days, weeks, or longer. Their observations have yielded some unexpected insights. Competition between synapses arising from different motor neurons does not involve the active displacement of the "losing" input by the eventual "winner." Instead, it appears that the neurotransmitter receptors beneath the input that will eventually be eliminated are first reduced in number. The loss reduces the synaptic strength of the input, which causes a further loss of postsynaptic receptors, leading to further reduction in the strength of the input. This downward spiral of synaptic efficacy eventually results in withdrawal of the presynaptic terminal. The remaining terminals do not grow to displace the endings that have withdrawn, but continue to enlarge and strengthen in place as the endplate region expands during muscle growth.

References

BALICE-GORDON, R. J. AND J. W. LICHTMAN (1994) Long-term synapse loss induced by focal blockade of postsynaptic receptors. Nature 372: 519–524.

BALICE-GORDON, R. J., C. K. CHUA, C. C. NELSON AND J. W. LICHTMAN (1993) Gradual loss of synaptic cartels precedes axon withdrawal at developing neuromuscular junctions. Neuron 11: 801–815.

LICHTMAN, J.W., L. MAGRASSI AND D. PURVES (1987) Visualization of neuromuscular junctions over periods of several months in living mice. J. Neurosci. 7: 1215-1222.

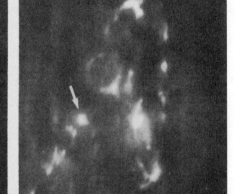

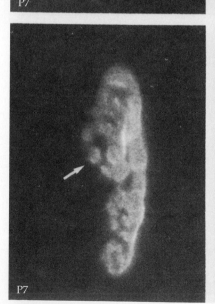

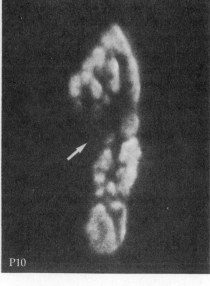

Changes in presynaptic terminals and postsynaptic receptors at a neuromuscular junction undergoing the transition from multiple to single innervation. In the top two panels, the presynaptic terminals were labeled with a green fluorescent dye and imaged in a living mouse on postnatal day 7 (P7) and again on postnatal day 10 (P10). At P7, two inputs are present (indicated by the arrows); only one remains on P10 (single arrow). The bottom panels show the corresponding changes at this synapse in the distribution of acetylcholine receptors, which have been labeled with a red fluorescent dye. As indicated by the arrow, a large portion of receptor-rich area present on P7, corresponding to the location of the eliminated presynaptic input, has also disappeared by P10. (From Balice-Gordon and Lichtman, 1993.)

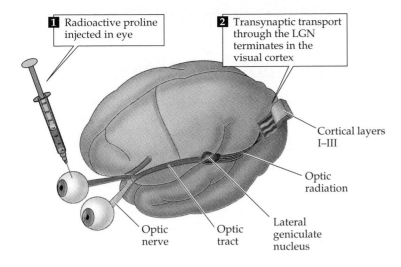

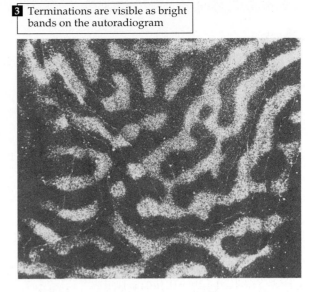

1 Radioactive proline injected in eye

2 Transynaptic transport through the LGN terminates in the visual cortex

3 Terminations are visible as bright bands on the autoradiogram

Cortical layers I–III

Optic radiation

Lateral geniculate nucleus

Optic tract

Optic nerve

Figure 22.2 Ocular dominance columns (which in most anthropoid primates are really stripes or bands) in layer IV of the primary visual cortex of an adult macaque monkey. Diagram indicates the labeling procedure (see also Box C); following transsynaptic transport, the pattern of geniculocortical terminations related to that eye is visible as a series of bright stripes in this autoradiogram of a section in the plane of the cortex (that is, as if looking down on the cortical surface). The dark areas are the zones occupied by geniculocortical terminals related to the other eye. The pattern of human ocular dominance column is shown in Figure 11.10. (From LeVay, Wiesel and Hubel, 1980.)

early postnatal life irreversibly altered neuronal connections in the visual cortex. These experiments provided the first evidence that the brain translates the effects of early experience (that is, neural activity) into permanently altered wiring.

To understand these experiments and their implications, it is important to review the organization and development of the mammalian visual system. Information from the two eyes is first integrated in the primary visual (striate) cortex (see Chapter 11), where most cortical afferents from the lateral geniculate nucleus (LGN) of the thalamus terminate. In some mammals—carnivores, anthropoid primates, and humans—the afferent terminals form an alternating series of eye-specific domains in cortical layer IV called **ocular dominance columns** (Figure 22.2). Ocular dominance columns can be visualized by injecting radioactive amino acids, such as proline, into one eye; the tracer is then transported along the visual pathway to specifically label the geniculocortical terminals corresponding to that eye (Box C). In the adult macaque monkey, the domains representing the two eyes are stripes of about equal width (0.5 mm) that occupy roughly equal areas of the primary visual cortex. Electrical recordings confirm that the cells within layer IV of macaques respond almost exclusively to stimulation of either the left or the right eye, although neurons in layers above and below layer IV integrate inputs from the left and right eyes, responding to binocular visual stimuli. Ocular dominance is thus apparent in two related phenomena: the degree to which individual cortical neurons are driven by stimulation of one eye or the other, and domains (stripes) in cortical layer IV in which the majority of neurons are driven exclusively by one eye.

Box C
TRANSNEURONAL LABELING WITH RADIOACTIVE AMINO ACIDS

Unlike most brain structures, ocular dominance columns cannot be made easily visible by conventional histology. Thus, the striking cortical patterns evident in cats and monkeys was not seen until the early 1970s, when the technique of anterograde tracing using radioactive amino acids was introduced. In this approach, an amino acid commonly found in proteins (usually proline) is radioactively tagged and injected into the area of interest. Neurons in the vicinity take up the label from the extracellular space and incorporate it into newly made proteins. Some of these proteins are involved in the maintenance and function of the neuron's synaptic terminals; thus, they are shipped via anterograde transport from the cell body to nerve terminals, where they accumulate. After a suitable interval the tissue is fixed, sections made, placed on glass slides, and coated with a sensitive photographic emulsion. The radioactive decay of the labeled amino acids in the proteins causes silver grains to form in the emulsion. After several months of exposure, a heavy concentration of silver grains accumulates over the regions that contain synapses originating from the injected site. For example, injections into the eye will heavily label the terminal fields of retinal ganglion cells in the lateral geniculate nucleus.

Transneuronal transport takes this process a step further. After tagged proteins reach the axon terminals, a fraction is actually released into the extracellular space, where the proteins are degraded into amino acids or small peptides that retain their radioactivity. An even smaller fraction of this pool of labeled amino acids is taken up by the postsynaptic neurons, incorporated again into proteins, and transported to synaptic terminals of the second set of neurons. Because the label passes from the presynaptic terminals of one set of cells to the postsynaptic target cells, the process is called transneuronal transport. By such transneuronal labeling, the chain of connections originating from a particular structure can be visualized. In the case of the visual system, proline injections into one eye label appropriate layers of the lateral geniculate nucleus (as well as other retinal ganglion cell targets such as the superior colliculus), and subsequently the terminals in the visual cortex of the geniculate neurons receiving inputs from that eye. Thus, when sections of the visual cortex are viewed with dark-field illumination to make the silver grains glow a brilliant white against the unlabeled background, ocular dominance columns in layer IV are easily seen (see Figure 22.2).

References

COWAN, W. M., D. I. GOTTLIEB, A. HENDRICKSON, J. L. PRICE AND T. A. WOOLSEY (1972) The autoradiographic demonstration of axonal connections in the central nervous system. Brain Res. 37: 21–51

GRAFSTEIN, B. (1971) Transneuronal transfer of radioactivity in the central nervous system. Science 172: 177–179.

GRAFSTEIN, B. (1975) Principles of anterograde axonal transport in relation to studies of neuronal connectivity. In *The Use of Axonal Transport for Studies in Neuronal Connectivity*, W. M. Cowan and M. Cuénod (eds.). Amsterdam: Elsevier, pp. 47–68.

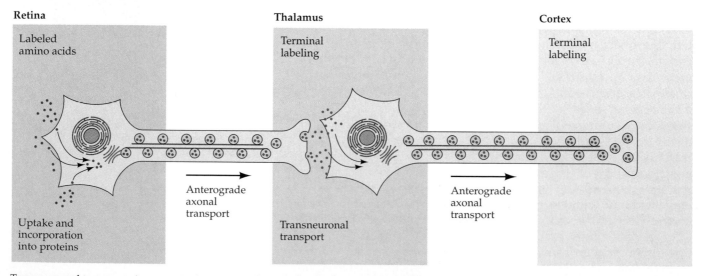

Transneuronal transport. A neuron in the retina is shown taking up a radioactive amino acid, incorporating it into proteins, and moving the proteins down the axons and across the extracellular space between neurons. This process is repeated in the thalamus and eventually label accumulates in the thalamocortical terminals in layer IV of the primary visual cortex.

■ EFFECTS OF VISUAL DEPRIVATION ON OCULAR DOMINANCE

If an electrode is passed at a shallow angle through the cortex while the responses of individual neurons to stimulation of one or the other eye are being recorded, detailed assessment of ocular dominance can be made at the level of individual cells (see Figure 11.13). In such studies, Hubel and Wiesel assigned neurons to one of seven ocular dominance categories. Group 1 cells were defined as being driven only by stimulation of the contralateral eye; group 7 cells are driven entirely by the ipsilateral eye. Binocular neurons driven equally well by either eye were assigned to group 4. Using this approach, they found that the ocular dominance distribution is roughly Gaussian in a normal adult cat; examining cells in all cortical layers, most had binocular responses, with roughly equal numbers of cells influenced by the right and left eyes (Figure 22.3A).

Hubel and Wiesel then asked whether this normal distribution of binocularity could be altered by visual experience. When they simply closed

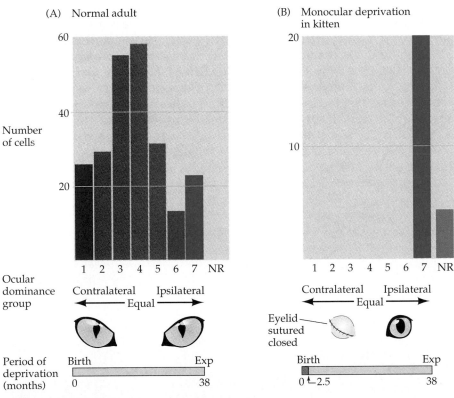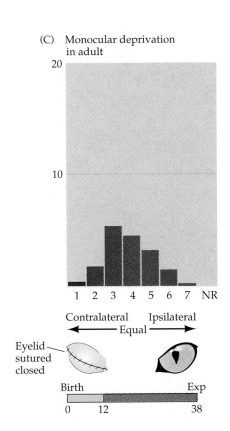

Figure 22.3 Effect of early closure of one eye on the distribution of cortical neurons driven by stimulation of each eye. (A) Ocular dominance distribution of recordings from a large number of neurons in the primary visual cortex of normal adult cats. Cells in group 1 were activated exclusively by the contralateral eye, cells in group 7 by the ipsilateral eye. (B) Following closure of one eye from 1 week after birth until 2½ months of age (indicated by the bar underneath the graph), no cells could be activated by the deprived (contralateral) eye. Some cells could not be activated by either eye (NR). Note that the closed eye is opened at the time of the experimental observations, and that the recordings are not restricted to any particular cortical layer. (C) A much longer period of monocular deprivation in an adult cat has little effect on ocular dominance. In this case, the contralateral eye was closed from 12 to 38 months of age. (A after Hubel and Weisel, 1962; B after Wiesel and Hubel, 1963; C after Hubel and Wiesel, 1970.)

one eye of a kitten early in life and let the animal mature to adulthood (which takes about 6 months), a remarkable change was observed. Electrophysiological recordings now showed that very few cells could be driven from the deprived eye; that is, the ocular dominance distribution had shifted such that all cells were driven by the eye that had remained open (Figure 22.3B). Recordings from the retina and lateral geniculate layers related to the deprived eye indicated that these more peripheral stations in the visual pathway worked quite normally. Thus, the absence of cortical cells that responded to stimulation of the closed eye was not a result of retinal degeneration or a loss of retinal connections to the thalamus. Rather, the deprived eye had been functionally disconnected from the visual cortex. Consequently, such animals are behaviorally blind in the deprived eye. This "cortical blindness," or **amblyopia**, is permanent. Even if the formerly deprived eye is subsequently left open for years, little or no recovery occurs.

Remarkably, the same manipulation—closing one eye—has no effect on the responses of cells in the visual cortex of an adult cat. Thus, even if one eye of a mature cat is closed for a year or more, both the electrophysiology of the visual cortex and the animal's visual behavior are indistinguishable from normal when tested through the reopened eye (Figure 22.3C). Thus, sometime between the time a kitten's eyes open (about a week after birth) and a year of age, visual experience determines how the visual cortex is wired with respect to eye dominance. In fact, further experiments showed that eye closure is effective only if the deprivation occurs during the first 3 months of life. Hubel and Wiesel called this period of susceptibility to visual deprivation the **critical period** for the development of ocular dominance. During the height of the critical period (about 4 weeks of age in the cat), as little as 3 to 4 days of eye closure profoundly alters the ocular dominance profile of the striate cortex (Figure 22.4). Similar experiments in the monkey have shown that the same phenomenon occurs in primates, although the critical period is longer (up to about 6 months of age).

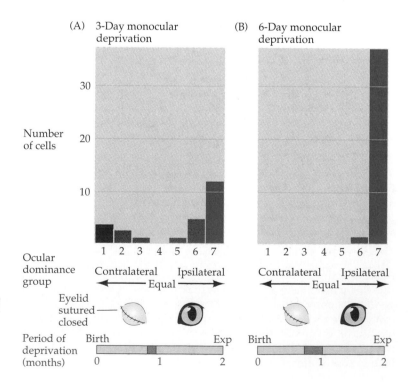

Figure 22.4   The consequences of a short period of monocular deprivation at the height of the critical period in the cat. Just 3 days of deprivation in this example (A) produced a significant shift of cortical innervation in favor of the nondeprived eye; 6 days of deprivation (B) produced an almost a compete shift. Bars below each histogram indicate the period of deprivation, as in Figure 22.3. (After Hubel and Wiesel, 1970.)

(A)

(B) Normal development

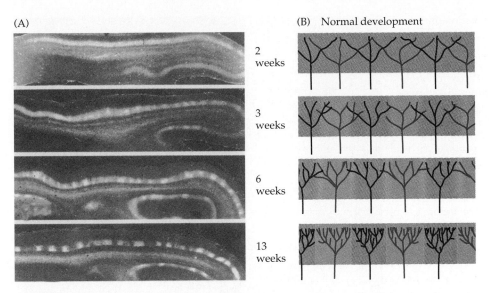

2 weeks

3 weeks

6 weeks

13 weeks

Figure 22.5   The postnatal development of ocular dominance columns in the cat. (A) Proline injections into one eye at 2 weeks of age label a continuous band in layer IV, demonstrating that the geniculocortical afferents have not yet formed stripes. (Coronal sections are shown here, in contrast to Figure 22.2 in which columns are shown in the plane of the cortical surface.) At 3 and 6 weeks, fluctuations in labeling intensity are apparent; by 13 weeks, the pattern of cortical striping is similar to that seen in the adult cat. Interestingly, recent experiments in developing monkeys have shown that ocular dominance columns are already discernable at birth. (B) Diagram of the modification of geniculocortical inputs during normal development. (A from LeVay, Stryker and Shatz, 1978.)

Anatomical experiments using transneuronal transport (see Box C) have provided additional insight into the events underlying the wiring changes in the cortex arising from visual deprivation. Injections of radiolabeled tracers into one eye early in development (prenatally in monkeys or during the first 2 postnatal weeks in cats) do not produce the distinct pattern of alternating eye-specific stripes seen in the adult (see Figure 22.2). Rather, a continuous band of labeling is observed in layer 4 (Figure 22.5). This result indicates that inputs from the two eyes initially overlap, rather than being limited to alternating right eye/left eye columns. In confirmation of this interpretation, electrophysiological studies during this period indicate that neurons in layer IV that will eventually receive thalamic inputs representing only one eye or the other initially receive inputs related to both eyes. Tracer injections made at progressively later developmental stages begin to reveal periodic fluctuations in the density of labeling in layer IV as the geniculocortical afferents grow and sort into increasingly well defined left and right eye zones. It is during this period of ocular dominance column formation that the visual system is most susceptible to visual deprivation.

If even brief periods of eye closure result in a rewiring of neurons normally activated by the closed eye, what happens to the anatomical pattern of ocular dominance stripes? Not surprisingly, animals in which one eye is closed during the critical period develop abnormal patterns of ocular dominance stripes in the visual cortex (Figure 22.6; compare with Figure 22.2). The open-eye stripes are wider than normal, whereas the stripes representing the deprived eye are severely shrunken. These anatomical observations imply that the absence of cortical neurons that respond to the deprived eye

Figure 22.6 Effect of monocular deprivation on ocular dominance columns in the macaque monkey. In normal monkeys, ocular dominance columns gradually form alternating stripes of roughly equal width (see Figure 22.2). The picture is quite different after monocular deprivation. This dark-field autoradiograph shows a reconstruction of several sections through layer IV of the primary visual cortex of a monkey whose right eye was sutured shut from 2 weeks of age to 18 months, when the animal was sacrificed. Two weeks before death, the normal (left) eye was injected with radiolabeled amino acids. The columns related to the nondeprived eye (white stripes) are much wider than normal, whereas as those related to the deprived eye are shrunken. (From Hubel, Wiesel and LeVay, 1977.)

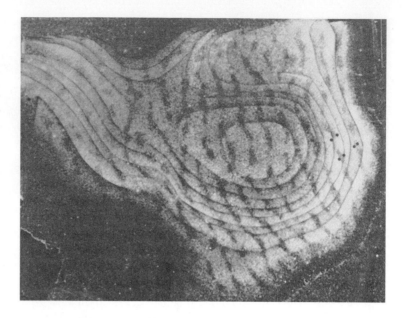

in electrophysiological studies is not due to relatively inactive inputs withering away altogether. (If this were the case, one would expect to see areas of layer IV devoid of any thalamic innervation.) Instead, inputs from the active (open) eye take over some of the territory that formerly belonged to the inactive (closed) eye. Hubel and Wiesel interpreted these results as demonstrating a competitive interaction between the two eyes during early development. In a normal animal, both eyes experience roughly comparable levels of visual stimulation, and therefore come to occupy roughly equal areas of the cortex. When an imbalance in visual experience is induced by monocular deprivation, the active eye has a competitive advantage and replaces many of the synaptic inputs from the closed eye, such that few if any neurons can be driven by the deprived eye (see Figure 22.3B).

The idea that a competitive imbalance underlies the altered distribution of inputs after deprivation has been confirmed by experiments in which *both* eyes are closed shortly after birth. In this case, all the visual cortical neurons are equally deprived of normal experience. The arrangement of ocular dominance recorded some months later is, by either electrophysiological or anatomical criteria, much more normal. Although several peculiarities in the response properties of cortical cells are apparent, roughly normal proportions of neurons representing the two eyes are present. Because there is no imbalance in the visual activity of the two eyes (both sets of related cortical inputs being deprived), both eyes evidently retain their territory in the cortex. If disuse atrophy of the closed-eye inputs were the main effect of deprivation, then binocular deprivation would cause the visual cortex to be largely unresponsive.

Experiments using techniques that label individual thalamic axons terminating in layer IV have shown in greater detail what happens to the arborizations of individual neurons after visual deprivation (Figure 22.7). As noted, monocular deprivation causes a loss of cortical territory related to the deprived eye, with a concomitant expansion of the open eye's territory. At the level of single axons, these changes are reflected in an increased extent and complexity of the arborizations related to the open eye, and a decrease in the size and complexity of the arborizations related to the deprived eye.

(A) SHORT-TERM
     MONOCULAR DEPRIVATION

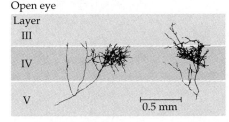

(B) LONG-TERM
     MONOCULAR DEPRIVATION

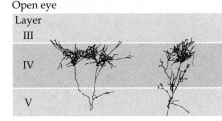

Figure 22.7 Terminal arborizations of lateral geniculate nucleus axons in the visual cortex can change rapidly in response to monocular deprivation during the critical period. (A) After only a week of monocular deprivation, axons from the deprived eye have greatly reduced numbers of branches compared with those from the open eye. (B) Deprivation for longer periods does not result in appreciably larger changes. Numbers on the left of each figure indicate cortical layers. (After Antonini and Stryker, 1993.)

Individual arborizations can be substantially altered after as little as 1 week of deprivation, and perhaps even less. This latter finding highlights the ability of the developing cortex to rapidly remodel connections—actually making and breaking synapses—in response to environmental changes.

Taken together, these observations show that the wiring of the visual system is not fully specified at birth, but requires visual experience to generate the adult pattern. The eventual diminishment of this capacity to remodel cortical (and subcortical) connections is presumably the cellular basis of critical periods in a variety of neural systems.

■ HOW NEURONAL ACTIVITY AFFECTS THE DEVELOPMENT OF NEURAL CIRCUITS

How do patterns of neuronal activity—whether spontaneously generated (Box D) or evoked by visual experience—modify patterns of cortical circuitry? In 1949, the psychologist D. O. Hebb hypothesized that coordinated activity of a presynaptic terminal and a postsynaptic neuron would strengthen the synaptic connections between them (see Chapter 23). Hebb's postulate, as it has come to be known, was originally formulated to explain the cellular basis of learning and memory, but has been widely applied to situations that involve long-term modifications in synaptic strength, including those that occur during cortical development. In this context, Hebb's postulate implies that synaptic terminals strengthened by correlated activity will be retained or sprout new branches, whereas those that are persistently weakened by uncorrelated activity will eventually lose their hold on the postsynaptic cell (Figure 22.8). In the visual system, the action potentials in the thalamocortical inputs related to one eye are probably better correlated with each other than with the activity related to the other eye—at least in layer IV. If sets of correlated inputs tend to exclude uncorrelated inputs from their territory, patches of cortex occupied exclusively by inputs representing one eye could arise. In this scenario, ocular dominance column formation in layer IV is generated by cooperation between inputs carrying *similar* patterns of activity, and competition between inputs carrying *dissimilar* patterns.

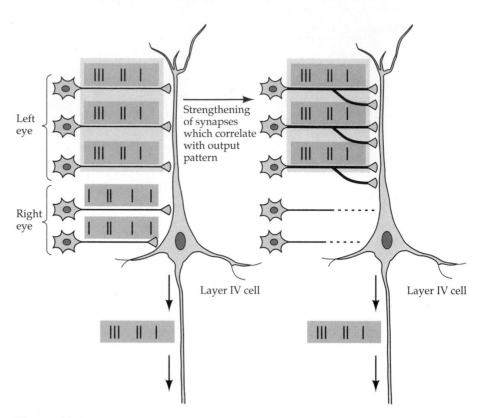

Figure 22.8 A representation of Hebb's postulate as it might operate during development of the visual system. The cell represents a postsynaptic neuron in layer IV of the primary visual cortex. Early in development, inputs from the two eyes converge on single postsynaptic cells. The two sets of presynaptic inputs, however, have different patterns of electrical activity (represented by the short vertical bars). In the example here, the three left eye inputs are better able to fire the postsynaptic cell; as a result, their activity is highly correlated with the postsynaptic cell's activity. According to Hebb's postulate, these synapses are therefore strengthened. The inputs from the right eye carry a different pattern of activity which is less well correlated with the majority of the activity elicited in the postsynaptic cell. These synapses gradually weaken and are eventually eliminated (right-hand side of figure), while the correlated inputs form additional synapses.

The idea that patterns of activity drive the competitive postnatal rearrangement of cortical connections is supported by two further experiments. If one of the extraocular muscles is cut, the two eyes can no longer be aligned, a condition called **strabismus**. The major consequence of being strabismic is that corresponding points on the two retinas no longer image the same location in visual space at the same time. As a result, differences in the visually evoked patterns of activity between the two eyes are far greater than normal. Unlike monocular deprivation, however, the overall amount of activity in each eye remains roughly the same; only the correlation of activity arising from corresponding retinal points is changed. The effects of surgically induced strabismus during cortical development are striking. The anatomical pattern of ocular dominance columns in layer IV of cats is sharper than normal, implying that the asynchronous patterns of activity have accentuated the normal separation of inputs from the two eyes. In addition, the ocular asynchrony prevents the binocular convergence that normally occurs in cells above and below layer IV: ocular dominance histograms from

Box D
PRENATAL ACTIVITY AND CIRCUIT FORMATION

Hubel and Wiesel's early investigations clearly implicated postnatal visual experience as a major determinant of the development of ocular dominance columns in cats. More recently, however, it has been shown that ocular dominance columns in monkeys are already well formed at birth. Thus, visual experience cannot be the sole cause of ocular dominance column formation, since monkeys in utero don't have anything much to look at. Evidently, spontaneous electrical activity in the developing visual system can also guide the formation of neuronal connectivity.

In 1988, L. Galli and L. Maffei performed the impressive technical feat of recording the electrical activity of retinal ganglion cells in embryonic rats. Even before photoreceptors developed, most cells were spontaneously active, firing short, high-frequency bursts of action potentials separated by up to 30 seconds of silence. Nearby cells fired these bursts in rough synchrony with one another. Based on this work, Carla Shatz and her colleagues used an innovative approach to confirm that spontaneous activity in the retina is indeed organized in coherent patterns. To do this, they placed small sections of retina, about half a mil-

limeter on a side, onto a flat grid containing 60 microelectrodes to record the activity of up to 100 cells at the same time. Such recordings showed that every minute or so, a wave of activity begins at one locus and spreads rapidly across a portion of the retina.

Previous work by Shatz and others had shown that the eye-specific layers of the lateral geniculate nucleus (see Chapter 11) emerge prenatally by the segregation of overlapping ganglion cell inputs from the two eyes. Like the formation of ocular dominance columns, this process requires activity. Studies by Pasko Rakic and more recent work by Jonathan Horton have shown further that ocular dominance columns can also develop prenatally. Because these phenomena occur before birth, segregation in the geniculate and the primary visual cortex is not simply a result of visual experience.

The experiments using the multielectrode array provide some insight into the organization of prenatal activity in the visual system. Since the spontaneous waves of retinal activity occur rarely, last only a few seconds, and are random in direction as they sweep across the retina, the chances that the

two retinas will generate identical patterns of waves is small. As described in the text, uncorrelated patterns of activity stimulate developing axons to sort into eye-specific sets; the presence of retinal waves of activity suggests that such patterns may also contribute to the prenatal formation of geniculate layers, and perhaps ocular dominance columns.

References

GALLI, L. AND L. MAFFEI (1988) Spontaneous impulse activity of rat retinal ganglion cells in prenatal life. Science 242: 90–91.

HORTON, J. C. AND D. R. HOCKING (1996) An adult-like pattern of ocular dominance columns in striate cortex of newborn monkeys prior to visual experience. J. Neurosci. 16: 1791–1807.

MEISTER, M., R. O. L. WONG, D. A BAYLOR AND C. J. SHATZ (1991) Synchronous bursts of action potentials in ganglion cells of the developing mammalian retina. Science 252: 939–943.

WONG, R. O., A. CHERNJAVSKY, S. J. SMITH AND C. J. SHATZ (1995) Early functional neural networks in the developing retina. Nature 374: 716–718.

WONG, R. O., S. J. SMITH AND C. J. SHATZ (1993) Transient period of correlated bursting activity during development of the mammalian retina. Neuron 11: 923–938.

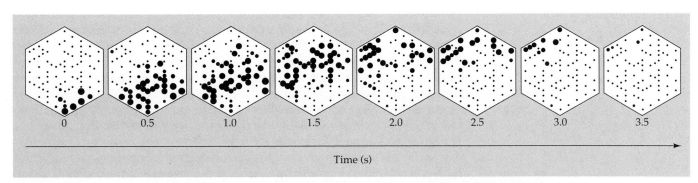

Correlated activity in an in vitro preparation of the developing ferret retina, recorded using a hexagonal grid of electrodes. Each dot represents the position of an active cell; the size of the dot corresponds to the average firing rate of the neuron during each 0.5 second interval. During the 4-second recording period shown, a wave of spontaneous activity begins in the lower right of the retina and propagates toward the upper left. (After Meister et al., 1991.)

Figure 22.9 Ocular dominance histograms obtained by electrophysiological recordings in normal adult cats (A) and adults cats in which strabismus was induced during the critical period (B). The data in (A) is the same as that shown in Figure 22.3A. The number of binocular cells is sharply decreased as a consequence of strabismus; most of the cells are driven exclusively by stimulation of one eye or the other. This enhanced segregation of the inputs presumably results from the greater discrepancy in the patterns of activity between the two eyes as a result of surgically interfering with normal conjugate vision. (After Hubel and Wiesel, 1965.)

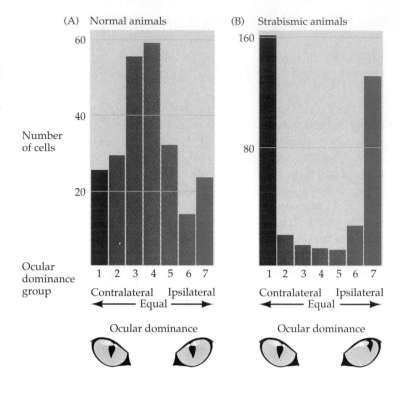

strabismic animals show that most cells in *all* layers are driven exclusively by one eye *or* the other (Figure 22.9). Evidently, strabismus not only accentuates the competition between the two sets of thalamic inputs in layer 4, but also prevents binocular interactions in the other layers, which are mediated by local connections originating from cells in layer 4. These observations on experimental animals have important implications for children with strabismus (see below). Unless the ocular deviation is corrected during the critical period (by patching the good eye or surgically altering the mechanics of the extraocular muscles), a strabismic child may ultimately have poor binocular fusion, diminished depth perception, and degraded acuity; in other words, they will become amblyopic.

A second study that supports the idea that competing activity patterns provide an essential cue for the formation of cortical circuitry involves directly stimulating the two optic nerves in a correlated or an uncorrelated manner (Figure 22.10). In these experiments, carried out by Michael Stryker and his colleagues, activity was eliminated by intraocular injection of tetrodotoxin into the eyes of a kitten, blocking sodium channels and therefore the generation of retinal ganglion cell action potentials. Stimulating electrodes were then implanted in each of the two optic nerves. In some kittens, the optic nerves were stimulated in synchrony for a few hours a day for several weeks. In other animals, the nerves were stimulated asynchronously The status of ocular dominance was then assessed by single-unit recordings in the visual cortex. The results were clear: when the two nerves were stimulated in synchrony, almost all neurons remained strongly binocular. In contrast, when the nerves were asynchronously activated, the ocular dominance profile resembled that of a strabismic animal (Figure 22.9B).

Evidently, neurons in the developing visual cortex do not distinguish inputs from the two eyes on the basis of intrinsic visual differences, but on the basis of patterns of neural activity.

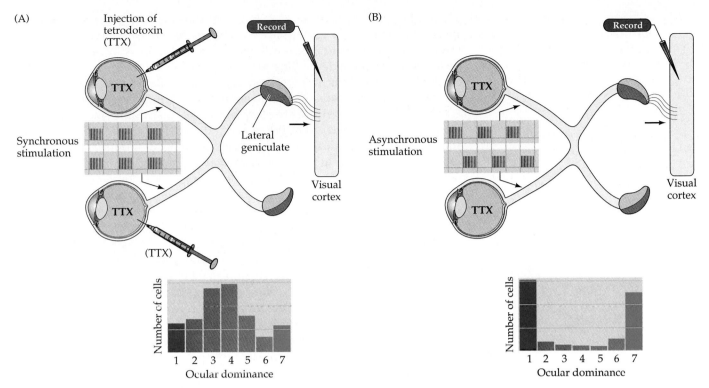

Figure 22.10 Further evidence that patterns of neural activity influence ocular dominance. In this experiment, all activity in the eyes was blocked by application of the sodium channel blocker tetrodotoxin (see Chapter 6). (A) Synchronous activity patterns were imposed on the left and right optic nerves by stimulating electrodes. In subsequent electrophysiological recordings in visual cortex, most cells were strongly binocular. (B) In contrast, when asynchronous stimulation was applied to the nerves, the ocular dominance profile resembled that of a strabismic animal. (After Stryker and Strickland, 1984.)

■ MOLECULAR MECHANISMS OF ACTIVITY-DRIVEN COMPETITION

Although the nature of the "reward" in the competition evident in the developing visual cortex is still highly speculative, one plausible candidate is the neurotrophins. By analogy with the peripheral nervous system, it is attractive to suppose that trophic molecules required for synaptic maintenance in the visual system are secreted in limited quantities by active postsynaptic neurons. According to this theory, presynaptic thalamic (or other) terminals that are activated at the same time as their postsynaptic target neurons are able to acquire trophic support, whereas asynchronously activated terminals are not. If early in development there existed local imbalances in the number of synapses representing each eye, then sets of co-active terminals, such as those originating from the same eye, would be more likely to activate a given cortical neuron. Such endings would eventually be stabilized and form additional synaptic connections with a target neuron, while asynchronous inputs—those originating from the other eye—would be less able to cause secretion of or respond to trophic factors. As a result, the asynchrous inputs to a particular neuron that were in the minority would gradually weaken and eventually be lost (see Figure 22.8).

Although a conclusive link between the neurotrophins and synaptic rearrangements in the cortex has yet to be made, experimental observations strongly implicate the activity-dependent regulation of these (or similar) growth factors in the developing brain. First, the levels of messenger RNA for specific neurotrophins and their receptors are regulated by levels of cortical activity. Second, infusion of specific neurotrophins into the developing visual cortex prevents or attenuates the formation of ocular dominance columns in the presence of normal visual experience. Finally, local sources of certain neurotrophins can forestall some of the effects of monocular deprivation.

Another possibility is that correlated neural activity affects synaptic strength by mechanisms similar to those proposed for **long-term potentiation** and **long-term depression** (**LTP** and **LTD**; see Chapter 23). One component in the induction of long-term synaptic enhancement (such as that which occurs when afferents from one eye are selectively activated) is a specific class of glutamate receptor, the N-methyl-D-aspartate (NMDA) receptor. This receptor exhibits an unusual form of voltage and ligand dependency. At the resting membrane potential, the channel is blocked by the presence of magnesium ions. When the cell is sufficiently depolarized via transmitters affecting other receptors, the magnesium block is relieved, and calcium ions can enter the cell. Calcium in turn activates intracellular cascades that could translate neural activity into more permanent changes in the efficacy or number of synapses. Weak inputs to a given cell (such as those related to a closed eye) would be relatively ineffective in removing the magnesium block; such inputs could be progressively weakened by LTD. Stronger inputs (like those from a group of highly correlated afferents) could remove the block and be potentiated. Whether these mechanisms actually influence cortical development has not yet been determined.

■ THE SIGNIFICANCE OF CRITICAL PERIODS

Activity-driven refinements of neural circuitry—and the profound behavioral consequences that follow—typically occur during a critical period in early life. The concept of critical periods originated from observations by some of the pioneers in the study of animal behavior. Niko Tinbergen, Konrad Lorenz, and others found that some surprisingly complex behaviors are innate, not requiring any experience for their expression in young animals (Box E). Other behaviors, however, are only expressed if animals have certain specific experiences during a restricted period in early postnatal (or posthatching) development. A dramatic example is **imprinting** in ducks and geese. Newly hatched birds will follow the first large moving object they encounter, which is normally their mother. However, other stimuli can readily substitute for the mother bird. These include other animals (not excepting humans) and even inanimate objects. Once an attachment is formed during the first few days after hatching, the birds will follow the "imprinted" object to the exclusion of all others; subsequent experience does little to alter the behavior. Nor are these effects of early experience confined to birds. Experiments carried out by Harry Harlow and his colleagues on the development of primate behavior showed that early social experience (or a lack thereof) profoundly and irreversibly affects a monkey's ability to engage in adult social interactions (see Box E). As in the case of the visual system, similar deprivations in adults have little or no effect. Monkeys raised in isolation are, as adults, fearful of other monkeys and show little interest in the opposite sex. Unfortunately, no amount of experience in later life can eradicate these devastating effects of early deprivation.

Box E
BUILT-IN BEHAVIORS

The idea that animals already possess a set of behaviors appropriate for a world not yet experienced has always been difficult to accept. However, as ethologist Konrad Lorenz pointed out, the preeminence of instinctual responses is obvious to any biologist who looks at what animals actually do. Perhaps the most thoroughly studied examples occur in young birds. Hatchlings emerge from the egg with an elaborate set of innate behaviors. First, of course, is the complex behavior that allows the chick to escape from the egg. Having emerged, a variety of additional abilities indicate how much early behavior is "preprogrammed." Chicks of precocial species preen, peck, gape their beaks, follow a parent, and carry out many other complex acts almost immediately. As Lorenz and Niko Tinbergen showed, hatchlings of some species automatically crouch down in the nest when a hawk or other raptor passes overhead, but are oblivious to the overflight of an innocuous bird. Evidently, even knowledge about predators can be built into the nervous system.

The relevance of this work to primates was underscored in the 1950s by Harry Harlow and his colleagues at the University of Wisconsin. Harlow isolated monkeys within a few hours of birth and raised them in the absence of either a natural mother or a human substitute. In the best-known of these experiments, the baby monkeys had one of two maternal surrogates: a "mother" constructed of a wooden frame covered with wire mesh that supported a nursing bottle, or a similarly shaped object covered with terrycloth. When presented with this choice, the baby monkeys preferred the terrycloth mother and spent much of their time clinging to it, even if the feeding bottle was with the wire mother. Harlow took this to mean that newborn monkeys have a built-in need for maternal care and have at least some innate idea of what a mother should be like. More recently, a number of other endogenous behaviors have been carefully studied in infant monkeys. These include a naive monkey's fear reaction to the presentation of a snake, and the "looming" response (fear elicited by the rapid approach of any formidable object). Most of these built-in behaviors have analogues in human infants.

Taken together, these observations make plain that many complicated behaviors, emotional responses, and other predilections are well established in the nervous system prior to any significant experience. The neural substrates for these behaviors develop under the influence of developmental rules that have presumably evolved to give newborns a better chance of surviving in a predictably dangerous world.

References

HARLOW, H. F. (1959) Love in infant monkeys. Sci. Am. 2(9): 68–74.

HARLOW, H. F. AND R. R. ZIMMERMAN (1959) Affectional responses in the infant monkey. Science 130: 421–432.

LORENZ, K. (1970) *Studies in Animal and Human Behaviour.* Translated by R Martin. Cambridge MA: Harvard University Press.

TINBERGEN, N. (1953) *Curious Naturalists.* Garden City, NY: Doubleday.

These are just a few examples of the myriad behaviors whose expression demands a particular sort of experience during a limited temporal window in early life. This window should not be taken to imply any one time that is critical for all behaviors. Nevertheless, in most experimental animals that have been studied, critical periods for a wide range of behaviors begin soon after birth and end at puberty, or shortly thereafter. Several decades of work on the development of the visual system in cats and monkeys—as well as much other work—have now shown compellingly that brain circuits are not completely hard-wired by intrinsic developmental mechanisms; rather, they are modulated by experience. It is reasonable to assume that abnormal experience of any sort in young animals may induce abnormal patterns of brain circuitry that cannot be rectified later in life.

■ CRITICAL PERIODS FOR THE DEVELOPMENT OF HUMAN BEHAVIOR

These observations on critical periods in experimental animals have important clinical implications for the developing human visual system and, by extension, for the development of human behavior more generally. An example is congenital cataracts. A cataract is an opacity of the lens that prevents normal form vision, in much the same way that surgically closing the eye of a cat or monkey prevents the clear perception of form. When cataracts develop in adults and are removed, even after decades, the patient typically recovers normal vision. In contrast, infants with an untreated cataract in one eye never recover normal vision in that eye after cataract extraction in later life. This unfortunate outcome emphasizes the importance of early detection and treatment of visual deficits; failure to do so leads to permanent damage. A related example is the incorrect use of eye patches to correct "lazy" eyes. One treatment for improper binocular alignment in children (strabismus) is to place a patch over the good eye so that the child is forced to use the eye that is becoming amblyopic. Patching, however, is also a form of monocular deprivation; improperly used, this strategy may result in a significant loss of vision in the good eye.

Another vital domain of human behavior in which a critical period plays an obvious part is language. The fact that there is a critical period for the acquisition of language illustrates the importance of early experience in a more cognitive aspect of mental function. During the first few months of life, human infants can perceive phonemes (the elementary components of speech) of all languages. But by 1 year of age, prior to the emergence of spoken language, babies begin to lose the ability to distinguish closely related phonemes that they do not hear spoken. Unless a child is exposed to the sounds of a language well before puberty, this ability is permanently lost (see Chapter 25). Persistent auditory deprivation can also lead to deficits in language acquisition. For example, recurrent middle ear infections in children can chronically muffle speech sounds and cause problems in learning language. A few extreme examples of deprivation in which a normal, hearing child was never exposed to a significant amount of language have unfortunately been reported. In one infamous case, a girl was raised by deranged parents until the age of 13 under conditions of almost total language deprivation. Despite intense training, she never learned more than a rudimentary level of communication.

The development of the visual system and the consequences of visual deprivation can thus be thought of as a paradigm for deprivation effects in other systems whose neural circuitry is less well understood.

■ SUMMARY

Once the broad outlines of connectivity have been established in the developing brain, neuronal activity begins to play an increasingly important role in determining the detailed arrangement of neural circuitry. In many developing circuits, ranging from the neuromuscular system to the neocortex, neurons or other target cells initially receive synapses from inputs that do not innervate them in maturity. Competition between sets of inputs with different patterns of activity is a central theme of such rearrangements. Although the basis for this struggle between different inputs is not fully understood, some aspects of the process appear to involve competition for limited amounts of trophic molecules presented by the postsynaptic target cells. Because neuronal activity is elicited by interactions with the outside world, activity-dependent modification of circuitry provides a means by which experience can influence the number and pattern of synaptic connections, and ultimately an animal's perceptual, emotional, and behavioral repertoire. The susceptibility of specific circuits to the influences of experience is particularly obvious early in life, defining a critical period for the normal development of each neural system. If not reversed before the end of this interval, the structural alterations in brain circuitry induced during the critical period are difficult or impossible to change. This general strategy of neural development evidently enables the maturing brain to store vast amounts of information in the circuitry formed during early life, and to retain it thereafter.

Additional Reading

Reviews

PURVES, D. AND J. W. LICHTMAN (1980) Elimination of synapses in the developing nervous system. Science 210: 153–157.

PURVES, D., W. D. SNIDER AND J. T. VOYVODIC (1988) Trophic regulation of nerve cell morphology and innervation in the autonomic nervous system. Nature 336: 123–128.

SHATZ, C. J. (1990) Impulse activity and the patterning of connections during CNS development. Neuron 5: 745–756.

SHERMAN, S. M. AND P. D. SPEAR (1982) Organization of visual pathways in normal and visually deprived cats. Physiol. Rev. 62: 738–855.

WIESEL, T. N. (1982) Postnatal development of the visual cortex and the influence of environment. Nature 299: 583–591.

Important Original Papers

ANTONINI, A. AND M. P. STRYKER (1993) Rapid remodeling of axonal arbors in the visual cortex. Science 260: 1819–1821.

BROWN, M. C., J. K. S. JANSEN AND D. VAN ESSEN (1976) Polyneuronal innervation of skeletal muscle in new-born rats and its elimination during maturation. J. Physiol. (London) 261: 387–422.

CABELLI, R. J, A. HOHN AND C. J. SHATZ (1995) Inhibition of ocular dominance column formation by infusion of NT-4/5 or BDNF. Science 267: 1662–1666.

HUBEL, D. H. AND T. N. WIESEL (1965) Binocular interaction in striate cortex of kittens reared with artificial squint. J. Neurophysiol. 28: 1041–1059.

HUBEL, D. H. AND T. N. WIESEL (1970) The period of susceptibility to the physiological effects of unilateral eye closure in kittens. J. Physiol. 206: 419–436.

HUBEL, D. H., T. N. WIESEL AND S. LEVAY (1977) Plasticity of ocular dominance columns in monkey striate cortex. Phil. Trans. R. Soc. Lond. B. 278: 377–409.

LEVAY, S., M. P. STRYKER AND C. J. SHATZ (1978) Ocular dominance columns and their development in layer IV of the cat's visual cortex: a quantitative study. J. Comp. Neurol. 179: 223–244.

LEVAY, S., T. N. WIESEL AND D. H. HUBEL (1980) The development of ocular dominance columns in normal and visually deprived monkeys. J. Comp. Neurol. 191: 1–51.

LICHTMAN, J. W. (1977) The reorganization of synaptic connexions in the rat submandibular ganglion during post-natal development. J. Physiol. (Lond.) 273: 155–177.

RAKIC, P. (1977) Prenatal development of the visual system in the rhesus monkey. Phil. Trans. R. Soc. Lond. B. 278: 245–260.

STRYKER, M. P. AND W. HARRIS (1986) Binocular impulse blockade prevents the formation of ocular dominance columns in cat visual cortex. J. Neurosci. 6: 2117–2133.

STRYKER, M. P. AND STRICKLAND, S. L. (1984) Physiological segregation of ocular dominance columns depends on the pattern of afferent electrical activity. Invest. Opthalmol. Vis. Sci [Suppl.] 25: 278.

WIESEL, T. N. AND D. H. HUBEL (1965) Comparison of the effects of unilateral and bilateral eye closure on cortical unit responses in kittens. J. Neurophysiol. 28: 1029–1040.

Books

CURTISS, S. (1977) Genie: A Psycholinguistic Study of a Modern-Day "Wild Child". New York: Academic Press.

HUBEL, D. H. (1988) Eye, Brain, and Vision. Scientific American Library Series. New York: W.H. Freeman.

PURVES, D. (1994) Neural Activity and the Growth of the Brain. Cambridge, UK: Cambridge University Press.

CHAPTER 23

■ OVERVIEW

The capacity of the nervous system to change—often referred to as neural plasticity—is especially prominent during development. Clearly, however, the ability to learn new skills and establish new memories continues throughout life. How does the adult nervous system mediate such changes? Although understanding the mechanisms responsible for learning and other plastic changes in the adult brain remains one of the central challenges of modern neuroscience, there is a consensus that these phenomena are based on carefully regulated changes in the strength of extant synapses. This strategy differs from developmental changes, which are largely based on changing the wiring between cells (that is, making new connections or removing existing ones). Experiments carried out in a variety of animals have shown that synaptic strength can be altered over periods ranging from milliseconds to months. The cellular mechanisms underlying these changes are transient modifications of neurotransmission and, in the case of longer-lasting alterations, changes in gene expression. Even the ability to alter the wiring of the cortex is not completely abolished in maturity, although as the phenomenon of critical periods makes plain, it is much diminished.

■ SHORT-TERM SYNAPTIC PLASTICITY

Some changes in synaptic efficacy arise acutely as a result of activity during the preceding few milliseconds to minutes (Figure 23.1). **Facilitation**, a transient increase in synaptic strength, occurs when two or more action potentials invade the presynaptic terminal in close succession, resulting in more neurotransmitter release with each succeeding action potential. As a result, the voltage change engendered in the postsynaptic cell increases progressively. In the 1950s, Bernard Katz and his colleagues at University College, London provided evidence that facilitation at the neuromuscular junction can be explained by elevated levels of calcium in motor nerve terminals following activity. Recall that the initial event that triggers synaptic vesicle release at synapses is the influx of calcium into the presynaptic terminal (see Chapters 5 and 6). Although the entry of Ca^{2+} after the invasion of the presynaptic terminal by an action potential occurs within a millisecond or two, the mechanisms that return calcium to resting levels are much slower. Thus, when action potentials occur close together, calcium levels in the presynaptic terminal tend to build up. Consequently, more neurotransmitter is released by a subsequent presynaptic action potential.

Synaptic transmission at the neuromuscular junction can also be gradually depressed following repeated use of a synapse. In **synaptic depression**, many action potentials in rapid succession release so much neurotransmitter that the mechanisms for vesicle reuptake and recharging are overwhelmed. As a result, excessive activity leads to progressive depletion of the pool of synaptic vesicles available for fusion and transmitter release, and the strength of the synapse declines until this pool can be replenished. At the same time, such intensive activity overwhelms the calcium-buffering capacity of the nerve terminal, resulting in prolonged elevation of calcium levels at

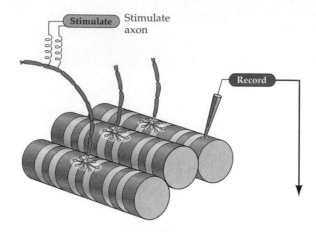

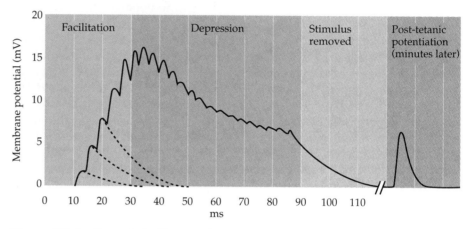

Figure 23.1   Synaptic facilitation, depression, and post-tetanic potentiation at the neuromuscular junction. Electrical recording in a single muscle fiber shows postsynaptic potentials elicited by electrical stimulation of the presynaptic nerve. Facilitation of the postsynaptic response occurs during the beginning of the stimulus train. As the nerve continues to be stimulated, this transient facilitation is followed by depression of the evoked response. After the train (tetanus) ends, the postsynaptic responses may again increase in size and become significantly larger than the initial response, a phenomenon called post-tetanic potentiation. As a result of facilitation, depression, and post-tetanic potentiation, the postsynaptic effect of neuromuscular synapses can vary substantially. (After Katz, 1966.)

the synapse. This elevation activates calcium-dependent processes that make more vesicles available for transmitter release. Thus, after an initial period of depression, the invasion of the terminal by a single action potential can again result in enhanced transmitter release. This form of enhancement, called **post-tetanic potentiation (PTP)**, persists for some minutes after a high-frequency burst of action potentials (called a tetanus).

■ LONG-TERM POTENTIATION

Whereas facilitation, depression, and post-tetanic potentiation can modify synaptic transmission briefly at the neuromuscular junction (and at many central synapses), such mechanisms are unlikely to provide the basis for memories or learned behaviors that persist for weeks, months, or years.

More enduring mechanisms of synaptic plasticity must be involved in such phenomena.

The most thoroughly studied example of a prolonged change in synaptic strength is **long-term potentiation (LTP)** in the mammalian hippocampus (Box A). In the early 1970s, Tim Bliss and his colleagues discovered that a few seconds of high-frequency electrical stimulation of a fiber pathway in the rabbit hippocampus enhances synaptic transmission between the stimulated axons and postsynaptic cells, and that this change persists for weeks. Accordingly, this phenomenon has been called "long-term" potentiation. Although LTP, as it is now universally known, was first observed in intact experimental animals, progress in understanding its cellular basis has relied on in vitro brain slice preparations (Box B). In brain slices taken from the hippocampus, the cell bodies of the pyramidal neurons lie in a single densely-packed layer that is readily apparent. This layer is divided into several distinct regions, the major ones being **CA1** and **CA3** (Figure 23.2). The dendrites of pyramidal cells in the CA1 region form a thick layer of neuropil (the stratum radiatum), where they receive synapses from the **Schaffer collaterals**, a major input pathway originating from pyramidal cells in the CA3 region. Much of the work on LTP has focused on the synaptic connections between CA3 and CA1 pyramidal cells, in part because they show a particularly robust form of LTP, but also because damage to the CA1 region has dramatic effects on human memory (see Chapter 29).

Electrical stimulation of the Schaffer collaterals generates excitatory postsynaptic potentials (EPSPs) in the postsynaptic CA1 cells. EPSPs can be recorded in individual cells with intracellular electrodes or as a "population" response (the summed effect from many postsynaptic neurons) recorded by

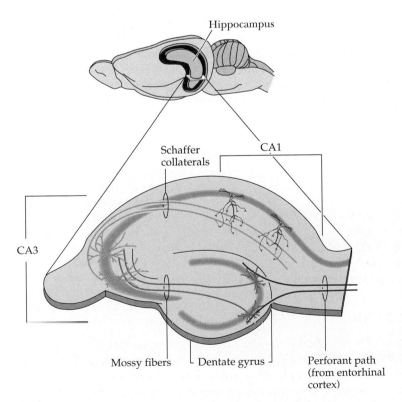

Figure 23.2 Diagram of a section through the rodent hippocampus showing the major excitatory pathways and their synaptic connections. Long-term potentiation of synaptic transmission has been observed in response to stimulation of each of the three pathways shown (the perforant path; the mossy fiber pathway; and the Schaffer collateral pathway).

Box A
WHY THE HIPPOCAMPUS?

Early efforts to understand the neurobiological basis of learned behavior in the brain led to some frustrating and ambiguous conclusions. Consider, for example, the work of Karl Lashley, who in the 1950s attempted to determine the repository of memory, or, as he termed it, the "engram." Lashley trained rats in behavioral tasks such as running a maze and subsequently removed different areas of the brain to determine where the memory was stored or held. To his surprise, Lashley found that remarkably large amounts of the cortex had to be ablated before the animal's maze-running ability was impaired. He reluctantly concluded that memories are distributed rather than located in a specific brain area.

It is now clear, however, that some brain areas are especially important in the formation and/or retrieval of memory.

The hippocampus in particular is clearly involved in the normal formation of longterm declarative memory (see Chapter 29). Thus, functional imaging of the normal human brain (using either fMRI or PET; see Box E in Chapter 1) shows that the hippocampus is activated during certain kinds of memory tasks. Moreover, bilateral damage to the hippocampus results in an inability to form new declarative memories, whereas memories established before the injury remain intact, implying that the hippocampus is not required for retrieval of long-term memories. Finally, hippocampal damage prevents rats from developing proficiency in spatial learning tasks (see Figure); in monkeys as well, hippocampal lesions cause consistent deficits in learning and memory. Recordings from rats navigating simple mazes show that some neurons in

the hippocampus fire action potentials only at certain locations. Such "place cells" may represent components of a neural system for encoding spatial memories. Although many other brain areas are involved in the complex process of memory formation, storage, and retrieval, these several observations have led many investigators to study adult plasticity in the hippocampus.

References

ALVAREZ, P., S. ZOLA-MORGAN AND L.R. SQUIRE (1995) Damage limited to the hippocampal region produces long-lasting memory impairment in monkeys. J. Neurosci. 15: 3796–3807.

LASHLEY, K. S. (1950) In search of the engram. Symp. Soc. Exp. Biol. 4: 454–482.

O'KEEFE, J. (1990) A computational theory of the hippocampal cognitive map. Prog. Brain Res. 83: 301–312.

SQUIRE, L. R., J. G. OJEMANN, F. M MIEZEN, S. E. PETERSEN, T. O. VIDEEN AND M. E. RAICHLE (1995) Activation of the hippocampus in normal humans: a functional anatomical study of memory. Proc. Natl. Acad. Sci. USA 89: 1837–1841.

(A) First trial — Control rat — Hidden platform — Cues in the surrounding environment — After 10 trials

(B) With hippocampus lesioned

Spatial learning in rats. (A) Rats are placed in a circular arena (about the size and shape of a child's wading pool) filled with cloudy water. The arena itself is featureless, but the surrounding environment contains the usual positional cues (windows, doors, light fixtures, and so on). A small platform is located just below the surface. As rats search for this resting place, the pattern of their swimming (indicated by the traces in the figure) is monitored by a video camera. After a few trials, normal rats swim directly to the platform on each trial. (B) The swimming patterns of rats with impaired spatial memories—induced by hippocampal lesions—indicate a seeming inability to remember where the platform is located. (After Schenk and Morris, 1985.)

an extracellular electrode (Figure 23.3). If the Schaffer collaterals are stimulated two or three times a minute, the size of the evoked EPSP in the CA1 neurons remains constant. The delivery of a brief, high-frequency train of stimuli to the same axons, however, causes a long-lasting increase in EPSP amplitude (that is, LTP). Thus, synaptic transmission in the hippocampus is enhanced when the synapse is repetitively activated.

A different way of eliciting LTP, termed **pairing**, further implicates the state of the postsynaptic cell in determining whether long-term enhancement takes place (Figure 23.4). If a single CA1 cell is impaled with an intracellular

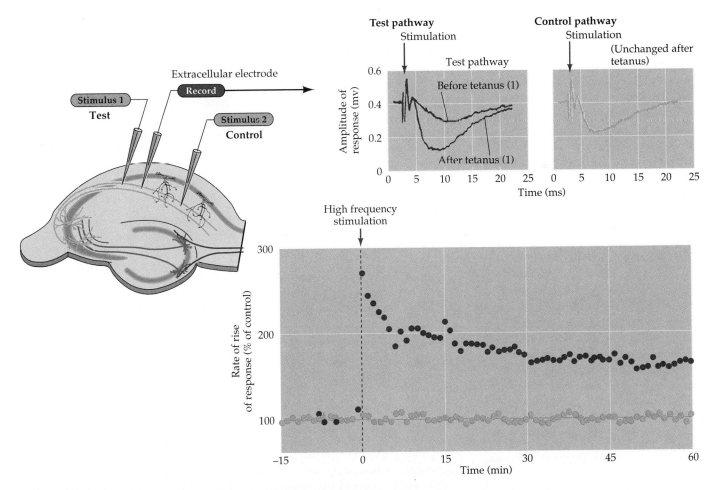

Figure 23.3 Long-term potentiation of the Schaffer collateral-CA1 synapses in a hippocampal slice. The diagram on the left indicates the position of the stimulating and recording electrodes. Stimulating electrodes 1 and 2 each activate separate populations of Schaffer collaterals, thus providing a "test" pathway and a "control" pathway. The panels on the upper right show the population response recorded in the CA1 region following a single stimulus applied separately to electrodes 1 and 2 before and some minutes after a train of high-frequency stimulation to the test pathway (stim 1). The size of the response elicited by electrode 1 is changed; a single stimulus now evokes a larger response. The response to activation of the control pathway (stim 2) is unchanged. In the lower panel, the initial slope of the population response (measured as a percentage increase over the normal rate of rise) was evaluated every 30 seconds before and after high-frequency stimulation (arrow). The response is enhanced for several hours in the stimulated pathway (purple dots); the response in the control pathway (orange dots) remains constant. (After Nicoll, Malenka and Kauer, 1988.)

Box B
BRAIN SLICES

Until the early 1970s, sorting out the mechanisms of synaptic transmission relied on a few preparations with large, accessible neurons such as the squid giant synapse, the synapses in mammalian autonomic ganglia, and the vertebrate neuromuscular junction. Although basic synaptic mechanisms are likely to be similar across different preparations and species, understanding the brain ultimately requires investigation of synapses in the mammalian central nervous system. Unfortunately, brain tissue has a very high demand for glucose and oxygen; simply placing a piece of brain in a dish does not give it adequate access to either resource, and the tissue quickly dies. In the late 1960s, however, C. Yamamoto and H. McIlwain introduced the use of brain slices—thin sections of living brain tissue that can be maintained in vitro. Because oxygen and nutrients rapidly diffuse into sufficiently thin pieces of tissue, the slices remain viable for many hours. As a result, slice preparations have become an extraordinarily valuable tool for understanding the behavior of mammalian brain cells. Nowhere is this fact more evident than in studies of

long-term potentiation (LTP); indeed, the cellular basis of LTP has been determined primarily through slice preparations of the hippocampus.

Brain slices have been prepared from almost all regions of the mammalian central nervous system, including the cortex, thalamus, spinal cord, hypothalamus, and cerebellum. The approach is similar in all these areas. After the animal is killed, the brain is rapidly removed and the area of interest dissected free. A variety of devices—tissue choppers, vibrating razor blades—are used to section the tissue into slices about half a millimeter thick. The slices are then placed in a solution of artificial cerebrospinal fluid in a warm, oxygenated chamber; under these conditions the neurons in the slices remain alive and electrically active for up to 24 hours.

A further advantage is that tissue slices are thick enough to preserve many of the circuits present in a given brain area. Thus, cell layers, fiber tracts, and other major features are readily apparent, and stimulating and recording electrodes can be guided to desired positions using a simple dissecting microscope. Intracellular and extracellular recordings are straightforward and quite stable, since there is no heartbeat or respiratory movement to disturb the placement of the electrodes. All this has made it possible to examine the ionic permeabilities that dictate the firing properties of many classes of central neurons, the neurotransmitters involved in the communication between them, and the way patterns of synaptic activity are translated into long-term synaptic modifications.

In more recent applications, investigators have combined slice recordings with other methods that would also be

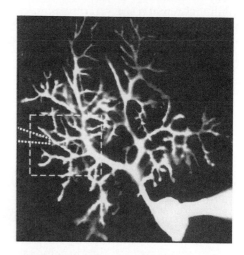

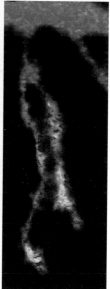

Purkinje cell in a slice of rat cerebellum. The neuron in the upper panel was filled with a calcium indicator dye via an intracellular electrode and viewed with confocal microscopy. The lower panels show the fluorescence of the dye in a portion of the dendritic arborization before (left) and after (right) stimulation of the parallel fibers in the slice. The increased dye fluorescence corresponds to an increased local concentration of intracellular calcium. Such changes may contribute to long-term depression (LTD) in the cerebellum (see Box C). (From Eilers et al., 1995.)

Brain slice from the hippocampus of a rat. The various cell layers are evident as relatively clear areas (compare with Figure 23.2); fiber tracts appear white. Synaptic connections in such slices undergo long-term potentiation and can be studied for many hours. (Courtesy of J. Kauer.)

difficult to employ in vivo. With imaging techniques, for example, it has been possible to monitor the changes in intracellular calcium levels in individual dendrites in cerebellar and hippocampal slices, and to transfect specific populations of neurons in a slice with genes whose products may be involved in long-term synaptic plasticity. While limited in some respects (many fiber tracts are of course severed by the sectioning procedure), brain slices offer a remarkably easy way to explore the cellular physiology of neurons in the mammalian brain.

References

BLANTON, M. J., J. J. LO TURCO AND A. R. KRIEGSTEIN (1989) Whole cell recording from neurons in slices of reptilian and mammalian cerebral cortex. J. Neurosci. Methods 30: 203–210.

DINGLEDINE, R. (ed.) (1984) *Brain Slices*. New York: Plenum Press.

TANK, D. W., M. SUGIMORI, J. A. CONNOR AND R. R. LLINÁS (1988) Spatially resolved calcium dynamics of mammalian Purkinje cells in cerebellar slice. Science 242: 773–777.

YAMAMOTO, C. AND H. MCILWAIN (1966) Potentials evoked in vitro in preparations from the mammalian brain. Nature 210: 1055–1056.

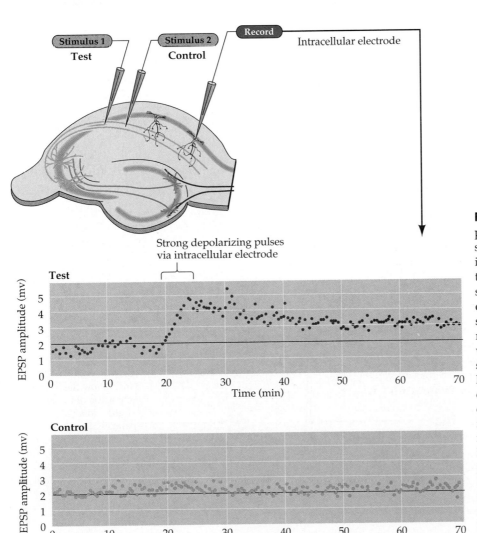

Figure 23.4 Pairing presynaptic and postsynaptic activity causes LTP. Two separate stimulation electrodes are used in conjunction with an intracellular electrode in a CA1 pyramidal cell. Single shocks were applied to each stimulating electrode every 10 seconds and the postsynaptic response (EPSP) of the CA1 cell recorded. Ordinarily, these stimuli would not elicit any change in synaptic strength (as in the graph labeled stimulus 2). However, if strong depolarizing current pulses are applied to the CA1 cell body in conjunction with the shocks for several minutes (during the time indicated in the graph of stimulus 1), a large, persistent increase in the size of the postsynaptic potential ensues. Responses to stimuli delivered via the control electrode (stimulus 2), which are not paired with intracellular depolarization remain unchanged. (After Gustafsson et al., 1987.)

electrode, current can be injected to elicit (or prevent) action potentials. In the pairing paradigm, a single electrical shock to the Schaffer collaterals (which would not normally elicit LTP) is paired with a strong depolarization of a single postsynaptic cell. After pairing occurs a number of times, the size of the intracellular EPSP is increased in the paired pathway, whereas the EPSP in the central pathway remains unchanged. The pairing paradigm works only if the activity of the presynaptic and postsynaptic cells is tightly linked in time, such that the strong postsynaptic depolarization occurs within about 100 ms of presynaptic transmitter release. Recall that the requirement for coincident activation of presynaptic and postsynaptic elements is the hallmark of the Hebbian postulate described in Chapter 22.

LTP has other features that are consistent with a mechanism involved in memory formation. One such characteristic is **specificity** (Figure 23.5A). A single CA1 pyramidal cell receives thousands of excitatory synaptic inputs via the Schaffer collaterals, each originating from a different cell in the CA3 region. If activation of one set of synapses led to all other synapses—even inactive ones—being potentiated, it would be impossible to selectively enhance one set of inputs, as is presumably required for learning and memory. In fact, when LTP is induced by the stimulation of one pathway, it does not occur in other, inactive inputs that contact the same neuron. Thus, LTP is specific to activated synapses rather than to all the synapses on a given cell.

Another important property of LTP from the standpoint of learning and memory is **associativity** (Figure 23.5B). As noted, weak stimulation of a pathway will not by itself trigger LTP. However, experiments using electrical activation in vivo show that if one pathway is weakly activated at the same time that a neighboring pathway is strongly activated, the weak pathway is potentiated. This behavior—selective enhancement of conjointly activated sets of inputs—might be expected in a network of neurons designed to associate two distinct pieces of information. Although there is an admittedly large gap between real memories and such experiments, this mechanism of cellular conjunction provides at least a plausible basis for associative memory functions.

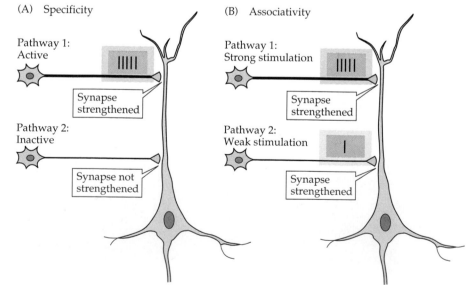

Figure 23.5 Properties of LTP in hippocampal slices. The cells represent CA1 pyramidal neurons receiving synaptic inputs from two independent sets of axons. (A) Strong activity initiates LTP at active synapses (pathway 1) without initiating LTP at nearby inactive synapses (pathway 2). (B) Weak stimulation of pathway 2 alone does not trigger LTP. However, when the same weak stimulus to pathway 2 is activated together with strong stimulation of pathway 1, both sets of synapses are strengthened.

■ THE MOLECULAR BASIS OF LTP

Despite the fact that LTP was first observed in the early 1970s, little progress was made in understanding its molecular basis until the mid-1980s, when the unique properties of a specific type of glutamate receptor, the NMDA receptor, were described. (This receptor is named for the glutamate analogue, *N*-methyl-D-aspartate, that selectively activates it; see Chapters 6 and 7). The first indication that this receptor is essential for LTP induction came from experiments using compounds that selectively block glutamate binding to NMDA receptors. Applying receptor antagonists to a hippocampal slice has no effect on the synaptic response evoked by low-frequency stimulation of the Schaffer collaterals; however, NMDA antagonists prevent LTP.

The biophysical properties of the NMDA and non-NMDA channels provide an elegant means of explaining the mechanics of LTP induction (Figure 23.6). The key elements of this mechanism are a voltage-dependent blockade of the NMDA channel by physiological concentrations of magnesium, and the unusual permeability of the NMDA channel to calcium ions. During low-frequency synaptic transmission, glutamate released from the axon terminals of the Schaffer collaterals binds to both NMDA and non-NMDA glutamate receptors. If the postsynaptic neuron is at its normal negative membrane potential, however, the NMDA channels are blocked by magnesium ions; consequently, no current flows through these ligand-gated channels. During low-frequency stimulation, then, the current that produces the EPSP flows almost exclusively through the non-NMDA channels, which are permeable to Na$^+$. The magnesium blockade of the NMDA channel is, however, voltage-dependent: although the channel is blocked at resting membrane potential, the occlusion disappears when the cell is strongly depolarized. The function of the synapse therefore changes markedly when the postsynaptic cell is significantly depolarized, as happens during high-frequency stimulation, or when the cell is depolarized directly, as in the pairing paradigm discussed earlier. Magnesium is then expelled from the NMDA channel, and current flows through both these and the non-NMDA channels. Compared to non-NMDA receptor channels, the pore of the NMDA channel is far more permeable to calcium, an important second messenger that is critical to the establishment of LTP. The NMDA receptor thus behaves like a molecular "and"

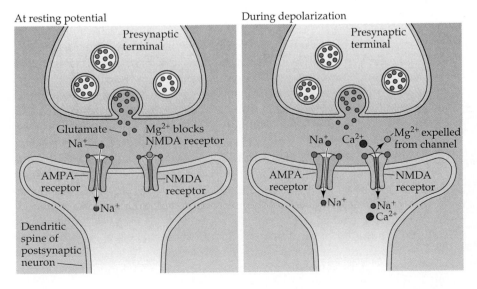

Figure 23.6 The NMDA receptor channel can open only during depolarization of the postsynaptic neuron from its normal resting level. Depolarization expels Mg^{2+} from the NMDA channel, allowing current to flow into the postsynaptic cell. Since the NMDA channel is permeable to Ca^{2+}, there is significant Ca^{2+} entry into the cell, which in turn triggers LTP. (After Nicoll, Malenka and Kauer, 1988.)

gate: the channel opens only (and therefore LTP can occur only) when glutamate is bound to NMDA receptors *and* the postsynaptic cell is depolarized to relieve the Mg^{2+} block of the NMDA channel.

These properties of the NMDA receptor presumably account for both the specificity and associativity of LTP. When only one pathway is strongly stimulated, LTP is confined to the active synapses because glutamate opens NMDA channels only at these sites. Associativity can also be understood in terms of NMDA receptor function. A weakly stimulated input releases glutamate but cannot sufficiently depolarize the postsynaptic cell to relieve the Mg^{2+} block; if neighboring inputs are strongly stimulated, they provide the "associative" depolarization necessary to relieve the block. LTP induced by the pairing of low-frequency input with single-cell depolarization may work similarly: the low-frequency input releases glutamate, while the depolarization relieves the Mg^{2+} block of the NMDA receptor.

The induction and maintenance of LTP depend on different mechanisms. The primary signal for LTP induction is calcium entering the cell through the NMDA receptor channel. In the absence of a sufficient increase in free intracellular calcium, LTP does not develop. Furthermore, injection of calcium chelators (compounds that bind free intracellular Ca^{2+}) blocks LTP induction, whereas rapid elevation of Ca^{2+} levels in postsynaptic neurons potentiates neurotransmission. A likely scenario for LTP induction, then, is that calcium ions entering through NMDA receptors stimulate one or more Ca^{2+}-activated enzymes in the postsynaptic neuron.

The involvement of Ca^{2+}-activated enzymes in the early stages of LTP is supported by experiments in which specific protein kinases (enzymes that add phosphate groups to target proteins) are blocked pharmacologically. At least two kinases have been implicated in LTP induction: protein kinase C and Ca^{2+}/calmodulin-dependent protein kinase (CaMKII). CaMKII may play an especially important part in LTP since it is the most abundant protein in the spines of hippocampal pyramidal cells, which is where Schaffer collateral synapses and NMDA receptors are found.

Although the mechanism that triggers the events leading to LTP are reasonably clear, the mechanism(s) responsible for maintaining potentiation remains controversial. The debate has centered on the locus of the changes that keep synaptic transmission persistently enhanced. One school of thought is that LTP maintenance is based on changes in the sensitivity of the postsynaptic cell to glutamate, either by rapid addition of new receptors to the potentiated synapses, or by increased current flow through receptors already present. However, another body of evidence implies a sustained increase in transmitter release by presynaptic terminals, perhaps by modification of the proteins involved in exocytosis (see Chapter 6). This presynaptic modification hypothesis requires some sort of retrograde signal from the affected postsynaptic region to the presynaptic terminals, which directs that more glutamate be released (Figure 23.7). Which of these schemes is correct remains to be seen; indeed, elements of both could operate.

Because the mechanics of LTP have been worked out almost exclusively in the in vitro slice preparation, little is known about the processes that account for LTP in vivo, where the potentiated synaptic response can persist for days or weeks. In experimental animals, blocking protein synthesis prevents LTP measured several hours after a tetanizing stimulus. Moreover the number and size of synaptic contacts may increase after tetanic stimulation of the hippocampus in vivo, although this issue has been difficult to decide. In any event, the mechanisms that account for LTP in an intact animal may trigger changes in gene expression and subsequent protein syn-

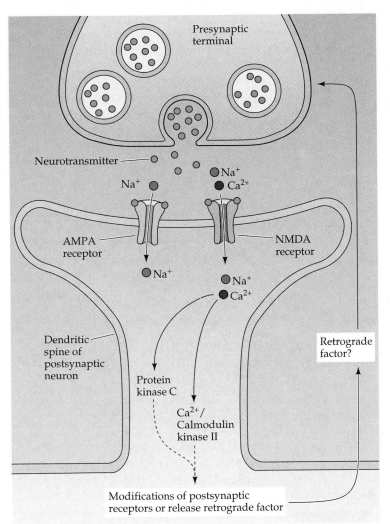

Figure 23.7 Mechanisms underlying LTP induction and maintenance. During glutamate release, the NMDA channel opens only if the postsynaptic cell is sufficiently depolarized. The calcium ions that enter the cell through the channel activate postsynaptic kinases. Through either kinase activation or a separate Ca^{2+}-dependent mechanism, a retrograde signal may be released that acts on the presynaptic transmitter release machinery. Alternatively, Ca^{2+} and/or kinases may act postsynaptically to increase sensitivity to glutamate (without altering transmitter release).

thesis that lead to more permanent alterations of synaptic function (see below).

■ LONG-TERM DEPRESSION

If synapses in the hippocampus simply continued to increase in strength as a result of LTP or similar phenomena, before long all synapses would reach some level of maximum efficacy. At this point it would be difficult or impossible, at least in principle, to encode new information; moreover, active synapses would be so plentiful that even a small stimulus would excite many cells, resulting in pathologies like epilepsy (see the next section). To make synaptic strengthening a useful mechanism for encoding information, other processes must selectively weaken specific sets of synapses. **Long-term depression (LTD)** in the hippocampus may be an example of such processes.

Hippocampal LTD, like LTP, occurs at the synapses between the Schaffer collaterals and the CA1 pyramidal cells. Whereas LTP requires brief, high-frequency stimulation, LTD occurs when the Schaffer collaterals are stimulated at a low rate—about 1 Hz—for long periods (10–15 minutes). This pattern of activity depresses the baseline intracellular EPSP or population response for several hours; it can also erase the increase in EPSP size due to LTP (Figure 23.8). Another form of long-term synaptic depression that depends on quite different synaptic and biochemical mechanisms is described in Box C.

Surprisingly, LTP and LTD share several key elements. Both require activation of NMDA-type glutamate receptors, and both entail calcium entry into the CA1 pyramidal cell. The major determinant of whether calcium entry via the NMDA receptor causes LTP or LTD appears to be the amount of unbound calcium inside the postsynaptic cell. Small increases in intracellular calcium set off a series of events that leads to depression, whereas large increases trigger potentiation. The different effects of low and high levels of calcium may result from the differential activation of calcium-dependent protein kinases and phosphatases. Recall that in LTP, high levels of calcium increase the activity of a specific calcium-dependent protein kinase, CaMKII, which in its active form results in the modification of target proteins by phosphorylation. Low levels of calcium, on the other hand, appear to acti-

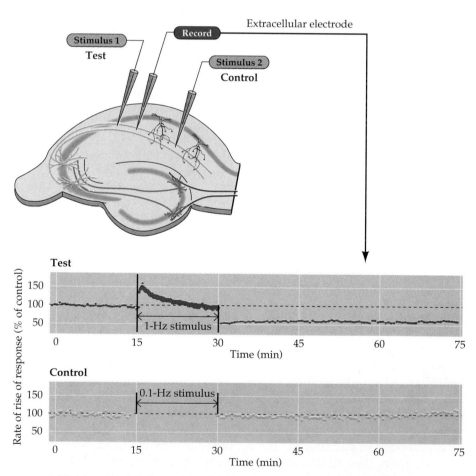

Figure 23.8 Long-term synaptic depression in the hippocampus. In this experiment two stimulating and one extracellular recording electrode are used. However, instead of the high-frequency stimulation used to elicit LTP, low-frequency stimulation (1 per second) is applied to the test pathway (stimulus electrode 1). After 15 minutes, the size and slope of the extracellular population response (see Figure 23.3) is depressed by about 50% (upper panel), while the control pathway (which was stimulated once every 10 seconds) remains unchanged (lower panel). The depression persists for the duration of the experiment. (After Mulkey and Malenka, 1992.)

vate one or several calcium-dependent phosphatases—enzymes that cleave phosphate groups from target molecules. Phosphatase inhibitors prevent LTD, but have no effect on LTP. An attractive speculation, therefore, is that LTP and LTD act together in a "push-pull" manner, phosphorylating and dephosphorylating the same set of regulatory proteins.

■ LTP AND EPILEPSY

The mechanisms that normally support LTP may also contribute to pathological conditions such as **epilepsy** (Box D). The hippocampus is particularly susceptible to epileptogenic activity, and is frequently the source of epileptic seizures. Seizures activate NMDA receptors and strengthen connections between the excited neurons. Thus, LTP may actually promote ongoing epileptic activity, and could explain why seizure activity at a particular site often spreads to involve other synaptically connected cortical regions.

The involvement of LTP or other mechanisms of synaptic plasticity in epilepsy is supported by an animal model of seizure production called **kindling**. To induce kindling, a stimulating electrode is implanted in the brain, often in the amygdala (a component of the limbic system that makes and receives connections with the cortex, thalamus, and other limbic structures, including the hippocampus; see Chapter 27). At the beginning of a typical kindling experiment, weak electrical stimulation, in the form of a low-amplitude train of electrical pulses, has no discernible effect on the animal's behavior or on the pattern of electrical activity in the brain (laboratory rats have typically been used for such studies). As this weak stimulation is repeated once a day for several weeks, it begins to produce behavioral and electrical indications of seizures. By the end of the experiment (usually 2 weeks later), the same weak stimulus that initially had no effect now causes full-blown seizures. This phenomenon is essentially permanent; even after an interval of a year, the same weak stimulus will again trigger a seizure. Thus, repetitive, weak activation produces long-lasting changes in the excitability of the brain that time cannot reverse (the word *kindling* is therefore quite appropriate: a single match can start a large and devastating fire). Many of the changes in the electrical patterns of brain activity detected in "kindled" animals resemble those in human epilepsy.

■ MECHANISMS OF SYNAPTIC PLASTICITY IN "SIMPLE" NERVOUS SYSTEMS

Despite suggestive relationships between the hippocampus, LTP, and memory (see Box A), a fundamental problem has hindered efforts to nail down the synaptic alterations related to stored information: Where, in the vastness of a human (or animal) brain, are specific memories stored? The answer to this question is simply not known. Thus, unambiguously associating a given behavioral change with a change in the strength or number of connections among specific sets of neurons remains a difficult problem.

One way of circumventing this dilemma is to examine learning and memory in nervous systems far simpler than the mammalian brain. The assumption in this strategy is that learning and memory are such fundamental processes that their cellular and molecular bases should share at least some features among very different organisms. The most successful example of this approach has been a series of studies carried out by Eric Kandel and his colleagues using the marine mollusk *Aplysia californica*. This animal has only about 20,000 neurons in its central nervous system, compared to the hundreds of millions of neurons in even a small mammalian brain. Further-

Box C
LONG-TERM DEPRESSION IN THE CEREBELLUM

The cerebellum, like the hippocampus, comprises principal cells—the Purkinje neurons—that receive two distinct types of excitatory input: climbing fibers and parallel fibers (see Chapter 18). Only one climbing fiber innervates each Purkinje cell, wrapping around the dendritic shafts and forming numerous excitatory contacts. In contrast, parallel fibers run perpendicular to the layer of Purkinje cells. A single Purkinje cell receives synapses from a huge number of parallel fibers—up to 100,000—on its distal dendrites. When climbing fibers and parallel fibers are activated at the same time, the strength of synaptic transmission in the parallel fiber pathway (but not the climbing fiber pathway) is depressed (Figure A). Such long-term depression (LTD) in the cerebellum was first observed in vivo; as with hippocampal LTP, however, the cellular mechanisms involved in LTD have been unraveled using brain slice (and to some extent tissue culture) preparations.

Activity of climbing fiber and parallel fiber synapses results in two distinct intracellular responses in the postsynaptic Purkinje cell. In the first pathway, glutamate released from the parallel fiber terminals binds at two types of glutamate receptors, the AMPA and metabotropic receptors (see Chapter 7). Glutamate binding to the AMPA receptor results in Na^+ influx and membrane depolarization, while binding to the metabotropic receptor initiates a second-messenger cascade that eventually activates an intracellular enzyme, protein kinase C (PKC). The second pathway is initiated by climbing fiber activation, which causes a large increase in intracellular Ca^{2+}. The conjoint activation of these two intracellular pathways leads to LTD; through an as-yet unknown mechanism, Ca^{2+} interacts with PKC to decrease the postsynaptic response of AMPA receptors to glutamate at the parallel fiber synapses (Figure B).

A number of observations support the idea that LTD requires activation of the metabotropic glutamate receptor on distal dendrites (mediated by parallel fibers) during a rise in intracellular Ca^{2+} (mediated by climbing fibers). Metabotropic receptor agonists can substitute for the parallel fiber stimulation required to trigger LTD, whereas compounds that activate other glutamate receptors cannot. Protein kinase C inhibitors prevent induction of LTD, and direct activation of the kinase produces an LTD-like effect. Coupling parallel fiber activation with artificial depolarization of the Purkinje cell (which mimics climbing fiber activation by allowing Ca^{2+} entry through voltage-gated Ca^{2+} channels) triggers LTD, while intracellular injection of Ca^{2+} chelators, which block the rise in intracellular calcium induced by climbing fiber activation, prevents LTD. In contrast to LTD in the hippocampus, then, cerebellar LTD requires the activity of a protein kinase, not a phosphatase, and does not involve calcium entry through the NMDA type of glutamate receptor.

Not surprisingly, different regions of the vertebrate brain utilize distinct mechanisms to adjust synaptic strength. Model systems like hippocampal LTP and cerebellar LTD highlight the diversity of possible mechanisms that can be employed to accomplish the important business of synaptic change.

References

Ito, M. (1989) Long-term depression. Annu. Rev. Neurosci. 12: 85–102.

Lev-Ram, V., L. R. Makings, P. F. Keitz, J. P. Kao and R. Y. Tsien (1995) Long-term depression in cerebellar Purkinje neurons results from coincidence of nitric oxide and depolarization-induced Ca^{2+} transients. Neuron 15: 407–415.

Linden, D. J., M. H. Dickinson, M. Smeyne and J. A. Connor (1991) A long-term depression of AMPA currents in cultured cerebellar Purkinje neurons. Neuron 7: 81–89.

more, specific cells can be identified by their size and location. It is thus possible to map the neuronal circuits involved in behaviors such as withdrawal of the animal's gill in response to noxious stimuli. Kandel's work has taken advantage of this stereotyped neural organization to correlate changes in a reflex behavior with synaptic changes of the relevant neurons.

An elementary form of learning in mollusks and many other species is **sensitization**, in which an animal learns to generalize an aversive response elicited by a noxious stimulus to a variety of other, non-noxious stimuli. In *Aplysia*, a light touch to the animal's siphon results in gill withdrawal, a response that gradually habituates (becomes less strong) with repeated stimulation. After several repetitions, the animal no longer bothers to withdraw the gill after being touched, presumably having identified this minor irritant

(A)

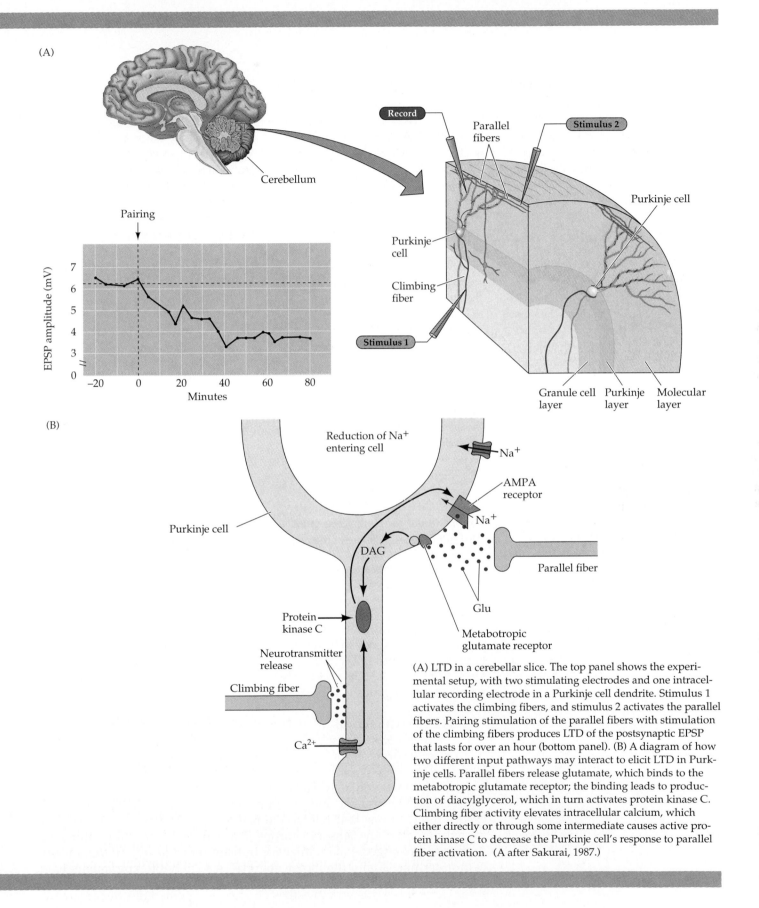

(B)

(A) LTD in a cerebellar slice. The top panel shows the experimental setup, with two stimulating electrodes and one intracellular recording electrode in a Purkinje cell dendrite. Stimulus 1 activates the climbing fibers, and stimulus 2 activates the parallel fibers. Pairing stimulation of the parallel fibers with stimulation of the climbing fibers produces LTD of the postsynaptic EPSP that lasts for over an hour (bottom panel). (B) A diagram of how two different input pathways may interact to elicit LTD in Purkinje cells. Parallel fibers release glutamate, which binds to the metabotropic glutamate receptor; the binding leads to production of diacylglycerol, which in turn activates protein kinase C. Climbing fiber activity elevates intracellular calcium, which either directly or through some intermediate causes active protein kinase C to decrease the Purkinje cell's response to parallel fiber activation. (A after Sakurai, 1987.)

Box D
EPILEPSY

Epilepsy is a brain disorder characterized by periodic and unpredictable seizures caused by the rhythmic firing of large groups of neurons. Epileptic seizures range from mild twitching in the extremities to loss of consciousness and uncontrollable convulsions. Although many highly accomplished people have been epileptics (Alexander the Great, Julius Caesar, Napoleon, Dostoevsky, and Van Gogh, to name a few), untreated epilepsy can be a debilitating condition, with seizures of sufficient intensity and frequency to preclude participating in many aspects of daily life. To make matters worse, uncontrolled convulsions can lead to degenerative changes in the brain, presumably as a result of excitotoxicity (see Box B in Chapter 6). Up to 1% of the population is afflicted, making epilepsy one of the most common neurological problems.

The first monograph on epilepsy, written in 400 B.C., concluded that seizures were caused by specific Greek gods, and the Christian Gospels refer to curing epilepsy by casting out evil spirits. Modern thinking about the causes

(and possible cures) of epilepsy has focused on where seizures originate and the mechanisms that make the affected region hyperexcitable. Most of the evidence suggests that small areas of the cerebral cortex (called foci) provide the "spark" that triggers a seizure by spreading to synaptically connected regions. Accordingly, the manifestations of seizures are determined by the particular brain areas involved. For example, a seizure originating in the thumb area of the right motor cortex will first be evident as uncontrolled movement of the left thumb. A seizure originating in the visual association cortex of the right hemisphere may be heralded by complex hallucinations in the left visual field. The behavioral manifestations of seizures therefore provide important clues for the neurologist seeking to pinpoint the abnormal region of cerebral cortex.

Epileptic seizures can be caused by a variety of acute or congenital factors, including cortical damage from trauma, stroke, neoplasm, congenital cortical dysgenesis (failure of the cortex

to grow properly) and congenital vascular malformations. One rare form of epilepsy, Rasmussen's encephalitis, has an autoimmune etiology (the body produces antibodies against components of certain glutamate receptors). Some forms of epilepsy are heritable; at least eight distinct single-locus defects are linked to unusual forms of epilepsy. Most forms of familial epilepsy (such as juvenile myoclonic epilepsy and petit mal epilepsy) are polygenic disorders; that is, they are caused by the simultaneous inheritance of several mutant genes.

No effective prevention or cure exists. In a small fraction of patients, the epileptogenic region can be surgically excised. In extreme cases, physicians must resort to cutting the corpus callosum to prevent the spread of seizures (most of the "split-brain" subjects described in Chapter 25 are patients with intractable epilepsy). Pharmacological therapies are based on drugs that inhibit seizures, often by acting as agonists of the inhibitory neurotransmitter GABA (see Chapter 6). Commonly used antiseizure medications include carba-

---

as unimportant. If, however, touching the siphon is paired with a single strong electrical shock to the animal's tail, the same light touch to the siphon now elicits a rapid gill withdrawal, and this reflex remains enhanced for at least an hour after the aversive tail shock (short-term sensitization). With repeated training, this behavior can be altered for days or weeks (long-term sensitization).

The gill withdrawal circuit comprises sensory neurons that innervate the siphon, motor neurons that control gill movement, and interneurons that receive inputs from sensory neurons in the tail (Figure 23.9). Short-term sensitization involves a cascade of neurotransmitters, second messengers, and ion channels that ultimately leads to increased efficacy of neurotransmission between the sensory and motor components of the circuit. The tail shock excites interneurons that release the neurotransmitter serotonin onto the axon terminals of the sensory neurons. The actions of serotonin ultimately produce a prolonged enhancement of transmitter release from the sensory neuron terminals onto the motor neurons. A variety of experiments have

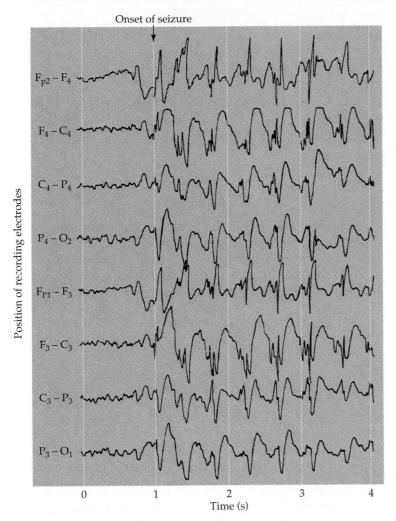

Onset of seizure

Position of recording electrodes

$F_{p2} - F_4$

$F_4 - C_4$

$C_4 - P_4$

$P_4 - O_2$

$F_{P1} - F_3$

$F_3 - C_3$

$C_3 - P_3$

$P_3 - O_1$

0   1   2   3   4

Time (s)

Electroencephalogram (EEG) recorded from a patient during a seizure. The traces show rhythmic activity that persisted much longer than the duration of this record; this abnormal pattern reflects the synchronous firing of large numbers of cortical neurons. (The designations are various positions of the electrodes on the head; see Box A in Chapter 26 for additional information about EEG recordings.) (After Dyro, 1989.)

mazepine, phenobarbital, phenytoin (Dilantin®), and valproic acid. These agents, which must be taken daily, successfully inhibit seizures in 60–70% of patients. A widespread hope is that understanding the cellular and molecular mechanisms of neuronal hyperexcitability and plasticity will lead to new and more effective approaches to this age-old problem.

References

McNamara, J. O. (1994) Cellular and molecular basis of epilepsy. J. Neurosci. 14: 3413–3425.

Engel, J. E. (1989) Epilepsy. In *Seizures and Epilepsy*, Contemporary Neurology Series No. 31. Philadelphia: F.A. Davis.

suggested a plausible cellular and molecular explanation of sensitization of the gill withdrawal reflex that is quite different from the events underlying LTP (Figure 23.10).

The short-term learning induced by a single tail shock is retained for about an hour and depends on covalent protein modification by phosphorylation. In contrast, long-term sensitization of the same pathway, which lasts up to several weeks, requires both protein synthesis and changes in gene expression. With repeated training (that is, additional tail shocks), the serotonin-activated, cAMP-dependent protein kinase, which phosphorylated K^+ channels during short-term sensitization, now phosphorylates—and thereby activates—a specific set of transcription factors (proteins that act within the nucleus to control gene expression; see Chapter 20). These transcription factors bind to specific DNA sequences, termed *cAMP responsive elements* (**CREs**); as a result, the transcription rates of genes located near these CREs are greatly increased. The transcription factors are known as **CREBs**, for *cAMP response element binding* proteins (see also Chapter 20). The depen-

Figure 23.9 Short-term sensitization of the *Aplysia* gill withdrawal reflex. (A) Diagram of the animal showing the relevant features. (B) Many cell bodies in the abdominal ganglion of this animal can be recognized by their size, shape, and position. Several neurons have been implicated in the gill withdrawal reflex circuit. (C) Stimulation of the facilitating interneurons in the gill withdrawal circuit by a shock to the animal's tail causes release of the neurotransmitter serotonin onto the presynaptic terminals of sensory neurons from the skin; the serotonin causes the presynaptic terminals of the sensory neurons to release more neurotransmitter onto the motor neurons innervating the gill, resulting in sensitization. (After Kandel, 1979.)

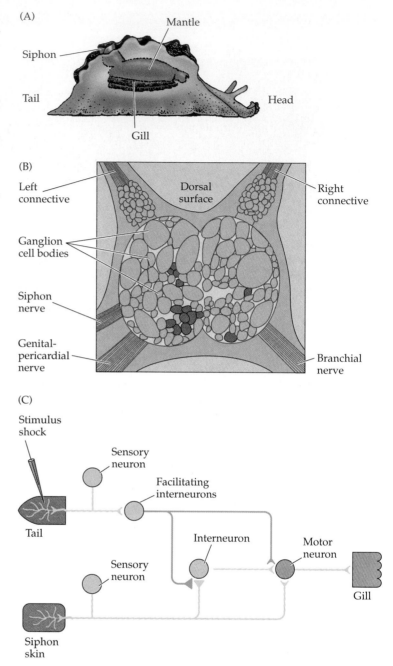

dence of long-term sensitization on interactions between CREs and CREB was shown by an experiment in which excess synthetic DNA containing CREs was injected into sensory neuron nuclei. This manipulation blocked long-term, but not short-term, sensitization. A possible explanation is that the excess synthetic CRE bound most of the CREB transcription factor molecules activated by training, effectively preventing CREB from binding to and activating the genuine CREs in the sensory neuron's own DNA. CREB has also been implicated in learning in other organisms, including fruit flies and mice. Mutant flies deficient in certain forms of CREB cannot learn to avoid odors paired with electrical shocks, and mice missing a mammalian form of CREB do poorly in certain learning tasks (and may have deficiencies in LTP).

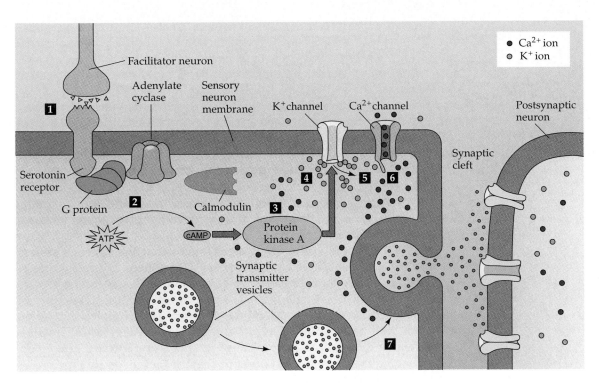

Figure 23.10 Events underlying short-term sensitization. Serotonin released from facilitating interneurons (1) activates adenylate cyclase, which raises cyclic AMP (cAMP) levels in the sensory cells. (2) Cyclic AMP turns on a protein kinase (3), whose activity reduces some of the voltage-sensitive K⁺ currents in the sensory cell (4). Decreasing K⁺ currents prolongs the sensory cell's action potentials, which increases the amount of Ca^{2+} entering the cell during each action potential (5, 6). The additional calcium in the presynaptic terminal presumably enhances neurotransmitter release onto the motor neurons (7), in much the same way that increased Ca^{2+} levels result in facilitation at the neuromuscular junction (see Figure 23.1) (After Klein and Kandel, 1978.)

The changes in genes and gene products that follow CRE activation have been difficult to sort out. Nonetheless, two consequences of gene activation in *Aplysia* have been identified. First, some cAMP-dependent protein kinases no longer require serotonin to be activated, but remain persistently active. Gene activation also results in an increase in the number of synapses between the sensory and the motor neurons. Such increases are not seen following short-term sensitization and may represent a long-lasting change in overall strength of the relevant connections. This and other observations suggest that neuronal growth, in addition to its importance in development, may also serve as a mechanism for long-lasting information storage in the adult nervous system.

■ COMMON THEMES IN ADULT PLASTICITY

Long term plastic changes in various systems share certain attributes. Synaptic plasticity, whether in the form of LTP, LTD, or sensitization in *Aplysia*, relies on second-messengers to activate protein kinases. In hippocampal LTP and LTD, and in cerebellar LTD (see Box C), Ca^{2+} acts as a second messenger, activating a host of kinases; in sensitization in *Aplysia*, cyclic AMP serves this function, activating cAMP-dependent protein kinase. These changes all involve the covalent modification of proteins by phosphorylation, and can be rendered more permanent by changes in gene expression and protein synthesis, which may eventually elicit structural as well as functional changes at the synapse.

There are also some obvious differences between the mechanisms neurons employ for these several forms of long-term plasticity. Most importantly, sensitization in *Aplysia* is non-associative, whereas cerebellar and hippocampal LTD and LTP result from associative activity between the presynaptic and postsynaptic cells. Furthermore, sensitization is apparently initiated by events at presynaptic terminals, whereas LTP and LTD are triggered postsynaptically by activation of glutamate receptors and Ca^{2+} entry. Induction of LTD and LTP requires nothing from the presynaptic terminal other than transmitter release.

LTP and LTD are examples of mechanisms that the nervous system uses to alter the strength of a particular set of synapses in adult animals. Models of neural networks propose, logically enough, that encoding memories requires not just a single strengthened synapse, but an assembly of synaptically linked neurons. (Memories are typically complex and often involve multiple associations.) Despite substantial advances in understanding the cellular and molecular bases of some forms of plasticity, how selectively changing the strength of a subset of synapses might encode memories or other complex information is simply not known.

■ CORTICAL PLASTICITY IN ADULT ANIMALS

Until recently, plasticity of the adult cerebral cortex received relatively little attention compared to developmental rewiring or phenomena like LTP in the hippocampus or LTD in the cerebellum, largely because the wiring of the mature cortex was assumed to be stable. Findings in the adult somatic sensory system and visual system, however, indicate that at least some aspects of neuronal circuitry, such as the size and arrangement of receptive fields, can change in response to altered circumstances. Although the functional significance (if any) of these adult rearrangements remains unclear, it has become apparent that some degree of cortical reorganization remains possible throughout life.

One approach to adult cortical plasticity has been to examine the malleability of somatic sensory maps. As described in Chapter 8, the body surface is mapped topographically onto the primary somatic sensory cortex. Moreover, each cortical neuron has a receptive field—a restricted area of the body surface to which that neuron preferentially responds. John Kaas, Michael Merzenich and their colleagues have used closely spaced microelectrode penetrations to determine the receptive fields of cortical neurons in monkeys, thereby creating a detailed map of the area of the somatic sensory cortex representing the hand (Figure 23.11). Merzenich then investigated what happened to the map if a single digit was surgically removed. When he and his collaborators recorded from the same region of cortex several months later, they found that the neurons within the cortical area occupied by the missing digit now responded to stimulation of the adjacent, intact digits.

Further experiments showed that such rearrangements in the somatotopic map do not necessarily require amputation. If a monkey is trained to use a specific digit for a particular task that is repeated many thousands of times, then the functional representation of that digit expands at the expense

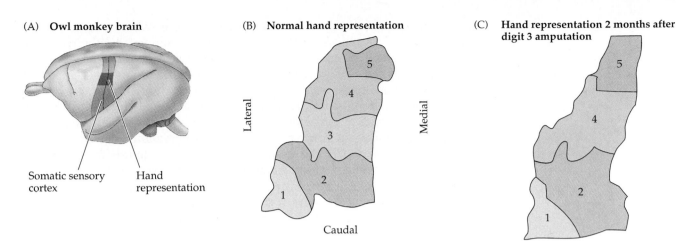

(A) **Owl monkey brain**

Somatic sensory cortex

Hand representation

(B) **Normal hand representation**

Lateral

Medial

Caudal

(C) **Hand representation 2 months after digit 3 amputation**

Figure 23.11 Functional changes in the somatic sensory cortex of an owl monkey following amputation of a digit. (A) Diagram of the somatic sensory cortex in the owl monkey, showing the approximate location of the hand representation. (B) The hand representation in the animal before amputation; the numbers correspond to different digits. (C) The cortical map determined in the same animal two months after amputation of digit 3. The map has changed substantially; neurons in the area formerly responding to stimulation of digit 3 now respond to stimulation of digits 2 and 4. (After Merzenich et al., 1984.)

of the other digits (Figure 23.12). Recent studies using functional magnetic resonance imaging (ƒMRI; see Box E in Chapter 1) are consistent with such a practice effect.

These results in the somatic sensory system could be explained by functional changes of synaptic efficancy at any of several subcortical sites. Related experiments in the visual system, however, indicate that at least some plasticity occurs in the cortex itself. As in the somatic sensory cortex, each neuron in the visual cortex has a receptive field—a circumscribed area of visual space to which a particular cortical neuron responds (see Chapter 11). Receptive field

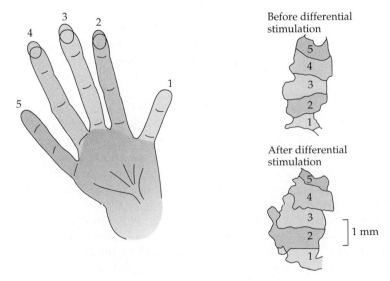

Before differential stimulation

After differential stimulation

1 mm

Figure 23.12 Functional expansion of a cortical representation by a repetitive behavioral task. An owl monkey was trained in a task that required heavy usage of digits 2, 3, and occasionally 4. The map of the digits in the primary somatic sensory cortex prior to training is shown. After several months of "practice," a larger region of the cortex contained neurons activated by the digits used in the task. Note that the specific arrangements of the digit representations are somewhat different from the monkey shown in Figure 23.11, indicating the variability of the cortical representation in particular animals. (After Jenkins et al., 1990.)

properties have generally been considered an immutable characteristic of each adult neuron. But several studies have now shown that the receptive fields of cortical neurons, and indeed the retinotopic map in the visual cortex, are also capable of some reorganization in adult cats and monkeys. If a small lesion is made at corresponding locations in the two retinas, an appreciable unresponsive region is produced in the striate cortex. What happens to the receptive fields of neurons within the deprived region of visual cortex? (Note that the question is similar to the one asked by Merzenich in amputating a digit.) Immediately after the lesion, the neurons are silent because they no longer receive stimulation from either eye. Within hours, however, neurons in the deprived zone begin reacting to stimulation of the retinal regions adjacent to the lesion (Figure 23.13). As in the somatic sensory system, there are several possible sites for the underlying change: the retina, the lateral geniculate nucleus in the thalamus, as well as the cortex. If the changes were retinal or thalamic, however, one would expect the size of the silent zone in the thalamus to shrink over time, in parallel with the cortical changes. This is not the case; the size of the thalamic silent zone remains unchanged after the lesion, implying that the locus of this plasticity is cortical.

The most likely explanation of these results in the visual and somatosensory cortices is that although cortical neurons normally respond best to stimulation of a small portion of the related sensory surface, they also receive weaker, but still significant synaptic connections from other nearby regions. When the principal drive to one set of neurons is removed (as in the case of retinal lesions or digit amputation), these inputs are unmasked and strengthened, so that now they provide the major excitatory drive to the deprived neurons. These weaker inputs probably arise from pyramidal neurons (the major excitatory cell type in the cortex), which make rich local connections—called **horizontal connections**—that span considerable distances.

These observations on the malleability of cortical cells suggest that their functional properties may continue to change in adulthood, even though the basic features of cortical organization—such as ocular dominance columns and the broader topographical organization of inputs from the thalamus—must remain largely fixed. The dramatic and essentially permanent changes that occur during critical periods reflect the actual growth and retraction of synaptic connections, particularly those originating from thalamic relay nuclei (see Chapter 22). Although the mechanisms involved in the more reversible rearrangements in the maps seen in the adult cortex are not yet understood, most of these changes probably reflect either enhancement or weakening of synapses already present, rather than substantial rewiring of the cortex. In any event, the changes that do take place after the phase of critical periods are certainly much less florid than the rewiring that occurs dur-

Figure 23.13 Functional changes in the visual cortex resulting from small, corresponding retinal lesions made in both eyes of an adult cat. The visual field is represented as a grid superimposed on the retina, with the corresponding maps shown on surface views of the primary visual cortex. The lesion (representing about 5° of visual space) is indicated; only one retina is shown. The retinal damage initially silenced an area of cortex about 10 mm in diameter. Over the ensuing two months, however, the receptive fields of cortical neurons reorganized such that the representation of the areas around the lesion increased, while the representation of the lesioned area decreased. (After Gilbert and Wiesel, 1992.)

Retina

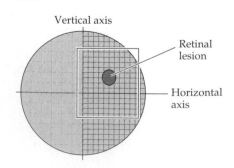

Visual cortex

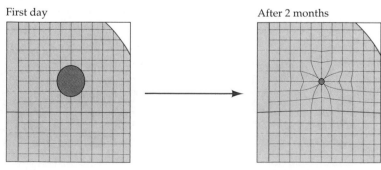

ing development. This fact presumably explains why no amount of effort or retraining in adults can fully reverse the effects of early experience.

■ SUMMARY

Changes in synaptic strength can occur over a broad temporal range and are presumably responsible for the nervous system's ability to modify behavioral responses in adult animals. At the shortest times (seconds to minutes), facilitation and depression provide rapid, transient modifications based on altered calcium levels at recently active synapses. These changes do not, however, trigger longer-lasting modifications. In other forms of synaptic plasticity, calcium and other second messengers trigger intracellular events, including protein phosphorylation and changes in gene expression, that may eventually lead to more permanent changes in synaptic strength. Several proximate mechanisms can cause such changes in the strength of connections between pre- and postsynaptic partners, including increasing or decreasing transmitter release from presynaptic terminals, changing the number or sensitivity of postsynaptic receptors, and adding (or subtracting) synapses in the relevant circuits. Different brain regions evidently use one or more of these strategies to learn new behaviors and acquire new memories. Although such synaptic changes are not in themselves learning or memory, they apparently provide the cellular substrate for these phenomena.

Additional Reading

Reviews

BLISS, T. V. P. AND G. L. COLLINGRIDGE (1993) A synaptic model of memory: Long-term potentiation in the hippocampus. Nature 361: 31–39.

JESSELL, T.. M. AND E. R. KANDEL (1993) Synaptic transmission: a bidirectional and self-modifiable form of cell-cell communication. Neuron 10: 1–30.

MADISON, D. V., R. C.MALENKA AND R. A. NICOLL (1991) Mechanisms underlying long-term potentiation of synaptic transmission. Annu. Rev. Neurosci. 14: 1379–1397

MALENKA, R. C. (1994) Synaptic plasticity in the hippocampus: LTP and LTD. Cell 78: 535–538.

MERZENICH, M. M., G. H. RECANZONE, W. M. JENKINS AND K. A. GRAJSKI (1990) Adaptive mechanisms in cortical networks underlying cortical contributions to learning and nondeclarative memory. Cold Spring Harbor Symp. Quant. Biol. 55: 873–887.

Important Original Papers

BLISS, T. V. P. AND A. R. GARDNER-MEDWIN (1973) Long-lasting potentiation of synaptic transmission in the dentate area of the unanaesthetized rabbit following stimulation of the perforant path. J. Physiol. 232: 357–374.

BLISS, T. V. P. AND T. LOMO (1973) Long-lasting potentiation of synaptic transmission in the dentate area of the anaesthetized rabbit following stimulation of the perforant path. J. Physiol. 232: 331–356.

COLLINGRIDGE, G. L., S. J. KEHL AND H. McLENNAN (1983) Excitatory amino acids in synaptic transmission in the Schaffer collateral-commissural pathway of the rat hippocampus. J. Physiol. 334: 33–46.

DASH, P. K., B. HOCHNER AND E. R. KANDEL (1990) Injection of the cAMP-responsive element into the nucleus of Aplysia sensory neurons blocks long-term facilitation. Nature 345: 718–721.

GILBERT, C. D. AND T. N. WIESEL (1992) Receptive field dynamics in adult primary visual cortex. Nature 356: 150–152.

JENKINS, W. M., M. M. MERZENICH, M. T. OCHS, E. ALLARD AND T. GUIC-ROBLES (1990) Functional reorganization of primary somatosensory cortex in adult owl monkeys after behaviorally controlled tactile stimulation. J. Neurophysiol. 63: 82–104.

KATZ, B. AND R. MILEDI (1968) The role of calcium in neuromuscular facilitation. J. Physiol. (Lond.) 195: 481–492.

KELSO, S. R. AND T. H. BROWN (1986) Differential conditioning of associative synaptic enhancement in hippocampal brain slices. Science 232: 85–87.

LINDEN, D. J., M. H. DICKINSON, M. SMEYNE AND J. A. CONNOR (1991) . A long-term depression of AMPA currents in cultured cerebellar Purkinje neurons. Neuron 7: 81–89.

MALENKA, R. C., J. A. KAUER, R. S. ZUCKER AND R. A. NICOLL (1988) Postsynaptic calcium is sufficient for potentiation of hippocampal synaptic transmission. Science 242: 81–84.

MAYER, M. L., G. L. WESTBROOK AND P. B. GUTHRIE (1984) Voltage-dependent block by Mg^{2+} of NMDA responses in spinal cord neurones. Nature 309: 261–263.

MERZENICH, M. M., R. J. NELSON, M. P. STRYKER, M. S. CYNADER, A. SCHOPPMANN AND J. M. ZOOK (1984) Somatosensory cortical map changes following digit amputation in adult monkeys. J. Comp. Neurol. 224: 591–605.

MULKEY, R. M., C. E. HERRON AND R.C. MALENKA (1993) An essential role for protein phosphatases in hippocampal long-term depression. Science 261: 1051–1055.

NOWAK, L., P. BREGESTOVSKI AND P. ASCHER (1984) Magnesium gates glutamate-activated channels in mouse central neurones. Nature 307: 462–465.

SCHACHER, S., V. F. CASTELLUCCI AND E. R. KANDEL (1988) cAMP evokes long-term facilitation in Aplysia sensory neuron that requires new protein synthesis. Science 240: 1667–1668.

SILVA, A. J., R. PAYLOR, J. M. WEHNER AND S. TONEGAWA (1992) Impaired spatial learning in alpha-calcium-calmodulin kinase II mutant mice. Science 257: 206–211.

Books

BAUDRY, M. AND J. D. DAVIS (1991) Long-Term Potentiation: A Debate of Current Issues. Cambridge, MA: MIT Press.

LANDFIELD, P. W. AND S. A. DEADWYLER (eds.) (1988) Long-Term Potentiation: From Biophysics to Behavior. New York: A. R. Liss.

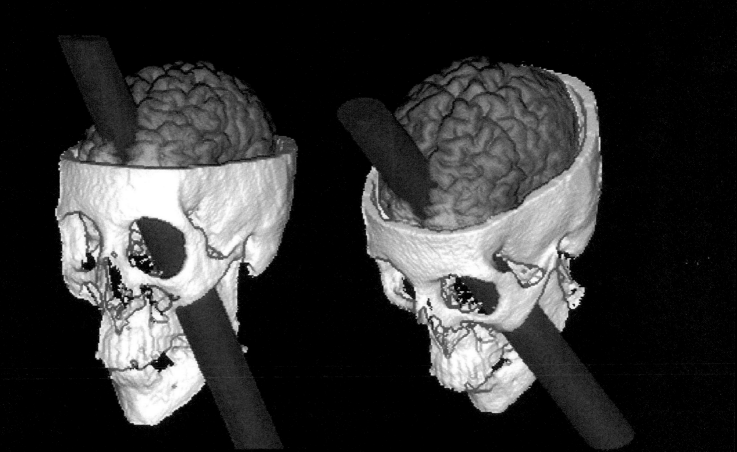

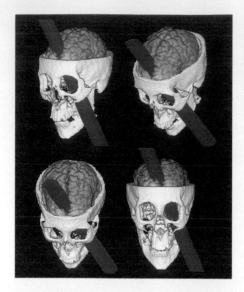

UNIT V
COMPLEX BRAIN FUNCTIONS

The function of the frontal cortex was first suggested by a dramatic nineteenth century accident in which a tamping rod was driven through the frontal part of the brain of a railroad worker named Phineas P. Gage. Remarkably, Gage survived, and his behavioral deficits stimulated much early thinking about complex brain functions. The illustration here is a reconstruction of the trajectory of the rod based on Gage's skull, which is now housed in the Warren Museum at Harvard Medical School (Courtesy of H. Damasio.)

Cognition, language, sleep and wakefulness, the emotions, the neurology of human sexuality, and memory—these are, for the majority of us, the most intriguing properties of the human brain. In fact, much of the brain is devoted to these functions. The intrinsic interest of these complex aspects of human behavior is unfortunately equaled by the difficulties involved—both technical and conceptual—in unraveling their neurobiological basis. Nonetheless, a good deal of progress has been made in deciphering the structural and functional organization of the relevant brain regions. Especially important has been the steady accumulation of human case studies documenting the signs and symptoms that result from damage to various brain regions. The advent of in vivo imaging and metabolic mapping has provided another important means of understanding the organization of these abilities in humans. Finally, complementary experiments in non-human primates have begun to indicate the cellular bases for some of these phenomena.

Based on this growing evidence, there is little doubt that a broad spectrum of neurological and psychiatric disorders will eventually be understood as cellular and molecular pathologies of the brain regions that subserve these more complex aspects of human brain function.

CHAPTER 24

■ OVERVIEW

The association cortices of the parietal, temporal, and frontal lobes account for approximately 75% of all human brain tissue. These cortical regions are responsible for much of the information processing that goes on between sensory input and motor output. The diverse functions of the association cortices are often referred to as cognition, which literally means the process by which we come to know the world. For neurobiologists, cognition refers more specifically to the ability to attend, identify, and plan meaningful responses to external stimuli or internal motivation. In keeping with these functions, the association cortices receive and integrate information from a variety of sources and can influence a broad range of behaviors. Inputs to the association cortices include connections from the primary and secondary sensory and motor cortices, the thalamus, and the brainstem. Outputs from the association cortices reach the hippocampus, the basal ganglia and cerebellum, the thalamus, and other association cortices. Insight into how the association areas work has come mainly from observations of human patients with specific brain lesions. Functional mapping at neurosurgery, noninvasive imaging of normal subjects, and electrophysiological analysis in comparable brain regions of subhuman primates have generally confirmed these clinical impressions. Together, these studies indicate that the parietal association cortex is essential for attending to complex stimuli in the external and internal environment, that the temporal association cortex is essential for identifying such stimuli, and that the frontal association cortex is essential for planning the appropriate behavioral responses.

■ THE ASSOCIATION CORTICES

The preceding chapters have considered in some detail the parts of the brain responsible for encoding sensory information and commanding and executing movements (Figure 24.1). These regions occupy perhaps 25% of the brain. What, then, does the other 75% do? Selectively attending to a particular stimulus, recognizing and identifying the relevant stimulus features, and planning and experiencing the response are some of the processes mediated by the rest of the human brain. Collectively, these abilities are referred to as cognition, and it is the association cortices in the parietal, temporal, and frontal lobes that make cognition possible. (Note that the occipital lobe is missing from this list; its functions appear to be more exclusively concerned with vision, although some of these are arguably cognitive as well.)

The most important component of the cortex in each of the cerebral lobes is the neocortex, defined as cortex that has six layers, or laminae. Each layer comprises a distinctive population of cells having different sizes, shapes, inputs, and outputs. (The organization and connectivity of the human cerebral cortex are summarized in Table 24.1 and Figure 24.2.) When the details of the thickness, cell-packing density, and connections of the six layers are cataloged across the cortical mantle, regional differences are apparent (Box A; Figure 24.2B). The structure of the neocortical layers and cellular morphology within them give identity to several dozen subdivisions

Figure 24.1 Lateral and medial views of the human brain, showing the extent of the association cortices. The primary sensory and motor regions of the neocortex are shaded in color. Notice that these regions occupy only a small fraction of the total area of the cortex. The remainder of the neocortex—defined by exclusion as the association cortices—is the seat of human cognitive ability. The term *association* refers to the fact that these regions of the cortex integrate (associate) information derived from other brain regions.

Primary sensory and motor areas

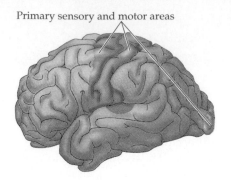
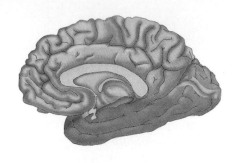

of the cerebral cortex referred to as **cytoarchitectonic areas**. Over the years, a small army of neuroanatomists has painstakingly mapped these areas in both humans and laboratory animals. At first, cytoarchitectonic regions were identified with little or no knowledge of their functional significance. Eventually, careful studies of patients with lesions of specific cortical areas, supplemented by electrophysiological mapping techniques in both laboratory animals and neurosurgical patients, showed that the regions anatomists had identified based on histological features are, in fact, functionally distinct. These distinctions are explained by the different connections of each area and the different morphological and physiological properties of the neurons within them. Gradually it became evident that the primary sensory and motor cortices occupy a relatively limited portion of the cortical mantle, the majority of the cortex being devoted to what were presumed to be more integrative functions. By default (at least initially), these other areas of largely unknown function were called **association cortices**.

TABLE 24.1
Some Basic Organizational Features of the Cerebral Cortex

Lobes	Cortical inputs	Targets of cortical output	Neocortical layers
Frontal lobe Temporal lobe Parietal lobe Occipital lobe	Thalamus Brainstem nuclei Other cortical regions	Thalamus Spinal cord Brainstem: Cranial motor nuclei Pontine nuclei (to cerebellum) Caudate and putamen	**Layer I:** Few cells; primarily axons, dendrites, and synapses **Layers II and III:** Pyramidal neurons that project to and receive projections from other cortical regions **Layer IV:** Stellate cells that receive most of thalamic input and project locally to other laminae **Layers V and VI:** Pyramidal neurons that project to subcortical regions such as the thalamus, brainstem, and spinal cord, and to other cortical areas

(A)

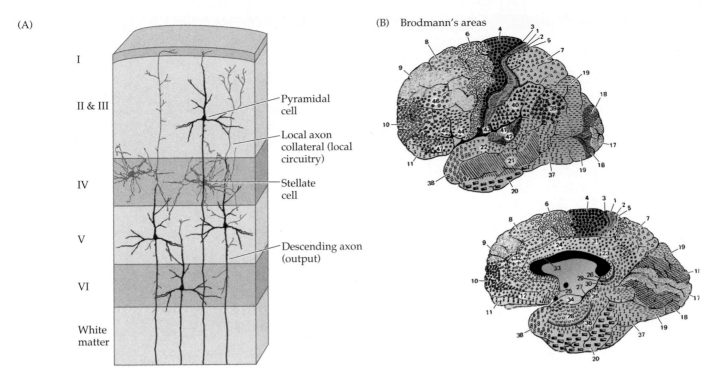

- Pyramidal cell
- Local axon collateral (local circuitry)
- Stellate cell
- Descending axon (output)

I

II & III

IV

V

VI

White matter

(B) Brodmann's areas

Figure 24.2 The structure of the human neocortex, including the association cortices. (A) A summary of the cellular composition of the six layers of the neocortex, showing the basic cellular composition of each of the six layers. (See Table 24.1 and Box A for additional detail.) (B) Based on variations in the thickness, cell density, and other histological features of the six neocortical laminae, the human brain can be divided into cytoarchitectonic areas, in this case those recognized by the neuroanatomist Korbinian Brodmann in his seminal monograph in 1909. Red indicates the primary motor cortex, blue the primary somatic sensory, green the primary auditory, and yellow the primary visual cortex. All other Brodmann areas are considered association cortex.

■ DIFFERENCES BETWEEN THE ASSOCIATION CORTICES AND THE REST OF THE NEOCORTEX

In light of what is now known about cortical connectivity, the association cortices are indeed quite different from primary and secondary sensory and motor cortices. Perhaps most apparent is a difference in the input and output of the association cortices. Three distinct thalamic nuclei provide input to the association cortex of the parietal, temporal, and frontal lobes. They are, respectively, the **pulvinar**, the **lateral posterior nuclei**, and the **medial dorsal nuclei**. The detailed projections and anatomy of these nuclei are not particularly relevant to the present discussion. However, unlike the thalamic projections to primary sensory cortex, most of the inputs to these thalamic nuclei come from other regions of the cortex. Thus, the signals coming into the association cortices via the thalamus reflect highly processed sensory and motor information from multiple areas of the cerebral cortex that is then fed back to the association regions. The primary sensory cortices, in contrast, receive thalamic information that is directly related to peripheral sense organs (see Unit 2), and much of the thalamic input to motor cortex is derived from the thalamic nuclei related to the basal ganglia and cerebellum (see Unit 3, Chapters 17 and 18).

Box A
A MORE DETAILED LOOK AT CORTICAL LAMINATION

Much current knowledge about the cerebral cortex is based on descriptions of differences in cell number and density throughout the cortical mantle. Nerve cells bodies, because of their high metabolic rate, are rich in basophilic substances (DNA and RNA, for instance) and therefore tend to stain darkly with reagents such as cresyl violet acetate. These so-called Nissl stains (after F. Nissl, who first described this technique when he was a medical student in the nineteenth century) provide a dramatic picture of brain structure at the histological level. The most striking feature revealed in this way is the distinctive lamination of the cortex in humans and other mammals. In humans, there are three to six cortical layers, which are usually designated by numerals, with letters for laminar subdivisions (layers IVa, b,

and c in the visual cortex, for example). Each of the cortical laminae has characteristic functional and anatomical features (see also Table 24.1). Thus, cortical layer IV is typically rich in stellate neurons with locally ramifying axons; in the primary sensory cortices, these neurons receive input from the thalamus, the major sensory relay from the periphery. Layer V, and to a lesser degree layer VI, contains pyramidal neurons whose axons generally leave the cortex. The neurons in layers II and III have primarily corticocortical connections, and layer I contains mainly neuropil. Korbinian Brodmann, who early in this century devoted his career to an analysis of brain regions distinguished in this way, described 40 to 50 distinct cortical regions, or cytoarchitectonic fields. These structural features of the cerebral cortex con-

tinue to figure importantly in most discussions of brain function.

Another cortical structure of the cerebrum is the hippocampus. The hippocampus lies deep in the temporal lobe and has been implicated in the control of emotional behavior and in the acquisition of memories (see Chapters 27 and 29). The shape evidently reminded the early neuroanatomist who named it of a sea horse (*hippocampus* is Greek for "sea horse"). The cortex of the hippocampus differs from the rest of the six-layered cerebral cortex in having only three laminae. This type of cortex is regarded as more primitive evolutionarily and is sometimes called archicortex to distinguish it from the six-layered neocortex.

(A)

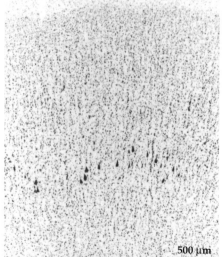

500 µm

(B)

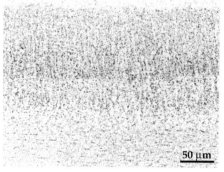

50 µm

(C)

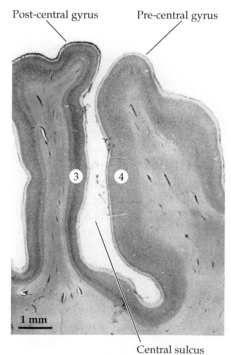

Post-central gyrus   Pre-central gyrus

3   4

1 mm

Central sulcus

Cytoarchitectonic differences in the structure of the human cerebral cortex in the region of the central sulcus. (A) Cytoarchitectonic distinctions based on nerve cell density in both the vertical dimension and the plane of the cortical surface allow different regions to be distinguished on a descriptive basis. This photomicrograph shows the primary motor cortex (Brodmann's area 4). (B) The somatic sensory cortex within the central sulcus (Brodmann's area 3) has an obviously different pattern of lamination. (C) A lower-power view of the same section, indicating the location of the primary motor cortex (area 4) and somatic sensory cortex (area 3). (Courtesy of Len White and Dale Purves.)

The other major source of innervation to the association cortices is direct projections from other cortical areas. Indeed, these connections are the largest source of input to the association cortices. Ipsilateral **corticocortical connections** arise from primary sensory, secondary sensory, and other association cortices within the same hemisphere. Such connections can also arise from the same (or different) cortical regions in the opposite hemisphere via the corpus callosum and anterior commissure. This special class of corticocortical connections is referred to as **interhemispheric connections**. Finally, the association areas receive diffuse inputs from the dopaminergic, noradrenergic, and serotonergic nuclei in the brainstem, as well as cholinergic nuclei in the brainstem and basal forebrain. These widespread inputs project to different cortical layers and, among other functions, establish mental status across a spectrum that ranges from deep sleep to high alert (see Chapter 26). A general wiring plan for the association cortices that summarizes the major pathways into and out of these regions is given in Figure 24.3 Although extensive, these connections certainly do not warrant the idea that everything is connected to everything else. On the contrary, each association

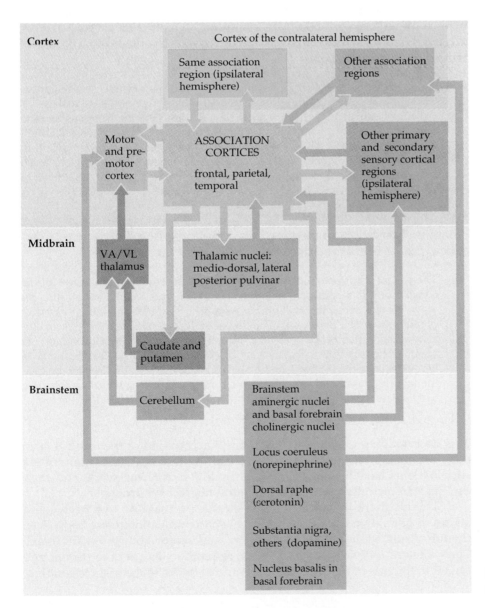

Figure 24.3 Summary of the inputs and outputs of the association cortices.

region is defined by a distinct, if overlapping, subset of thalamic, corticocortical and subcortical connections.

Whereas the connections of each association region suggest distinct circuitry subserving specific functions, it is impossible to conclude much about the role of different cortical areas based solely on connectivity (this information is, in any event, rather limited for the human association cortices; most of the evidence comes from comparisons with anatomical tracing studies in nonhuman primates and the limited pathway tracing that can be done in human brain tissue postmortem). As a result, inferences about the function of human association areas continue to depend quite heavily on observation of patients with cortical lesions. The three regions that have been studied most intensively are the association cortices of the parietal, temporal, and frontal lobes. Damage to each results in specific cognitive deficits that indicate their general functions.

■ LESIONS OF THE PARIETAL LOBE: ATTENTION DEFICITS

In 1941, the British neurologist W. R. Brain (really!) reported a series of patients with unilateral parietal lobe lesions accompanied by varying degrees of perceptual difficulty. Brain described the most profoundly affected of these individuals as follows:

> Three patients exhibited quite a different form of visual disorientation. Thought not suffering from a loss of topographical memory or an inability to describe familiar routes, they nevertheless got lost in going from one room to another in their own homes, always making the same error of choosing a right turning instead of a left, or a door on the right instead of one on the left. In each case there was a massive lesion in the right parieto-occipital region, and it is suggested that this ... resulted in an inattention to or neglect of the left half of external space.
>
> W. R. Brain, 1941 (*Brain* 64: p. 257)

From his observations of these patients and subsequent postmortem examination, Brain summarized the features of this syndrome:

> The patient who is thus cut off from the sensations which are necessary for the construction of a body scheme may react to the situation in several different ways. He may remember that the limbs on his left side are still there, or he may periodically forget them until reminded of their presence. He may have an illusion of their absence, i.e. they may 'feel absent' although he knows that they are there; he may believe that they are absent but allow himself to be convinced by evidence to the contrary; or, finally, his belief in their absence may be unamenable to reason and evidence to the contrary and so constitute a delusion. ... His state [is] thus comparable with an amnesia for the left half of the body.
>
> (Ibid, p. 264)

This description is considered the first full accounting of the link between parietal lobe lesions and deficits in perception and attention. Many more such patients have been studied since Brain's pioneering work, and their deficits are now referred to as **contralateral neglect syndrome**.

The hallmark of contralateral neglect is an inability to perceive and attend to one's own body or to objects in space in relation to the body, even though visual acuity, somatic sensation, and motor ability remain intact. Thus, affected individuals fail to report, respond to, or orient to stimuli presented to the side of the body (or visual space) opposite the brain lesion (Fig-

(A) "Draw a clock"

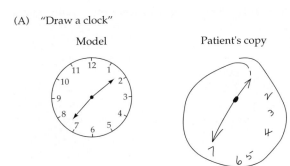

Model   Patient's copy

(B) "Draw a house"

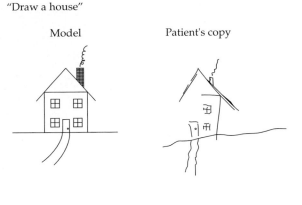

Model   Patient's copy

Figure 24.4 Characteristic performance on a visuospatial task by an individual suffering from contralateral neglect syndrome. The patient was asked to draw a clock (A) and a house (B) by copying the figures on the left; on the right is the subject's imitation. In (C), the patient was asked to bisect the line. (Adapted from Posner and Raichle, 1994.)

(C) "Bisect the line"

ure 24.4). They may also have difficulty performing complex motor tasks on the neglected side, including dressing themselves, reaching for objects, writing, drawing, and, to a lesser extent, orienting to sounds. The evidence of neglect can be as mild as a temporary lack of attention that resolves as the patient recovers, or it can be a complete and permanent denial of the side opposite the lesion. As Brain pointed out, such patients may even deny that the limbs on the neglected side are actually theirs.

Since the original description of contralateral neglect and its relationship to lesions of the parietal lobe, it has been generally accepted that the parietal cortex, particularly the inferior parietal lobe, is the primary cortical region responsible for attention (Figure 24.5).

Figure 24.5 The location of the underlying lesions in eight patients diagnosed with contralateral neglect syndrome. The site of damage was ascertained from CT scans (see Box C in Chapter 1). The lesions shown in 1–5 include parietal cortical areas. The lesions in 6–8, however, are restricted to the temporal lobe in the right hemisphere. These cases may reflect disruption of the pathways to and from the parietal cortex to other centers. (After Heilman and Valenstein, 1985.)

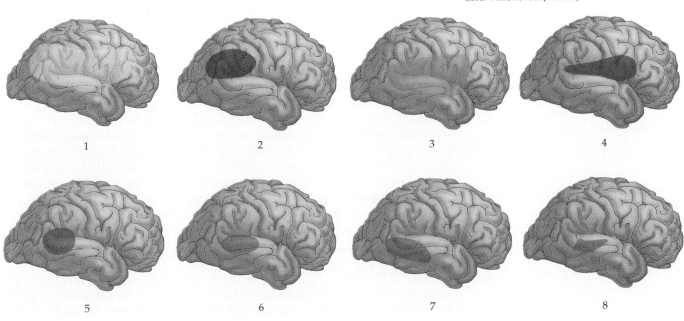

Figure 24.6 In confirmation of the impressions derived from neurological patients with parietal lobe damage, the right parietal cortex of normal subjects is highly active during tasks requiring attention. At the top, a subject has been asked to attend to objects in the left visual field, thus activating the right parietal cortex (arrow). Even when attention is shifted from the left visual field to the right, as shown at the bottom, the right parietal cortex remains active. (From Posner and Raichle, 1994.)

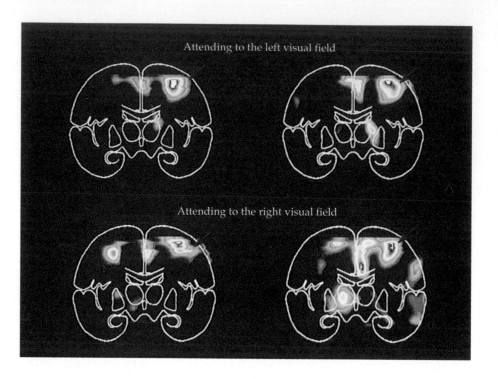

The contralateral neglect syndrome is specifically associated with damage to the parietal cortex of the right hemisphere. The unequal distribution of this particular cognitive function between the hemispheres is thought to arise because the right parietal cortex mediates attention to *both* left and right halves of the body and extrapersonal space, whereas the left hemisphere only attends to the right. Thus, left parietal lesions tend to be compensated by the intact right hemisphere. In contrast, when the right parietal cortex is damaged, there is no compensatory capacity in the left hemisphere for attention to the left side.

The use of imaging techniques such as PET scanning during specific attention tasks performed by normal subjects supports this view (Figure 24.6). Thus, blood flow is increased in the right parietal cortex when subjects are asked to perform tasks in the left visual field requiring selective attention to distinct aspects of a visual stimulus such as its shape, velocity, and color. When a similar challenge is presented to the right visual field, there is activation of both left and right parietal cortices. Moreover, when subjects are asked to maintain general alertness, the right parietal cortex is joined by increased activity in the right frontal cortex. These studies suggest that to some extent regions outside the parietal lobe also contribute to attentive behavior, and perhaps to the pathology of neglect syndromes (see Figure 24.5). Nevertheless, metabolic mapping confirms the fact that contralateral neglect typically arises from a right parietal lesion, and supports the notion that there is some degree of hemispheric specialization for attention, just as there is for a number of other cognitive functions (see Chapter 25).

■ LESIONS OF THE TEMPORAL LOBE: DEFICITS OF RECOGNITION

Damage to the association cortex of either temporal lobe can result in problems recognizing, identifying, and naming familiar objects. These disorders, collectively called **agnosias** (from the Greek for "not knowing"), are quite different from the neglect syndromes. In neglect, affected individuals deny awareness of sensory information in the affected field, even though sensation remains intact (an individual with contralateral neglect syndrome responds when his left arm is pinched, even though he may deny the arm's existence). Patients with agnosia acknowledge the presence of a stimulus, but are unable to report exactly what it is. These disorders can have both a **lexical** aspect—a mismatching of verbal or cognitive symbols with sensory stimuli (see Chapter 25)—and a **mnemonic** aspect—a failure to recall stimuli when confronted with them again (see Chapter 29).

One of the most intriguing agnosias following damage to the temporal association cortex in humans is the inability to recognize and identify faces. This disorder, called **prosopagnosia** (*prosopo*, from the Greek for "face" or "person") was recognized by neurologists as early as the late nineteenth century. After damage to the temporal lobe, which can be unilateral or bilateral, such patients are unable to identify familiar individuals by their facial characteristics, and in some cases cannot recognize a face at all. Nonetheless, these individuals are perfectly aware that some sort of visual stimulus is present and can describe particular aspects of it.

An example of the importance of the temporal cortex in facial recognition is the case of L.H., described by N. L. Etcoff and colleagues in 1991. (The use of initials to identify important neurological patients is standard practice, as will be evident in subsequent chapters.) This 40-year-old minister and social worker sustained a severe head injury as the result of an automobile accident when he was 18. After his initial recovery, L.H. could not recognize familiar faces, report that they were familiar, or answer questions about faces from memory. He was nonetheless able to conduct a fairly normal and productive life. He could identify other common objects, could discriminate subtle shape differences, and could recognize the sex, age, and "likability" of faces. Moreover, he could recognize particular people by a number of nonfacial cues such as voice, body shape, and gait. The only other category of visual stimuli he had trouble recognizing was animals and their expressions, though these impairments were not as severe as for human faces. L.H.'s prosopagnosia derived from a large lesion of the right temporal lobe (Figure 24.7).

Prosopagnosia is a specific instance of a broad range of functional deficits that have as their hallmark the inability to recognize a complex sensory stimulus as familiar, and to identify and name that stimulus as a meaningful object in the environment. Depending upon the location and size of

(A)

(B)

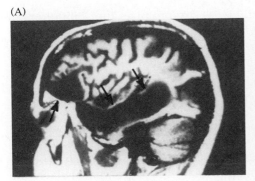
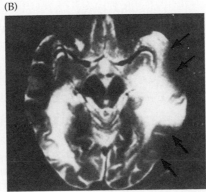

Figure 24.7 MRI images of the right hemisphere from patient L.H. in the sagittal (A) and horizontal (B) plane. The lesion (dark area of the scan, indicated by the arrows) includes most of the association cortex of the right temporal lobe. In addition, the right anterior frontal lobe has been damaged (arrow at left in A). (From Etcoff et al., 1991.)

the lesion, agnosias can be specific as is that for human faces, or as general as an inability to name most familiar objects. These deficits are also distinct from those that arise after damage to the specific regions of the left temporal lobe that are crucial for the lexical aspects of language (see Chapter 25).

■ LESIONS OF THE FRONTAL LOBE: PLANNING DEFICITS

The functional deficits that result from damage to the human frontal lobe are diverse and devastating, particularly if both hemispheres are involved. The diversity stems from the fact that the frontal cortex includes a wider repertoire of functional capacities than any other neocortical region (consistent with the fact that frontal cortex comprises a wide variety of cytoarchitectonic zones). The devastating nature of the behavioral deficits after frontal lobe damage reflects the role of this part of the brain in maintaining what we normally think of as an individual's "personality." The frontal cortex amalgamates complex perceptual information from sensory and motor cortices, as well as from the parietal and temporal association cortices. The result is an appreciation of the individual in relation to the world that allows behaviors to be planned and executed normally. When this cognitive capacity is compromised, the afflicted patient often has difficulty carrying out complex behaviors that are temporally, spatially, and even socially appropriate. These deficiencies are evident in the inability to match ongoing behavior to present or future demands and are often interpreted as a change in the patient's "character." Such deficits can be detected with specific neuropsychological tests (Box B), as well as through changes in personality and social behavior in affected individuals.

A well documented example of frontal lobe deficits is a patient studied by R. M. Brickner for almost 20 years. A, as Brickner referred to him, was a stockbroker who underwent bilateral frontal lobe resection because of a large tumor at age 39. After the operation, A had no obvious sensory or motor deficits; he could speak and understand verbal communication and was aware of people, objects, and temporal order in his environment. He acknowledged his illness and retained a high degree of intellectual power, as judged from an ongoing ability to play an expert game of checkers. Nonetheless, A's personality had undergone a dramatic change. Once a restrained, modest man, he became boastful of professional, physical, and sexual prowess. He showed little restraint in conversation and was unable to match the appropriateness of what he said to his audience. His ability to plan for the future was largely lost, as was much of his earlier initiative and creativity. Even though he retained the ability to learn complex procedures, he was unable to return to work, and relied on his family for support and care.

Thus, the damage to the frontal lobes compromises the ability to plan behavior in relationship to the environment and to use memories to guide the appropriateness of behavior in various situations. By inference, these cognitive abilities reside primarily in the frontal lobes.

■ "ATTENTION NEURONS" IN THE PARIETAL CORTEX

The complexity of cognitive functions, together with the inability to carry out cellular studies in humans, has made a more mechanistic understanding of the human association cortices difficult. A number of informative observations have been made in nonhuman primates, however, particularly the macaque (rhesus) monkey (Figure 24.8A). These animals have cognitive abilities mediated by the parietal, frontal, and temporal cortices that are in many ways similar to those in humans (Figure 24.8B).

Because electrophysiological recordings can be made from single neurons in the brains of awake, behaving monkeys, it is possible to assess the physiological responses of individual cells in the association cortices as cognitive tasks are performed (Figure 24.8). Such studies of neurons in the parietal cortex have often taken advantage of the fact that eye movements provide an excellent indicator of attentive behavior in primates. Thus, the fixation of the eyes on an interesting target can be used to identify attention-sensitive neurons in this part of the cortex (Figure 24.9). Neurons in the parietal areas thought to be responsible for spatial attention in the rhesus monkey (Areas 7A, B, and M) increase their rate of firing when the animal fixates on a target of interest, such as food or a novel object. The neurons maintain their activity for the duration of the fixation. Such responses are also observed when the monkey fixates on a test stimulus that has been associated with a food reward. When attention to the stimulus flags, eye move-

(A)

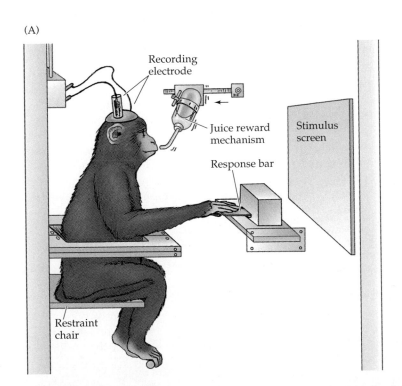

(B)

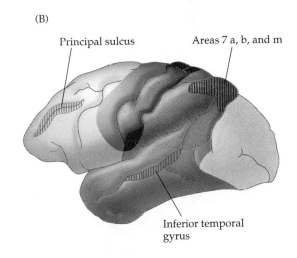

Figure 24.8 Recording from single neurons in the brain of an awake, behaving rhesus monkey. (A) The animal is seated in a chair and gently restrained. Several months before data collection begins, a recording well is placed through the skull using a sterile surgical technique. For electrophysiological recording experiments, a tungsten microelectrode is inserted through the dura and arachnoid, and into the cortex. The screen and the response bar in front of the monkey are for behavioral testing. In this way, individual neurons can be monitored while the monkey performs specific cognitive tasks. (B) Lateral view of the rhesus monkey brain showing the primary sensory and motor areas (shaded darker green and red) and the remaining parietal (pink), temporal (blue), and frontal (light green) association cortices. The occipital cortex is shaded yellow. The hatching in each region indicates the cortical areas where neurons with the specific response properties described here have been found.

Box B
NEUROPSYCHOLOGICAL TESTING

Long before PET scanning and functional MRI began to be used to evaluate normal and abnormal cognitive function, several "low-tech" methods proved to be reliable means of assessing these abilities in human subjects. From the late 1940s onward, psychologists and neurologists developed a battery of behavioral tests—generally called neuropsychological tests—to evaluate the integrity of cognitive function and to localize lesions. One of the most frequently used measures was the Wisconsin Card Sorting Task. In this test, the examiner places four cards with symbols that differ in number, shape, or color before the subject, who is given a set of response cards with similar symbols on them. The subject is then asked to place an appropriate response card in front of the stimulus card based upon a sorting rule established, but not stated, by the examiner (i.e., sort by color, number, or shape; see Figure). The examiner then indicates whether the response is "right" or "wrong," which is the only

Results of the Wisconsin Card Sorting Test				
Location of surgical excision	*Number of cases*	*Total errors*	*Repeated errors*	*Single errors*
Dorsolateral frontal	7	78.2	68.1	10.1
Unilateral temporal	12	43.6	31.7	11.9
Bilateral temporal	1	24	19	5
Parietotemporal	1	4	1	3
Inferior frontal	1	13	4	9

From Milner, 1963.

feedback provided. After 10 consecutive correct responses, the examiner changes the sorting rule simply by saying "wrong." The subject must then ascertain the new sorting rule and perform 10 correct trials. The sorting rule is then changed again, until six cycles have been completed.

In 1963, the neuropsychologist Brenda Milner at the Montreal Neurological Institute showed that patients with frontal lobe lesions have a remarkable deficit in performing the Wisconsin Card Sorting Task. By comparing patients with known brain lesions as a

result of surgery for epilepsy or tumor, Milner was able to demonstrate that this impairment is fairly specific for the frontal lobes (see table). A widely accepted explanation for the sensitivity of the Wisconsin Card Sorting Task for frontal lobe function is the "planning" aspect of this test. To respond correctly, the subject must retain information about the previous trial, which is then used to guide behavior on future trials. Particularly striking is the tendency for frontal lobe patients to make repeated errors. This perseveration is thought to reflect the inability to use previous infor-

Sort by color

Sort by shape

Sort by number

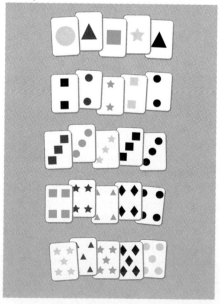

mation to guide subsequent behavior. Processing this sort of information is characteristic of frontal lobe function.

A variety of other neuropsychological tests have been devised to evaluate the functional integrity of other cortical regions and cognitive functions. These include tasks in which a patient is asked to identify familiar faces in a series of pictures, and others in which "distractors" interfere with the patient's ability to attend to salient stimulus features. An example of the latter is the Stroop Interference Test, in which patients are asked to read the names of colors presented in color-conflicting print (for example, the word *green* printed in red ink). This sort of challenge evaluates both attention and identification abilities.

The simplicity, economy, and accumulated experience with such tests continue to make them a valuable means of evaluating cognitive functions.

References

BERG, E. A. (1948) A simple objective technique for measuring flexibility in thinking. J. Gen. Psychol. 39: 15–22.

LEZAK, M. D. (1995) *Neuropsychological Assessment*, 3rd Ed. New York: Oxford University Press.

MILNER, B. (1963) Effects of different brain lesions on card sorting. Arch. Neurol. 9: 90–100.

MILNER, B. AND M. PETRIDES (1984) Behavioural effects of frontal-lobe lesions in man. Trends Neurosci. 4: 403–407.

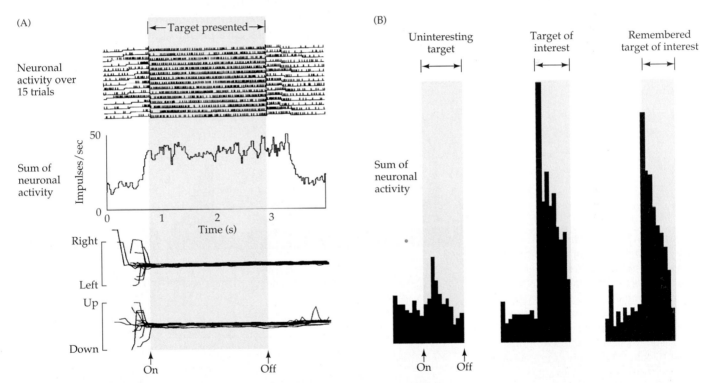

Figure 24.9 Selective activation of neurons in the parietal cortex of a rhesus monkey during the fixation of a significant visual target (in this case, a spot of light associated with a food reward). (A) When the target is presented, saccadic eye movements cease (indicated by the absence of deviations from 0 to the left and right or up and down in the lower traces). During the time the target is presented, the neuron increases its rate of firing almost tenfold. (B) Another parietal neuron fires only during fixation on a target of interest. As a control (uninteresting target), the monkey fixates randomly on objects that have no particular significance (that is, no reward has been paired with the stimulus). Fixation on such insignificant stimuli does not alter the firing of attention-selective neurons. In contrast, when a monkey is presented with a target of interest, or trained to fixate on a point where an interesting target had been presented (a remembered target of interest), another parietal neuron fires selectively. Thus, when a monkey attends to a relevant stimulus, whether present or remembered, neurons in the parietal cortex respond specifically during the attentive period. (After Lynch et al., 1977.)

ments resume and the firing of these neurons in the parietal cortex falls to baseline levels. Thus, the monkey parietal cortex contains neurons that respond specifically during attention to a behaviorally meaningful stimulus.

■ "RECOGNITION NEURONS" IN THE TEMPORAL CORTEX

In accord with human recognition deficits following temporal lobe lesions, neurons with responses that correlate with the recognition of specific stimuli are found in the temporal cortex of rhesus monkeys. The behavior of these neurons in the vicinity of the inferior temporal gyrus (see Figure 24.8B) is generally consistent with one of the major functions ascribed to the temporal cortex—namely, the recognition and identification of complex stimuli. For example, some neurons in the inferior temporal gyrus of the rhesus monkey respond specifically to the presentation of a monkey face. These cells are generally quite selective, some responding only to the frontal view of a face and others only to profiles (Figure 24.10). Furthermore, the cells are not easily tricked. When parts of faces or generally similar objects are presented, the cells fail to respond. Such neurons have therefore been called **face cells**. So far, no face cells have been found that are selective for a particular face; rather they respond to an entire class of faces. In principal, it is unlikely that there are cells that are tuned to specific faces or objects in an

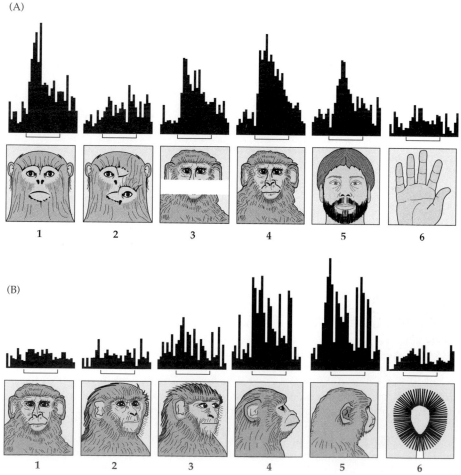

Figure 24.10 Selective activation of face cells in the inferior temporal cortex of a rhesus monkey. (A) This cell responds selectively to faces seen from the front. Scrambled parts of faces (stimulus 2) or faces with parts omitted (stimulus 3) do not elicit a maximal response. The cell responds best to different monkey faces, as long as they are complete and viewed from the front (stimulus 4); it also responds to a bearded human face (stimulus 5), although not quite as robustly. An irrelevant stimulus (in this case a hand; stimulus 6) does not elicit a response. (B) This cell responds to profiles of faces. A face viewed from the front (stimulus 1), 30° (stimulus 2), or 60° (stimulus 3) is not as effective as a true profile (stimulus 4). The cell will respond to profiles of different monkeys (stimulus 5), but is unresponsive to an irrelevant stimulus (in this case a brush; stimulus 6). (After Desimone et al., 1984.)

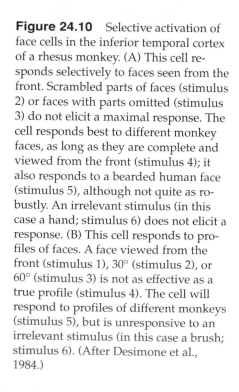

animal's environment (the infamous "grandmother cell" hypothesis). However, a population of cells, each differentially responsive to various categories of objects, could act in concert to enable the recognition of such complex sensory stimuli.

■ "PLANNING NEURONS" IN THE FRONTAL CORTEX

Neurons that appear to be specifically involved in planning have been identified in the frontal cortices of rhesus monkeys. Again, the response properties of these cells in the nonmotor areas of the frontal lobes accord with the function of the frontal association cortex deduced from studies of human patients. The behavioral challenge used to study the frontal cortex cells in monkeys is called the **delayed response task** (Figure 24.11). Variants of this task are used to assess frontal lobe function in a wide variety of situations, including the clinical evaluation of frontal lobe function in humans (see Box B). In the delayed response task, the monkey watches an experimenter place a food morsel in one of two wells; both wells are then covered by lids that can be easily removed. Subsequently, a screen is lowered for an interval of a few seconds to several minutes (the delay). When the screen is raised, the monkey gets only one chance to uncover the well containing food and receive the reward. Thus, the animal must decide that he wants the food, remember where it is placed, recall that the cover must be removed to obtain it, and keep all this information available during the delay so that it can be used to get the reward. The monkey's ability to carry out this task is abolished if the area anterior to the motor region of the frontal cortex—the **prefrontal cortex** (particularly that around the principal sulcus in monkeys)—is destroyed bilaterally (in accord with clinical findings in human patients).

Some neurons in the prefrontal cortex, particularly those around the principal sulcus (see Figure 24.8B), generate a response that is correlated with the delayed response task; that is, they are maximally active during the period of the delay, as if their firing represented the information maintained

Figure 24.11 Delayed response task. The experimenter randomly varies the well in which the food morsel is placed. The monkey watches the morsel being covered, and then the screen is lowered for a standard time. When the screen is raised, the monkey is allowed to uncover only one well to retrieve the food morsel. Normal monkeys learn this task quickly, usually performing at a level of 90% correct after less than 500 training trials.

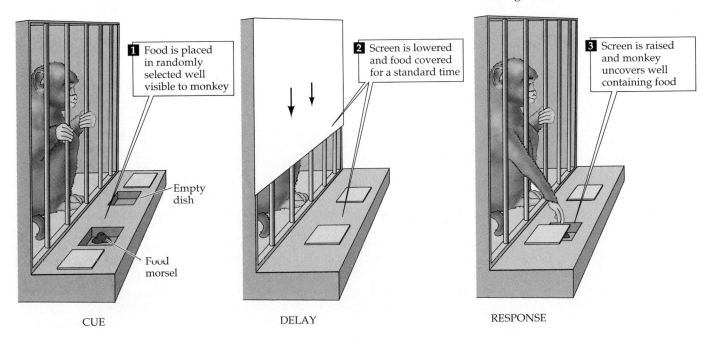

1 Food is placed in randomly selected well visible to monkey

Empty dish

Food morsel

2 Screen is lowered and food covered for a standard time

3 Screen is raised and monkey uncovers well containing food

CUE

DELAY

RESPONSE

(A) Stimulus (food morsel) presented

Cue   Delay   Response

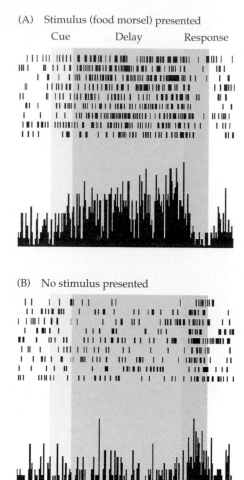

(B) No stimulus presented

Figure 24.12 A delay-specific neuron in the prefrontal cortex of a rhesus monkey recorded during the delayed response task shown in Figure 24.11. The rasters indicate the firing pattern of the cell over 9 consecutive trials. The histograms summarize the frequency of response during the cue, delay, and response periods. (A) The neuron begins firing when the screen is lowered and remains active throughout the delay period. (B) When no stimulus is presented, but the screen is still lowered and raised, the neuron is silent. (After Goldman-Rakic, 1987.)

from the presentation part of the trial (that is, the cognitive information needed to guide behavior when the screen is raised; Figure 24.12). Such neurons return to a low level of activity during the actual motor phase of the behavior, confirming that they represent short-term memory and planning rather that the actual movement itself. Delay-specific, or "planning" neurons in the prefrontal cortex are also active in monkeys that have been trained to perform a variant of the delayed response task in which the response is to internally generated memories. Evidently, these neurons are equally capable of using stored information to guide current behavior. Thus, if a monkey is trained to associate eye movements to a particular target with a delayed reward, the delay-associated neurons in the prefrontal cortex will fire during the delay, even if the monkey moves his eyes to the appropriate region of the visual field in the absence of the target.

The existence of delay-specific neurons in the frontal cortex of rhesus monkeys, as well as attention-specific cells in the parietal cortex and recognition-specific cells in the temporal cortex, supports the functional identity of these cortical areas inferred from clinical evidence. Thus, the concept of functional localization and cellular specificity that underlies distinct functions is as cogent for higher-order processing as for primary sensory and motor functions. Nonetheless, functional localization, whether inferred by examining human patients or by recording single neurons in monkeys, is an imprecise business. The observations summarized here serve only as a rudimentary guide to thinking about how complex cognitive information is represented and processed in the brain, and how the relevant brain areas and their constituent neurons contribute to such important but still ill-defined qualities as personality or intelligence (Box C).

■ SUMMARY

The majority of the human brain is devoted to tasks that transcend encoding primary sensations or commanding motor actions. Collectively, the association cortices mediate the cognitive functions of the brain—the ability to identify, order, and act meaningfully in response to external stimuli or internal motivation. The regions of the cerebral cortex devoted to cognition are found in the parietal, temporal, and frontal lobes. The association cortices in each of these regions have distinct input and output pathways that underlie their function. Descriptions of patients with cortical lesions, functional brain imaging of normal subjects, and behavioral and electrophysiological studies of nonhuman primates have established the general purpose of each of the major association cortices. Parietal cortical regions are involved in attention and awareness of the body and the stimuli that act on it; temporal cortical regions are involved in the recognition and identification of highly processed sensory information; and frontal cortical regions are involved in guiding complex behavior by planning responses to ongoing stimulation (or remembered information), thus matching such behaviors to the demands of a particular situation.

Box C
BRAIN SIZE AND INTELLIGENCE

The fact that so much of the brain is occupied by the association cortices raises a fundamental question: does more of it provide individuals with greater cognitive ability? Humans and other animals clearly vary in their talents and predispositions for a wide range of cognitive behaviors. Does a particular talent imply a greater amount of neural space in the service of that function?

Historically, the most popular approach to the issue of neural space and behavior in humans has been to relate the size of the brain to a broad index of performance, conventionally measured in humans by "intelligence" tests. This way of studying the relationship between brain and behavior has caused considerable trouble. In general terms, the idea that the size of brains from different species reflects intelligence represents a simple and apparently valid idea. The ratio of brain weight to body weight for fish is 1:5000; for reptiles it is about 1:1500; for birds, 1:220; for most mammals, 1:180, and for humans, 1:50. If intelligence is defined as the full scope of cognitive performance, surely no one would dispute that a human is more intelligent than a mouse, or that this difference is explained in part by the 3000-fold difference in the size of the brains of these species. Does it follow, however, that relatively small differences in the size of the brain among related species, strains, genders, or individuals—which often persist even after correcting for differences in body size—are also a valid measure of cognitive abilities? Certainly no issue in neuroscience has provoked a more heated debate than the notion that alleged differences in brain size among races—or the demonstrable differences in brain size between men and women—reflect differences in performance. The passion attending this controversy has

been generated not only by the scientific issues involved, but also by the spectre of racist or misogynist mischief.

Nineteenth-century enthusiasm for brain size as a simple measure of human performance was championed by some remarkably astute scientists (including Francis Galton and Paul Broca), as well as others whose motives and methods are now suspect (see Gould, 1978 for an interesting commentary). Broca, one of the great neurologists of his day and a gifted observer, not only thought that brain size reflected intelligence, but was of the opinion (as was just about every other nineteenth-century scientist) that white European males had larger and better-developed brains than anyone else. Based on what was known about the human brain in the late nineteenth century, it was reasonable for Broca to consider it an organ comparable to the liver or the lung; that is, a structure with a largely homogeneous function. Ironically, it was Broca himself who laid the groundwork for the modern view that the brain is a heterogeneous collection of highly interconnected but functionally discrete systems (see Chapter 25). Nonetheless, the nineteenth-century approach to brain size and intelligence has persisted in some quarters.

There are at least two reasons why measures such as brain weight or cranial capacity are not easily interpretable indices of intelligence, even though small observed differences may be statistically valid. First is the obvious difficulty of defining and accurately measuring intelligence among animals, particularly among humans with different educational and cultural backgrounds. Second is the functional diversity and connectional complexity of the brain. Imagine assessing the relationship between body size and athletic ability,

which might be considered the somatic analogue of intelligence. Body weight, or any other global measure of somatic phenotype, would be a woefully inadequate index of athletic ability. Although the evidence would presumably indicate that bigger is better in the context of sumo wrestling, more subtle somatic features would be correlated with extraordinary ability in ping pong, gymnastics, or figure skating. The diversity of somatic function vis-à-vis athletic ability confounds the interpretation of any simple measure such as body size.

The implications of this analogy for the brain are straightforward. Any program that seeks to relate brain weight, cranial capacity, or some other measure of overall brain size to individual performance ignores the reality of the brain's functional diversity. Thus, quite apart from the political or ethical probity of attempts to measure "intelligence" by brain size, by the yardstick of modern neuroscience (or simple common sense) this approach will inevitably generate more heat than light.

References

BROCA, P. (1861) Sur le volume et al forme du cerveau suivant les individus et suivant les races. Bull. Soc. Anthrop. 2: 139–207, 301–321.

GALTON, F. (1883) *Inquiries into Human Faculty and Its Development.* London: Macmillan.

GOULD, S. J. (1978) Morton's ranking of races by cranial capacity. Science 200: 503–509.

GOULD, S. J. (1981) *The Mismeasure of Man.* New York: W.W. Norton and Company.

GROSS, B. R. (1990) The case of Phillipe Rushton. Acad. Quest. 3: 35–46.

SPITZKA, E. A. (1907) A study of the brains of six eminent scientists and scholars belonging to the American Anthropometric Society, together with a description of the skull of Professor E. D. Cope. Trans. Amer. Phil. Soc. 21: 175–308.

WALLER, A. D. (1891) *Human Physiology.* London: Longmans, Green.

Additional Reading

Reviews

ANDERSEN, A. R. (1989) Visual and eye movement functions of the posterior parietal cortex. Annu. Rev. Neurosci. 12: 377–404.

DAMASIO, A. R. (1985) The frontal lobes. In *Clinical Neuropsychology*, 2nd Ed. K. H. Heilman and E. Valenstein (eds.). New York: Oxford University Press.

DAMASIO, A. R., H. DAMASIO AND G. W. VAN HOESEN (1982) Prosopagnosia: anatomic basis and behavioral mechanisms. Neurology 32: 331–341.

DESIMONE, R. (1991) Face-selective cells in the temporal cortex of monkeys. J. Cog. Neurosci. 3: 1–8.

FILLEY, C. M. (1995) *Neurobehavioral Anatomy*. Chapter 8: Right hemisphere syndromes. Boulder: University of Colorado Press.

GOLDMAN-RAKIC, P. S. (1987) Circuitry of the prefrontal cortex and the regulation of behavior by representational memory. *Handbook of Physiology 5*, Part 1, Chapter 9, pp. 373–417.

POSNER, M. I. AND S. E. PETERSEN (1990) The attention system of the human brain. Annu. Rev. Neurosci. 13: 25–42.

Important Original Papers

BRICKNER, R. M. (1952) Brain of patient A after bilateral frontal lobectomy: status of frontal lobe problem. Arch. Neurol. Psychiatr. 68: 293–313.

BRAIN, W. R. (1941) Visual disorientation with special reference to lesions of the right cerebral hemisphere. Brain 64: 224–272.

DESIMONE, R., T. D. ALBRIGHT, C. G. GROSS AND C. BRUCE (1984) Stimulus-selective properties of inferior temporal neurons in the macaque. J. Neurosci. 4: 2051–2062.

ETCOFF, N. L., R. FREEMAN AND K. R. CAVE (1991) Can we lose memories of faces? Content specificity and awareness in a prosopagnosic. J. Cog. Neurosci. 3: 25–41.

FUNAHASHI, S., C. J. BRUCE AND P. S. GOLDMAN-RAKIC (1993) Dorsolateral prefrontal lesions and oculomotor delayed response performance: Evidence for mnemonic "scotomas." J. Neurosci. 13: 1479–1497.

FUNAHASHI, S., M. V. CHAFEE AND P. S. GOLDMAN-RAKIC (1993) Prefrontal neuronal activity in rhesus monkeys performing a delayed antisaccade task. Nature 365: 753–756.

FUSTER, J. M. (1973) Unit activity in prefrontal cortex during delayed-response performance: neuronal correlates of transient memory. J. Neurophysiol. 36: 61–78.

GESCHWIND, N. (1965) Disconnexion syndromes in animals and man. Parts I and II. Brain 88: 237–294.

MISHKIN M. AND K. PRIBRAM (1956) Analysis of the effects of frontal lesions in monkeys II: Variations of delayed response. J. Comp. Physiol. Psychol. 49: 36–40.

MOUNTCASTLE, V. B., J. C. LYNCH, A. GEORGOPOULOUS, H. SAKATA AND C. ACUNA (1975) Posterior parietal association cortex of the monkey: Command function from operations within extrapersonal space. J. Neurophys. 38: 871–908.

Books

DAMASIO, A. R. (1994) *Descartes' Error: Emotion, Reason and the Human Brain*. New York: Grosset/Putnam.

DEFELIPE, J. AND E. G. JONES (1988) *Cajal on the Cerebral Cortex: An Annotated Translation of the Complete Writings*. New York: Oxford University Press.

GAREY, L. J. (1994) *Brodmann's 'Localisation in the Cerebral Cortex'*. London: Smith-Gordon. (Translation of K. Brodmann's 1909 book, Leipzig: Verlag von Johann Ambrosius Barth.)

HEILMAN, H. AND E. VALENSTEIN (1985) *Clinical Neuropsychology*, 2nd Ed. New York: Oxford University Press, Chapters 8, 10, 12.

POSNER, M. I. AND M. E. RAICHLE (1994) *Images of Mind*. New York: Scientific American Library.

CHAPTER 25

LANGUAGE AND LATERALIZATION

■ OVERVIEW

One of the most remarkable features of human cognition is the ability to associate arbitrary symbols with specific meanings to report thoughts and emotions—in a word, language. Studies of patients with damage to specific cortical regions have shown that the linguistic abilities of the human brain reside in several specialized areas of the association cortices in the temporal and frontal lobes. In most people, these major language functions are located in the *left* hemisphere; thus, the sensory representation of words and symbols is found primarily in the left temporoparietal cortex, and the representation of the motor commands that organize the production of meaningful speech is primarily in the left frontal cortex. Despite this left-sided predominance, the emotional (affective) content of language is governed largely by the right hemisphere. Nor are the cortical areas specialized for language concerned solely with words. Studies of congenitally deaf individuals have shown that the brain areas devoted to sign language are the same as those that organize spoken and heard communication. Such regions are therefore specialized for symbolic representation and communication rather than for spoken language as such. Understanding functional localization and hemispheric lateralization of language is especially important to the practice of neurology and neurosurgery. The loss of language is such a devastating blow to humans that every effort is made to identify and spare those cortical areas that are involved in its comprehension and production.

■ LANGUAGE IS BOTH LOCALIZED AND LATERALIZED

It has been known for more than a century that two distinct regions in the temporoparietal and frontal association cortices of the left cerebral hemisphere are especially important for normal human language. That language abilities are localized is expected; ample evidence of the localization of other cognitive functions was reviewed in Chapter 24. A novel aspect, however, is the unequal representation of language functions in the two cerebral hemispheres (although a hemispheric inequality was implied by the apparent lateralization of attention in hemineglect syndrome). This functional asymmetry is referred to as **hemispheric lateralization**, a phenomenon that has given rise to the misleading idea that one hemisphere in humans is actually "dominant" over the other—namely, the hemisphere in which the major capacity for language resides. The true significance of lateralization, however, lies in the efficient subdivision of complex functions between the hemispheres, rather than in any superiority of one hemisphere over the other.

The representation of language in the brain is distinct from the circuitry concerned with the motor control of the mouth, tongue, larynx, and pharynx, the structures that produce speech sounds; it is also distinct from the circuits underlying the auditory perception of spoken words and the visual perception of written words (Figure 25.1). The neural substrate for language transcends these essential motor and sensory functions because it is concerned with a system of symbols—spoken and heard, written and read (or, in the case of sign language, gestured and seen). The essence of language,

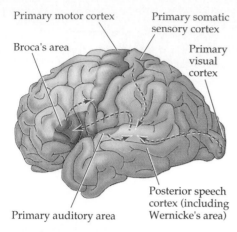

Primary motor cortex

Primary somatic sensory cortex

Broca's area

Primary visual cortex

Posterior speech cortex (including Wernicke's area)

Primary auditory area

Figure 25.1 Diagram of the major brain areas involved in the comprehension and production of language. The primary sensory, auditory, visual, and motor cortices are indicated to show the relation of Broca's and Wernicke's language areas to the less specialized areas that are nonetheless involved in the comprehension and production of speech.

then, is symbolic representation. Syntax, grammar, and intonation, are all recognizable regardless of the particular mode of representation, comprehension, and expression.

■ APHASIAS

The distinction between language and the related sensory and articulatory capacities on which it depends was first apparent in patients with damage to particularly relevant brain regions. The ability to move the muscles of the mouth, tongue, larynx, and pharynx can be compromised without abolishing the ability to use spoken language to communicate (even though the motor deficit may make communication difficult). Conversely, damage to other regions of the brain can compromise essential language functions while leaving the sensory and motor components of verbal communication intact. These latter syndromes, collectively referred to as the **aphasias**, diminish or abolish the ability to comprehend or to produce language, while sparing the ability to perceive verbal stimuli and to produce words. Missing in these patients is the capacity to recognize or employ the symbolic value of words, thus depriving them of the syntax, grammar, and intonation that distinguishes language from nonsense.

The localization of language function to a specific region (and hemisphere) of the cerebrum is usually attributed to the French neurologist Paul Broca and the German neurologist Carl Wernicke, both of whom made their seminal observations in the late 1800s. Broca and Wernicke examined the brains of individuals who had become aphasic and later died. Based on these postmortem correlations, Broca suggested that language abilities were localized in the ventroposterior region of the frontal lobe. More importantly, he observed that the loss of the ability to produce meaningful language—as opposed to the ability to move the mouth and produce words—was usually associated with damage to the left hemisphere. "On parle avec l'hemisphere gauche," Broca concluded. The preponderance of aphasic syndromes associated with damage to the left hemisphere amply supports his claim that one speaks with the left hemisphere.

Although Broca was basically correct, he failed to grasp the limitations of thinking about language as a unitary function localized in a single cortical region. This issue was better appreciated by Wernicke, who distinguished between patients who had lost the ability to comprehend language and those who could no longer produce language. Wernicke recognized that some aphasic patients do not understand language but retain the ability to produce utterances with reasonable syntactic, grammatical, and symbolic content. He concluded that lesions of the left temporal lobe tend to result in this sort of deficit. In contrast, other aphasic patients continue to comprehend language but lack the ability to control the linguistic content of their response. Thus, they produce nonsense syllables, transposed words, and generally utter syntactically or grammatically incomprehensible phrases. These deficits are associated with damage to the left frontal lobe.

As a consequence of these early observations, two rules about the localization of language have been taught ever since. The first is that lesions of the left frontal lobe in a region referred to as **Broca's area** (see Figure 25.1) affect the ability to produce language. This deficiency is called **motor** or **production aphasia**, and is also known as **Broca's aphasia**. Such aphasias must be distinguished from **dysarthria**, which is the inability to move the muscles of the face and tongue that mediate speaking. Note that the deficient motor-planning aspects of production aphasia accord with the complex motor functions of the posterior frontal lobe (see Chapters 15 and 24). The second rule is that damage to the left temporal lobe causes difficulty understanding spoken language, a deficiency referred to as **sensory** or **receptive aphasia**. (Deficits of reading and writing—alexias and agraphias—are separate disorders that arise from damage to other brain areas.) Sensory aphasia, also known as **Wernicke's aphasia**, generally reflects damage to the auditory and visual association cortices in the temporal and adjacent parietal lobe, a region often referred to as **Wernicke's area**. A final broad category of language syndromes is **conduction aphasia**, which arises from lesions to the pathways connecting the relevant temporal and frontal regions, such as the arcuate fasiculus that links Broca's and Wernicke's areas (see Figure 25.1). Interruption of this pathway leads to an inability to produce appropriate responses to heard communication, even though the communication is understood.

Despite the validity of Broca's and Wernicke's original observations, the classification of language disorders is more complex. An effort to refine the nineteenth-century categorization of aphasias was undertaken by the American neurologist Norman Geschwind during the 1950s and early 1960s. The revision was especially useful because an ongoing debate about the evidence for functional localization had created an atmosphere of uncertainty. Based on data from a large number of patients and a thorough understanding of cortical connectivity gleaned from studies in animals, Geschwind argued that several regions of the parietal, temporal, and frontal cortices were critically involved in human linguistic capacities. His clarification of the definitions of language disorders (Table 25.1) remains the basis for much contemporary clinical work on the aphasias.

As discussed in Chapter 24, a multitude of histologically distinct regions, referred to as cytoarchitectonic areas or fields, has been described in the human cerebral cortex. Primary sensory and motor functions are localized to some of these areas, whereas general cognitive functions like attention, identification, and planning appear to encompass several regions in a particular cortical lobe. The language functions described here are primarily associated with three of the cytoarchitectonic areas defined by Brodmann: area 22, at the junction of the parietal and temporal lobes (Wernicke's area); and areas 44 and 45, in the ventral and posterior region of the frontal lobe (Broca's area; Figure 25.2).

■ A DRAMATIC CONFIRMATION OF LANGUAGE LATERALIZATION

Until the 1960s, observations about language localization and lateralization were based primarily on patients with brain lesions of variable severity, location and etiology. This fact allowed skeptics to suggest that language function (or other complex cognitive functions) might not be localized or lateralized at all. Definitive evidence for lateralization came from studies of patients whose corpus callosum and anterior commissure had been severed as a treatment for medically intractable epileptic seizures. In these patients, the investigators could assess the function of the two cerebral hemispheres independently, since the major axon tracts that connect them had been interrupted. The pioneering studies of these so-called **split-brain patients** were carried out by Roger Sperry and his colleagues in the 1960s and 1970s and established the hemispheric lateralization of language beyond any doubt; these studies also demonstrated many other functional differences between the left and right hemispheres and continue to stand as an extraordinary contribution to understanding the organization of the brain.

To evaluate the functional capacity of each hemisphere in split-brain patients, it is essential to provide information to one side of the brain only.

Figure 25.2 The relationship of the major language areas to the classical cytoarchitectonic map of the cerebral cortex. Shown here is Brodmann's map (see Chapter 24); the cytologically distinct areas associated with language (areas 22, 44, and 45) are highlighted.

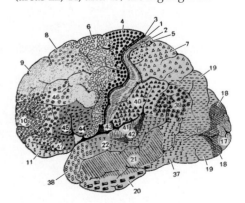

TABLE 25.1
Categorization of Aphasias

Type	Verbal output	Ability to repeat sentences	Comprehension	Naming of objects	Associated clinical signs	Lesion includes
Broca's	Deficient	Impaired	Normal	Marginally impaired	Right hemiparesis apraxia of the left limbs and face	
Wernicke's	Fluent	Impaired	Impaired	Impaired	± Right homonymous hemianopsia	
Conduction	Fluent	Impaired	Normal	Impaired	± Right hemisensory defect, apraxia of all limbs and face	
Global	Deficient	Impaired	Impaired	Impaired	Right hemiparesis, Right hemisensory defect, Right homonymous hemianopsia	
Anomic	Fluent	Normal (but difficulty naming objects)	Normal	Impaired	None	
Transcortical motor	Deficient	Normal	Normal	Impaired	Right hemiparesis	
Sensory	Fluent	Normal	Impaired	Impaired	± Right homonymous hemianopsia	

Sperry and others devised several simple ways to do this, the most straight-forward of which was to ask the subject to use each hand independently to identify objects without any visual assistance (Figure 25.3A). Recall from Chapter 8 that the somatic sensory information from the right hand is processed by the left hemisphere and vice versa. By asking the subject to

Figure 25.3 Confirmation of the linguistic specialization of the left hemisphere in the vast majority of humans by studying individuals in whom the connections between the right and left hemispheres have been surgically divided. (A) Single-handed, vision-independent stereognosis can be used to evaluate the language capabilities of each hemisphere in "split-brain" patients. Objects held in the right hand, which provides somatic sensory information to the left hemisphere, are easily named; remarkably, objects held in the left hand can not be identified by name in these patients. (B) Schematic representation of some of the different functional abilities that reside in the left and right hemispheres, deduced from a variety of behavioral tests in split-brain patients. Visual instructions can also be given independently to the right or left hemisphere in these individuals. Since the left visual field is perceived by the right hemisphere (and vice versa; see Chapter 11), a briefly presented (tachistoscopic) instruction in the left visual field is appreciated only by the right brain (assuming that the individual maintains his or her gaze straight ahead). A wide range of functions can be evaluated using this tachistoscopic method, even in normal subjects.

(A)

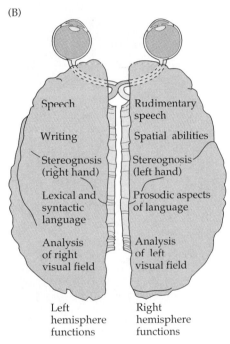

(B)

Left hemisphere functions	Right hemisphere functions
Speech	Rudimentary speech
Writing	Spatial abilities
Stereognosis (right hand)	Stereognosis (left hand)
Lexical and syntactic language	Prosodic aspects of language
Analysis of right visual field	Analysis of left visual field

describe an item being manipulated by one hand or the other, the language capacity of the relevant hemisphere could be examined. Such testing showed clearly that the two hemispheres differ in their language ability. Using the left hemisphere, the split-brain patients were able to name objects held in the right hand without difficulty. In contrast, and quite amazingly, an object held in the left hand could not be named! Using the right hemisphere, the subjects could only produce an indirect description of the object that relied on rudimentary words and phrases rather than the precise lexical symbol for the object (for instance, "a round thing" instead of "a ball"). Observations using special techniques to present visual information to the hemispheres independently (a technique called tachistoscopic presentation) showed further that the left hemisphere can respond to written commands linguistically, whereas the right hemisphere can respond to such stimulation only by nonverbal means (or by rudimentary linguistic expression). These linguistic distinctions reflect broader hemispheric differences that can be summarized by the statement that, among other things, the left hemisphere in most humans is specialized for processing verbal and symbolic material, whereas the right hemisphere has a greater capacity for visuospatial and emotional functions (Figure 25.3B).

Sperry's ingenious work on split-brain patients put an end to the century-long controversy about language lateralization; in most individuals, the left hemisphere is unequivocally the seat of the major language functions. It would be wrong to imagine, however, that the right hemisphere has no language capacity. As mentioned, it can produce rudimentary words and phrases, and contributes emotional context to language (see Chapter 27 and below). Moreover, the right hemisphere obviously understands language, since it can respond to written (or spoken) commands. Consequently, Broca's conclusion that we speak with our left brain is not strictly correct; it would be more accurate to say that the left hemisphere speaks very much better than the right.

■ ANATOMICAL ASYMMETRIES BETWEEN THE RIGHT AND LEFT HEMISPHERES

The clear differences in language function between the left and right hemispheres have inspired neurologists and neuropsychologists to find a structural correlate of this behavioral asymmetry. One apparent hemispheric difference was identified in the late 1960s when Norman Geschwind and his colleagues found an asymmetry in the superior aspect of the temporal lobe known as the **planum temporale** (Figure 25.4). This area was significantly larger on the left side in about two-thirds of the human subjects studied postmortem. A similar difference has been described in higher apes (Box A), but not in other primates. A smaller asymmetry has also been noted in the operculum of the frontal lobe, the region that includes Broca's area. Because the planum is near (although not congruent with) the regions of the temporal lobe that contain cortical areas essential to language, it has been widely

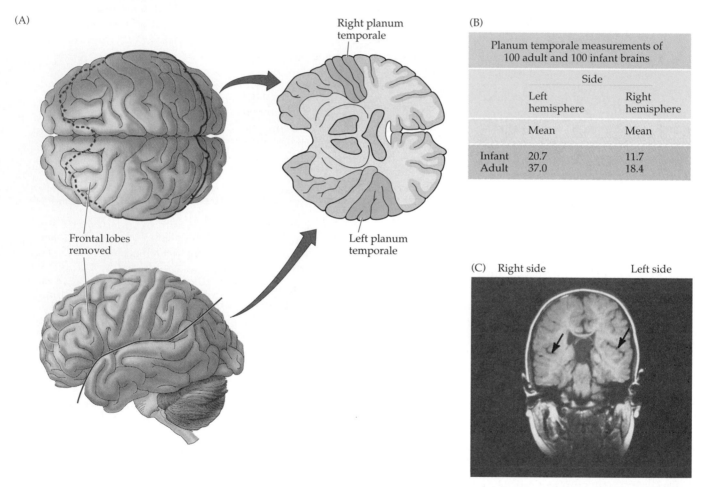

(B)

Planum temporale measurements of 100 adult and 100 infant brains	
Side	
Left hemisphere	Right hemisphere
Mean	Mean
Infant 20.7	11.7
Adult 37.0	18.4

Figure 25.4 Asymmetry of the right and left human temporal lobes. (A) The superior portion of the brain has been removed as indicated to reveal the dorsal surface of the temporal lobes in the righthand diagram (which presents a dorsal view of the horizontal plane). A region of the surface of the temporal lobe called the planum temporale is significantly larger in the left hemisphere of most (but far from all) individuals. (B) Measurements of the planum temporale in 100 adult and infant brains. The size of the planum temporale is expressed in arbitrary planimetric units to get around the difficulty of measuring the curvature of the gyri within the planum. The asymmetry is evident at birth and persists in adults at roughly the same magnitude (on average, the left planum is 57% larger in infants and 50% larger in adults). (C) An MRI image in the frontal plane showing this asymmetry in a normal adult subject.

Box A
DO APES HAVE LANGUAGE?

Over the centuries, theologians, natural philosophers, and even some modern neuroscientists have argued that language is uniquely human, this extraordinary behavior setting us qualitatively apart from our fellow animals. The gradual accumulation of evidence demonstrating highly sophisticated systems of communication in species as diverse as whales, birds, and bees has made this point of view untenable, at least in a broad sense. Despite these examples in "lower animals," human language *does* appear unique in its semantic aspect (that is, the ability to associate specific meanings with arbitrary symbols). This latter distinction raises fascinating questions in the context of both evolution and comparative neurology. How did language arise in hominids, and, perhaps more pointedly, what are the functions of those parts of the brains of higher apes that in humans instantiate language? Several approaches have been taken to these issues.

Perhaps the most controversial strategy has been to probe the ability of higher apes to learn human language by teaching aspects of language to developing chimpanzees (our closest extant relative), much as human infants are taught. Although some of these efforts were patently misguided (older attempts to teach chimpanzees to speak were without merit simply because the animal

lacks the necessary vocal apparatus), modern attempts to explore this issue have been quite revealing. If chimpanzees are given the means to communicate symbolically, they demonstrate some surprising talents. Although techniques have varied, most psychologists who study primates have used some form of manipulable symbols that can be arranged to express ideas in an interpretable manner. For example, chimpanzees can be trained to manipulate tiles or other symbols such as the gestures of sign language that represent words and syntactical constructs, allowing the animals to communicate simple demands, questions, and even spontaneous expressions. As a result of these studies, a consensus has gradually emerged that chimpanzees *are* capable of considerable symbolic thinking and that this ability is arguably a forerunner of human language. (The sign language studies mentioned in the text are obviously pertinent here.) It remains to be demonstrated that the regions of the temporal, parietal, and frontal cortices in the human brain that support language also serve these symbolic functions in chimpanzees, but that goal is certainly feasible. At the same time, ethologists studying chimpanzees in the wild have described extensive social communication based on gestures, the manipulation of objects, and facial expressions. The

intricate social interactions among chimpanzees imply the expression and interpretation of information that can be considered an antecedent of human language; one need only think of the importance of gestures and facial expressions as ancillary aspects of our own speech to appreciate this point. Although much uncertainty remains, only someone given to extraordinary anthropocentrism could now argue that symbolic communication is uniquely human.

References

GOODALL, J. (1990) *Through a Window: My Thirty Years with the Chimpanzees of Gombe.* Boston: Houghton Mifflin Company.

GRIFFIN, D. R. (1992) *Animal Minds.* Chicago: The University of Chicago Press.

HELTNE, P. G. AND L. A. MARQUARDT (EDS.) (1989) *Understanding Chimpanzees.* Cambridge, MA: Harvard University Press.

MILES, H. L. W. AND S. E. HARPER (1994) "Ape language" studies and the study of human language origins. In *Hominid Culture in Primate Perspective*, D. Quiatt and J. Itani (eds.). Niwot, CO: University Press of Colorado, pp. 253–278.

TERRACE, H. S. (1983) Apes who "talk": Language or projection of language by their teachers? In *Language in Primates: Perspectives and Implications*, J. de Luce and H. T. Wilder (eds.). New York: Springer-Verlag, pp. 19–42.

WALLMAN, J. (1992) *Aping Language.* New York: Cambridge University Press.

assumed that this leftward asymmetry reflects the greater involvement of the left hemisphere in language. Nonetheless, these anatomical differences in the two hemispheres of the brain, which are recognizable before birth, have by no means been proven to be an anatomical reflection of the lateralization of language function. The fact that a detectable planum asymmetry is present in only 67% of human brains, whereas the preeminence of language in the left hemisphere is evident in 97% of the population (see below), argues for regarding this association with some caution.

■ MAPPING LANGUAGE FUNCTION

The pioneering work of Broca and Wernicke, and later Geschwind and Sperry, clearly established differences in hemispheric function. Several techniques have since been developed that allow hemispheric function to be assessed in neurological patients with an intact corpus callosum, and even in normal subjects. One such method for the clinical assessment of language lateralization was devised in the 1960s by Juhn Wada at the Montreal Neurological Institute. Wada injected a short-acting anesthetic, sodium amytal, into one carotid artery; this procedure transiently anesthetizes one hemisphere and thus tests the functional capabilities of the affected half of the brain. For example, if the left hemisphere is anesthetized by left carotid artery injection, the patient becomes transiently aphasic. Less invasive ways to test the cognitive abilities of the two hemispheres in intact subjects include positron emission tomography (PET), functional magnetic resonance imaging (fMRI) (see Box E in Chapter 1), and the sort of tachistoscopic presentation used so effectively by Sperry. (Even when the hemispheres are normally connected, subjects show delayed verbal responses and other differences when the right hemisphere receives the instruction.) Inferences about lateralization can also be made by testing hand preferences (Box B). These techniques have all confirmed earlier evidence of hemispheric lateralization. More importantly, they have provided valuable diagnostic tools to determine which hemisphere is "eloquent" in preparation for neurosurgery: although most individuals have the major language functions in the left hemisphere, a few—about 3% of the population—do not.

By the 1930s, the efforts of the neurosurgeon Wilder Penfield had already made possible a more detailed localization of cortical capacities (see Chapter 8). Electrical stimulation and mapping techniques adapted from neurophysiological work in animals provided a way to delineate the language areas of the cortex prior to the neurosurgical removal of brain tissue to treat tumors or epilepsy. Such intraoperative mapping guaranteed that the cure would not be worse than the disease, and has been widely used ever since. As a result, considerable new information about language localization emerged. Penfield's observations, together with recent studies performed by George Ojemann and his colleagues at the University of Washington, confirm the conclusions inferred from postmortem correlations and other approaches: a large region of the perisylvian cortex of the left hemisphere is clearly involved in language production and comprehension (Figure 25.5). A surprise, however, was the variability in language localization from patient to patient. Thus, Ojemann found that the brain regions involved in language are only approximately those indicated by most textbook treatments and that their exact location differed unpredictably among individuals. Indeed, bilingual patients do not necessarily use the same cortical area for storing the names of the same objects in two different languages!

Despite these advances, neurosurgical stimulation studies are complicated by their intrinsic difficulty and the fact that the patients in whom they are carried out are far from normal. The advent of positron emission tomography in the 1980s allowed studies of the language regions in normal subjects by noninvasive measurement of regional cerebral blood flow. Recall that PET reveals the regions of the brain that are active during a particular task because the related electrical activity increases local blood flow (see Box E in Chapter 1). The results of this approach, particularly in the hands of Marcus Raichle, Steve Peterson, and their colleagues at Washington University, have challenged excessively rigid views of localization and lateralization of linguistic function. Although high levels of activity occur in the expected re-

(A)

(B)

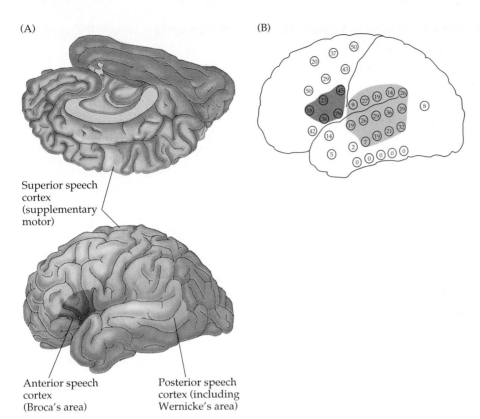

Superior speech cortex (supplementary motor)

Anterior speech cortex (Broca's area)

Posterior speech cortex (including Wernicke's area)

Figure 25.5 Cortical mapping of the language areas in the left cerebral cortex during neurosurgery. (A) Some of the key areas in the left hemisphere that caused interference with language production or comprehension when focally stimulated by small electrical currents. (B) Evidence for the variability of language representation among individuals. This diagram summarizes data from 117 patients whose language areas were electrically mapped at surgery. The number in each circle indicates the percentage of the patients who showed interference with language in response to stimulation at that site. Note also that many of the sites that elicited interference fall outside the classic language areas. (A after Roberts and Penfield, 1959; B after Ojemann et al., 1989.)

gions, large areas of both hemispheres are activated in word recognition or production tasks (Figure 25.6).

Since the same cortical areas exist in both hemispheres, a puzzling issue remains. What do the regions in the right hemisphere that are homologous to the left hemisphere language areas actually do? Perhaps the answer lies in the fact that subtle language deficits *do* occur following damage to the right hemisphere. Most obvious is an absence of the normal emotional and tonal components—called **prosodic** elements—that impart additional meaning to

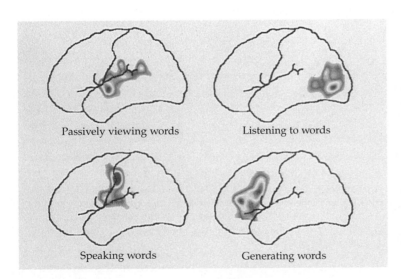

Passively viewing words

Listening to words

Speaking words

Generating words

Figure 25.6 Language-related regions of the left hemisphere mapped by positron emission tomography (PET) in a normal human subject. Language tasks such as listening to words and generating words elicit activity in Broca's and Wernicke's areas, as expected. However, there is also activity in primary and association sensory and motor areas for both active and passive language tasks. These observations indicate that language processing involves cortical regions other than the classic language areas. (After Posner and Raichle, 1994.)

Box B
HANDEDNESS

Approximately 9 out of 10 people are right-handed, a proportion that appears to have been stable over thousands of years and across all cultures in which handedness has been examined. Handedness is usually assessed by having individuals answer a series of questions about preferred manual behaviors, such as "Which hand do you use to write?," "Which hand do you use to throw a ball?," or "Which hand do you use to brush your teeth?" Each answer is given a value, depending on the preference indicated, providing a quantitative measure of the inclination toward right- or left-handedness. Anthropologists have determined the incidence of handedness in ancient cultures by examining artifacts; the shape of a flint ax, for example, can indicate whether it was made by a right- or left-handed individual. Handedness in antiquity has also been assessed by examining the incidence of figures in artistic representations who are using one hand or the other. Based on this evidence, our species appears always to have been a right-handed one. Moreover, handedness is probably not peculiar to humans; many studies have demonstrated paw preference in animals ranging from mice to monkeys that is, at least in some ways, similar to human handedness.

Whether an individual is right-handed or left-handed has a number of interesting consequences. As will be obvious to left-handers, the world of man-made objects is in many respects a right-handed one. Implements such as scissors, knives, coffee pots and power tools, are constructed for the right-handed majority. Books and magazines are also designed for right-handers (try turning these pages with your left hand), as are golf clubs and guitars. Perhaps as a consequence of this bias, the accident rate for left-handers in all categories (sports, work, home, and so on) is higher than for right-handers. Indeed, the traffic fatality rate of left-handers is several times that of right-handers. However, there are also some advantages to being left-handed. For example, an inordinate number of international fencing champions have been left-handed. The reason for this fact is obvious if you think about it. The majority of any individual's opponents will be right-handed; thus, the average fencer, right or left-handed, is less prepared to parry thrusts from left-handers.

One of the most hotly debated questions about the consequences of handedness in recent years has been whether being left-handed entails a diminished life expectancy. No one disputes the fact that there is currently a surprisingly small number of left-handers among the elderly (see figure). These data have come from studies of the general population and have been supported by information gleaned from *The Baseball Encyclopedia*, in which longevity and other characteristics of a large number of healthy left- and right-handers have been recorded because of interest in the national pastime. Two explanations of this peculiar finding have been put forward. Stanley Coren and his collaborators have argued that these statistics reflect a higher mortality rate among left-handers, partly as a result of increased accidents, but also because of other data that show that left-handedness is associated with a variety of pathologies. In this regard, Coren and others have suggested that left-handedness may arise because of developmental problems in the pre- or perinatal period. If true, then a further rationale for decreased longevity would have been identified. An alternative explanation, however, is that the diminished number of left-handers among the

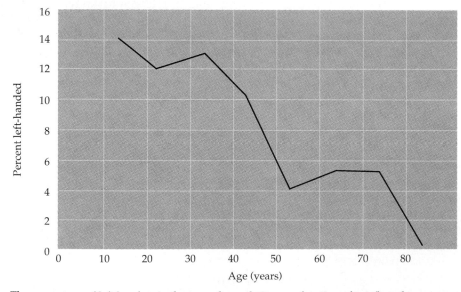

The percentage of left handers in the normal population as a function of age (based on more than 5000 individuals). Taken at face value, these data indicate that right-handers live longer than left-handers. A more subtle analysis, however, suggests that the paucity of elderly left-handers at present may simply reflect changes over the decades in the social pressures on children to become right-handed. (From Coren, 1992.)

elderly is primarily a reflection of sociological factors—namely, a greater acceptance of left-handed children today compared to earlier in the twentieth century. In this view, there are fewer older left-handers now because parents, teachers, and others encouraged right-handedness in earlier generations. Although this controversy continues, the weight of the evidence favors the sociological explanation.

The relationship between handedness and other lateralized functions—language in particular—has long been a source of confusion. It is unlikely that there is any direct relationship between language and handedness, despite much speculation to the contrary. The most straightforward evidence on this point

comes from the results of the Wada test (the injection of sodium amytal into one carotid artery to determine the hemisphere in which language function is located; see text). A large number of such tests carried out for clinical purposes indicate that about 97% of humans, including the majority of left-handers, have their major language functions in the left hemisphere. Since most left-handers have language function on the side of the brain opposite the control of their preferred hand, it is hard to argue for any strict relationship between these two lateralized functions.

In all likelihood, handedness, like language, is an example of the advantage of having any specialized function

on one side of the brain or the other to make maximum use of the available neural circuitry in a brain of limited size.

References

BAKAN, P. (1975) Are left-handers brain damaged? New Scientist 67: 200-202.

COREN, S. (1992) *The Left-Hander Syndrome: The Causes and Consequence of Left-Handedness.* New York: The Free Press.

DAVIDSON, R. J. AND K. HUGDAHL (EDS.) (1995) *Brain Asymmetry.* Cambridge, MA: MIT Press.

SALIVE, M. E., J. M. GURALNIK AND R. J. GLYNN (1993) Left-handedness and mortality. Amer. J. Pub. Health 83: 265-267.

---

verbal communication. These deficiencies, referred to as **aprosodias**, are associated with right-hemisphere lesions of the cortical areas that correspond to Broca's and Wernicke's areas in the left hemisphere. The aprosodias emphasize that although one hemisphere (or distinct cortical regions within that hemisphere) may figure prominently in the comprehension and production of language, many regions are needed to provide the full richness of everyday language.

In short, whereas the classically defined regions of the left hemisphere are activated in a way consistent with their assumed function, a variety of contemporary studies have shown that other left- and right-hemisphere areas clearly make a significant contribution to language function.

■ THE DEVELOPMENT OF LANGUAGE SKILLS

The development of language abilities in infants can be assessed by observing behaviors like sucking rate that give an indication of the child's attentiveness. Infants as young as 4 months respond to novel stimuli—simple language sounds, for example—by sucking more frequently on a pacifier, then gradually ceasing this excess activity as they habituate to the stimulus. Using this behavioral test, developmental psychologists have asked whether an infant discriminates between various language stimuli by allowing habituation to occur and then presenting a novel stimulus. If the infant recognizes a difference in the sounds, it responds by an elevated sucking rate. Remarkably, infants a few months old can discriminate distinct consonant sounds in spoken English, although similar nonlinguistic sounds are ineffective (Figure 25.7). For example, the sounds *b* and *p*, known respectively as voiced and voiceless end consonants, can readily be discriminated. This ability implies that the neural machinery for the reception and processing of language sounds is in place at an extremely early age.

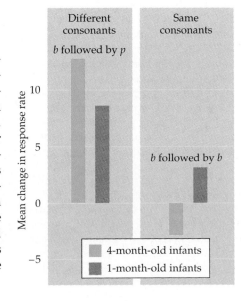

Figure 25.7 Evidence that differences between speech sounds are perceived by infants. The ability to discriminate normal language sounds was shown here by measuring sucking responses after presenting the babies with sequential consonant sounds. These results indicate that infants can readily perceive extraordinarily subtle differences in speech sounds. (After Eimas, 1971.)

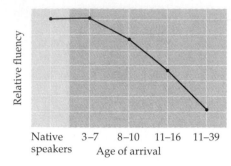

Figure 25.8 A critical period for learning language is shown by the decline in language ability (fluency) of non-native speakers of English as a function of their age upon arrival in the United States. The ability to score well on tests of English grammar and vocabulary declines from approximately age 7 onward. (After Johnson and Newport, 1989.)

Even though some of the basic machinery for language is present in infants, language is obviously learned. The acquisition of most learned behaviors occurs during a period when the brain is especially plastic (see Chapter 22). This critical period phenomenon has been effectively studied for language by examining the ability of non-native speakers to learn English as a function of age (Figure 25.8). When non-native speakers are tested for lexical and syntactic ability, performance suffers if they arrived in the English-speaking environment after about 7 or 8 years of age. Thus, although some of the circuitry for language comprehension and production is in place at a very early stage of human development, the language system of the brain apparently follows the same rules of developmental plasticity and use-dependent change as other systems (the visual system, for example; see Chapter 22). Like other behaviors, the acquisition of language depends on innate capacities that can be modified by experience for a limited period during maturation.

■ SIGN LANGUAGE

Does language localization and lateralization simply reflect specialization for hearing and speaking, or are the language regions of the brain more broadly organized for processing symbols? The answer to this important question has come from studies by Ursula Bellugi and her colleagues of sign language in individuals deaf from birth.

Bellugi's initial analysis showed that American Sign Language has all the components and constraints (such as grammar and syntax) of spoken

Figure 25.9 Signing deficits in congenitally deaf individuals who had learned sign language from birth and later suffered lesions of the language areas in the left hemisphere. Lesions of the left hemisphere produced signing problems in these patients analogous to the aphasias seen after comparable lesions in hearing, speaking patients. In this example, the patient (lower panels) is expressing the sentence "We arrived in Jerusalem and stayed there." Compared to a normal control (upper panels), he cannot properly control the spatial orientation of the signs. The direction of the correct signs and the aberrant direction of the "aphasic" signs are indicated in the upper left-hand corner of each panel. (After Bellugi et al., 1989.)

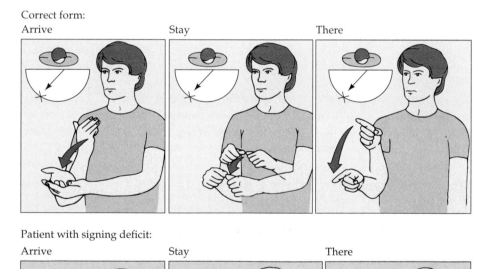

and heard language. She then examined the localization of sign language in patients who had suffered localized lesions of either the left or right hemisphere; all these individuals were prelingually deaf, had been signing throughout their lives, had deaf spouses, were members of the deaf community, and were right-handed. The patients with left-hemisphere lesions, which in each case involved the language areas of the frontal and/or temporal lobes, had measurable deficits in sign production and comprehension when compared to normal signers of similar age (Figure 25.9). In contrast, the patients with lesions in comparable areas in the right hemisphere did not have sign "aphasias." Instead, as predicted from other studies, visuospatial and other abilities were impaired. Although the number of subjects is necessarily small (deaf signers with lesions of the language areas are understandably difficult to identify and study), the capacity for signed and seen communication is evidently represented predominantly in the left hemisphere in the same areas as spoken language. These results indicate that the language regions of the brain are specialized for the representation of symbolic communication rather than heard and spoken language per se.

The capacity for seen and signed communication, like its heard and spoken counterpart, also emerges in early infancy. Careful observation of babbling in hearing (and eventually, speaking) infants shows the production of a predictable pattern of sounds related to the ultimate acquisition of spoken language. Thus, babbling represents an early behavior that prefigures true language, reflecting an innate capacity for language imitation. The congenitally deaf offspring of deaf, signing parents "babble" with their hands in gestures that are apparently the forerunners of signs (Figure 25.10). The amount of signed "babbling" increases with age until the child begins to form accurate, meaningful signs. These observations indicate that the strategy for acquiring the rudiments of symbolic communication from parental or other cues—regardless of the means of expression—is similar in deaf and hearing individuals.

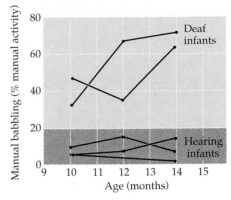

Figure 25.10 Manual "babbling" in deaf infants raised by deaf, signing parents compared to spoken babble in hearing infants. Babbling was judged by scoring hand positions and shapes that showed some resemblance to the components of American Sign Language. In deaf infants, meaningful hand shapes increase as a percentage of manual activity between ages 10 and 14 months. Hearing children raised by hearing, speaking parents do not produce similar hand shapes. (After Petitto and Marentette, 1991.)

■ SUMMARY

Neurological, psychological, and physiological methods have all been used to localize linguistic function in the human brain. This effort began in the nineteenth century by correlating clinical signs and symptoms with the location of brain lesions determined postmortem. In this century, studies of "split-brain" patients, mapping at neurosurgery, sodium amytal anesthesia of a single hemisphere, and noninvasive imaging techniques such as positron emission tomography have greatly extended knowledge about the localization of language. Taken together, these various approaches show that the perisylvian cortices of the left hemisphere are especially important for normal language. The right hemisphere also contributes to language, most obviously by giving it emotional meaning. The similarity of the deficits after comparable brain lesions in congenitally deaf patients and their speaking counterparts supports the idea that the cortical representation of language is independent of the means of its expression or perception (spoken and heard versus gestured and seen). The specialized language areas that have been identified to date are evidently the major components of a widely distributed set of brain regions that allow us to communicate effectively by means of symbols.

Additional Reading

Reviews

BELLUGI, U., H. POIZNER AND E. S. KLIMA (1989) Language, modality and the brain. Trends Neurosci. 12: 380–388.

DAMASIO, A. R. (1992) Aphasia. New Eng. J. Med. 326: 531–539.

DAMASIO, A. R. AND H. DAMASIO (1992) Brain and language. Sci. Amer. 267: 89–95.

DAMASIO, A. R. AND N. GESCHWIND (1984) The neural basis of language. Annu. Rev. Neurosci. 7: 127–147.

LENNEBERG, E. H. (1967) Language in the context of growth and maturation. In: *Biological Foundations of Language.* New York: John Wiley and Sons, Chapter 4, pp. 125–395.

OJEMANN, G. A. (1983) The intrahemispheric organization of human language, derived with electrical stimulation techniques. Trends Neurosci. 4: 184–189.

OJEMANN, G. A. (1991) Cortical organization of language. J. Neurosci. 11: 2281–2287.

SPERRY, R. W. (1974) Lateral specialization in the surgically separated hemispheres. In *The Neurosciences: Third Study Program,* F. O. Schmitt and F. G. Worden (eds.). Cambridge: The MIT Press, pp. 5–19.

SPERRY, R. W. (1982) Some effects of disconnecting the cerebral hemispheres. Science 217: 1223–1226.

Important Original Papers

EIMAS, P. D., E. R. SIQUELAND, P. JUSCZYK AND J. VIGORITO (1971) Speech perception in infants. Science 171: 303–306.

GAZZANIGA, M. S. AND R. W. SPERRY (1967) Language after section of the cerebral commissures. Brain 90: 131–147.

GESCHWIND, N. AND W. LEVITSKY (1968) Human brain: left-right asymmetries in temporal speech region. Science 161: 186–187.

JOHNSON, J. S. AND E. I. NEWPORT (1989) Critical period effects in second language learning: The influences of maturational state on the acquisition of English as a second language. Cogn. Psychol. 21: 60–99.

MIYAWAKI, M., W. STRANGE, R. VERBRUGGE, A. LIBERMAN, J. J. JENKINS AND O. FUJIMURA (1975) An effect of linguistic experience: The discrimination of [r] and [l] by native speakers of Japanese and English. Percep. and Psychophys. 18: 331–340.

OJEMANN, G. A. AND H. A. WHITAKER (1978) The bilingual brain. Arch. Neurol. 35: 409–412.

PETERSEN, S. E., P. T. FOX, M. I. POSNER, M. MINTUN AND M. E. RAICHLE (1988) Positron emission tomographic studies of the cortical anatomy of single-word processing. Nature 331: 585–589.

PETTITO, L. A. AND P. F. MARENTETTE (1991) Babbling in the manual mode: Evidence for the ontogeny of language. Science 251: 1493–1496.

WADA, J. A., R. CLARKE AND A. HAMM (1975) Cerebral hemispheric asymmetry in humans: cortical speech zones in 100 adult and 100 infant brains. Arch. Neurol. 32: 239–246.

Books

MILLER, G. A. (1991) *The Science of Words.* New York: Scientific American Library.

POSNER, M. I. AND M. E. RAICHLE (1994) *Images of Mind.* New York: Scientific American Library.

CHAPTER 26

SLEEP AND WAKEFULNESS

■ OVERVIEW

Sleep—which is defined behaviorally by the suspension of normal consciousness and electrophysiologically by specific brain wave criteria—consumes fully a third of our lives. Sleep occurs in all mammals, and probably all vertebrates. We crave sleep when deprived of it, and, to judge from animal studies, continued sleep deprivation can ultimately be fatal. Surprisingly, however, this peculiar state is not the result of a simple diminution of brain activity; rather, sleep is a series of precisely controlled brain states, and in some of these the brain is as active as it is when we are awake. The sequence of sleep states is governed by a group of brainstem nuclei that project widely throughout the rest of the brain and modulate overall levels of brain activity. Despite many advances, major aspects of sleep are still incompletely understood. The reason for high levels of brain activity during some phases of sleep, the significance of dreaming, and the basis of the restorative effect of sleep are all tantalizing puzzles that continue to motivate much ongoing research.

■ SLEEP AS AN ACTIVE STATE

For centuries—indeed up until the 1950s—most people who thought about sleep considered it a unitary phenomenon whose physiology was essentially passive and whose purpose was largely restorative. In 1953, however, Nathaniel Kleitman and Eugene Aserinksy showed, by means of electroencephalographic recordings from normal subjects, that sleep actually comprises two quite different components. The first is called **non-rapid eye movement (non-REM) sleep**; its most prominent feature is a phase called **slow-wave sleep**. Slow-wave sleep has the key characteristic that had always been assigned to sleep: the apparent generation of neurological "rest." Kleitman and Aserinsky's surprising finding was that much of the night is spent in a wholly different form of sleep called **rapid eye movement (REM) sleep**. REM sleep is characterized by the same high-frequency, low-voltage electroencephalographic activity that provides the electrical signature of the waking brain (Figure 26.1).

■ NON-REM SLEEP

Humans descend into sleep in stages that succeed each other over the first hour or so after retiring (Figure 26.2). The characteristic stages of non-REM sleep are defined primarily by electroencephalographic criteria (Box A). Initially, during the period that we are simply drowsy, the frequency spectrum of the electroencephalogram (EEG) is shifted toward lower values and the amplitude of the cortical waves increases somewhat. This period, called **stage 1 sleep**, eventually gives way to light sleep (**stage 2 sleep**), which is characterized by a further decrease in the frequency of the EEG waves and an increase in their amplitude, together with intermittent high-frequency spike clusters called **sleep spindles** (see Figure 26.1). Sleep spindles are periodic bursts of activity at about 10–12 Hz that generally last 1 or 2 seconds;

Figure 26.1 Electroencephalographic recordings from a normal volunteer during descent into sleep (see Box A and Figure 26.2). The waking state is characterized by high-frequency, low-amplitude activity, whereas descent into non-REM sleep is characterized by decreasing frequency and increasing amplitude of EEG waves. The typical frequency range of EEG activity in the waking state with the eyes open is 14–60 Hz (beta activity); with the eyes closed most people have prominent oscillations in the 8–13 Hz range (alpha waves), primarily over the occipital region. In deep sleep (stage 4), slow waves (also called delta waves) appear at a characteristic frequency of 0.5–2 Hz. Finally, rapid eye movement sleep is characterized by the reappearance of low-voltage, high-frequency activity that is remarkably similar to the EEG activity of individuals who are awake. The PGO waves that mark the onset of REM sleep are not shown here.

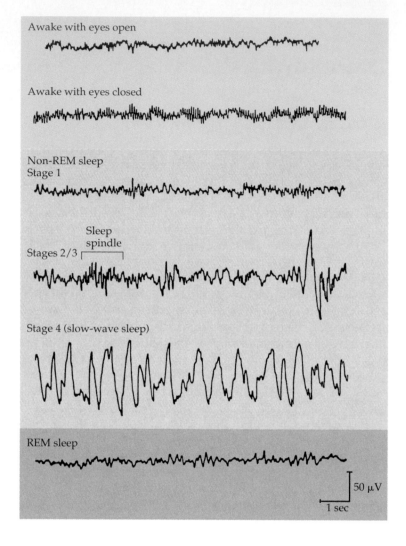

they arise as a result of incompletely understood interactions between thalamic and cortical neurons. In **stage 3 sleep**, which represents moderate to deep sleep, the number of spindles decreases whereas the amplitude of low-frequency waves increases still more. In the deepest level of sleep, **stage 4 sleep**, the predominant EEG activity consists of low-frequency, high-amplitude fluctuations called **delta waves**, the characteristic slow waves for which this phase of sleep is named. The delta waves during slow-wave sleep are

Figure 26.2 A typical night's sleep. This diagram shows the characteristic descent into stage 4 slow-wave sleep, followed by ascent into rapid eye movement sleep. An abbreviated version of this cycle is normally repeated four or five times during the night. (After Biddle and Oaster, 1990.)

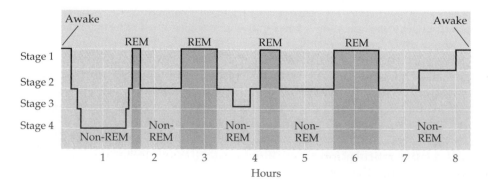

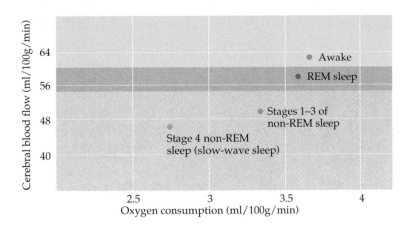

Figure 26.3 Decreased cerebral metabolism during sleep. The metabolic rate, measured in this case by both cerebral blood flow and oxygen consumption per minute, was determined in normal volunteers during wakefulness and various stages of sleep. (After Madsen and Vorstrup, 1991.)

thalamocortical oscillations that occur in the absence of activating, cholinergic inputs from the midbrain to the thalamus. The entire sequence from drowsiness to deep stage 4 sleep usually takes about an hour (Figure 26.2).

The evidence that non-REM sleep is, broadly speaking, restorative comes from several observations. First, the metabolism of the brain during slow-wave sleep, as measured by cerebral blood flow or oxygen consumption, is reduced by as much as 45% (Figure 26.3). In accord with reduced energy consumption, body temperature drops during non-REM sleep. Second, total sleep deprivation has drastic consequences that eventually lead to a breakdown of homeostatic function and ultimately to the death of experimental animals chronically deprived of slow-wave sleep. Since REM sleep deprivation has only minor effects (see below), the consequences of total sleep deprivation presumably arise from the absence of non-REM sleep.

Despite the fact that the brain is relatively quiescent during non-REM sleep, the body, as monitored by muscle recordings (electromyography), is remarkably active during this period, and most reflexes are intact. Thus, sleepwalking occurs during slow-wave sleep. This observation has led to the aphorism that non-REM sleep is characterized by an inactive brain in an active body, whereas REM sleep (described in the following section) is characterized by an active brain in an inactive body.

■ REM SLEEP

Up until this point in the sleep cycle, the activity of the brain recorded electroencephalographically generally accords with our commonsense experience. Kleitman and Aserinsky's findings in the 1950s, however, permanently changed the perception of sleep as a unitary, passive phenomenon. Their initial discovery was that slow-wave sleep is punctuated several times during the night by a completely different condition in which the eyes move rapidly; hence the name rapid eye movement sleep. During REM sleep, electroencephalographic recordings again show the high-frequency, low-amplitude activity of normal wakefulness (see Figure 26.1). Moreover, the body, which in non-REM sleep remains responsive to a variety of stimuli, is inhibited from responding. For these several reasons, rapid eye movement sleep is also called "paradoxical sleep." In short, REM sleep belies any simplistic view of sleep as a "turning off" of the nervous system.

The onset of REM sleep is characterized by EEG waves that originate in the pontine reticular formation and propagate through the lateral geniculate nucleus of the thalamus to the occipital cortex. These **pontine-geniculo-**

Box A
ELECTROENCEPHALOGRAPHY

Although electrical activity recorded from the exposed cerebral cortex of a monkey was reported in 1875, it was not until 1929 that Hans Berger, a psychiatrist at the University of Jena, first made scalp recordings of this activity in humans. Since then, the electroencephalogram, or EEG, has received a mixed press, touted by some as a unique opportunity to understand human thinking and denigrated by others as too complex and poorly resolved to allow anything more than a superficial glimpse of what the brain is actually doing. The truth probably lies somewhere in between. Certainly no one disputes that electroencephalography has provided a valuable tool to both researchers and clinicians, particularly in the fields of sleep physiology and epilepsy.

The major advantage of electroencephalography, which involves the application of a set of electrodes to standard positions on the scalp, is its great simplicity. Its most serious limitation is its poor spatial resolution, allowing localization of an active site only to within several centimeters. Four basic EEG phenomena have been defined in humans (albeit somewhat arbitrarily). The alpha rhythm is typically recorded in awake subjects with their eyes closed. By definition, the frequency of the alpha rhythm is 8–13 Hz, with an amplitude that is typically 10–50 μV. Lower-amplitude beta activity is defined by frequencies of 14–60 Hz and is indicative of mental activity and attention. The theta and delta waves, which are characterized by frequencies of 4–7 Hz and less than 4 Hz, respectively, imply drowsiness, sleep, or one of a variety of pathological conditions. Far and away the most obvious component of these various oscillations is the alpha rhythm. Its prominence in the occipital region—and its modulation by eye opening and closing—implies that it is somehow linked to visual processing, as was first pointed out in 1935 by the British physiologist E. D. Adrian.

In the 1940s, E. W. Dempsey and R. S. Morrison showed that these EEG rhythms depend in part on activity in the thalamus, since thalamic lesions can reduce or abolish the oscillatory cortical discharge (although some oscillatory activity remains even after the thalamus has been inactivated). At about the same time, H. W. Magoun and G. Moruzzi showed that the reticular activating system in the brainstem is also important in modulating EEG activity (see text). For example, activation of the reticular formation changes the cortical alpha rhythm to beta activity, in association with greater behavioral alertness. In the 1960s, P. Andersen and his colleagues further advanced these studies by showing that virtually all areas of the cortex participate in these oscillatory rhythms, which reflect a feedback loop between neurons in the thalamus and cortex.

The cortical origin of EEG activity has been clarified by animal studies, which have shown that the current source of the fluctuating scalp potential is neurons and their synaptic connections in the deeper layers of the cortex. (This conclusion was reached by noting the

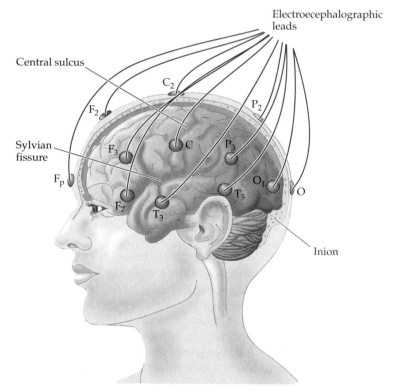

The electroencephalogram represents the voltage recorded between two electrodes applied to the scalp. Typically, pairs of electrodes are placed in 19 standard positions distributed over the head, as shown here. The recording obtained from each pair of electrodes is somewhat different because each samples the activity of a population of neurons in a different brain region.

location of electrical field reversal upon passing an electrode vertically through the cortex from surface to white matter.) In general, oscillations come about either because of pacemaker cells, whose membrane voltage fluctuates spontaneously, or the reciprocal interaction of excitatory and inhibitory neurons in circuit loops. The oscillations of the EEG are thought to arise from the latter mechanism.

Finally, evidence from very large numbers of subjects suggests that at least several different regions of the brain have their own characteristic rhythms; for example, within the alpha band (8–13 Hz), one rhythm, the classic alpha rhythm, is associated with visual cortex, one (the mu rhythm) with the sensorimotor cortex around the central sulcus, and yet another (the kappa rhythm) with the auditory cortex.

Psychologists have had some suc-cess in demonstrating a cycle of excita-tion associated with each cycle in the ongoing alpha rhythm. For example, reaction time is somewhat improved if a visual cue falls at the peak of the alpha wave cycle. Moreover, there are several perceptual phenomena that have a peri-odicity of approximately 10 Hz, the aver-age frequency of the alpha rhythm. Thus, a flashing light appears brighter at a repetition rate of 10 Hz than at fre-quencies above or below this rate, and epileptic attacks are sometimes induced in susceptible people by flashing lights at about 15 Hz. Despite these intriguing observations, the functional significance of the alpha (or any other) cortical rhythm is not known. The purpose of the brain's remarkable oscillatory activ-ity is a puzzle that has now defied elec-troencephalographers and neurobiolo-gists for more than 60 years.

References

ADRIAN, E. D. AND K. YAMAGIWA (1935) The origin of the Berger rhythm. Brain 58: 323-351.

ANDERSEN, P. AND S. A. ANDERSSON (1968) *Physiological Basis of the Alpha Rhythm.* Meredith Corporation.

CATON, R. (1875) The electrical currents of the brain. Brit. Med. J. 2: 278.

DA SILVA, F. H. AND W. S. VAN LEEUWEN (1977) The cortical source of the alpha rhythm. Neurosci. Letters 6: 237–241.

DEMPSEY, E. W. AND R. S. MORRISON (1943) The electrical activity of a thalamocortical relay system. Amer. J. Physiol. 138: 283–296.

NIEDERMEYER, E. AND F. L. DA SILVA (1993) *Electroencephalography. Basic Principles, Clinical Applications, and Related Fields.* Maryland: Williams & Wilkins.

NUÑEZ, P. L. (1981) *Electric Fields of the Brain. The Neurophysics of EEG.* New York: Oxford University Press.

occipital (**PGO**) **waves** therefore provide a useful marker for the beginning of REM sleep. Although the specific purpose of the PGO waves is not clear, they signify brainstem activation of the cortex.

The first period of REM sleep that follows the descent through the four stages of non-REM sleep typically lasts about 20 minutes, followed by another episode of non-REM sleep. This sequence—descent into slow-wave sleep, a quick ascent into REM sleep, followed by descent again into non-REM sleep—repeats itself four or five times during the night (see Figure 26.2). In total then, the typical 8 hours of sleep is divided into about 1.5 to 2 hours of rapid eye movement sleep and about 6 hours of repeated descent into non-REM sleep and subsequent ascent to REM sleep. Only 1 to 2 hours of this time is actually spent in stage 4 slow-wave sleep.

Interestingly, the overall duration of REM sleep varies as a function of age (Figure 26.4). Infants (and fetuses) spend a great deal of time sleeping, and a high proportion of fetal and infant sleep is REM sleep. This fraction decreases during childhood, remains steady during much of adult life, then diminishes in old age. The meaning of these changes has provoked a good deal of speculation (a link between REM sleep and learning has been suggested) but remains obscure.

■ DREAMING AND THE POSSIBLE FUNCTIONS OF REM SLEEP

Despite the wealth of descriptive information about the stages of sleep, the functional purposes of the various sleep states are not yet known. Whereas

Figure 26.4 Differences in the daily amount of REM and non-REM sleep in humans as a function of age. The amount of time spent sleeping falls with increasing age, the amount of REM sleep declining relatively more than the amount of non-REM sleep. (After Roffwarg et al., 1966.)

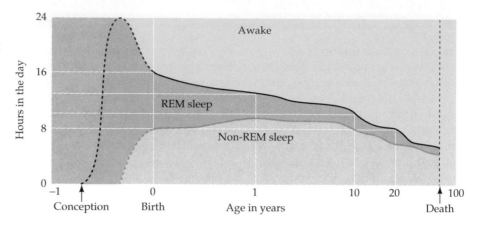

most sleep researchers accept the idea that the purpose of non-REM sleep is at least in part restorative, the function of REM sleep remains a matter of considerable controversy. A possible clue about the purposes of REM sleep is the prevalence of dreams during these epochs of the sleep cycle. The occurrence of dreams can be tested by waking volunteers during either non-REM or REM sleep and asking them if they were dreaming. Subjects awakened during non-REM sleep rarely report detailed dreams; in contrast, elaborate dreams are almost always described by volunteers awakened from REM sleep. Evidently, the dreams that occur during REM sleep are quickly forgotten, since subjects awakened only minutes after the end of a period of REM sleep remember few, if any, of the dreams that they presumably had during the preceding REM epoch.

Dreams have been studied in a variety of ways, perhaps most notably within the psychoanalytic framework of revealing unconscious thought processes considered to be at the root of neuroses. Sigmund Freud's *The Interpretation of Dreams*, published in 1900, speaks eloquently to the complex relationship between conscious and unconscious mentation. It is by no means agreed upon, however, that dreams have the deep significance that Freud and others have given them. Although there is little scientific evidence on the matter, the psychoanalytic interpretation of dreams has recently fallen into disfavor (nevertheless, most people probably give some credence to the significance of dream content, at least privately).

An imaginative hypothesis about dreams in REM sleep has been advanced by Francis Crick (of DNA fame) and Grahame Mitchison. In Crick and Mitchison's theory, the function of dreams is to act as an "unlearning" mechanism, whereby certain modes of neural activity are erased by random activation. Their theory is based on the idea that the human brain represents information by the activity of sets of neuronal networks that are widely distributed and overlapping. Computer simulations of neural networks imply, among other things, that such systems are subject to parasitic modes, meaning that some patterns of activity become established that degrade rather than enhance the information content of the system. Roughly speaking, one can think of these parasitic modes of activity as unwanted thoughts or erroneous information, which, if not expunged, might become the basis for obsession, paranoia, or other thought pathology.

At first glance, the comparative neurology of REM sleep appears consistent with this sort of role for REM sleep. Although REM sleep occurs in nearly all mammals, adult birds lack it, as do reptiles and amphibians. Since mammals (both placental and marsupial) have undergone extraordinary evolutionary development of the neocortex, REM sleep is arguably a characteristic of large, intelligent brains. On the other hand, a few mammals lack REM sleep. An example is the spiny anteater, which has a remarkably large cerebral cortex that does not show any signs of REM sleep. Crick and Mitchison have argued that the anomalously large size of the cerebral cortex in this species may reflect a compensation for an inability to expunge parasitic modes of thought. Nevertheless, the example of the anteater raises the possibility that other species with more ordinary amounts of neocortex will also be found to lack REM sleep when someone gets around to studying them in detail. Indeed, the bottle-nosed dolphin has also been reported to lack REM sleep, and the male (but not the female) rabbit shows very little.

■ SLEEP DEPRIVATION

Adding to uncertainty about the purposes of REM sleep (and arguing for some skepticism about existing theories) is the fact that deprivation of REM sleep in humans for as much as 2 weeks has little or no effect on behavior. Such studies have been done by waking volunteers whenever their EEG recordings showed the characteristic signs of REM sleep. The minds of such individuals are not, as far as can be discovered, disrupted by any pathological modes of thought; and although the volunteers compensate for the lack of REM sleep by experiencing more of it after the period of deprivation has ended, they suffer no obvious adverse effects. Similarly, patients taking certain antidepressants (MAO inhibitors; see Box D in Chapter 27) have little or no REM sleep, yet show no obvious ill effects, even after months or years of treatment.

The apparent innocuousness of REM sleep deprivation contrasts markedly with the effects of total sleep deprivation. The longest documented period of voluntary sleeplessness is 288 hours (approximately 12 days), a record chalked up by a 23-year-old Californian who accomplished this feat in the showroom of a San Diego waterbed store using coffee as his only source of pharmacological stimulation. He recovered after a few days during which he slept a great deal, and seemed none the worse for wear. As most people know from their own experience, however, total sleep deprivation has dramatic effects on the efficiency of mental (and motor) functioning. Sleep deprivation has not been carried further in humans for good reason: if sleep is prevented for several weeks in rats, the animals invariably die as a result of failed homeostatic mechanisms (Figure 26.5). The implication of these several findings is that we can get along without REM sleep, but need non-REM sleep in order to survive.

■ STYLES OF SLEEP AMONG DIFFERENT SPECIES

Sleep comprising non-REM and REM phases is largely restricted to mammals. A wide variety of animals, however, have a rest-activity cycle that often (but not always) occurs in a daily **circadian** rhythm (*circa* means "around," *dia* means "day"). Even among mammals, the organization of sleep depends very much on the lifestyle of the species in question. As a general rule, predatory animals can indulge, as we do, in long, uninterrupted periods of sleep that can be nocturnal or diurnal, depending on the time of day when the animal acquires food, mates, cares for its young, and deals with life's other necessities. The survival of animals that are preyed upon, however, depends much more critically on continued vigilance. Such

(A) Experimental setup

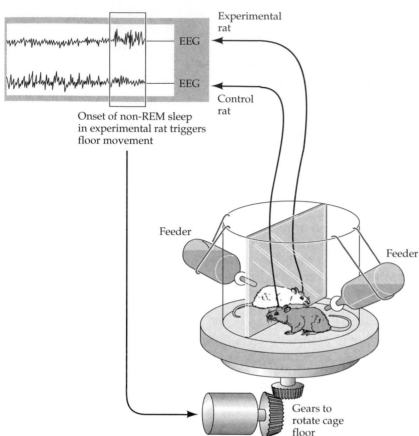

Experimental
rat

EEG

Onset of non-REM sleep
in experimental rat triggers
floor movement

Control
rat

EEG

Feeder

Feeder

Gears to
rotate cage
floor

Motor

(B) Experimental animal

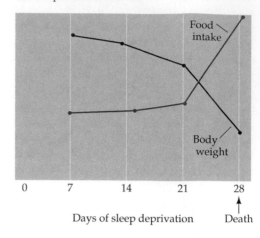

Food
intake

Body
weight

0   7   14   21   28

Days of sleep deprivation   Death

Figure 26.5 The consequences of total sleep
deprivation in an experimental animal. (A) In
this apparatus, the experimental rat is kept
awake because the onset of sleep (detected electroencephalographically) triggers movement of
the cage floor. The control rat can thus sleep
intermittently. (B) After 2 to 3 weeks of sleep
deprivation, the experimental animals begin to
lose weight, fail to control their body temperature, and eventually die. (After Bergmann et al.,
1989.)

species—as diverse as rabbits and giraffes—sleep during short intervals that
usually last no more than a few minutes. Shrews, the smallest mammals,
hardly sleep at all. An especially remarkable solution to the problem of maintaining vigilance during sleep is shown by dolphins and seals, in whom sleep
alternates between the two cerebral hemispheres. Thus, one hemisphere can
exhibit the electroencephalographic signs of wakefulness, while the other
shows the characteristics of sleep (Figure 26.6). In short, although periods of
rest are evidently essential to the proper functioning of the brain, and more
generally to normal homeostasis, the manner in which rest is obtained
depends on the particular needs of each species.

■ BIOLOGICAL CLOCKS

Human sleep occurs with circadian periodicity. Biologists interested in circadian rhythms have explored a number of interesting and important questions
about this daily cycle. What happens, for example, if individuals are prevented from sensing the cues they normally have about night and day? Will
the subjects continue to coordinate sleep and wakefulness with the light-dark
cycle they were used to, or will some other rhythm supervene? Such questions can be answered by placing volunteers in an environment without
external cues about time (caves or bunkers have sometimes been used), while
otherwise allowing them to proceed with a more or less normal life.

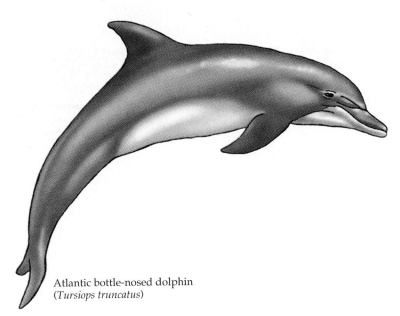

Figure 26.6 Some animals can sleep one hemisphere at a time. These EEG tracings were taken simultaneously from left and right cerebral hemispheres of a dolphin. Slow-wave sleep is apparent in the left hemisphere (recording sites 1–3); the right hemisphere, however, shows low-voltage, high frequency waking activity (sites 4–6). (After Mukhametov, Supin and Polyakova, 1977.)

Atlantic bottle-nosed dolphin
(*Tursiops truncatus*)

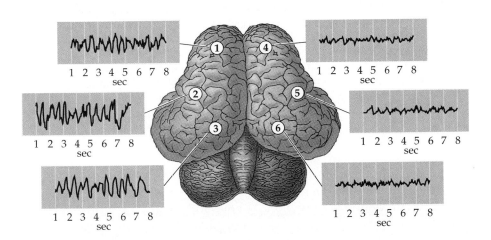

In the absence of cues about day and night, the normal circadian rhythm is maintained but loses its usual relationship to the actual time (Figure 26.7A). Moreover, the cycle of sleep and wakefulness gradually lengthens to about 25 hours instead of the normal 24. Thus, humans (and many other mammals) have an internal clock that continues to operate in the absence of any external information about the time of day. Presumably, this circadian clock evolved to maintain appropriate periods of sleep and wakefulness in spite of the variable amount of daylight and darkness in different seasons and at different places on the planet.

Animal experiments have shown that the location of this clock is the **suprachiasmatic nucleus (SCN)** of the hypothalamus (Figure 26.7B). Not surprisingly, this region receives projections from the retina (presumably to provide information about light and dark), and is connected to the brainstem systems important in controlling sleep and wakefulness (see below). Perhaps the most convincing evidence of the SCN's role as a biological clock is that its removal in experimental animals abolishes their circadian rhythm of sleep and waking. Although the overall amounts of REM and non-REM

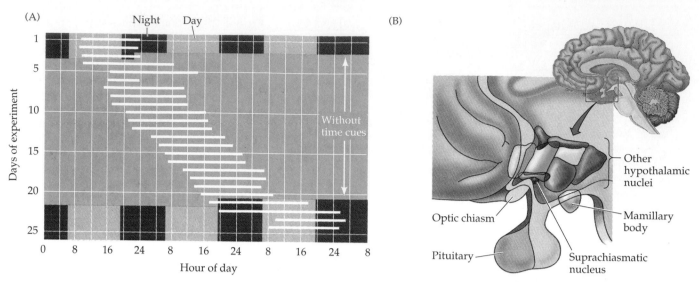

Figure 26.7 The circadian cycle of sleep and wakefulness in humans. (A) When a human volunteer was isolated from cues about time by living in an underground bunker, his daily activity rhythm changed from a 24-hour cycle (days 1–3) to one of about 26 hours (days 4–21). The heavy lines represent waking hours; these run in a diagonal direction because the subject awoke and went to sleep later each day (that is, his internal clock became "free running"). The circadian pattern returned to the normal 24 hour cycle when the volunteer was again exposed to a natural day/night cycle (days 22–26). (B) Diagram of the hypothalamus, showing the location of the suprachiasmatic nucleus, which is thought to be the primary "biological clock" in mammals. The name derives from its location just above the optic chiasm. (A after Aschoff, 1965.)

sleep remain about the same as before, in the absence of the SCN, these sleep epochs occur randomly during the 24-hour cycle of day and night. Whether the influence of this biological clock is mediated by a hormonal mechanism or by synaptic connections to other brain centers is not clear. The ability to reestablish circadian rhythms by implantation of the SCN from a donor into the brain of a host animal (rats have often been used) suggests that the mechanism is a hormonal one. Nor is it known how the clock tells time, although a variety of molecular hypotheses are being explored (Box B).

■ BRAINSTEM MECHANISMS OF SLEEP AND WAKEFULNESS

Among the most controversial aspects of sleep has been the structural and chemical basis of non-REM and REM sleep. This ongoing debate reflects the fact that both the anatomy and the pharmacology of sleep are exceedingly complex, and admit quite different interpretations (Box C). Nonetheless, a consensus has emerged on a number of important points.

Sleep, whether non-REM or REM sleep, is not something elicited, as the nineteenth-century psychologist William James suggested, by a diminished ability to perceive sensations. Quite the contrary, sleep and wakefulness are now understood to be generated by the activation of specific neural centers. This counterintuitive conclusion was first suggested by Horace Magoun and Giuseppe Moruzzi in 1949, when they found that electrical

Box B
MOLECULAR MECHANISMS OF BIOLOGICAL CLOCKS

Understanding the molecular basis of biological clocks in humans and other mammals is a difficult problem. Happily, a simple animal model has provided some powerful clues about how at least some neurons are able to keep time. As noted in the text, virtually all animals have periods of rest and activity, often in a circadian pattern; the fruit fly *Drosophila*, the animal of choice for many genetic analyses, is no exception. In the early 1970s, Ron Konopka and Seymour Benzer, working at the California Institute of Technology, discovered three mutant strains of flies whose circadian rhythms were abnormal. Further analysis showed the mutants to be alleles of a single locus, which Konopka and Benzer called the *period* or *per* gene. In the absence of normal environmental cues (that is, in constant light or dark), wild-type flies have periods of activity geared to a 24-hour cycle; per^s mutants have 19 hour rhythms, per^l mutants have 29-hour rhythms, and per^0 mutants have no apparent rhythm at all.

Working independently, Michael Young at Rockefeller University and Jeffrey Hall and Michael Rosbash at Brandeis University cloned the *per* gene in the early 1980s. Cloning a gene does not necessarily reveal its function, and so it was in this case. Nonetheless, the gene product Per, a nuclear protein, is found in many *Drosophila* cells that are pertinent to the production of the fly's circadian rhythms. Moreover, normal flies show a circadian variation in the amount of *per* mRNA and Per protein, whereas per^0 flies, which lack a circadian rhythm, do not show this circadian rhythmicity of gene expression. There is evidently a feedback loop, with the *per* gene product regulating its own mRNA levels; this loop could represent the basic elements of a circadian pacemaker at the molecular level.

A mammalian homologue of *per* has not been found. Nevertheless, similar mechanisms may be involved in mammalian circadian pacemaker function, and several mutant strains of rodents with abnormal circadian rhythms have recently been isolated.

Although many uncertainties remain, it appears that cells in the neural regions that serve as biological clocks in a wide variety of species contain oscillatory molecular mechanisms that control the transcription and translation of specific gene products, effectively creating a timing mechanism for a range of circadian functions.

References

DUNLAP, J. C. (1993) Genetic analysis of circadian clocks. Annu. Rev. Physiol. 55: 683-728.

HARDIN, P. E., J. C. HALL AND M. ROSBASH (1990) Feedback of the *Drosophila period* gene product on circadian cycling of its messenger RNA levels. Nature 348: 536-540.

TAKAHASHI, J. S. (1992) Circadian clock genes are ticking. Science 258: 238-240.

TAKAHASHI, J. S. (1995) Molecular neurobiology and genetics of circadian rhythms in mammals. Annu. Rev. Neurosci. 18: 531-538.

VITATERNA, M. H., D. P. KING, A. M. CHANG, J. M. KORNHAUSER, P. L. LOWREY, J. D. McDONALD, W. F. DOVE, L. H. PINTO, F. W. TUREK AND J. S. TAKAHASHI (1994) Mutagenesis and mapping of a mouse gene, *clock*, essential for circadian behavior. Science 264: 719-725.

stimulation of the midbrain reticular formation (Figure 26.8) causes a state of wakefulness and arousal (the name **reticular activating system** was therefore given to this region of the brainstem). This observation implied that wakefulness requires a special mechanism, not just the presence of adequate sensory experience.

Most workers now agree that a key component of the reticular activating system is a group of cholinergic nuclei near the pons-midbrain junction. Many of the neurons in these nuclei have high discharge rates during waking and in REM sleep; conversely, they are quiescent during non-REM sleep. Moreover, when stimulated, the neurons of the cholinergic nuclei cause desynchronization of the electroencephalogram (that is, a shift of EEG activity from high-amplitude, synchronized waves to lower-amplitude, higher-frequency, desynchronized ones). These several features imply that activity of cholinergic cells in the reticular activating system is a primary cause of wakefulness and REM sleep, and that their inactivity is important for producing non-REM sleep. However, activity of these neurons is not the only

Box C
SLEEP FACTORS

A quite different idea about the mechanism of sleep is that one or more molecules builds up in the brain during wakefulness and induce sleep (presumably by influencing the neural systems described in the text). This theory had long been attractive to some sleep physiologists, and the possibility was finally taken on by J. Pappenheimer, M. Karnowsky, and their colleagues in the 1960s and 70s. Pappenheimer first confirmed in detail an earlier observation that the transfer of cerebrospinal fluid from a sleep-deprived animal into a normal recipient induces sleep. In practice, the transfer was accomplished by collecting cerebrospinal fluid from sleep-deprived goats. Pappenheimer found that small amounts of this material, when injected into normal rabbits, increased non-REM sleep over the next 6 to 8 hours by as much as 40%.

The robust nature of this result allowed Pappenheimer and Karnowsky to identify the relevant factor in the spinal fluid of the goats, then in the brains of sleep-deprived rabbits, and eventually in human urine. This heroic effort was sparked by the vision of ameliorating human sleep disorders through understanding this potentially critical aspect of sleep pharmacology. After several years of work, the sleep factor isolated from 15,000 rabbit brains—and subsequently from vast quantities of human urine—turned out to be a relatively small molecule identified as a muramyl dipeptide. Surprisingly, and no doubt disappointingly for the principals, muramyl peptides are not synthesized by mammalian cells, but are components of bacterial cell walls. Introduction of the isolated peptide and its synthetic analogues left no doubt that it did indeed induce sleep in experimental animals, but its bacterial provenance raised questions about its physiological relevance.

Since the mid-1980s, this field has remained more or less in limbo, skeptics taking the view that all the effort had turned up a bacterial contaminant that happens to have sleep-inducing effects. The proponents have argued that the nay-sayers should be more open-minded and recognize that bacterial products do serve some important human purposes (in the gut, for instance, bacteria are a source of vitamin K). Moreover, sleep deprivation decreases the efficiency of the immune system, suggesting how a bacterial peptide might find its way into the cerebrospinal fluid of sleep-deprived animals. It remains to be seen how this controversy will be resolved.

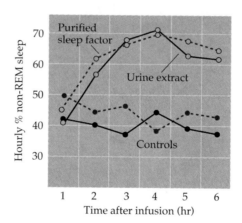

Evidence for a sleep-inducing factor in normal human urine. Graph shows the effect of the injection of this material on the amount of non-REM sleep in rabbits. (After Pappenheimer, 1983.)

References

KRUEGER, J. M., J. R. PAPPENHEIMER AND M. L. KARNOVSKY (1982) The composition of sleep-promoting factor isolated from human urine. J. Biol. Chem. 257: 1664-1669.

KRUEGER, J. M. AND L. JOHANNSEN (1988) Bacterial products, cytokines and sleep. In *Molecular Memory in Health and Disease*. Elsevier Science Publishers, pp. 35-46.

MARTIN, S. A., M. L. KARNOVSKY, J. M. KRUEGER, J. R. PAPPENHEIMER AND K. BIEMANN (1984) Peptidoglycans as promoters of slow-wave sleep. J. Biol. Chem. 259: 12652-12658.

PAPPENHEIMER, J. R. (1979) "Nature's soft nurse": A sleep-promoting factor isolated from brain. Johns Hopkins Med. J. 145: 49-56.

PAPPENHEIMER, J. R. (1983) Induction of sleep by muramyl peptides. J. Physiol. 336: 1-11.

cellular basis of wakefulness; noradrenergic neurons of the **locus coeruleus** and serotonergic neurons of the **raphe nuclei** are also involved. As summarized in Table 26.1, increased activity of these cells is also associated with wakefulness, and decreased activity with the onset of sleep. These effects on mental status are evidently achieved largely by modulating the rhythmicity of interactions between the thalamus and the cortex. The activity of these several ascending systems decreases both the rhythmic bursting of the neurons in the thalamic nuclei and the related synchronized activity of cortical neurons.

The cholinergic neurons in the brainstem that generate the EEG activity characteristic of REM sleep are also responsible for the associated rapid

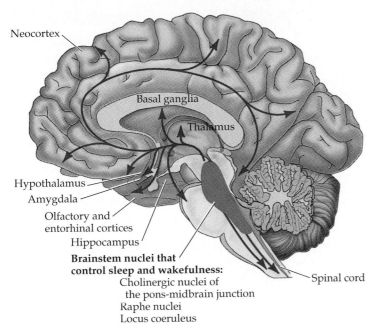

Neocortex

Basal ganglia

Thalamus

Hypothalamus

Amygdala

Olfactory and
entorhinal cortices

Hippocampus

**Brainstem nuclei that
control sleep and wakefulness:**
    Cholinergic nuclei of
      the pons-midbrain junction
    Raphe nuclei
    Locus coeruleus

Spinal cord

Figure 26.8 The brainstem systems that generate wakefulness and sleep. A variety of brainstem nuclei using several different neurotransmitters determine mental status on a continuum that ranges from deep sleep to a high level of alertness. These nuclei, which include the cholinergic nuclei of the pons-midbrain junction, the locus coeruleus, and the raphe nuclei, all have widespread ascending and descending connections to other regions, which explain their numerous effects.

eye movements, PGO waves, and muscle atonia that are the hallmarks of this phase of sleep. In accord with this conclusion, administration of cholinergic agonists desynchronizes the EEG and elicits REM sleep in experimental animals. Furthermore, levels of secreted acetylcholine increase in the cortex of experimental animals during REM sleep. Inactivation of the serotonergic neurons of the raphe nuclei has also been implicated in the induction of REM sleep (see Table 26.1). Turning off REM sleep depends on a different group of neurons that are noradenergic. These cells have an opposite discharge pattern from the cholinergic neurons that turn on REM sleep and

TABLE 26.1
Summary of the Cellular Mechanisms of Sleep and Wakefulness

Brainstem nuclei responsible	Neurotransmitter involved	Activity state of the relevant brainstem neurons
Wakefulness		
Cholinergic nuclei of pons-midbrain junction	Acetylcholine	Active
Locus coeruleus	Norepinephrine	Active
Raphe nuclei	Serotonin	Active
Non-REM sleep		
Cholinergic nuclei of pons-midbrain junction	Acetylcholine	Inactive
Locus coeruleus	Norepinephrine	Inactive
Raphe nuclei	Serotonin	Inactive
REM sleep on		
Cholinergic nuclei of pons-midbrain junction	Acetylcholine	Active (PGO waves)
Raphe nuclei	Serotonin	Inactive
REM sleep off		
Locus coeruleus	Norepinephrine	Active

Figure 26.9 Summary diagram of the interactions between the brainstem, thalamus, and cortex that determine mental status on the continuum from deep sleep to alert wakefulness.

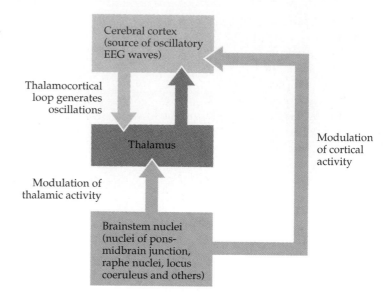

Box D
CONSCIOUSNESS

Quite apart from consciousness in the context of sleep and wakefulness, philosophers, psychologists, and some neurobiologists have been greatly interested in the phenomenon of consciousness as such. Considerable intellectual effort notwithstanding, there remains deep uncertainty about the definition of consciousness, its evolutionary origins, and its relation to the nuts and bolts of brain function. At one extreme in such arguments are the disciples of artificial intelligence, people who prefer to think of brains as glorified computers. In the view of John McCarthy, for example, any feedback device has the essential quality that leads eventually to consciousness in humans. (In one famous instance, he argued that even a thermostat would qualify.) At the other extreme are those who argue that consciousness is a uniquely human property, not being present in "lower" animals. Biologists generally feel that this assertion could

only be made by those who haven't bothered to get to know the behavior and brain structure of other animals.

One of the most engaging contemporary philosophers who works on consciousness is John Searle. Searle offers an argument to rebut those who imagine that a computer, because its operation in some ways resembles mental processes, can be conscious. His famous Chinese Room analogy describes a cubicle in which workers are handed English letters that they then translate into Chinese characters. The workers themselves have no knowledge of English or Chinese, but simply a set of rules that enables the characters to be translated. The output of the workers in the room consists of sensible statements in Chinese. Yet the workers have no knowledge of the meaning of the information they are dealing with, or of the room's larger purpose. Searle uses this image to emphasize that meaningful output from a computer, however sophis-

ticated, cannot provide evidence for consciousness within it.

Perhaps the most frustrating problem in thinking about consciousness is how to rationalize the "internal observer" that we routinely think of as our "self"; this riddle has been discussed for centuries without resolution (other than the consensus that this perspective must be wrongheaded, since it entails an infinite regress). As a result of this protracted and so far fruitless argument, many neurobiologists have come to feel that consciousness is best regarded as an emergent property of the brain, rather than a specific entity that can be studied in its own right.

References

CHURCHLAND, P. M. AND P. S. CHURCHLAND (1990) Could a machine think? Sci. Amer. 262(1): 32–37.

SEARLE, R. W. (1992) *The Rediscovery of the Mind.* Cambridge, MA: MIT Press.

are located primarily in the locus coeruleus. In short, turning on REM sleep—and turning it off—involves several nuclei using different neurotransmitters.

These extraordinarily complex interactions can be summarized as follows (Figure 26.9). The control of sleep and wakefulness depends on brainstem modulation of the thalamus and cortex. (Box D considers consciousness in a broader context.) It is this thalamocortical loop that normally generates the EEG signature of mental function along the continuum of deep sleep to high alert. The major components of the brainstem modulatory system are the cholinergic nuclei of the pons-midbrain junction, the noradrenergic cells of the locus coeruleus in the pons, and the serotonergic raphe nuclei. All of these nuclei are characterized by projections to both the cortex and the thalamus, where they have direct as well as indirect effects on cortical function.

■ SUMMARY

All animals have a restorative cycle of rest following activity, but only mammals divide the period of rest into distinct phases of non-REM and REM sleep. Why mammals (or other animals) need a restorative phase of suspended consciousness accompanied by decreased metabolism and lowered body temperature is not known, although this phenomenon is such an integral part of our lives that we readily accept the need for sleep as making good sense. Even more mysterious is why the brain is periodically active during sleep at levels not appreciably different from the waking state (that is, the neural activity during REM sleep). The highly organized sequence of human sleep states is actively generated by nuclei in the brainstem (most importantly the cholinergic nuclei of the pons-midbrain junction, the noradrenergic cells of the locus coeruleus, and the serotonergic neurons of the raphe nuclei). The activity of these cell groups controls the degree of mental alertness on a continuum from deep sleep to waking attentiveness. These systems are in turn influenced by a circadian clock located in the suprachiasmatic nucleus of the hypothalamus; the clock adjusts periods of sleep and wakefulness to appropriate durations during the 24-hour cycle of light and darkness that is fundamental to life on earth.

Additional Reading

Reviews

ALLISON, T. H. AND H. VAN TWYVER (1970) The evolution of sleep. Natural History 79: 56–65.

HOBSON, J. A. (1990) Sleep and dreaming. J. Neurosci. 10: 371–382.

McCARLEY, R. W. (1995) Sleep, dreams and states of consciousness. In *Neuroscience in Medicine*, P. M. Conn (ed.). Philadelphia: J.B. Lippincott, pp. 535–554.

McCORMICK, D. A. (1989) Cholinergic and noradrenergic modulation of thalamocortical processing. Trends Neurosci. 12: 215–220.

McCORMICK, D. A. (1992) Neurotransmitter actions in the thalamus and cerebral cortex. J. Clin. Neurophysiol. 9: 212–223.

POSNER, M. I. AND S. DEHAENE (1994) Attentional networks. Trends Neurosci. 17: 75–79.

SAPER, C. B. AND F. PLUM (1985) Disorders of consciousness. In *Handbook of Clinical Neurol-ogy, Volume 1 (45): Clinical Neuropsychology.* Amsterdam: Elsevier Science Publishers, pp. 107–128.

STERIADE, M. (1992) Basic mechanisms of sleep generation. Neurol. 42: 9–18.

STERIADE, M., D. A. McCORMICK AND T. J. SEJNOWSKI (1993) Thalamocortical oscillations in the sleeping and aroused brain. Science 262: 679–685.

Important Original Papers

ALLISON, T., H. VAN TWYVER AND W. R. GOFF (1972) Electrophysiological studies of the echidna, *Tachyglossus aculeatus.* Arch. Ital. Biol. 110: 145–184.

ALLISON, T. AND D. V. CICCHETTI (1976) Sleep in mammals: ecological and constitutional correlates. Science 194: 732–734.

ASERINSKY, E. AND N. KLEITMAN (1953) Regularly occurring periods of eye motility, and concomitant phenomena, during sleep. Science 118: 273–274.

ASCHOFF, J. (1965) Circadian rhythms in man. Science 148: 1427–1432.

CRICK, F. AND G. MITCHISON (1983) The function of dream sleep. Science 304: 111–114.

DEMENT, W. C. AND N. KLEITMAN (1957) Cyclic variation in EEG during sleep and their relation to eye movements, body motility and dreaming. Electroenceph. Clin. Neurophysiol. 9: 673–690.

MORUZZI, G. and H. W. MAGOUN (1949). Brain stem reticular formation and activation of the EEG. Electroenceph. Clin. Neurophysiol. 1: 455–473.

ROFFWARG, H. P., J. N. MUZIO AND W. C. DEMENT (1966) Ontogenetic development of the human sleep-dream cycle. Science 152: 604–619.

Book

HOBSON, J. A. (1989) *Sleep.* New York: Scientific American Library.

CHAPTER 27

■ OVERVIEW

The subjective feelings known as emotions are an essential feature of normal human experience; moreover, some of the most devastating psychiatric problems involve emotional (affective) disorders. Although everyday emotions are as varied as happiness, surprise, anger, fear, and sadness, some characteristics are common to all of them. Thus, all emotions are expressed through both physiological changes and stereotyped motor responses, especially of the facial muscles. These responses accompany subjective experiences that are not easily described, but which are much the same in all human cultures. Expression of the emotions is closely tied to the autonomic nervous system and therefore entails the activity of certain brainstem nuclei, the hypothalamus, and the amygdala, as well as the preganglionic neurons in the spinal cord, the autonomic ganglia and peripheral effectors. The centers that coordinate emotional responses have been grouped under the rubric of the limbic system. At the cortical level, the two hemispheres differ in their governance of the emotions, the right hemisphere being more critically involved than the left—yet another example of hemispheric specialization (see Chapters 24 and 25).

■ PHYSIOLOGICAL CHANGES ASSOCIATED WITH EMOTION

The most obvious signs of emotional arousal involve changes in the activity of the autonomic nervous system (see Box A in Chapter 1). Thus, increases or decreases in sweating, heart rate, cutaneous blood flow (blushing or turning pale), piloerection, and gastrointestinal motility can all accompany various emotions. These responses are brought about by changes in activity in the sympathetic, enteric, and parasympathetic components of the autonomic nervous system, which govern cardiac muscle, smooth muscle, and glands throughout the body. Walter B. Cannon, who made many fundamental contributions to understanding the autonomic control of physiological processes, argued that intense activity of the sympathetic division prepares the animal to fully utilize metabolic and other resources for emergencies (that is, preparation for, in his words, "fight or flight"). Conversely, activity of the parasympathetic division (and the enteric division) of the autonomic system promotes a building up of metabolic reserves. Cannon further suggested that the natural opposition of the expenditure and storage of resources is reflected in a parallel opposition of the emotions associated with these different physiological states. As Cannon put it, "The desire for food and drink, the relish of taking them, all the pleasures of the table are naught in the presence of anger or great anxiety."

For many years activation of the autonomic nervous system, particularly of the sympathetic division, was considered an all-or-nothing process. Once effective stimuli engaged the system, it was argued, there was a diffuse and widespread discharge of all of its components. More recent studies have led to the current view that the responses of the neurons of the autonomic nervous system are actually quite specific, with different patterns of activation characterizing different situations and their associated emotions. It is interesting in this regard that emotion-specific expressions produced voluntarily can engender distinct patterns of autonomic activity. In one study, subjects were given muscle-by-muscle instructions that resulted in facial expressions recognizable as anger, disgust, fear, happiness, sadness, or surprise without being told which emotion they were simulating. During this exercise, indices of autonomic activity, such as heart rate, skin conductance, and temperature were measured; surprisingly, it was found that each pattern of facial muscle activity was accompanied by specific and reproducible differences in autonomic activity. Moreover, autonomic responses were strongest when the facial expressions were judged to most closely resemble actual emotional expression; and, even more remarkably, the muscular expression of a particular emotion often led to the subjective experience of that emotion! One interpretation of these findings is that when voluntary facial expressions are produced, signals in the brain engage not only the motor cortex but also some of the circuits that produce emotional responses. Perhaps this relationship helps explain how good actors can be so convincing.

Activity in the autonomic nervous system is controlled by inputs from a number of sources. One important input is sensory drive from the internal organs (see Box A in Chapter 1); this source forms the sensory limb of reflex circuitry that allows rapid physiological changes in response to altered conditions. However, not all activity in the autonomic nervous system arises in response to sensory stimuli. Physiological responses may also be elicited by complex and idiosyncratic stimuli, which achieve their significance only through activity in the forebrain. For example, an anticipated tryst with a lover, a suspenseful episode in a novel or film, stirring patriotic or religious music, or dishonest statements about one's activities can all lead to autonomic activation. The neural activity evoked by such complex stimuli is relayed from the forebrain to autonomic and somatic motor nuclei via the brainstem reticular formation and hypothalamus, the major structures that coordinate emotional behavior.

■ THE INTEGRATION OF EMOTIONAL BEHAVIOR

In 1928, Phillip Bard reported the results of a series of experiments that pointed to the hypothalamus (Box A) as a critical center for coordination of the autonomic and somatic components of emotional behavior. Bard removed both cerebral hemispheres (including the cortex, underlying white matter, and basal ganglia) in a series of cats. When the surgical anesthesia had worn off, the animals behaved as if they were very angry. The angry behavior occurred spontaneously and included the usual autonomic correlates of this emotion: increased blood pressure and heart rate, retraction of the nictitating membranes (thin connective tissue sheets associated with feline eyelids), dilation of the pupils, and erection of the hairs on the back and tail. The cats also exhibited somatic motor components of anger, such as arching the back, extending the claws, lashing the tail, and snarling. This behavior was called "sham rage" because it had no obvious target (unlike the rage exhibited by a normal cat). Bard showed that a complete rage response was obtained as long as the caudal hypothalamus was intact (Figure 27.1). Sham rage could not be elicited, however, when the brain was transected at the junction of the hypothalamus and midbrain, although some uncoordinated components of the response were still apparent. Bard suggested that whereas the subjective experience of emotion might depend on an intact cerebral cortex, coordinated emotional behavior does not necessarily depend on cortical processes. He

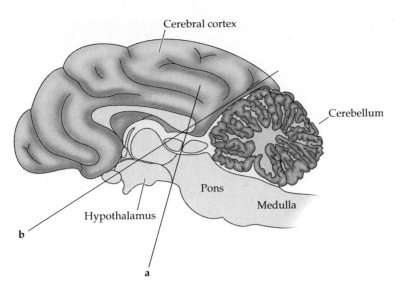

Figure 27.1 Midsagittal view of a cat's brain, illustrating the regions sufficient for the expression of emotional behavior. Transection through plane **a** abolishes integrated emotional responses. Emotional behavior survives removal of the cerebral hemispheres (plane **b**), as long as the hypothalamus remains intact. (After LeDoux, 1987.)

emphasized that emotional behaviors are often directed toward self-preservation (a point also made by Charles Darwin in his book on the evolution of emotion) and that emotions are shared by many vertebrate species. The prevalence of emotions among various animals implies the involvement of phylogenetically older parts of the nervous system.

Complementary results were reported by Walter Hess, who showed that electrical stimulation of discrete sites in the hypothalamus of awake, freely moving cats could also lead to a rage response and even to subsequent attack behavior. Stimulation of other sites in the hypothalamus caused a defensive posture that resembled fear. In 1949, a share of the Nobel Prize in physiology or medicine was awarded to Hess "for his discovery of the functional organization of the interbrain [hypothalamus] as a coordinator of the activities of the internal organs." Experiments like those of Bard and Hess led to the important conclusion that the basic circuits for organized behaviors accompanied by emotion exist in the diencephalon and the parts of the brainstem connected to it. Furthermore, their work showed that the control of the autonomic nervous system is not entirely separable from the control of other effector pathways. Stimulation that elicits autonomic responses almost invariably elicits somatic motor responses as well.

The routes by which the hypothalamus (and other forebrain structures) influence the autonomic and somatic motor systems are complex. Major targets of the hypothalamus lie in the reticular formation, the tangled web of cells and fibers in the core of the brainstem. This structure contains over 100 identifiable cell groups, including some of the nuclei that control the brain states associated with sleep and wakefulness described in the previous chapter. Other important circuits in the reticular formation are those that control cardiovascular function, respiration, urination, vomiting, and swallowing. The reticular neurons receive hypothalamic input and feed into both somatic and autonomic effector systems in the brainstem and spinal cord. Their activity can therefore produce widespread autonomic and somatic responses, often overriding reflex function and sometimes involving almost every organ in the body.

Unified emotional behavior is thus achieved through the convergence in the brainstem reticular formation of hypothalamic and other descending

Box A
THE HYPOTHALAMUS

The hypothalamus is located at the base of the forebrain, bounded by the optic chiasm rostrally, and the midbrain tegmentum caudally. This critical structure forms the floor and ventral walls of the third ventricle and is continuous through the infundibular stalk with the posterior pituitary. Because of its central position in the brain and its proximity to the pituitary, it is not surprising that the hypothalamus integrates information from the forebrain, brainstem, spinal cord, and various endocrine systems.

The hypothalamus is made up of a large number of small but distinct nuclei, each with its own complex pattern of connections and functions. The nuclei do not function separately from each other, however, since there are intricate interconnections among them. The hypothalamic nuclei are grouped in three longitudinal regions called the periventricular, medial, and lateral zones. Nuclei in the periventricular zone include the paraventricular and supraoptic nuclei, which contain neurosecretory neurons whose axons extend into the posterior pituitary. With appropriate stimulation, these neurons secrete oxytocin or vasopressin (also known as antidiuretic hormone) into the bloodstream. Scattered neurons in the periventricular zone (as well as in the other zones) manufacture peptides known as releasing or inhibiting factors that control secretion of hormones by the anterior pituitary. The axons of these neurons project to the median eminence, a region at the junction of the hypothalamus and pituitary stalk, where they secrete their peptides into the portal circulation that supplies the anterior pituitary. Other neurons in the paraventricular nucleus project to the brainstem and spinal cord, where they innervate preganglionic autonomic neurons. The

periventricular zone also contains the suprachiasmatic nucleus, which receives direct retinal input and drives circadian rhythms, both visceral and behavioral (see Chapter 26). The periventricular zone nuclei receive massive inputs from the other hypothalamic zones. The medial zone nuclei, including the dorsomedial and ventromedial nuclei, and the nuclei of the mammillary bodies are involved in feeding, reproductive and parenting behavior (see Chapter 28), thermoregulation, and water balance. A

major goal of current studies of the hypothalamus and associated structures is the identification of specific neuronal circuits that control such functions. The medial zone nuclei receive inputs from structures of the limbic system and from visceral sensory nuclei of the brainstem. The lateral zone of the hypothalamus is best considered a rostral continuation of the midbrain reticular formation. Thus, the neurons of the lateral zone are not grouped into nuclei, but are scattered among the fibers of the medial forebrain

Diagram of the human hypothalamus showing some of the nuclei in its medial (periventricular) zone.

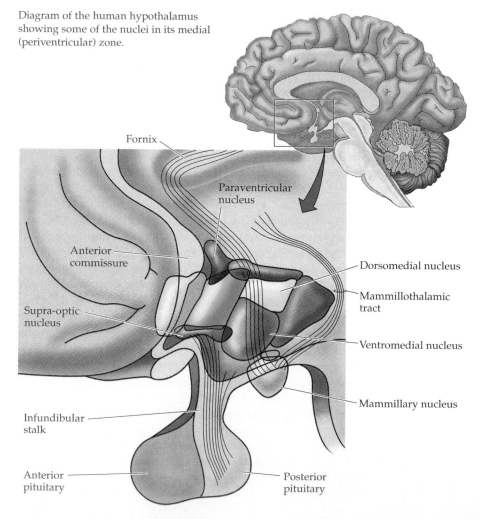

Fornix

Paraventricular nucleus

Anterior commissure

Dorsomedial nucleus

Mammillothalamic tract

Supra-optic nucleus

Ventromedial nucleus

Mammillary nucleus

Infundibular stalk

Anterior pituitary

Posterior pituitary

bundle, which runs through the lateral hypothalamus. These cells control behavioral arousal and shifts of attention, especially as related to reproductive activities and homeostasis.

In summary, the hypothalamus regulates a wide range of physiological and behavioral activities, including control of body temperature, sexual activity, reproductive endocrinology, and attack-and-defense (aggressive) behavior. The hypothalamic functions discussed in this chapter are therefore just one aspect of its many roles.

References

SWANSON, L. W. (1987) The hypothalamus. In *Handbook of Chemical Neuroanatomy*, Vol. 5, *Integrated Systems of the CNS*, Part I, *Hypothalamus, Hippocampus, Amygdala, Retina*. A Björklund and T Hökfelt (eds.), Amsterdam: Elsevier, pp. 1–124.

pathways that control motor neurons (Figure 27.2). A common clinical observation underscores the point that motor neurons can be activated by more than one pathway from the forebrain. Patients with unilateral facial paralysis due to damage of descending pathways from the motor cortex (upper motor neuron syndrome; see Chapter 16) are unable to move their lower facial muscles on one side, either voluntarily or in response to commands. Many of these individuals, however, produce completely symmetrical *involuntary* facial movements when they laugh, frown, or cry in response to amusing or distressing stimuli. In such patients, pathways from regions of the forebrain other than the motor cortex (the hypothalamus, for example) remain available to activate motor behavior in response to stimuli with emotional significance. In short, emotional behavior is effected through the activity of a

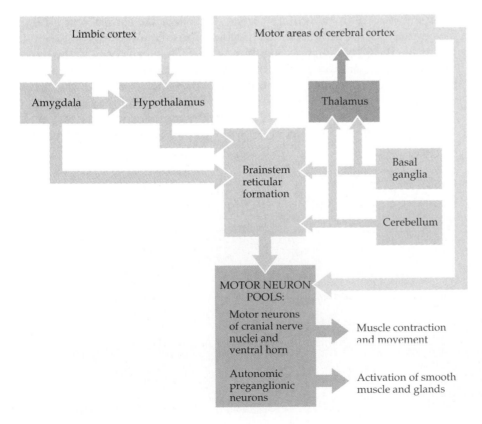

Figure 27.2 Diagram of the motor system (see Chapter 15), to which modulatory pathways discussed in this chapter have been added. In addition to direct projections from the motor cortex, motor neurons in the spinal cord and cranial nerve nuclei receive indirect inputs from parts of the limbic cortex via the hypothalamus, the amygdala, and the brainstem reticular formation. These connections mediate the expression of emotion.

final common pathway—the somatic and autonomic motor neurons—that integrate inputs from a variety of sources.

■ THE LIMBIC SYSTEM

"Higher" cortical processes (our awareness of an embarrassing situation, for example) obviously influence the emotions. The search for a link between the cerebral cortex and the effector systems that control emotional behavior has a long history. In 1937, James Papez first proposed that specific brain circuits are devoted to emotional experience and expression (much as the occipital cortex is devoted to vision). Seeking parts of the brain that might serve this function, he began to explore a region of the cortex known as the **limbic lobe**. In the 1850s, Paul Broca had popularized the term limbic lobe to refer to the part of the cerebral cortex that forms a rim around the corpus callosum on the medial aspect of the hemispheres (Figure 27.3). Two prominent components of the limbic lobe are the cingulate gyrus, which lies above the corpus callosum, and the hippocampus, which lies in the medial temporal lobe. For many years, the structures of the limbic lobe, along with the olfactory bulbs, were called the rhinencephalon; as the name suggests, they were thought to be devoted to the sense of smell (*rhino* means "of the nose").

Papez speculated that the function of the limbic lobe might be more interesting than the term *rhinencephalon* implies. His argument was based largely on neuroanatomy. He knew from the work of Bard and Hess that the hypothalamus influences the expression of emotion. Further, he knew that emotions reach consciousness and that higher cognitive functions affect emotional behavior. Ultimately, Papez showed that the cortex and hypothalamus are interconnected through pathways that came to be known as **Papez' circuit** (Figure 27.4). In this circuit, the hypothalamus (specifically, the **mammillary bodies**) projects to the **anterior nucleus of the dorsal thalamus**, which projects in turn to the **cingulate cortex**. The cingulate cortex (and a lot of other cortex as well) projects to the **hippocampus**. Finally, the hippocampus projects via the **fornix** (a large fiber bundle; see Chapter 1) back to the hypothalamus. Papez postulated that these pathways provided the connections necessary for cortical control of emotional expression. Over time, the circuit acquired some new elements and came to be called the **limbic system** (Figure 27.5). One of the most prominent of these newer components is the **amygdala** (Box B), a nuclear mass buried in the white matter of the temporal lobe, ros-

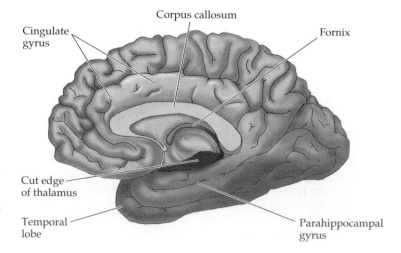

Figure 27.3 The limbic lobe, to use the initial terminology for this component of the brain, comprises the cortex on the medial aspect of the cerebral hemisphere that forms a rim around the corpus callosum. Thus, the limbic lobe includes the cingulate gyrus (lying above the corpus callosum) and the hippocampus, which is normally hidden by the parahippocampal gyrus, itself a part of the limbic lobe. These cortical areas are richly interconnected.

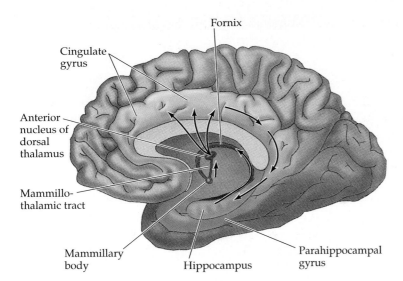

Figure 27.4 Papez's circuit. While not all his assignments of function to the various components of the circuit are accepted, Papez deserves credit as the first person to consider emotional functions in terms of their cerebral location; he was also correct in proposing that cortical structures could influence emotional behavior via these pathways.

tral to the hippocampus. Some of the structures that Papez originally described (the hippocampus, for example) appear to have little to do with emotional behavior; ironically, the amygdala, which was hardly mentioned by Papez, is now known to play a major role in emotional control.

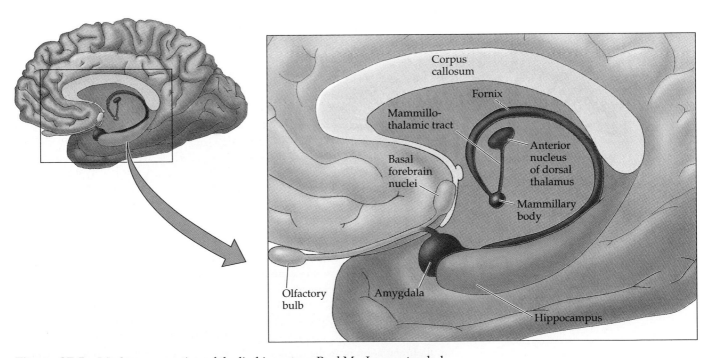

Figure 27.5 Modern conception of the limbic system. Paul MacLean extended Papez' suggestion that the limbic lobe was involved in emotional expression, noting that the elements of Papez' circuit were interconnected with many other parts of the central nervous system, some of which are illustrated here. These interconnected structures, together with the cortex of the cingulate and parahippocampal gyri, have come to be known as the limbic system. An especially important component of the limbic system is the amygdala, which has extensive interconnections with both the hypothalamus and the cerebral cortex. Although some investigators consider the concept of a limbic system to be outmoded, no newer idea has emerged to replace it; thus the concept of the limbic system remains an important one in discussions of emotion.

Box B
THE AMYGDALA

The amygdala occupies the rostral pole of the temporal lobe. It is made up of a mass of gray matter buried in the cerebral hemisphere, as well as some associated cortex on the medial aspect of the hemispheric surface. The amygdala (or amygdaloid complex, as it is often called) comprises three subdivisions, each of which has a unique set of connections with other parts of the brain. The corticomedial group of nuclei has extensive connections with the olfactory bulb and the olfactory cortex. The basolateral group, which is especially large in

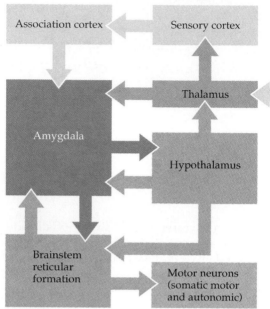

The major connections of the amygdala. Linkages with the association areas of cortex, the phylogenetically older limbic cortex, and the visceral sensory and effector nuclei of the brainstem allow the amygdala to coordinate many aspects of behavior, thus organizing the responses associated with emotions.

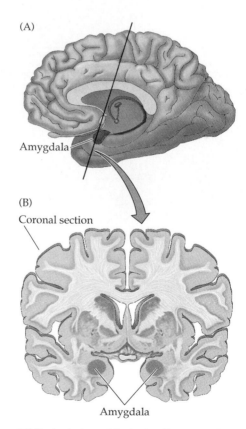

(A) Sagittal view of the brain, illustrating the location of the amygdala in the temporal lobe. The line indicates the level of the section in (B). (B) Coronal section through the forebrain at the level of the amygdala.

humans, has its main connections with the cerebral cortex, especially the sensory association areas. The central and anterior group of nuclei is characterized by connections with the brainstem and hypothalamus and with viscerosensory areas such as the nucleus of the solitary tract. Two major projection pathways carry information into and out of the amygdala: the ventral amygdalofugal system, a diffuse collection of fibers, and the stria terminalis, a more compact fiber bundle.

The amygdala links those cortical regions that process sensory information with hypothalamic and brainstem effector systems. Cortical inputs provide information about highly processed visual, somatic sensory and auditory stimuli. Innervation from the insula, a cortical area that processes visceral sensory input, has also been demonstrated. These cortical inputs distinguish the amygdala from the hypothalamus, which receives only relatively unpro-

cessed sensory inputs. The amygdala also receives sensory input directly from some thalamic nuclei and from the olfactory bulb and the nucleus of the solitary tract in the brainstem. Physiological studies have confirmed this convergence of sensory information. Thus, many neurons in the amygdala respond to visual, auditory, somatic sensory, gustatory, and olfactory stimuli. Moreover, highly complex stimuli (faces, for instance) are often required to evoke a response. Projections from the amygdala to the hypothalamus and brainstem (and possibly as far as the spinal cord) allow it to control activity in both the autonomic and somatic motor systems.

Reference

PRICE, J. L., F. T. RUSSCHEN AND D. G. AMARAL (1987) The limbic region II: The amygdaloid complex. In *Handbook of Chemical Neuroanatomy*, Vol. 5, *Integrated Systems of the CNS*, Part I, *Hypothalamus, Hippocampus, Amygdala, Retina*. A Björklund and T Hökfelt (eds.), Amsterdam: Elsevier, pp. 279–388.

About the same time that Papez proposed that the limbic lobe and associated structures were important for the integration of emotional behavior, Heinrich Klüver and Paul Bucy were carrying out a series of experiments on rhesus monkeys in which they removed a large part of both medial temporal lobes, thus destroying much of Papez' circuit. They reported a set of abnormalities in the behavior of these animals that is now known as the Klüver-Bucy syndrome (Box C). Among the most prominent changes they observed was visual agnosia: the animals appeared to be unable to recognize objects, although they were not blind. This deficit is similar to that seen in certain human patients following lesions of the temporal cortex (see Chapter 24). In addition, the monkeys displayed bizarre oral behaviors. For instance, these animals would put all sorts of objects into their mouths, including things a normal monkey would have nothing to do with. They also exhibited hyperactivity and hypersexuality, approaching and making physical contact with virtually everything in their environment. Finally, there were marked changes in the emotional behavior of the monkeys. Because they had been caught in the wild, these monkeys typically reacted with hostility and fear to humans before their surgery. Postoperatively, however, they were tame. Motor and vocal reactions generally associated with anger or fear were no longer elicited by the approach of humans, and the monkeys now showed little or no excitement when the experimenters handled them. Nor did they show fear when presented with a snake—a strong aversive stimulus for a normal rhesus monkey.

Klüver and Bucy concluded that this marked change in behavior was at least partly due to the interruption of the pathways described by Papez. A similar syndrome has been described in humans who have suffered bilateral damage of the temporal lobes. When it was later demonstrated that the emotional disturbances of the Klüver-Bucy syndrome could be elicited by removal of the amygdala alone, attention turned more specifically to the role of this structure in the control of emotional behavior.

∎ THE INFLUENCE OF THE AMYGDALA ON EMOTIONAL BEHAVIOR

Experiments performed by J. L. Downer at University College, London vividly demonstrate the importance of the amygdala in aggressive behavior. Downer removed one amygdala in rhesus monkeys, at the same time transecting the optic chiasm and all of the commissures that link the two hemispheres (see Chapter 25). In so doing, he produced an animal with a single amygdala that had access only to visual inputs from the eye on the same side. Downer found that behavior depended on which eye was used to view the world. When the eye on the side of the intact amygdala was covered, the animals behaved very much like the monkeys described by Klüver and Bucy; that is, they were quite placid in the presence of humans, even approaching the experimenter to take raisins from his hand. If, however, the eye on the side of the remaining amygdala was allowed to see, the animals reverted to their normal fearful and aggressive behavior. As Downer reported, "When observers looked into its cage, the animal would bare its teeth and dash forward, jumping at the door and attempting to bite and claw." Thus, in the absence of the amygdala, a monkey does not seem to interpret the significance of the visual stimulus presented by an approaching human in the same way as does an intact animal. Only visual stimuli presented to the eye on the side of the ablation produced this abnormal state; if the animal was *touched* on either side, a full aggressive reaction occurred, implying that somatic

Box C
THE REASONING BEHIND AN IMPORTANT DISCOVERY*

Paul Bucy explains why he and Heinrich Klüver removed the temporal lobes in monkeys.

When we started out, we were not trying to find out what removal of the temporal lobe would do, or what changes in behavior of the monkeys it would produce. What we found out was completely unexpected! Heinrich had been experimenting with mescaline. He had even taken it himself and had experienced hallucinations. He had written a book about mescaline and its effects. Later Heinrich gave mescaline to his monkeys. He gave everything to his monkeys, even his lunch! He noticed that the monkeys acted as though they experienced paraesthesias in their lips. They licked, bit and chewed their lips. So he came to me and said, "Maybe we can find out where mescaline has its actions in the brain." So I said, "OK."

We began by doing a sensory denervation of the face, but that didn't make any difference to the mescaline-induced behavior. So we tried motor denervation. That didn't make any difference, either. Then we had to sit back and think hard about where to look. I said to Heinrich, "This business of licking and chewing the lips is not unlike what you see in cases of temporal lobe epilepsy. Patients chew and smack their lips inordinately. So, let's take out the uncus." Well, we could just as well take out the whole temporal lobe, including the uncus. So we did.

We were especially fortunate with our first animal. This was an older female. ... She had become vicious—absolutely nasty. She was the most vicious animal you ever saw; it was dangerous to go near her. If she didn't hurt you, she would at least tear your cloth-

ing. She was the first animal on which we operated. I removed one temporal lobe. ... The next morning my phone was ringing like mad. It was Heinrich, who asked, "Paul, what did you do to my monkey? She is tame!" Subsequently, in operating on non-vicious animals, the taming effect was never so obvious.

That stimulated our getting the other temporal lobe out as soon as we could evaluate her. When we removed the other temporal lobe, the whole syndrome blossomed.

*Excerpt from an interview of Bucy by K. E. Livingston in 1981. K. E. Livingston (1986) Epilogue: Reflections on James Wenceslas Papez, According to Four of his Colleagues. In *The Limbic System: Functional Organization and Clinical Disorders*. B. K. Doane and K. E. Livingston, New York: Raven Press.

sensory information about both sides of the body was reaching the remaining amygdala. These results indicate that the amygdala forms a critical link in the processes that invest sensory experience with emotional significance.

To better understand the role of the amygdala in evaluating stimuli and to define more precisely the specific circuits and mechanisms involved, several other animal models of emotional behavior have been developed. One of the most useful of these is based on conditioned fear in rats. Conditioned fear develops when an initially neutral stimulus is repeatedly paired with an aversive one. Over time, the animal begins to respond to the neutral stimulus with behaviors similar to those elicited by the threatening stimulus. That is, the animal learns to attach a new meaning to the stimulus. Studies of the parts of the brain involved in the development of conditioned fear in such animals have begun to shed some light on the processes underlying human anxiety. Indeed, a rat's response to threatening stimuli is remarkably similar to the responses of people placed in anxiety-provoking situations (Table 27.1). Joseph LeDoux and his colleagues trained rats to associate a tone with a foot shock delivered shortly after onset of the sound. To assess the animals' responses, they measured blood pressure and the length of time the animals crouched without moving (a behavior called "freezing"). Before

TABLE 27.1
Signs of Fear in Animals Compared with Signs and Symptoms of Anxiety in Humans[a]

Signs of fear in animals	Signs and symptoms of anxiety in humans
Increased heart rate and stroke volume	Increased heart rate and stroke volume
Decreased salivation	Dry mouth
Gastric ulcers	Upset stomach
Altered respiratory rate	Increased respiratory rate
Scanning and vigilance	Scanning and vigilance
Increased startle response	Jumpiness, easily startled
Frequent urination	Frequent urination
Frequent defecation	Diarrhea
Increased grooming	Fidgeting
Periods of immobility ("freezing")	Apprehension (expectation that something bad is going to happen)

(After Davis, 1992.)

[a]List of measures in animals typically used as indices of fear compared with the criteria that characterize generalized anxiety in humans as summarized in the standard manual of psychiatric diagnosis, *The Diagnostic and Statistical Manual of Mental Disorders* (DSM-IV).

training, the rats did not react to the tone, nor did their blood pressure change when the tone was presented. After training, however, the onset of the tone caused a marked increase in blood pressure and prolonged periods of behavioral freezing. Using this paradigm, LeDoux determined the neural circuitry that established the association between the tone and fear (Figure 27.6). First, he demonstrated that the medial geniculate nucleus is necessary for the development of the conditioned fear response. This result is not surprising, since all auditory information that reaches the forebrain travels through the medial geniculate nucleus of the dorsal thalamus (see Chapter 12). He went on to show, however, that the responses were still elicited if the

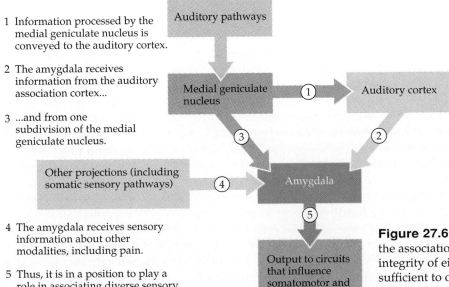

1 Information processed by the medial geniculate nucleus is conveyed to the auditory cortex.

2 The amygdala receives information from the auditory association cortex...

3 ...and from one subdivision of the medial geniculate nucleus.

4 The amygdala receives sensory information about other modalities, including pain.

5 Thus, it is in a position to play a role in associating diverse sensory inputs, leading to new behavioral and autonomic responses.

Figure 27.6 Pathways in the rat brain that mediate the association of auditory and aversive stimuli. The integrity of either auditory pathway to the amygdala is sufficient to obtain fear conditioning; if both paths are damaged, however, conditioned fear to auditory stimuli cannot be learned.

connections between the medial geniculate and auditory cortex were severed, leaving only a projection between the medial geniculate and the amygdala. Furthermore, if the part of the medial geniculate that projected to the amygdala was also destroyed, the fear responses were abolished. Subsequent work in LeDoux's laboratory established that projections from the amygdala to the midbrain reticular formation are important for the expression of freezing behavior, and that projections from the amygdala to the hypothalamus control the rise in blood pressure.

Since the amygdala is a site where neural activity produced by both tones and shocks can be processed (because of the convergence of multiple sensory pathways there; see Box C), it is reasonable to suppose that it is also the site where learning about fearful stimuli occurs. The discovery that long-term potentiation (LTP; see Chapter 23) can be evoked in the amygdala has strengthened the suggestion that associative learning takes place there. Indeed, the acquisition of conditioned fear in rats is blocked by infusion into the amygdala of NMDA antagonists (which block LTP). These results among others, have led to the broader hypothesis that the amygdala participates in establishing associations between stimuli and rewards, thus helping the animal evaluate the emotional significance of events in its environment.

■ THE NEOCORTEX AND EMOTIONAL EXPRESSION

As Papez' work on the limbic system suggested, understanding the neural basis of emotions requires an appreciation of the role of the cerebral cortex. In animals like the rat, most behavioral responses are highly stereotyped. In more complex brains, however, individual experience is increasingly influential in determining responses to particular stimuli. In humans, a stimulus that evokes fear or sadness in one person may have no effect on the emotions of another (Box D). Although the neuronal circuitry and mechanisms of such idiosyncratic responses are not yet understood, clinical observations and the results of psychological experimentation have already provided some insight into higher order emotional processing. Thus, damage to specific parts of the human cerebral hemispheres can result in mood disorders; and, as in many other aspects of cortical function, the two hemispheres make different contributions to the governance of emotion (see also Chapters 24 and 25).

Emotionality appears to be lateralized in the cerebral hemispheres in at least two ways. First, as discussed in Chapter 25, parts of the right hemisphere are especially important for the expression and comprehension of the affective aspects of speech. Thus, patients with damage to the supra-Sylvian portions of the posterior frontal and anterior parietal lobes on the right side may lose the ability to express emotion by modulation of their speech patterns. (Such a loss of emotional expression is referred to as **aprosody**; a similar lesion on the left side would give rise to a Broca's aphasia.) Patients exhibiting aprosody tend to speak in a monotonous voice, no matter what the circumstances or meaning of what is said. One such patient, a teacher, was no longer able to maintain classroom discipline by modulating the sound of her voice. Because her pupils (and even her own children) couldn't tell when she was angry or upset, she had to resort to adding phrases such as "I am angry and mean it" to the end of her remarks. The wife of another patient insisted that her husband no longer loved her because he could not imbue his speech with cheerfulness or affection. Although such patients cannot express emotion in speech, they experience normal emotional feelings.

A second body of evidence about asymmetrical hemispheric function in emotionality relates to the establishment of mood. Both clinical and

Box D
AFFECTIVE DISORDERS

Whereas some degree of disordered emotion is present in virtually all psychiatric problems, in affective (mood) disorders, the essence of the disease is an abnormal regulation of the feelings of sadness and happiness that are part of everyone's life. The most severe of these afflictions are major depression and manic depression. (Manic depression is also called bipolar disorder, since such patients experience alternating episodes of depression and euphoria.) Depression, the most common of the major psychiatric disorders, has a lifetime prevalence of about 5–8% in the population; if one includes bipolar disorder, the lifetime prevalence of affective disorders rises to nearly 10%! For clinical purposes, depression (as distinct from bereavement or neurotic unhappiness) is defined by a set of standard criteria, one or more of which must be present to tender the diagnosis. In addition to an abnormal sense of sadness, despair, and bleak feelings about the future (depression itself), these criteria include disordered eating and weight control; disordered sleeping (insomnia or hypersomnia); and diminished sexual interest. The overwhelming quality of major depression has been compellingly described by patient/authors such as William Styron, and by afflicted psychologists such as Kay Jamison. But the depressed patient's profound sense of despair has been nowhere better expressed than by Abraham Lincoln, who during a period of depression said:

> I am now the most miserable man living. If what I feel were equally distributed to the whole human family, there would not be one cheerful face on earth. Whether I shall ever be better, I cannot tell; I awfully forebode I shall not. To remain as I am is impossible. I must die or be better, it appears to me.

Indeed, about half the suicides in this country occur in individuals with clinical depression.

In earlier times, depression and mania were considered disorders that arose from circumstances or the neurotic inability to cope with normal problems. It is now universally accepted that these conditions are neurobiological disorders. Among the strongest lines of evidence for this consensus are studies of the inheritance of these diseases. For example, the concordance of affective disorders is very high in monozygotic compared to dizygotic twins.

Despite evidence for a genetic predisposition to affective disorders, the cause remains unknown. The efficacy of a large number of drugs that influence catecholaminergic and serotonergic neurotransmission strongly implies that the basis of the disorder is ultimately neurochemical. The majority of patients (about 70%) can be effectively treated with one of a variety of drugs (including tricyclic antidepressants, monoamine oxidase inhibitors and selective serotonin reuptake inhibitors), which are among the most widely prescribed agents worldwide.

Pharmaceutical companies have now succeeded in synthesizing drugs that selectively block the uptake of serotonin without affecting the uptake of other neurotransmitters. Three such inhibitors—fluoxetine (Prozac®), sertraline (Zoloft®), and paroxetine (Paxil®)—are effective in treating depression and have few of the side effects of the older, less specific drugs. Perhaps the best indicator of the success of these drugs has been their wide acceptance: although Prozac® was approved for clinical use only in the late 1980s, it is now the second-largest selling drug in the United States.

Most depressed patients who use drugs such as Prozac® report that they lead fuller lives and are much more energetic and organized. Based on such information, Prozac® is now used not only to combat depression but also to "treat" individuals who have no definable psychiatric disorder. This abuse raises important social questions; Prozac® has sometimes been compared to "Soma," the mythical drug routinely administered to bring pharmacological bliss to the inhabitants of Aldous Huxley's *Brave New World*.

References

BREGGIN, P. R. (1994) *Talking Back to Prozac: What Doctors Won't Tell You about Today's Most Controversial Drug*. New York: St. Martin's Press.

FREEMAN, P. S., D. R. WILSON AND F. S. SIERLES (1993) Psychopathology. In *Behavior Science for Medical Students*, F. S. Sierles (ed.). Baltimore: Williams and Wilkins, pp. 239–277.

GREENBERG, P. E., L. E. STIGLIN, S. N. FINKELSTEIN AND E. R. BERNDT (1993) The economic burden of depression in 1990. J. Clin. Psychiatry 54: 405–424.

JAMISON, K. R. (1995) *An Unquiet Mind*. New York: Alfred A. Knopf.

JEFFERSON, J. W. AND J. H. GRIEST (1994) Mood disorders. In *Textbook of Psychiatry*, J. A. Talbott, R. E. Hales and S. C. Yudofsky (eds.). Washington: American Psychiatric Press, pp. 465–494.

ROBINS, E. (1981) *The Final Months. A Study of the Lives of 134 Persons Who Committed Suicide*. New York: Oxford University Press.

STYRON, W. (1990) *Darkness Visible. A Memoir of Madness*. New York: Random House.

WONG, D. T. AND F. P. BYMASTER (1995) Development of antidepressant drugs: Fluoxetin (Prozac®) and other selective serotonin uptake inhibitors. Adv. Exp. Med. Biol. 363: 77–95.

WONG, D. T., F. P. BYMASTER AND E. A. ENGLEMAN (1995) Prozac® (fluoxetine, Lilly 110140), the first selective serotonin uptake inhibitor and an antidepressant drug: Twenty years since its first publication. Life Sci. 57(5): 411–441.

WURTZEL, E. (1994). *Prozac Nation: Young and Depressed in America*. Boston: Houghton-Mifflin.

Figure 27.7 Smiles on some famous faces. Studies of normal subjects indicate that facial expressions are more quickly and fully expressed by the left facial musculature compared to the right, as is apparent in these examples. Since the left lower face is governed by the right hemisphere, some psychologists have suggested that the majority of humans are "left-faced," in the same general sense that most of us are right-handed. (After Moscovitch and Olds, 1982.)

experimental studies support the idea that the left hemisphere is more concerned with what can be thought of as positive emotions, whereas the right hemisphere is more concerned with negative emotions. For example, the incidence and severity of depression is significantly higher in patients with lesions of the left anterior hemisphere compared to any other location. In contrast, patients with lesions of the right anterior hemisphere are often described as unduly cheerful. These observations suggest that lesions in the left hemisphere may result in the loss of positive feelings, leading to depression, whereas lesions of the right hemisphere may result in the loss of negative feelings, leading to inappropriate optimism.

Hemispheric asymmetry related to mood is also observable in normal individuals. For instance, listening experiments that introduce sound into one ear or the other indicate a right-hemisphere superiority in detecting the emotional nuances of speech (or even nonspeech sounds). Moreover, when facial expressions are presented tachistoscopically to either the right or the left hemifield, the emotions depicted are more readily and accurately identified from the information in the left hemifield (that is, the hemifield perceived by the right hemisphere; see Chapters 11 and 25). Kinematic studies of facial expressions also show that most right-handers more quickly and fully express emotions with the left facial musculature than with the right (recall that the left lower face is controlled by the right hemisphere, and vice versa) (Figure 27.7). Taken together, this evidence is consistent with the idea that the right hemisphere is more intimately involved with both the perception and expression of emotions than is the left hemisphere, although as in the case of other lateralized behaviors (language, for instance) both hemispheres participate.

■ SUMMARY

The word *emotion* covers a wide range of states that have in common the association of physiological responses, expressive behavior, and distinct subjective feelings. The physiological responses are mediated largely by the autonomic nervous system, which is itself regulated by inputs from many other parts of the brain. The organization of the expressive behavior associated with emotion appears to be carried out by circuits in the limbic system, which includes the hypothalamus, and the amygdala, as well as several other cortical and subcortical structures. Although a good deal is known about the neuroanatomy and transmitter chemistry of the different parts of the limbic system, there is still a dearth of information about how this circuitry mediates specific emotional functions. On a broader scale, a variety of evidence indicates that the two hemispheres are differently specialized for the governance of emotion, the right hemisphere being the more important in this regard. The prevalence and social significance of human emotions and their disorders (depression, for example) ensures that knowledge about the neurobiology of emotions will continue to grow rapidly.

Additional Reading

Reviews

APPLETON, J. P. (1993) The contribution of the amygdala to normal and abnormal emotional states. Trends Neurosci. 16: 328–333.

CAMPBELL, R. (1986) Asymmetries of facial action: Some facts and fancies of normal face movement. In *The Neuropsychology of Face Perception and Facial Expression*, R Bruyer (ed.). Hillsdale, NJ: Erlbaum, pp. 247–267.

DAVIS, M. (1992) The role of the amygdala in fear and anxiety. Annu. Rev. Neurosci. 15: 353–375.

LeDOUX, J. E. (1987) Emotion. In *Handbook of Physiology. Section 1. The Nervous System*, Vol. 5. American Physiological Society, pp. 419–459.

SMITH, O. A. AND J. L. DeVITO (1984) Central neural integration for the control of autonomic responses associated with emotion. Annu. Rev. Neurosci. 7: 43–65.

Important Original Papers

BARD, P. (1928) A diencephalic mechanism for the expression of rage with special reference to the sympathetic nervous system. Am. J. Physiol. 84: 490–515.

DOWNER, J. L. DE C. (1961) Changes in visual agnostic functions and emotional behaviour following unilateral temporal pole damage in the 'split-brain' monkey. Nature 191: 50–51.

EKMAN, P., R. W. LEVENSON AND W. V. FRIESEN (1983) Autonomic nervous system activity distinguishes among emotions. Science 221: 1208–1210.

KLÜVER, H. AND P. C. BUCY (1939) Preliminary analysis of functions of the temporal lobes in monkeys. Arch. Neurol. and Psychiat. 42: 979–1000.

MACCLEAN, P. D. (1964) Psychosomatic disease and the "visceral brain": Recent developments bearing on the Papez theory of emotion. In *Basic Readings in Neuropsychology*, R. L. Isaacson (ed.). New York: Harper & Row, Inc., pp. 181–211.

PAPEZ, J. W. (1937) A proposed mechanism of emotion. Arch. Neurol. Psychiat. 38: 725–743.

ROSS, E. D. AND M.-M. MESULAM (1979) Dominant language functions of the right hemisphere? Prosody and emotional gesturing. Arch. Neurol. 36: 144–148.

Books

APPLETON, J. P. (ed.) (1992) *The Amygdala. Neurobiological Aspects of Emotion, Memory and Mental Dysfunction*. New York: Wiley-Liss.

CORBALLIS, M. C. (1991) *The Lopsided Ape: Evolution of the Generative Mind*. New York: Oxford University Press.

DARWIN, C. (1890) *The Expression of Emotion in Man and Animals*, 2nd Ed. In *The Works of Charles Darwin*, Vol. 23, 1989. London: William Pickering.

HELLIGE, J. P. (1993) *Hemispheric Asymmetry: What's Right and What's Left*. Cambridge, MA: Harvard University Press.

LOEWY, A. D. AND K. M. SPYER (1990) *Central Regulation of Autonomic Functions*. New York: Oxford University Press.

CHAPTER 28

SEX, SEXUALITY, AND THE BRAIN

◼ OVERVIEW

"Vive la difference." "Isn't that just like a (wo)man?" "It's on the Y chromosome." These are all expressions used to denote pleasure (or displeasure) with behavioral differences associated with the sexes. While some of these distinctions may be rooted in societal expectations or individual biases, they also arise because the brains of females and males are in some respects different. These differences in central nervous system structure are called sexual dimorphisms. In the rat, the animal in which most experimental work has been done, several brain structures of females and males differ in size and number of constituent neurons. In humans and other primates, however, structural differences are less pronounced and, as a result, more controversial. In both rodents and humans, sexually dimorphic brain structures tend to cluster around the third ventricle in the anterior hypothalamus, although they are also discernible in some other cerebral structures. The functional consequences of sexual dimorphisms in rodents are reasonably well understood. In humans, the significance of such brain dimorphisms is less clear; nonetheless, they provide an increasingly plausible basis for a variety of behaviors that are different in the two sexes.

◼ SEXUALLY DIMORPHIC BEHAVIOR

A wide range of animal behaviors are **sexually dimorphic** (*dimorphic* means "two forms"). Some of these behaviors are reflex in nature, while others require a high level of cognitive activity. In rodents, examples of sexually dimorphic reflex behaviors are often related to the sex act: priming of the genitalia for sexual intercourse, a stereotypical position assumed while having sex, and the drive for sex. Somewhat more complex sexual behaviors are those involved in the outcome of sex, such as taking care of the young and other parental behaviors. In humans, still more complex examples are behaviors associated with sexual affiliation and the choice of sexual partner. Other human behaviors vary according to sex but are not related directly to sexual or reproductive function; examples are spatial thinking and the use of language.

All of these behaviors, from simple reflexes to complex mentation, are to some extent sexually dimorphic. Because all behaviors are based on the details of the underlying neural circuitry, neurobiologists have long looked for differences between the brains of females and males that might explain sexually dimorphic behaviors. Just as body structure differs between females and males, so do certain brain structures. In both rodents and primates, including humans, various brain nuclei and fiber tracts differ in size, number of neurons, and synaptic architecture according to sex. These brain differences, like behavioral differences, have been called sexually dimorphic. Bear in mind, however, that while brain differences in rodents often take two distinct forms in females and males, in humans these differences usually vary along a continuum.

Box A
THE DEVELOPMENT OF MALE AND FEMALE PHENOTYPES

The presence of two X chromosomes or an X and Y chromosome in the cells of an embryo sets in motion events that ultimately affect the development of the brain. These neural effects are determined by the production of estrogens or androgens, which depends in turn on the presence of either female or male gonads.

The early stages of human embryonic development follow a plan that produces common precursors for the gonads. At about the sixth week of gestation, the primordial gonads have formed from somatic mesenchyme tissue, near the developing kidneys. Cells in the gonads differentiate into supporting and hormone-producing cells; the cells that will become the gametes (germ cells that give rise to ova and sperm) migrate from elsewhere in the embryo (the yolk sac). Attached to the primordial gonads are two sets of tubes that are the progenitors of the internal genitalia. These are the Müllerian and Wolffian ducts. Developing simultaneously is an undifferentiated structure called the urogenital groove, the progenitor of the external genitalia.

The primary genetic influence on the development of male gonads is the sex determining region on the Y chromosome. In males, this region of the Y chromosome (the *Sry* gene) becomes activated. The *Sry* gene produces a protein called testicular determining factor, or TDF, which instructs the testes to begin development. It is not clear whether the switch that turns on the *Sry* gene is on the X or Y chromosome, but if it is activated, the primordial gonads develop into testes. In the absence of TDF (that is, in XX embryos), the indifferent gonad differentiates into an ovary, the Wolffian ducts degenerate, and the Müllerian ducts develop into the oviducts, uterus, and cervix. Tissue

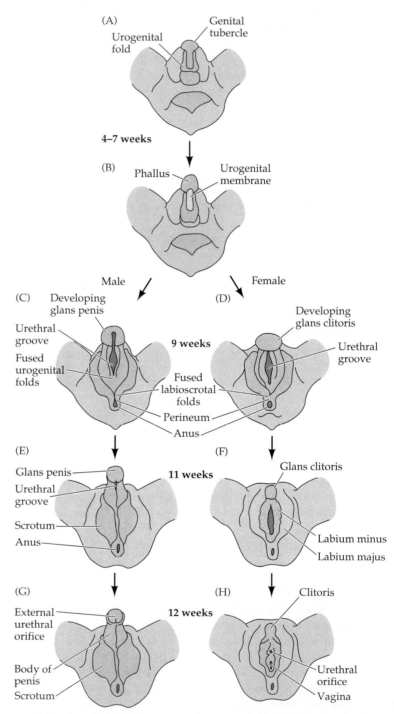

Development of female and male external genitalia. A and B show the indifferent stage during weeks 4–7 of gestation. D, F, and H show differentiation in the female genitalia at weeks 9, 11, and 12, respectively. C, E, and G show differentiation into male genitalia at the same intervals. (After Moore, 1977.)

around the urogenital groove becomes the clitoris, labia, and vagina.

In the presence of TDF (in XY embryos), the undifferentiated gonad becomes testes, whose cells secrete testosterone and Müllerian-inhibiting hormone (MIH). MIH prevents the Müllerian ducts from developing and allows the Wolffian ducts to develop into the epididymis, vas deferens, and seminal vesicles. Tissue around the urogenital groove becomes the penis and scrotum. In short, development of the female phenotype depends on the absence of TDF and the consequent absence of androgens during perinatal life. The early absence of androgens leads to the differentiation of the female body and brain. The presence of TDF and the consequent production of androgens early in life lead to the differentiation of the male body and brain.

References

JOHNSON, M. H. AND B. J. EVERITT (1988) *Essential Reproduction*, 3rd Ed. Oxford: Blackwell Scientific, pp. 1–34.

KOOPMAN, P., J. GUBBAY, N. VIVIAN, P. GOODFELLOW AND R. LOVELL-BADGE (1991) Male development of chromosomally female mice transgenic for *Sry*. Nature 351: 117–121.

SINCLAIR, A. H., P. BERTA, M. S. PALMER, J. R. HAWKINS, B. L. GRIFFITHS, M. J. SMITH, J. W. FOSTER, A. M. FRISCHAUF, R. LOVELL-BADGE AND P. N. GOODFELLOW (1990) A gene from the human sex-determining region encodes a protein with homology to a conserved DNA-binding motif. Nature 346: 240–242.

■ DEFINITIONS OF SEX

In keeping with the apparently continuous variation of sexually relevant brain structures, human sexual behaviors are also less stereotyped than those in rodents. As a result, the word *sex* has several different connotations: genotypic sex, phenotypic sex, and gender identification.

Genotypic sex is determined by the two sex chromosomes, X and Y. Most people have either two X chromosomes or an X and a Y chromosome; XX is a genotypic female, XY a genotypic male. **Phenotypic sex** is determined by the development of the internal and external genitalia (Box A). If everything goes according to plan during development, the XX genotype leads to a person with ovaries, oviducts, uterus, cervix, clitoris, labia, and vagina—a phenotypic female. By the same token, the XY genotype leads to a person with testicles, epididymis, vas deferens, seminal vesicles, penis, and scrotum—a phenotypic male. **Gender identification** is determined by the subjective perception of one's sex, which is often harder to define and can be influenced by societal expectations. Gender identity entails self-appraisal according to the traits most often associated with one sex or the other, called gender traits. For purposes of understanding the neurobiological issues related to sex, it is helpful to think of genotypic sex as being largely immutable, phenotypic sex as being modifiable (by developmental processes, hormones, or surgery), and gender identification as being a societal construct that an individual may or may not accept.

Once categorized in this way, it is clear that genotypic sex, phenotypic sex, and gender identification need not always be aligned. A mismatch can lead to medical problems, psychological problems, and sexual dysfunction. For instance, genotypic and phenotypic males or females may marry someone of the opposite sex, have children, and exhibit gender-typical behavior. However, some of these individuals may also suffer greatly because of discrepant gender identification, regarding themselves as members of the opposite sex. People with this sort of mismatch are called **transsexuals**. Some transsexuals perceive the discrepancy so strongly that they resort to surgery and hormone treatment to make their phenotypic sex more closely match their gender identification.

Other individuals are genotypically XY but phenotypically female, due to a defective gene for the androgen receptor, a condition called **androgen insensitivity syndrome** or **testicular feminization**. The receptor deficiency leads to the development of the internal genitalia of a male and the external genitalia of a female. Thus, people with androgen insensitivity syndrome describe themselves as female even though they have a Y chromosome; they are generally not aware of their condition until puberty, when they fail to menstruate. In this case, gender identity matches the external sexual phenotype but not genotype; the only major problem of individuals with androgen insensitivity syndrome is sterility.

Another sexual mismatch is genotypic males who are phenotypic females early in life, but whose sexual phenotype changes at puberty. As infants and children, these individuals are phenotypic females because they lack an enzyme, 5-α-reductase, that converts testosterone to dihydrotestosterone, the agent that promotes the early development of male genitalia. Such individuals develop somewhat ambiguous but generally female-appearing genitalia (they have labia with an enlarged clitoris and undescended testes). As a result, they are usually raised as females and their early gender identification is female. At puberty, however, when the testicular secretion of androgen becomes high, the clitoris develops into a penis and the testes descend, changing these individuals into phenotypic males. In the Dominican Republic, where this congenital syndrome has been thoroughly studied in a particular pedigree, this condition is referred to colloquially as "testes-at-twelve." Such individuals generally change their gender behavior at puberty, and most eventually assume a male role.

■ HORMONAL INFLUENCES ON SEXUAL DIMORPHISM

The development of sexual dimorphisms in the central nervous system is ultimately an outcome of genotypic sex. Normally, genotype determines the phenotype of the gonads; the gonads, in turn, are responsible for producing most of the circulating sex hormones. Because the gonads are dimorphic, the production of sex steroids during development is itself dimorphic; the testes begin making androgens and the ovaries begin making estrogens. The resulting differences in circulating hormones lead to a variety of differential effects on the development of females and males, including the development of the brain.

Hormonally generated sexual differences in brain development have been documented in experimental animals by administering testosterone to females, or by depriving males of testosterone by castrating them just after birth. For example, Geoffrey Raisman and Pauline Field, working at Oxford University, found a greater number of synapses on spines in the preoptic region of the hypothalamus in normal female rats compared to the equivalent region in males. Castrating males within 12 days of birth increased the density of these synapses to female levels, whereas administration of testosterone to developing females led to a reduction of preoptic spine synapses to male levels. Subsequently, Roger Gorski and his colleagues at the University of California at Los Angeles discovered a nucleus in the rodent hypothalamus that is sexually dimorphic, which (logically enough) they called the sexually dimorphic nucleus (SDN). The SDN is small in females and large in males. Its development is also under the influence of hormones; Gorski found that the SDN in male rats could be reduced in size to that of the female by castration within the first 2 weeks after birth. Similarly, the size of the female SDN could be increased to that of the male by early administration of androgens. These studies, as well as many others, show that the development of sexually dimorphic structures in the rodent brain is under the control of circulating sex hormones.

In general, the establishment of such brain dimorphisms in rodents is generated by different levels of hormones at different times in females and males (males have an early surge of testosterone, and females a later surge of estrogens). The active agent in both cases is actually **estradiol**. Although testosterone is popularly considered the "male" hormone and estrogen the "female" hormone, testosterone is converted to estradiol once it has entered the relevant neurons (Figure 28.1). Thus, it is an estrogen that acts inside neurons to stimulate sexually dimorphic patterns of neuronal circuitry as a result of its higher concentration in developing males. Female gonads do not produce a fetal surge of estradiol similar to the male surge of testosterone; were levels higher, estradiol would act in females to produce the neural circuitry of the male brain. Maternal blood is also rich in estrogens produced by the mother's gonads and placenta, which could affect the developing fetal brain. To mitigate this potential problem, many mammals have a circulating estrogen-binding protein called **α-fetoprotein**. In fetal blood and cerebrospinal fluid, α-fetoprotein binds circulating estrogens from any source. In

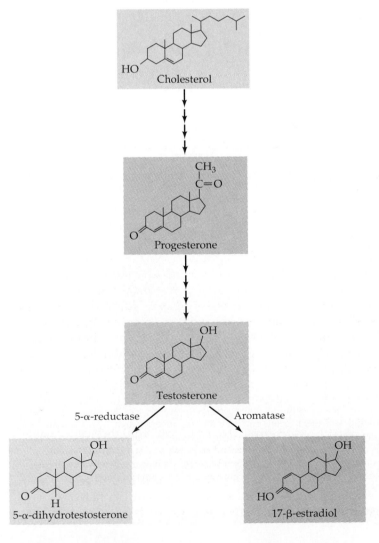

Figure 28.1 All sex steroids are synthesized from cholesterol. Cholesterol is first converted to progesterone, the common precursor, by four enzymatic reactions (represented by the four arrows). Progesterone can then be converted into testosterone via another series of enzymatic reactions; testosterone in turn is converted to 5-α-dihydrotestosterone via 5-α-reductase, or to 17-β-estradiol via an aromatase. 17-β-estradiol mediates most known hormonal effects in the brains of both female and male rodents.

Box B
THE ACTIONS OF SEX HORMONES

Sex hormones, which include progestagens, androgens, and estrogens, are all steroids derived from a common precursor, cholesterol (see Figure 28.1). Despite the tendency to speak of the these hormones as female or male, it is not really correct to think of estrogens as female and androgens as male; females and males synthesize *both* estrogens and androgens. What is important is the ratio of these two types of steroids in the circulation at any time, and the receptors available to bind them.

Because sex steroids are lipids, they do not need special membrane receptors to enter cells; they simply diffuse through the lipid bilayer of the membrane. However, neurons and other cells have the capacity to select, concentrate, and retain specific steroids by means of receptors and binding proteins in the cytoplasm and nucleus. Different areas of the adult brain have different steroid receptor patterns, with overlapping distributions of receptor types. Thus, particular brain regions can be targets for the actions of different classes of steroids (Figure A). For instance, estradiol receptors are sparsely distributed in the neocortex of the rat, but are prevalent in preoptic and hypothalamic areas and the anterior pituitary. Conversely, whereas receptors for 5-α-dihydrotestosterone (5-DHT) are found only in certain nuclei in the septum and hypothalamus, both estradiol and 5-DHT receptors are abundant in the frontal, prefrontal, and cingulate areas of the cortex. Some neurons express receptors for more than one steroid. Thus, all neurons with progesterone receptors also express estrogen receptors. As a result, hormones can have a synergistic effect; for example certain reproductive behaviors brought on by progesterone can be enhanced if estrogen is given first.

Steroids can have a "direct" effect

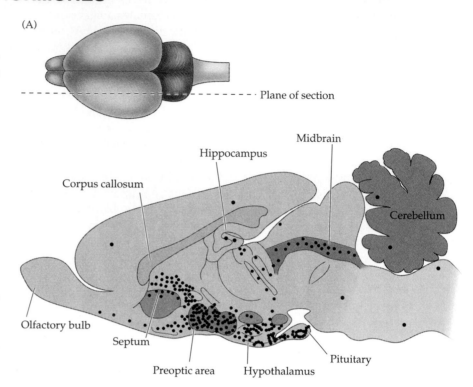

(A)

Plane of section

Midbrain
Hippocampus
Corpus callosum
Cerebellum
Olfactory bulb
Septum
Preoptic area
Hypothalamus
Pituitary

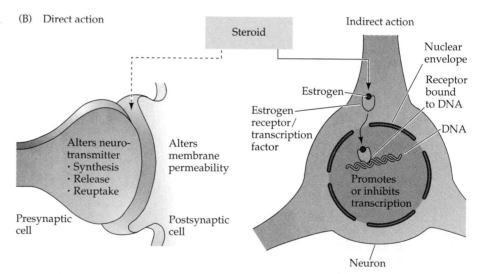

(B)   Direct action

Steroid

Indirect action

Nuclear envelope
Receptor bound to DNA
Estrogen
Estrogen receptor/ transcription factor
DNA
Alters neurotransmitter
· Synthesis
· Release
· Reuptake
Alters membrane permeability
Promotes or inhibits transcription
Presynaptic cell
Postsynaptic cell
Neuron

(A) Distribution of estradiol-sensitive neurons in a sagittal section of the rat brain. Animals were given radioactively labeled estradiol; dots represent regions where the label accumulated. In the rat, most estradiol-sensitive neurons are located in the preoptic area, hypothalamus, and amygdala. (B) Steroids have direct and indirect effects on neurons. Dashed line shows direct effects of hormones on the pre- or postsynaptic membrane, which alters neurotransmitter release, and affects neurotransmitter receptors. Solid line shows indirect effects of hormones, which act at the level of the nucleus to alter protein synthesis. (A after McEwen, 1976; B after McEwen et al., 1978.)

on neural activity by altering the permeability of the membrane to neurotransmitters and their precursors, or altering the functioning of the neurotransmitter receptors (Figure B). This type of effect has a latency to onset of seconds to minutes. As a consequence of these actions, sex steroids can modulate the efficacy of neural signaling.

Sex steroids can also have an indirect effect on neural activity by forming noncovalent bonds with the receptors, causing a conformational change that allows the receptor to bind to specific DNA recognition elements called hormone responsive elements. Consequently, hormones can alter gene expression, leading to changes in the synthesis of specific proteins (Figure B). Such hormonal actions are said to be "indirect," and their effects have a latency to onset of minutes to hours.

Most sexually dimorphic differences in the brains of females and males are thought to arise by the *indirect* actions of hormones on gene expression.

References

McEwen, B. S., P. G. Davis, B. S. Parsons and D. W. Pfaff (1979) The brain as a target for steroid hormone action. Annu. Rev. Neurosci. 2: 65–112.

Tsai, M.-J. and B. W. O'Malley (1994) Molecular mechanisms of action of steroid/thyroid receptor superfamily members. Annu. Rev. Biochem. 63: 451–486.

contrast, fetal testosterone has free access to steroid-sensitive neurons, where it is aromatized to estradiol.

These developmental studies in rats give some insight into the behavior of humans with congenital conditions that affect circulating levels of sex hormones. For example, women with **congenital adrenal hyperplasia** have overactive adrenals during development, which cause abnormally high levels of circulating androgens. Such women exhibit "tomboyish" behavior as children and tend to form homosexual relationships as adults. By analogy with the rodent studies, high levels of circulating androgens may stimulate sexually dimorphic brain circuitry to have a male rather than female organization, leading to more aggressive play and the eventual choice of a female sexual partner.

In short, the way that estrogens—the ultimate agents of both female and male sexual differentiation—mediate their effects on the brain is fairly well understood, at least in rodents. Estradiol and testosterone diffuse across the neuronal membrane where estradiol—or testosterone aromatized to estradiol—are bound by estrogen receptors. Estrogen receptors are transcription factors activated by estrogen binding. Thus, the receptors can influence gene transcription, and ultimately the development of sexually dimorphic neural circuits (see Box B).

(A)

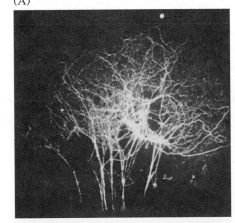

(B)

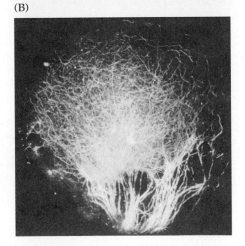

■ THE EFFECT OF ESTROGENS ON NEURAL CIRCUITRY

Early in life, estradiol can produce brain dimorphisms by increasing neuronal size, nuclear volume, dendritic length, dendritic branching, dendritic spine density, and number of synapses. One of the first demonstrations of such effects was provided by Dominique Toran-Allerand of Columbia University, who observed these phenomena after adding estrogens to fetal hypothalamic explants (Figure 28.2). Later in life, estrogens can continue to

Figure 28.2 Estrogen causes exuberant outgrowth of neurites in hypothalamic explants from newborn mice. (A) Control explant showing only a few silver-impregnated processes growing from the explant. (B) An estradiol-treated explant has many more neurites growing from its center. (From Toran-Allerand, 1978.)

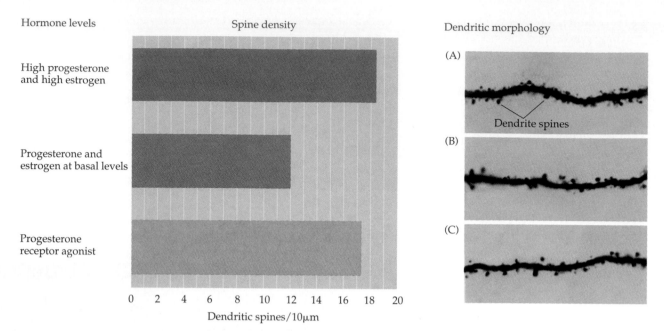

Hormone levels   Spine density   Dendritic morphology

High progesterone
and high estrogen

Progesterone and
estrogen at basal levels

Progesterone
receptor agonist

0   2   4   6   8   10   12   14   16   18   20

Dendritic spines/10μm

(A)

Dendrite spines

(B)

(C)

Figure 28.3   Changes in the dendrites of rat hippocampal neurons following various hormonal regimes. *Left*: Dendritic spine density under each of the indicated conditions (recall that dendritic spines, which are small extensions from the dendritic shaft, are sites of synapses). *Right*: Tracings of representative apical dendrites from hippocampal pyramidal neurons. (A) After administration of progesterone and estrogen in high dosage. (B) After administration of progesterone and estrogen at basal levels. (C) After administration of a progesterone receptor antagonist. (After Woolley and McEwen, 1992.)

modify selected neurons by acting as a growth factor. Thus, even in the adult, estrogens can stimulate some neurons to increase their somal and nuclear size, as well as the number of synaptic contacts they receive. For example, during periods of high circulating estrogen in the estrous cycle of female rodents—or after administration of estrogen to castrated males—there is an increase in the number of dendritic spines in the ventromedial hypothalamus, an area involved in parental behavior. This same response to estrogen occurs in the hippocampus, where high levels of estrogen and progesterone cause an increase in the density of spines and synapses on the apical dendrites of pyramidal neurons (Figure 28.3). Such changes during the estrous cycle in rodents may be pertinent to understanding the course of catemenial epilepsy, a disorder that occurs only in women. The frequency of seizures in catemenial epilepsy is linked to certain phases of the estrous cycle and may be explained by heightened neuronal excitability due to an increased number of dendritic spines on hippocampal pyramidal neurons.

■ CENTRAL NERVOUS SYSTEM DIMORPHISMS RELATED TO RELATIVELY SIMPLE SEXUAL BEHAVIORS

The actions of sex hormones on neurons provide powerful mechanisms for the production of behavioral differences between females and males. Sexual dimorphisms in the central nervous system can influence a wide range of behaviors, from simple reflexes to cognition. Perhaps the best example of a dimorphism related to an essentially reflex sexual behavior is the difference in size between females and males of a nucleus in the lumbar segment of the rat spinal cord called the **spinal nucleus of the bulbocavernosus**. The motor neurons of this nucleus innervate two striated muscles of the perineum, the bulbocavernosus and ischiocavernosus (Figure 28.4A). In males, the bulbocavernosus is under voluntary control and is used to eject the last drops of urine; both the bulbocavernosus and the ischiocavernosus attach to the penis and also play a role in penile erection. In females, the bulbocavernosus is used to constrict the opening of the vagina; both the bulbocavernosus and the ischiocavernosus attach to the base of the clitoris. Marc Breedlove and his colleagues showed that the spinal nucleus containing the motor neurons that innervate the bulbocavernosus in rodents is absent in females but quite large in males (Figure 28.4B, C), and that the development of this dimorphism in the spinal cord depends on the maintenance of target muscles by circulating

(A) Male rat pelvis

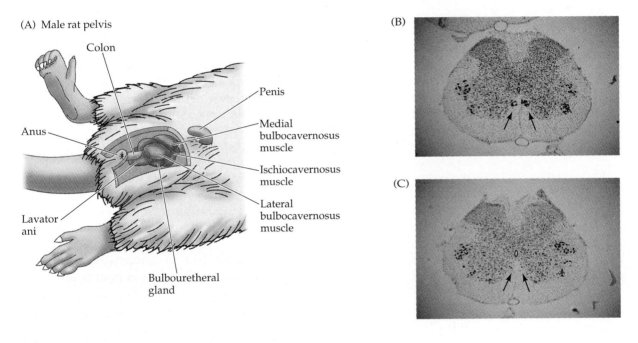

(D)

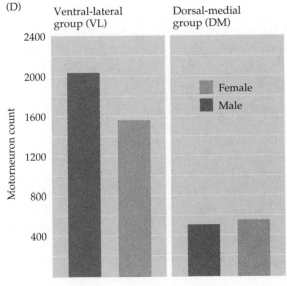

Figure 28.4 The number of spinal motor neurons related to the perineal muscles is different in female and male rodents. (A) Diagram of the perineal region of a male rat. (B) A histological cross section through the fifth lumbar segment of the male. Arrows indicate the spinal nucleus of the bulbocavernosus of the male. (C) Same region of the spinal cord in the female rat. There is no equivalent grouping of densely stained neurons. (D) Histograms showing motor neuron counts in the VL and DM groups of Onuf's nucleus in human females and males. (A after Breedlove and Arnold, 1984; B and C from Breedlove and Arnold, 1983; D after Forger and Breedlove, 1986.)

androgens. Since developing males have high levels of circulating sex steroids whereas females do not, these muscles degenerate in developing female rats, leaving the motor neurons to atrophy in the absence of trophic support (see Chapter 22). In humans, the corresponding spinal cord structure is called **Onuf's nucleus.** Onuf's nucleus consists of two cell groups in the sacral cord: the dorsal medial and the ventral lateral groups. Although the dorsal medial group is not sexually dimorphic, human females have fewer neurons in the ventral lateral group than males (Figure 28.4D). The female perineal muscles remain throughout life, but are smaller than in the male. The difference in nuclear size presumably reflects, as in rodents, the difference in the number of muscle fibers the motor neurons must innervate.

Since a variety of sexual behaviors are governed by the hypothalamus, it has been widely assumed that structural dimorphisms of the anterior hypothalamus could underlie other behavioral differences between females and males (Figure 28.5). In fact, a number of dimorphic hypothalamic nuclei exist in both rodents (the sexually dimorphic nucleus already mentioned, for example) and humans; and, as with the spinal nucleus of the bulbocavernosus and Onuf's nucleus, the differences between females and males are more obvious in rodents than in humans. In the rodent, lesioning the sexually dimorphic nucleus (or not allowing it to develop) makes male rats much less interested in sexual intercourse with females; indeed, males tend to assume sexual postures commonly assumed by females. In rhesus monkeys, physiological recordings from hypothalamic neurons during sexual activity show that neurons of the medial preoptic area of the anterior hypothalamus fire during specific sexual behaviors. Such recordings have been carried out on male monkeys sitting in a flexible restraining chair that allows the male to gain access to a receptive female by pressing a bar, which brings the female close enough to allow mounting by the male. In this way, the responses of hypothalamic neurons can be correlated with "desire" (number of bar presses) and mating behavior (contact, mounting, intromission, thrusting). Such studies show that neurons in the medial preoptic area of the male hypothalamus fire rapidly before sexual behavior, but decrease firing upon contact with the female and mating (Figure 28.6). In contrast, neurons in the dorsal anterior hypothalamus begin firing at the onset of mating and continue to fire vigorously during intercourse. Thus, in monkeys, some medial preoptic hypothalamic neurons are involved in arousal, the initiation of copulation, and copulation itself.

In humans, the linkage of nuclear dimorphisms in the hypothalamus to sexual behavior is based on analogies to rodents and nonhuman primates. The most thoroughly documented examples of sexually dimorphic hypothalamic nuclei in humans have been described by Laura Allen and Roger Gorski at UCLA as well as Dick Swaab and E. Fliers at the Netherlands Institute for Brain Research. Within the anterior hypothalamus there are four cell groupings, collectively called the **interstitial nuclei of the anterior hypothalamus** (INAH); the nuclei are numbered 1–4 from dorsolateral to ventral medial (Figure 28.7). Nuclei 1–3 of the INAH are often more than twice as large in males as they are in females. Moreover, at least two of these, INAH-1 and 2, change in size over time. INAH-1 is the same size in females and males up until 2–4 years of age; it then becomes larger in males until approximately 50 years of age, when it decreases in size in both sexes. Although generally larger in males, INAH-2 is larger in females of childbearing age than in prepubescent and postmenopausal females. Such changes in size with age suggest that in humans, as in rodents, hypothalamic nuclear dimorphisms may be related to levels of circulating sex steroids.

(A)

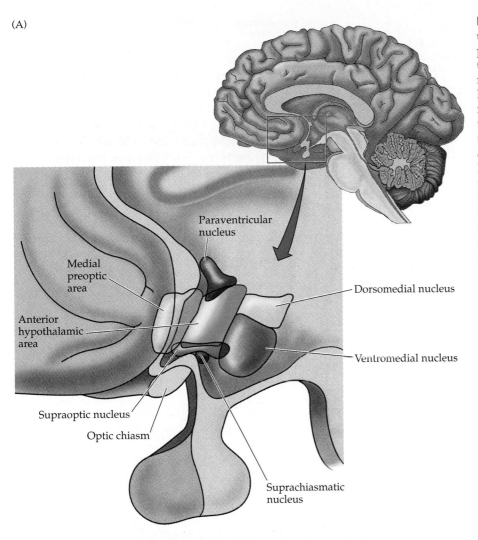

Figure 28.5 Organization of the sexual hypothalamus. (A) The human hypothalamus, illustrating the location of the anterior hypothalamic area and other nuclei in which sexual dimorphisms have been observed in either humans or rodents. (B) Diagram of the major relationships of the rodent anterior hypothalamus with other brain regions. Blue arrows denote neural connections; yellow arrows denote hormonal links; and purple arrow denotes a combination of hormonal and neural connections. Although similar information is not available for humans, it is reasonable to assume that these interactions are characteristic of mammals.

(B)

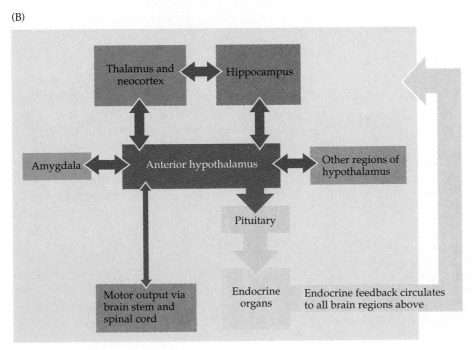

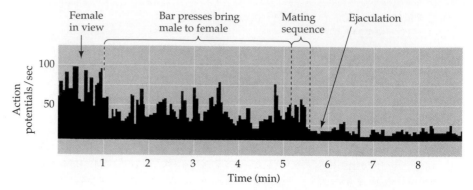

Figure 28.6 Neurons in different regions of the primate hypothalamus are actively associated with sexual behavior. Plots of neuronal activity recorded in the medial preoptic area, showing that firing increases prior to intromission and ejaculation. (After Oomura et al., 1983.)

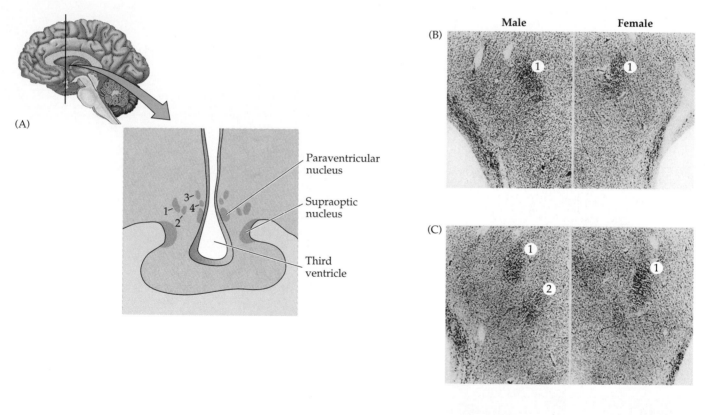

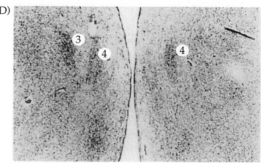

Figure 28.7 Sexual dimorphisms in the interstitial nuclei of the human anterior hypothalamus (INAH). (A) Diagrammatic coronal section through the anterior hypothalamus. The four interstitial nuclei of the anterior hypothalamus (red) are indicated by the numbers 1–4. The micrographs below show the interstitial nuclei from a male (left column) and a female (right column). The male examples were taken from the left side of the brain, female examples from the right side at the same level. (B) INAH-1. (C) INAH-1 and 2. Note that INAH-2 is less compact in the female. (D) INAH-3 and 4. INAH-4 is well represented in both the male and female, whereas INAH-3 is less distinct in the female. (From Allen et al., 1989.)

■ BRAIN DIMORPHISMS RELATED TO MORE COMPLEX SEXUAL BEHAVIORS

Although copulation may seem complicated enough, other aspects of human sexual behavior are even more complex. The choice of a sexual partner is an example. In addition to heterosexual behavior, some humans express sexual behaviors toward both females and males (**bisexuality**), some toward members of their own phenotypic sex (**homosexuality**), and others toward the opposite sex but with a gender identity at odds with their phenotypic sex (**transsexuality**). Based on experimental work in animals and evidence that relatively simple sexual behaviors are influenced by brain dimorphisms, explaining these more complex behaviors in the same general way has been an attractive possibility for workers in this field. To pursue this issue, Simon LeVay, then working at the Salk Institute, compared the INAH nuclei of females, heterosexual males, and homosexual males. LeVay first confirmed Allen and Gorski's findings that of the four INAH nuclei, at least two are sexually dimorphic. He went on to discover that one of these nuclei—the INAH-3—is more than twice as large in male heterosexuals as in male homosexuals (Figure 28.8A). LeVay suggested that this difference is related to sexual orientation.

(A) INAH-3

Heterosexual male

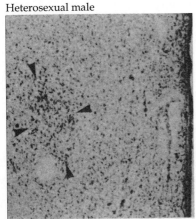

Homosexual male

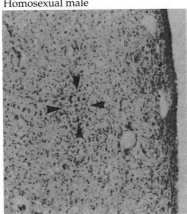

(B) Suprachiasmatic nucleus

Volume

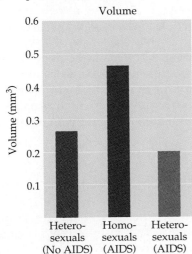

Number of neurons

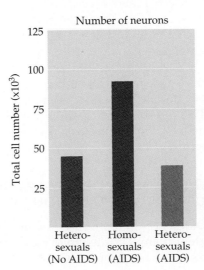

Figure 28.8 Brain dimorphisms in heterosexual and homosexual human males. (A) Micrographs showing difference in INAH-3 between heterosexual and homosexual males. Arrowheads outline the nucleus. (B) The suprachiasmatic nucleus may also differ between homosexual and heterosexual males. Note that the suprachiasmatic nucleus of homosexual males appears to be larger (left histogram) and to contain more neurons (right histogram) than that of heterosexual males with or without AIDS. (A from LeVay, 1991; B after Swaab and Hofman, 1990.)

Other researchers have also concluded that dimorphisms of hypothalamic nuclei are related to sexual orientation and gender identity. Dick Swaab and Michel Hofman examined the suprachiasmatic nucleus of the hypothalamus, which lies just above the optic chiasm in both rodents and humans and generates circadian rhythms (see Figure 28.5A and Chapter 26). This nucleus is also involved in reproduction. In examining the suprachiasmatic nuclei of females, heterosexual males, and homosexual males, Swaab and Hofman found the volume of the suprachiasmatic nucleus to be almost twice as large in the brains of male homosexuals, as compared with the brains of male heterosexuals (Figure 28.8B). They found no difference, however, between the size of the suprachiasmatic nucleus in females and heterosexual males. Like LeVay, they concluded that the difference in size between homosexual and heterosexual men might be related to sexual orientation. This same group has also reported a dimorphism that may be related to gender identity. In comparing male-to-female transsexuals to heterosexual males, they found that another hypothalamic structure, the bed nucleus of the stria terminalis, is smaller in transsexual males, being closer in size to that of females.

Taken together, this evidence suggests that one way to explain the continuum of human sexuality is that small differences in brain structure generate significant differences in sexual behavior. In analogy to the rodent, such brain dimorphisms could be established by the early influence of hormones acting on the brain nuclei that mediate aspects of sexuality as complex as sexual orientation and gender identification. Low levels of circulating androgens in a male early in life could lead to a relatively feminine brain in genotypic males, whereas high levels of circulating androgen in females could lead to a masculinized brain in genotypic females.

As attractive as this hypothesis appears, the development of sexuality in humans is probably a good deal more convoluted. Although LeVay's findings support the idea that homosexuality is related to "feminization" of the male brain (INAH-3 in homosexual males is smaller than in heterosexual males), Swaab and Hofman's data on the size of the suprachiasmatic nucleus undermine the interpretation that the male homosexual's brain is simply "feminized" by a lack of androgens early in development. Whereas there was a difference in the volume of the suprachiasmatic nucleus between homosexual and heterosexual males, in contrast to LeVay they found no difference in the volume of the nucleus between females and heterosexual males. In addition, the development of the INAH-1 dimorphism occurs between 2 and 4 years of age—long after the first testosterone surge in human males. These several discrepancies suggest that the development of sexually dimorphic nuclei in humans does not depend solely on early hormone levels.

Despite these uncertainties (much more will have to be done before these essentially preliminary findings can be fully accepted), work over the last decade has placed human sexuality in a much more biological context. This is a welcome advance over the not too distant past when unusual sexual behavior was commonly explained in Freudian or, worse yet, moralistic terms.

■ BRAIN DIMORPHISMS RELATED TO COGNITIVE FUNCTION

Differences in brain structure may also influence the ability to attend to and identify important stimuli and plan meaningful responses to them (that is, cognitive behavior; see Chapter 24). Evidence for sexually dimorphic cognitive abilities comes mainly from clinical observations. For example, neurologists had reported that females suffer aphasia less often than males after damage to the left hemisphere. This observation led to the suggestion that language functions are to some degree differently represented in females and males. To explore this issue in more detail, Doreen Kimura at the University of Western Ontario studied language function in right-handed patients with unilateral lesions of the left cerebral cortex. She found that females were more likely to suffer aphasia if the damage was to the anterior left hemisphere, while males were more likely to suffer aphasia if damage was located posteriorly (Table 28.1). Kimura concluded that language areas of the female brain are more anteriorly represented and thus less vulnerable to stroke.

Perhaps the most famous (or infamous) structure that might influence the cognitive abilities of females and males is the corpus callosum, the large fiber bundle that connects the right and left cerebral hemispheres (see Chapter 25). In the most complete study to date, Laura Allen and her colleagues found that the corpus callosum, while not differing in size between females and males, does differ in shape: a region called the splenium is more bulbous in females than in males (Figure 28.9 A,B). The functional significance of this difference, if any, is unclear. Another fiber bundle that connects the cerebral hemispheres, the anterior commissure, is also sexually dimorphic. The midsagittal surface area of the anterior commissure in females is, on average, about 12% larger than in males (Figure 28.9C). Moreover, the size of the anterior commissure of homosexual males is the same as that of females. Since the anterior commissure, like the corpus callosum, mediates the interhemispheric transfer of visual, auditory, and olfactory information, these findings raise the possibility that there may be some differences between females and males—and perhaps between gays and straights—in the way information is integrated across the hemispheres.

The differential effects of unilateral stroke in females and males, together with anatomical differences of the cerebral commissures, lend some support to the idea that there may be sex differences in lateralized cognitive functions. To study this issue more directly, tasks that depend primarily on one hemisphere or the other have been given to females and males. One such test for visuospatial differences concerns how well girls and boys are able to identify shapes after palpating them with the right or left hand. Both sexes perform equally well with either hand up to about 6 years of age. Thereafter, boys start scoring better when they use their left hand, whereas

TABLE 28.1
Incidence of Aphasia in Females and Males with Damage Restricted to the Left Hemisphere

	Total sample	Number aphasic	Average age	Number nonaphasic	Average age
Anterior lesions					
Females	13	8 (**62%**)	58	5	58
Males	15	6 (**40%**)	54	9	46
Posterior lesions					
Females	19	2 (**11%**)	49	17	39
Males	34	14 (**41%**)	56	20	39

After Kimura, 1983

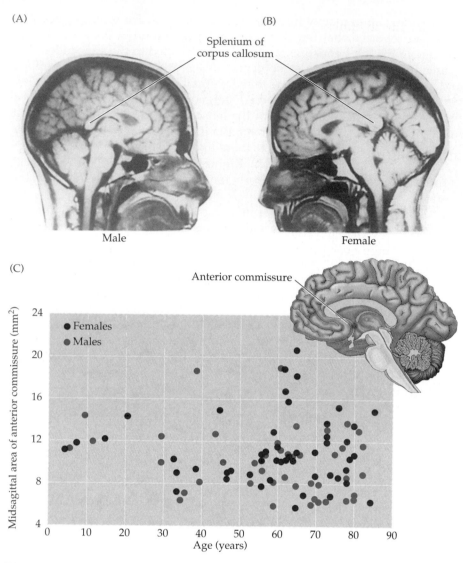

(A) (B)

Splenium of
corpus callosum

Male

Female

(C)

Anterior commissure

Figure 28.9 Structural dimorphisms in the human brain that may be related to cognitive behaviors in females and males. The magnetic resonance images show the body and splenium of the human corpus callosum taken from the midsagittal plane of a representative male (A) and female (B) brain. (C) Size differences of the anterior commissures of human females and males measured at autoposy. Although the differences in the means of the two populations are statistically significant, the distributions overlap substantially. (A and B from Allen et al., 1991; C after Allen and Gorski, 1991.)

girls continue to score equally well with either hand up to 13 years of age, when they also begin to do better with their left hand. This study suggests that boys develop right hemispheric lateralization of visuospatial skills earlier than girls. The idea that females and males develop lateralized functions at different rates is supported by studies of the development of the prefrontal cortex of nonhuman primates. Removing the prefrontal cortex before 15 to 18 months of age does not affect motor-planning functions in female rhesus monkeys. However, the same lesion in male monkeys at this age impedes motor-planning skills.

Perhaps in humans, as in these animal models, differences in levels of circulating hormones early in development and throughout life influence brain structures related to cognitive function.

■ ONGOING CHANGES IN SEX-RELATED BRAIN CIRCUITS

One of the problems associated with studies of differences in female and male brain structures is that the relevant regions of the brain are usually studied long after the behaviors are established. Thus, determining cause and effect is difficult. In LeVay's study, for example, the size difference of INAH-3 in homosexuals might be generated by homosexual behavior, rather than being a cause of the behavior.

There is growing evidence, however, that some brain circuits can continue to change over the course of an individual's life, providing a way around this dilemma. (Recall that estrogens can influence the structure of developing neurons; see Figure 28.2.) For instance, neural connections among cells in the rat hypothalamus that secrete vasopressin and oxytocin change markedly during parturition. In females prior to pregnancy, these neurons are isolated from each other by thin astrocytic processes. Under the influence of the hormones prevailing during birth and lactation, the glial processes retract and the oxytocin- and vasopressin-secreting neurons become electrically coupled by gap junctions (Figure 28.10). Whereas these neurons fire independently before the female gives birth, during lactation they fire synchronously, releasing pulses of oxytocin into the maternal circulation. These surges of oxytocin cause the contraction of smooth muscles in the mammary glands, and hence milk ejection.

Another change in the brain circuits of adult rats associated with parenting behavior is the altered representation of the ventrum (chest) in the somatic sensory cortex of virgin and lactating female rats. As determined by electrophysiology, the representation of the ventrum is approximately twice as large in nursing females as in nonlactating controls. Moreover, the receptive fields of the neurons representing the skin of the ventrum in lactating females are decreased in size by about a third (Figure 28.11). Both the increase

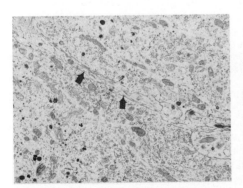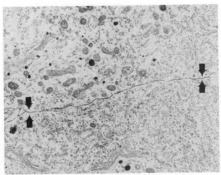

Figure 28.10 Changes in neurons of the rat supraoptic nucleus during lactation. *Left*: Before birth; the relevant neurons and their dendrites are isolated from each other by astrocytic processes (arrows). *Right*: During nursing of the young, the astrocytic processes withdraw and the neurons and their dendrites show the close apposition (arrow pairs) that allows electrical synapses to form between adjacent neurons. (From Modney and Hatton, 1990.)

(A) Female rat

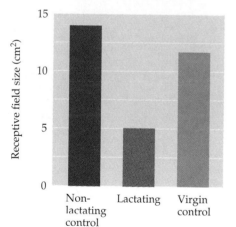

Location of nipples on ventrum

Primary somatic sensory cortex

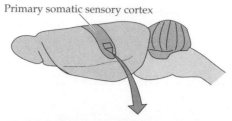

(B) Nonlactating rat (18 days postpartum)

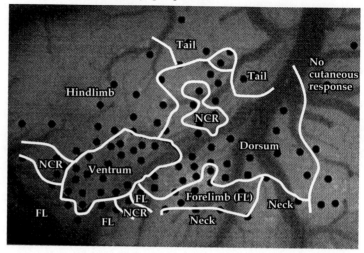

(C) Lactating rat (19 days postpartum)

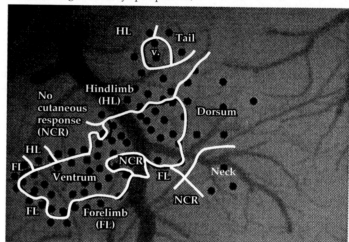

(D)

Receptive field size (cm²)

Non-lactating control | Lactating | Virgin control

Figure 28.11 Changes in the cortical representation of the skin of the ventrum in the rat primary somatic sensory cortex during lactation. (A) Ventrum of the female rat; dots mark the position of nipples. (B) Computer image of somatic sensory cortex in a nonlactating control rat, showing the amount of cortex normally activated by stimulation of the ventrum. Dots mark electrode penetrations; heavy lines delineate the estimated representation. (C) Similar image from a 19-day postpartum, lactating rat. Note the expansion of the representation of the ventrum. NCR = not cutaneous; FL = forelimb; HL = hindlimb. (D) Histogram of receptive field sizes of single neurons in nonlactating control, lactating, and virgin control rats. The receptive field sizes of neurons in lactating mothers are decreased. (From Xerri et al., 1994.)

in cortical representation and the decrease in receptive field size show that parenting behavior can be reflected in changes of cortical circuitry in adult animals.

■ SUMMARY

Differences in female and male behaviors ranging from copulation to cognition are linked to differences in brain structure. Although the neural basis for these sexual dimorphisms is much clearer in experimental animals, the evidence for sex-related differences in the human brain is now substantial. The region of the brain in which the most clear-cut structural dimorphisms occur is the anterior hypothalamus. In rats and monkeys, the nuclei in this region play a role not only in the mechanics of sex, but also in desire, parenting, and sexual orientation. In the rodent, sexual dimorphisms develop primarily as a result of hormonal action on neurons during perinatal development. On the strength of this knowledge about sexual development in experimental animals, neurobiological explanations for human behaviors such as homosexuality and transsexuality have been proposed. Such models remain controversial at present because neither the sexual dimorphisms of the human brain nor their functional significance are fully established. Nonetheless, from what is now known it seems likely that a deeper understanding of how sex hormones influence developing (and to some extent adult) brain circuitry will eventually explain much about the continuum of human sexual behavior.

Additional Readings

Reviews

MacLusky, N. J. and F. Naftolin (1981) Sexual differentiation of the central nervous system. Science 211: 1294–1302.

Swaab, D. F. (1992) Gender and sexual orientation in relation to hypothalamic structures. Horm. Res. 38 (Suppl. 2): 51–61.

Swaab, D. F. and M. A. Hofman (1984) Sexual differentiation of the human brain: A historical perspective. In Progress in Brain Research, Vol. 61. G. J. De Vries (ed.). Amsterdam: Elsevier, pp. 361–374.

Important Original Papers

Allen, L. S., M. Hines, J. E. Shryne and R. A. Gorski (1989) Two sexually dimorphic cell groups in the human brain. J. Neurosci. 9: 497–506.

Allen, L. S., M. F. Richey, Y. M. Chai and R. A. Gorski (1991) Sex differences in the corpus callosum of the living human being. J. Neurosci. 11: 933–942.

Breedlove, S. M. and A. P. Arnold (1981) Sexually dimorphic motor nucleus in the rat lumbar spinal cord: Response to adult hormone manipulation, absence in androgen-insensitive rats. Brain Res. 225: 297–307.

Gorski, R. A., J. H. Gordon, J. E. Shryne and A. M. Southam (1978) Evidence for a morphological sex difference within the medial preoptic area of the rat brain. Brain Res. 143: 333–346.

LeVay, S. (1991) A difference in hypothalamic structure between heterosexual and homosexual men. Science 253: 1034–1037.

Meyer-Bahlburg, H. F. L., A. A. Ehrhardt, L. R. Rosen and R. S. Gruen (1995) Prenatal estrogens and the development of homosexual orientation. Dev. Psych. 31:12–21.

Modney, B. K. and G. I. Hatton (1990) Motherhood modifies magnocellular neuronal interrelationships in functionally meaningful ways. In Mammalian Parenting, N. A. Krasnegor and R. S. Bridges (eds.). New York: Oxford University Press, pp. 306–323.

Raisman, G. and P. M. Field (1973) Sexual dimorphism in the neuropil of the preoptic area of the rat and its dependence on neonatal androgen. Brain Res. 54: 1–29.

Woolley, C. S. and B. S. McEwen (1992) Estradiol mediates fluctuation in hippocampal synapse density during the estrous cycle in the adult rat. J. Neurosci. 12: 2549–2554.

Xerri, C., J. M. Stern and M. M. Merzenich (1994) Alterations of the cortical representation of the rat ventrum induced by nursing behavior. J. Neurosci. 14: 1710–1721.

Zhou, J.-N., M. A. Hofman, L. J. G. Gooren and D. F. Swaab (1995) A sex difference in the human brain and its relation to transsexuality. Nature 378: 68–70.

Books

Fausto-Sterling, A. (1992) Myths of Gender: Biological Theories About Women and Men, 2nd Ed. New York: Basic Books.

Goy, R. W. and B. S. McEwen (1980) Sexual Differentiation of the Brain. Cambridge, MA: MIT Press.

LeVay, S. (1993) The Sexual Brain. Cambridge, MA: MIT Press.

CHAPTER 29

HUMAN MEMORY

■ OVERVIEW

Perhaps the most intriguing property of the brain is its ability to store information and retrieve much of it at will. Equally fascinating is the normal—and sometimes abnormal—ability to forget stored information. Three related terms are fundamental to any discussion of these phenomena: learning, memory, and forgetting. Learning is the process by which new information is acquired by the nervous system. Memory is the storage and/or retrieval of that information, and forgetting is the process whereby stored information is lost over time. Amnesia, in contrast, is the pathological inability to learn new information or to retrieve information that has already been acquired. Because of the great importance of memory in human affairs, understanding these issues is widely perceived as one of the major challenges of modern neuroscience. Mechanisms of synaptic change that may provide some of the cellular and molecular bases for information storage were considered in Chapter 23. The focus here is on observations that give insight into the broader organization of human memory.

■ QUALITATIVE CATEGORIES OF HUMAN MEMORY

Humans demonstrate at least two qualitatively different types of information storage: **declarative memory** and **procedural memory** (Figure 29.1). Declarative memory is the storage (and retrieval) of material that is available to the conscious mind and which can therefore be encoded in symbols and expressed by language (hence, declarative). Examples of declarative memory are the ability to remember a telephone number, a birthday, a shopping list, or the image of an important event. Procedural memory, on the other hand, is not available to the conscious mind, at least not in any detail. Such memories involve skills and associations that are, by and large, acquired and retrieved at an unconscious level. Remembering how to ride a bicycle or play a musical instrument, or the associations made by classical conditioning are all examples of memories that fall in this category. It is difficult to say how we do these things. When we perform complex motor actions, for instance, we are not conscious of a particular memory—or at least we need not be. In fact, thinking about such acts (a golf swing comes to mind) may actually inhibit the ability to perform them smoothly and efficiently.

■ TEMPORAL CATEGORIES OF MEMORY

In addition to the types of memory defined by the nature of what is remembered, memory can also be categorized according to the time over which it is effective. Although the details are still debated by both psychologists and neurobiologists, three classes of memory have been described (Figure 29.2). The first of these is **immediate memory**. By definition, immediate memory is the routine ability to hold an experience in mind for a few seconds. The capacity of this register is very large, involves all modalities (visual, verbal, tactile, and so on), and provides our ongoing sense of the present.

    Short-term memory, the second temporal category, is the ability to hold information in mind for periods of seconds to minutes once the present

Figure 29.1 The major qualitative categories of human memory. Declarative memory includes those memories that can be brought to consciousness and expressed as remembered events, images, sounds, and so on. Procedural memory includes motor skills, cognitive skills, simple classical conditioning, priming, and other information that is acquired and retrieved without having to "think about it."

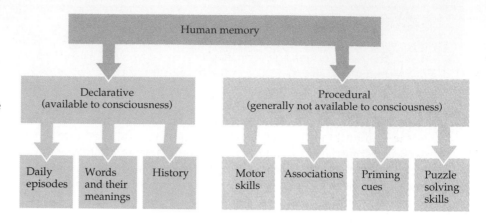

moment has passed. A conventional way of testing short-term memory is to present a string of digits, which the subject is then asked to repeat. Operationally, this evaluation tests short-term declarative memory. A special sort of short-term memory is called **working memory**. Working memory refers to the ability to hold things in mind long enough to carry out sequential actions. An example is searching for a lost object; working memory allows the hunt to proceed efficiently, avoiding places already inspected. A particular advantage of working memory is that it can be readily tested in animals (see Chapter 24).

Long-term memory—the retention of information for days, weeks, or even a lifetime—entails the transfer of information initially acquired to a more permanent form of storage. There is general agreement that the so-called **engram** (a term that refers to the physical embodiment of the memory in neural machinery) underlying this more permanent form of memory depends on changes in synapses, either long-term changes in the efficacy of existing synapses, or the actual reordering of synaptic connections. There is, in fact, evidence for both these varieties of synaptic change (see Chapters 22 and 23).

The subtle transfer of information from short-term to long-term storage is apparent in the phenomenon of **priming**. Priming is demonstrated by presenting a series of sentences such as "The rattle hung over the baby's cradle." Much later, subjects are asked to complete three-letter strings to make the first word that comes to mind. Typically, the words presented earlier (for example, "cradle" to complete "cra___" or "rattle" to complete "rat___") are given at a much higher frequency than expected by chance, despite the subject's inability to recall directly the initial words or sentences. Clearly, the temporal boundaries of short-term and long-term memory are indistinct, and some workers prefer to use the phrase **intermediate-term memory** for phenomena like priming.

Figure 29.2 The major temporal categories of human memory.

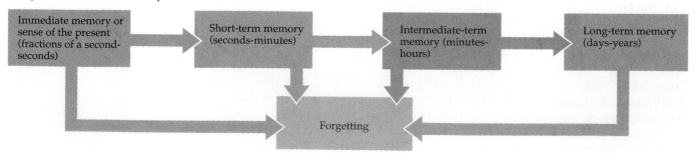

■ FORGETTING

Some years ago, two psychologists at the University of Washington found that 84% of their colleagues agreed with the statement that "everything we learn is permanently stored in the mind, although sometimes particular details are not accessible." The 16% who thought otherwise are more likely to be correct. Common sense indicates that were it not for forgetting, our brains would be impossibly burdened with a welter of useless information. In fact, the human brain is very good at forgetting, and there are mechanisms, again poorly understood, that cause information to be lost at every stage of the memory storage process, including long-term memories. For example, our memory of the appearance of a penny (an icon seen thousands of times) is uncertain at best, and people gradually forget the TV shows they have seen (Figure 29.3).

These and many other observations indicate that we readily forget things that have no particular importance. Indeed, the ability to forget unimportant information is probably as critical for normal mentation as retaining information that we judge to be significant. This conclusion is underscored by individuals who have difficulty with the normal "erasure" of information. Perhaps the best-known case is a subject studied over several decades by the Russian neuropsychologist A.R. Luria; Luria referred to this man simply as S (his name was Shereshevskii). Luria's description of an early encounter (A. R. Luria, 1987, *The Mind of a Mnemonist*) gives some idea why S, then a newspaper reporter, was so interesting:

> I gave S a series of words, then numbers, then letters, reading them to him slowly or presenting them in written form. He read or listened attentively and

(A)

(B)

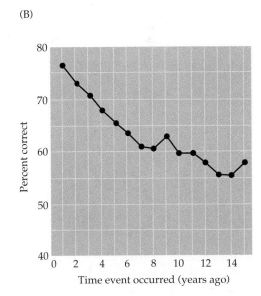

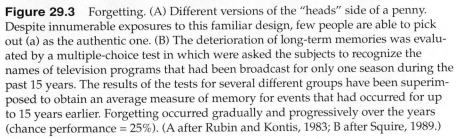

Figure 29.3 Forgetting. (A) Different versions of the "heads" side of a penny. Despite innumerable exposures to this familiar design, few people are able to pick out (a) as the authentic one. (B) The deterioration of long-term memories was evaluated by a multiple-choice test in which were asked the subjects to recognize the names of television programs that had been broadcast for only one season during the past 15 years. The results of the tests for several different groups have been superimposed to obtain an average measure of memory for events that had occurred for up to 15 years earlier. Forgetting occurred gradually and progressively over the years (chance performance = 25%). (A after Rubin and Kontis, 1983; B after Squire, 1989.)

then repeated the material exactly as it had been presented. I increased the number of elements in each series, giving him as many as thirty, fifty, or even seventy words or numbers, but this too, presented no problem for him. He did not need to commit any of the material to memory; if I gave him a series of words or numbers, which I read slowly and distinctly, he would listen attentively, sometimes ask me to stop and enunciate a word more clearly, or, if in doubt whether he had heard a word correctly, would ask me to repeat it. Usually during an experiment he would close his eyes or stare into space, fixing his gaze on one point; when the experiment was over, he would ask that we pause while he went over the material in his mind to see if he had retained it. Thereupon, without another moment's pause, he would reproduce the series that had been read to him. (pp. 9–10)

S's phenomenal memory, however, did not always serve him well. He had difficulty ridding his mind of the trivial information that he tended to focus on, sometimes to the point of incapacitation. As Luria put it:

Thus, trying to understand a passage, to grasp the information it contains (which other people accomplish by singling out what is most important) became a tortuous procedure for S, a struggle against images that kept rising to the surface in his mind. Images, then, proved an obstacle as well as an aid to learning in that they prevented S from concentrating on what was essential. Moreover, since these images tended to jam together, producing still more images, he was carried so far adrift that he was forced to go back and rethink the entire passage. Consequently, a simple passage—a phrase, for that matter—would turn out to be a Sisyphean task. (p. 113)

Clearly, forgetting is an essential and quite normal mental process.

Pathological forgetting is called **amnesia**, which may be evident as an inability to fully establish new memories or an inability to retrieve old ones. The inability to establish new memories is called **anterograde amnesia**, whereas difficulty retrieving previously established memories is called **retrograde amnesia**. Anterograde and retrograde amnesia are often present together, but can be dissociated under various circumstances. Some of the major causes of amnesia are listed in Table 29.1. The best-studied of these are Korsakoff's syndrome, amnesia following head trauma, amnesia associated with vascular accidents, and, perhaps most revealingly, amnesia following bilateral lesions of the temporal lobe and diencephalon (see below).

TABLE 29.1
Some Clinical Causes of Amnesia

Causes	Examples	Site of damage
Vascular (occlusion of both posterior cerebral arteries)	Patient R.B.	Bilateral medial temporal lobe, the hippocampus in particular
Tumors	—	Medial thalamus bilaterally (hippocampus and other related structures if tumor is large enough)
Trauma	Patient N.A.	Bilateral medial temporal lobe
Surgery	Patient H.M.	Bilateral medial temporal lobe
Infections	Herpes simplex encephalitis	Bilateral medial temporal lobe
Vitamin B1 deficiency	Korsakoff's syndrome	Medial thalamus and mammillary bodies
Electroconvulsive therapy (ECT) for depression	—	Uncertain

■ THE HUMAN CAPACITY FOR INFORMATION STORAGE

The capacity of the short-term memory buffer for relatively meaningless information is quite limited, being a string of about seven or eight numbers or other arbitrary items for normal individuals. This capacity can, however, be increased dramatically with practice. For example, a college student who for more than a year spent an hour each day practicing the task of remembering numbers increased his reiterative ability up to a string of about 80 digits (Figure 29.4). He did this by making subsets of the string of numbers he was given signify dates, times at track meets (he was a competitive runner), and so on—in essence, relating meaningless items to a meaningful context. This same strategy is used by most professional "mnemonists," who amaze audiences by their apparently prodigious feats of memory. Another compelling example is that a good chess player can remember the position of many more pieces on a briefly examined chess board than a poor player (Figure 29.5). Thus, short-term memory capacity depends on the significance of the information in question.

Despite a limited capacity for storing meaningless data, all of us have a remarkable ability to remember information that we care about. An especially remarkable example is the case of Arturo Toscanini, the late conductor of the NBC Philharmonic Orchestra, who allegedly kept in his head the complete orchestral scores of more than 250 works, as well as the music and librettos for about 100 operas. Once, just before a concert in St. Louis, the first bassoonist approached Toscanini in some consternation because he had just discovered that the lowest note on his bassoon was broken. After a minute or two of deep concentration, the story goes, Toscanini turned to the alarmed bassoonist and informed him that there was no need for concern, since that note did not appear in any of the bassoon parts for the evening's program. An example of a remarkable quantitative memory is the mathematician

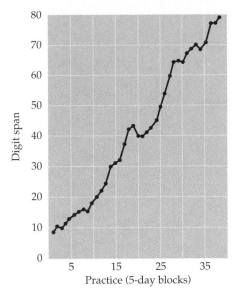

Figure 29.4 Increasing the digit span by a learning strategy. During many months involving 1 hour of practice a day for 3 to 5 days a week, this subject increased his digit span from 7 to 79 numbers. Random digits were read to him at the rate of 1 per second. If a sequence was recalled correctly, 1 digit was added to the next sequence. (After Ericsson et al., 1980.)

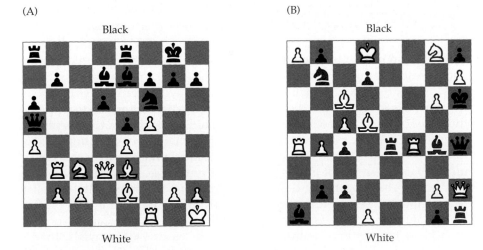

Figure 29.5 The retention of briefly presented information depends on past experience, context, and its perceived importance. (A) Board position after white's twenty-first move in game 10 of the 1985 World Chess Championship between A. Karpov (white) and G. Kasparov (black). (B) A random arrangement of the same 28 pieces. After briefly viewing the board from the real game, master players reconstruct the positions of the pieces with much greater accuracy than beginning or average players. With a randomly arranged board, experts and beginners perform at about the same level.

Alexander Aitken. After an undistinguished career in elementary school, Aitken was, at the age of 13, greatly taken with the manipulation of numbers. For the next four years he undertook, as a personal challenge, to master mental calculation. He began by memorizing the value of π to 1000 places, and could soon do calculations in his head with such facility that he became a local celebrity. When asked for the squares of three-digit numbers, he was able to give these almost instantly. The square roots for each were produced to five significant digits in 2 to 3 seconds; the squares of four-digit numbers took him about 5 seconds. Aitken went on to become a professor of mathematics at Edinburgh and was eventually elected a Fellow of the Royal Society for his contributions to numerical mathematics, statistics, and matrix algebra. At the age of 30 or so, he began to lose his enthusiasm for "mental yoga," as he called his penchant. In part, his waning enthusiasm stemmed from the realization that the advent of mechanical calculators was making his prowess obsolete (it was then 1930). He also discovered that the last 180 digits of π that he had memorized as a boy were wrong; he had taken the values from the published work of another mental calculator, who erred in an era when there were no computers to readily determine the correct value. But in middle life, Aitken decided to memorize π once more:

> I amused myself again by learning the correct value as far as 1000 places, and once again found it no trouble, except that I needed to "fix" the join where [the] error had occurred. The secret, to my mind, is relaxation, the complete antithesis of concentration as usually understood.

Aitken's mental processes, into which he obviously had much insight, were not rote, but more in the nature of the fascination that aficionados bring to chess, music, or language. Although few of us can boast the mnemonic prowess of Aitken or Toscanini, our ability to remember the things that deeply interest us—whether baseball statistics, soap opera plots, or the details of brain structure—is remarkable.

■ BRAIN SYSTEMS UNDERLYING DECLARATIVE AND PROCEDURAL MEMORIES

Three extraordinary clinical cases have revealed a great deal about the brain systems responsible for the short-term storage of declarative information and are now familiar to neurobiologists and clinicians as patients H.M., N.A., and R.B. (Box A). Taken together, they provide dramatic evidence of the importance of **midline diencephalic** and **medial temporal structures**— the hippocampus, in particular—in establishing new declarative memories (Figure 29.6). These patients also demonstrate that there is a different anatomical substrate for anterograde and retrograde amnesia. In each of these individuals, memory for events prior to the precipitating event was reasonably good. The devastating deficiency is (or was) the inability to establish new memories. Retrograde amnesia—the loss of memory for events preceding an injury or illness—is more typical of the generalized lesions associated with head trauma and neurodegenerative diseases (Alzheimer's, for instance; see Box B). Although a degree of retrograde amnesia can occur with the focal lesions that cause anterograde amnesia, the long-term storage of memories is presumably distributed throughout the brain. Thus, the hippocampus and related diencephalic structures indicated in Figure 29.6 form and consolidate declarative memories that are ultimately stored elsewhere. Finally, it is important to emphasize that H.M., N.A., and R.B. had no problems establishing (or recalling) *procedural* memories. This fact indicates that

(A) Brain areas associated with declarative memory disorders

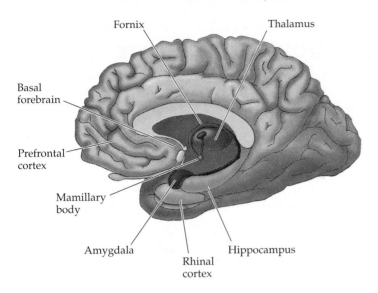

(B) Ventral view of hippocampus and related structures with part of temporal lobes removed

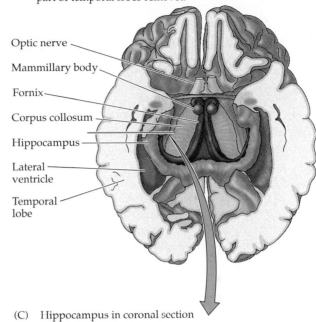

(C) Hippocampus in coronal section

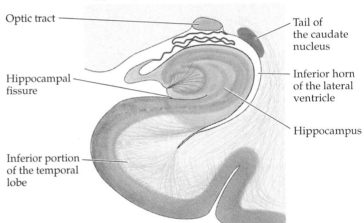

Figure 29.6 Brain areas that, when damaged, tend to give rise to declarative memory disorders. By inference, declarative memory is normally based on the physiological activity of these structures. (A) Studies of amnesic patients have shown that the formation of declarative memories depends on the integrity of a subset of limbic circuits (see Chapter 27), particularly those of the hippocampus and its subcortical connections to the mammillary bodies and dorsal thalamus. (B) Diagram showing the location of the hippocampus in a cutaway view in the horizontal plane. (C) The hippocampus as it would appear in a histological section in the coronal plane, at approximately the level indicated by the line in (B).

procedural memories must also have a different anatomical basis than declarative memories.

There are, of course, many causes of amnesia in addition to hippocampal damage (see Table 29.1). **Korsakoff's syndrome**, for example, can occur in chronic alcoholics as a result of thiamine (vitamin B_1) deficiency. In such cases, loss of brain tissue occurs bilaterally in the mammillary bodies and the medial thalamus. The cause of this loss is not well understood. Typically, these patients have memory dysfunction, and often confabulate (make up tall stories) in the acute stage of the disease. Korsakoff's patients are difficult to fit into a neat pathophysiological framework for amnesia because chronic alcoholism gives rise to a variety of neurological disorders.

Electroconvulsive therapy (ECT) can also cause amnesia. Patients with severe depression are often treated by the passage of enough electrical current through the brain to cause the equivalent of a full-blown seizure. Indeed, this therapy was discovered because depression in epileptics was sometimes

Box A
CLINICAL CASES THAT REVEAL THE ANATOMICAL SUBSTRATE FOR DECLARATIVE MEMORIES

The Case of H.M

At the age of 27, H.M., who had suffered minor seizures since age 10 and major seizures since age 16, underwent surgery to correct his increasingly debilitating epilepsy. A high school graduate, H.M. had been working as a technician in a small electrical business until shortly before the time of his operation. His attacks involved generalized convulsions with tongue biting, incontinence, and loss of consciousness (all this typical of grand mal seizures). Despite a variety of medications, the seizures remained uncontrolled and increased in severity. A few weeks before his surgery, H.M. became unable to work and had to quit his job.

On September 1, 1953, a bilateral medial temporal lobe resection was carried out in which the amygdala, uncus, hippocampal gyrus, and anterior two-thirds of the hippocampus were removed. At the time, it was unclear that bilateral surgery of this kind would cause a profound memory defect. Severe amnesia was evident, however, upon H.M.'s recovery from the operation, and his life was changed radically.

The first formal psychological exam of H.M. was conducted nearly 2 years after the operation, at which time a profound memory defect was still obvious. Just before the examination, for instance, H.M. had been talking to the psychologist; yet he had no recollection of this experience a few minutes later, denying that anyone had spoken to him. He gave the date as March 1953 and seemed oblivious to the fact that he had undergone an operation, or that he had become incapacitated as a result. Nonetheless, his score on the Wechsler-Bellevue Intelligence Scale was 112, a value not significantly different from his pre-operative IQ. Various psychological tests

failed to reveal any deficiencies in perception, abstract thinking, or reasoning; moreover, he seemed highly motivated and, in the context of casual conversation, normal. Importantly, he also performed well on tests of the ability to learn new skills, such as mirror writing

or puzzle solving (that is, his ability to form procedural memories was intact). Moreover, his early memories were easily recalled, showing that the structures removed during H.M.'s operation are not a permanent repository for such information. On the Wechsler Memory

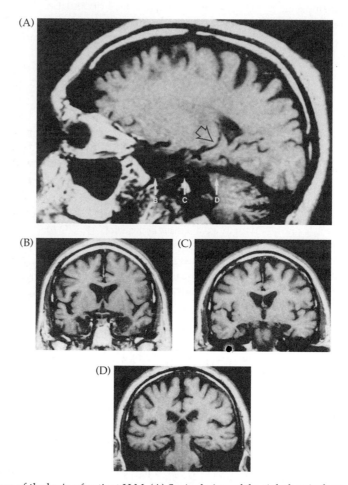

MRI images of the brain of patient H.M. (A) Sagittal view of the right hemisphere; the area of the anterior temporal lobectomy is indicated by the filled arrows. The intact posterior hippocampus is the banana-shaped object indicated by the open arrow. (B–D) Coronal sections at approximately the levels indicated by the arrows in (A). Image (B) is the most rostral and is located at the level of the amygdala. The amygdala and the underlying entorhinal cortex are entirely missing. Image (C) is at the level of the rostral hippocampal formation; again, this structure and the underlying entorhinal cortex have been removed. Image (D) is at the caudal level of the hippocampus; the posterior hippocampus appears intact, although somewhat shrunken. (From Corkin et al., 1996.)

Scale (a specific test of declarative memory), however, he performed very poorly, and could not recall a preceding test-set once he had turned his attention to another part of the exam. These formal deficits, along with his obvious inability to recall events in his daily life, all indicate a profound loss of short-term declarative memory function.

During the subsequent decades, H.M. has been studied extensively, primarily by Brenda Milner and her colleagues at the Montreal Neurological Institute. His memory deficiency has continued unabated, and, according to Milner, he has little idea who she is in spite of their acquaintance for more than 40 years. Sadly, he has gradually come to appreciate his predicament. "Every day is alone," H.M. reports, "whatever enjoyment I've had and whatever sorrow I've had."

The Case of N.A.

N.A. was born in 1938 and grew up with his mother and stepfather, attending public schools in California. After a year of junior college, he joined the Air Force. In October of 1959 he was assigned to the Azores as a radar technician and remained there until December 1960, when a bizarre accident made him a celebrated neurological case. N.A. was assembling a model airplane in his barracks room while his roommate, unbeknownst to him, was making thrusts and parries with a miniature fencing foil behind N.A.'s chair. N.A. turned suddenly and was stabbed through the right nostril. The foil penetrated the cribriform plate (the structure through which the olfactory nerve enters the brain) and took an upward course into the left forebrain. N.A. lost consciousness within a few minutes (presumably because of bleeding in the region of brain injury) and was taken to a hospital. There he exhibited a right-sided weakness and paralysis of the right eye muscles innervated by the third cranial nerve. Exploratory surgery

was undertaken and the dural tear repaired. Gradually he recovered and was sent home to California. After some months, his only general neurological deficits were some weakness of upward gaze and mild double vision. He retained, however, a severe anterograde amnesia for declarative memories. MRI studies first carried out in 1986 showed extensive damage to the thalamus and the medial temporal lobe, mostly on the right side; the mammillary bodies also appeared to be missing bilaterally. The exact extent of his lesion, however, is not known, as N.A. remains alive and well.

N.A.'s memory from the time of his injury over 35 years ago to the present has remained impaired, and like H.M. he fails badly on formal tests of new learning ability. His IQ is 124 and he shows no defects in language skills, perception, or other measures of intelligence. He can also learn new procedural skills quite normally. His amnesia is not as dense as that of H.M. and is more verbal than spatial. He can, for example, draw accurate diagrams of material presented to him earlier. Nonetheless, he loses track of his possessions, forgets what he has done, and tends to forget who has come to visit him. He has only vague impressions of the political, social, and sporting events that have occurred since his injury. Even watching television is difficult because he tends to forget the storyline during commercials. On the other hand, his memory for events prior to 1960 is extremely good; indeed, his lifestyle, according to his physicians, tends to reflect the 1950s.

The Case of R.B.

At the age of 52, R.B. suffered an ischemic episode during cardiac bypass surgery. Following recovery from anesthesia, a profound amnesic disorder was apparent. As in the cases of H.M. and N.A., his IQ was normal (111) and he showed no evidence of cognitive defects other than memory impairment. R.B. was tested extensively for the next 5

years, and while his amnesia was not as severe as that of H.M. or N.A., he consistently failed the standard tests of the ability to establish new declarative memories. When R.B. died in 1983 of congestive heart failure, a detailed examination of his brain was carried out. The only significant finding was bilateral lesions of the hippocampus—specifically, cell loss in the CA1 region that extended the full rostral-caudal length of the hippocampus on both sides. The amygdala, thalamus, and mammillary bodies, as well as the structures of the basal forebrain, were normal. R.B.'s case is particularly important because it suggests that hippocampal lesions by themselves can result in profound anterograde amnesia for declarative memory.

References

CORKIN, S. (1984) Lasting consequences of bilateral medial temporal lobectomy: Clinical course and experimental findings in H.M. Semin. Neurol. 4: 249–259.

CORKIN, S., D. G. AMARAL, R. G. GONZÁLEZ, K. A. JOHNSON AND B. T. HYMAN (1996) H. M.'s medial temporal lobe lesion: Findings from MRI. J. Neurosci. (in press)

HILTS, P. J. (1995) Memory's Ghost: The Strange Tale of Mr. M. and the Nature of Memory. New York: Simon and Schuster.

MILNER, B., S. CORKIN AND H.-L. TEUBER (1968) Further analysis of the hippocampal amnesic syndrome: A 14-year follow-up study of H.M. Neuropsychologia 6: 215–234.

SCOVILLE, W. B. AND B. MILNER (1957) Loss of recent memory after bilateral hippocampal lesions. J. Neurol. Neurosurg. Psychiat. 20: 11–21.

SQUIRE, L. R., D. G. AMARAL, S. M. ZOLA-MORGAN, M. KRITCHEVSKY AND G. PRESS (1989) Description of brain injury in the amnesic patient N.A. based on magnetic resonance imaging. Exp. Neurol. 105: 23–35.

TEUBER, H. L., B. MILNER AND H. G. VAUGHN (1968) Persistent anterograde amnesia after stab wound of the basal brain. Neuropsychologia 6: 267–282.

ZOLA-MORGAN, S., L. R. SQUIRE AND D. AMARAL (1986) Human amnesia and the medial temporal region: Enduring memory impairment following a bilateral lesion limited to the CA1 field of the hippocampus. J. Neurosci. 6: 2950–2967.

alleviated following a seizure. The use of this therapy over the decades has left no doubt about its efficacy in alleviating depression (see Chapter 27). However, patients undergoing ECT often suffer both anterograde and retrograde amnesia. Patients typically do not remember the treatment itself or the events of the preceding days; their recall of events of the previous 1–3 years can also be affected. Animal studies (rats tested for maze learning, for example) have confirmed the amnesic consequences of ECT. The amnesia usually clears over a period of weeks to months; however, to mitigate this worrisome side effect (which may be the result of excitotoxicity; see Box B in Chapter 6), ECT is often delivered to only one hemisphere at a time.

■ THE LONG-TERM STORAGE OF INFORMATION

As revealing as they have been, clinical studies of amnesic patients have provided relatively little insight into the long-term storage of information in the brain (other than to indicate that such information is *not* stored in the midline diencephalic structures that are so important in anterograde amnesias). In this aspect of memory, the cerebral cortex appears to be the major repository. Different cortical regions have clearly different cognitive functions (see Chapter 24). Not surprisingly, at least some of these are sites of long-term information storage that accords with their particular role in mentation. The lexicon that determines our vocabulary, for instance, is located in the association cortex of the superior temporal lobe and nearby parietal lobe (Wernicke's area; see Chapter 25). Presumably, the widespread connections of the hippocampus to the language areas serves to route declarative information to such destinations (Figure 29.7). By the same token, the clinical picture of patients with temporal lobe lesions suggests that memories of objects and faces are located there; and frontal lobe syndromes imply that memories about social context and future plans reside in the frontal cortex.

With respect to procedural learning, the motor skills that we gradually acquire through practice are evidently stored in the basal ganglia, cerebel-

(A)   Afferent connections of the hippocampal region

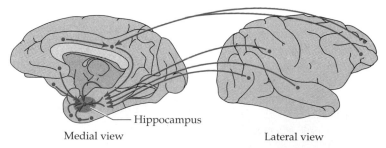

Medial view   Lateral view

Figure 29.7   Connections between the region of the hippocampus and possible memory storage sites; the rhesus monkey brain is shown because these connections are much better documented in monkeys than in humans. Projections to this region are shown in (A) and efferent projections in (B). Projections from numerous cortical areas converge on the structures known to be involved in human memory; most of these sites also send projections to the same cortical areas. Medial and lateral views are shown, the latter rotated 180° for clarity. (After Van Hoesen, 1982.)

(B)   Efferent connections of the hippocampal region

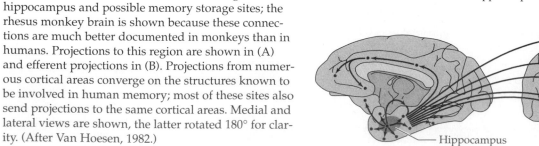

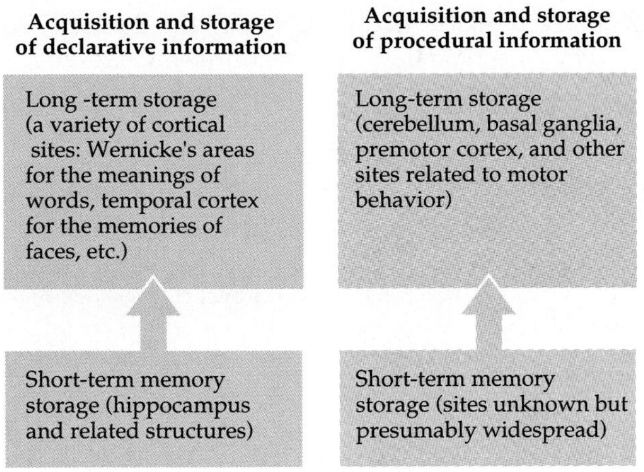

Acquisition and storage of declarative information
Acquisition and storage of procedural information
Long -term storage (a variety of cortical sites: Wernicke's areas for the meanings of words, temporal cortex for the memories of faces, etc.)
Long-term storage (cerebellum, basal ganglia, premotor cortex, and other sites related to motor behavior)
Short-term memory storage (hippocampus and related structures)
Short-term memory storage (sites unknown but presumably widespread)

Figure 29.8 Summary diagram of the acquisition and storage of declarative versus procedural information.

lum, and premotor cortex (see Chapters 17 and 18). Learned associations, another form of procedural memory, presumably involve the appropriate association cortices as well (Chapter 23). This general scheme is diagrammed in Figure 29.8.

■ MEMORY AND AGING

Although our outward appearance inevitably changes with age, we tend to imagine that our brains resist the ravages of time. The evidence suggests that this optimistic view is unjustified. From early adulthood onward, the average weight of the normal human brain, as determined at autopsy, steadily decreases (Figure 29.9). In many elderly individuals, this effect can be observed with noninvasive imaging as a slight but perceptible shrinkage of the brain. Counts of synapses generally decrease in old age, although the number of neurons probably does not change very much, suggesting that it is mainly neuropil that is lost (the uncertainty of this statement reflects the difficulty of making such determinations). These observations accord with the subjective experience of decreased memory function in older individuals

1.8
1.6
1.4
1.2
1.0
0.8
0.6
0.4
0.2
Brain weight (kg)
Males
Females
0–5 1 3 6–7 10–12 16–18 22–30 41–50 56–60 66–70 81–85 86+
Age (years)

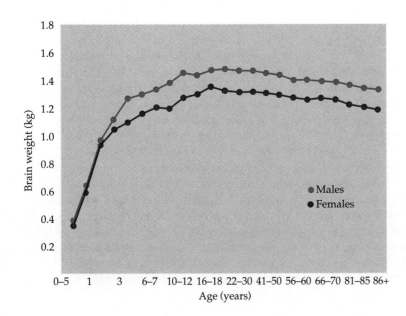

Figure 29.9 Brain size as a function of age. The human brain reaches its maximum size (measured by weight in this case) in early adult life and decreases progressively thereafter. This gradual decrease presumably represents the loss of some neural circuitry in the aging brain, which may help explain the diminished memory function in most older individuals. (After Dekaban and Sadowsky, 1978.)

Box B
ALZHEIMER'S DISEASE

Dementia refers to a pathological failure of memory and other cognitive functions. Alzheimer's disease (AD) is the most common dementia, afflicting 5–10% of the population over the age of 65 and an even higher percentage of the population over 85. The earliest manifestation is typically an impairment of recent memory function and attention; this deficit is followed by a deterioration of language skills, visuospatial orientation, abstract thinking, and judgment, and alterations of personality. There is no effective therapy.

The diagnosis of AD is established by these characteristic clinical features in association with a distinctive histopathology characterized by intraneuronal cytoskeletal filaments (neurofibrillary tangles) and deposits of an abnormal extracellular matrix material called amyloid (senile) plaques. There is also an accompanying loss of neurons. These histopathological findings are most apparent in the neocortex, in limbic structures (particularly the hippocampus, amygdala, and entorhinal cortex), and in selected nuclei (especially basal fore-

brain nuclei, locus coeruleus, and raphe nuclei).

Although the cause of AD is not known, the identification of a mutant gene in a few families with an early-onset form of the disease has recently provided an opportunity to understand some aspects of the the problem. AD is familial in approximately 5% of cases. Investigators had long suspected that the mutant gene responsible for familial AD resided on chromosome 21, in part because Down's syndrome, which is caused by an extra copy of chromosome 21, shows similar clinical and neuropathological features (plaques, tangles, and the early onset of dementia). The prominence of amyloid in the histopathology of AD further suggested the mutation of a gene encoding amyloid precursor protein (APP). The gene for APP was cloned by D. Goldgaber and his colleagues and found to be on chromosome 21. This observation eventually led to the identification of missense mutations of the APP gene in families with the early-onset form of AD.

The apparent importance of these

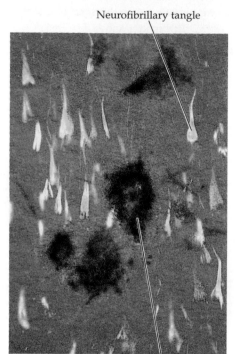

Neurofibrillary tangle

Amyloid plaque

Histological section of the cerebral cortex from a patient with Alzheimer's disease, showing characteristic amyloid plaques and neurofibrillary tangles. (From Roses, 1995; courtesy of Gary W. Van Hoesen.)

and declining scores on tests of memory. The normal loss of some memory function with age means that there is a large "gray zone" between individuals undergoing normal aging and patients suffering from age-related dementias such as Alzheimer's disease (Box B).

■ SUMMARY

Memory is sometimes referred to as a monolithic phenomenon to be explained by a particular cellular or molecular mechanism. This idea is misleading, since human memory involves a number of strategies and anatomical substrates. Primary among these are a system for memories that can be expressed by means of language and are available to the conscious mind (declarative memory); a separate system deals with skills and associations that are essentially prelinguistic, operating at a largely unconscious level (procedural memory). Based on evidence from amnesic patients and normal

APP gene mutations in families with the early-onset form of the disease implied that APP might be involved in other forms of AD. To test this idea, biochemists W. Strittmatter and G. Salvesen immobilized a derivative of APP on nitrocellulose; then, using cerebrospinal fluid from patients with AD, they searched for proteins that bound with high affinity. One of the proteins that bound with the APP derivative was apolipoprotein E (apoE), a protein that chaperons cholesterol in blood serum.

The significance of this discovery became apparent when M. Pericak-Vance and A. Roses found that affected members in some families with the late-onset form of AD exhibited an association with DNA markers on chromosome 19, which harbors a gene encoding an isoform of apoE, the E4 allele. This result led to the investigation of other alleles for apoE in affected members of families with late-onset AD. The frequency of the E3 allele in the general population is 0.78, the frequency of the E4 allele 0.14. Roses, Strittmatter, Pericak-Vance and their colleagues found that the frequency of allele E4 in the late-onset familial AD patients was 0.52, almost four times higher than expected. They therefore concluded that

inheritance of the E4 allele was a risk factor for late-onset AD. In fact, people homozygous for E4 are eight times more likely to develop AD compared to subjects homozygous for E3. Among individuals in late-onset Alzheimer's families with no copies of E4, only 20% developed AD by age 75, in contrast to 90% of individuals with two copies of E4. An increased association of the E4 allele was also shown in late-onset nonfamilial AD—an especially important discovery because this is the most common form of the disease.

A question arising from these observations is whether the E4 allele for ApoE is itself responsible for the increased risk, or whether a closely linked gene on chromosome 19 is the culprit. The fact that ApoE binds to plaques in the brains of early-onset AD patients favors the idea that the E4 allele is itself the problem. Conversely, a mutation of the E4 allele is probably not directly responsible for late-onset AD; rather, inheriting this particular form of the gene somehow increases the risk of developing AD. Importantly, some individuals with early-onset forms of familial AD due to APP mutations do not have the E4 allele, indicating consider-

able heterogeneity of the disease.

Thus an ongoing challenge is to determine how the E4 allele is related to increased risk for developing AD. The discovery of this risk factor has provided a valuable set of tools to help dissect the molecular pathogenesis of the most common form of dementia.

References

CORDER, E. H., A. M. SAUNDERS, W. J. STRITTMATTER, D. E. SCHMECHEL, P. C. GASKELL, G. W. SMALL, A. D. ROSES, J. L. HAINES AND M. A. PERICAK-VANCE (1993) Gene dose of apolipoprotein E type 4 allele and the risk of Alzheimer's disease in late onset families. Science 261: 921–923.

GOLDGABER, D., M. I. LERMAN, O. W. MCBRIDE, U. SAFFIOTTI AND D. C. GAJDUSEK (1987) Characterization and chromosomal localization of a cDNA encoding brain amyloid of Alzheimer's disease. Science 235: 877–880.

MURRELL, J., M. FARLOW, B. GHETTI AND M. D. BENSON (1991) A mutation in the amyloid precursor protein associated with hereditary Alzheimer's disease. Science 254: 97–99.

ROSES, ALLEN D. (1995) Apolipoprotein E and Alzheimer disease. Sci. Amer. Sci. Med. 2: 16–25.

YANKNER, B. A. AND M. M. MESULAM (1991) β-amyloid and the pathogenesis of Alzheimer's disease. N. Engl. J. Med. 325: 1849–1857.

patterns of neural connections, the hippocampus and associated midline diencephalic and medial temporal lobe structures are clearly important in laying down new declarative memories, but not in storing them. In contrast, procedural memory for motor (and other) skills depends on the integrity of the cortex, basal ganglia, and cerebellum and is not affected by lesions that impair the declarative memory system. Given the importance of retaining the lessons of experience, it is not surprising that multiple systems and mechanisms operate to encode information in the brain.

Additional Reading

Reviews

MISHKIN, M. AND T. APPENZELLER (1987) The anatomy of memory. Sci. Amer. 256(6): 80–89.

PETRI, H. AND M. MISHKIN (1994) Behaviorism, cognitivism and the neuropsychology of memory. Amer. Sci. 82: 30–37.

SMITH, S. B. (1983) *The Great Mental Calculators. The Psychology, Methods, and Lives of Calculating Prodigies, Past and Present.* New York: Columbia University Press.

SQUIRE, L. R. (1992) Memory and hippocampus: A synthesis from findings with rats, monkeys, and humans. Psych. Rev. 99: 195–231.

THOMPSON, R. F. (1986) The neurobiology of learning and memory. Science 223: 941–947.

ZOLA-MORGAN, S. M. AND L. R. SQUIRE (1993) Neuroanatomy of memory. Annu. Rev. Neurosci. 16: 547–563.

Important Original Papers

SCOVILLE, W. B. AND B. MILNER (1957) Loss of recent memory after bilateral hippocampal lesions. J. Neurol. Neurosurg. Psychiat. 20: 11–21.

SQUIRE, L. R. (1989) On the course of forgetting in very long-term memory. J. Exp. Psychol. 15: 241–245.

ZOLA-MORGAN, S. M. AND L. R. SQUIRE (1990) The primate hippocampal formation: Evidence for a time-limited role in memory storage. Science 250: 288–290.

Books

BADDELEY, A. (1982) *Your Memory: A User's Guide.* New York: Macmillan.

LURIA, A. R. (1987) The Mind of a Mnemonist (Transl. Lynn Solotaroff). Cambridge, MA: Harvard University Press.

NEISSER, U. (1982) *Memory Observed: Remembering in Natural Contexts.* San Francisco: W.H. Freeman.

PENFIELD, W. AND L. ROBERTS (1959) *Speech and Brain Mechanisms.* Princeton, NJ: Princeton University Press.

SAPER, C. B. AND F. PLUM (1985) *Handbook of Clinical Neurology,* Vol. 1(45): *Clinical Neuropsychology,* P. J. Vinken, G. S. Bruyn and H. L. Klawans (eds.). New York: Elsevier, pp. 107–128.

SQUIRE, L. R. (1987) *Memory and Brain.* New York: Oxford University Press, pp. 202–223.

ZECHMEISTER, E. B. AND S. E. NYBERG (1982) *Human Memory: An Introduction to Research and Theory.* Monterey, CA: Brooks/Cole Publishing.

Glossary

acetylcholine Neurotransmitter at motor neuron synapses, in autonomic ganglia and a variety of central synapses; binds to two types of receptors—ligand-gated ion channels (nicotinic receptors), and G-protein-coupled receptors (muscarininc receptors).

action potential The electrical signal conducted along axons (or muscle fibers) by which information is conveyed from one place to another in the nervous system.

activation The time-dependent opening of ion channels in response to a stimulus, typically membrane depolarization.

adenylyl cyclase Membrane-bound enzyme that can be activated by G-proteins to catalyze the synthesis of cyclic AMP from ATP.

adhesion molecules *see* cell adhesion molecules.

adrenaline *see* epinephrine.

adrenergic Refers to synaptic transmission mediated by the release of norepinephrine or epinephrine.

adult The mature form of an animal, usually defined by the ability to reproduce.

afferent An axon that conducts action potentials from the periphery toward the central nervous system.

agnosia The inability to name objects.

alpha motor neurons Neurons in the ventral horn of the spinal cord that innervate skeletal muscle.

amacrine cells Retinal neurons that mediate lateral interactions between bipolar cell terminals and the dendrites of ganglion cells.

amblyopia Diminished visual acuity as a result of the failure to establish appropriate visual cortical connections in early life.

amnesia The pathological inability to remember or establish memories; retrograde amnesia is the inability to recall existing memories, whereas anterograde amnesia is the inability to lay down new memories.

amphetamine A synthetically produced central nervous system stimulant with cocaine-like effects; drug abuse may lead to dependence.

ampullae The juglike swellings at the base of the semicircular canals that contain the hair cells and cupulae (*see also* cupulae).

amygdala A nuclear complex in the temporal lobe that forms part of the limbic system; its major functions concern autonomic, emotional and sexual behavior.

androgen insensitivity syndrome A condition in which, due to a defect in the gene that codes for the androgen receptor, testosterone cannot act on its target tissues.

anencephaly A congenital defect of neural tube closure, in which much of the brain fails to develop.

anosmia Loss of the sense of smell.

anterior Toward the front; sometimes used as a synonym for rostral, and sometimes as a synonym for ventral.

anterior commissure A small midline fiber tract that lies at the anterior end of the corpus callosum; like the callosum, it serves to connect the two hemispheres.

anterior hypothalamus Region of the hypothalamus containing nuclei that mediate sexual behaviors; not to be confused with region in rodent called the medial preoptic area, which lies anterior to hypothalamus and also contains nuclei that mediate sexual behavior (most notably the sexually dimorphic nucleus).

anterograde A movement or influence acting from the neuronal cell body toward the axonal target.

anterolateral pathway (anterolateral system) Ascending sensory pathway in the spinal cord and brainstem that carries information about pain and temperature to the thalamus.

antiserum Serum harvested from an animal immunized to an agent of interest.

aphasia The inability to comprehend and/or produce language as a result of damage to the language areas of the cerebral cortex (or their white matter interconnections).

apoptosis Cell death resulting from a programmed pattern of gene expression; also known as "programmed cell death."

aprosodia The inability to infuse language with its normal emotional content.

arachnoid mater One of the three coverings of the brain that make up the meninges; lies between the dura mater and the pia mater.

areflexia Loss of reflexes.

association cortex Defined by exclusion as those neocortical regions that are not involved in primary sensory or motor processing.

associativity In the hippocampus, the enhancement of a weakly activated group of synapses when a nearby group is strongly activated.

astrocytes One of the three major classes of glial cells found in the central nervous system; important in regulating the ionic milieu of nerve cells, and in some cases transmitter reuptake.

astrotactin A cell-surface molecule that causes neurons to adhere to radial glial fibers during neuronal migration.

athetosis Slow, writhing movements seen primarily in patients with disorders of the basal ganglion.

ATPase pumps Membrane pumps that use the hydrolysis of ATP to translocate ions against their electrochemical gradients.

attention   The selection of a particular sensory stimulus from a more complex stimulus array for further analysis.

auditory space map   Topographic representation of sound source location, as occurs in the inferior colliculus.

autonomic nervous system   The components of the nervous system (peripheral and central) concerned with the regulation of smooth muscle, cardiac muscle, and glands.

axon   The neuronal process that carries the action potential from the nerve cell body to a target.

axoplasmic transport   The process by which materials are carried from nerve cell bodies to their terminals (anterograde transport), or from nerve cell terminals to the neuronal cell body (retrograde transport).

basal ganglia   A group of nuclei lying deep in the subcortical white matter of the frontal lobes that organize motor behavior. The caudate and putamen and the globus pallidus are the major components of the basal ganglia; the subthalamic nucleus and substantia nigra are often included.

basal lamina (basement membrane)   A thin layer of extracellular matrix material (primarily collagen, laminin, and fibronectin) that surrounds muscle cells and Schwann cells. Also underlies all epithelial sheets.

basilar membrane   The membrane that forms the floor of the cochlear duct, upon which the cochlear hair cells are located.

basket cells   Inhibitory interneurons in the cerebellar cortex whose cells bodies are located within the Purkinje cell layer and whose axons make basketlike terminal arbors around Purkinje cell bodies.

biogenic amines   The bioactive amine neurotransmitters; includes the catecholamines (epinephrine, norepinephrine, dopamine), serotonin and histamine.

bipolar cells   Retinal neurons that provide a direct link between photoreceptor terminals and ganglion cell dendrites.

bisexuality   Sexual attraction to members of both the opposite and the same phenotypic sex.

blastomere   A cell produced when the egg undergoes cleavage.

blastula   An early embryo during the stage when the cells are typically arranged to form a hollow sphere.

blind spot   The region of visual space that falls on the optic disk; due to the lack of photoreceptors in the optic disk, objects that lie completely within the blind spot are not perceived.

blood-brain barrier   A diffusion barrier between the brain vasculature and the substance of the brain formed by tight junctions between capillary endothelial cells.

bouton (synaptic bouton)   A swelling specialized for the release of neurotransmitter that occurs along or at the end of an axon.

bradykinesia   Pathologically slow movement.

brainstem   The portion of the brain that lies between the diencephalon and the spinal cord; comprises the midbrain, pons and medulla.

Broca's aphasia   Difficulty producing speech as a result of damage to Broca's area in the left frontal lobe.

Broca's area   An area in the left frontal lobe specialized for the production of language.

CA1   A region of the hippocampus that shows a robust form of long-term potentiation.

CA3   A region of the hippocampus containing the neurons that form the Schaffer collaterals.

cadherins   A family of calcium-dependent cell adhesion molecules found on the surfaces of growth cones and the cells over which they grow.

calcarine sulcus   The major sulcus on the medial aspect of the occipital lobe; the primary visual cortex lies largely within this sulcus.

cAMP response element binding protein (CREB)   A protein activated by cyclic AMP that binds to specific regions of DNA, thereby increasing the transcription rates of nearby genes.

cAMP response elements (CREs)   Specific DNA sequences that bind transcription factors activated by cAMP (*see* cAMP response element binding protein).

catecholamine   A term referring to molecules containing a catechol ring and an amino group; examples are the neurotransmitters epinephrine, norepinephrine, and dopamine.

cauda equina   The collection of segmental ventral and dorsal roots that extend from the caudal end of the spinal cord to their exit from the spinal canal.

caudal   Posterior, or "tailward."

caudate nucleus   One of the three major components of the basal ganglia (the other two are the globus pallidus and putamen).

cell adhesion molecules   A family of molecules on cell surfaces that cause them to stick to one another (*see also* laminin and fibronectin).

central nervous system   The brain and spinal cord of vertebrates (by analogy, the central nerve cord and ganglia of invertebrates).

central pattern generator   Oscillatory spinal cord or brainstem circuits responsible for programmed, rhythmic movements such as locomotion.

central sulcus   A major sulcus on the lateral aspect of the hemispheres that forms the boundary between the frontal and parietal lobes. The anterior bank of the sulcus contains the primary motor cortex; the posterior bank contains the primary sensory cortex.

cerebellar ataxia   A pathological inability to make coordinated movements; associated with lesions to the cerebellum. Movements show a jerky tremor when attempted.

cerebellar cortex   The superficial gray matter of the cerebellum.

cerebellar peduncles   The three bilateral pairs of axon tracts (inferior, middle, and superior cerebellar peduncles) that carry information to and from the cerebellum.

cerebellum   Prominent hindbrain structure concerned with motor coordination, posture, and balance. Composed of a three-layered cortex and deep nuclei; attached to the brainstem by three sets of cerebellar peduncles.

cerebral achromatopsia Loss of color vision as a result of damage to extrastriate visual cortex.

cerebral aqueduct The portion of the ventricular system that connects the third and fourth ventricles.

cerebral cortex The superficial gray matter of the cerebral hemispheres.

cerebral peduncles The major fiber bundles that connect the brainstem to the cerebral hemispheres.

cerebro-cerebellum The part of the cerebellar cortex that receives input from the cerebral cortex via axons from the pontine relay nuclei.

cerebrum The largest and most rostral part of the brain in humans and other mammals, consisting of the two cerebral hemispheres.

chemical synapses Synapses that transmit information via the secretion of chemical signals (neurotransmitters).

chemoaffinity (chemoaffinity hypothesis) The idea that nerve cells bear chemical labels that determine their connectivity.

chemotaxis The movement of a cell up (or down) the gradient of a chemical signal.

chemotropism The growth of a part of a cell (axon, dendrite, filopodium) up (or down) a chemical gradient.

chimera An experimentally generated embryo (or organ) comprising cells derived from two or more species (or other genetically distinct sources).

cholinergic Referring to synaptic transmission mediated by the release of acetylcholine.

chorea Jerky, involuntary movement of the face or extremities associated with damage to the basal ganglia.

choreoathetosis The combination of jerky, ballistic, and writhing movements that characterizes the late stages of Huntington's Disease.

choroid plexus Specialized epithelium in the ventricular system that produces cerebrospinal fluid.

chromosome Nuclear organelle that bears the genes.

cingulate gyrus Prominent gyrus on the medial aspect of the hemisphere, lying just superior to the corpus callosum; forms a part of the limbic system.

cingulate sulcus Prominent sulcus on the medial aspect of the hemisphere.

circadian rhythms Variations in physiological functions that occur on a daily basis.

circle of Willis Arterial anastomosis on the ventral aspect of the midbrain; connects the posterior and anterior cerebral circulation.

cisterns Large, cerebrospinal-fluid-filled spaces that lie within the subarachnoid space.

class A taxonomic category subordinate to phylum; comprises animal orders.

climbing fibers Axons that originate in the inferior olive, ascend through the inferior cerebellar peduncle, and make terminal arborizations that invest the dendritic tree of Purkinje cells.

clone The progeny of a single cell.

co-transmitters Two or more types of neurotransmitters within a single synapse; may be packaged into separate populations of synaptic vesicles or co-localized within the same synaptic vesicles.

cochlea The coiled structure within the inner ear where vibrations caused by airborne sound are transduced into neural impulses.

cognition The ability of the central nervous system to attend, identify, and act on complex stimuli.

collapsin A molecule that causes collapse of growth cones; a member of the semaphorin family of signaling molecules.

colliculi The two paired hillocks that characterize the dorsal surface of the midbrain; the superior colliculi concern vision, the inferior colliculi audition.

competition The struggle among nerve cells, or nerve cell processes, for limited resources essential to survival or growth.

conduction aphasia Difficulty producing speech as a result of damage to the connection between Wernicke's and Broca's language areas.

conduction velocity The speed at which an action potential is propagated along an axon.

cone opsins The three distinct photopigments found in cones; the basis for color vision.

cones Photoreceptors specialized for high visual acuity and the perception of color.

congenital adrenal hyperplasia Genetic deficiency that leads to overproduction of androgens and a resultant masculinization of external genitalia in genotypic females.

conjugate The paired movements of the two eyes in the same direction, as occurs in the vestibulo-ocular reflex (*see also* vestibulo-ocular reflex and vergence movements).

conspecific Fellow member of a species.

contralateral On the other side.

contralateral neglect syndrome Neurological condition in which the patient does not acknowledge or attend to the left visual hemifield or the left half of the body. The syndrome typically results from a lesion of the right parietal cortex.

convergence Innervation of a target cell by axons from more than one neuron.

coronal Referring to a plane through the brain that runs parallel to the coronal suture (the mediolateral plane).

corpus callosum The large midline fiber bundle that connects the cortices of the two cerebral hemispheres.

cortex The gray matter of the cerebral hemispheres and cerebellum, where most of the neurons in the brain are located.

cortico-cortical connections Connections made between cortical areas in the same hemisphere or betweem the two hemispheres via the cerebral commissures (the corpus callosum and the anterior commissure).

cranial nerve ganglia The sensory ganglia associated with the cranial nerves; these correspond to the dorsal root ganglia of the spinal segmental nerves.

cranial nerve nuclei Nuclei in the brainstem that contain the neurons related to cranial nerves III–XII.

cranial nerves The 12 pairs of nerves that carry sensory information toward (and sometimes motor information away from) the brain.

crista The hair cell-containing sensory epithelium of the semicircular canals.

critical period A restricted developmental period during which organisms are particularly sensitive to the effects of experience.

cuneate nuclei Sensory relay nuclei that lie in the lower medulla; they contain the second-order sensory neurons that relay mechanosensory information from peripheral receptors in the upper body to the thalamus.

cupulae Gelatinous structures in the semicircular canals in which the hair cell bundles are embedded.

cytoarchitectonic areas Distinct regions of the neocortical mantle identified by differences in cell size, packing density, and laminar arrangement.

decerebrate rigidity Excessive tone in extensor muscles as a result of damage to descending motor pathways at the level of the brainstem.

declarative memory Memories available to consciousness that can be expressed by language.

decussation A crossing of fiber tracts in the midline.

deep cerebellar nuclei The nuclei at the base of the cerebellum that relay information from the cerebellar cortex to the thalamus.

delayed response task A behavioral paradigm used to test cognition and memory.

delta waves Slow (<4 Hz) electroencephalographic waves that characterize stage IV (slow-wave) sleep.

dendrite A neuronal process that receives synaptic input; usually branches near the cell body and is typically unable to support an action potential.

denervation Removal of the innervation to a target.

dentate gyrus A region of the hippocampus; shaped like a tooth.

depolarization The displacement of a cell's membrane potential toward a less negative value.

dermatome The area of skin supplied by the sensory axons of a single spinal nerve.

determination Commitment of a developing cell or cell group to a particular fate.

diencephalon Portion of the brain that lies just rostral to the midbrain; comprises the thalamus and hypothalamus.

differentiation The progressive specialization of developing cells.

dihydrotestosterone A more potent form of testosterone that masculinizes the external genitalia.

disinhibition Arrangement of inhibitory and excitatory cells in a circuit that generates excitation by the transient inhibition of a tonically active inhibitory neuron.

disjunctive eye movements Movements of the two eyes in opposite directions (*see also* vergence movements).

distal Farther away from a point of reference (the opposite of proximal).

divergence The branching of an axon to innervate multiple target cells.

dopamine A catecholamine neurotransmitter.

dorsal Referring to the back.

dorsal column nuclei Second-order sensory neurons in the lower medulla that relay mechanosensory information from the spinal cord to the thalamus; comprises the cuneate and gracile nuclei.

dorsal columns Major ascending tracts in the spinal cord that carry mechanosensory information from the first-order sensory neurons in dorsal root ganglia to the dorsal column nuclei; also called the posterior funiculi.

dorsal horn The dorsal portion of the spinal cord gray matter; contains neurons that process sensory information.

dorsal root ganglia The segmental sensory ganglia of the spinal cord; contain the first-order neurons of the dorsal column/medial lemniscus and spinothalamic pathways.

dorsal roots The bundle of axons that runs from the dorsal root ganglia to the dorsal horn of the spinal cord, carrying sensory information from the periphery.

dura mater The thick external covering of the brain and spinal cord; one of the three components of the meninges, the other two being the pia mater and arachnoid mater.

dynorphins A class of endogenous opioid peptides.

dysarthria Difficulty producing speech as a result of damage to the primary motor centers that govern the muscles of articulation; distinguished from aphasias, which result from cortical damage.

dysmetria Inaccurate movements due to faulty judgment of distance.

dystonia Lack of muscle tone.

early inward current The initial electrical current, measured in voltage clamp experiments, that results from the voltage-dependent entry of a cation such as Na^+ or Ca^{2+}; produces the rising phase of the action potential.

ectoderm The most superficial of the three embryonic germ layers; gives rise to the nervous system and epidermis.

Edinger-Westphal nucleus Midbrain nucleus containing the autonomic neurons that constitute the efferent limb of the pupillary light reflex.

efferent An axon that conducts information away from the central nervous system.

electrical synapses Synapses that transmit information via the direct flow of electrical current at gap junctions.

electrochemical equilibrium The condition in which no net ionic flux occurs across a membrane because ion concentration gradients and opposing transmembrane potentials are in exact balance.

electrogenic Capable of generating an electrical current; usually applied to membrane pumps that create electrical currents while translocating ions.

embryo The developing organism before birth or hatching.

end plate current (EPC) Postsynaptic current produced by neurotransmitter release and binding at the motor end plate.

end plate potential (EPP) Depolarization of the membrane potential of skeletal muscle fiber, caused by the action of the transmitter acetylcholine at the neuromuscular synapse.

endocytosis A budding off of vesicles from the plasma membrane, which allows uptake of materials in the extracellular medium.

endoderm The deepest of the three embryonic germ layers.

endogenous opioids Peptides in the central nervous system that have the same pharmacological effects as morphine and other derivatives of opium.

endolymph The potassium-rich fluid filling both the cochlear duct and the membranous labyrinth; bathes the apical end of the hair cells.

endorphins One of a group of neuropeptides that are agonists at opioid receptors, virtually all of which contain the sequence Tyr-Gly-Gly-Phe.

endplate The complex postsynaptic specialization at the site of nerve contact on skeletal muscle fibers.

engram The term used to indicate the physical basis of a stored memory.

enkephalins A general term for endogenous opioid peptides.

ependyma The epithelial lining of the canal of the spinal cord and the ventricles.

ependymal cells Epithelial cells that line the ventricular system.

epidermis The outermost layer of the skin; derived from the embryonic ectoderm.

epigenetic Referring to influences on development that arise from factors other than genetic instructions.

epinephrine (adrenaline) Catecholamine hormone and neurotransmitter that binds to *a*- and *b*-adrenergic G-protein-coupled receptors.

epineurium The connective tissue surrounding axon fascicles of a peripheral nerve.

epithelium Any continuous layer of cells that covers a surface or lines a cavity.

equilibrium potential The membrane potential at which there is electrochemical equilibrium for a given ion.

estradiol One of the biologically important C_{18} class of steroid hormones capable of inducing estrous in females.

eukaryote An organism that contains cells with nuclei.

excitatory postsynaptic potential (EPSP) Neurotransmitter-induced postsynaptic potential changes that depolarize the cell, and hence increase the likelihood of initiating a postsynaptic action potential.

exocytosis A form of cell secretion resulting from the fusion of the membrane of a storage organelle, such as a synaptic vesicle, with the plasma membrane.

explant A piece of tissue maintained in culture medium.

external segment A subdivision of the globus pallidus.

extracellular matrix A matrix composed of collagen, laminin, and fibronectin that surrounds most cells (*see also* basal lamina)

extrafusal muscle fibers Fibers of skeletal muscles; a term that distinguishes ordinary muscle fibers from the specialized intrafusal fibers associated with muscle spindles.

face cells Neurons in the temporal cortex of rhesus monkeys that respond specifically to faces.

facilitation The increased transmitter release produced by an action potential that follows closely upon a preceding action potential.

family A taxonomic category subordinate to order; comprises genera.

fasciculation The aggregation of neuronal processes to form a nerve bundle; also refers to the spontaneous discharge of motor units after muscle denervation.

***a*-fetoprotein** A protein that actively sequestors circulating estrogens.

fetus The developing mammalian embryo at relatively late stages when the parts of the body are recognizable.

fibrillation Spontaneous contractile activity of denervated muscle fibers.

fibroblast growth factor (FGF) A peptide growth factor, originally defined by its mitogenic effects on fibroblasts; also acts as an inducer during early brain development.

fibronectin A large cell adhesion molecule that binds integrins.

filopodium Slender protoplasmic projection, arising from the growth cone of an axon or dendrite, that explores the local environment.

fissure A deep cleft in the brain; distinguished from sulci, which are shallower cortical infoldings.

flexion reflex Polysynaptic reflex mediating withdrawal from a painful stimulus.

folia The name given to the gyral formations of the cerebellum.

forebrain The anterior portion of the brain that includes the cerebral hemispheres (the telencephalon and diencephalon).

fornix An axon tract, best seen from the medial surface of the divided brain, that interconnects the hypothalamus and hippocampus.

fourth ventricle The ventricular space that lies between the pons and the cerebellum.

fovea Area of the retina specialized for high acuity; contains a high density of cones and few rods.

foveola Capillary and rod-free zone in the center of the fovea.

frontal lobe One of the four lobes of the brain; includes all the cortex that lies anterior to the central sulcus and superior to the lateral fissure.

G-proteins Term for two large groups of proteins—the heterotrimeric G-proteins and the small molecule G-proteins—that can be activated by exchanging bound GDP for GTP.

G-protein-coupled receptors (metabotropic receptors) A large family of neurotransmitter or hormone receptors, characterized by seven transmembrane domains; the binding of these receptors by agonists leads to the activation of intracellular G-proteins.

gamma motor neurons Class of spinal motor neurons specifically concerned with the regulation of muscle spindle length; these neurons innervate the intrafusal muscle fibers of the spindle.

ganglion (plural, ganglia) Collections of hundreds to thousands of neurons found outside the brain and spinal cord, along the course of peripheral nerves.

ganglion cell A neuron located in a ganglion.

gap junction A specialized type of intercellular contact formed by channels that directly connect the cytoplasm of two cells.

gastrula The early embryo during the period when the three embryonic germ layers are formed; follows the blastula stage.

gastrulation The cell movements (invagination and spreading) that transform the embryonic blastula into the gastrula.

gender identification Self-perception of one's alignment with the traits associated with being a phenotypic female or male in a given culture.

gene A hereditary unit located on the chromosomes; genetic information is carried by linear sequences of nucleotides in DNA that code for corresponding sequences of amino acids.

genome The complete set of an animal's genes.

genotype The genetic makeup of an individual.

genotypic sex Sexual characterization according to the complement of sex chromosomes; XX is a genotypic female and XY is a genotypic male.

genus A taxonomic division that comprises a number of closely related species within a family.

germ cell The egg or sperm (or the precursors of these cells).

germ layers The three primary layers of the developing embryo from which all adult tissues arise: ectoderm, mesoderm, and endoderm.

glia (neuroglial cells) The support cells associated with neurons (astrocytes, oligodendrocytes, and microglia in the central nervous system; Schwann cells in peripheral nerves; and satellite cells in ganglia).

globus pallidus One of the three nuclei that make up the basal ganglia; relays information from the caudate and putamen to the thalamus.

glomeruli Characteristic collections of neuropil in the olfactory bulbs; formed by dendrites of mitral cells and terminals of olfactory receptor cells, as well as processes from local interneurons.

glutamate-glutamine cycle A metabolic cycle of glutamate release and re-synthesis involving both neuronal and glial cells.

G_{olf} A G-protein found uniquely in olfactory epithelium.

Golgi tendon organs Receptors located in muscle tendons that provide mechanosensory information to the central nervous system about muscle tension.

gracile nuclei Sensory nuclei in the lower medulla; these second-order sensory neurons relay mechanosensory information from the lower body to the thalamus.

gradient A systematic variation of the concentration of a molecule (or some other agent) that influences cell behavior.

granule cell layer The layer of the cerebellar cortex where granule cell bodies are found. Also used to refer to cell-rich layers in neocortex and hippocampus.

gray matter General term that describes regions of the central nervous system rich in neuronal cell bodies and neuropil; includes the cerebral and cerebellar cortices, the nuclei of the brain, and the central portion of the spinal cord.

growth cone The specialized end of a growing axon (or dendrite) that generates the motive force for elongation.

gyri The ridges of the infolded cerebral cortex (the valleys between these ridges are called sulci).

hair cells The sensory cells within the inner ear that transduce mechanical displacement into neural impulses.

helicotrema The opening at the apex of the cochlea that joins the scala vestibuli and scala tympani.

Hensen's node *see* primitive pit.

heterotrimeric G-proteins A large group of proteins consisting of three subunits (*a*, *b*, and *g*) that can be activated by exchanging bound GDP with GTP resulting in the liberation of two signaling molecules—*a*GTP and the *bg*-dimer.

higher-order neurons Neurons that are relatively remote from peripheral targets.

hindbrain *see* rhombencephalon.

hippocampus A cortical structure in the medial portion of the temporal lobe; in humans concerned with short-term declarative memory, among many other functions.

histamine A biogenic amine neurotransmitter derived from the amino acid histidine.

homeobox genes A set of master control genes whose expression establishes the early body plan of developing organisms (*see also* homeotic mutant).

homeotic mutant A mutation that transforms one part of the body into another (e.g., insect antennae into legs). Affects homeobox genes.

homologous Technically, referring to structures in different species that share the same evolutionary history; more generally, referring to structures or organs that have the same general anatomy and perform the same function.

homosexuality Sexual attraction to an individual of the same phenotypic sex.

horizontal cells Retinal neurons that mediate lateral interactions between photoreceptor terminals and the dendrites of bipolar cells.

horseradish peroxidase   A plant enzyme widely used to stain nerve cells (after injection into a neuron, it generates a visible precipitate by one of several histochemical reactions).

Huntington's disease   An autosomal dominant genetic disorder in which a single gene mutation results in personality changes, progressive loss of the control of voluntary movement, and eventually death. Primary target is the basal ganglia.

hyperalgesia   Increased perception of pain.

hyperkinesis   Excessive movement.

hypokinesis   A paucity of movement.

hypothalamus   A collection of small but critical nuclei in the diencephalon that lies just inferior to the thalamus; governs reproductive, homeostatic and circadian functions.

imprinting   A rapid and permanent form of learning that occurs in response to early experience.

in vitro   Referring to any biological process studied outside of the organism (literally, "in glass").

in vivo   Referring to any biological process studied in an intact living organism. (Literally "in life.")

inactivation   The time-dependent closing of ion channels in response to a stimulus, such as membrane depolarization.

inducers   Chemical signals originating from one set of cells that influence the differentiation of other cells.

induction   The ability of a cell or tissue to influence the fate of nearby cells or tissues during development by chemical signals.

inferior colliculi (singular, colliculus)   Paired hillocks on the dorsal surface of the midbrain; concerned with auditory processing.

inferior olive (inferior olivary nucleus)   Prominent nucleus in the medulla; a major source of input to the cerebellum.

infundibulum   The connection between the hypothalamus and the pituitary gland; also known as the pituitary stalk.

inhibitory postsynaptic potential (IPSP)   Neurotransmitter-induced postsynaptic potential change that tends to decrease the likelihood of a postsynaptic action potential.

innervate   Establish synaptic contact with a target.

innervation   Referring to all the synaptic contacts made with a target.

input   The innervation of a target cell by a particular axon; more loosely, the innervation of a target.

input elimination   The developmental process by which the number of axons innervating some classes of target cells is diminished.

instructive   A developmental influence that dictates the fate of a cell rather than simply permitting differentiation to occur.

insula   The portion of the cerebral cortex that is buried within the depths of the lateral fissure.

integral membrane proteins   Proteins that possess hydrophobic domains that are inserted into membranes.

integration   The summation of excitatory and inhibitory synaptic conductance changes by postsynaptic cells.

integrins   A family of receptor molecules found on growth cones that bind to cell adhesion molecules such as laminin and fibronection.

intention tremor   Tremor that occurs while performing a voluntary motor act. Characteristic of cerebellar pathology.

internal arcuate tract   Mechanosensory pathway in the brainstem that runs from the dorsal column nuclei to form the medial lemniscus.

internal capsule   Large white matter tract that lies between the diencephalon and the basal ganglia; contains, among others, sensory axons that run from the thalamus to the cortex and motor axons that run from the cortex to the brainstem and spinal cord.

interneuron   Technically, a neuron in the pathway between primary sensory and primary effector neurons; more generally, a neuron that branches locally to innervate other neurons.

interstitial nuclei of the anterior hypothalamus (INAH)   Four cell groups located slightly lateral to the third ventricle in the anterior hypothalamus of primates; thought to play a role in sexual behavior.

intrafusal muscle fibers   Specialized muscle fibers found in muscle spindles.

invertebrate   An animal without a backbone (includes about 97% of extant animals).

ion channels   Integral membrane proteins possessing pores that allow certain ions to diffuse across cell membranes, thereby conferring selective ionic permeability.

ion exchangers   Membrane pumps that translocate one or more ions against their concentration gradient by using the electrochemical gradient of other ions as an energy source.

ion pumps   Integral membrane proteins that use a cellular energy source to establish ion concentration gradients across cell membranes.

ionotropic   *see* ligand-gated ion channels.

ipsilateral   On the same side of the body.

ischemia   Insufficient blood supply.

kinocilium   A true ciliary structure which, along with the stereocilia, comprises the hair bundle of vestibular and fetal cochlear hair cells in mammals (it is not present in the adult mammalian cochlear hair cell).

Korsakoff's syndrome   An amnesic syndrome seen in chronic alcoholics.

labyrinth   The elaborate set of interconnected canals that forms the main peripheral component of the vestibular system.

lamellipodia   The leading edge of a motile cell or growth cone, which is rich in actin filaments.

laminae (singular, lamina)   Cell layers that characterize the neocortex, hippocampus, and cerebellar cortex. The gray matter of the spinal cord is also arranged in laminae.

laminin   A large cell adhesion molecule that binds integrins.

late outward current The delayed electrical current, measured in voltage clamp experiments, that results from the voltage-dependent efflux of a cation such as K^+. Produces the repolarizing phase of the action potential.

lateral columns The lateral regions of spinal cord white matter that convey motor information from the brain to the spinal cord.

lateral (Sylvian) fissure The cleft on the lateral surface of the brain that separates the temporal and frontal lobes.

lateral geniculate nucleus (LGN) A nucleus in the thalamus that receives the axonal projections of retinal ganglion cells in the primary visual pathway.

lateral olfactory tract The projection of mitral cells from the olfactory bulbs to higher olfactory centers.

lateral posterior nucleus A thalamic nucleus that receives its major input from sensory and association cortices and projects in turn to association cortices, particularly in the parietal and temporal lobes.

lateral superior olive (LSO) The auditory brainstem structure that processes interaural intensity differences and, in humans, mediates sound localization for stimuli greater than 3 kHz.

learning The acquisition of novel behavior through experience.

lexical The quality of associating a symbol (e.g., a word) with a particular object, emotion, or idea.

lexicon Dictionary. Sometimes used to indicate region of brain that stores the meanings of words.

ligand-gated ion channels Term for a large group of neurotransmitter receptors that combine receptor and ion channel functions into a single molecule.

limb bud The limb rudiment of vertebrate embryos.

limbic lobe Cortex that lies superior to the corpus callosum on the medial aspect of the cerebral hemispheres; forms the cortical component of the limbic system.

limbic system Term that refers to those cortical and subcortical structures concerned with the emotions; the most prominent components are the cingulate gyrus, the hippocampus, and the amygdala.

lobes The four major divisions of the cerebral cortex (frontal, parietal, occipital, and temporal).

long-term Lasting weeks, months, or longer.

long-term depression A persistent weakening of synapses based on past patterns of activity.

long-term memory Memories that last weeks, months, years, or a lifetime.

long-term potentiation (LTP) A persistent strengthening of synapses based on past patterns of activity.

lower motor neuron Spinal motor neuron; directly innervates muscle (also referred to as alpha or primary motor neuron).

lower motor neuron syndrome Signs and symptoms that result from damage to alpha motor neurons; these include paralysis or paresis, muscle atrophy, areflexia, and fibrillations.

macroscopic currents Ionic currents flowing through large numbers of ion channels distributed over a substantial area of membrane.

macula The central region of the retina that contains the fovea (the term derives from the yellowish appearance of this region in ophtholmoscopic examination); also, the sensory epithelia of the otolith organs.

mammal An animal the embyros of which develop in a uterus and the young of which begin to suckle at birth (technically, a member of the class Mammalia).

mammillary bodies Small prominences on the ventral surface of the diencephalon; functionally, part of the caudal hypothalamus.

map The ordered projection of axons from one region of the nervous system to another, by which the organization of the body is reflected in the organization of the nervous system.

medial Located nearer to the midsagittal plane of an animal (the opposite of lateral).

medial dorsal nucleus A thalamic nucleus that receives its major input from sensory and association cortices and projects in turn to association cortices, particularly in the frontal lobe.

medial geniculate complex The major thalamic relay for auditory information.

medial lemniscus Axon tract in the brainstem that carries mechanosensory information from the dorsal column nuclei to the thalamus.

medial longitudinal fasciculus Axon tract that carries excitatory projections from the abducens nucleus to the contralateral oculomotor nucleus; important in coordinating conjugate eye movements.

medial superior olive (MSO) The auditory brainstem structure that processes interaural time differences and serves to compute the horizontal location of a sound source.

medium spiny neuron The principal projection neuron of the caudate and putamen.

medulla The caudal portion of the brainstem, extending from the pons to the spinal cord.

Meissner's corpuscles Encapsulated cutaneous mechanosensory receptors specialized for the detection of fine touch and pressure.

membrane conductance An electrical term representing the reciprocal of membrane resistance. Changes in membrane conductance result from, and are used to describe, the opening or closing of ion channels.

meninges The external covering of the brain; includes the pia, arachnoid, and dura mater.

Merkel's disks Encapsulated cutaneous mechanosensory receptors specialized for the detection of fine touch and pressure.

mesencephalon *see* midbrain.

mesoderm The middle of the three germ layers; gives rise to muscle, connective tissue, skeleton, and other structures.

mesopic   Light levels at which both the rod and cone systems are active.

metabotropic receptors   *see* G-protein-coupled receptors.

Meyer's loop   That part of the optic radiation that runs under the caudal portion of the temporal lobe.

microglial cells   One of the three main types of central nervous system glia; concerned primarily with repairing damage following neural injury.

microscopic currents   Ionic currents flowing through single ion channels.

midbrain (mesencephalon)   The most rostral portion of the brainstem; identified by the superior and inferior colliculi on its dorsal surface, and the cerebral penduncles on its ventral aspect.

middle cerebellar peduncle   Large white matter tract that carries axons from the pontine relay nuclei to the cerebellar cortex.

miniature end plate potential (MEPP)   Small, spontaneous depolarization of the membrane potential of skeletal muscle cells, caused by the release of a single quantum of acetylcholine.

mitral cells   The major output neurons of the olfactory bulb.

mnemonic   Having to do with memory.

modality   A category of function. For example, vision, hearing, and touch are different sensory modalities.

molecular layer   The layer of the cerebellar cortex containing the apical dendrites of Purkinje cells, parallel fibers from granule cells, a few local circuit neurons and the synapses between these elements.

monoclonal antibody   An antibody molecule raised from a clone of transformed lymphocytes.

morphine   A plant alkaloid that gives opium its analgesic properties.

morphogen   A molecule that influences morphogenesis.

morphogenesis   The generation of animal form.

morphology   The study of the form and structure of organisms; or, more commonly, the form and structure of an animal or animal part.

motor   Pertaining to movement.

motor cortex   The region of the cerebral cortex lying anterior to the central sulcus concerned with motor behavior; includes the primary motor cortex in the precentral gyrus, and associated cortical areas in the frontal lobe.

motor neuron   By usage, a nerve cell that innervates skeletal muscle. Also called primary or *a* motor neuron.

motor neuron pool   The collection of motor neurons that innervates a single muscle.

motor system   A broad term used to describe all the central and peripheral structures that support motor behavior.

motor unit   A motor neuron and the skeletal muscle fibers it innervates; more loosely, the collection of skeletal muscle fibers innervated by a single motor neuron.

muscarinic receptors   A group of G-protein-coupled acetylcholine receptors activated by the plant alkaloid muscarine.

muscle spindle   Highly specialized sensory organ found in most skeletal muscles; provides mechanosensory information about muscle length.

muscle tone   The normal, ongoing tension in a muscle; measured by resistance of a muscle to passive stretching.

myelin   The multilaminated wrapping around many axons formed by oligodendrocytes or Schwann cells. Myelination serves to increase axonal conduction velocity.

myelination   Process by which glial cells wrap axons to form multiple layers of glial cell membrane that electrically insulate the axon and, thereby, speed up conduction of action potentials.

myotome   The part of each somite that contributes to the development of skeletal muscles.

Na$^+$/K$^+$ pump (or Na$^+$ pump)   A type of ATPase pump in the plasma membrane of most cells that is responsible for accumulating intracellular K$^+$ and extruding intracellular Na$^+$.

near reflex   Reflexive responses induced by changing binocular fixation to a closer target; includes convergence, accommodation, and pupillary constriction.

neocortex   The six-layered cortex that covers the bulk of the cerebral hemispheres.

Nernst equation   A mathematical relationship that predicts the equilibrium potential across a membrane that is permeable to only one ion.

nerve   A collection of peripheral axons that are bundled together and travel a common route.

nerve growth factor   A small dimeric protein that acts as a trophic agent in several components of the nervous system.

netrins   A family of diffusible molecules that act as attractive or repulsive cues to guide growing axons.

neural cell adhesion molecule (N-CAM)   Molecule that is structurally related to immunoglobin; helps bind axons together and is widely distributed in the developing nervous system.

neural crest   A group of progenitor cells that forms along the dorsum of the neural tube and gives rise to peripheral neurons and glia (among other derivatives).

neural plate   The thickened region of the dorsal ectoderm of a neurula that gives rise to the neural tube.

neural tube   The primordium of the brain and spinal cord; derived from the neural ectoderm.

neurite   A neuronal branch (usually used when the process in question could be either an axon or a dendrite, such as the branches of isolated nerve cells in tissue culture).

neuroblast   A dividing cell, the progeny of which develop into neurons.

neurogenesis   The development of the nervous system.

neuroglial cells   *see* glial cells.

neuroleptics   A group of antipsychotic agents that cause indifference to stimuli by blocking brain dopamine receptors.

neuromuscular junction   The synapse made by a motor axon on skeletal muscle fiber.

neuron Cell specialized for the conduction and transmission of electrical signals in the nervous system.

neuron-glia cell adhesion molecule (Ng-CAM) A cell adhesion molecule, structurally related to immunoglobin molecules, that promotes adhesive interactions between neurons and glia.

neuronal geometry The spatial arrangement of neuronal branches.

neuropeptides A general term describing a large number of peptides that function as neurotransmitters or neurohormones.

neuropil The dense tangle of axonal and dendritic branches, and the synapses between them, that lies between neuronal cell bodies in the gray matter of the brain and spinal cord.

neurotransmitter Substance released by synaptic terminals for the purpose of transmitting information from one nerve cell to another.

neurotrophic factors A general term for molecules that promote the growth and survival of neurons.

neurotrophic hypothesis The idea that developing neurons compete for a limited supply of trophic factors secreted by their targets.

neurotrophins A family of trophic factor molecules that promote the growth and survival of several different classes of neurons.

neurula The early vertebrate embryo during the stage when the neural tube forms from the neural plate; follows the gastrula stage.

neurulation The process by which the neural plate folds to form the neural tube.

nociceptors Cutaneous and subcutaneous receptors (usually free nerve endings) specialized for the detection of harmful (noxious) stimuli.

nodes of Ranvier Periodic gaps in the myelination of axons.

non-rapid eye movement (non-REM) sleep Collectively, those phases of sleep characterized by the absence of rapid-eye movements.

norepinephrine (noradrenaline) Catecholamine hormone and neurotransmitter that binds to *a*- and *b*-adrenergic receptors, both of which are G-protein-coupled receptors.

notochord A transient, cylindrical structure of mesodermal cells underlying the neural plate (and later the neural tube) in vertebrate embryos. Source of important inductive signals for spinal cord.

nucleus (plural, nuclei) Collection of nerve cells in the brain that are anatomically discrete, and which typically serve a particular function.

nucleus proprius Region of the dorsal horn of the spinal cord that receives information from nociceptors.

occipital lobe The posterior lobe of the cerebral hemisphere; primarily devoted to vision.

ocular dominance columns The segregated termination patterns of thalamic inputs representing the two eyes in primary visual cortex of some mammalian species.

olfactory bulb Olfactory relay station that receives axons from the first (olfactory) cranial nerve, and transmits this information via the olfactory tract to higher centers.

olfactory epithelium Pseudostratified epithelium that contains olfactory receptor cells, supporting cells and mucous secreting glands.

olfactory receptor neurons Bipolar neurons in olfactory epithelium that contain receptors for odors.

olfactory tracts *see* lateral olfactory tracts.

oligodendrocytes One of three classes of central neuroglial cells; their major function is to elaborate myelin.

ontogeny The developmental history of an individual animal; also used as a synonym for development.

Onuf's nucleus Sexually dimorphic nucleus in the human spinal cord that innervates striated perineal muscles mediating contraction of the bladder in males, and vaginal constriction in females.

opioid Any natural or synthetic drug that has pharmacological actions similar to those of morphine.

optic chiasm The junction of the two optic nerves on the ventral aspect of the diencephalon, where axons from the nasal parts of each retina cross the midline.

optic cup *see* optic vesicle.

optic disk The region of the retina where the axons of retinal ganglion cells exit to form the optic nerve.

optic nerve The nerve (cranial nerve II) containing the axons of retinal ganglion cells; extends from the eye to the optic chiasm.

optic radiation Portion of the internal capsule that contains the axons of lateral geniculate neurons that carry visual information to the striate cortex.

optic tectum The first central station in the visual pathway of many vertebrates (analogous to the superior colliculus in mammals).

optic tract The axons of retinal ganglion cells after they have passed through the region of the optic chiasm en route to the lateral geniculate nucleus of the thalamus.

optic vesicle The evagination of the forebrain vesicle that generates the retina and induces lens formation in the overlying ectoderm.

optokinetic eye movements Movements of the eyes that compensate for head movements; the stimulus for optokinetic movements is large scale motion of the visual field.

optokinetic nystagmus Repeated reflexive responses of the eyes to ongoing large-scale movements of the visual scene.

order A taxonomic category subordinate to class; comprises animal families.

orientation selectivity A property of many neurons in visual cortex; such neurons respond selectively to edges presented over a narrow range of orientations.

oscillopsia An inability to fixate visual targets while the head is moving as a result of vestibular damage.

otoconia The calcium carbonate crystals that rest on the otolithic membrane overlying the hair cells of the sacculus and utricle.

otolithic membrane The gelatinous membrane on which the otoconia lie and in which the tips of the hair bundles are embedded.

outer segment Portion of photoreceptors made up of membranous disks that contain the photopigment responsible for initiating phototransduction.

oval window Site where the middle ear ossicles transfer vibrational energy to the cochlea.

overshoot phase The peak, positive-going phase of an action potential, caused by high membrane permeability to a cation such as Na^+ or Ca^{2+}.

oxytocin A 9-amino-acid neuropeptide that is both a putative neurotransmitter and neurohormone.

Pacinian corpuscle Encapsulated mechanosensory receptor specialized for the detection of high frequency vibrations.

parallel fibers The bifurcated axons of cerebellar granule cells that synapse upon dendritic spines of Purkinje cells.

paralysis Complete loss of voluntary motor control.

paramedian pontine reticular formation (PPRF) Neurons in the reticular formation of the pons that coordinate the actions of motor neurons in the abducens and oculomotor nuclei to generate horizontal movements of the eyes; also known as the horizontal gaze center.

parasympathetic nervous system A division of the peripheral autonomic nervous system comprising cholinergic ganglion cells located near target organs.

paresis Partial loss of voluntary motor control; weakness.

parietal lobe The lobe of the brain that lies between the frontal lobe anteriorly, and the occipital lobe posteriorly.

Parkinson's disease A neurodegenerative disease of the substantia nigra that results in a characteristic tremor at rest and a general paucity of movement.

passive current flow Current flow across neuronal membranes that does not entail the action potential mechanism.

patch clamp An extraordinarily sensitive voltage-clamp method that permits the measurement of ionic currents flowing through individual ion channels.

periaqueductal gray matter Region of brainstem gray matter that contains, among others, nuclei associated with the modulation of pain perception.

perilymph The potassium-poor fluid that bathes the basal end of the cochlear hair cells.

perineurium The connective tissue that surrounds a nerve fascicle in a peripheral nerve.

peripheral nervous system All nerves and neurons that lie outside the brain and spinal cord.

permissive An influence during development that permits differentiation to occur but does not specifically instruct cell fate.

phasic Transient firing of action potentials in response to a prolonged stimulus; the opposite of tonic.

phenotype The visible (or otherwise discernible) characteristics of an animal that arise during development.

phenotypic sex The visible body characteristics associated with sexual behaviors.

phospholipase A2 A G-protein-activated enzyme that hydrolizes membrane phospholipids at the inner leaflet of the plasma membrane to release fatty acids such as arachadonic acid.

phospholipase C A G-protein-activated enzyme that hydrolizes membrane phospholipids at the inner leaflet of the plasma membrane to release a diacylglycerol and an inositol phosphate such as inositol-trisphosphate (IP³).

photopic vision Vision at high light levels that is mediated entirely by cones.

phylogeny The evolutionary history of a species or other taxonomic category.

phylum A major division of the plant or animal kingdom that includes classes having a common ancestry.

pia mater The innermost of the three layers of the meninges, which is closely applied to the surface of the brain.

pituitary gland Endocrine structure comprising an anterior lobe made up of many different types of hormone-secreting cells, and a posterior lobe that secretes neuropeptides produced by neurons in the hypothalamus.

placebo An inert substance that when administered may, because of the circumstances, have physiological effects.

planum temporale Region on the superior surface of the temporal lobe, posterior to Heschl's gyrus; notable because it is larger in the left hemisphere in about two-thirds of humans.

plasticity Term that refers to structural or functional changes in the nervous system, usually in the adult.

polarity Referring to a continually graded organization along one of the major embryonic axes.

polyneuronal innervation A state in which developing neurons or muscle fibers receive synaptic inputs from multiple, rather than single axons.

pons One of the three components of the brainstem, lying between the midbrain rostrally and the medulla caudally.

pontine relay nuclei Collections of neurons in the pons that receive input from the cerebral cortex, and send their axons across the midline to the cerebellar cortex via the middle cerebellar peduncle.

pontine-geniculate-occipital (PGO) waves Characteristic encephalographic waves that signal the onset of rapid-eye movement sleep.

pore A structural feature of membrane ion channels that allows ions to diffuse through the channel by virtue of aqueous continuity.

pore loop An extracellular domain of amino acids, found in certain ion channels, that lines the channel pore and allows only certain ions to pass through.

postcentral gyrus The hemispheric gyrus that lies just posterior to the central sulcus; contains the primary somatic sensory cortex.

posterior Toward the back; sometimes used as a synonym for caudal and/or dorsal.

postsynaptic current (PSC) The current produced in a postsynaptic neuron by the binding of neurotransmitter released from a presynaptic neuron.

postsynaptic potential (PSP) The potential change produced in a postsynaptic neuron by the binding of neurotransmitter released from a presynaptic neuron.

postsynaptic Referring to the component of a synapse specialized for transmitter reception; downstream at a synapse.

post-tetanic potentiation (PTP) An enhancement of synaptic transmission resulting from high-frequency trains of action potentials.

precentral gyrus The hemispheric gyrus that lies just anterior to the central sulcus; contains the primary motor cortex.

prefrontal cortex Cortical regions in the frontal lobe that are anterior to the primary and association motor cortices; thought to be involved in planning complex cognitive behaviors, and in the expression of personality and appropriate social behavior.

premotor areas Cortical areas in the frontal lobe anterior to primary motor cortex; thought to be involved in planning or programming of voluntary movements.

pre-proproteins The first protein translation products synthesized in a cell. These polypeptides are usually much larger than the final, mature peptide, and often contain signal sequences that target the peptide to the lumen of the endoplasmic reticulum.

presynaptic Referring to the component of a synapse specialized for transmitter release; upstream at a synapse.

pretectum A group of nuclei located at the junction of the thalamus and the midbrain; these nuclei are important in the pupillary light reflex, relaying information from the retina to the Edinger-Westphal nucleus.

primary auditory cortex The major cortical target of the neurons in the medial geniculate nucleus.

primary motor cortex The major source of descending projections to motor neurons in the the spinal cord and cranial nerve nuclei; located in the precentral gyrus (Brodmann's area 4) and essential for the voluntary control of movement.

primary neuron A neuron that directly links muscles, glands, and sense organs to the central nervous system.

primary sensory cortex Any one of several cortical areas receiving the thalamic input for a particular sensory modality.

primary visual cortex *see* striate cortex.

primate An order of mammals that includes lemurs, tarsiers, marmosets, monkeys, apes, and humans (technically, a member of this order).

priming A phenomenon in which the memory of an initial exposure is expressed by improved performance at a later time.

primitive pit The thickened anterior end of the primitive streak; an important source of inductive signals during early development.

procedural memory Unconscious memories such as motor skills and associations.

production aphasia Aphasia that derives from cortical damage to those centers concerned with the motor aspects of speech.

proproteins Partially processed forms of proteins containing peptide sequences that play a role in the correct folding of the final protein.

prosencephalon The part of the brain that includes the diencephalon and telencephalon (derived from the embryonic forebrain vesicle).

prosody (adjective, prosodic) The emotional tone or quality of speech.

prosopagnosia The inability to recognize faces; usually associated with lesions to the right temporal cortex.

proteoglycan Molecule consisting of a core protein to which one or more long, linear carbohydrate chains (glycosaminoglycans) are attached.

proximal Closer to a point of reference (the opposite of distal).

psychotropic drugs Drugs that alter behavior, mood, and perception.

pulvinar A thalamic nucleus that receives its major input from sensory and association cortices and projects in turn to association cortices, particularly in the parietal lobe.

pupillary light reflex The decrease in the diameter of the pupil that follows stimulation of the retina with light.

Purkinje cell The large principal projection neuron of the cerebellar cortex that has as its defining characteristic an elaborate apical dendrite.

putamen One of the three major nuclei that make up the basal ganglia.

pyramidal tract White matter tract that lies on the ventral surface of the medulla and contains axons descending from motor cortex to the spinal cord.

radial glia Glial cells that contact both the luminal and pial surfaces of the neural tube, providing a substrate for neuronal migration.

rapid eye movement (REM) sleep Phase of sleep characterized by low voltage, high frequency electroencephalographic activity accompanied by rapid eye movements.

receptive field Region of the body surface that causes a sensory nerve cell (or axon) to respond.

receptor Membrane protein containing an extracellular binding site for a neurotransmitter or hormone, as well as intracellular or transmembrane domains for signaling the agonist-bound state to the interior of the cell. Also used to refer to cells specialized for sensory transduction.

5-a-Reductase Enzyme that converts testosterone to dihydrotestosterone.

reflex A stereotyped (involuntary) motor response elicited by a defined stimulus.

refractory period A brief time period after the generation of an action potential during which the generation of a second action potential is difficult or impossible.

remodeling Change in the anatomical arrangement of neural connections.

reserpine An antihypertensive drug that is no longer used due to side effects such as behavioral depression.

resting potential The inside-negative electrical potential that is normally recorded across cell membranes.

reticular activating system Region in the brainstem tegmentum that, when stimulated, causes arousal; involved in modulating sleep and wakefulness.

reticular formation A network of neurons and fibers that occupies the core of the brainstem, giving it a reticulated appearance in myelin-stained material; its major functions include control of respiration and heart rate, posture, and state of consciousness.

retinoic acid A derivative of vitamin A that acts as an inducer during early brain development.

retinotectal system The pathway between ganglion cells in the retina and the optic tectum of vertebrates.

retrograde A movement or influence acting from the axon terminal toward the cell body.

rhodopsin The photopigment found in rods.

rhombencephalon The part of the brain that includes the pons, cerebellum, and medulla (derived from the embryonic hindbrain vesicle).

rising phase The initial, depolarizing, phase of an action potential, caused by the regenerative, voltage-dependent influx of a cation such as Na^+ or Ca^{2+}.

rods Photoreceptors specialized for operating at low light levels.

rostral Anterior, or "headward."

rostral interstitial nucleus Neurons in the midbrain reticular formation that coordinate the actions of neurons in the oculomotor nuclei to generate vertical movements of the eye; also known as the vertical gaze center.

saccades Ballistic, conjugate eye movements that change the point of foveal fixation.

sacculus The otolith organ that detects linear accelerations and head tilts in the vertical plane.

sagittal Referring to the anterior-posterior plane of an animal.

saltatory conduction Mechanism of action potential propagation in myelinated axons; so named because action potentials "jump" from one node of Ranvier to the next due to generation of action potentials only at these sites.

Scarpa's ganglion The ganglion containing the bipolar cells that innervate the semicircular canals and otolith organs.

Schaffer collaterals The axons of cells in the CA3 region of hippocampus that form synapses in the CA1 region.

Schwann cells Neuroglial cells in the peripheral nervous system that elaborate myelin (named after the nineteenth-century anatomist and physiologist Theodor Schwann).

scotopic vision Vision at low light levels that is mediated entirely by rods.

second order neurons Projection neurons in a sensory pathway that lie between the primary receptor neurons and the third order neurons.

segment One of a series of more or less similar anterior-posterior units that make up segmental animals.

segmentation The anterior-posterior division of animals into roughly similar repeating units.

semaphorins A family of diffusible, growth-inhibiting molecules (*see also* collapsin).

semicircular canals The vestibular end-organs within the inner ear that sense rotational accelerations of the head.

sensitization Increased sensitivity to stimuli in an area surrounding an injury. Also, a generalized aversive response to an otherwise benign stimulus when it is paired with a noxious stimulus.

sensory Pertaining to sensation.

sensory aphasia Difficulty communicating that derives from cortical damage to those areas concerned with the comprehension of speech.

sensory ganglia *see* dorsal root ganglia

sensory system Term sometimes used to describe all the components of the central and peripheral nervous system concerned with sensation.

sensory transduction Process by which energy in the environment is converted into electrical signals by sensory receptors.

serotonin A biogenic amine neurotransmitter derived from the amino acid tryptophan.

sexually dimorphic Having two different forms depending on genotypic or phenotypic sex.

short-term memory Memories that last from seconds to minutes.

silver stain A classical method for visualizing neurons and their processes by impregnation with silver salts (the best-known technique is the Golgi stain, developed by the Italian anatomist Camillo Golgi in the late nineteenth century).

size principle The orderly recruitment of motor neurons by size to generate increasing amounts of muscle tension.

sleep spindles Bursts of electroencephalographic activity, at a frequency about 10–14 Hz and lasting a few seconds, that characterize the initial descent into non-REM sleep.

small molecule neurotransmitters Referring to the nonpeptide neurotransmitters such as acetylcholine, the amino acids glutamate, aspartate, GABA, and glycine, as well as the biogenic amines.

smooth pursuit movements Slow, tracking movements of the eyes designed to keep a moving stimulus aligned with the fovea.

soma (plural, somata) A nerve cell body.

somatic cells Referring to the cells of an animal other than its germ cells.

somatic sensory cortex That region of the hemispheric cortex concerned with processing sensory information from the body surface, subcutaneous tissues, muscles, and joints; located primarily in the posterior bank of the central sulcus and on the postcentral gyrus.

somatic sensory system Denoting those parts of the nervous system involved in processing sensory information about the mechanical forces active on both the body surface and on deeper structures such as muscles and joints.

somatotopic maps Cortical or subcortical arrangements of sensory pathways that reflect the organization of the body.

somites Segmentally arranged masses of mesoderm that lie alongside the neural tube and give rise to skeletal muscle, vertebrae, and dermis.

species A taxonomic category subordinate to genus; members of a species are defined by extensive similarities, including the ability to interbreed.

spina bifida A developmental defect in which the neural tube fails to close at its posterior end.

spinal cord The portion of the central nervous system that extends from the lower end of the brainstem (the medulla) to the cauda equina.

spinal ganglia *see* dorsal root ganglia.

spinal nucleus of the bulbocavernosus Sexually dimorphic collection of neurons in the lumbar region of the rodent spinal cord that innervate striated perineal muscles.

spinal shock The initial flaccid paralysis that accompanies damage to descending motor pathways.

spinal trigeminal tract Brainstem tract carrying fibers from the trigeminal nerve to the spinal nucleus of the trigeminal complex (which serves as the relay for painful stimulation of the face).

spino-cerebellum Region of the cerebellar cortex that receives input form the spinal cord, particularly Clarke's column in the thoracic spinal cord.

spinothalamic pathway *see* anterolateral pathway.

spinothalamic tract Ascending white matter tract carrying information about pain and temperature from the spinal cord to the VP nuclear complex in the thalamus; also referred to as the anterolateral tract.

split-brain patients Individuals who have had the cerebral commissures divided in the midline to control epileptic seizures.

sporadic Cases of a disease that occur at random in a population; contrasts with familial.

stereocilia The actin-rich processes which, along with the kinocilium, form the hair hundle extending from the apical surface of the hair cell; site of mechanotransduction.

stereopsis The perception of depth that results from the fact that the two eyes view the world from slightly different angles.

strabismus Misalignment of the two eyes, such that normal binocular vision is compromised.

stria vascularis Specialized epithelium lining the cochlear duct that maintains the high potassium concentration of the endolymph.

striate cortex Primary visual cortex in the occipital lobe (also called Brodmann's area 17). So named because the prominence of layer IV in myelin-stained sections gives this region a striped appearance.

striatum (neostriatum) Generic name for the caudate and putamen; the numerous axon bundles in this region give it a striped appearance.

striola A line found in both the sacculus and utricle that divides the hair cells into two populations with opposing hair bundle polarities.

subarachnoid space The cerebrospinal fluid filled space over the surface of the brain that lies between the arachnoid and the pia.

substance P An 11-amino acid neuropeptide; the first neuropeptide to be characterized.

substantia nigra Nucleus at the base of the midbrain that receives input from a number of cortical and subcortical structures. The dopaminergic cells of the substantia nigra send their output to the caudate/putamen, while the GABAergic cells send their output to the thalamus.

subthalamic nucleus A nucleus in the ventral diencephalon that receives input from the caudate/putamen and participates in the modulation of motor control.

sulci (singular, sulcus) The infoldings of the cerebral hemisphere that form the valleys between the gyral ridges.

superior colliculus Laminated structure that helps form the roof of the midbrain; plays an important role in orienting movements of the head and eyes.

suprachiasmatic nucleus Hypothalamic nucleus lying just above the optic chiasm that receives direct input from the retina; involved in light entrainment of circadian rhythms.

Sylvian fissure *see* lateral fissure.

sympathetic nervous system A division of the peripheral autonomic nervous system in vertebrates comprising, for the most part, adrenergic ganglion cells located relatively far from the related end organs.

synapse Specialized apposition between a neuron and its target cell for transmission of information by release and reception of a chemical transmitter agent.

synaptic cleft The space that separates pre- and postsynaptic neurons at chemical synapses.

synaptic depression A short-term decrease in synaptic strength resulting from the depletion of synaptic vesicles at active synapses.

synaptic vesicle recycling A sequence of budding and fusion reactions that occurs within presynaptic terminals and produces synaptic vesicles.

synaptic vesicles Spherical, membrane-bound organelles in presynaptic terminals that store neurotransmitters.

syncytium A group of cells in protoplasmic continuity.

target (neural target) The object of innervation, which can be either non-neuronal targets, such as muscles, glands, and sense organs, or other neurons.

taste buds Onion-shaped structures in the mouth and pharynx that contain taste cells.

tectorial membrane The fibrous sheet overlying the apical surface of the cochlear hair cells; produces a shearing motion of the stereocilia when the basilar membrane is displaced.

tectum A general term referring to the dorsal region of the brainstem (*tectum* means "roof").

tegmentum   A general term that refers to the central gray matter of the brainstem.

telencephalon   The part of the brain derived from the anterior part of the embryonic forebrain vesicle; includes the cerebral hemispheres.

temporal lobe   The hemispheric lobe that lies inferior to the lateral fissure.

terminal   A presynaptic (axonal) ending.

tetraethylammonium   A quaternary ammonium compound that selectively blocks voltage-sensitive K^+ channels; eliminates the delayed K^+ current measured in voltage clamp experiments.

tetrodotoxin   An alkaloid neurotoxin, produced by certain puffer fish, tropical frogs and salamanders, that selectively blocks voltage-sensitive Na^+ channels; eliminates the initial Na^+ current measured in voltage clamp experiments.

thalamus   A collection of nuclei that forms the major component of the diencephalon. Although its functions are many, a primary role of the thalamus is to relay sensory information from lower centers to the cerebral cortex.

threshold   The level of membrane potential at which an action potential is generated.

tight junction   A specialized junction between epithelial cells that seals them together, preventing most molecules from passing across the cell sheet.

tip links   The filamentous structures that link the tips of adjacent stereocilia; thought to mediate the gating of the hair cell's transduction channels.

tonic   Sustained activity in response to an ongoing stimulus; the opposite of phasic.

tonotopy   The regular ordering of best frequency responses along one axis of an auditory structure.

transducin   G-protein involved in the phototransduction cascade.

transforming growth factor (TGF)   A class of peptide growth factors that acts as an inducer during early development.

transmitter   *see* neurotransmitter.

transsexuality   Gender identification with the opposite phenotypic sex.

tricyclic antidepressants   A class of antidepressant drugs named for their three-ringed molecular structure, and thought to act by blocking the reuptake of biogenic amines.

trigeminal ganglion   The sensory ganglion associated with the trigeminal (fifth) cranial nerve.

Trk receptors   The receptors for the neurotrophin family of growth factors.

trophic   The ability of one tissue or cell to support another; usually applied to long-term interactions between pre- and postsynaptic cells.

trophic factor (agent)   A molecule that mediates trophic interactions.

trophic interactions   Referring to the long-term interdependence of nerve cells and their targets.

trophic molecules   *see* trophic factor.

tropic   An influence of one cell or tissue on the direction of movement (or outgrowth) of another.

tropic molecules   Molecules that influence the direction of growth or movement.

tropism   Orientation of growth in response to an external stimulus.

tympanic membrane   The eardrum.

undershoot phase   The final, hyperpolarizing phase of an action potential, typically caused by the voltage-dependent efflux of a cation such as K^+.

upper motor neurons   The neurons that give rise to the descending projections that control the activity of lower motor neurons in the brainstem and spinal cord.

upper motor neuron syndrome   Signs and symptoms that result from damage to descending motor systems; these include paralysis, spasticity, and a positive Babinski sign.

utricle   The otolith organ that senses linear accelerations and head tilts in the horizontal plane.

vasopressin   A 9-amino-acid neuropeptide that acts as a neurotransmitter, as well as a neurohormone.

ventral   Referring to the belly. The opposite of dorsal.

ventral horn   The ventral portion of the spinal cord gray matter; contains the primary motor neurons.

ventral posterior complex   Group of thalamic nuclei that receives the somatic sensory projections from the dorsal column nuclei and the trigeminal nuclear complex.

ventral posterior lateral nucleus   Component of the ventral posterior complex of thalamic nuclei that receives brainstem projections carrying somatic sensory information from the body excluding the face.

ventral posterior medial nucleus   Component of the ventral posterior complex of thalamic nuclei that receives brainstem projections related to somatic sensory information from the face.

ventral roots   The collection of nerve fibers containing motor axons that exit ventrally from the spinal cord, and contribute the motor component of each segmental spinal nerve.

ventricles   The spaces in the vertebrate brain that represent the lumen of the embryonic neural tube.

ventricular zone   The sheet of cells closest to the ventricles in the developing neural tube.

vergence movements   Disjunctive movements of the eyes (convergence or divergence) that align the fovea of each eye with targets located at different distances from the observer.

vertebrate   An animal with a backbone (technically, a member of the subphylum Vertebrata).

vesicle   Literally, a small sac. Used to refer to the organelles that store and release transmitter at nerve endings. Also used to refer to any of the three dilations of the anterior end of the neural tube that give rise to the three major subdivisions of the brain.

vestibulo-cerebellum The part of the cerebellar cortex that receives direct input from the vestibular nuclei or vestibular nerve.

vestibulo-ocular reflex Involuntary movement of the eyes in response to displacement of the head. This reflex allows retinal images to remain stable while the head is moved.

visceral nervous system The autonomic nervous system.

vital dye A reagent that stains cells only when they are alive.

voltage clamp A method that uses electronic feedback to control the membrane potential of a cell, simultaneously measuring transmembrane currents that result from the opening and closing of ion channels.

voltage-gated Term used to describe ion channels whose opening and closing is sensitive to membrane potential.

Wallerian degeneration The process by which the distal portion of a damaged axon segment degenerates; named after Augustus Waller, a nineteenth-century neuroanatomist.

Wernicke's aphasia Difficulty comprehending speech as a result of damage to Wernicke's language area.

Wernicke's area Region of cortex in the superior and posterior region of the left temporal lobe that helps mediate language comprehension. Named after the nineteenth-century neurologist, Karl Wernicke.

white matter A general term that refers to large axon tracts in the brain and spinal cord; the phrase derives from the fact that axonal tracts have a whitish cast when viewed in the freshly cut material.

working memory Memories held briefly in mind that enable a particular task to be accomplished (e.g., efficiently searching a room for a lost object).

Illustration Credits

Chapter 1 The Organization of the Nervous System
Figure 1.3 JONES, E. G. AND M. W. COWAN (1983) The nervous tissue. In *The Structural Basis of Neurobiology*, E. G. Jones (ed.). New York: Elsevier, Chapter 8. Figure 1.4 PETERS, A., S. L. PALAY AND H. DE F. WEBSTER (1970) *The Fine Structure of the Nervous System: The Cells and Their Processes*. New York: Harper and Row. Figure 1.21 HASSLER, O. (1967) Arterial pattern of human brain stem. Normal appearance and deformation in expanding supratentorial conditions. Neurology 17: 368–375. Box A GOLDSTEIN, G. W. AND A. L. BETZ (1986) The blood-brain barrier. Sci. Amer. 255: 77–83. Box D PETERS, A., S. L. PALAY AND H. DE F. WEBSTER (1991) *The Fine Structure of the Nervous System: The Cells and Their Processes*, 3rd Ed. New York: Oxford University Press.

Chapter 2 Electrical Signals of Nerve Cells
Figure 2.5 HODGKIN, A. L. AND B. KATZ (1949) The effect of sodium ions on the electrical activity of the giant axon of the squid. J. Physiol. (Lond.) 108: 37–77. Figure 2.6 HODGKIN, A. L. AND B. KATZ (1949) The effect of sodium ions on the electrical activity of the giant axon of the squid. J. Physiol. (Lond.) 108: 37–77.

Chapter 3 Voltage-Dependent Membrane Permeability
Figures 3.1, 3.2, & 3.3 HODGKIN, A. L., A. F. HUXLEY AND B. KATZ (1938) Measurements of current voltage relations in the membrane of the giant axon of *Loligo*. J. Physiol. 116: 424–448. Figure 3.4 HODGKIN, A. L. AND A. F. HUXLEY (1952a) Currents carried by sodium and potassium ions through the membrane of the giant axon of *Loligo*. J. Physiol. 116: 449–472. Figure 3.5 ARMSTRONG, C. M. AND L. BINSTOCK (1965) Anomalous rectification in the squid giant axon injected with tetraethylammonium chloride. J. Gen. Physiol. 48: 859–872. MOORE, J. W., M. P. BLAUSTEIN, N. C. ANDERSON AND T. NARAHASHI (1967) Basis of tetrodotoxin's selectivity in blockage of squid axons. J. Gen. Physiol. 50: 1401–1411. Figures 3.6 & 3.7 HODGKIN, A. L. AND A. F. HUXLEY (1952b) The components of membrane conductance in the giant axon of *Loligo*. J. Physiol. 116: 473–496. Figure 3.8 HODGKIN, A. L. AND A. F. HUXLEY (1952d) A quantitative description of membrane current and its application to conduction and excitation in nerve. J. Physiol. 116: 507–544. Figure 3.10 HODGKIN, A. L. AND W. A. RUSHTON (1938) The electrical constants of a crustacean nerve fibre. Proc. R. Soc. Lond. 133: 444–479.

Chapter 4 Channels and Pumps
Figure 4.1B & C BEZANILLA, F. AND A. M. CORREA (1995) Single-channel properties and gating of Na^+ and K^+ channels in the squid giant axon. In *Cephalopod Neurobiology*, N. J. Abbott, R. Williamson and L. Maddock (eds.). New York: Oxford University Press, pp. 131–151. Figure 4.1D VANDERBERG, C. A. AND F. BEZANILLA (1991) A sodium channel model based on single channel, macroscopic ionic, and gating currents in the squid giant axon. Biophys. J. 60: 1511–1533. Figure 4.1E CORREA, A. M. AND F. BEZANILLA (1994) Gating of the squid sodium channel at positive potentials. II. Single channels reveal two open states. Biophys. J. 66: 1864–1878. Figure 4.2 PEROZO, E., D. S. JONG AND F. BEZANILLA (1991) Single-channel studies of the phosphorylation of K^+ channels in the squid giant axon. II. Nonstationary conditions. J. Gen. Physiol. 98: 19–34. Figure 4.2B & C HILLE, B. (1992) *Ionic Channels of Excitable Membranes*, 2nd Ed. Sunderland, MA: Sinauer Associates. Figure 4.2D AUGUSTINE, C. K. AND F. BEZANILLA (1990) Phosphorylation modulates potassium conductance and gating current of perfused giant axons of squid. J. Gen. Physiol. 95: 245–271. Figure 4.2E PEROZO, E., D. S. JONG AND F. BEZANILLA (1991) Single-channel studies of the phosphorylation of K^+ channels in the squid giant axon. II. Nonstationary conditions. J. Gen. Physiol. 98: 19–34. Figure 4.4A CATTERALL, W. A. (1988) Structure and function of voltage-sensitive ion channels. Science 242: 50–61. Figure 4.4D HILLE, B. (1992) *Ionic Channels of Excitable Membranes*, 2nd Ed. Sunderland, MA: Sinauer Associates. Figure 4.5 WEI, A. M., A. COVARRUBIAS, A. BUTLER, K. BAKER, M. PAK AND L. SALKOFF (1990) K^+ current diversity is produced by an extended gene family conserved in *Drosophila* and mouse. Science 248: 599–603. Figure 4.7A HODGKIN, A. L. AND R. D. KEYNES (1955) Active transport of cations in giant axons from *Sepia* and *Loligo*. J. Physiol. 128: 28–60. Figure 4.7B LINGREL, J. B., J. VAN HUYSSE, W. O'BRIEN, E. JEWELL-MOTZ, R. ASKEW AND P. SCHULTHEIS (1994) Structure-function studies of the Na, K-ATPase. Kidney Internat. 45: S32–S39. Figure 4.8 RANG, H. P. AND J. M. RITCHIE (1968) On the electrogenic sodium pump in mammalian non-myelinated nerve fibres and its activation by various external cations. J. Physiol. 196: 183–221. Figure 4.9 LINGREL, J. B., J. VAN HUYSSE, W. O'BRIEN, E. JEWELL-MOTZ, R. ASKEW AND P. SCHULTHEIS (1994) Structure-function studies of the Na, K-ATPase. Kidney Internat. 45: S32–S39.

Chapter 5 Synaptic Transmission
Figure 5.2 HALL, Z. (1992) *An Introduction to Molecular Neurobiology*. Sunderland, MA: Sinauer Associates. FURSHPAN, E. J. AND D. D. POTTER (1959) Transmission at the giant motor synapses of the crayfish. J. Physiol. (Lond.) 145: 289–325. Figure 5.4 FATT, P. AND B. KATZ (1952) Spontaneous subthreshold activity at motor nerve endings. J. Physiol. (Lond.) 117: 109–128. Figure 5.5 BOYD, I. A. AND A. R. MARTIN (1955) Spontaneous subthreshold activity at mammalian neuromuscular junctions. J. Physiol. 132: 61–73. Figure 5.6 HEUSER, J. E., T. S. REESE, M. J. DENNIS, Y. JAN, L. JAN AND L. EVANS (1979) Synaptic vesicle exocytosis captured by quick freezing and correlated with quantal transmitter release. J. Cell Biol. 81: 275–300. Figure 5.7 HEUSER, J. E. AND T. S. REESE (1973) Evidence for recycling of synaptic vesicle membrane during transmitter release at the frog neuromuscular junction. J. Cell Biol. 57: 315–344. Figure 5.8 AUGUSTINE, G. J. AND R. ECKERT (1984) Divalent cations differentially support transmitter release at the squid giant synapse. J. Physiol. 346: 257–271. Figure 5.9A SMITH, S. J., J. BUCHANAN, L. R. OSSES, M. P. CHARLTON AND G. J. AUGUSTINE (1993) The spatial distribution of calcium signals in squid presynaptic terminals. J. Physiol. (Lond.) 472: 573–593. Figure 5.9B MILEDI, R. (1973) Transmitter release induced by injection of calcium

ions into nerve terminals. Proc. R. Soc. Lond. B 183: 421–425. Figure 5.9C ADLER, E., M. ADLER, G. J. AUGUSTINE, M. P. CHARLTON AND S. N. DUFFY (1991) Alien intracellular calcium chelators attenuate neurotransmitter release at the squid giant synapse. J. Neurosci. 11: 1496–1507. Figure 5.10 JESSELL, T. M. AND E. R. KANDEL (1993) Synaptic transmission: A bidirectional and self-modifiable form of cell-cell communication. Cell 72/Neuron 10: (Supplement): 1–30.

Chapter 7 Neurotransmitter Receptors and Their Effects
Figure 7.3 TAKEUCHI, A. AND N. TAKEUCHI (1960) On the permeability of end-plate membrane during the action of transmitter. J. Physiol. 154: 52–67. Figure 7.8 TOYOSHIMA, C. AND N. UNWIN (1990) Three-dimensional structure of the acetylcholine receptor by cryoelectron microscopy and helical image reconstruction. J. Cell Biol. 111: 2623–2635.

Chapter 8 The Somatic Sensory System
Figure 8.3 DARIAN-SMITH, I. (1984) The sense of touch: Performance and peripheral neural processes. In Handbook of Physiology: The Nervous System, Vol. III., J. M. Brookhart and V. B. Mountcastle (eds.). Bethesda, MD: American Physiological Society, pp. 739–788. Figure 8.4 WEINSTEIN, S. (1969) Neuropsychological studies of the phantom. In Contributions to Clinical Neuropsychology, A. L. Benton (ed.). Chicago: Aldine Publishing Company, pp. 73–106. Figure 8.5 MATTHEWS, P. B. C. (1964) Muscle spindles and their motor control. Physiol. Rev. 44: 219–288. Figure 8.7 BRODAL, P. (1992) The Central Nervous System: Structure and Function. New York: Oxford University Press, p. 151. Figure 8.7B JONES, E. G. AND D. P. FRIEDMAN (1982) Projection pattern of functional components of thalamic ventrobasal complex on monkey somatosensory cortex. J. Neurophysiol. 48: 521–544. Figure 8.8 PENFIELD, W. AND T. RASMUSSEN (1950) The Cerebral Cortex of Man: A Clinical Study of Localization of Function. New York: Macmillan, p. 14. CORSI, P. (1991) The Enchanted Loom: Chapters in the History of Neuroscience, P. Corsi (ed.). New York: Oxford University Press. Figure 8.9 KAAS, J. H. (1989) The functional organization of somatosensory cortex in primates. Ann. Anat. 175: 509–518.

Chapter 9 Pain
Figure 9.1 FIELDS, H. L. (1978) Pain. New York: McGraw-Hill. Figure 9.2

FIELDS, H. L. (ed.) (1990) Pain Syndromes in Neurology. London: Butterworths. Box B SOLONEN, K. A. (1962) The phantom phenomenon in amputed Finnish war veterans. Acta. Orthop. Scand. Suppl. 54: 1–37.

Chapter 10 Vision: The Eye
Figure 10.3A-C HILFER, S. R. AND J.-J. W. YANG (1980) Accumulation of CPC-precipitable material at apical cell surfaces during formation of the optic cup. Anat. Rec. 197: 423–433.

Chapter 11 Central Visual Pathways
Figure 11.10B J. HORTON (1992) The central visual pathways. In Adler's Physiology of the Eye: Clinical Application, 9th Ed. St. Louis: Mosby Year Book. Figure 11.15A MAUNSELL, J. H. R. (1992) Functional visual streams. Curr. Opin. Neurobiol. 2: 506–510. Figure 11.15B FELLEMAN, D. J. AND D. C. VAN ESSEN (1991) Distributed hierarchial processing in primate cerebral cortex. Cereb. Cortex 1: 1–47. Figure 11.16 ZEKI, S. (1990) La construction des images par le cerveau. La Recherche (Societe d'editions scientifiques) 21: 712–721. Box B, Figure B WANDELL, B. A. (1995) Foundations of Vision. Sunderland, MA: Sinauer Associates. Box B, Figure C Super Stereogram (1994) San Francisco: Cadence Books, p. 40.

Chapter 12 The Auditory System
Figure 12.5 DALLOS, P. (1992) The active cochlea. J. Neurosci. 12: 4575–4585. Figure 12.6 VON BÈKÈSY, G. (1960) Experiments in Hearing. New York: McGraw-Hill. Figure 12.7A LINDEMAN, H. H. (1973) Anatomy of the otolith organs. Adv. Otorhinolaryngol. 20: 405–433. Figure 12.7B HUDSPETH, A. J. (1983) The hair cells of the inner ear. Sci. Amer. 248: 54–64. Figure 12.7C PICKLES, J. O., S. D. COMIS AND M. P. OSBORNE (1984) Crosslinks between stereocilia in the guinea pig organ of Corti, and their possible relation to sensory transduction. Hear. Res. 15: 103–112. Figure 12.8A LEWIS, R. S. AND A. J. HUDSPETH (1983) Voltage- and ion-dependent conductances in solitary vertebrate hair cells. Nature 304: 538–541. Figure 12.8B PALMER, A. R., AND I. J. RUSSELL (1986) Phase-locking in the cochlear nerve of the guinea-pig and its relation to the receptor potential of inner hair cells. Hear. Res. 24: 1–15. Figure 12.9 KIANG, N. Y. AND E. C. MOXON (1972) Physiological considerations in artificial stimulation of the inner ear. Ann. Otol. Rhinol. Laryngol. 81: 714–730. Figure 12.10 KIANG, N. Y. S. (1984) Peripheral neural processing of auditory

information. In Handbook of Physiology: A Critical, Comprehensive Presentation of Physiological Knowledge and Concepts, Section 1: The Nervous System, Vol. III. Sensory Processes, Part 2, J. M. Brookhart, V. B. Mountcastle, I. Darian-Smith and S. R. Geiger (eds.). Bethesda, MD: American Physiological Society, pp. 639–674. Figure 12.12 JEFFRESS, L. A. (1948) A place theory of sound localization. J. Comp. Physiol. Psychol. 41: 35–39.

Chapter 13 The Vestibular System
Figure 13.3 LINDEMAN, H. H. (1973) Anatomy of the otolith organs. Adv. Otorhinolaryngol. 20: 405–433. Figure 13.6 GOLDBERG, J. M. AND C. FERNÁNDEZ (1976) Physiology of peripheral neurons innervating otolith organs of the squirrel monkey, Parts 1, 2, 3. J. Neurophysiol. 39: 970–1008. Figure 13.9 GOLDBERG, J. M. AND C. FERNÁNDEZ (1971) Physiology of peripheral neurons innervating semicircular canals of the squirrel monkey, Parts 1, 2, 3. J. Neurophysiol. 34: 635–684.

Chapter 14 The Chemical Senses
Figure 14.2 PELOSI, P. (1994) Odorant-binding proteins. Crit. Rev. Biochem. Mol. Biol. 29: 199–228. Figure 14.3 CAIN, W. S. AND J. F. GENT (1986) Use of odor identification in clinical testing of olfaction. In Clinical Measurement of Taste and Smell, H. L. Meiselman and R. S. Rivlin (eds.). New York: Macmillan, pp. 170–186. Figure 14.4 MURPHY, C. (1986) Taste and smell in the elderly. In Clinical Measurement of Taste and Smell, H. L. Meiselman and R. S. Rivlin (eds.). New York: Macmillan, pp. 343–371. Figure 14.5 FIRESTEIN, S., F. ZUFALL AND G. M. SHEPHERD (1991) Single odor-sensitive channels in olfactory receptor neurons are also gated by cyclic nucleotides. J. Neurosci. 11: 3565–3572. Figure 14.7A FIRESTEIN, S. AND G. M. SHEPHERD (1992) Neurotransmitter antagonists block some odor responses in olfactory receptor neurons. NeuroReport 3: 661–664. Figure 14.7B GETCHELL, T. V. AND G. M. SHEPHERD (1978b) Responses of olfactory receptor cells to step pulses of odour at different concentrations in the salamander. J. Physiol. 282: 521–540. Figure 14.8A LaMANTIA, A.-S., S. L. POMEROY AND D. PURVES (1992) Vital imaging of glomeruli in the mouse olfactory bulb. J. Neurosci. 12: 976–988. Figure 14.8B POMEROY, S. L., A.-S. LaMANTIA AND D. PURVES (1990) Postnatal construction of neural activity in the mouse olfactory bulb. J. Neurosci. 10: 1952–1966. Figure 14.8D AXEL, R. (1995) The molecular

logic of smell. Sci. Amer. 273: 154–159. Figure 14.8E SHEPHERD, G. M. (1994) Discrimination of molecular signals by the olfactory receptor neuron. Neuron 13: 771–790. Figure 14.10 ROSS M. H., L. J. ROMMELL AND G. I. KAYE (1995) *Histology, A Text and Atlas.* Baltimore: Williams and Wilkins. Figure 14.11A SCHIFFMAN, S. S., E. LOCKHEAD AND F. W. MAES (1983) Amiloride reduces taste intensity of salts and sweeteners. Proc. Natl. Acad. Sci. USA 80: 6136–6140. Figure 14.11B AVANET, P. AND B. LINDEMANN (1988) Amiloride-blockable sodium currents in isolated taste receptor cells. J. Mem. Biol. 105: 245–255. Figure 14.13 SMITH, D. V. AND M. E. FRANK (1993) Sensory coding by peripheral taste fibers. In *Mechanisms of Taste Transduction,* S. Simon and S. Roper (eds.). Boca Raton, FL: CRC Press, pp. 295–338. Figure 14.15 COMETTO-MUNIZ, J. E. AND W. S. CAIN (1990) Thresholds for odor and nasal pungency. Physiol. Behav. 48: 719–725.

Chapter 15 Spinal Cord Circuits and Motor Control
Figure 15.2 BURKE, R. E., P. L. STRICK, K. KANDA, C. C. KIM AND B. WALMSLEY (1977) Anatomy of medial gastrocnemius and soleus motor nuclei in cat spinal cord. J. Neurophysiol. 40: 667–680. Figure 15.5 BURKE, R. E., D. N. LEVINE, M. SALCMAN AND P. TSAIRIS (1974) Motorunits in cat soleus muscle: Physiological, histochemical and morphological characteristics. J. Physiol. (Lond.) 238: 503–514. Figure 15.6 WALMSLEY, B., J. A. HODGSON AND R. E. BURKE (1978) Forces produced by medical gastrocnemius and soleus muscles during locomotion in freely moving cats. J. Neurophysiol. 41: 1203–1216. Figure 15.8 MONSTER, A. W. AND H. CHAN (1977) Isometric force production by motor units of extensor digitorum communis muscle in man. J. Neurophysiol. 40: 1432–1443. Figure 15.10 HUNT, C. C. AND S. W. KUFFLER (1951) Stretch receptor discharges during muscle contraction. J. Physiol. (Lond.) 113: 298–315. Figure 15.11B PATTON, H. D. (1965) Reflex regulation of movement and posture. In *Physiology and Biophysics,* 19th Ed., T. C. Ruch and H. D. Patton (eds.). Philadelphia: Saunders, pp. 181–206. Figure 15.14 PEARSON, K. (1976) The control of walking. Sci. Amer. 235: 72–86.

Chapter 16 Descending Control of Spinal Cord Circuitry
Figure 16.4 NASHNER, L. M. (1979) Organization and programming of motor activity during posture control. In *Reflex Control of Posture and Movement,* R. Granit and O. Pompeiano (eds.). Prog. Brain Res. 50: 177–184. Figure 16.8 ROLAND, P. E., B. LARSEN, N. A. LASSEN AND E. SKINHOF (1980) Supplementary motor area and other cortical areas in organization of voluntary movements in man. J. Neurophysiol. 43: 118–136.

Chapter 17 Modulation of Movement by the Basal Ganglia and Cerebellum
Figure 17.3A & B BRADLEY, W. G., R. B. DAROFF, G. M. FENICHEL AND C. D. MARSDEN (eds.) (1991) *Neurology in Clinical Practice.* Boston: Butterworth-Heinemann, Figures 29–77. Figure 17.3C VICTOR, M., R. D. ADAMS AND E. L. MANCALL (1959) A restricted form of cerebellar cortical degeneration occurring in alcoholic patients. Arch. Neurol. 1: 579–688.

Chapter 18 Mechanisms of Motor Modulation
Figure 18.5 DELONG, M. R. AND P. L. STRICK (1974) Relation of basal ganglia, cerebellum, and motor cortex units to ramp and ballistic movements. Brain Res. 71: 327–335. Figure 18.6 DELONG, M. R. (1990) Primate models of movement disorders of basal ganglia origin. Trends Neurosci. 13: 281–285. Figure 18.7 THACH, W. T. (1968) Discharge of Purkinje and cerebellar nuclear neurons during rapidly alternating arm movements in the monkey. J. Neurophysiol. 31: 785–797. Figure 18.8 STEIN, J. F. (1986) Role of the cerebellum in the visual guidance of movement. Nature 323: 217–221. Figure 18.10 HIKOSAKA, O. AND R. H. WURTZ (1989) The basal ganglia. Rev. Oculomot. Res. 3: 257–281.

Chapter 19 Eye Movements and Sensory-Motor Integration
Figure 19.1 YARBUS, A. L. (1967) *Eye Movements and Vision.* Basil Haigh (trans.) New York: Plenum Press. Figures 19.4 & 19.5 FUCHS, A. F. (1967) Saccadic and smooth pursuit eye movements in the monkey. J. Physiol. (Lond.) 191: 609–631. Figure 19.6 FUCHS, A. F. AND E. S. LUSCHEI (1970) Firing patterns of abducens neurons of alert monkeys in relationship to horizontal eye movements. J. Neurophysiol. 33: 382–392. Figure 19.8 SCHILLER, P. H. AND M. STRYKER (1972) Single unit recording and stimulation in superior colliculus of the alert rhesus monkey. J. Neurophysiol. 35: 915–924. Figure 19.10 OPTICAN, L. M. AND D. A. ROBINSON (1980) Cerebellar-dependent adaptive control of primate saccadic system. J. Neurophysiol. 44: 1058–1076. Figure 19.11 HIKOSAKA, O. AND R. H. WURTZ (1989) The basal ganglia. Rev. Oculomot. Res. 3: 257–281. Box A PRITCHARD, R. M. (1961) Stabilized images on the retina. Sci. Amer. 204: 72–78.

Chapter 20 Early Brain Development
Figure 20.2 SANES, J. R. (1989) Extracellular matrix molecules that influence neural development. Annu. Rev. Neurosci. 12: 491–516. Figure 20.4A INGHAM, P. (1988) The molecular genetics of embryonic pattern formation in *Drosophila.* Nature 335: 25–34. Figure 20.4B GILBERT, S. F. (1994) *Developmental Biology,* 4th Ed. Sunderland, MA: Sinauer Associates. Figure 20.6 RAKIC, P. (1974) Neurons in rhesus monkey visual cortex: Systematic relation between time of origin and eventual disposition. Science 183: 425–427. Figure 20.8 GALILEO, S. S., GRAY, G. E., OWENS, G. C., MAJORS, J., AND SANES, J. R. (1990) Neurons and glia arise from a common progenitor in chicken optic tectum: Demonstration with two retroviruses and cell type-specific antibodies. Proc. Natl. Acad. Sci. USA 87: 458–462. Figure 20.10 RAKIC, P. (1974) Neurons in rhesus monkey visual cortex: Systematic relation between time of origin and eventual disposition. Science 183: 425–427.

Chapter 21 Construction of Neural Circuits
Figure 21.2A & B GODEMONT, P., L. C. WANG AND C. A. MASON (1994) Retinal axon divergence in the optic chiasm: Dynamics of growth cone behavior at the midline. J. Neurosci. 14: 7024–7039. Figure 21.2C SRETAVAN, D. W., L. FENG, E. PURE AND L. F. REICHARDT (1994) Embryonic neurons of the developing optic chiasm express L1 and CD44, cell surface molecules with opposing effects on retinal axon growth. Neuron 12: 957–975. Figure 21.3 REICHARDT, L. F. AND K. J. TOMASELLI (1991) Extracellular matrix molecules and their receptors: Functions in neural development. Annu. Rev. Neurosci. 14: 531–570. Figure 21.4A SERAFINI, T., T. E. KENNEDY, M. J. GALKO, C. MIRZAYAN, T. M. JESSELL, M. TESSIER-LAVIGNE (1994) The netrins define a family of axon outgrowth-promoting proteins homologous to *C. elegans* UNC-6. Cell 78: 409–424. Figure 21.4B, C, & D KENNEDY, T. E., T. SERAFINI, J. R. DE LA TORRE AND M. TESSIER-LAVIGNE (1994) Netrins are diffusible chemotropic factors for commissural axons in the embryonic spinal cord. Cell 78: 425–435. Figure 21.5 MESSERSMITH, E. K., E. D. LEONARDO, C. J. SHATZ, M. TESSIER-LAVI-

GNE, C. S. GOODMAN AND A. L. KOLODKIN (1995) Semaphorin III can function as a selective chemorepellent to pattern sensory projections in the spinal cord. Neuron. 14: 949–959. Figure 21.6A SPERRY, R. W. (1963) Chemoaffinity in the orderly growth of nerve fiber patterns and connections. Proc. Natl. Acad. Sci. USA 50: 703–710. Figure 21.6B PURVES, D. (1988) *Body and Brain: A Trophic Theory of Neural Connections.* Cambridge, MA: Harvard University Press. Figure 21.7A WALTER, J., S. HENKE-FAHLE AND F. BONHOEFFER (1987) Avoidance of posterior tectal membranes by temporal retinal axons. Development. 101: 909–913. Figure 21.7B CHENG, H. J., M. NAKAMOTO, A. D. BERGEMANN AND J. G. FLANAGAN (1995) Complementary gradients in expression and binding of ELF-1 and Mek4 in development of the topographic retinotectal projection map. Cell 82: 371–381. Figure 21.9 HOLLYDAY, M. AND V. HAMBURGER (1976) Reduction of the naturally occurring motor neuron loss by enlargement of the periphery. J. Comp. Neurol. 170: 311–320 . HOLLYDAY, M. AND V. HAMBURGER (1958) Regression versus peripheral controls of differentiation in motor hypoplasia. Am. J. Anat. 102: 365–409. HAMBURGER, V. (1977) The developmental history of the motor neuron. The F. O. Schmitt Lecture in Neuroscience, 1970, Neurosci. Res. Prog. Bull. 15, Suppl. III: 1–37. Figure 21.10 PURVES, D. AND J.W. LICHTMAN (1985) *Principles of Neural Development.* Sunderland, MA: Sinauer Associates, Inc. Figure 21.11B McDONALD, N. Q., R. LAPATTO, J. MURRAY-RUST, J. GUNNING, A. WLODAWER AND T. L. BLUNDELL (1991) New protein fold revealed by a 2.3-Å resolution crystal structure of nerve growth factor. Nature 354: 411–414. Figure 21.12 CHUN, L. L. AND P. H. PATTERSON (1977) Role of nerve growth factor in the development of rat sympathetic neurons in vitro. III; Effect on acetylcholine production. J. Cell Biol. 75: 712–718. Figure 21.13 MAISONPIERRE, P.C., L. BELLUSCIO, S. SQUINTO, N.Y. IP, M.E. FURTH, R.M. LINDSAY AND G.D. YANCOPOULOS (1990) Neurotrophin-3: A neurotrophic factor related to NGF and BDNF. Science 247: 1446–1451.

Chapter 22 Neural Activity and the Modification of Developing Brain Circuits
Figure 22.1 PURVES, D. AND J. W. LICHTMAN (1980) Elimination of synapses in the developing nervous system. Science 210: 153–157. Figure 22.2 LEVAY, S., T. N. WIESEL AND D. H. HUBEL (1980) The development of ocular dominance col-
umns in normal and visually deprived monkeys. J. Comp. Neurol. 191: 1–51. Figure 22.3A & B WIESEL, T. N. AND D. H. HUBEL (1965) Comparison of the effects of unilateral and bilateral eye closure on cortical unit responses in kittens. J. Neurophysiol. 28: 1029–1040. Figures 22.3C & 22.4 HUBEL, D. H. AND T. N. WIESEL (1970) The period of susceptibility to the physiological effects of unilateral eye closure in kittens. J. Physiol. 206: 419–436. Figure 22.5A LEVAY, S., M. P. STRYKER AND C. J. SHATZ (1978) Ocular dominance columns and their development in layer IV of the cat's visual cortex: A quantitative study. J. Comp. Neurol. 179: 223–224. Figures 22.5B, C & 22.6 HUBEL, D. H., T. N. WIESEL AND S. LEVAY (1977) Plasticity of ocular dominance columns in monkey striate cortex. Phil. Trans. R. Soc. Lond. B. 278: 377–409. Figure 22.7 ANTONINI, A. AND M. P. STRYKER (1993) Rapid remodeling of axonal arbors in the visual cortex. Science 260: 1819–1821. Figure 22.9 HUBEL, D. H. AND T. N. WIESEL (1965) Binocular interaction in striate cortex of kittens reared with artificial squint. J. Neurophysiol. 28: 1041–1059. Figure 22.10 STRYKER, M. P. AND S. L. STRICKLAND (1984) Physiological segregation of ocular dominance columns depends on the pattern of afferent electrical activity. Invest. Ophthalmol. Vis. Sci. (Suppl.) 25: 278.

Chapter 23 Plasticity in the Adult Nervous System
Figure 23.1 KATZ, B. (1966) *Nerve, Muscle and Synapse.* New York: McGraw Hill. Figure 23.3 NICOLL, R. A., J.A. KAUER AND R. C. MALENKA (1988) The current excitement in long-term potentiation. Neuron. 1: 97–103. Figure 23.4 GUSTAFSSON, B., H. WIGSTROM, W. C. ABRAHAM AND Y. Y. HUANG (1987) Long-term potentiation in the hippocampus using depolarizing current pulses as the conditioning stimulus to single volley synaptic potentials. J. Neurosci. 7: 774–780. Figure 23.5 NICOLL, R. A., J. A. KAUER AND R. C. MALENKA (1988) The current excitement in long-term potentiation. Neuron. 1: 97–103. Figure 23.6 NICOLL, R. A., R. C. MALENKA AND J. A. KAUER (1988) The role of calcium in long-term potentiation. Ann. N. Y. Acad. Sci. 568: 166–170. Figure 23.8 MULKEY, R. M. AND R. C. MALENKA (1992) Mechanisms underlying induction of homosynaptic long-term depression in area CA1 of the hippocampus. Neuron 9: 967–975. Figure 23.9 KANDEL, E. R. (1979) Small systems of neurons. In *The Brain.* San Francisco: W. H. Freeman and Company. Figure 23.10 KLEIN, M. AND E. R. KANDEL (1978) Presynaptic modulation of a voltage dependent Ca^{2+} current: Mechanism for behavioral sensitization in *Aplysia californica.* Proc. Natl. Acad. Sci. USA 75: 3512–3516. Figure 23.11 MERZENICH, M. M., R. J. NELSON, M. P. STRYKER, M. S. CYNADER, A. SCHOPPMAN AND J. M. ZOOK (1984) Somatosensory cortical map changes following digit amputation in adult monkeys. J. Comp. Neurol. 224: 591–605. Figure 23.12 JENKINS, W. M., M. M. MERZENICH, M. T. OCHS, E. ALLARD AND T. GUIC-ROBLES (1990) Functional reorganization of primary somatosensory cortex in adult owl monkeys after behaviorally controlled tactile stimulation. J. Neurophysiol. 61: 82–104. Figure 23.13 GILBERT, C. D. (1992) Horizontal integration and cortical dynamics. Neuron 9: 1–13. Box A MORRIS, R. G., J. J. HAGAN AND J. N. RAWLINS (1986) Allocentric spatial learning by hippocampectomised rats: A further test of the "spatial mapping" and "working memory" theories of hippocampal function. Q. J. Exp. Psychol. 38: 365–395. Box B EILERS, J., G. J. AUGUSTINE AND A. KONNERTH (1995) Subthreshold synaptic Ca^{2+} signalling in fine dendrites and spines of cerebellar Purkinje neurons. Box C SAKURAI, M. (1990) Calcium is an intracellular mediator of the climbing fiber in induction of cerebellar long-term depression. Proc. Natl. Acad. Sci. USA 87: 3383–3385. Box D DYRO, F. M. (1989) *The EEG Handbook.* Boston: Little, Brown and Company.

Chapter 24 Cognition
Figure 24.5 HEILMAN, H. AND E. VALENSTEIN (1985) *Clinical Neuropsychology,* 2nd Ed. New York: Oxford University Press, Figures 8, 10, and 12. Figure 24.6 POSNER, M. I. AND M. E. RAICHLE (1994) *Images of Mind.* New York: Scientific American Library. Figure 24.7 ETCOFF, N. L., R. FREEMAN AND K. R. CAVE (1991) Can we lose memories of faces? Content specificity and awareness in a prosopagnosic. J. Cog. Neurosci. 3: 25–41. Figure 24.9 LYNCH, J. C., V. B. MOUNTCASTLE, W. H. TALBOT AND T. C. YIN (1977) Parietal lobe mechanisms for directed visual attention. J. Neurophysiol. 40: 362–369.

Chapter 25 Language and Lateralization
Figure 25.5A PENFIELD, W. AND L. ROBERTS (1959) *Speech and Brain Mechanisms.* Princeton: Princeton University Press. Figure 25.5B OJEMANN, G. A. (1991) Cortical organization of language. J. Neurosci. 11: 2281–2287. Figure 25.6 POSNER, M. I. AND M. E. RAICHLE

(1994) *Images of Mind*. New York: Scientific American Library. Figure 25.7 EIMAS, P. D., E. R. SIQUELAND, P. JUSCZYK AND J. VIGORITO (1971) Speech perception in infants. Science 171: 303–306. Figure 25.8 JOHNSON, J. S. AND E. I. NEWPORT (1989) Critical period effects in second language learning: The influences of maturational state on the acquisition of English as a second language. Cogn. Psychol. 21: 60–99. Figure 25.9 BELLUGI, U., H. POIZNER AND E. S. KLIMA (1989) Language, modality and the brain. Trends Neurosci. 12: 380–388. Figure 25.10 PETTITO, L. A. AND P. F. MARENTETTE (1991) Babbling in the manual mode: Evidence for the ontogeny of language. Science 251: 1493–1496.

Chapter 26 Sleep and Wakefulness
Figure 26.2 BIDDLE, C. AND T. R. F. OASTER (1990) The nature of sleep. AANA J. 58: 36–44. Figure 26.3 LUNDMADSEN, P. AND S. VORSTRUP (1991) Cerebral blood flow and metabolism during sleep. Cerebrovasc. Brain Metab. Rev. 3: 281–296. Figure 26.4 ROFFWARG, H. P., J. N. MUZIO AND W. C. DEMENT (1966) Ontogenetic development of the human sleep-dream cycle. Science 152: 604–619. Figure 26.5 BERGMANN, B. M., C. A. KUSHIDA, C. A. EVERSON, M. A. GILLILAND, W. OBERYMEYER AND A. RECHTSCHAFFEN (1989) Sleep deprivation in the rat: II. Methodology. Sleep 12: 5–12. Figure 26.6 MUKHAMETOV, L. H., A. Y. SUPIN AND I. G. POLYAKOVA (1977) Interhemispheric asymmetry of the electroencephalographic sleep patterns in dolphins. Brain Res. 134: 581–584. Figure 26.7 ASCHOFF, J. (1965) Circadian rhythms in man. Science 148: 1427–1432.

Chapter 27 Emotions
Figure 27.1 LEDOUX, J. E. (1987) Emotion. In *Handbook of Physiology: A Critical, Comprehensive Presentation of Physiological Knowledge and Concepts*. Section I: The Nervous System, Vol. V. Higher Functions of the Brain, Part 1, V. B. Mountcastle, F. Plum and S. R. Geiger (eds.). Bethesda, MD: American Physio-

logical Society, pp. 419–459. Figure 27.7 MOSCOVITCH, M. AND J. OLDS (1982) Asymmetries in spontaneous facial expression and their possible relation to hemispheric specialization. Neuropsych. 20: 71–81.

Chapter 28 Sex, Sexuality, and the Brain
Figure 28.2 TORAN-ALLERAND, C. D. (1978) Gonadal hormones and brain development. Cellular aspects of sexual differentiation. Amer. Zool. 18: 553–565. Figure 28.3 WOOLLEY, C. S. AND B. S. MCEWEN (1992) Estradiol mediates fluctuation in hippocampal synapse density during the estrous cycle in the adult rat. J. Neurosci. 12: 2549–2554. Figure 28.4A BREEDLOVE, S. M. AND A. P. ARNOLD (1984) Sexually dimorphic motor nucleus in the rat lumbar spinal cord: Response to adult hormone manipulation, absence in androgen-insensitive rats. Brain Res. 225: 297–307. Figure 28.4B & C BREEDLOVE, S. M. AND A. P. ARNOLD (1983) Hormonal control of a developing neuromuscular system. II. Sensitive periods for the androgen-induced masculinization of the rat spinal nucleus of the bulbocavernosus. J. Neurosci. 3: 424–432. Figure 28.4D FORGER, N. G. AND S. M. BREEDLOVE (1986) Sexual dimorphism in human and canine spinal cord: Role of early androgen. Proc. Natl. Acad. Sci. USA 83: 7527–7531. Figure 28.6 OOMURA, Y., H. YOSHIMATSU AND S. AOU (1983) Medial preoptic and hypothalamic neuronal activity during sexual behavior of the male monkey. Brain Res. 266: 340–343. Figure 28.7 ALLEN, L. S., M. HINES, J. E. SHYRNE AND R. A. GORSKI (1989) Two sexually dimorphic cell groups in the human brain. J. Neurosci. 9: 497–506. Figure 28.8A LEVAY, S. (1991) A difference in hypothalamic structure between heterosexual and homosexual men. Science 253: 1034–1037. Figure 28.8B SWAAB, D. F. AND M. A. HOFFMAN (1984) Sexual differentiation of the human brain: A historical perspective. In *Progress in Brain Research*, Vol. 61, G. J.

De Vries (ed.). Amsterdam: Elsevier, pp. 361–374. Figure 28.9A & B ALLEN, L. S., M. F. RICHEY, Y. M. CHAI AND R. A. GORSKI (1991) Sex differences in the corpus callosum of the living human being. J. Neurosci. 11: 933–942. Figure 28.9C ALLEN, L. S. AND R. A. GORSKI (1991) Sexual dimorphism of the anterior commissure and massa intermedia of the human brain. J. Comp. Neurol. 312: 97–104. Figure 28.10 MODNEY, B. K. AND G. I. HATTON (1990) Motherhood modifies magnocellular neuronal interrelationships in functionally meaningful ways. In *Mammalian Parenting*, N. A. Krasnegor and R. S. Bridges (eds.). New York: Oxford University Press, pp. 306–323. Figure 28.11 XERRI, C., J. M. STERN AND M. M. MERZENICH (1994) Alterations of the cortical representation of the rat ventrum induced by nursing behavior. J. Neurosci. 14: 1710–1721. Box A MOORE, K. L. (1977) *The Developing Human*, 2nd Ed. Philadelphia: W. B. Saunders, p. 219. Box B MCEWEN, B. S. (1976) Interactions between hormones and nerve tissue. Sci. Amer. 235: 48–58; MCEWEN, B. S., P. G. DAVIS, B. PARSONS AND D. W. PFAFF (1979) The brain as a target for steroid hormone action. Annu. Rev. Neurosci. 2: 65–112.

Chapter 29 Human Memory
Figure 29.3A RUBIN, D. C. AND T. C. KONTIS (1983) A schema for common cents. Mem. Cog. 11: 335–341. Figure 29.3B SQUIRE, L. R. (1989) On the course of forgetting in very long-term memory. J. Exp. Psychol. 15: 241–245. Figure 29.4 ERICSSON, K. A., W. G. CHASE, AND S. FALOON (1980) Acquisition of a memory skill. Science. 208: 1181–1182. Figure 29.7 DAMASIO, A. R., H. DAMASIO AND G. W. VAN HOESEN (1982) Prosopagnosia: Anatomic basis and behavioral mechanisms. Neurology 31: 331–341. Figure 29.9 DEKABAN, A. S. AND D. SADOWSKY (1978) Changes in brain weights during the span of human life: Relation of brain weights to body heights and body weights. Ann. Neurol. 4: 345–356.

Index

About the Book

Editor: Andrew D. Sinauer

Project Manager: Carol J. Wigg

Copy Editor: Janet Greenblatt

Production Manager: Christopher Small

Book Layout and Production: Michele Ruschhaupt

Art Editing and Illustration Program: J/B Woolsey Associates

Book Design: Rodelinde Graphic Design

Cover Design: Mickey Boisvert

Color Separations: Vision Graphics, Inc.

Cover Manufacturer: Henry Sawyer Company, Inc.

Book Manufacturer: The Courier Companies, Inc.